Lecture Notes in Electrical Engineering

Volume 998

The book series *Lecture Notes in Electrical Engineering* (LNEE) publishes the latest developments in Electrical Engineering—quickly, informally and in high quality. While original research reported in proceedings and monographs has traditionally formed the core of LNEE, we also encourage authors to submit books devoted to supporting student education and professional training in the various fields and applications areas of electrical engineering. The series cover classical and emerging topics concerning:

- Communication Engineering, Information Theory and Networks
- Electronics Engineering and Microelectronics
- Signal, Image and Speech Processing
- Wireless and Mobile Communication
- Circuits and Systems
- Energy Systems, Power Electronics and Electrical Machines
- Electro-optical Engineering
- Instrumentation Engineering
- Avionics Engineering
- Control Systems
- Internet-of-Things and Cybersecurity
- Biomedical Devices, MEMS and NEMS

For general information about this book series, comments or suggestions, please contact leontina.dicecco@springer.com.

To submit a proposal or request further information, please contact the Publishing Editor in your country:

China

Jasmine Dou, Editor (jasmine.dou@springer.com)

India, Japan, Rest of Asia

Swati Meherishi, Editorial Director (Swati.Meherishi@springer.com)

Southeast Asia, Australia, New Zealand

Ramesh Nath Premnath, Editor (ramesh.premnath@springernature.com)

USA, Canada

Michael Luby, Senior Editor (michael.luby@springer.com)

All other Countries

Leontina Di Cecco, Senior Editor (leontina.dicecco@springer.com)

** **This series is indexed by EI Compendex and Scopus databases.** **

Pradeep Singh · Deepak Singh · Vivek Tiwari ·
Sanjay Misra
Editors

Machine Learning and Computational Intelligence Techniques for Data Engineering

Proceedings of the 4th International Conference MISP 2022, Volume 2

Springer

Editors
Pradeep Singh
Department of Computer Science and Engineering
National Institute of Technology Raipur
Raipur, Chhattisgarh, India

Deepak Singh
Department of Computer Science and Engineering
National Institute of Technology Raipur
Raipur, Chhattisgarh, India

Vivek Tiwari
Department of Computer Science and Engineering
International Institute of Information Technology
Naya Raipur, Chhattisgarh, India

Sanjay Misra
Østfold University College
Halden, Norway

ISSN 1876-1100 ISSN 1876-1119 (electronic)
Lecture Notes in Electrical Engineering
ISBN 978-981-99-0049-7 ISBN 978-981-99-0047-3 (eBook)
https://doi.org/10.1007/978-981-99-0047-3

This Springer imprint is published by the registered company Springer Nature Singapore Pte Ltd.
The registered company address is: 152 Beach Road, #21-01/04 Gateway East, Singapore 189721, Singapore

Contents

About the Editors

Dr. Pradeep Singh received a Ph.D. in Computer science and Engineering from the National Institute of Technology, Raipur, and an M.Tech. in Software engineering from the Motilal Nehru National Institute of Technology, Allahabad, India. Dr. Singh is an Assistant Professor in the Computer Science & Engineering Department at the National Institute of Technology. He has over 15 years of experience in various government and reputed engineering institutes. He has published over 80 refereed articles in journals and conference proceedings. His current research interests areas are machine learning and evolutionary computing and empirical studies on software quality, and software fault prediction models.

Dr. Deepak Singh completed his Bachelor of Engineering from Pt. Ravi Shankar University, Raipur, India, in 2007. He earned his Master of Technology with honors from CSVTU Bhilai, India, in 2011. He received a Ph.D. degree from the Department of Computer Science and Engineering at the National Institute of Technology (NIT) in Raipur, India, in 2019. Dr. Singh is currently working as an Assistant Professor at the Department of Computer Science and Engineering, National Institute of Technology Raipur, India. He has over 8 years of teaching and research experience along with several publications in journals and conferences. His research interests include evolutionary computation, machine learning, domain adaptation, protein mining, and data mining.

Dr. Vivek Tiwari is a Professor in Charge of the Department of Data Science and AI and Faculty of Computer Science and Engineering at IIIT Naya Raipur, India. He received B.Eng. from the Rajiv Gandhi Technical University, Bhopal, in 2004 and M.Tech. from SATI, Vidisha (MP), in 2008. He obtained a Ph.D. degree from the National Institute of Technology, Bhopal (MA-NIT), India, in 2015 in data mining and warehousing. Dr. Tiwari has over 65 research papers, 2 edited books, and one international patent published to his credit. His current research interest is in machine/deep learning, data mining, pattern recognition, business analytics, and data warehousing.

Dr. Sanjay Misra Sr. Member of IEEE and ACM Distinguished Lecturer, is Professor at Østfold University College (HIOF), Halden, Norway. Before coming to HIOF, he was Professor at Covenant University (400-500 ranked by THE (2019)) for 9 years. He holds a Ph.D. in Information & Knowledge Engineering (Software Engineering) from the University of Alcala, Spain, and an M.Tech. (Software Engineering) from MLN National Institute of Tech, India. Total around 600 articles (SCOPUS/WoS) with 500 co-authors worldwide (-130 JCR/SCIE) in the core & appl. area of Software Engineering, Web engineering, Health Informatics, Cybersecurity, Intelligent systems, AI, etc.

A Review on Rainfall Prediction Using Neural Networks

Sudipta Mandal, Saroj Kumar Biswas, Biswajit Purayastha, Manomita Chakraborty, and Saranagata Kundu

1 Introduction

Rain plays the most vital function in human life during all types of meteorological events [1]. Rainfall is a natural climatic phenomenon that has a massive impact on human civilization and demands precise forecasting [2]. Rainfall forecasting has a link with agronomics, which contributes remarkably to the country's providence [3, 4]. There are three methods for developing rainfall forecasting: (i) Numerical, (ii) Statistical, and (iii) Machine Learning.

Numerical Weather Prediction (NWP) forecasts using computer power [5, 6]. To forecast future weather, NWP computer models process current weather observations. The model's output is formulated on current weather monitoring, which digests into the model's framework and is used to predict temperature, precipitation, and lots of other meteorological parameters from the ocean up to the top layer of the atmosphere [7].

Statistical forecasting entails using statistics based on historical data to forecast what might happen in the future [8]. For forecasting, the statistical method employs

S. Mandal (✉) · S. K. Biswas · B. Purayastha
Computer Science and Engineering, National Institute of Technology, Silchar, Assam, India
e-mail: sudiptamandal1998@gmail.com

S. K. Biswas
e-mail: saroj@cse.nits.ac.in

B. Purayastha
e-mail: biswajit@nits.ac.in

M. Chakraborty
Computer Science and Engineering, VIT-AP Campus, Amaravati, Andhra Pradesh, India
e-mail: Manomita.c@vitap.ac.in

S. Kundu
Computer Science and Engineering, The University of Burdwan, Bardhaman, West Bengal, India

P. Singh et al. (eds.), *Machine Learning and Computational Intelligence Techniques for Data Engineering*, Lecture Notes in Electrical Engineering 998,
https://doi.org/10.1007/978-981-99-0047-3_1

linear time-series data [9]. Each statistical model comes with its own set of limitations. The statistical model, Auto-Regressive (AR), regresses against the series' previous values. The AR term simply informs how many linearly associated lagged observations are there, therefore it's not suitable for data with nonlinear correlations. The Moving Average (MA) model uses the previous error which is used as a describing variable. It keeps track of the past of distinct periods for each anticipated period, and it frequently overlooks intricate dataset linkages. It does not respond to fluctuations that occur due to factors such as cycles and seasonal effects [10]. The ARIMA model (Auto-Regressive Integrated Moving Average) is a versatile and useful time-series model that combines the AR and MA models [11]. Using stationary time-series data, the ARIMA model can only forecast short-term rainfall. Because of the dynamic nature of climatic phenomena and the nonlinear nature of rainfall data, statistical approaches cannot be used to forecast long-term rainfall.

Machine learning can be used to perform real-time comparisons of historical weather forecasts and observations. Because the physical processes which affect rainfall occurrence are extremely complex and nonlinear, some machine learning techniques such as Artificial Neural Network (ANN), Support Vector Machine (SVM), Random Forest Regression Model, Decision Forest Regression, and Bayesian linear regression models are better suited for rainfall forecasting. However, among all machine learning techniques, ANNs perform the best in terms of rainfall forecasting. The usage of ANNs has grown in popularity, and ANNs are one of the most extensively used models for forecasting rainfall. ANNs are a data-driven model that does not need any limiting suppositions about the core model's shape. Because of their parallel processing capacity, ANNs are effective at training huge samples and can implicitly recognize complex nonlinear correlations between conditional and non-conditional variables. This model is dependable and robust since it learns from the original inputs and their relationships with unseen data. As a result, ANNs can estimate the approximate peak value of rainfall data with ease.

This paper presents the different rainfall forecasting models proposed using ANNs and highlights some special features observed during the survey. This study also reports the suitability of different ANN architectures in different situations for rainfall forecasting. Besides, the paper finds the weather parameters responsible for rainfall and discusses different issues in rainfall forecasting using machine learning. The paper has been assembled with the sections described as follows. Section 2 discusses the literature survey of papers using different models. Section 3 discusses the theoretical analysis of the survey and discussion and, at last, the conclusion part is discussed in Sect. 4. Future scope of this paper is also discussed in Sect. 4.

2 Literature Survey

Rainfall prediction is one of the most required jobs in the modern world. In general, weather and rainfall are highly nonlinear and complex phenomena, which require the latest computer modeling and recreation for their accurate prediction. An Artificial

Neural Network (ANN) can be used to foresee the behavior of such nonlinear systems. Soft computing hands out with estimated models where an approximation answer and result are obtained. Soft computing has three primary components; those are Artificial Neural Network (ANN), Fuzzy logic, and Genetic Algorithm. ANN is commonly used by researchers in the field of rainfall prediction. The human brain is highly complex and nonlinear. On the other hand, Neural Networks are simplified models of biological neuron systems. A neural network is a massively parallel distributed processor built up of simple processing units, which has a natural tendency for storing experiential knowledge and making it available for use. Many researchers have attempted to forecast rainfall using various machine learning models. In most of the cases, ANNs are used to forecast rainfall. Table 1 shows some types of ANNs like a Backpropagation Neural Network (BPNN) and Convolutional Neural Network (CNN) that are used based on the quality of the dataset and rainfall parameters for better understanding and comprehensibility. Rainfall accuracy is measured using accuracy measures such as MSE and RMSE.

One of the most significant advancements in neural networks is the Backpropagation learning algorithm. For complicated, multi-layered networks, this network remains the most common and effective model. One input layer, one output layer, and at least one hidden layer make up the backpropagation network. The capacity of a network to provide correct outputs for a given dataset is determined by each layer's neuron count and the hidden layer's number. The raw data is divided into two portions, one for training purpose and the other for testing purpose of the model. Vamsidhar et al. [1] have proposed an efficient rainfall prediction system using BPNN. They created a 3-layered feedforward neural network architecture and initialized the weights of the neural network. A 3-layered feedforward neural network architecture was created by initializing the weights of the neural network by random values between -1.0 and 1.0. Monthly rainfall from the year 1901 to the year 2000 was used here. Using humidity, dew point, and pressure as input parameters, they obtained an accuracy of 99.79% in predicting rainfall and 94.28% for testing purposes. Geeta et al. [2] have proposed monthly monsoon rainfall for Chennai, using the BPNN model. Chennai's monthly rainfall data from 1978 to 2009 were taken for the desired output data for training and testing purposes. Using wind speed, mean temperature, relative humidity, and aerosol values (RSPM) as rainfall parameter, they got a prediction of 9.96 error rate. Abhishek et al. [3] have proposed an Artificial Neural Network system-based rainfall prediction model. They concluded that when there is an increase in the number of hidden neurons in ANN, then MSE of that model decreases. The model was built by five sequential steps: 1. Input and the output data selection for supervised backpropagation learning. 2. Input and the output data normalization. 3. Normalized data using Backpropagation learning. 4. Testing of fit of the model. 5. Comparing the predicted output with the desired output. Input parameters were the average humidity and the average wind speed for the 8 months of 50 years for 1960–2010. Back Propagation Algorithm (BPA) was implemented in the nftools, and they obtained a minimum $MSE = 3.646$. Shrivastava et al. [4] have proposed a rainfall prediction model using backpropagation neural network. They used rainfall data from Ambikapur region of Chhattisgarh, India. They concluded that

Table 1 Different rainfall forecasting models using neural network

Year	Authors	Region used	Dataset	Model used	Parameters used	Accuracy
2010	E. Vamsidhar et al. [1]	Rainfall data of time period 1901–2000	Website (www.tyndall.ac.uk)	Backpropagation neural network	Humidity, Dew point, and Pressure	Obtained 99.79% of accuracy and 94.28% in testing
2011	G. Geetha et al. [2]	Case study of Chennai	32 years of monthly mean data	Backpropagation neural network	Wind speed, mean temperature, relative humidity, aerosol values (RSPM)	Prediction of error was 9.96% only
2012	K. Abhishek et al. [3]	Udipi district of Karnataka	50 years, 1960–2010	Backpropagation Algorithm (BPA)	Average humidity and the average wind speed	Accuracy measure—MSE (Mean Squared Error) MSE = 3.646
2013	G. Shrivastava et al. [4]	Ambikapur region of Chhattisgarh, India	1951 to 2012	Backpropagation Neural (BPN)	Not mentioned	94.4% of L.P.A
2015	A. Sharma et al. [5]	Region of Delhi (India)	Year not given, 365 samples	Backpropagation Neural (BPN)	Temperature, humidity, wind speed, pressure, dew point	Accuracy Graph plotted with NFTOOL. MSE = 8.70
2015	A. Chaturvedi [6]	Region of Delhi (India)	Monsoon period from May to September	Backpropagation Feedforward Neural Network	Humidity, wind speed	MSE = 8.70
2018	Y. A. Lesnussaa et al. [7]	Ambon City	Monthly data, 2011–2015	Backpropagation Feedforward Neural Network	Air temperature, wind speed, air pressure	MSE = 0.022
1998	S Lee et al. [8]	Switzerland	367 locations based on the daily rainfall at nearby 100 locations	Radial basis function neural network	Rainfall data of 4 regions	RMSE = 78.65 Relative Error—0.46 Absolute Error—55.9

(continued)

Table 1 (continued)

Year	Authors	Region used	Dataset	Model used	Parameters used	Accuracy
2006	C. Lee et al. [9]	Taipei City in Taiwan	Daily rainfall-runoff of Taipei	Radial basis function neural network	Rainfall frequency, the amount of runoff, the water continuity, and the reliability	Success Rate—98.6%
2015	Liu et al. [10]	Handan city	Monthly rainfall data on 39-year in Handan city	Radial basis function neural network	Rainfall Data	IMF Signal Graph
2017	M. Qiu et al. [10]	Manizales city	Daily accumulated rainfall data	Convolutional Neural Networks	Rain gauges	RMSE = 11.253 (Root Mean Square Error)
2018	A. Haider et al. [12]	Bureau of Meteorology	Not mentioned	One-dimensional Deep Convolutional Neural Network	Mean minimum temperature (MinT) and the mean maximum temperature (MaxT)	RMSE = 15.951
2018	S. Aswin et al. [13]	Not Given	468 months	Convolutional Neural Networks	Precipitation	RMSE = 2.44
2020	C. Zhang et al. [14]	Shenzhen, China,	2014 to 2016 (March to September),	Deep convolutional neural network	Gauge rainfall and Doppler radar echo map	RMSE = 9.29
2019	R. Kaneko. et al. [15]	Kyushu region in Japan	From 2016 to 2018	2-layer stacked LSTMs	Wind direction and wind velocity, temperature, precipitation, pressure, and relative humidity	RMSE = 2.07
2020	A. Pranolo et al. [16]	Tenggarong of East Kalimantan-Indonesia	1986 to 2008	A Long Short-Term Memory	Not mentioned	RMSE = 0.2367

(continued)

Table 1 (continued)

Year	Authors	Region used	Dataset	Model used	Parameters used	Accuracy
2020	I. Salehin et al. [17]	Bangladesh Meteorological Department	2020 (1 Aug to 31 Aug)	Long Short-Term Memory	Temperature, dew point, humidity, wind properties (pressure, speed, and direction)	76% accuracy
2020	A. Samad et al. [18]	Albany, Walpole, and Witchcliffe	2007–2015 for training, 2016 for testing	Long Short-Term Memory	Temperature, pressure, humidity, wind speed, and wind direction	MSE, RMSE, and MAE RMSE = 5.343
2020	D. Zatusiva et al. [19]	East Java, Indonesia	December 29, 2014 to August 4, 2019	Long Short-Term Memory	El Nino Index (NI) and Indian Ocean Dipole (IOD)	MAAPE = 0.9644
2019	S. Poornima et al. [20]	Hyderabad, India	Hyderabad region starting from 1980 until 2014	Intensified Long Short-Term Memory	Maximum temperature, minimum temperature, maximum relative humidity, minimum relative humidity, wind speed, sunshine and	Accuracy—87.99%

BPN is suitable for the identification of internal dynamics of high dynamic monsoon rainfall. The performance of the model was evaluated by comparing Standard Deviation (SD) and Mean Absolute Deviation (MAD). Based on backpropagation, they were able to get 94.4% accuracy. Sharma et al. [5] have proposed a rainfall prediction model on backpropagation neural network by using Delhi's rainfall data. The input and target data had to be normalized because of having different units. By using temperature, humidity, wind speed, pressure, and dew point as input parameters of the prediction model, MSE was approximately 8.70 and accuracy graph was plotted with NFTools. Chaturvedi [6] has proposed rainfall prediction using backpropagation neural network. He took 70% of data for training purpose, 15% for testing, and other 15% for validation purpose. The input data for the model consisted of 365 samples within that testing purpose of 255 samples, 55 samples for testing, and the rest samples are for validation purpose. He plotted a graph using NFTools among the predicted value and the target values which showed a minimized MSE of 8.7. He also concluded increase in the neuron number of the network shows a decrease in MSE of the model. Lessnussaa et al. [7] have proposed a rainfall prediction using backpropagation neural network in Ambon city. The researchers have used monthly rainfall data from 2011 to 2015 and considered weather parameters such as air temperature, air velocity, and pressure. They got a result of accuracy 80% by using alpha 0.7, iteration number (in terms of epoch) 10,000, and also MSE value 0.022.

Radial Basis Function Networks are a class of nonlinear layered feedforward networks. It is a different approach which views the design of neural network as a curve fitting problem in a high-dimensional space. The construction of a RBF network involves three layers with entirely different roles: the input layer, the only hidden layer, and the output layer. Lee et al. [8] have proposed rainfall prediction using an artificial neural network. The dataset has been taken from 367 locations based on the daily rainfall at nearly 100 locations in Switzerland. They proposed a divide-and-conquer approach where the whole region is divided into four sub-areas and each is modeled with a different method. For two larger areas, they used radial basis function (RBF) networks to perform rainfall forecasting. They achieved a result of RMSE of the whole dataset: 78.65, Relative and Absolute Errors, Relative Error—0.46, Absolute Error—55.9 from rainfall prediction. For the other two smaller sub-areas, they used a simple linear regression model to predict the rainfall. Lee et al. [9] have proposed "Artificial neural network analysis for reliability prediction of regional runoff utilization". They used artificial neural networks to predict regional runoff utilization, using two different types of artificial neural network models (RBF and BPNN) to build up small-area rainfall–runoff supply systems. A historical rainfall for Taipei City in Taiwan was applied in the study. As a result of the impact variances between the results used in training, testing, and prediction and the actual results, the overall success rates of prediction are about 83% for BPNN and 98.6% for RBF. Liu Xinia et al. [10] have proposed Filtering and Multi-Scale RBF Prediction Model of Rainfall Based on EMD Method, a new model based on empirical mode decomposition (EMD) and the Radial Basis Function Network (RBFN) for rainfall prediction. They used monthly rainfall data for 39 years in Handan city. Therefore, the results obtained were evidence of the fact that the RBF network can be successfully applied to determine the relationship between rainfall and runoff.

Convolutional Neural Network (ConvNet/CNN) is a well-known deep learning algorithm which takes an input image, sets some relevance (by using learnable weights and biases) to different aspects/objects in the image, and discriminates among them. CNN is made up of different feedforward neural network layers, such as convolution, pooling, and fully connected layers. CNN is used to predict rainfall for time-series rainfall data. Qiu et al. [11] have proposed a multi-task convolutional neural networks-based rainfall prediction system. They evaluated two real-world datasets. The first one was the daily collected rainfall data from the meteorological station of Manizales city. Another was a large-scale rainfall dataset taken from the observation sites of Guangdong province, China. They got a result of RMSE = 11.253 in their work. Halder et al. [12] have proposed a one-dimensional Deep Convolutional Neural Network based on a monthly rainfall prediction system. Additional local attributes were also taken like Mint and MaxT. They got a result of RMSE = 15.951 in their work. Aswin et al. [13] have proposed a rainfall prediction model using a convolutional neural network. They used precipitation as an input parameter and using 468 months of precipitation as an input parameter, they got an RMSE accuracy of 2.44. Ghang et al. [14] have proposed a rainfall prediction model using deep convolutional neural network. They collected this rainfall data from the meteorological observation center in Shenzhen, China, for the years 2014 to 2016 from March to September. They got RMSE = 9.29 for their work. They have concluded that Tiny-RainNet model's overall performance is better than fully connected LSTM and convolutional LSTM.

Recurrent Neural Network is an abstraction of feedforward neural networks that possess intrinsic memory. RNN is recurring as it brings about the same function for every input data and the output depends on the past compilation. After finding the output, it is copied and conveyed back into the recurrent network unit.

LSTM is one of the RNNs that has the potential to forecast rainfall. LSTM is a component of the Recurrent Neural Network (RNN) layer, which is accustomed to addressing the gradient problem by forcing constant error flow. A LSTM unit is made up of three primary gates, each of which functions as a controller for the data passing through the network, making it a multi-layer neural network. Kaneko et al. [15] have proposed a 2-layer stacked RNN-LSTM-based rainfall prediction system with batch normalization. The LSTM model performance was compared with MSM (Meso Scale Model by JMA) from 2016 to 2018. The LSTM model successfully predicted hourly rainfall and surprisingly some rainfall events were predicted better in the LSTM model than MSM. RMSE of the LSTM model and MSM were 2.07 mm h-1 and 2.44 mm h-1, respectively. Using wind direction and wind velocity, temperature, precipitation, pressure, and relative humidity as rainfall parameters, they got an RMSE of 2.07. Pranolo et al. [16] have proposed a LSTM model for predicting rainfall. The data consisted of 276 data samples, which were subsequently separated into 216 (75%) training datasets for the years 1986 to 2003, and 60 (25%) test datasets for the years 2004 to 2008. In this study, the LSTM and BPNN architecture included a hidden layer of 200, a maximum epoch of 250, gradient threshold of 1, and learning rate of 0.005, 0.007, and 0.009. These results clearly indicate the advantages of the LSTM produced good accuracy than the BPNN algorithm. They got a result of RMSE = 0.2367 in their work. Salehin et al. [17] have proposed a LSTM and Neural Network-based rainfall prediction system. Time-series forecasting with LSTM is a modern approach to building a rapid model of forecasting. After analyzing all data using LSTM, they found 76% accuracy in this work. LSTM networks are suitable for time-series data categorization, processing, and prediction. So, they concluded that LSTM gives the most controllability and thus better results were obtained. Samad et al. [18] have proposed a rainfall prediction model using Long Short-Term memory. Using temperature, pressure, humidity, wind speed, and wind direction as input parameters on the rainfall data of years 2007–2015, they got an accuracy of RMSE 5.343. Haq et al. [19] have proposed a rainfall prediction model using long short-term memory based on El Nino and IOD Data. They used 60% training data with variation in the hidden layer, batch size, and learn rate drop periods to achieve the best prediction results. They got an accuracy of MAAPE = 0.9644 in their work. S. Poornima [20] has proposed an article named "Prediction of Rainfall Using Intensified LSTM-Based Recurrent Neural Network with Weighted Linear Units". This paper presented Intensified Long Short-Term Memory (Intensified LSTM)-based Recurrent Neural Network (RNN) to predict rainfall. The parameters considered for the evaluation of the performance and the efficiency of the proposed rainfall prediction model were Root Mean Square Error (RMSE), accuracy, number of epochs, loss, and learning rate of the network. The initial learning rate was fixed to 0.1, and no momentum was set as default, with a batch size of 2500 undergone for 5 iterations since the total number of rows in the dataset is 12,410 consisting of 8 attributes. The accuracy achieved by the Intensified LSTM-based rainfall prediction model is 87.99%.

For prediction, all of these models use nearly identical rainfall parameters. Humidity, wind speed, and temperature are important parameters for backpropagation [21]. Temperature and precipitation are important factors in convolutional neural networks (Covnet). Temperature, wind speed, and humidity are all important factors for Recurrent Neural Network (RNN) and Long Short-Term Memory (LSTM) networks. In most of the cases, accuracy measures such as MSE, RMSE, and MAE are used. With temperature, air pressure, humidity, wind speed, and wind direction as input parameters, BPNN has achieved an accuracy of 2.646, CNN has achieved an accuracy of 2.44, and LSTM has achieved a better accuracy of RMSE = 0.236. As a result, from this survey it can be said that LSTM is an effective model for rainfall forecasting.

3 Theoretical Analysis of Survey and Discussion

High variability in rainfall patterns is the main problem of rainfall forecasting. Data inefficiency and absence of the records like temperature, wind speed, and wind directions can affect prediction [22, 23]. So, data preprocessing is required for compensating the missing values. As future data is unpredictable, models have to use estimated data and assumptions to predict future weather [24]. Besides massive deforestation, abrupt changes in climate conditions may prove the prediction false. In the case of the yearly rainfall dataset, there is no manageable procedure to determine rainfall parameters such as wind speed, humidity, and soil temperature. In some models, researchers have used one hidden layer, and for that large number of hidden nodes are required and performance gets minimized. To compensate this, 2 hidden layers are used. More than 2 hidden layers give the same results. Either a few or more input parameters can influence the learning or prediction capability of the network [25]. The model simulations use dynamic equations which demonstrate how the atmosphere will respond to changes in temperature, pressure, and humidity over time. Some of the frequent challenges while implementing several types of ANN architecture for modeling weekly, monthly, and yearly rainfall data are such as hidden layer and node count, and training and testing dataset division. So, prior knowledge about these methods and architectures is needed. As ANNs are prone to overfitting problems, this can be reduced by early stopping or regularizing methods. Choosing accurate performance measures and activation functions for simulation are also an important part of rainfall prediction implementation.

4 Conclusions

This paper considers a study of various ANNs used by researchers to forecast rainfall. The survey shows that BPN, CNN RNN, LSTM, etc. are suitable to predict rainfall than other forecasting techniques such as statistical and numerical methods. Moreover, this paper discussed the issues that must be addressed when using ANNs for rainfall forecasting. In most cases, previous daily data of rainfall and maximum and minimum temperature, humidity, and wind speed are considered. All the models provide good prediction accuracy, but as the models progress from neural networks to deep learning, the accuracy improves, implying a lower error rate. Finally, based on the literature review, it can be stated that ANN is practical for rainfall forecasting because several ANN models have attained significant accuracy. RNN shows better accuracy as there are memory units incorporated, so it can remember the past trends of rainfall. Depending on past trends, the model gives a more accurate prediction. Accuracy can be enhanced even more if other parameters are taken into account. Rainfall prediction will be more accurate as ANNs progress, making it easier to understand weather patterns.

From this research work after analyzing all the results from these mentioned research papers, it can be concluded that neural networks perform better, so, for further works, rainfall forecasting implementation will be done by using neural networks. If RNN and LSTM are used, then forecasting would be better for their additional memory unit. So, for the continuation of this paper, rainfall forecasting of a particular region will be done using LSTM. And additionally, there will be a comparative study with other neural networks for a better understanding of the importance of artificial neural networks in rainfall forecasting.

References

1. Vamsidhar E, Varma KV, Rao PS, Satapati R (2010) Prediction of rainfall using back propagation neural network model. Int J Comput Sci Eng 02(04):1119–1121
2. Geetha G, Samuel R, Selvaraj (2011) Prediction of monthly rainfall in Chennai using back propagation neural network model. Int J Eng Sci Technol 3(1):211–213
3. Abhishek K, Kumar A, Ranjan R, Kumar S (2012) A rainfall prediction model using artificial neural network. IEEE Control Syst Graduate Res Colloquium. https://doi.org/10.1109/ICSGRC.2012.6287140
4. Shrivastava G, Karmakar S, Kowar MK, Guhathakurta P (2012) Application of artificial neural networks in weather forecasting: a comprehensive literature review. IJCA 51(18):0975–8887. https://doi.org/10.5120/8142-1867
5. Sharma A, Nijhawan G (2015) Rainfall prediction using neural network. Int J Comput Sci Trends Technol (IJCST) 3(3), ISSN 2347–8578
6. Chaturvedi A (2015) Rainfall prediction using back propagation feed forward network. Int J Comput Appl (0975 – 8887) 119(4)
7. Lesnussa YA, Mustamu CG, Lembang FK, Talakua MW (2018) Application of backpropagation neural networks in predicting rainfall data in Ambon city. Int J Artif Intell Res 2(2). ISSN 2579–7298
8. Lee S, Cho S, Wong PM (1998) Rainfall prediction using artificial neural network. J Geog Inf Decision Anal 2:233–242
9. Lee C, Lin HT (2006) Artificial neural network analysis for reliability prediction of regional runoff utilization. S. CIB W062 symposium 2006
10. Xinia L, Anbing Z, Cuimei S, Haifeng W (2015) Filtering and multi-scale RBF prediction model of rainfall based on EMD method. ICISE 2009:3785–3788
11. Qiu M, Zha P, Zhang K, Huang J, Shi X, Wa X, Chu W (2017) A short-term rainfall prediction model using multi-task convolutional neural networks. In: IEEE international conference on data mining. https://doi.org/10.1109/ICDM.2017.49
12. Haidar A, Verma B (2018) Monthly rainfall forecasting using one-dimensional deep convolutional neural network. Project: Weather Forecasting using Machine Learning Algorithm, UNSW Sydney. https://doi.org/10.1109/ACCESS.2018.2880044
13. Aswin S, Geetha P, Vinayakumar R (2018) Deep learning models for the prediction of rainfall. In: International conference on communication and signal procesing. https://doi.org/10.1109/ICCSP.2018.8523829,2018
14. Zhang CJ, Wang HY, Zeng J, Ma LM, Guan L (2020) Tiny-RainNet: a deep convolutional neural network with bi-directional long short-term memory model for short-term rainfall prediction. Meteorolog Appl 27(5)
15. Kaneko R, Nakayoshi M, Onomura S (2019) Rainfall prediction by a recurrent neural network algorithm LSTM learning surface observation data. Am Geophys Union, Fall Meeting

16. Pranolo A, Mao Y, Tang Y, Wibawa AP (2020) A long short term memory implemented for rainfall forecasting. In: 6th international conference on science in information technology (ICSITech). https://doi.org/10.1109/ICSITech49800.2020.9392056
17. Salehin I, Talha IM, Hasan MM, Dip ST, Saifuzzaman M, Moon NN (2020) An artificial intelligence based rainfall prediction using LSTM and neural network. https://doi.org/10.1109/WIECON-ECE52138.2020.9398022
18. Samad A, Gautam V, Jain P, Sarkar K (2020) An approach for rainfall prediction using long short term memory neural network. In: IEEE 5th international conference on computing communication and automation (ICCCA) Galgotias University, GreaterNoida,UP, India. https://doi.org/10.1109/ICCCA49541.2020.9250809
19. Haq DZ, Novitasari DC, Hamid A, Ulinnuha N, Farida Y, Nugraheni RD, Nariswari R, Rohayani H, Pramulya R, Widjayanto A (2020) Long short-term memory algorithm for rainfall prediction based on El-Nino and IOD Data. In: 5th international conference on computer science and computational intelligence
20. Poornima S, Pushpalatha M (2019) Prediction of rainfall using intensified LSTM based recurrent neural network with weighted linear units. Comput Sci Atmos. 10110668
21. Parida BP, Moalafhi DB (2008) Regional rainfall frequency analysis for Botswana using L-Moments and radial basis function network. Phys Chem Earth Parts A/B/C 33(8). https://doi.org/10.1016/j.pce.2008.06.011
22. Dubey AD (2015) Artificial neural network models for rainfall prediction in Pondicherry. Int J Comput Appl (0975–8887). 10.1.1.695.8020
23. Biswas S, Das A, Purkayastha B, Barman D (2013) Techniques for efficient case retrieval and rainfall prediction using CBR and Fuzzy logic. Int J Electron Commun Comput Eng 4(3):692–698
24. Basha CZ, Bhavana N, Bhavya P, Sowmya V (2020) Proceedings of the international conference on electronics and sustainable communication systems. IEEE Xplore Part Number: CFP20V66-ART; ISBN: 978-1-7281-4108-4.
25. Biswas SK, Sinha N, Purkayastha B, Marbaniang L (2014) Weather prediction by recurrent neural network dynamics. Int J Intell Eng Informat Indersci, 2(2/3):166–180 (ESCI journal)

Identifying the Impact of Crime in Indian Jail Prison Strength with Statical Measures

Sapna Singh kshatri and Deepak Singh

1 Introduction

The use of machine learning algorithms to forecast any crime is becoming commonplace. This research is separated into two parts: the forecast of violent crime and its influence in prison, and the prediction of detainees in jail. We are using data from a separate source. The first two datasets used are violent crime and total FIR data from the police department, followed by data on prisoners and detainees sentenced for violent crimes from the Jail Department.

A guide for the correct use of correlation in crime and jail strength is needed to solve this issue. Data from the NCRB shows how correlation coefficients can be used in real-world situations. As shown in Fig. 1, a correlation coefficient will be used for the forecast of crime and the prediction of jail overcrowding.

Regression and correlation and are two distinct yet complementary approaches. In general, regressions are used to make predictions (which do not extend beyond the data used in the research), whereas correlation is used to establish the degree of link. There are circumstances in which the x variable is neither fixed nor readily selected by the researcher but is instead a random covariate of the y variable [1]. In this article, the observer's subjective features and the latest methodology are used. The beginnings and increases of crime are governed by age groups, racial backgrounds, family structure, education, housing size [2], employed-to-unemployed ratio, and cops per capita. Rather than systematic classification to categorize under the impressionistic

S. S. kshatri (✉)
Department of Computer Science and Engineering (AI), Shri Shankaracharya Institute of Professional Management and Technology, Raipur, C.G., India
e-mail: sapna.singh@ssipmt.com

D. Singh
Department of Computer Science and Engineering, National Institute of Technology, Raipur, C.G., India
e-mail: dsingh.cs@nitrr.ac.in

P. Singh et al. (eds.), *Machine Learning and Computational Intelligence Techniques for Data Engineering*, Lecture Notes in Electrical Engineering 998,
https://doi.org/10.1007/978-981-99-0047-3_2

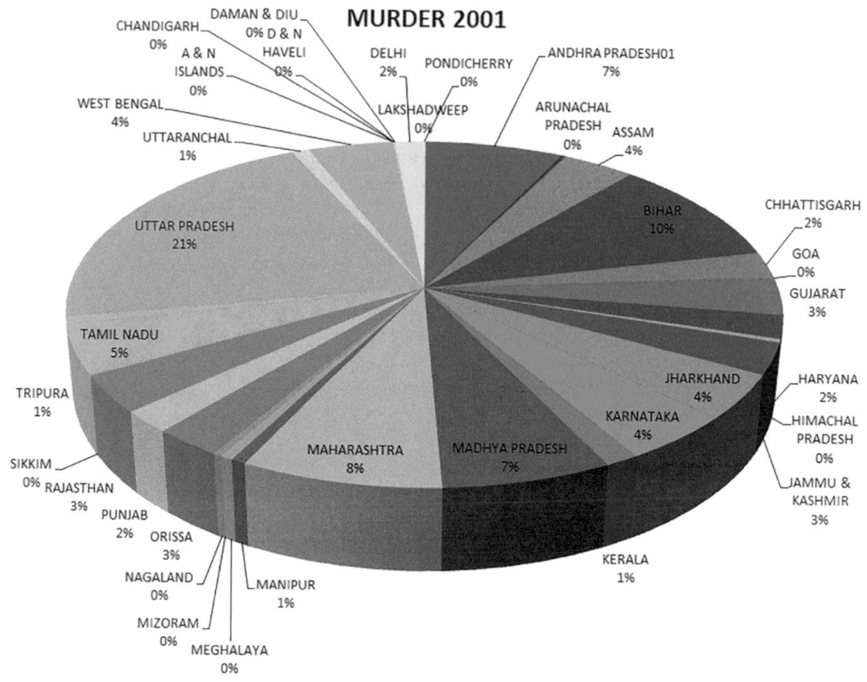

Fig. 1 Violence Crime state-wide in 2001

literary paradigm, we present a quick overview of a few of these measures in this section. This is not a thorough assessment; rather, it emphasizes measurements that demonstrate the many techniques that have been offered. Logical associates use the term correlation to refer to affiliation, association, or, on the other hand, any partnership, relation, or correspondence. This extensional everyday meaning now means that researchers in science are misusing the factual word "correlation." Correlation is so widely used that some analysts say they never invented it [3]. Crime percentages are widely used to survey the threat of crime by adjusting for the population at risk of this property, which defines the crime percentage as one of the most well-known indicators of crime investigation [4]. Is the rate of various crimes a result of the progress in the number of prisoners in jails?

The correlation coefficient is the statistical extent of similarity between two continuous variables. There are several types of correlation coefficients, but the most common is the Pearson correlation coefficient, denoted by the symbol.

A correlation matrix is a simple way; it summarizes the correlation between all variables in a dataset. We have a dataset of statistics about crime rates and jail population staring at us. Fortunately, a correlation matrix helps us understand in quickly comprehending the relationship between each variable. A key principle of multiple linear regression is that no independent variables in the model are highly associated with one another. Correlation describes the link between defined values of one variable (the independent, explanatory, regressor, exogenous, carrier, or predictor) and the means of all corresponding values of the second variable (the dependent, outcome, response variable, or variable being explained). On the other hand, regression expresses the link between specific values of one variable (the independent, predictor, explanatory, exogenous, regressor, or carrier) and the means of all corresponding values. We could argue that investigating interdependence leads to an examination of correlations in general. The concept of regression is derived from the study of dependency. When the x variable is a random covariate to the y variable, we are more concerned with establishing the linear relationship's strength than with prediction. That is, x and y vary together (continuous variables), and the sample correlation coefficient, r_{xy} (r), is the statistics used for this purpose.

The purpose of this study is to compare crime and prison data in order to determine the relationship so that we can reduce the stench of the prison. We have divided the rest of the work as follows: in the following section, the literature on artificial intelligence and deep learning method, which is one of the crime predictions researched, is discussed. The third section discusses the dataset's content and defines the problem. Following a discussion of the problem, the fourth section explains the methodology of the correlation employed in the study. Findings are presented in the fifth section.

2 Related Work

The classification paradigm was the clinical/theoretical approach, which began to include philosophy and surveys. The justice system uses forecasting models to predict future crime and recidivism, allocate resources effectively, and provide risk-based and criminogenic treatment programs [5]. A novel path-building operator is provided as a single population, which is integrated with MMAS and ACS to produce a three-colony Ant Colony Optimization method. The method uses the Pearson correlation coefficient as the evaluation criterion, choosing two colonies with the highest similarity, and rewarding the parameters of the standard route in the two colonies to accelerate convergence. We may adaptively manage the frequency of information transmission between two colonies based on the dynamic feedback of the diversity among colonies to ensure the ant colony algorithm finds better solutions [6].

Urban crime is an ongoing problem that concerns people all around the globe. The article examines the impact of overlapping noises (local outliers and irregular waves) in actual crime data and suggests a DuroNet encoder-decoder network for capturing deep crime patterns. We compare two real-world crime databases from New York and Chicago. DuroNet model outperforms previous approaches in high accuracy and robustness [7]. The current work developed and implemented machine learning models using machine learning algorithms (ensemble and simile), namely SMO, J48, and Naive Bayes in an ensemble SVM-bagging, SVM-Random-forest, and SVM-stacking (C4.5, SMO, J48). Each preset element is included in a dataset for training on violent crime (murder, rape, robbery, etc.). After successfully training and verifying six models, we came to a significant conclusion [8]. J. Jeyaboopathiraja et al. present the police department used big data analytics (BDA), support vector machine (SVM), artificial neural networks (ANNs), K-means algorithm Naive Bayes. AI (machine learning) and DL (deep learning approaches). This study's aim is to research the most accurate AI and DL methods for predicting crime rates and the application of data approaches in attempts to forecast crime, with a focus on the dataset.

Modern methods based on machine learning algorithms can provide predictions in circumstances where the relationships between characteristics and outcomes are complex. Using algorithms to detect potential criminal areas, these forecasts may assist politicians and law enforcement to create successful programs to minimize crime and improve the nation's growth. The goal of this project is to construct a machine learning system for predicting a morally acceptable output value. Our results show that utilizing FAMD as a feature selection technique outperforms PCA on machine learning classifiers. With 97.53 percent accuracy for FAMD and 97.10 percent accuracy for PCA, the naive Bayes classifier surpasses other classifiers [9]. Retrospective models employ past crime data to predict future crime. These include hotspot approaches, which assume that yesterday's hotspots are likewise tomorrow's. Empirical research backs this up: although hotspots may flare up and quiet down quickly, they tend to stay there over time [10].

Prospective models employ more than just historical data to examine the causes of major crime and build a mathematical relationship between the causes and levels of crime. Future models use criminological ideas to anticipate criminal conduct. As a consequence, these models should be more relevant and provide more "enduring" projections [11]. Previous models used either socioeconomic factors (e.g., RTM [15]) or near-repeat phenomena (e.g., Promap [12]; PredPol [13]). The term "near-repeat" refers to a phenomenon when a property or surrounding properties or sites are targeted again shortly after the initial criminal incident.

Another way, Drones may also be used to map cities, chase criminals, investigate crime scenes and accidents, regulate traffic flow, and search and rescue after a catastrophe. In Ref. [14], legal concerns surrounding drone usage and airspace allocation are discussed. The public has privacy concerns when the police acquire power and influence. Concerns concerning drone height are raised by airspace dispersal. These include body cameras and license plate recognition. In Ref. [15], the authors state that face recognition can gather suspect profiles and evaluate them from various databases. A license plate scanner may also get data on a vehicle suspected of committing a crime. They may even employ body cameras to see more than the human eye can perceive, meaning the reader sees and records all a cop sees. Normally, we cannot recall the whole picture of an item we have seen. The influence of body cameras on officer misbehaviors and domestic violence was explored in Ref. [16]. Patrol personnel now have body cameras. Police misconduct protection. However, wearing a body camera is not just for security purposes but also to capture crucial moments during everyday activities or major operations.

While each of these ways is useful, they all function separately. While the police may utilize any of these methods singly or simultaneously, having a device that can combine the benefits of all of these techniques would be immensely advantageous. Classification of threats, machine learning, deep learning, threat detection, intelligence interpretation, voice print recognition, natural language processing Core analytics, Computer linguistics, Data collection, Neural networks Considering all of these characteristics is critical for crime prediction.

3 Methods and Materials

3.1 *Dataset*

The dataset in this method consists of 28 states and seven union territories. As a result, the crime has been split into parts. We chose some of the pieces that were held in the category of Violence Crime for our study. A type for the Total Number of FIRs has also been included. The first dataset was gathered from the police and prison departments, and it is vast. Serious sequential data are typically extensive in size, making it challenging to manage terabytes of data every day from various crimes. Time series modeling is performed by using classification models, which simplify the data and enable it to model to construct an outcome variable. Data from 2001 to 2015 was plotted in an excel file in a state-by-state format. The most common crime datasets are chosen from a large pool of data. Within the police and jail departments, violent offenses are common—one proposed arduous factor for both departments. Overcrowding refers to a problem defining and describing thoughts, and it is also linked to externally focused thought. This study aimed to investigate the connection between violent crime, FIR, and strength in jail. There are some well-documented psychological causes of aggression. For example, both impulsivity and anger have indeed been linked to criminal attacks [17].

The frequent crime datasets are selected from huge data [17]. A line graph is used to analyze the total IPC crimes for each state (based on districts) from 2001 to 2015. The attribute "States? UT" is used to generate the data, as compared to the attribute "average FIR." The supervised and unsupervised data techniques are used to predict crime accurately from the collected data [18, 19] (Table 1).

3.2 *Experimental Work*

The imported dataset is pictured with the class attribute being STATE/UT. The representation diagram shows the distribution of attribute STATE/UT with different attributes in the dataset; each shade in the perception graph represents a specific state. The imported dataset is pictured; the representation diagram shows the circulation of crime as 1–5 levels specific attributes with class attributes which are people captured during the year.

Table 1 Types of violence crime

Number	Types of violence crime
1	Chnot amounting to murder
2	Riots
3	Rape
4	Murder
5	Prep
6	Dowry death
7	Robbery
8	Dacoity
9	Kidnapping and abduction
10	Assembly for dacoity
11	Arson
12	Attempt to murder

The blue region in the chart represents high crime like murder, and the pink area represents low crime like the kidnapping of a particular attribute in the dataset. Police data create a label for murder, attempt to murder, and dowry death as 1—the rape, 2—attempt to rape, 3—dacoity, assembly to dacoity, and, likewise, up to 5 (Fig. 2).

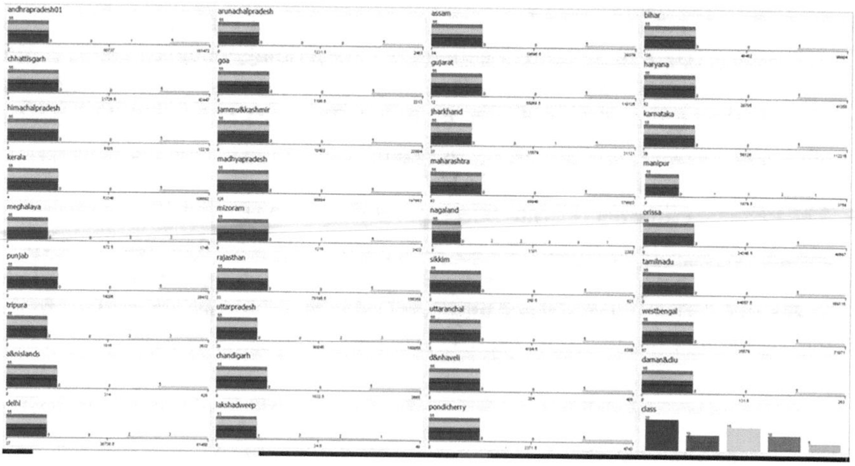

Fig. 2 Violence crime visualizations of 2021 of different states of india

3.3 Correlation Coefficient Between Two Random Variables

The dataset with crime rates and the prison population is gazing at us. Fortunately, a correlation matrix can assist us in immediately comprehending the relationship between each variable. One basic premise of multiple linear regression is that no independent variable in the model is substantially associated with another variable.

Numerous numerical techniques are available to help you understand how effectively a regression equation relates to the data, in addition to charting. The sample coefficient of determination, R^2, of a linear regression fit (with any number of predictors) is a valuable statistic to examine. Assuming a homoscedastic model ($w_i = 1$), R^2 is the ratio between SS_{Reg} and S_{yy}, the sum of squares of deviations from the mean (S_{yy},) accounted for by regression [1].

The primary objective behind relapse is to demonstrate and examine the relationship between the dependent and independent variables. The errors are proportionally independent and normally distributed with a mean of 0 and variance σ. By decreasing the error or residual sums of squares, the βs are estimated:

$$S(\beta_0, \beta_{1,\dots\dots\beta_m}) = \sum\nolimits_{i=1}^{n} \left(Y_i - (\beta_0 + \sum\nolimits_{j=1}^{k} \beta_j X_{ij}) \right) \tag{1}$$

To locate the base of (2) regarding β, the subsidiary of the capacity in (2), as for each of the βs, is set to zero and tackled. This gives the accompanying condition:

$$\frac{\delta s}{\delta\beta|\hat{\beta}_0, \hat{\beta}_1 \dots \hat{\beta}_m} = -2\sum_{i=1}^{n} \left(Y_i - \left(\hat{\beta}_0 + \sum_{j=1}^{k} \hat{\beta}_j X_{ij} \right) \right) = 0, j = 0, 1, 2 \dots k \tag{2}$$

And

$$\frac{\delta s}{\delta\beta|\hat{\beta}_0, \hat{\beta}_1 \dots \hat{\beta}_m} = -2\sum_{i=1}^{n} \left(Y_i - \left(\hat{\beta}_0 + \sum_{j=1}^{k} \hat{\beta}_j X_{ij} \right) \right) = 0, j = 0, 1, 2 \dots k \tag{3}$$

The $\hat{\beta}$s, the answers for (3) and (4), are the least squares appraisals of the βs.

It is helpful to communicate both the n conditions in (1) and the k + 1 condition in (3) and (4) (which depend on straight capacity of the βs) in a lattice structure. Model (1) can be communicated as

$$y = X\beta + \epsilon \tag{4}$$

where y is the nx1 vector of perception, X is a nx(k + 1) network of autonomous factors (and an additional section of 1 s for the intercept $\hat{\beta}_0$, β is a (k + 1)X_i vector

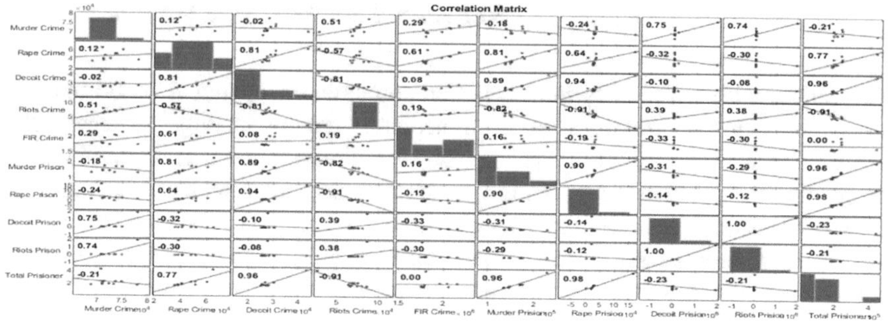

Fig. 3 Correlation matrix between crime data and prison strength

of coefficients, and ε is a X_i vector of free and indistinguishably circulated mistakes related to (1).

At the point when two free variables are exceptionally corresponded, this ends in issue regression. Looking at a grid and outwardly checking whether any of the elements are profoundly connected is one of the most accessible techniques to collinearity outwards containing to recognize a potential multicollinearity issue.

The estimations of the correlation coefficient can extend from −1 to + 1. The closer it is to + 1 or −1, the more intently the two factors are connected. The positive sign means the heading of the relationship; for example, on the off chance that one of the factors expands, the other variable is additionally expected to increment as shown in Fig. 3, correlation matrixes between crime information and jail population. Every cell in the table offers a connection between two explicit factors. For instance, the featured cell beneath indicates that the relationship between "Assault cases in FIR" and "Assault cases in prison" is 0.81, alluded to as multicollinearity. It can make it challenging to decipher the consequences of the that they're unequivocally emphatically connected. Each piece of information is coordinating, however, 0.57, which demonstrates that they're feebly adversely related.

A correlation value of precisely r = 1.000 indicates that the two variables have a completely positive association. When one variable rises, the other rises with it. A correlation value of precisely r = −1.000 indicates that the two variables have a completely negative connection. The value of one variable drops as the value of the other rises.

4 Result and Discussion

The crime expands, and the measure of detainees doesn't diminish; this shows a negative relationship and would, by expansion, have a negative connection coefficient. A positive correlation coefficient would be the relationship between crime and prisoners' strength; as crime increases, so does the prison crowd. As we can see in the correlation matrix, there is no relation between crime and prison strength, so we say there is a weak correlation coefficient in Prison strength. There is no impact on the prison strength of the crime rate, to solve this problem.

The estimations of the correlation coefficient can extend from −1 to + 1. The closer it is to + 1 or −1, the more intently the two factors are connected. The positive sign means the heading of the relationship; for example, on the off chance that one of the factors expands, the other variable is additionally expected to increment as shown in Fig. 3.10, correlation matrixes between crime information and jail quality. Every cell in the table offers a connection between two explicit factors. For instance, the featured cell beneath indicates that the relationship between "Assault cases in FIR" and "Assault cases in prison" is 0.81, alluded to as multicollinearity. It can make it challenging to decipher the consequences of the that they're unequivocally emphatically connected. However, each piece of information is coordinating 0.57, which demonstrates that they're feebly adversely related.

Furthermore, the matrix reveals that the relationship between "dacoity crime" and "dacoity prison" is −0.32, indicating that they are unrelated. Similarly, the relationship coefficients along the incline of the table are mainly equal to 1 since each element is entirely correlated with itself. These cells are not helpful for comprehension.

5 Conclusion

The research moves on to depictions of the survey structure, information estimates, and scaling. Following that, the discussion focuses on information-gathering tactics, crime prediction, and resolving prison overcrowding. Standard statistical metrics and measures are designed specifically for use in criminology research. However, there is no consensus on the best way to measure and compare predictive model outcomes. It is difficult to find the correlation between crime and prison strength crime. We observed that could be improved research in many ways. To enhance crime prediction, we can say that few crimes impact prison strength. This study shows no association between FIR and violent crime in prisoners' strength when established correlations of all violent crimes such as murder, robbery, dacoity, and kidnapping are taken into account.

In future, we can create a model for every correlated crime, not only violent but even with other crimes that directly impact jail.

References

1. Asuero AG, Sayago A, González AG (2006) The correlation coefficient: an overview. Crit Rev Anal Chem 36(1):41–59. https://doi.org/10.1080/10408340500526766
2. Wang Z, Lu J, Beccarelli P, Yang C (2021) Neighbourhood permeability and burglary: a case study of a city in China. Intell Build Int 1–18. https://doi.org/10.1080/17508975.2021.1904202
3. Mukaka MM (2012) Malawi Med J 24, no. September:69–71. https://www.ajol.info/index.php/mmj/article/view/81576

4. Andresen MA (2007) Location quotients, ambient populations, and the spatial analysis of crime in Vancouver, Canada. Environ Plan A Econ Sp 39(10):2423–2444. https://doi.org/10.1068/a38187
5. Clipper S, Selby C (2021) Crime prediction/forecasting. In: The encyclopedia of research methods in criminology and criminal justice, John Wiley & Sons, Ltd, 458–462
6. Zhu H, You X, Liu S (2019) Multiple ant colony optimization based on pearson correlation coefficient. IEEE Access 7:61628–61638. https://doi.org/10.1109/ACCESS.2019.2915673
7. Hu K, Li L, Liu J, Sun D (2021) DuroNet: a dual-robust enhanced spatial-temporal learning network for urban crime prediction. ACM Trans Internet Technol 21, 1. https://doi.org/10.1145/3432249
8. Kshatri SS, Singh D, Narain B, Bhatia S, Quasim MT, Sinha GR (2021) An empirical analysis of machine learning algorithms for crime prediction using stacked generalization: an ensemble approach. IEEE Access 9:67488–67500. https://doi.org/10.1109/ACCESS.2021.3075140
9. Albahli S, Alsaqabi A, Aldhubayi F, Rauf HT, Arif M, Mohammed MA (2020) Predicting the type of crime: intelligence gathering and crime analysis. Comput Mater Contin 66(3):2317–2341. https://doi.org/10.32604/cmc.2021.014113
10. Spelman W (1995) The severity of intermediate sanctions. J Res Crime Delinq 32(2):107–135. https://doi.org/10.1177/0022427895032002001
11. Caplan JM, Kennedy LW, Miller J (2011) Risk terrain modeling: brokering criminological theory and GIS methods for crime forecasting. Justice Q 28(2):360–381. https://doi.org/10.1080/07418825.2010.486037
12. Johnson SD, Birks DJ, McLaughlin L, Bowers KJ, Pease K (2008) Prospective crime mapping in operational context: final report. London, UK Home Off. online Rep., vol. 19, no. September, pp. 07–08. http://www-staff.lboro.ac.uk/~ssgf/kp/2007_Prospective_Mapping.pdf
13. Wicks M (2016) Forecasting the future of fish. Oceanus 51(2):94–97
14. McNeal GS (2014) Drones and aerial surveillance: considerations for legislators, p 34. https://papers.ssrn.com/abstract=2523041.
15. Fatih T, Bekir C (2015) Police Use of Technology To Fight, Police Use Technol. To Fight Against Crime 11(10):286–296
16. Katz CM et al (2014) Evaluating the impact of officer worn body cameras in the Phoenix Police Department. Centre for Violence Prevention and Community Safety, Arizona State University, December, pp 1–43
17. Krakowski MI, Czobor P (2013) Depression and impulsivity as pathways to violence: implications for antiaggressive treatment. Schizophr Bull 40(4):886–894. https://doi.org/10.1093/schbul/sbt117
18. Kshatri SS, Narain B (2020) Analytical study of some selected classification algorithms and crime prediction. Int J Eng Adv Technol 9(6):241–247. https://doi.org/10.35940/ijeat.f1370.089620
19. Osisanwo FY, Akinsola JE, Awodele O, Hinmikaiye JO, Olakanmi O, Akinjobi J (2017) Supervised machine learning algorithms: classification and comparison. Int J Comput Trends Technol 48(3):128–138. https://doi.org/10.14445/22312803/ijctt-v48p126

Visual Question Answering Using Convolutional and Recurrent Neural Networks

Ankush Azade, Renuka Saini, and Dinesh Naik

1 Introduction

"Visual Question Answering" is a topic that inculcates the input as an image and a set of questions corresponding to a particular image which when fed to neural networks and machine learning models generate an answer or multiple answers. The purpose of building such systems is to assist the advanced tasks of computer vision like object detection and automatic answering by machine learning models when receiving the data in the form of images or in even advanced versions, receiving as video data. This task is very essential when we consider research objectives in artificial intelligence. In recent developments of AI [1], the importance of image data and integration of tasks involving textual and image forms of input is huge. Visual question-answering task will sometimes be used to answer open-ended questions, otherwise multiple choice, or close-ended answers. In our methodology, we have considered the formulation of open-ended answers instead of close-ended ones because in the real world, we see that most of the human interactions involve non-binary answers to questions. Open-ended questions are a part of a much bigger pool of the set of answers, when compared to close-ended, binary, or even multiple choice answers.

Some of the major challenges that VQA tasks face is computational costs, execution time, and the integration of neural networks for textual and image data. It is practically unachievable and inefficient to implement a neural network that takes into account both text features and image features and learns the weights of the network to

A. Azade (✉) · R. Saini · D. Naik
National Institute of Technology Karnataka, Surathkal 575025, India
e-mail: azadeankush8@gmail.com

R. Saini
e-mail: renukasaini.202it021@nitk.edu.in

D. Naik
e-mail: din_nk@nitk.edu.in

P. Singh et al. (eds.), *Machine Learning and Computational Intelligence Techniques for Data Engineering*, Lecture Notes in Electrical Engineering 998,
https://doi.org/10.1007/978-981-99-0047-3_3

make decisions and predictions. For the purposes of our research, we have considered the state-of-the-art dataset which is publically available. The question set that could be formed using that dataset is very wide. For instance one of the questions for an image containing multiple 3-D shapes of different colors can be "How many objects of cylinder shape are present?" [1]. As we can see this question pertains to a very deep observation, similar to human observation. After observing, experimenting, and examining the dataset questions we could see that each answer requires multiple queries to converge to an answer. Performing this task requires knowledge and application of natural language processing techniques in order to analyze the textual question and form answers. In this paper, we discuss the model constructed using Convolutional Neural Network layers for processing image features and Recurrent Neural Network based model for analyzing text features.

2 Literature Survey

A general idea was to take features from a global feature vector by convolution network and to basically extract and encode the questions using a "lstm" or long short term memory networks. These are then combined to make out a consolidated result. This gives us great answers but it fails to give accurate results when the answers or questions are dependent on the specific focused regions of images.

We also came across the use of stacked attention networks for VQA by Yang [3] which used extraction of the semantics of a question to look for the parts and areas of the picture that related to the answer. These networks are advanced versions of the "attention mechanism" that were applied in other problem domains like image caption generation and machine translation, etc. The paper by Yang [3] proposed a multiple-layer stacked attention network.

This majorly constituted of the following components: (1) A model dedicated to image, (2) a separate model dedicated to the question, which can be implemented using a convolution network or a Long Short Term Memory (LSTM) [8] to make out the semantic vector for questions, and (3) the stacked attention model to see and recognize the focus and important part and areas of the image. But despite its promising results, this approach had its own limitations.

Research by Yi et al. [4] in 2018 proposed a new model, this model had multiple parts or components to deal with images and questions/answers. They made use of a "scene parser", a "question parser" and something to execute. In the first component, Mask R-CNN was used to create segmented portions of the image. In the second component meant for the question, they used a "seq2seq" model. The component used for program execution was made using modules of python that would deal with the logical aspects of the questions in the dataset.

Focal visual-text attention for visual question-answering Liang et al. [5] This model (Focal Visual Text Attention) combines the sequence of image features generated by the network, text features of the image, and the question. Focal Visual Text Attention used a hierarchical approach to dynamically choose the modalities and snippets in the sequential data to focus on in order to answer the question, and so can not only forecast the proper answers but also identify the correct supporting arguments to enable people to validate the system's results. Implemented on a smaller dataset and not tested against more standard datasets.

Visual Reasoning Dialogs with Structural and Partial Observations Zhu et al. [7] Nodes in this Graph Neural Network model represent dialog entities (title, question and response pairs, and the unobserved questioned answer) (embeddings). The edges reflect semantic relationships between nodes. They created an EM-style inference technique to estimate latent linkages between nodes and missing values for unobserved nodes. (The M-step calculates the edge weights, whereas the E-step uses neural message passing (embeddings) to update all hidden node states.)

3 Dataset Description

The CLEVR10("A Diagnostic Dataset for Compositional Language and Elementary Visual Reasoning") [2] dataset was used, which includes a 70,000-image training set with 699,989 questions, a 15,000-image validation set with 149,991 questions, a 15,000-image test set with 14,988 questions, and responses to all train and val questions. Refer Dataset-1 statistics from Table 1 and a sample image from Fig. 1.

For Experminet-2 we have used a dataset titled easy-VQA which is publically available. This dataset is a simpler version of the CLEVR dataset, it mainly contains 2-Dimensional images of different shapes with different colors and positions. Dataset Statistics can be referred from Table 2 and a sample image from the easy-VQA dataset from Fig. 2.

4 Proposed Method

After reading about multiple techniques and models used to approach VQA task, we have used CNN+LSTM as the base approach for the model and worked our way up. CNN-LSTM model, where Image features and language features are computed separately and combined together and a multi-layer perceptron is trained on the combined features. The questions are encoded using a two-layer LSTM, while the visuals are encoded using the last hidden layer of CNN. After that, the picture features are l2 normalized. Then the question and image features are converted to a common

Table 1 Dataset-1 statistics

	Train	Validation	Test
Image	70,000	15,000	15,000
Question	699,989	149,991	14,988

Q: Are there an equal number of large things and metal spheres? **Q:** What size is the cylinder that is left of the brown metal thing that is left of the big sphere? **Q:** There is a sphere with the same size as the metal cube; is it made of the same material as the small red sphere? **Q:** How many objects are either small cylinders or metal things?

Fig. 1 Sample image from Dataset-1

Table 2 Dataset-2 statistics

	Train	Test
Image	4,000	1,000
Question	38,575	9,673
Binary questions	28,407	7,136

Fig. 2 Sample image from Dataset-2

space and we have taken an element-wise multiplication to obtain an answer. As a part of another approach we have used CNN-based model architecture for image feature extraction and for text features extraction bag of words technique has been

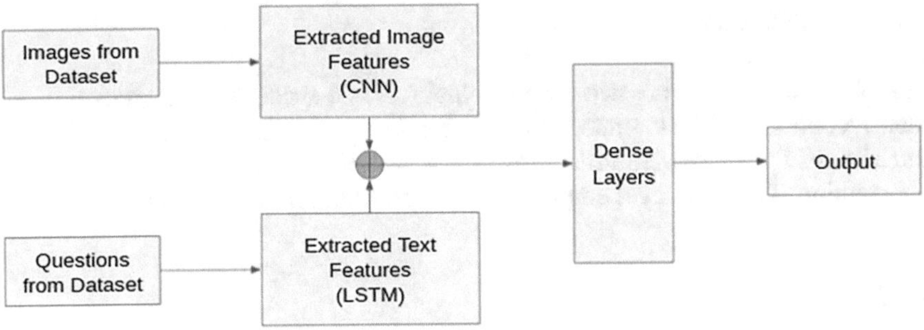

Fig. 3 Proposed model

used to form a fixed length vector and simple feed forward network to extract the features. Refer Fig. 3 for the proposed model.

4.1 Experiment 1

4.1.1 CNN

A CNN takes into account the parts and aspects of an input fed to the network as an image. The importance termed as weights and biases in neural networks is assigned based on the relevance of the aspects of the image and also points out what distinguishes them. A ConvNet requires far less pre-processing than other classification algorithms. CNN model is shown in Fig. 4. We have used mobilenetv2 in our CNN model. MobileNetV2 is a convolutional neural network design that as the name suggests is portable and in other words "mobile-friendly". It is built on an inverted residual structure, with residual connections between bottleneck levels. MobileNetV2 [9] is a powerful feature extractor for detecting and segmenting objects. The CNN model consists of the image input layer, mobilenetv2 layer, and global average pooling layer.

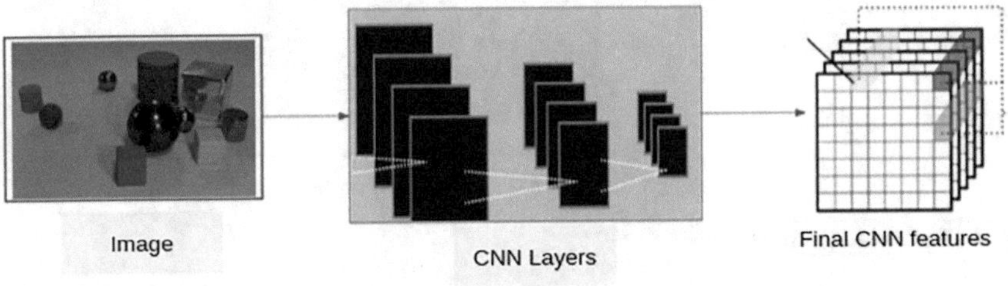

Fig. 4 Convolutional neural network

4.1.2 MobileNetV2

In MobileNetV2, there are two types of blocks. A one-stride residual block is one of them. A two-stride block is another option for downsizing. Both sorts of blocks have three levels. 1×1 convolution using ReLU6 is the initial layer, followed by depthwise convolution. The third layer employs a 1×1 convolution with no non-linearity.

4.1.3 LSTM

In sequence prediction problems, LSTM networks are a type of recurrent neural network that can learn order dependency. Given time lags of varying lengths, LSTM is ideally suited to identifying, analyzing, and forecasting time series. The model is trained via back-propagation. Refer to Fig. 5.

LSTM model consists of the text input layer, one embedding layer and three bidirectional layers consisting of LSTM layers.

After implementation of CNN and the LSTM model, we take their outputs and concatenate them.

$$Out = Multiply([x1, x2]) \tag{1}$$

where
x1 = Output from CNN,
x2 = Output from LSTM,
Out = Concatenation of x1 and x2.

After this, we will create a dense layer consisting of a softmax activation function with the help of TensorFlow. Then we will give CNN output, LSTM output, and the concatenated dense layer to the model. Refer Fig. 6 for overall architecture. The adam optimizer and sparse categorical cross-entropy loss were used to create this model. For merging the two components, we have used element-wise multiplication and fed it to the network to predict answers.

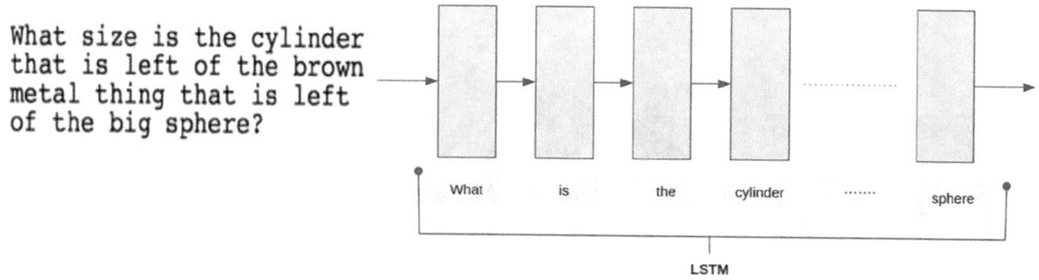

Fig. 5 Recurrent neural network

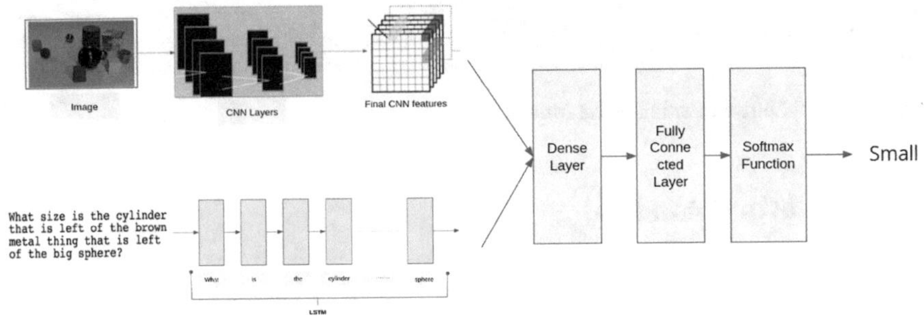

Fig. 6 Visual question-answering

4.2 *Experiment 2*

As a first step, we have preprocessed both the image data and the text data, i.e., the questions given as input. For this experiment, we have used a CNN model for extracting features from the image dataset. In Fig. 8, we have represented the model architecture used in the form of block representation. The input image of 64 * 64 is given as the input shape and fed to further layers. Then through a convolution layer with eight 3 × 3 filters using "same" padding, the output of this layer results in 64 × 64 × 8 dimensions. Then we used a maxpooling layer to reduce it to 32 × 32 × 8, further the next convolution layer uses 16 filters and generates in 32 × 32 × 16. Again with the use of maxpooling layer, it cuts the dimension down to 16 × 16 × 16. And finally, we flatten it to obtain the output of the 64 × 64 image in form of 4096 nodes. Refer Fig. 7.

In this experiment instead of using a complex RNN architecture to extract the features from the text part that is the questions. We have used the bag of words technique to form a fixed length vector and simple feedforward network to extract the features refer to Fig. 8. The figure below represents the process. Here, we have passed the bag of words to two fully connected layers and applied "tanh" activation function to obtain the output. Both these components have been merged using the element-wise multiplication as discussed in the previous section as well.

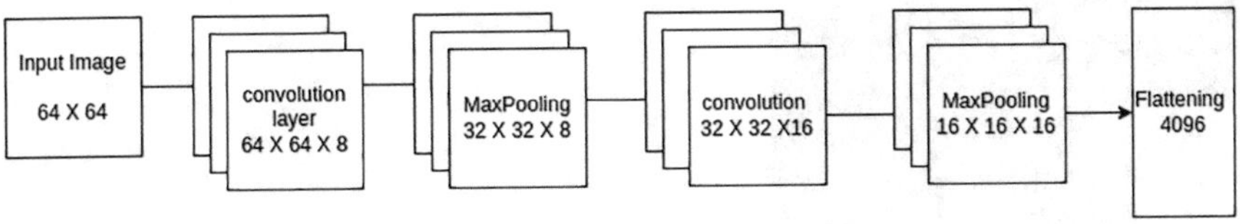

Fig. 7 CNN—Experiment 2

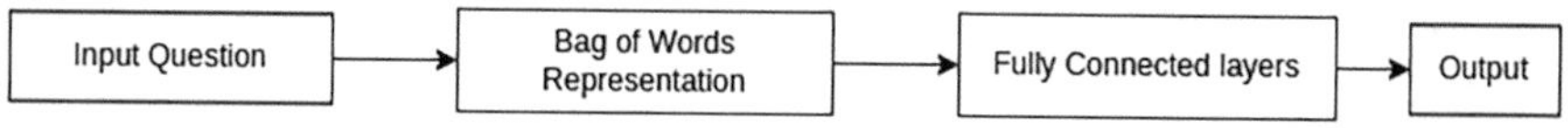

Fig. 8 Text feature extraction—Experiment 2

5 Results and Analysis

Following are the results for the Experiment 1 and Experiment 2.

5.1 *Experiment 1*

From Figs. 9 and 10 we can see that in the given image there are few solid and rubber shapes having different colors. For this respective image, we have a question "What number of small rubber balls are there". For this question we have an actual answer as 1. and our model also predicts the value as 1 which is correct.

5.2 *Experiment 2*

In the second experiment, we have considered a simpler form of the CLEVR dataset. And as explained in the methodology uses different models and variations of the approach. In Fig. 11 we can see that we have given an image and for that image we have a question "Does this image not contain a circle?" and our model predicted the correct answer as "No".

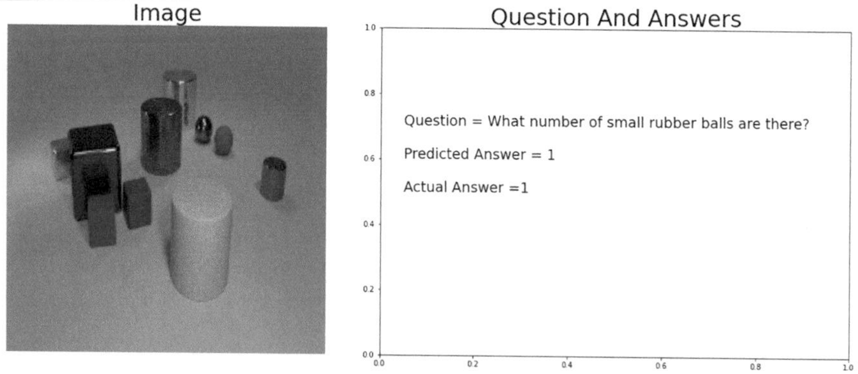

Fig. 9 Results of Experiment 1a

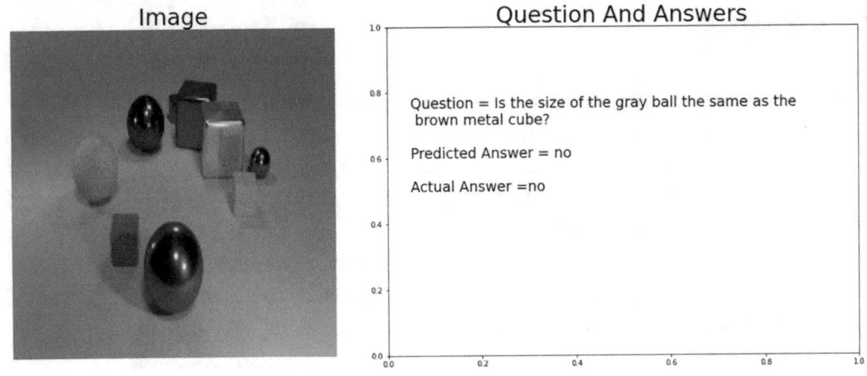

Fig. 10 Results of Experiment 1b

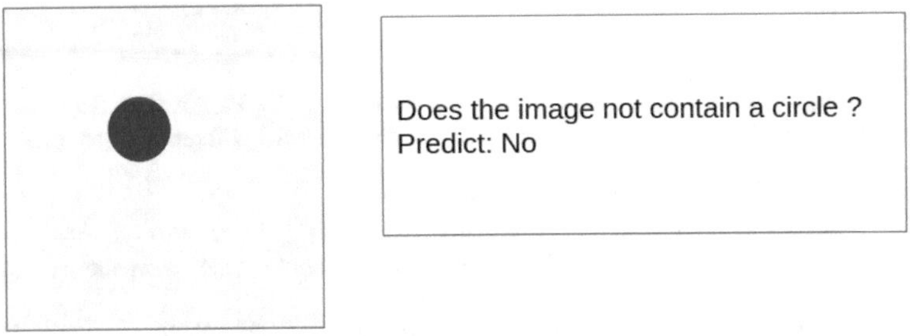

Fig. 11 Results of Experiment 2

Table 3 Train and test accuracy

Epoch	Train accuracy	Test accuracy
1	67.79%	72.69%
2	74.68%	76.89%
3	76.55	77.20%
4	77.77	77.87%
5	79.10	79.09%
6	82.17	81.82%
7	85.28	83.32%
8	87.02	83.60
9	88.40%	84.23%
10	90.01%	85.5%

Observing the gradual increase in accuracy with each epoch with positive changes shows us that there is learning happening in our model at each step. Since calculating the accuracy for a VQA task is not objective because of open-ended nature of the questions. We have achieved a training accuracy of 90.01% and test accuracy of 85.5% Table 3, this is a decent result when compared to the existing methodologies [1]. These results were observed on easy-VQA dataset.

6 Conclusion

Visual question-answering result analysis is a subjective task. We used two-component approaches which after performing separate extractions, merged their findings to obtain a consolidated result and predict the open-ended answers. It can be concluded that the approach performed well and that the use of CNN network is very essential for image feature extraction. And also the use of natural language processing techniques is essential for question feature extraction. Compared to baseline models the strategy is similar with tweaks discussed in the methodology section proved to be working well for a visual question-answering system.

References

1. Antol S, Agrawal A, Lu J, Mitchell M, Batra D, Zitnick CL, Parikh D (2015) Vqa: Visual question answering. In: Proceedings of the IEEE international conference on computer vision, pp 2425–2433
2. Dataset: https://visualqa.org/download.html
3. Yang Z, He X, Gao J, Deng L, Smola A (2016) Stacked attention networks for image question answering. In: Proceedings of the IEEE conference on computer vision and pattern recognition, pp 21–29
4. Yi K, Wu J, Gan C, Torralba A, Kohli P, Tenenbaum J (2018) Neural-symbolic vqa: disentangling reasoning from vision and language understanding. Adv Neural Inf Process Syst 31
5. Liang J, Jiang L, Cao L, Li LJ, Hauptmann AG (2018) Focal visual-text attention for visual question answering. In: Proceedings of the IEEE conference on computer vision and pattern recognition, pp 6135–6143
6. Wu C, Liu J, Wang X, Li R (2019) Differential networks for visual question answering. Proc AAAI Conf Artif Intell 33(01), 8997–9004. https://doi.org/10.1609/aaai.v33i01.33018997
7. Zheng Z, Wang W, Qi S, Zhu SC (2019) Reasoning visual dialogs with structural and partial observations. In: Proceedings of the IEEE/CVF conference on computer vision and pattern recognition, pp 6669–6678
8. https://www.analyticsvidhya.com/blog/2017/12/fundamentals-of-deep-learning-introduction-to-lstm/
9. https://towardsdatascience.com/review-mobilenetv2-light-weight-model-image-classification-8febb490e61c
10. Liu Y, Zhang X, Huang F, Tang X, Li Z (2019) Visual question answering via attention-based syntactic structure tree-LSTM. Appl Soft Comput 82, 105584. https://doi.org/10.1016/j.asoc.2019.105584, https://www.sciencedirect.com/science/article/pii/S1568494619303643
11. Nisar R, Bhuva D, Chawan P (2019) Visual question answering using combination of LSTM and CNN: a survey, pp 2395–0056
12. Kan C, Wang J, Chen L-C, Gao H, Xu W, Nevatia R (2015) ABC-CNN, an attention based convolutional neural network for visual question answering
13. Sharma N, Jain V, Mishra A (2018) An analysis of convolutional neural networks for image classification. Procedia Comput Sci 132, 377–384. ISSN 1877-0509. https://doi.org/10.1016/j.procs.2018.05.198, https://www.sciencedirect.com/science/article/pii/S1877050918309335
14. Staudemeyer RC, Morris ER (2019) Understanding LSTM–a tutorial into long short-term memory recurrent neural networks. arXiv:1909.09586

15. Zabirul Islam M, Milon Islam M, Asraf A (2020) A combined deep CNN-LSTM network for the detection of novel coronavirus (COVID-19) using X-ray images. Inform Med Unlocked 20, 100412. ISSN 2352-9148. https://doi.org/10.1016/j.imu.2020.100412
16. Boulila W, Ghandorh H, Ahmed Khan M, Ahmed F, Ahmad J (2021) A novel CNN-LSTM-based approach to predict urban expansion. Ecol Inform 64. https://doi.org/10.1016/j.ecoinf.2021.101325, https://www.sciencedirect.com/science/article/pii/S1574954121001163

Brain Tumor Segmentation Using Deep Neural Networks: A Comparative Study

Pankaj Kumar Gautam, Rishabh Goyal, Udit Upadhyay, and Dinesh Naik

1 Introduction

In a survey conducted in 2020, in USA about 3,460 children were diagnosed with the brain tumor having age under 15 years, and around 24,530 adults [1]. Tumors like Gliomas are most common, they are less threatening (lower grade) in a case where the expectancy of life is of several years or more threatening (higher grade) where it is almost two years. One of the most common medications for tumors is brain surgery. Radiation and Chemotherapy have also been used to regulate tumor growth that cannot be separated through surgery. Detailed images of the brain can be obtained using Magnetic resonance imaging (MRI). Brain tumor segmentation from MRI can significantly impact improved diagnostics, growth rate prediction, and treatment planning.

There are some categories of tumors like gliomas, glioblastomas, and meningiomas. Tumors such as meningiomas can be segmented easily, whereas the other two are much harder to locate and segment [2]. The scattered, poorly contrasted, and extended arrangements make it challenging to segment these tumors. One more difficulty in segmentation is that they can be present in any part of the brain with nearly

P. Kumar Gautam (✉) · R. Goyal · U. Upadhyay · D. Naik
Department of Information Technology, National Institute of Technology Karnataka, Surathkal, Karnataka, India
e-mail: pnkjgautam.202it018@nitk.edu.in

R. Goyal
e-mail: rishabhgoyal.202it022@nitk.edu.in

U. Upadhyay
e-mail: uditupadhyay.202it031@nitk.edu.in

D. Naik
e-mail: din_nk@nitk.edu.in

P. Singh et al. (eds.), *Machine Learning and Computational Intelligence Techniques for Data Engineering*, Lecture Notes in Electrical Engineering 998,
https://doi.org/10.1007/978-981-99-0047-3_4

any size-shape. Depending on the type of MRI machine used, the identical tumor cell may vary based on gray-scale values when diagnosed at different hospitals.

There are three types of tissues that form a healthy brain: white matter, gray matter, and cerebro-spinal fluid [3]. The tumor image segmentation helps in determining the size, position, and spread [4]. Since glioblastomas are permeated, the edges are usually blurred and tough to differentiate from normal brain tissues. T1-contrasted (T1-C), T1, T2 (spin-lattice and spin-spin relaxation, respectively) pulse sequences are frequently utilized as a solution [5]. Every sort of brain tissue receives a nearly different perception due to the differences between the modalities.

Segmenting brain tumors using the 2-pathway CNN design has already been proven to assist achieve reasonable accuracy and resilience [6, 7]. The research verified their methodology on MRI scan datasets of BRATS 2013 and 2015 [7]. Previous investigations also used encoder-decoder-based CNN design that uses autoencoder architectures. The research attached a different path to the end of the encoder section to recreate the actual scan image [8]. The purpose of adopting the autoencoder path was to offer further guidance and regularization to the encoder section because the size of the dataset was restricted [9]. In the past, the Vgg and Resnet designs were used to transfer learning for medical applications such as electroencephalograms (EEG). "EEG is a method of measuring brainwaves that have been often employed in brain-computer interface (BCI) applications" [10].

In this research, segmentation of tumors in the brain using two different CNN architectures is done. Modern advances in Convolutional Neural Network designs and learning methodologies, including Max-out hidden nodes and Dropout regularization, were utilized in this experiment. The BRATS-13 [11] dataset downloaded from the SMIR repository is available for educational use. This dataset was used to compare our results with the results of previous work [6]. In pre-processing, the one percent highest and lowest intensity levels were removed to normalize data. Later, the work used CNN to create a novel 2-pathway model that memorizes local brain features and then uses a two-stage training technique which was observed to be critical in dealing with the distribution of im-balanced labels for the target variable [6]. Traditional structured output techniques were replaced with a unique cascaded design, which was both effective and theoretically superior. We proposed a U-net machine learning model for further implementation, which has given extraordinary results in image segmentation [12].

The research is arranged as follows. Section 2 contains the methodology for the research, which presents two different approaches for the segmentation of brain tumor, i.e., Cascade CNN and U-net. Section 3 presents empirical studies that include a description of data, experimental setup, and performance evaluation metrics. Section 4 presents the visualization and result analysis, while Sect. 5 contains the conclusion of the research.

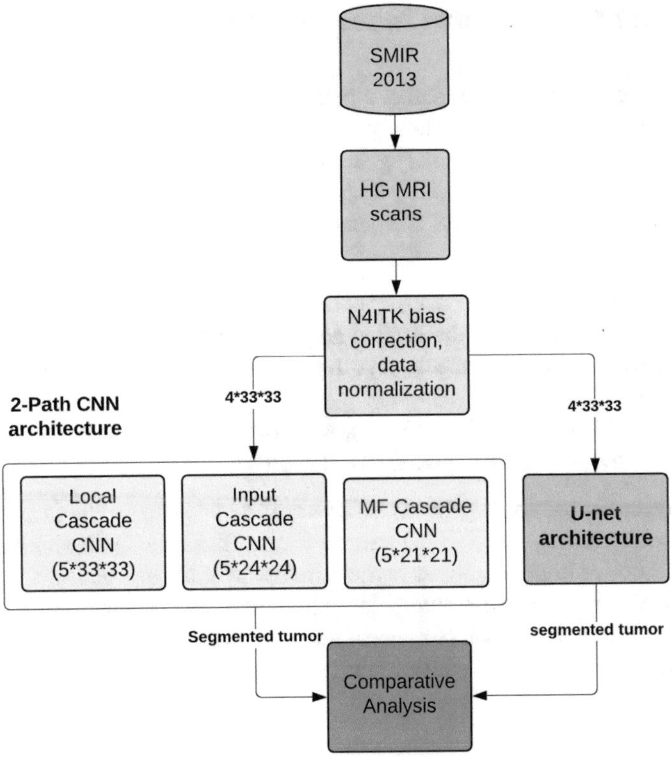

Fig. 1 Proposed methodology flow diagram

2 Methodology

This section presents the adopted methodology based on finding the tumor from the MRI scan of the patients by using two different architectures based on Convolutional Neural Networks (CNN). (A) Cascaded CNN [6], and (B) U-net [12]. First, we modeled the CNN architecture based on the cascading approach and then calculated the F1 score for all three types of cascading architecture. Then secondly, we modeled the U-net architecture and calculated the dice score and the dice loss for our segmented tumor output. Finally, we compared both these models based on dice scores. Figure 1 represents the adopted methodology in our research work. The research is divided into two parallel which represents the two approaches described above. The results of these two were then compared based on the F1 score and Dice loss.

2.1 2-Path Convolutional Neural Network

The architecture includes two paths: a pathway with 13 * 13 large receptive and 7 * 7 small receptive fields [6]. These paths were referred to as the global and local pathway, respectively as shown in Fig. 2. This architectural method is used to predict the pixel's label to be determined by 2 characteristics: (i) visible features of the area nearby the pixel, (ii) location of the patch. The structure of the two pathways is as follows:

1. Local: The 1st layer is of size (7, 7) and max-pooling of (4, 4), and the 2nd one is of size (3, 3). Because of the limited neighborhood and visual characteristics of the area around the pixel, the local path processes finer details because the kernel is smaller.
2. Global: The layer is of size (13, 13), Max-out is applied, and there is no max-pooling in the global path, giving (21, 21) filters.

Two layers for the local pathway were used to concatenate the primary-hidden layers of both pathways, with 3 * 3 kernels for the 2nd layer. This signifies that the effective receptive field of features in the primary layer of each pathway is the same. Also, the global pathway's parametrization models feature in that same region more flexibly. The union of the feature maps of these pathways is later supplied to the final output layer. The "Softmax" activation is applied to the output activation layer.

2.2 Cascaded Architecture

The 2-Path CNN architecture was expanded using a cascade of CNN blocks. The model utilizes the first CNN's output as added inputs to the hidden layers of the secondary CNN block.

This research implements three different cascading designs that add initial convolutional neural network results to distinct levels of the 2nd convolutional neural network block as described below [6]:

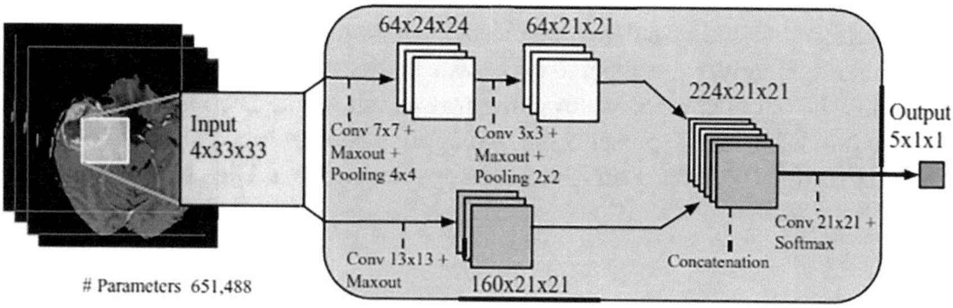

Fig. 2 2-Path CNN architecture [6]

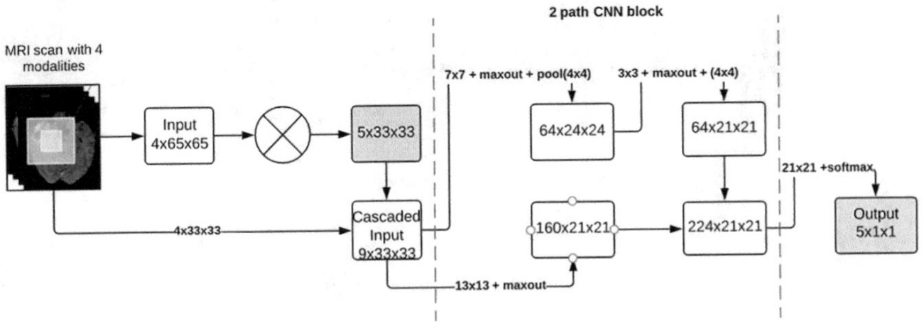

Fig. 3 Architecture for input cascade CNN

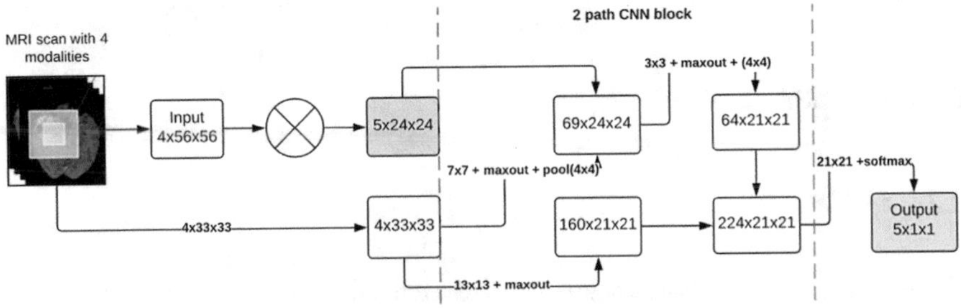

Fig. 4 Architecture for local cascade CNN

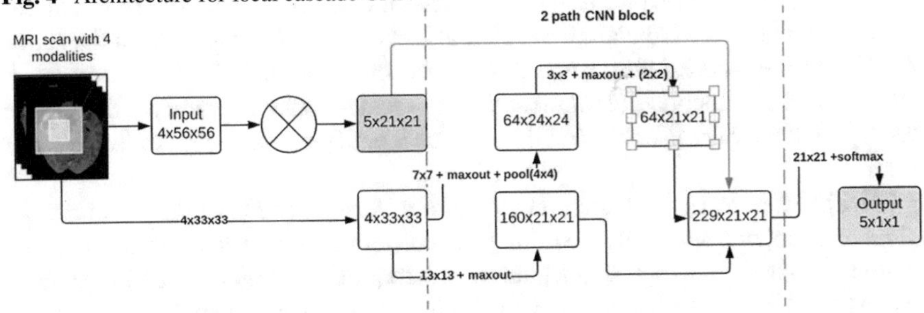

Fig. 5 Architecture for MF cascade CNN

1. Input cascade CNN: The first CNN's output is directly applied to the second CNN (Fig. 3). They are thus treated as additional MRI images scan channels of the input patch.
2. Local cascade CNN: In the second CNN, the work ascends up a layer in the local route and add to its primary-hidden layer (Fig. 4).
3. Mean-Field cascade CNN: The work now goes to the end of the second CNN and concatenates just before the output layer (Fig. 5). This method is similar to computations performed in Conditional random fields using a single run of mean-field inference.

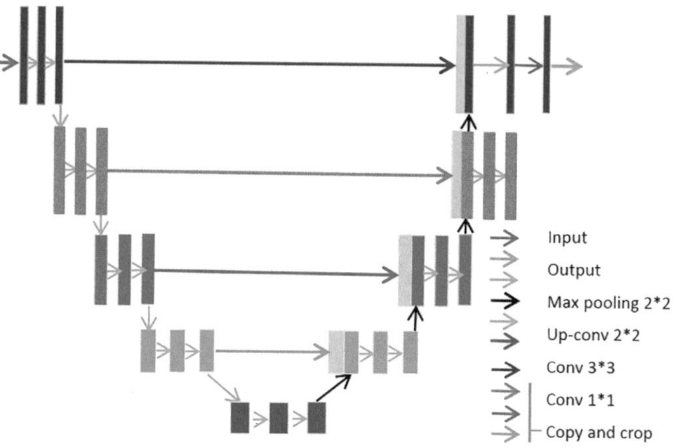

Fig. 6 U-net architecture [12]

2.3 U-Net

The traditional convolutional neural network architecture helps us predict the tumor class but cannot locate the tumor in an MRI scan precisely and effectively. Applying segmentation, we can recognize where objects of distinct classes are present in our image. U-net [13] is a Convolutional Neural Network (CNN) modeled in the shape of "U" that is expanded with some changes in the traditional CNN architecture. It was designed to semantically segment the bio-medical images where the target is to classify whether there is contagion or not, thus identifying the region of infection or tumor.

CNN helps to learn the feature mapping of an image, and it works well for classification problems where an input image is converted into a vector used for classification. However, in image segmentation, it is required to reproduce an image from this vector. While transforming an image into a vector, we already learned the feature mapping of the image, so we use the same feature maps used while contracting to expand a vector to a segmented image. The U-net model consists of 3 sections: Encoder, Bottleneck, and Decoder block as shown in Fig. 6. The encoder is made of many contraction layers. Each layer takes an input and employs two 3 * 3 convolutions accompanied by a 2 * 2 max-pooling. The bottom-most layer interferes with the encoder and the decoder blocks. Similarly, each layer passes the input to two convolutional layers of size 3 * 3 for the encoder, accompanied by a 2 * 2 up-sampling layer, which follows the same as encoder blocks.

To maintain symmetry, the amount of feature maps gets halved. The number of expansion and contraction blocks is the same. After that, the final mapping passes through another 3 * 3 convolutional layer with an equal number of features mapped as that of the number of segments.

In image segmentation, we focus on the shape and boundary of the segmented image rather than its colors, texture, and illumination. The loss function can measure how clean and meaningful boundaries were segmented from the original image. The loss is computed as the mean of per-pixel loss in cross-entropy loss, and the per-pixel loss is calculated discretely without knowing whether or not its nearby pixels are borders. As a result, cross-entropy loss only takes into account loss in a micro-region rather than the entire image, which is insufficient for medical image segmentation. As a result, the research uses the Dice loss function to train our U-net model.

3 Empirical Studies

This section presents the dataset description, experimental setup, data pre-processing, and metrics for performance evaluation.

3.1 Dataset

BRATS-13 MRI dataset [11] was used for the research. It consists of actual patient scans and synthetic scans created by SMIR (SICAS medical image repository). The size is around 1Gb and was stored in "Google drive" for further use. Dataset consists of synthetic and natural images. Each category contains MRI scans for high-graded gliomas (HG) and low-graded gliomas (LG). There are 25 patients with synthetic HG and LG scans and 20 patients with actual HG, and ten patients with actual LG scans. Dataset consists of four modalities (different types of scans) like T1, T1-C, T2, and FLAIR. For each patient and each modality, we get a 3-D image of the brain. We're concatenating these modalities as four channels slice-wise. Figure 7 shows tumors along with their MRI scan. We have used 126th slice for representation. For HG, the dimensions are (176, 216, and 160). Image in gray-scale represents the MRI scan, and that in blue-colored represents the tumor for their respective MRI scans.

3.2 Experiment Setup

The research was carried out using Google Colab, which provides a web interface to run Jupyter notebooks free of cost. "Pandas" and "Numpy" libraries were used for data pre-processing, CNN models were imported from "Keras" library for segmentation, and "SkLearn" is used for measuring different performance metrics like F1 score (3), and Dice loss (4). Also, the MRI scan data was the first download under the academic agreement and is then uploaded on Colab. For data pre-processing,

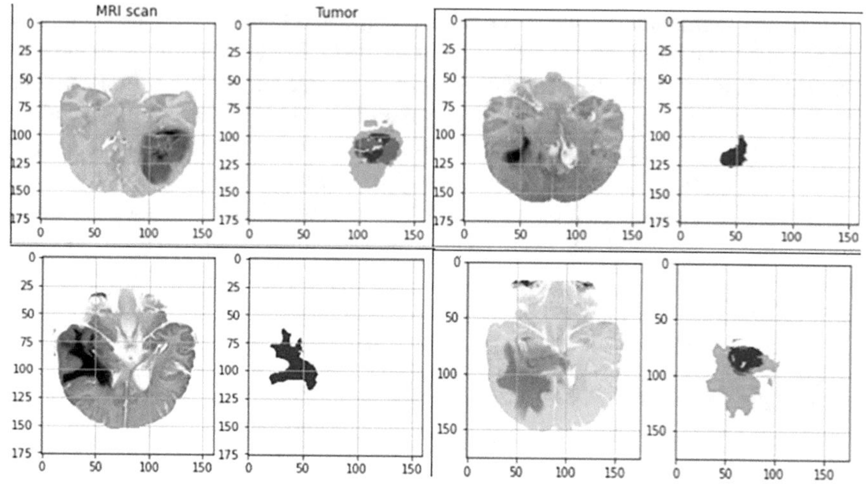

Fig. 7 MRI scan with their labeled data (tumor location)

multiple pre-processing steps have been applied to the dataset as presented in the next section. The data was split into 70:30 for training and testing data, respectively.

3.3 Data Preprocessing

First, slices of MRI scans where the tumor information was absent were removed from the original dataset. This will help us in minimizing the dataset without affecting the results of the segmentation. Then the one percent highest and lowest intensity levels were eliminated. Intensity levels for T1 and T1-C modalities were normalized using N4ITK bias field correction [14]. Also, the image data is normalized in each input layer by subtracting the average and then dividing it by the standard deviation of a channel. Batch normalization was used because of the following reasons:

1. Speeds up training makes the optimization landscape much smoother, producing a more predictive and constant performance of gradients, allowing quicker training.
2. In the case of "Batch Norm" we can use a much larger learning rate to get to the minima, resulting in fast learning.

3.4 Performance Evaluation Metrics

We have used various performance metrics for comparing both model performance. Precision, Recall, F1-Score, and Dice Loss were selected as our performance parameters. Precision (Pr) is the proportion between the True Positives and all the Positives.

The Recall (Re) is the measure of our model perfectly identifying True Positives. F1-Score is a function of Recall and Precision and is the harmonic mean of both. It helps in considering both the Recall and Precision values. Finally, accuracy is the fraction of predictions our model got correct.

$$\text{Precision(Pr)} = \frac{TP}{FP + TP} \tag{1}$$

$$\text{Recall(Re)} = \frac{TP}{FN + TP} \tag{2}$$

$$F1\,\text{Score} = 2 \times \frac{Pr \times Re}{Pr + Re} \tag{3}$$

where True Positive (TP) represents that the actual and the predicted labels co-respond to the same positive class. True Negative (TN) represents that the actual and the predicted label co-responds to the same negative class. False Positive (FP) tells that the actual label belongs to a negative class; however, the predicted label belongs to a positive class. It is also called the Type-I error. False Negative (FN) or the Type-II error tells that the actual labels belong to a positive class; however, the model predicted it into a negative class.

$$\text{Loss}_{\text{dice}} = \frac{2 \times \sum_i p_i \times g_i}{\sum_i (p_i{}^2 + g_i{}^2)} \tag{4}$$

Also, Dice loss ($\text{Loss}_{\text{dice}}$) (4) measures how clean and meaningful boundaries were calculated by the loss function. Here, p_i and g_i represent pairs of corresponding pixel values of predicted and ground truth, respectively. Dice loss considers the loss of information both locally and globally, which is critical for high accuracy.

4 Visualization and Result Analysis

This section presents the results after performing both the Cascaded and U-net architecture.

4.1 Cascaded CNN

The three cascading architectures were trained on 70% of data using the "cross-entropy loss" and "Adam Optimizer." Testing was done on the rest 30% of the data, and then the F1 score was computed for all the three types of cascading architecture (as shown in Table 1). The F1 score of Local cascade CNN is the highest for the

Table 1 F1 score comparison with [6]

Model type	[6]	Proposed work
Input cascaded	0.88	0.848
Local cascade	0.88	0.872
MF cascade	0.86	0.828

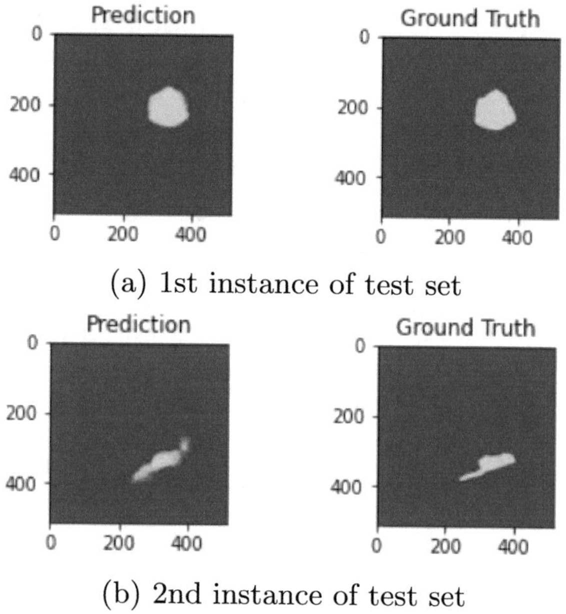

Fig. 8 Ground truth versus predicted segment of tumor mask

research; also, it is very similar to the previous work done by [6]. For Input Cascade and MF cascade, the model has a difference of around 4% compared to previous work.

Figure 8a, b shows the results for the segmentation on two instances of test MRI scan images. The segmented output was compared with the ground truth and was concluded that the model was able to get an accurate and precise boundary of the tumor from the test MRI scan image dataset.

4.2 U-Net

This deep neural network (DNN) architecture is modeled using the Dice Loss, which takes account information loss both globally and locally and is essential for high accuracy. Dice loss varies from [0, 1], where 0 means that the segmented output and the ground truth do not overlap at all, and 1 represents that both the segmented result

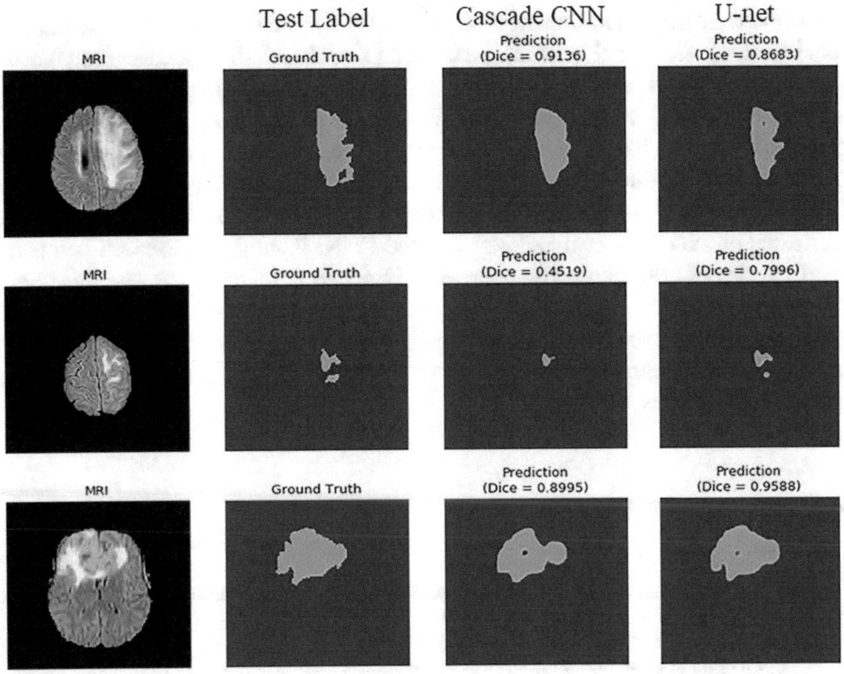

Fig. 9 F1 score: cascade CNN versus U-net architecture

and the ground truth image are fully overlapped. We achieved a dice loss of 0.6863 on our testing data, which means most of our segmented output is similar in terms of boundaries and region with ground truth images.

Figure 9 shows the results for the segmentation on three random instances of test MRI scan images. From left to right, we have the MRI scan, the Ground truth image, then we have segmented output from Cascade CNN, and finally, we have the segmented output for the U-net model. The segmented output was compared to ground truth, and the model was capable of obtaining an accurate and precise boundary of the tumor from the test MRI scan image dataset.

From Fig. 9 it was concluded that the U-net model performs better than the Cascaded architecture in terms of F1 score.

5 Conclusions

The research used convolutional neural networks (CNN) to perform brain tumor segmentation. The research looked at two designs (Cascaded CNN and U-net) and analyzed their performance. We test our findings on the BRAT 2013 dataset, which contains authentic patient images and synthetic images created by SMIR. Significant performance was produced using a novel 2-pathway model (which can represent the local features and global meaning), extending it to three different cascading models and represent local label dependencies by piling 2 convolutional neural networks. Two-phase training was followed, which allowed us to model the CNNs when the distribution has un-balanced labels efficiently. The model using the cascading architecture could reproduce almost similar results compared with the base paper in terms of F1 score. Also, in our research, we concluded that the Local cascade CNN performs better than the Local and MF cascade CNN. Finally, the research compared the F1 score of cascaded architecture and U-net model, and it was concluded that the overall performance of the semantic-based segmentation model, U-net performs better than the cascaded architecture. The Dice loss for the U-net was 0.6863, which describes that our model produces almost similar segmented images like that of the ground truth images.

References

1. ASCO: Brain tumor: Statistics (2021) Accessed 10 Nov 2021 from https://www.cancer.net/cancer-types/brain-tumor/statistics
2. Zacharaki EI, Wang S, Chawla S, Soo Yoo D, Wolf R, Melhem ER, Davatzikos C (2009) Classification of brain tumor type and grade using mri texture and shape in a machine learning scheme. Magn Reson Med: Off J Int Soc Magn Reson Med 62(6):1609–1618
3. 3T How To: Structural MRI Imaging—Center for Functional MRI - UC San Diego. Accessed 10 Nov 2021 from https://cfmriweb.ucsd.edu/Howto/3T/structure.html
4. Rajasekaran KA, Gounder CC (2018) Advanced brain tumour segmentation from mri images. High-Resolut Neuroimaging: Basic Phys Princ Clin Appl 83
5. Lin X, Zhan H, Li H, Huang Y, Chen Z (2020) Nmr relaxation measurements on complex samples based on real-time pure shift techniques. Molecules 25(3):473
6. Havaei M, Davy A, Warde-Farley D, Biard A, Courville A, Bengio Y, Pal C, Jodoin PM, Larochelle H (2017) Brain tumor segmentation with deep neural networks. Med Image Anal 35, 18–31
7. Razzak MI, Imran M, Xu G (2018) Efficient brain tumor segmentation with multiscale two-pathway-group conventional neural networks. IEEE J Biomed Health Inform 23(5):1911–1919
8. Myronenko, A (2018) 3d mri brain tumor segmentation using autoencoder regularization. In: International MICCAI brainlesion workshop. Springer, Berlin, pp 311–320
9. Aboussaleh I, Riffi J, Mahraz AM, Tairi H (2021) Brain tumor segmentation based on deep learning's feature representation. J Imaging 7(12):269
10. Singh D, Singh S (2020) Realising transfer learning through convolutional neural network and support vector machine for mental task classification. Electron Lett 56(25):1375–1378
11. SMIR: Brats—sicas medical image repository (2013) Accessed 10 Nov 2021 from https://www.smir.ch/BRATS/Start2013
12. Yang T, Song J (2018) An automatic brain tumor image segmentation method based on the u-net. In: 2018 IEEE 4th international conference on computer and communications (ICCC). IEEE, pp 1600–1604
13. Ronneberger O, Fischer P, Brox T (2015) U-net: Convolutional networks for biomedical image segmentation. In: Medical image computing and computer-assisted intervention (MICCAI). LNCS, vol 9351, pp 234–241. Springer, Berlin
14. Tustison NJ, Avants BB, Cook PA, Zheng Y, Egan A, Yushkevich PA, Gee JC (2010) N4itk: improved n3 bias correction. IEEE Trans Med Imaging 29(6), 1310–1320

Predicting Bangladesh Life Expectancy Using Multiple Depend Features and Regression Models

Fatema Tuj Jannat, Khalid Been Md. Badruzzaman Biplob, and Abu Kowshir Bitto

1 Introduction

The word "life expectancy" refers to how long a person can expect to live on average [1]. Life expectancy is a measurement of a person's projected average lifespan. Life expectancy is measured using a variety of factors such as the year of birth, current age, and demographic sex. A person's life expectancy is determined by his surroundings. Surrounding refers to the entire social system, not just society. In this study, our target area is the average life expectancy in Bangladesh, the nation in South Asia where the average life expectancy is 72.59 years. Research suggests that the average life expectancy depends on lifestyle, economic status (GDP), healthcare, diet, primary education, and population. The death rate in the present is indeed lower than in the past. The main reason is the environment. Lifestyle and Primary Education are among the many environmental surroundings. Lifestyle depends on primary education. If a person does not receive primary education, he will not be able to be health conscious in any way. This can lead to premature death from the damage to the health of the person. So that it affects the average life expectancy of the whole country. Indeed, the medical system was not good before, so it is said that both the baby and the mother would have died during childbirth. Many people have died because they did not know what medicine to take, or how much to take because they did not have the right knowledge and primary education. It is through this elementary education that economic status (GDP) and population developed. The average lifespan varies

F. Tuj Jannat · K. B. Md. B. Biplob · A. K. Bitto (✉)
Department of Software Engineering, Daffodil International University, Dhaka 1207, Bangladesh
e-mail: abu.kowshir777@gmail.com

F. Tuj Jannat
e-mail: jannat.fatema7940@gmail.com

K. B. Md. B. Biplob
e-mail: khalid@daffodilvarsity.edu.bd

P. Singh et al. (eds.), *Machine Learning and Computational Intelligence Techniques for Data Engineering*, Lecture Notes in Electrical Engineering 998,
https://doi.org/10.1007/978-981-99-0047-3_5

from generation to generation. We are all aware that our life expectancy is increasing year after year. Since its independence in 1971, Bangladesh, a poor nation in South Asia, has achieved significant progress in terms of health outcomes. There was the expansion of the economic sector. There were a lot of good things in the late twentieth century and amifications all around the globe.

In this paper, we used some features for a measure of life expectancies such as GDP, Rural Population Growth (%), Urban Population Growth (%), Services Value, Industry Value Food Production, Permanent Cropland (%), Cereal production (metric tons), Agriculture, forestry, and fishing value (%). We will measure the impact of these depending on features to predict life expectancy. Use various regression models to find the most accurate model in search of find life expectancy of Bangladesh with these depending on features. It will assist us in determining which feature aids in increasing life expectancy. This research aids a country in increasing the value of its features for life expectancy and also finding which regression model performs best for predicting life expectancy.

2 Literature Review

Several studies on life expectancy have previously been produced by several different researchers. As part of the literature review, we are reporting a few past studies to understand the previously identified factors.

Beeksma et al. [2] obtained data from seven different healthcare facilities in Nijmegen, the Netherlands, with a set of 33,509 EMRs dataset. The accuracy of their model was 29%. While clinicians overestimated life expectancy in 63 percent of erroneous prognoses, causing delays in receiving adequate end-of-life care, his model which was the keyword model only overestimated life expectancy in 31% of inaccurate prognoses. Another study by Nigri et al. [3] worked on recurrent neural networks with a long short-term memory, which was a new technique for projecting life expectancy, and lifespan discrepancy was measured. Their projections appeared to be consistent with the past patterns and offered a more realistic picture of future life expectancy and disparities. The LSTM model, ARIMA model, DG model, Lee-Carter model, CoDa model, and VAR model are examples of applied recurrent neural networks. It is shown that both separate and simultaneous projections of life expectancy and lifespan disparity give fresh insights for a thorough examination of the mortality forecasts, constituting a valuable technique to identify irregular death trajectories. The development of the age-at-death distribution assumes more compressed tails with time, indicating a decrease in longevity difference across industrialized nations. Khan et al. [4] analyzed gender disparities in terms of disabilities incidence and disability-free life expectancy (DFLE) among Bangladeshi senior citizens. They utilized the data from a nationwide survey that included 4,189 senior people aged 60 and above, and they employed the Sullivan technique. They collected the data from the Bangladeshi household income and expenditure survey (HIES)-2010, a large nationwide survey conducted by the BBS. The data-collecting procedure was a year-long program. There was a total of 12,240 households chosen, with 7,840 from rural regions and 4,400 from urban areas. For a total of 55,580 people, all members of chosen homes were surveyed. They discovered that at the age of 70, both men and women can expect to spend more than half of their lives disabled and have a significant consequence for the likelihood of disability, as well as the requirement for the usage of long-term care services and limitations, including, to begin with, the study's data is self-reported. Due to a lack of solid demographic factors, the institutionalized population was not taken into consideration. The number of senior individuals living in institutions is tiny, and they have the same health problems and impairments as the elderly in the general population.

Tareque, et al. [5] explored the link between life expectancy and disability-free life expectancy (DFLE) in the Rajshahi District of Bangladesh by investigating the connections between the Active Aging Index (AAI) and DFLE. Data were obtained during April 2009 from the Socio-Demographic status of the aged population and elderly abuse study project. They discovered that urban, educated, older men are more engaged in all parts of life and have a longer DFLE. In rural regions, 93 percent of older respondents lived with family members, although 45.9% of nuclear families and 54.1 percent of joint families were noted. In urban regions, however, 23.4 percent were nuclear families and 76.6 percent were joint families, and they face restrictions in terms of several key indicators, such as the types and duration of physical activity. For a post-childhood-life table, Preston and Bennett's (1983) estimate technique was used. Because related data was not available, the institutionalized population was not examined. Tareque et al. [6] multiple linear regression models, as well as the Sullivan technique, were utilized. They based their findings on the World Values Survey, which was performed between 1996 and 2002 among people aged 15 and above. They discovered that between 1996 and 2002, people's perceptions of their health improved. Males predicted fewer life years spent in excellent SRH in 2002 than females, but a higher proportion of their expected lives were spent in good SRH. The study has certain limitations, such as the sample size being small, and the institutionalized population was not included in the HLE calculation. The subjective character of SRH, as opposed to health assessments based on medical diagnoses, may have resulted in gender bias in the results. In 2002, the response category "very poor" was missing from the SRH survey. In 2002, there's a chance that healthy persons were overrepresented. Tareque et al. [6] investigated how many years older individuals expect to remain in excellent health, as well as the factors that influence self-reported health (SRH). By integrating SRH, they proposed a link between LE and HLE. The project's brief said that it was socioeconomic and demographic research of Rajshahi district's elderly population (60 years and over). They employed Sullivan's approach for solving the problem. For their work, SRH was utilized to estimate HLE. They discovered that as people became older, LE and anticipated life in both poor and good health declined. Individuals in their 60 s were anticipated to be in excellent health for approximately 40% of their remaining lives, but those in their 80 s projected just 21% of their remaining lives to be in good health, and their restrictions were more severe. The sample size is small, and it comes from only one district, Rajshahi; it is not indicative of the entire country. As a result, generalizing the findings of this study to the entire country of Bangladesh should be approached with caution. The institutionalized population was not factored into the HLE calculation.

Ho et al. [7] examine whether decreases in life expectancy happened across high-income countries from 2014 to 2016 with 18 nations. They conducted a demographic study based on aggregated data and data from the WHO mortality database, which was augmented with data from Statistics Canada and Statistics Portugal, and their contribution to changes in life expectancy between 2014 and 2015. Arriaga's decomposition approach was used. They discovered that in the years 2014–15, life expectancy fell across the board in high-income nations. Women's life expectancy fell in 12 of the 18 nations studied, while men's life expectancy fell in 11 of them. They also have certain flaws, such as the underreporting of influenza and pneumonia on death certificates, the issue of linked causes of death, often known as the competing hazards dilemma, and the comparability of the cause of death coding between nations. Meshram et al. [8] for the comparison of life expectancy between developed and developing nations, Linear Regression, Decision Tree, and Random Forest Regressor were applied. The Random Forest Regressor was chosen for the construction of the life expectancy prediction model because it had R^2 scores of 0.99 and 0.95 on training and testing data, respectively, as well as Mean Squared Error and Mean Absolute Error of 4.43 and 1.58. The analysis is based on HIV or AIDS, Adult Mortality, and Healthcare Expenditure, as these are the key aspects indicated by the model. This suggests that India has a higher adult mortality rate than other affluent countries due to its low healthcare spending.

Matsuo et al. [9] investigate survival predictions using clinic laboratory data in women with recurrent cervical cancer, as well as the efficacy of a new analytic technique based on deep-learning neural networks. Alam et al. [10] using annual data from 1972 to 2013 investigate the impact of financial development on Bangladesh's significant growth in life expectancy. The unit root properties of the variables are examined using a structural break unit root test. In their literature review, they mention some studies on the effects of trade openness and foreign direct investment on life expectancy. Using annual data from 1972 to 2013, investigate the impact of financial development on Bangladesh's significant growth in life expectancy. The unit root properties of the variables are examined using a structural break unit root test. In their literature review, they mention some studies on the effects of trade openness and foreign direct investment on life expectancy. Furthermore, the empirical findings support the occurrence of counteraction in long-run associations. Income disparity appears to reduce life expectancy in the long run, according to the long-run elasticities. Finally, their results provide policymakers with fresh information that is critical to improving Bangladesh's life expectancy. Husain et al. [11] conducted a multivariate cross-national study of national life expectancy factors. The linear and log-linear regression models are the first regression models. The data on explanatory factors comes from UNDP, World Bank, and Rudolf's yearly statistics releases (1981). His findings show that if adequate attention is paid to fertility reduction and boosting calorie intake, life expectancies in poor nations may be considerably enhanced.

3 Proposed Methodology

In any research project, we must complete numerous key stages, including data collecting, data preparation, picking an appropriate model, implementing it, calculating errors, and producing output. To achieve our aim, we use the step-to-step working technique illustrated in Fig. 1.

3.1 Data Preprocessing

Preprocessing, which includes data cleaning and standardization, noisy data filtering, and management of missing information, is necessary for machine learning to be done. Any data analysis will succeed if there is enough relevant data. The information was gathered from Trends Economics. The dataset contained data from 1960 to 2020. Combine all of the factors that are linked to Bangladesh's Life Expectancy. We replaced the null values using the mean values. We examined the relationship where GDP, Rural Population Growth (%), Urban Population Growth (%), Services Value, Industry Value Food Production, Permanent Cropland (%), Cereal production (metric tons), Agriculture, forestry, and fishing value (%) were the independent features and Life Expectancy (LE) being the target variable. We separated the data into two subsets to test the model and develop the model: A total of 20% of the data was used for testing, with the remaining 80% divided into training subsets.

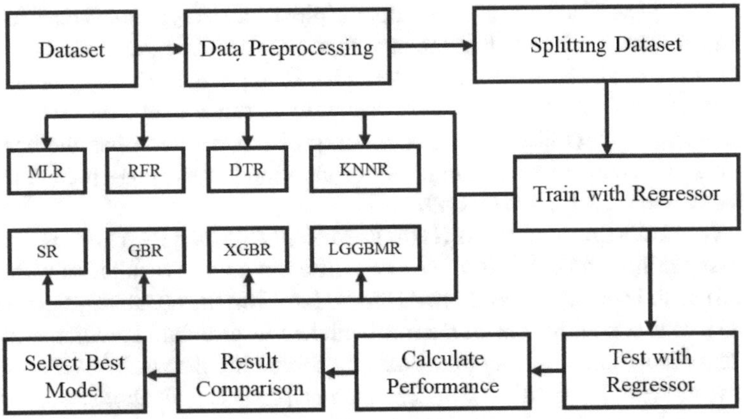

Fig. 1 Working procedure diagram

3.2 Regressor Relevant Theory

Multiple Linear Regression (MLR): A statistical strategy [12] for predicting the outcome of a variable using the values of two or more variables is known as multiple linear regression. Multiple regression is a type of regression that is an extension of linear regression. The dependent variable is the one we're trying to forecast, and the independent or explanatory elements are employed to predict its value. In the case of multiple linear regression, the formula is as follows in "(1)".

$$Y = \beta_0 + \beta_1 X_1 + \beta_2 X_2 + \ldots.. + \beta_n X_n + \in \tag{1}$$

K-Neighbors Regressor (KNNR): It's a non-parametric strategy for logically averaging data in the same neighborhood to approximate the link between independent variables and continuous outcomes [13]. To discover the neighborhood size that minimizes the mean squared error, the analyst must define the size of the neighborhood.

Decision Tree Regressor (DTR): A decision tree [14] is a hierarchical architecture that resembles a flowchart and is used to make decisions. In a supervised learning approach, the decision tree technique is categorized. It may be utilized with both categorical and continuous output variables. The Decision Tree method has become one of the most commonly used machine learning algorithms. The use of a Decision Tree can help with both classification and regression difficulties.

Random Forest Regressor (RFR): A Random Forest is an ensemble method for solving regression and classification problems that use several decision trees with the Bootstrap and Aggregation methodology. Rather than relying on individual decision trees to decide the outcome, the fundamental concept is to combine many decision trees. Random Forest employs several decision trees as a foundation learning paradigm.

Stacking Regressor (SR): The phrase "stacking" or "stacked" refers to the process of stacking objects. Each estimator's output is piled, and a regressor is used to calculate the final forecast. By feeding the output of each estimate into a final estimator, you may make use of each estimate's strengths. Using a meta-learning technique, it learns how to combine predictions from two or more fundamental machine learning algorithms. On a classification or regression problem, stacking has the benefit of combining the talents of several high-performing models to create predictions that surpass any one model in the ensemble.

Gradient Boosting Regressor (GBR): Gradient Boosting Regressor is a forward stage-wise additive model that allows any differentiable loss function to be optimized. At each level, a regression tree is fitted based on the negative gradient of the supplied loss function. It's one of the most efficient ways to build predictive models. It was feasible to build an ensemble model by combining the weak learners or weak predictive models. The gradient boosting approach can help with both regression and classification issues. The Gradient Boosting Regression technique is used to fit the model that predicts the continuous value.

Extreme Gradient Boosting Regressor (XGBR): Extreme Gradient Boosting is an open-source application that executes the gradient boosting approach efficiently and effectively. Extreme Gradient Boosting (EGB) is a machine learning technique that creates a prediction model from a set of weak prediction models, most frequently decision trees, for regression, classification, and other tasks. When a decision tree is a poor learner, the resulting technique is called gradient enhanced tree, and it often outperforms random forests.

Light Gradient Boosting Machine Regressor (LGBMR): Light Gradient Boosted Machine is an open-source toolkit that efficiently and effectively implements the gradient boosting approach. LightGBM enhances the gradient boosting approach by incorporating automated feature selection and focusing on boosting situations with larger gradients. This might result in a considerable boost in training speed as well as im- proved prediction accuracy. As a result, LightGBM has been the de facto technique for machine learning contests when working with tabular data for regression and classification predictive modeling tasks.

3.3 *Preformation Calculation*

On the basis of their prediction, error, and accuracy, the estimated models are compared and contrasted.

Mean Absolute Error (MAE): The MAE is a measure for evaluating regression models. The MAE of a model concerning the test set is the mean of all individual prediction errors on all occurrences in the test set. For each event, a prediction error is a difference between the true and expected value. Following is the formula in "(2)".

$$\mathrm{MAE} = \frac{1}{n}\sum_{i=1}^{n} |A_i - A| \tag{2}$$

Mean Squared Error (MSE): The MSE shows us how close we are to a collection of points. By squaring the distances between the points and the regression line, it achieves this. Squaring is required to eliminate any undesirable signs. Inequalities with greater magnitude are also given more weight. The fact that we are computing the average of a series of errors gives the mean squared error its name. The better the prediction, the smaller the MSE. The following is the formula in "(3)".

$$\mathrm{MSE} = \frac{1}{n}\sum_{i=1}^{n} |\mathrm{Actual} - \mathrm{Prediction}| \tag{3}$$

Root Mean Square Error (RMSE): The RMSE measures the distance between data points and the regression line, and the RMSE is a measure of how to spread out these residuals. The following is the formula in "(4)".

$$\text{RMSE} = \sqrt{\frac{1}{n}\sum_{i=1}^{n}|\text{Actual} - \text{Prediction}|} \quad (4)$$

4 Results and Discussions

The life expectancy of a nation is determined by several variables. Figure 2 depicted the pairwise association between life expectancy and a variety of independent characteristics such as GDP, Rural Population Growth (%), Urban Population Growth (%), Services Value, Industry Value Food Production, Permanent Cropland (%), Cereal production (metric tons), Agriculture, forestry, and fishing value (%).

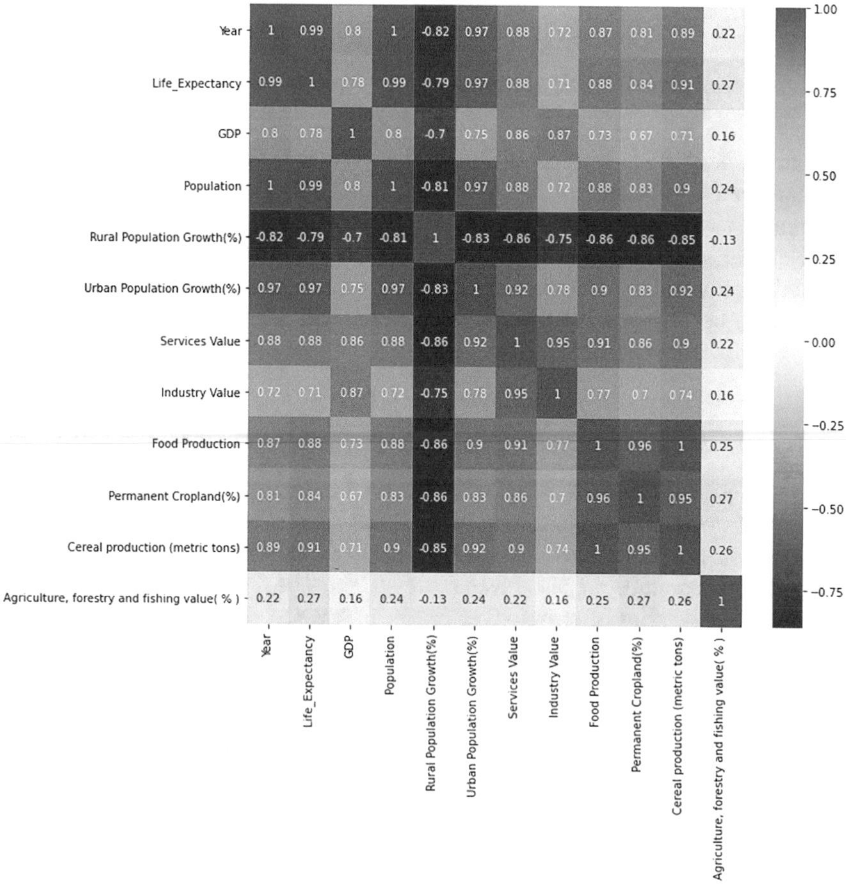

Fig. 2 Correlation matrix between features

Figure 3 shows that the data reveals the value of GDP that has risen steadily over time. As a consequence, GDP in 1960 was 4,274,893,913.49536, whereas GDP in 2020 was 353,000,000,000. It was discovered that the value of GDP had risen. The two factors of life expectancy and GDP are inextricably linked. The bigger the GDP, the higher the standard of living will be. As a result, the average life expectancy may rise. Life expectancy is also influenced by service value and industry value. The greater the service and industry values are, the better the quality of life will be. As can be seen, service value and industry value have increased significantly year after year, and according to the most recent update in 2020, service value has increased significantly and now stands at 5,460,000,000,000. And the industry value was 7,540,000,000,000, which has a positive impact on daily life. Food production influences life expectancy and quality of life. Our level of living will improve if our food production is good, and this will have a positive influence on life expectancy. From 1990 to 2020, food production ranged between 26.13 and 109.07. Agriculture, forestry, and fishing value percent are also shortly involved with life expectancy.

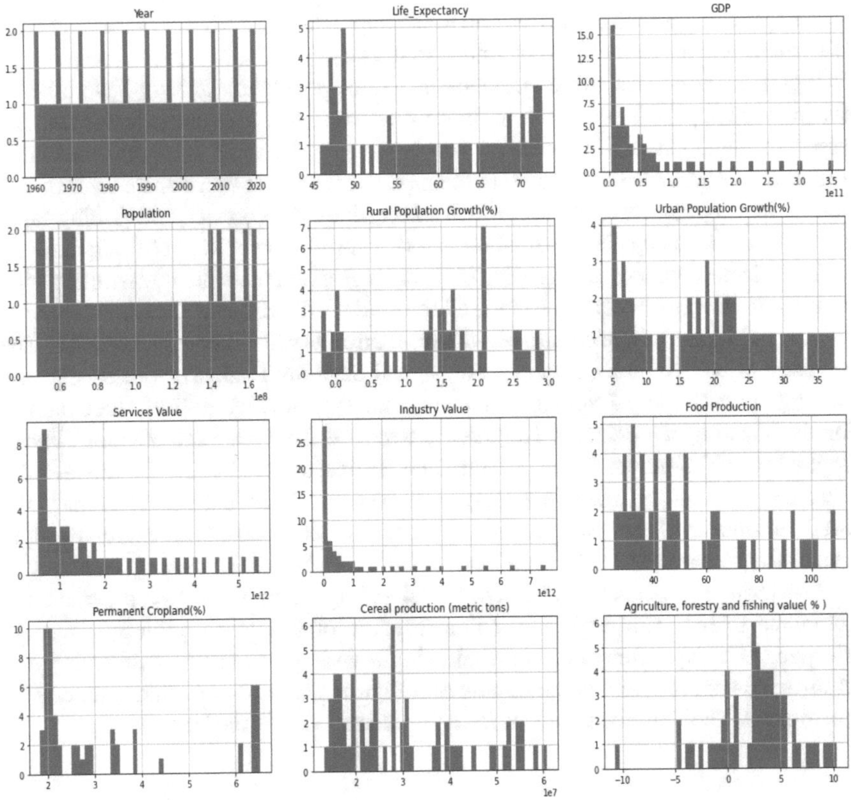

Fig. 3 Visualize all of the features

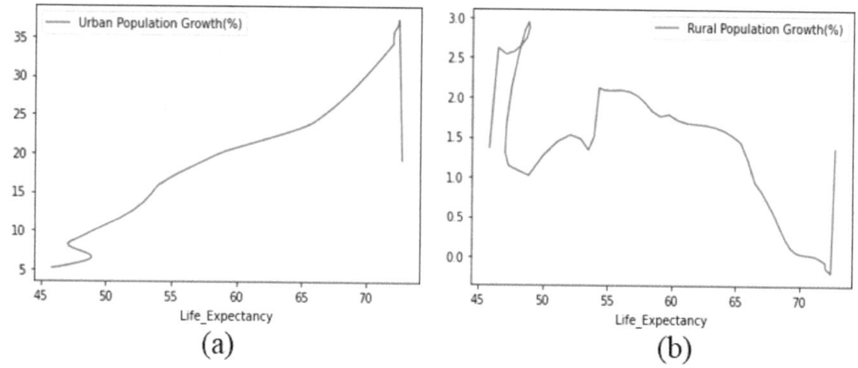

Fig. 4 Population growth of **a** Urban Area (%) and **b** Rural Area (%)

Figure 4a, b shows there are two types of population growth; rural and urban. In the 1990s century urban population percent was more than rural and year by year rural population growth decreased and urban population growth increased. The level of living improves as more people move to the city.

Figure 2 shows that life expectancy and rural population growth have a negative relationship. We can see how these characteristics are intertwined with life expectancy and have an influence on how we live our lives. Its worth has fluctuated over time. Its value has fluctuated in the past, increasing at times and decreasing at others. We drop Rural population growth and Agriculture, Forestry, and Fishing value as it was having a negative correlation and less correlation between life expectancy.

Table 1 shows that we utilize eight different regression models to determine which models are the most accurate. Among all the models, the Extreme Gradient Boosting Regressor has the best accuracy and the least error. It was 99 percent accurate. The accuracy of K-Neighbors, Random Forest, and Stacking Regressor was 94 percent. Among them, Slightly Stacking had the highest accuracy. We utilized three models for the stacking regressor: K-Neighbors, Gradient Boosting, and Random Forest Regressor, and Random Forest for the meta regressor. Among all the models, the Decision Tree has the lowest accuracy at 79 percent. With 96 percent accuracy, the Gradient Boosting Regressor comes in second. 88 percent and 87 percent for Multiple Linear Regression and Light Gradient Boosting Machine Regressor, respectively.

The term "life expectancy" refers to the average amount of time a person can anticipate to live. Life expectancy is a measure of a person's projected average lifespan. Life expectancy is calculated using a variety of factors such as the year of birth, current age, and demographic sex. Figure 5 shows the accuracy among all the models. The Extreme Gradient Boosting Regressor has the best accuracy.

Table 1 Error and accuracy comparison between all the regressor models

Models	MAE	MSE	RMSE	ACCURACY
Multiple linear regression	1.46	8.82	2.97	88.07%
K-Neighbors regressor	0.96	4.17	2.04	94.35%
Decision tree regressor	2.63	15.30	3.91	79.32%
Random forest regressor	1.06	4.28	2.06	94.21%
Stacking regressor	1.02	3.90	1.97	94.72%
Gradient boosting regressor	0.94	2.43	1.55	96.71%
Extreme gradient boosting regressor	0.58	0.44	0.66	99.39%
Light gradient boosting machine regressor	2.62	9.57	3.09	87.06%

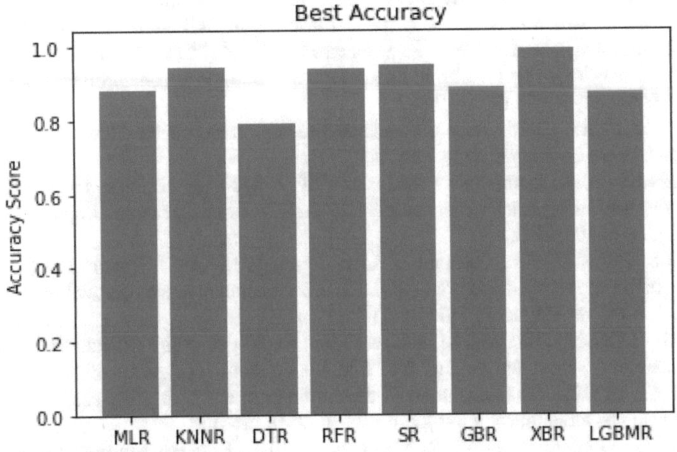

Fig. 5 Accuracy among all the models

5 Conclusion and Future Work

A country's life expectancy is affected by a variety of factors. The paper showed the pairwise relationship between life expectancy and several independent variables. We apply some machine learning models to make the prediction. Extreme Gradients Boosting Regressors in general forecast better than other regressors. Our findings lead us to conclude that life expectancy may be predicted using GDP, urban population growth (%), services value, industry value, food production, permanent cropland (%), and cereal output (metric tons). Larger datasets may result in more accurate predictions. In the ahead, additional data and newer machine learning methods would be used to improve the accuracy of forecasts.

References

1. Rubi MA, Bijoy HI, Bitto AK (2021) Life expectancy prediction based on GDP and population size of Bangladesh using multiple linear regression and ANN model. In: 2021 12th international conference on computing communication and networking technologies (ICCCNT), pp 1–6. https://doi.org/10.1109/ICCCNT51525.2021.9579594.
2. Beeksma M, Verberne S, van den Bosch A, Das E, Hendrickx I, Groenewoud S (2019) Predicting life expectancy with a long short-term memory recurrent neural network using electronic medical records. BMC Med Informat Decision Making 19(1):1–15.
3. Nigri A, Levantesi S, Marino M (2021) Life expectancy and lifespan disparity forecasting: a long short-term memory approach. Scand Actuar J 2021(2):110–133
4. Khan HR, Asaduzzaman M (2007) Literate life expectancy in Bangladesh: a new approach of social indicator. J Data Sci 5:131–142.
5. Tareque MI, Hoque N, Islam TM, Kawahara K, Sugawa M (2013) Relationships between the active aging index and disability-free life expectancy: a case study in the Rajshahi district of Bangladesh. Canadian J Aging/La Revue Canadienne du vieillissement 32(4):417–432
6. Tareque MI, Islam TM, Kawahara K, Sugawa M, Saito Y (2015) Healthy life expectancy and the correlates of self-rated health in an ageing population in Rajshahi district of Bangladesh. Ageing & Society 35(5):1075–1094
7. Ho JY, Hendi AS (2018) Recent trends in life expectancy across high income countries: retrospective observational study. bmj 362
8. Meshram SS (2020) Comparative analysis of life expectancy between developed and developing countries using machine learning. In 2020 IEEE Bombay Section Signature Conference (IBSSC), pp 6–10. IEEE
9. Matsuo K, Purushotham S, Moeini A, Li G, Machida H, Liu Y, Roman LD (2017) A pilot study in using deep learning to predict limited life expectancy in women with recurrent cervical cancer. Am J Obstet Gynecol 217(6):703–705
10. Alam MS, Islam MS, Shahzad SJ, Bilal S (2021) Rapid rise of life expectancy in Bangladesh: does financial development matter? Int J Finance Econom 26(4):4918–4931
11. Husain AR (2002) Life expectancy in developing countries: a cross-section analysis. The Bangladesh Development Studies 28, no. 1/2 (2002):161–178.
12. Choubin B, Khalighi-Sigaroodi S, Malekian A, Kişi Ö (2016) Multiple linear regression, multi-layer perceptron network and adaptive neuro-fuzzy inference system for forecasting precipitation based on large-scale climate signals. Hydrol Sci J 61(6):1001–1009
13. Kramer O (2013) K-nearest neighbors. In Dimensionality reduction with unsupervised nearest neighbors, pp 13–23. Springer, Berlin, Heidelberg
14. Joshi N, Singh G, Kumar S, Jain R, Nagrath P (2020) Airline prices analysis and prediction using decision tree regressor. In: Batra U, Roy N, Panda B (eds) Data science and analytics. REDSET 2019. Communications in computer and information science, vol 1229. Springer, Singapore. https://doi.org/10.1007/978-981-15-5827-6_15

A Data-Driven Approach to Forecasting Bangladesh Next-Generation Economy

Md. Mahfuj Hasan Shohug, Abu Kowshir Bitto, Maksuda Akter Rubi, Md. Hasan Imam Bijoy, and Ashikur Rahaman

1 Introduction

Although Bangladesh ranks 92nd in terms of landmass, it now ranks 8th in terms of people, showing that Bangladesh is the world's most populous country. After a 9-month length and deadly battle, in 1971, under the leadership of Banga Bandhu Sheikh Mujibur Rahman., the Father of the Nation, a war of freedom was waged. We recognized Bangladesh as an independent sovereign country. But Bangladesh, despite being a populous country, remains far behind the wealthy countries of the world [1] and the developed world, particularly in economic terms. Bangladesh is a developing country with a primarily agricultural economy. According to the United Nations, it is a least developed country. Bangladesh's per capita income was $12.5992 US dollars in March 2016. It increased to 2,084 per capita in August 2020. However, according to our population, this is far too low. The economy of

Md. M. H. Shohug · A. K. Bitto · A. Rahaman
Department of Software Engineering, Daffodil International University, Dhaka-1216, Bangladesh
e-mail: mahfuj.shohug@gmail.com

A. K. Bitto
e-mail: abu.kowshir777@gmail.com

A. Rahaman
e-mail: ashikur35-1575@diu.edu.bd

M. A. Rubi
Department of General Educational Development, Daffodil International University, Dhaka-1216, Bangladesh
e-mail: rubi.ged@diu.edu.bd

Md. H. I. Bijoy (✉)
Department of Computer Science and Engineering, Daffodil International University, Dhaka-1216, Bangladesh
e-mail: hasan15-11743@diu.edu.bd

P. Singh et al. (eds.), *Machine Learning and Computational Intelligence Techniques for Data Engineering*, Lecture Notes in Electrical Engineering 998,
https://doi.org/10.1007/978-981-99-0047-3_6

Bangladesh is described as a creating market economy. Its Gross Domestic Product (GDP) is dramatically expanding after freedom. Total GDP is a significant pointer of financial action and is regularly utilized by chiefs to design monetary strategy. It's a standard metric for determining the size of a country's level of economy. A country's gross domestic product (GDP) is the monetary value of a significant number of completed economic consumption produced inside its bounds over a period of time [2]. It addresses the total measurement of all financial actions. The exhibition of the economy can be estimated with the help of GDP. The issues of GDP have gotten the most worried among macroeconomic factors and statistics on GDP is displayed as the fundamental file for evaluating the public economical turn of events and for deciding about the working status of the macro-economy [3]. It is crucial to forecast microeconomic variables in the economic terminology. The main macroeconomic factors to gauge are the Gross Domestic Product (GDP), swelling, and joblessness. As a total proportion of absolute financial creation for a country, GDP is one of the essential markers used to gauge the nation's economy. Since significant monetary and political choices depend on conjectures of these macroeconomic factors, it is basic that they are just about as solid and exact as could be expected. Erroneous figures might bring about destabilizing strategies and a more unstable business cycle. GDP is possibly the main pointer of public financial exercises for a nation [4].

In this manner, the remainder of the study is in order. Section 2 of the paper is a review of the literature. The approach for forecasting GDP of Bangladesh is discussed in Sect. 3. The analysis and results are demonstrated in Sect. 4. Section 5 of the document, certainly, brings the whole thing to a conclusion.

2 Literature Review

Many papers, articles, and research projects focus on text categorization, text recognition, and categories, while some focus on particular points. Here are some of the work reviews that have been provided.

Hassan et al. [5] used the Box-Jenkins method to develop an ARIMA method for the Sudan GDP from 1960 to 2018 and evaluate the autoregressive and moving normal portions' elective ordering. The four phases of the Box-Jenkins technique are performed to produce an OK ARIMA model. They used MLE to evaluate the model. From the monetary year 1972 to the financial year 2010, Anam et al. [6] provide a period series model based on Agriculture's contribution to GDP. In this investigation, they discovered the ARIMA (1, 2, 1) methods to be a useful method for estimating Bangladesh's annual GDP growth rate. From 1972 to 2013, Sultana et al. [7] used univariate analysis to time series data on annual rice mass production in Bangladesh. The motivation of this study was to analyze the factors that influence the behavior of ARIMA and ANN. The backpropagation approach was used to create a simple ANN model with an acceptable amount of hubs or neurons in a single secret layer, variable edge worth, and swotting value [8]. The values of RMSE, MAE, and MAPE are used. The findings revealed that the ANN's estimated blunder is significantly larger than the selected ARIMA's estimated error. In this article, they considered the ARIMA model and the ANN model using univariate data.

Wang et al. [9] used Shenzhen GDP for time series analysis, and the methodology shows that the ARIMA method created using the B-J technique has more vaticination validity. The ARIMA (3, 3, 5) method developed in this focus superior addresses the principle of financial evolution and is employed to forecast the Shenzhen GDP over the medium and long term. In light of Bangladesh's GDP data from 1960 to 2017, Miah et al. [10] developed an ARIMA method and forecasted. The used method was ARIMA (autoregressive coordinated moving normal) (1, 2, 1). The remaining diagnostics included a correlogram, Q-measurement, histogram, and ordinariness test. For solidity testing, they used the Chow test. In Bangladesh, Awal et al. [11] develop an ARIMA model for predicting instantaneous rice yields. According to the review, the best-fitted models for short-run expecting Aus, Aman, and Boro rice generation were ARIMA (4,1,1), ARIMA (2,1,1), and ARIMA (2,2,3), respectively. Abonazel et al. [12] used the Box-Jenkins approach to create a plausible ARIMA technique for the Egyptian yearly GDP. The World Bank provided yearly GDP statistics figures for Egypt from 1965 to 2016. They show that the ARIMA method is superior for estimating Egyptian GDP (1, 2, 1). Lastly, using the fitted ARIMA technique, Egypt's GDP was front-projected over the next ten years.

From 2008–09 to 2012–13, Rahman et al. [13] used the ARIMA technique to predict the Boro rice harvest in Bangladesh. The ARIMA (0,1,2) model was shown to be excellent for regional, current, and absolute Boro rice proffering, respectively. Voumik et al. [14] looked at annual statistics for Bangladesh from 1972 to 2019 and used the ARIMA method to estimate future GDP per capita. ARIMA is the best model for estimating Bangladeshi GDP apiece, according to the ADF, PP, and KPSS tests (0, 2, 1). Finally, in this study, we used the ARIMA method (0,2,1) to estimate Bangladesh's GDP apiece for the following 10 years. The use of ARIMA demonstration techniques in the Nigeria Gross Domestic Product between 1980 and 2007 is depicted in this research study by Fatoki et al. [15]. Zakai et al. [16] examine the quality of the International Monetary Fund's (IMF) annual GDP statistics for Pakistan from 1953 to 2012. To display the GDP, a number of ARIMA methods are created using the Box-Jenkins approach. They discovered that by using the master modeler technique and the best-fit model, they were able to achieve ARIMA (1,1,0). Finally, using the best-fit ARIMA model, gauge values for the next several years have been obtained. According to their findings, they were in charge of test estimates from 1953 to 2009, and visual representation of prediction values revealed appropriate behavior.

To demonstrate and evaluate GDP growth rates in Bangladesh's economy, Voumik et al. [17] used the time series methods ARIMA and the method of exponential smoothing. World Development Indicators (WDI), a World Bank subsidiary, compiled the data over a 37-year period. The Phillips-Perron (PP) and Augmented Dickey-Fuller (ADF) trials were used to look at the fixed person of the features. Smoothing measures are used to guess the rate of GDP growth. Furthermore, the triple exceptional model outperformed all other Exponential Smoothing models in terms of the lowest Sum of Square Error (SSE) and Root Mean Square Error (RMSE).

Khan et al. [18] started the ball rolling. A time series model can assess the value-added of financial hypotheses in comparison to the pure evaluative capacity of the variable's prior actions; continuous improvements in the analysis of time series that suggest more current time series techniques might impart more precise standards for monetary techniques. From the monetary years 1979–1980 to 2011–2012, the characteristics of annual data on a modern commitment to GDP are examined. They used two strategies to create their informative index: Holt's straight Smoothing technique and the Auto-Regressive Integrated Moving Average (ARIMA).

3 Methodology

The main goal of our research is to develop a model to forecast the Growth Domestic Product (GDP) of Bangladesh. Our proposed model relevant theory is given below.

Autoregressive Model (AR): The AR model stands for the autoregressive model. An auto-backward model is created when a value from a time series is reverted on earlier gain against a comparable time series. This model has a request with the letter "p" in it. The documentation AR indicates that request "p" has an auto-backward model (p). In "(1)", the AR(p) model is depicted.

$$Y_t = \varphi_0 + \varphi_1 \times y_{t-1} + \varphi_2 \times y_{t-2} + \varphi_3 \times y_{t-3} \ldots\ldots + \varphi_m \times y_{t-m} \quad (1)$$

Here, T = 1, 2, 3............., t and Y_t = signifies Y as a function of time t, and ϕ_m = is in the autoregression coefficients.

Moving Average Model (MA): The moving normal model is a time series model that compensates for extremely low short-run autocorrelation. It demonstrates that the next impression is the normal of all previous perceptions. The request for the moving assert age model "q" may be decided in great part by the ACF plot of the time series. The documentation MA (q) refers to a moving normal model request "q". In "(2)", the MA(q) model is depicted.

$$Y_t = \sigma_0 + \sigma_1 \times \alpha_{t-1} + \sigma_2 \times \alpha_{t-2} + \sigma_3 \times \alpha_{t-3} \ldots\ldots\ldots\ldots + \sigma_k \times \alpha_{t-k} \quad (2)$$

where σ is the mean of the series, the parameters of the mode are $\sigma_0, \sigma_1, \sigma_3 \ldots\ldots\ldots \sigma_k$, and the white noise error terms are $\alpha_{t-1}, \alpha_{t-2}, \alpha_{t-3} \ldots \alpha_{t\text{-}k}$

Autoregressive Integrated Moving Average Model (ARIMA): The Autoregressive Integrated Moving Average model [19, 20] is abbreviated as ARIMA. In time series data, a type of model may catch a variety of common transitory occurrences. ARIMA models are factual models that are used to analyze and figure out time series data. In the model, each of these elements is clearly stated as a border. ARIMA (p, d, q) is a type of standard documentation in which the borders are replaced by numerical attributes in order to recognize the ARIMA method. We may suppose that the ARIMA (p, 1, q) method and the condition decide in "(3)" in this connected model.

$$\Delta Y_t = \varphi_0 + \varphi_1 \times \Delta y_{t-1} \ldots + \varphi_m \times \Delta y_{t-m} + \sigma_0 + \sigma_1 \times \Delta \alpha_{t-1} + \ldots. + \sigma_k \times \Delta \alpha_{t-k} \quad (3)$$

In this equation here, ΔY_t is defined as a combined of those (1) number and (2) number equations. Therefore, $\Delta Y_t = Y_t - Y_{t-k}$, to account for a linear trend in the data, a first difference might be utilized.

Seasonal Autoregressive Integrated Moving Average Exogenous Model: SARIMAX stands for Seasonal Autoregressive Integrated Moving Average Exogenous model. The SARIMAX method is created by stretching the ARIMA technique. This method has a sporadic component. As we've shown, ARIMA can make a non-fixed time series fixed by modifying the pattern. By removing patterns and irregularities, the SARIMAX model may be able to handle a non-fixed time series. SARIMAX grew as a result of the model's limitations (P, D, Q, s). They are described as follows:

P This denotes the autoregressive seasonality's order.
D This is the seasonal differentiation order.
Q This is the seasonality order of the moving average.
s This is mainly defining our season's number of periods.

Akaike Information Criterion (AIC): The Akaike Information Criterion (AIC) permits us to examine how good our model runs the enlightening record beyond overfishing it. Furthermore, the AIC score pursues a method that gains a maximum fairness of-fit rate and rebuffs them assuming they suit exorbitantly synthesis. With no one else, the AIC score isn't very useful except if we contrast it and the AIC score of a contending time series model. It relied on the model with the lower AIC score to find harmony between its capacity to fit the informational index and its capacity to try not to over-fit the informational index. The formula of AIC value:

$$AIC = 2m - 2\ln(\delta) \quad (4)$$

Here the parameters define that m = Number of model parameters. $\delta = \delta(\theta)$ = highest value of the possible function of the method. For my model here, θ = maximum likelihood.

Autocorrelation Function (ACF): It demonstrates how data values in a time series are correlated to the data values before them on the mean value.

Partial Autocorrelation Function (PACF): The theoretical PACF for an AR model "closes off" once the model is solicited. The articulation "shut off" implies that the partial auto-relationships are equivalent to 0 beyond that point on a fundamental level. In other words, the number of non-zero halfway autocorrelations provides the AR model with the request. The most ludicrous leeway of "Bangladesh GDP development rate" that is used as a pointer is referred to as "demand for the model.".

Mean Square Error: The mean square error (MSE) is another strategy for assessing an estimating method. Every error or leftover is squared. The quantity of perceptions then added and partitioned these. This method punishes enormous determining errors because the mistakes are squared, which is significant. The MSE is given by

$$MSE = \frac{1}{n}\sum_{i=1}^{n}\left(y_i - \overline{y}_i\right)^2 \tag{5}$$

Root Mean Square Error: Root mean square error is a commonly utilized fraction of the difference between allying rate (test and real) by a technique or assessor and the characteristics perceived. The RMSE is calculated as in "(6)":

$$RMSE = \sqrt{\frac{1}{n}\sum_{i=1}^{n}\left(y_i - \overline{y}_i\right)^2} \tag{6}$$

4 Analysis and Results

We use the target variable Bangladesh GDP growth rate (according to the percentage) from the year 1960 to 2021 collected from the World Bank database's official website. A portion of this typical data is shown in Table 1.

We showed the time series plots of the whole dataset from the year 1960 to 2021 on a yearly basis in Fig. 1 for both (a) GDP growth (annual%) data and (b) First Difference of GDP growth (annual%). It is observed that there is a sharp decrease in GDP growth from 1970 to 1972. After that, on an average, an upward trend is observed but has another decrease in 2020 because of the spread of Coronavirus infection.

In Fig. 2, decomposing of a time series involves the collection of level, trend, seasonality, and noise components. The Auto ARIMA system provided the AIC values for the several combinations of the p, d, and q values. The ARIMA model with minimum AIC value is chosen and also suggested SARIMAX (0, 1, 1) function

Table 1 The typical data of GDP Growth (annual %) of Bangladesh (Partial)

Year	GDP growth (Annual %)
01/01/1960	2.632
01/01/1961	6.058
01/01/1962	5.453
01/01/1963	−0.456
01/01/1964	10.953
01/01/1965	1.606

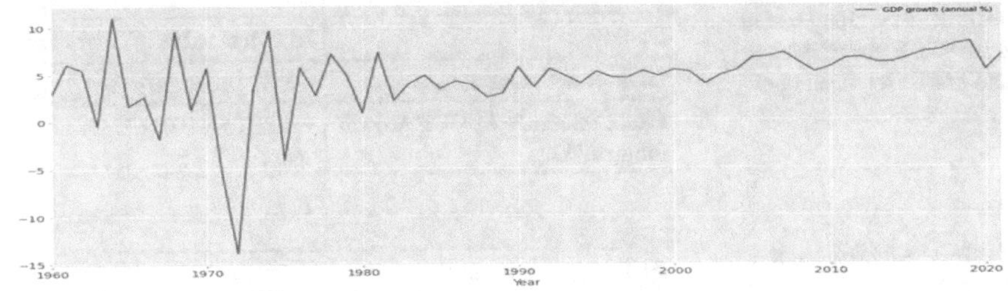

(a)

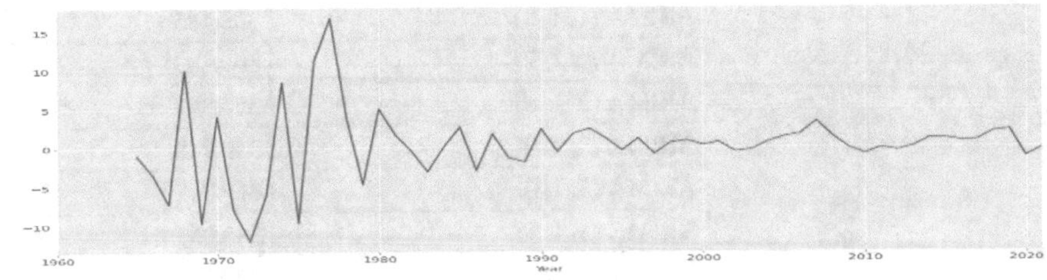

(b)

Fig. 1 The time series plots of yearly **a** GDP growth (annual %) data and **b** first difference of GDP growth (annual %)

to be used. According to this sequence, the ARIMA (p, d, q) which is ARIMA (0, 1, 1) and after that with auto ARIMA system, which is shown in Table 4. In this figure, here auto ARIMA system is defined as a SARIMAX (0, 1, 1) for creating seasonality. Here, this is dependent on the AIC value and makes the result in SARIMAX function for the next fitted ARIMA model through the train data (Tables 2 and 3).

After finding the function and ARIMA (p, d, q) values in this dataset for fitting the model, it divided the data into 80% as training data and the other 20% as test

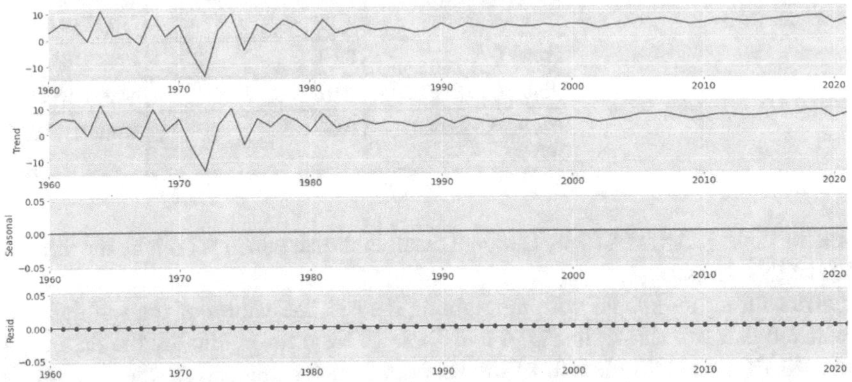

Fig. 2 Decomposition of GDP growth (annual %)

Table 2 Stationary test of actual GDP growth rate data and first differenced data

Data type	ADF test value	Stationary
GDP growth (annual %)	−1.87 (p > 0.10)	No
First Difference of GDP growth (annual %)	−4.80 (p < 0.01)	Yes

Table 3 ARIMA order selection

ARIMA (p, d, q) model	AIC
ARIMA (0, 1, 0)	371.792
ARIMA (1, 1, 0)	358.185
ARIMA (0, 1, 2)	369.803
ARIMA (2, 1, 0)	350.347
ARIMA (3, 1, 0)	346.948
ARIMA (5, 1, 0)	342.707
ARIMA (3, 1, 1)	340.057
ARIMA (0, 1, 1)	**334.750**
ARIMA (0, 1, 2)	336.632
ARIMA (0, 1, 0)	356.198
ARIMA (1, 1, 2)	338.457

Table 4 SARIMAX (0, 1, 1) model estimation

Model	No. observation: 62 Years	SARIMAX (0, 1, 1)
Running date	Tue, 17 Aug 2021	AIC: 334.750
ma.L1	Std err	0.075
ma.L1	P > |z|	0.000
sigma2	Std err	1.137
sigma2	P > |z|	0.000
Ljung-Box (L1) (Q)	0.22	Jarque–Bera (JB):188.41
Prob(Q)	0.64	Prob (JB): 0.00
Prob(H) (two-sided)	0.04	Skew:-1.91
Heteroskedasticity (H)	0.00	Kurtosis: 10.72

data. In Table 5, ARIMA (0, 1, 1) model result is shown which is built by the training data of this dataset.

After fitting the model with the training dataset, the values of the test data and predicted data are shown in Fig. 6 and Table 7. Here for predicting the data using SARIMAX seasonality is half of the year context; for this reason, the SARIMAX function is defined as SARIMAX (0, 1, 1, 6) as it has shown less error compared to the other seasonal orders. And this will be the best-fitted model, which is defined in

Table 5 Results of ARIMA (0, 1, 1) model converted from SARIMAX (0, 1, 1)

Model	No. observation: 48 years	ARIMA (0, 1, 1)
Running date	Tue, 17 Aug 2021	AIC: 274.666
Method	css-mle	S.D. of innovations: 3.816
Const	coef	0.0736
Const	std err	0.039
Const	P > |z|	0.056
ma.L1.D.GDP growth (annual %)	coef	−1.0000
ma.L1.D.GDP growth (annual %)	std err	0.062
ma.L1.D.GDP growth (annual %)	P > |z|	0.000
MA.1	Real	1.0000
MA.1	Imaginary	+ 0.0000j

the model evaluation. The RMSE value, MAE value, and model accuracy are given in Table 8, which suggested that SARIMAX (0, 1, 1, 6) model can be used as the best model for predicting the GDP growth rate (annual %).

Figure 7 depicted the forthcoming 10 years Bangladesh GDP growth rate plot after the model was evaluated. The built web application, GDP indicator [21] based on time series ARIMA model, and Fig. 8a, which introduces the GDB indicator application with the table of predicted GDP growth (%) values shown in Fig. 8b.

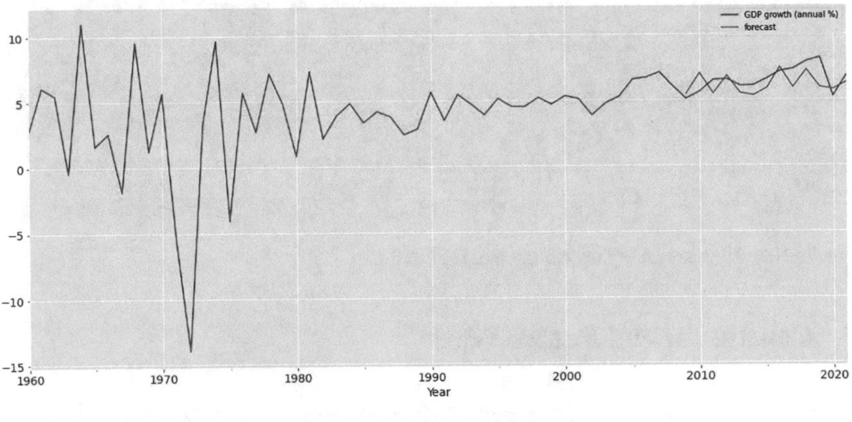

Fig. 6 Actual and predicted GDP growth rate (%)

Table 7 Actual and predicted GDP growth Rate (%) from the year 2009 to 2021

Year	GDP growth rate (%)	Forecast
2009–01-01	5.045	5.398161
2010–01-01	5.572	6.978662
2011–01-01	6.464	5.440322
2012–01-01	6.521	6.786356
2013–01-01	6.014	5.461063
2014–01-01	6.061	5.292917
2015–01-01	6.553	5.834626
2016–01-01	7.113	7.415128
2017–01-01	7.284	5.876787
2018–01-01	7.864	7.222821
2019–01-01	8.153	5.897529
2020–01-01	5.200	5.729383
2021–01-01	6.800	6.271092

Table 8 Evaluation parameter values for the SARIMAX (0, 1, 1, 6) model

Evaluation parameter for model SARIMAX (0, 1,1, 6)	Value
RMSE error value	0.991
MAE error value	0.827
Model accuracy	**87.51%**

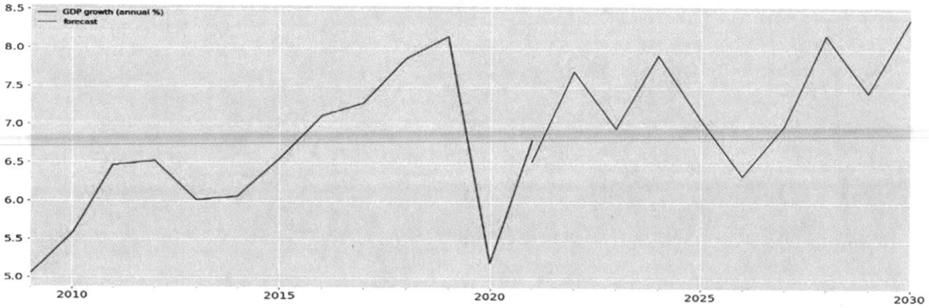

Fig. 7 Next 10 years GDP Growth (annual %) Prediction

5 Conclusion and Future Work

According to our study, we are successfully predicting the Bangladesh GDP Growth Rate with the machine learning time series ARIMA model with the order of (0, 1, 1). Here, in this model, we found this model performs 87.51% accurately. This model is verified with a minimum AIC value which is generated by the auto ARIMA function. In this model, auto ARIMA defines SARIMAX (0, 1, 1) model which is observed by

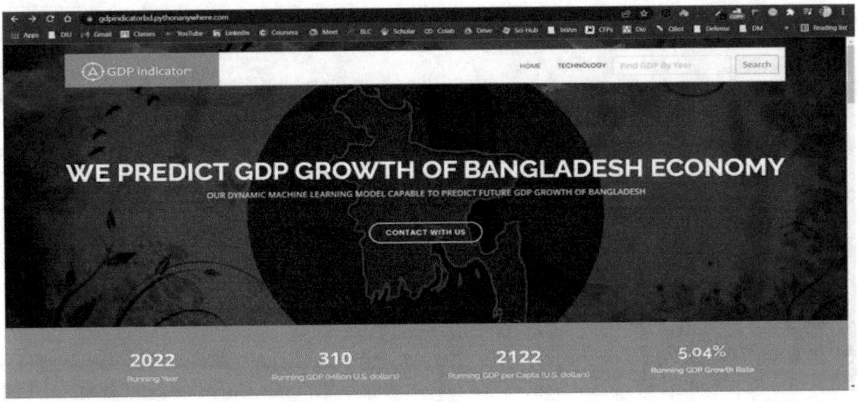

(a)

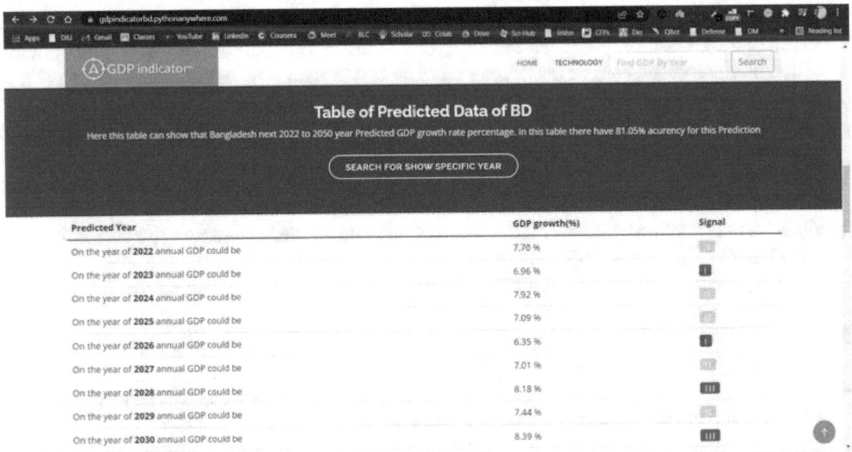

(b)

Fig. 8 User interface of GDB indicator **a** homepage and **b** predicted GDP growth rate for the next upcoming year (2022–2050)

the whole historical data. The half-yearly seasonality of this data observed this and, after that, this model predicts automatically for the upcoming year. We implement this Machine Learning time series ARIMA model on the web application as GDB indicator-BD. Here users can find Bangladesh's future GDP growth rate and they can observe that yearBangladesh's upcoming economy. In this dataset, we can also implement another machine learning or upgraded deep learning model, but we cannot implement this. So, we think that this is the gap in our research. In the future, we also work on this data with multiple features and implement other upcoming and upgraded models and I will show how it's performed in this dataset for future prediction.

References

1. Jahan N (2021) Predicting economic performance of Bangladesh using Autoregressive Integrated Moving Average (ARIMA) model. J Appl Finance Banking 11(2):129–148
2. Jamir I (2020) Forecasting potential impact of COVID-19 outbreak on India's GDP using ARIMA model. Available at SSRN 3613724
3. Chowdhury IU, Khan MA (2015) The impact of climate change on rice yield in Bangladesh: a time series analysis. Russian J Agric Socio-Econom Sci 40(4):12–28
4. Rubi MA, Bijoy HI, Bitto AK (2021) Life expectancy prediction based on GDP and population size of Bangladesh using multiple linear regression and ANN model. In: 2021 12th international conference on computing communication and networking technologies (ICCCNT), pp 1–6. https://doi.org/10.1109/ICCCNT51525.2021.9579594
5. Hassan HM (2020) Modelling GDP for Sudan using ARIMA. Available at SSRN 3630099
6. Anam S, Hossain MM (2012) Time series modelling of the contribution of agriculture to GDP of Bangladesh
7. Sultana A, Khanam M (2020) Forecasting rice production of Bangladesh using ARIMA and artificial neural network models. Dhaka Univ J Sci 68(2):143–147
8. Chowdhury S, Rubi MA, Bijoy MH (2021) Application of artificial neural network for predicting agricultural methane and CO_2 emissions in Bangladesh. In: 2021 12th international conference on computing communication and networking technologies (ICCCNT), pp 1–5. https://doi.org/10.1109/ICCCNT51525.2021.9580106
9. Wang T (2016) Forecast of economic growth by time series and scenario planning method—a case study of Shenzhen. Mod Econ 7(02):212
10. Miah MM, Tabassum M, Rana MS (2019) Modeling and forecasting of GDP in Bangladesh: an ARIMA approach. J Mech Continua Math Sci 14(3): 150–166
11. Awal MA, Siddique MAB (2011) Rice production in Bangladesh employing by ARIMA model. Bangladesh J Agric Res 36(1):51–62
12. Abonazel MR, Abd-Elftah AI (2019) Forecasting Egyptian GDP using ARIMA models. Reports on Economics and Finance 5(1): 35–47
13. Rahman NMF (2010) Forecasting of boro rice production in Bangladesh: an ARIMA approach. J Bangladesh Agric Univ 8(1):103–112
14. Rahman M, Voumik LC, Rahman M, Hossain S (2019) Forecasting GDP growth rates of Bangladesh: an empirical study. Indian J Econom Developm 7(7): 1–11
15. Fatoki O, Ugochukwu M, Abass O (2010) An application of ARIMA model to the Nigeria Gross Domestic Product (GDP). Int J Statistics Syst 5(1):63–72
16. Zakai M (2014) A time series modeling on GDP of Pakistan. J Contemporary Issues Business Res 3(4):200–210
17. Voumik LC, Smrity DY (2020) Forecasting GDP Per Capita In Bangladesh: using Arima model. Eur J Business Manag Res 5(5)
18. Khan T, Rahman MA (2013) Modeling the contribution of industry to gross domestic product of Bangladesh. Int J Econom Res 4: 66–76
19. Rahman A, Hasan MM (2017) Modeling and forecasting of carbon dioxide emissions in Bangladesh using Autoregressive Integrated Moving Average (ARIMA) models. Open J Statistics 7(4): 560–566
20. Khan MS, Khan U (2020) Comparison of forecasting performance with VAR vs. ARIMA models using economic variables of Bangladesh. Asian J Probab Stat 10(2): 33–47
21. Hasan MS, Bitto AK, Rubi MA, Hasan IB, Rahman A (2021) GDP Indicator BD. https://gdpindicatorbd.pythonanywhere.com/. Accessed 11 Feb 2022

A Cross Dataset Approach for Noisy Speech Identification

A. K. Punnoose

1 Introduction

Noisy speech poses a great challenge to a real-time, real-world speech recognition system. Speech recognition errors can be introduced at the phoneme level or at the word level depending on the type of noise. There are many ways to deal with noise in speech. One is to figure out whether the utterance is noisy before passing to the core recognition engine. This is suitable if the recognition engine is trained using clean speech data. Once the incoming utterance is identified as noisy, then appropriate mechanisms can be employed to deal with the noise. In the worst case, if the speech is too noisy, the utterance can be discarded. Here, the noisy speech detection algorithm works as a pre-processing step before the core speech recognition stage.

Another way is to simply ignore the noise in the training phase of the speech recognition engine. One advantage of this approach is that though the speech is noisy, the learned phoneme models are averaged with respect to noise. This increases the robustness of the core recognition engine. But during testing, unintended recognition error patterns like high precision for certain phonemes at the expense of other phonemes could be observed. This could require further benchmarking at the phoneme level, to be used as a general purpose recognition engine.

Another approach to deal with noisy speech identification is to add noise to the training data and train with the noise. The noise can be added through multiple noise class labels at the frame level but is a difficult task. Another way of dealing with the noisy speech identification problem is to focus on speech rather than noise. As speech is more organized compared to that of noise [1], spectral level patterns would be easily discernible for clean speech compared to that of noisy speech. At this point, it is worth noting the difference between noisy speech identification and voice activity detection (VAD). Noisy speech identification assumes a default speech

A. K. Punnoose (✉)
Flare Speech Systems, Bangalore, India
e-mail: punnoose@flarespeech.com

P. Singh et al. (eds.), *Machine Learning and Computational Intelligence Techniques for Data Engineering*, Lecture Notes in Electrical Engineering 998,
https://doi.org/10.1007/978-981-99-0047-3_7

recording and noise could be present. The task is to identify whether noise is present in the speech. On the other hand, in VAD the default is a noisy recording and speech could be present. The task is to identify whether speech is present in the recording. The techniques developed for VAD can be used interchangeably with noisy speech identification.

2 Problem Statement

Given an utterance, identify whether the utterance is noisy or not.

3 Prior Work

Noisy speech detection is covered extensively in the literature. Filters like Kalman filter [2, 3] and spectral subtraction [4, 5] have been used to remove noise in speech. But this requires an understanding of the nature of the noise, which is mostly infeasible. A more generic way is to estimate the signal-to-noise ratio(SNR) of the recording and use appropriate thresholding on SNR to filter out noisy recordings [6–9]. Voice activity detection is also extensively covered in the literature. Autocorrelation functions and their various derivatives have been used extensively for voice activity detection. Subband decomposition and suppression of certain sub-bands based on stationarity assumptions on autocorrelation function are used for robust voice activity detection [10]. Autocorrelation derived features like harmonicity, clarity, and periodicity provide more speech-like characteristics. Pitch continuity in speech has been exploited for robust speech activity detection [11]. For highly degraded channels, GABOR features along with autocorrelation derived features are also used [12]. Modulation frequency is also used in conjunction with harmonicity for VAD [13].

Another very common method is to use mel frequency cepstral features with classifiers like SVMs to predict speech regions [14]. Derived spectral features like low short-time energy ratio, high zero-crossing rate ratio, line spectral pairs, spectral flux, spectral centroid, spectral rolloff, ratio of magnitude in speech band, top peaks, and ratio of magnitude under top peaks are also used to predict speech/non-speech regions [15].

Sparse coding has been used to learn a combined dictionary of speech and noise and then, remove the noise part to get the pure speech representation [16, 17]. The correspondence between the features derived from the clean speech dictionary and the speech/non-speech labels can be learned using discriminative models like conditional random fields [18]. Along with sparse coding, acoustic-phonetic features are also explored for speech and noise analysis [19].

From the speech intelligibility perspective, vowels remain more resilient to noise [20]. Moreover, speech intelligibility in the presence of noise also depends on the listener's native language [21–24]. Any robust noisy speech identification system must take into consideration the inherent intelligibility of phonemes while scoring the sentence hypothesis. The rest of the paper is organized as follows. The experimental setup is first defined. Certain measures, that could be used to differentiate clean speech from noisy speech, are explored. A scoring function is defined to score the noisy speech. Simple thresholding on the scoring is used to differentiate noisy speech and clean speech.

4 Experimental Setup

60 h of Voxforge dataset is used to train the MLP. The rationale behind using Voxforge data is its closeness to real-world conditions, in terms of recording, speaker variability, noise, etc. ICSI Quicknet [25] is used for the training. Perceptual linear coefficients (plp) along with delta and double-delta coefficients are used as the input. Softmax layer is employed at the output. Cross entropy error is the loss function used. Output labels are the standard English phonemes.

For a 9 plp frame window given as the input, MLP outputs a probability vector with individual components corresponding to the phonemes. The phoneme which gets the highest probability is treated as the top phoneme for that frame. The highest softmax probability of the frame is termed as the top softmax probability of the frame. A set of consecutive frames classified as the same phoneme constitutes a phoneme chunk. The number of frames in a phoneme chunk is referred to as the phoneme chunk size.

For the subsequent stages, TIMIT training set is used as the clean speech training data. A subset of background noise data from the CHiME dataset [26] is mixed with the TIMIT training set and is treated as the noisy speech training data. We label this dataset as d_{train}. d_{train} is passed through the MLP to get the phoneme posteriors. From the MLP posteriors, the required measures and distributions needed to detect noisy speech recording are computed. A noisy speech scoring mechanism is defined. For testing, the TIMIT testing set is used as a clean speech testing dataset. TIMIT testing set mixed with a different subset of CHiME background noise is used as the noisy speech testing data. This data is labeled as d_{test}

We define 2 new measures, phoneme detection rate and softmax probability of clean and noisy speech. These measures are combined to get a recording level score, which is used to determine the noise level in a recording.

4.1 Phoneme Detection rate

For a phoneme p, let g be the ratio of the number of frames that got recognized as true positives to the number of frames that got recognized as false positives, for clean speech. Let h represent the same ratio for the noisy speech. The phoneme detection nature of clean speech and noisy speech can be broadly classified into three cases.

In the first category, both g and h are low. In the second case, g is high and h is low. In the third case, both g and h are high. A phoneme weighting function is defined as

$$f_1(p; g, h) = \begin{cases} x_1 \; g < 1 \text{ and } h < 1 \\ x_2 \; g > 1 \text{ and } h < 1 \\ x_3 \; g > 1 \text{ and } h > 1 \end{cases} \tag{1}$$

where $\sum x_i = 1$ and $x_i \in (0, 1]$. This is not a probability distribution function. The optimal values of x_1, x_2 and x_3 will be derived in the next section. Note that g and h are computed from the clean speech and noisy speech training data. x_3 corresponds to the most robust phoneme while x_1 corresponds to non-robust phoneme.

4.2 *Softmax Probability of Clean Speech and Noisy Speech*

Figure 1 plots the density of top softmax probability of the frames of true positive detections for the noisy speech. Figure 2 plots the same for false positive detections of clean speech. Any approach to identify noisy recordings must be able to take into account the subtle difference in these densities. As the plots are asymmetrical and skewed, we use gamma distribution to model the density. The probability density function of the gamma distribution is given by

$$f_2(x; \alpha, \beta) = \frac{\beta^{\alpha} x^{\alpha-1} e^{-\beta x}}{\Gamma(\alpha)} \tag{2}$$

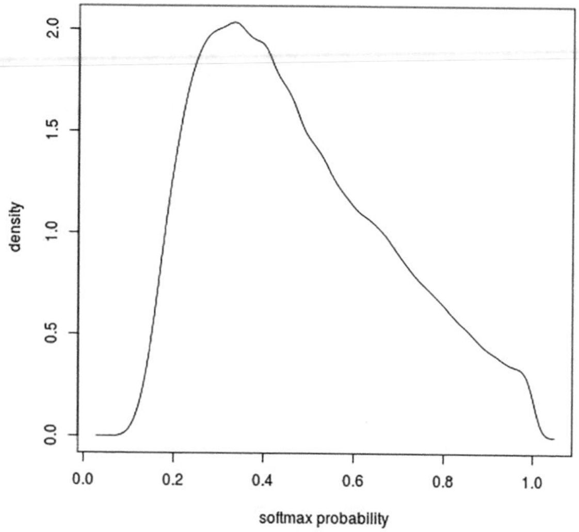

Fig. 1 Density of noisy speech true positive softmax probabilities

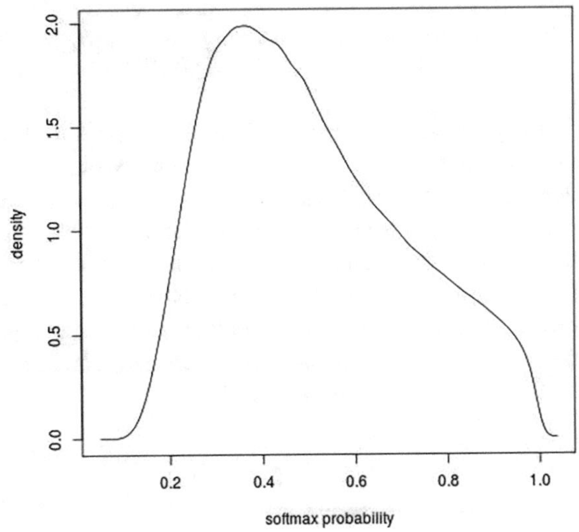

Fig. 2 Density of clean speech false positive softmax probabilities

where

$$\Gamma(\alpha) = (\alpha - 1)! \tag{3}$$

α and β are the shape and rate parameters.

4.3 Utterance Level Scoring

Given a sequence of top phonemes $[p_1 p_2 \dots p_N]$ along with the associated softmax probabilities $[q_1 q_2 \dots q_N]$, corresponding to a recording. To get the utterance level score, we first compute the geometric mean of the density ratio weighed by the phoneme probability, of all frames.

$$s = \sqrt[N]{\prod_{i=1}^{N} f_1(p_i) \frac{f_2(q_i; \alpha_+, \beta_+)}{f_2(q_i; \alpha_-, \beta_-)}} \tag{4}$$

where α_+ and β_+ are the shape and rate parameters of the true positive detection of noisy speech and α_- and β_- are the same for false positive detection of clean speech.

Using $w_i = f_1(p_i)$ and $A_i = \frac{f_2(q_i;\alpha_+,\beta_+)}{f_2(q_i;\alpha_-,\beta_-)}$, Eq. 4 can be rewritten as

$$s = \exp\left(\frac{1}{N} \sum_{i}^{N} \ln(w_i A_i)\right) \tag{5}$$

which implies

$$s \propto \frac{1}{N} \sum_{i}^{N} \ln(w_i A_i) \tag{6}$$

Equation 8 is the average of N terms, each term corresponding to a frame. We label these terms as factors. Note that f_2 is independent of phoneme and w_i is phoneme dependent. To increase the robustness of the overall recording level score, a set of conditions are introduced on these factors. Factors corresponding to frames where the phoneme detected is non-robust should be covered by 2 factors of frames where the phoneme detected is robust. Similarly, factors corresponding to frames where the phoneme detected is of intermediate confidence should be covered by 3 factors of frames where the phoneme is predicted with the highest confidence.

Define the max density ratio A as

$$A = \max_i \left[\frac{f_2(q_i; \alpha_+, \beta_+)}{f_2(q_i; \alpha_-, \beta_-)} \right] \tag{7}$$

and define the average density ratio B as,

$$B = \operatorname{avg}_i \left[\frac{f_2(q_i; \alpha_+, \beta_+)}{f_2(q_i; \alpha_-, \beta_-)} \right] \tag{8}$$

The conditions defined above can be expressed through appropriate values of the variables x_1, x_2 and x_3, which could be found by solving the following optimization problem.

$$\begin{aligned} \min_{x_1, x_2, x_3} \quad & \ln(Ax_1) + \ln(Ax_2) - 5\ln(Bx_3) \\ \text{s.t.} \quad & \ln(Ax_1) - 2\ln(Bx_3) > 0 \\ & \ln(Ax_2) - 3\ln(Bx_3) > 0 \\ & x_1 + x_2 + x_3 = 1 \\ & 0 < x_i \leq 1 \end{aligned}$$

The objective function ensures that the inequalities are just satisfied. The Hessian of the objective function is given by

$$H = \begin{bmatrix} \frac{-1}{x_1^2} & 0 & 0 \\ 0 & \frac{-1}{x_2^2} & 0 \\ 0 & 0 & \frac{5}{x_3^2} \end{bmatrix} \tag{9}$$

H is indefinite and the inequality constraints are not convex. Hence the standard convex optimization approaches can't be employed. In the training phase, the values of A and B have to be found. For a given A and B, the values of x_1, x_2 and x_3 which satisfy the inequalities have to be computed. As the optimization problem is in R^3 a grid search will yield the optimal solution.

4.3.1 Need for Inequalities

Assume the same w_i for all the frames, i.e., for every phoneme, the weightage is the same. Now consider the scenario where a set of noisy speech recordings with a roughly equal number of non-robust and robust frames are recognized, per recording. And assume that A_i values are high for non-robust phonemes, and low for robust phonemes. Then any threshold t, set for classification, will be dominated by the non-robust phoneme frames. While testing, assume a noisy speech recording with predominantly robust phonemes with low A_i values, then the recording level score s will be less than the required threshold value t, thus effectively reducing the recall of the system. To alleviate this issue, conditions are set on the weightage of phonemes based on their robustness.

5 Results

The variable values $A = 4.1$ and $B = 1.27$ are computed from d_{train}. The optimal variable values $x_1 = 0.175$, $x_2 = 0.148$, $x_3 = 0.677$ are obtained by grid search on the variable space. With the optimal variable values, testing is done for noisy speech recording identification on d_{test}. A simple thresholding on the recording level score s is used as the decision mechanism. In this context, a true positive refers to the identification of a noisy speech recording correctly. Figure 3 plots the ROC curve for noisy speech recording identification. Note that silence phonemes are excluded from all the computations.

Fig. 3 ROC curve

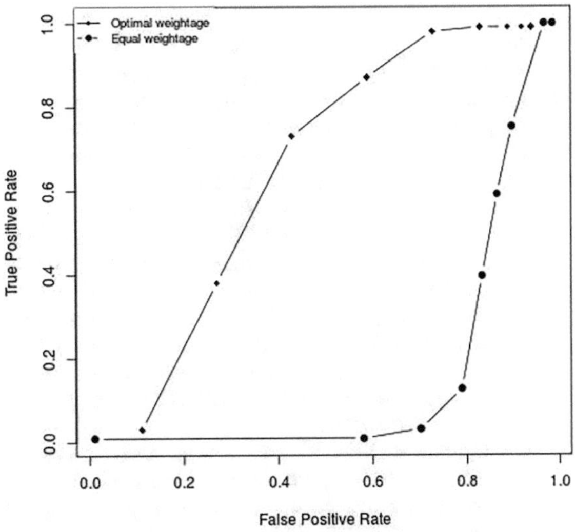

In the ROC curve, it is evident that the utterance level scoring with equal weightage for all the phonemes is not useful. But the differential scoring of phonemes based on their recognition capability makes the utterance level scoring much more meaningful.

6 Conclusion and Future Work

A computationally simple approach for detecting noisy speech recording is presented. The difference in the distribution of frame-level softmax probabilities of true positive detection of the noisy speech and false positive detection of the clean speech is demonstrated. A ratio-based scoring is defined, which is weighed by a framewise phoneme detection confidence score. To ensure robustness, a set of 2 conditions on framewise scores are imposed, which gets reflected in the values of the parameters of phoneme confidence scoring function. Grid search is done to obtain the optimal values of the phoneme confidence scoring function parameters. The geometric mean of the frame-level scores of a recording is considered as the recording level score for the noisy speech. ROC curve for various thresholds on the recording level score is plotted, from the testing dataset.

In the future, we plan to incorporate more features into this framework. Formant transitions and stylized pitch contours can be used to improve the predictive power of this framework. Other phoneme level features like plosives, voice bar, etc. can also be used for noisy speech recording identification.

References

1. Renevey P, Drygajlo A (2001) Entropy based voice activity detection in very noisy conditions. In: Proceedings of eurospeech, pp 1887–1890
2. Shrawankar U, Thakare V (2010) Noise estimation and noise removal techniques for speech recognition in adverse environment. In: Shi Z, Vadera S, Aamodt A, Leake D (eds) Intelligent information processing V. IIP 2010. IFIP advances in information and communication technology, vol 340. Springer, Berlin
3. Fujimoto M, Ariki Y (2000) Noisy speech recognition using noise reduction method based on Kalman filter. In: 2000 IEEE international conference on acoustics, speech, and signal processing. Proceedings (Cat. No.00CH37100), vol 3, pp 1727–1730. https://doi.org/10.1109/ICASSP.2000.862085
4. Boll S (1979) Suppression of acoustic noise in speech using spectral subtraction. IEEE Trans Acoust, Speech, Signal Process 27(2): 113–120. https://doi.org/10.1109/TASSP.1979.1163209
5. Mwema WN, Mwangi E (1996) A spectral subtraction method for noise reduction in speech signals. In: Proceedings of IEEE. AFRICON '96, vol 1, pp 382–385. https://doi.org/10.1109/AFRCON.1996.563142
6. Kim C, Stern R (2008) Robust signal-to-noise ratio estimation based on waveform amplitude distribution analysis. In: Proceedings of interspeech, pp 2598–2601
7. Papadopoulos P, Tsiartas A, Narayanan S (2016) Long-term SNR estimation of speech signals in known and unknown channel conditions. IEEE/ACM Trans Audio, Speech, Lang Process 24(12): 2495–2506

8. Papadopoulos P, Tsiartas A, Gibson J, Narayanan S (2014) A supervised signal-to-noise ratio estimation of speech signals. In: Proceedings of ICASSP, pp 8237–8241
9. Plapous C, Marro C, Scalart P (2006) Improved signal-to-noise ratio estimation for speech enhancement. IEEE Trans Audio, Speech, Lang Process 14(6): 2098–2108
10. Lee K, Ellis DPW (2006) Voice activity detection in personal audio recordings using autocorrelogram compensation. In: Proceedings of interspeech, pp 1970–1973
11. Shao Y, Lin Q (2018) Use of pitch continuity for robust speech activity detection. In: Proceedings of ICASSP, pp 5534–5538
12. Graciarena M, Alwan A, Ellis DPW, Franco H, Ferrer L, Hansen JHL, Janin AL, Lee BS, Lei Y, Mitra V, Morgan N, Sadjadi SO, Tsai T, Scheffer N, Tan LN, Williams B (2013) All for one: feature combination for highly channel-degraded speech activity detection. In: Proceedings of the annual conference of the international speech communication association, interspeech, pp 709–713
13. Chuangsuwanich E, Glass J (2011) Robust voice activity detector for real world applications using harmonicity and modulation frequency. In: Proceedings of interspeech, pp 2645–2648
14. Tomi Kinnunen, Evgenia Chernenko (2012) Tuononen Marko. Voice activity detection using MFCC features and support vector machine, Parsi Fr
15. Misra A (2012) Speech/nonspeech segmentation in web videos. In: Proceedings of interspeech, vol 3, pp 1975–1978
16. Deng S-W, Han J (2013) Statistical voice activity detection based on sparse representation over learned dictionary. In: Proceedings of digital signal processing, vol 23, pp 1228–1232
17. Ahmadi P, Joneidi M (2014) A new method for voice activity detection based on sparse representation. In: Proceedings of 7th international congress on image and signal processing CISP, pp 878–882
18. Teng P, Jia Y (2013) Voice activity detection via noise reducing using non-negative sparse coding. IEEE Signal Process Lett 20(5): 475–478
19. Ramakrishnan AG, Vijay GKV (2017) Speech and noise analysis using sparse representation and acoustic-phonetics knowledge
20. Julien M, Dentel L, Meunier F (2013) Speech recognition in natural background noise. PloS one 8(11):e79279
21. Jin S-H, Liu C (2014) Intelligibility of American English vowels and consonants spoken by international students in the US. J Speech, Lang, Hear Res 57
22. Rogers CL, Jennifer L, Febo DM, Joan B, Harvey A (2006) Effects of bilingualism, noise, and reverberation on speech perception by listeners with normal hearing. Appl Psycholinguist 27:465–485
23. Stuart A, Jianliang Z, Shannon S (2010) Reception thresholds for sentences in quiet and noise for monolingual english and bilingual mandarin-english listeners. J Am Acad Audiol 21:239–48
24. Van Engen K (2010) Similarity and familiarity: second language sentence recognition in first- and second-language multi-talker babble. Speech Commun 52:943–953
25. David J (2004) ICSI quicknet software package
26. Jon B, Ricard M, Vincent E, Shinji W (2016) The third 'CHiME' speech separation and recognition challenge: analysis and outcomes. Comput Speech Lang

A Robust Distributed Clustered Fault-Tolerant Scheduling for Wireless Sensor Networks (RDCFT)

Sandeep Sahu and Sanjay Silakari

1 Introduction

As daily demands, requests, and significance develop, the obligation is to strengthen and solidify the sensor network. These destinations contribute to the development of a fault-tolerant wireless sensor network. Additionally, WSNs are transported to far-flung check-in sites, segregated or hazardous zones, necessitating the use of a profoundly robust fault-tolerant component. As the organization's requirements and significance increase, it is necessary to strengthen its reliability. It resulted in the development of fault-tolerant wireless sensor networks. Typically, a WSN monitors or operates a remote or hazardous location that requires a highly reliable, fault-tolerant architecture. Weakness is a Latin term that refers to the characteristics of the framework, and portions of it can be easily modified and destroyed. The fundamental key concern in WSNs is whether the principal consequences of battery depletion and energy shortages are sensor failure and the exchange of erroneous information among sensors. As a result, enhanced fault tolerance capabilities in WSNs result in increased sensor residual lives. Due to its necessity, the sensor must withstand failure and transmit accurate data to the base station [1].

Even in the presence of a fault, a fault-tolerant framework will continue to administer itself. It is also capable of identifying errors and reviving the framework following failure. As a result, a fault-tolerant framework requires several conditions. In the realm of wireless sensor networks, fault tolerance mechanisms have

S. Sahu (✉)
Faculty, School of Computing Science & Engineering, VIT Bhopal University, Sehore (MP) 466114, India
e-mail: sandeep.sahu@vitbhopal.ac.in

S. Silakari
Professor, University Institute of Technology, Rajiv Gandhi Proudyogiki Vishwavidyalaya, Bhopal, Madhya Pradesh, India

P. Singh et al. (eds.), *Machine Learning and Computational Intelligence Techniques for Data Engineering*, Lecture Notes in Electrical Engineering 998,
https://doi.org/10.1007/978-981-99-0047-3_8

been extensively explored and discussed [2]. Some sensors fail to operate after their estimated battery life has expired, reducing the network's total lifespan and functionality. Numerous researchers have made significant contributions to fault-related obstacles such as sensor failures, coverage, connectivity issues, network partitioning, data delivery inaccuracy, and dynamic routing, among others [3, 4].

2 Literature Review

2.1 *Classification of Fault Levels*

The following are the two-level faults considered in [3–6].

Sensor Level: Faults might manifest in either the network's nodes' hardware or software components. A Fault at the **Sink/Cluster Head Level**: A fault at the sink node level will result in system failure. The active sensor's major source of energy is its limited battery capacity. After sensors are deployed in the R, they are not easily rechargeable, changeable, or, as we might say, nearly impossible to replace. Certain applications require a high coverage quality and an extended network lifetime. Due to the sensor network's inability to meet the required level of coverage and quality of service due to insufficient sensor scheduling, the network's operational life is cut short or reduced [7, 8].

2.2 *Redundancy Based Fault Tolerance in WSNs*

References [7, 8] authors use the sweep-line-based sensor redundancy check in WSNs. The authors proposed a distributed multilevel energy-efficient fault-tolerant scheduling approach for WSNs [8] based on coverage redundancy.

Clustering methods are most often used to reduce energy usage, but they may also be used to achieve various quality-driven goals like fault-tolerant capability in WSNs [9, 10]. As a network management challenge, clustering methods should tolerate malfunctioning nodes while maintaining connectivity and stability. Numerous factors may contribute to node failure in WSNs [8]. Battery depletion may result in comparable failures to physical components, such as transceiver and processor failures, susceptible to harm from external causes. Physical or environmental issues may also cause connectivity failures, rectified via topology management techniques. Failure of a node might result in a loss of connection or coverage.

Additionally, the researchers considered energy usage and the number of dead SNs. The results indicated that the new approach could dramatically minimize power usage and data loss. To overcome these concerns, [9] presented a novel technique for ensuring an energy efficient fault tolerant whale optimization based routing algorithm in WSNs. The recommended approach was utilized to ensure the network's coverage and connection. When a node fails, the "up to fail" node is evaluated and replaced before the entire network fails. However, if the "up to fail" node cannot be replaced, a quick rerouting method has been suggested to redirect the routed traffic initially through the "up to fail" node. The performance assessment of the proposed technique indicated that the number of nodes suitable for the "up to fail" node replacement is dependent on characteristics such as the node redundancy level threshold and network density [10].

Numerous researchers have examined different redundancy mechanisms in WSNs, including route redundancy, time redundancy or temporal redundancy, data redundancy, node redundancy, and physical redundancy [10]. These strategies maximize energy efficiency and assure WSNS's dependability, security, and fault tolerance. When the collector node detects that the central cluster head (CH) has failed, it sends data to the backup cluster head (CH) rather than simultaneously broadcasting data to the leading CH and backup CH. IHR's efficacy was compared to Dual Homed Routing (DHR) and Low-Energy Adaptive Clustering Hierarchy (LEACH) [11].

In this [12] paper, the authors offer a novel fault-tolerant sensor node scheduling method, named FANS (Fault-tolerant Adaptive Node Scheduling), that takes into account not only sensing coverage but also sensing level. The suggested FANS algorithm helps retain sensor coverage, enhance network lifespan, and achieve energy efficiency.

Additionally, it may result in data loss if sensors or CHs are affected (forwarders). Fault-tolerant clustering techniques can replace failed sensors with other redundant sensors and keep the network stable. These approaches allow for replacing failing sensors with other sensors, maintaining the network's stability.

We have extended the article proposed in [8] and proposed a robust distributed clustered fault-tolerant scheduling which is based on the redundancy check algorithm (sweep-line approach [7, 8]) that provides the number of redundant sensors for *R*. The proposed RDCFT determines the 1-coverage requirement precisely and fast while ensuring the sensor's redundancy eligibility criterion at a low cost and with better fault tolerance capability at sensor and cluster level fault detection and replacement. Additionally, we simulated and analyzed the suggested work's correctness and efficiency in various situations.

3 Proposed Work

3.1 Network Model, Preliminaries, and Assumptions

This paper discusses a distributed fault-tolerant strategy based on a clustering methodology to ensure that the whole network is wholly linked. *Clustering* is a distributed system that enables scalability in network management processes. By reducing communication messages to the sink or base station, this strategy enables us to build a fault-tolerant network with an energy-efficient network. Two distinct

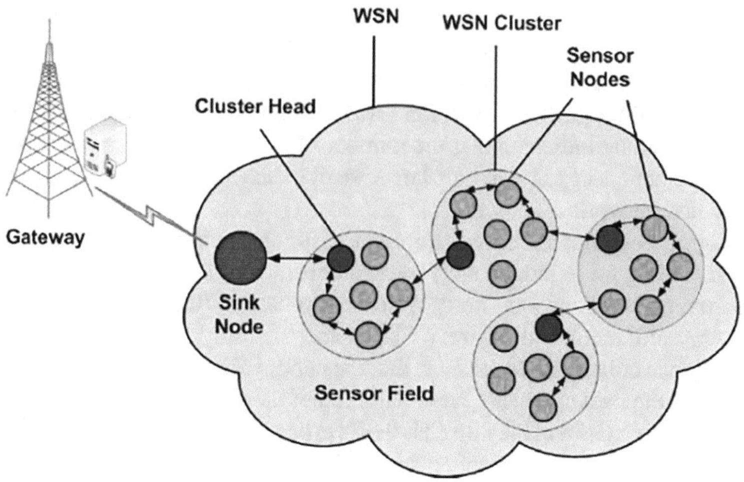

Fig. 1 Clustered architecture of A WSN

ways will be invoked if fault detection and recovery modes are necessary. Why it is called "two-way"? Because the proposed approach has two modes of execution: first, when all sensors are internally deployed, i.e., at the start of the first network round, and second, when all sensors are externally deployed (assuming a 100% energy level). Following that, the second method is when the remaining energy level of all sensors is 50% or less. The suggested process that we apply in our scheme consists of two phases: randomly selecting a cluster head (CH) and forming a collection of clusters. We should emphasize that we presume the WSNs employed in our method are homogeneous. Figure 1 shows the clustered architecture of WSNs that we consider in this section.

The failure of one or more sensors may disrupt connection and result in the network being divided into several discontinuous segments. It may also result in connection and coverage gaps in the surrounding region, which may damage the monitoring process of the environment. The only way to solve this issue is to replace the dead sensors with other redundant ones. Typically, the CH monitors the distribution, processing of data and making judgments. When a CH fails, its replacement alerts all sensors of its failure.

3.2 *Fault Detection and Recovery*

The first step of the fault management process is fault identification, the most critical phase. However, errors must be identified precisely and appropriately. One challenge is the specification of fault tolerance in WSNs; there is a trade-off between energy usage and accuracy. As a result, we use a cluster-based fault detection approach that conserves node energy and is highly accurate. Our technique for detecting faults is as follows [12–15]:

Detection of intra-cluster failures: If CH does not receive data from a node for a preset length of time, it waits for the next period. Due to the possibility of data loss due to interference and noise when the node is healthy, if CH does not receive a packet after the second period, this node is presumed to be malfunctioning. As a result, CH transmits a message to all surrounding CHs and cluster nodes, designating this node with this ID as faulty.

Intra-cluster error detection: When CH obtains data from nodes that are physically close together, it computes and saves a "median value" for the data. CH compares newly collected data to the per-request "median value." When the difference between the two values exceeds a predefined constant deviation, represented by CH detects an error and declares the node that generated the data faultily. Again, CH notified all surrounding CHs and nodes in his cluster that the node with this ID was faulty.

Detection of inter-cluster faults: CHs are a vital component of WSNs, and their failure must be identified promptly. As a result, we employ this method. CHs communicate with other CHs regularly. This packet contains information on the cluster's nodes. If a CH cannot receive this packet from an adjacent CH, it is deemed faulty.

3.3 Redundancy Check and Clustering in WSNs

The sensors are assumed to be arranged randomly and densely over an R-shaped dispersed rectangular grid. All sensors are identical in terms of sensing and communication ranges, as well as battery power consumption. Consider two sensors S_i and S_j, with a distance between them of R_s (S_i and $S_j \le R_s$). The sector apexed at Si with an angle of 2α can be used to approximate the fraction of S_i's sensing region covered by S_j, as illustrated in Fig. 2. As indicated in Eq. 1, the angle can be computed using the simple cosine rule, also explained in [6].

$$\cos\alpha = \frac{\overline{|S_i p|}^2 + \overline{|S_i S_j|}^2 - \overline{|S_j p|}^2}{2\overline{|S_i p|}^2\,\overline{|S_i S_j|}}.\ \text{Hence}\ \alpha = \text{arcos}\frac{\overline{|S_i S_j|}^2}{2R_S} \tag{1}$$

By their initial setup phase, each sensor creates a table of 1-hop detecting neighbors based on received HELLO messages. The contribution of a sensor's one-hop detecting neighbors is determined. A sensor S_j is redundant for full-covered if its 1-hop active sensing neighbors cover the complete 360° circle surrounding it. To put it another way, the union of the sectors contributed by sensors in its vicinity to cover the entire 360° is defined as a sensor S_j's redundant criterion for full-covered. As a result, the sensor S_j is redundant.

It is possible to accomplish this algorithmically by extending the sweep -ine-based algorithm for sensor redundancy checking. Assume an imaginary vertical line sweeps these intervals between 0° to 360°. If the sweep-line intersects k_p intervals from I_{Nj}

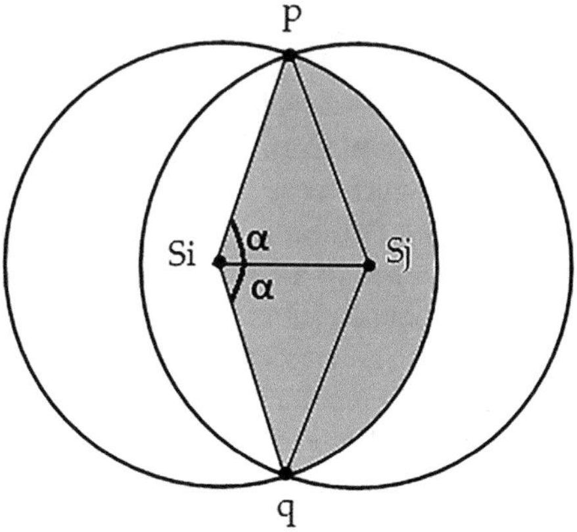

Fig. 2 Approximated region of S_i covered by S_j

and is in the interval ipi in I_{CQj}, the sensor Sj is redundant in ip. If this condition holds true for all intervals in I_{CQj}, then the sensor S_j is redundant, as illustrated in Fig. 3 as a flowchart for a sweep-line algorithm-based redundancy check of a sensor [7, 8]. We hold a variable CCQ for the current CQ and a sweep-line status l for the length of an interval from I_{CQj} intersected by the sweep-line. Only when the sweep-line crosses the left, or right terminus of an interval does its status change. As a result, the event queue Q retains the endpoints of the intervals I_{CQj} and I_{Nj}.

- If sweep-line crosses the event left endpoint of i_p in I_{CQj} then CCQ is k_p.
- If sweep-line crosses the event left endpoint of s_p in I_{Nj} then increment l.
- If sweep-line crosses the event right endpoint of s_p in I_{Nj} then decrement l.

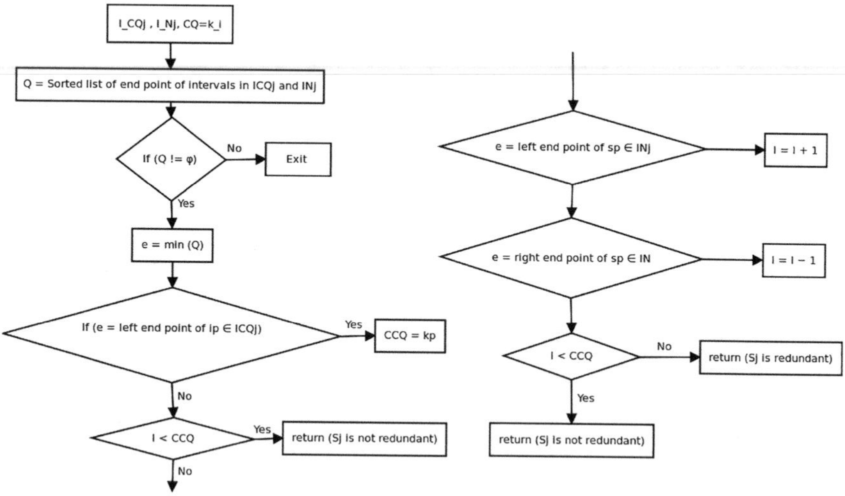

Fig. 3 Flowchart for Redundancy check of a sensor (sweep-line based [7, 8])

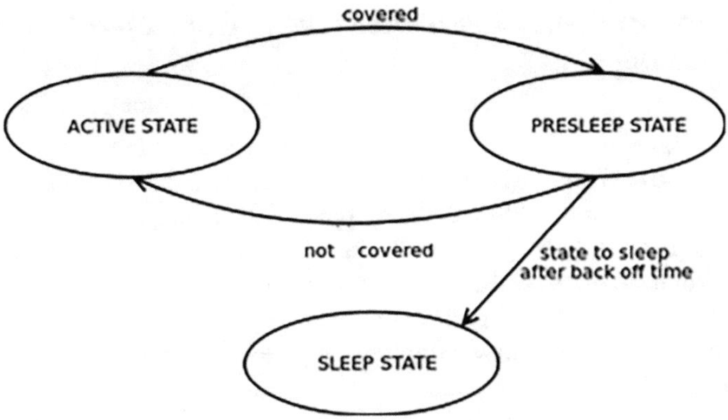

Fig. 4 State transition of a sensor

If the sweep-line status l remains greater than or equal to CCQ for the sweep duration, the sensor node S_j serves as a redundant sensor for the full-covered. Figure 4 shows the transition state of a sensor, i.e., it can be either one of the states viz., active, presleep, or sleep. This sleeping competition can be avoided using simple back-off time.

3.4 Selection of Cluster Head

The Cluster Head (CH) is chosen at random by the base station (BS) among the cluster's members. Then, CH will send a greeting message to all cluster members, requesting their energy levels. The CH will communicate the energy levels to the BS and then perform the hierarchy process. The BS will now establish a hierarchy based on the excess energy. A CH will be created according to the hierarchy that has been established. As the CH grows with each round, the hierarchy is built. The first round's CH will be the node with the most considerable energy storage capacity. The second round includes the node with the second-highest energy level.

Similarly, the third round's CH will comprise the three top nodes. Dynamically, CH is selected for each round. The initial round continues until the highest node's energy level meets the energy level of the second-highest node. The second round will continue until the first two nodes with the most significant energy levels reach the third level. Likewise, for the third round, the process is repeated. As a result, time allocation is also accomplished dynamically.

3.5 *Algorithm Phase: Distributed Clustered Fault-Tolerant Scheduling*

Sensors may be deterministically or randomly scattered at the target region for monitoring within the R. We propose a clustered fault-tolerant sensor scheduling consisting of a sequence of algorithms to effectively operate the deployed WSN. Each sensor executes the defined duties periodically in every round of the total network lifetime and periodically detects the faulty sensor nodes. The flowchart of the proposed mechanism is also shown in Fig. 5.

Our proposed clustered fault-tolerant sensor scheduling protocol has the following assumptions:

- All deployed sensor nodes are assigned a unique identifier ($sensor_{id}$).
- Sensors are homogeneous and all are locally synchronized.

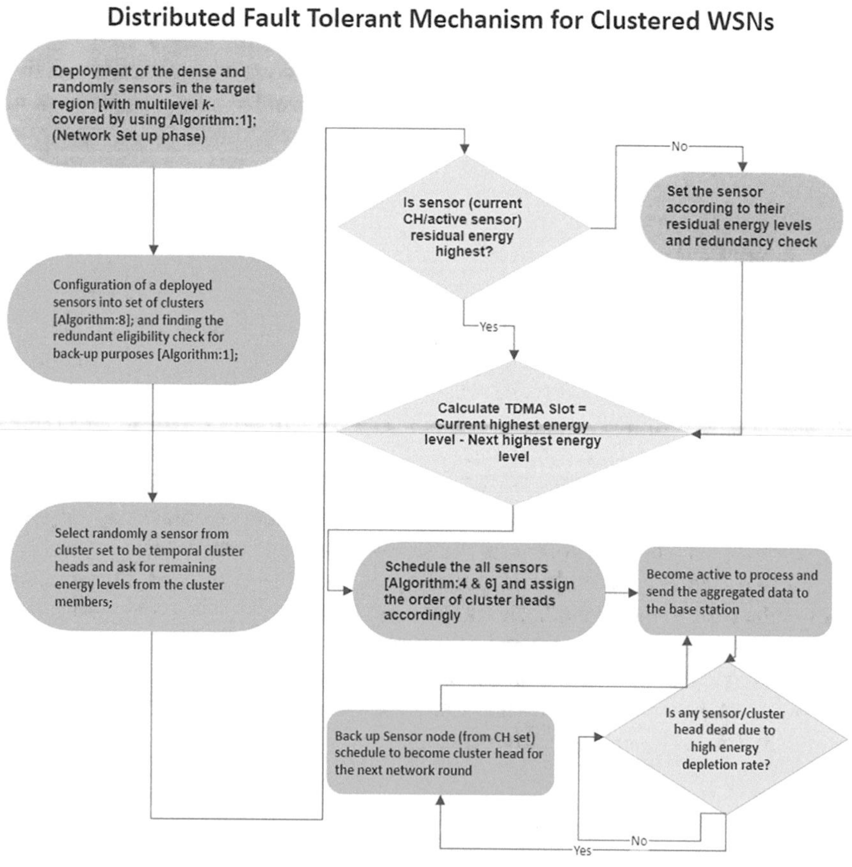

Fig. 5 Flowchart for the proposed scheme *RDCFT* for WSNs

- Sensors are densely deployed with static mode. (redundancy gives better fault tolerance capability in WSNs).
- The set of active/alive sensor nodes is represented by $\{S_{an}\} = \{S_{a1}\}, S_{a2}, S_{a3}, \ldots, S_{an}\}$.
- The set of cluster head nodes is represented by $\{CH_m\} = \{S_{CH1}, S_{CH2}, S_{CH3}, \ldots, S_{CHm}\}$.
- The set of faulty sensor nodes is represented by $\{S_{fn}\} = \{S_{f1}\}, S_{f2}, S_{f3}, \ldots, S_{fn}\}$.

PROPOSED ALGORITHM: *RDCFT*

Input: Set of randomly deployed sensor nodes $S_n = \{s_1, s_2, \ldots s_n\}$ in the target region, and these sensors arranged as Cluster Head set (CH) (CH_1, CH_2, CH_3 ..., CH_m) for all S_n;.

Process: Configure the sensors into set of clusters and also find their redundant eligibility for backup CH purpose;

Process: Random selection of temporal CH for the first network round and ask for remaining cluster members to check the energy levels;

Output: Selection of the sensor node as CH and Set of Cluster Head (CH_m), set of faulty sensors;

Initialize: Random selection of temporal CH for the first network round and ask for remaining cluster members to check the energy levels.

```
If {Current CH (residual energy) Er ≥ THRESHOLD (highest residual energy level)}
{
Eligible for the current round Tr ;
For each active CH: Calculate TDMA slot (TSactive);
        Schedule all CHs;
        Become active and perform assign tasks and send aggregated data to base station;
        Sensor node declares itself as a non-faulty;
        }
        Else
        {
        If Sensor node detected/observed low energy or as a faulty;
                Procedure Call: redundancy check;
                check redundancy eligibility and go for backup;
                Broadcast Ibackup message forall Sn;
}
Initialize Timer for all Sn T = 0;
While (current network round is over);
        If (transmission energy ≥ residual energy) or Is any current active sensor or CH
dead? ;
{
Yes, Go to backup set and activate alarm for all backup sensors;
Procedure Call: Fault-check;
        Else; remain active;
        End if;
        End for;
}
```

Table 1 Simulation parameters

Parameter	Value
Number of sensor nodes	100, 200, 300, & random
Network area (meter^2)	100 × 100
Clusters	Differs
Distributed subregion size	30 m x 30 m
Initial level of energy in each sensor	10 J
Energy consumption for transmission	0.02 J
Energy consumption for receiving	0.001 Jules
Communication & sensing ranges	4 m- 3 m
Threshold energy in each sensor	2 J
Simulation & round time	1000–1500 s and 200 s

3.6 Simulation Setup and Results

This section illustrates the experimental setup and the proposed algorithms' findings. We assessed the proposed algorithms' performance using a network simulator [16]. The proposed protocol for *RDCFT* is simulated using NS2, and the parameters utilized are shown in Table 1. *RDCFT* is being tested against existing methods LEACH and randomized scenarios using the mentioned standard metrics and is defined as follows in Table 1. The simulation is divided into the following steps:

(1) Specify the properties of the sensor node;
(2) Assign the x and y axes to a two-dimensional rectangular coordinate system (distributed region/subregion) for each sensor node in the *R*;
(3) Assign a transmission radius (a communication range denoted by the symbol R_c);
(4) Use the edges of the network graph to represent the connections between each sensor node and its one-hop neighbors.
(5) Begin the timer for the TDMA time slot.
(6) Using the energy calculation, redundancy and assign a cluster head randomly.
(7) Apply the scheduling *RDCFT* algorithm for faulty sensor node(s) detection of the network;
(8) Each sensor node uses energy when it generates, receives, or transmits a packet;
(9) Terminate the time slot timer (on a round basis) when the number of packets received in the sink matches the number of network nodes;
(10) Repeat steps until the sink have no live neighbors;
(11) Stop the TDMA time slot timer (round basis) when the network lifetime is over.

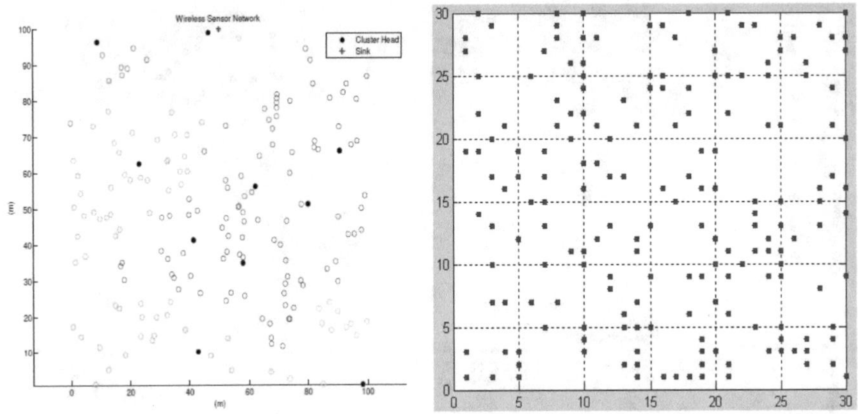

Fig. 6 Cluster Formation in *R* (total area 100 × 100 & 30 × 30 subregions)

The duration of an experiment is measured in time units (seconds in our simulator), and it is defined as the number of running steps required until no node can reach the CH. A node that lacks the energy necessary to send or relay a packet is a dead node. A disconnected node is also considered to be a dead node.

Figure 6 shows the deployment of sensor nodes for a subregion of RoI and Fig. 6 represents the cluster Formation in RoI (total area 100m_100m). As a simulation result, Fig. 7 represents the average number of alive sensors for RoI compared with the randomized method. Figure 6 represents the average number of faulty sensors (including member sensor nodes and CHs) for RoI. There are two types of faulty sensors that can be detected as faulty CH nodes and/or faulty, normal sensor nodes (members of a cluster). Subsequently, Fig. 7 represents the average number of backup, CH, and active sensors for RoI during several simulation rounds. Figure 7 shows the proposed RDCFT simulated, and the result shows there are more number of alive nodes in our proposed method than LEACH and randomized methods versus number of network rounds.

4 Conclusion and Future Remarks

The proposed approach is based on the redundancy of the sensor and CHs. This clustering approach maximizes the longevity of the network. We have extended the article proposed in [8] and the proposed method begins with detecting defects and can find the faulty sensor nodes using fault detection algorithm and replace the faulty sensor with redundant sensors for the same *R*. The fault detection process is carried out by scheduled communication messages exchanged between nodes and CHs is O (nlogn). Second, the approach commences a recovery period for CHs/common sensors that have been retrieved with the help of redundant sensors using the proposed

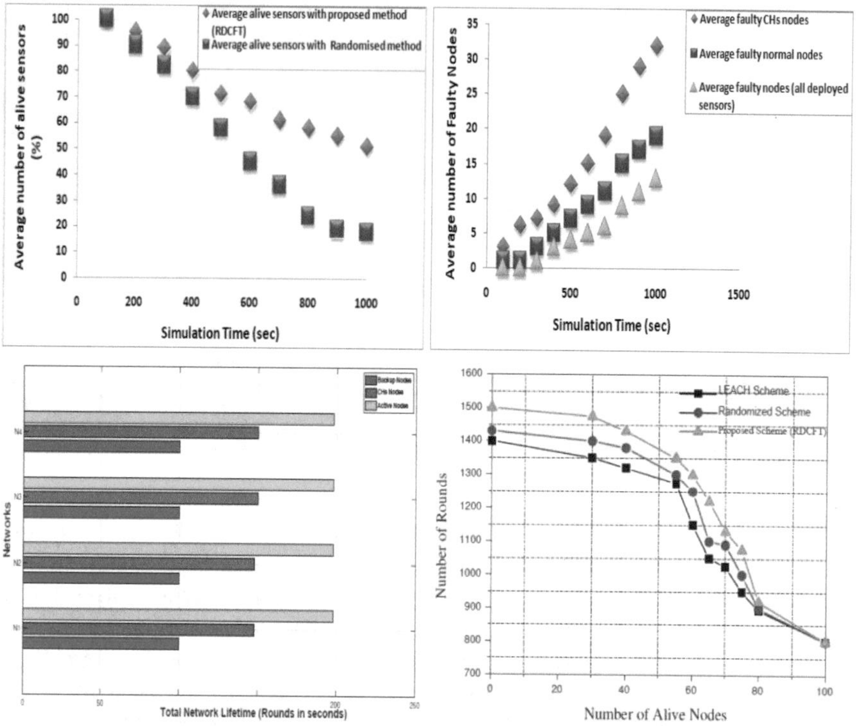

Fig. 7 Measurements of average alive, CHs, Backup and faulty sensors for *R*

algorithm. This is a novel and self-contained technique since the proposed method does not need communication with the BS/sink to work. Simulations are performed to evaluate the efficiency and validity of the overall proposed works in terms of energy consumption, coverage ratio, fault tolerance scheduling.

Our proposed efforts are based on WSNs' largely two-dimensional architecture. Future remarks should include 3D-based WSNs. Future studies will also address three critical challenges in 3D-WSNs, including energy, coverage, and faults, which may pave the way for a new approach to researching sustainable WSNs to optimize the overall network lifetime.

References

1. Yick J, Mukherjee B, Ghosal D (2008) Wireless sensor network survey. Comput Netw 52(12):2292–2330
2. Alrajei N, Fu H (2014) A survey on fault tolerance in wireless sensor networks. Sensor COMM 09366–09371
3. Sandeep S, Sanjay S (2021) Analysis of energy, coverage, and fault issues and their impacts on applications of wireless sensor networks: a concise survey. IJCNA 8(4):358–380

4. Kakamanshadi G, Gupta S, Singh S (2015) A survey on fault tolerance techniques in wireless sensor networks. Google Sch 168–173
5. Mitra S, Das A (2017) Distributed fault tolerant architecture for wireless sensor network. MR3650784 94A12 (68M15) 41(1), 47–55
6. Raj R, Ramesh M, Kumar S (2008) Fault-tolerant clustering approaches in wireless sensor network for landslide area monitoring, pp 107–113
7. Tripathi RN, Rao SV (2008) Sweep line algorithm for k-coverage in wireless sensor networks: In 2008 fourth international conference on wireless communication and sensor networks, pp 63–66
8. Sahu S, Silakari S (2022a) Distributed Multilevel k-Coverage Energy-Efficient Fault-Tolerant Scheduling for Wireless Sensor Networks. Wireless Personal Communications Springer, Vol 124 (4), 2893–2922. https://doi.org/10.1007/s11277-022-09495-3
9. Sahu S, Silakari S (2022b) A whale optimization-based energy-efficient clustered routing for wireless sensor networks. Soft Computing: Theories and Applications: Proceedings of SoCTA 2021, LNNS vol 425, 333–344, Springer, Singapore https://doi.org/10.1007/978-981-19-0707-4_31
10. Shahraki A, Taherkordi A, Haugen Ø, Eliassen F (2020) Clustering objectives in wireless sensor networks: a survey and research direction analysis. Comput Netw 180:107376
11. I. El Korbi, Y. Ghamri-Doudane, R. Jazi and L. A. Saidane: Coverage-connectivity based fault tolerance procedure in wireless sensor networks," 9th International Wireless Communications and Mobile Computing Conference (IWCMC), 2013, pp. 1540–1545 (2013).
12. Choi J, Hahn J, Ha R (2009) A fault-tolerant adaptive node scheduling scheme for wireless sensor networks. J Inf Sci Eng 25(1):273–287
13. Qiu M, Ming Z, Li J, Liu J, Quan G, Zhu Y (2013) Informer homed routing fault tolerance mechanism for wireless sensor networks. J Syst Archit 59:260–270
14. Heinzelman WR, Chandrakasan A, Balakrishnan H (2000) Energy-efficient communication protocol for wireless microsensor networks. In: Proceedings of the 33rd Annual Hawaii international conference on system sciences, vol 2, pp 10
15. Zhang Z, Mehmood A, Shu L, Huo Z, Zhang Y, Mukherjee M (2018) A survey on fault diagnosis in wireless sensor networks. IEEE Access 6:11349–11364
16. Issariyakul T, Hossain E, Issariyakul T, Hossain E (2012) Introduction to Network Simulator NS2, USA, Springer US

Audio Scene Classification Based on Topic Modelling and Audio Events Using LDA and LSA

J. Sangeetha, P. Umamaheswari, and D. Rekha

1 Introduction

With the rapidly increasing availability of digital media, the ability to efficiently process audio data has become very essential. Organizing audio documents with topic labels is useful for sorting, filtering and efficient searching to find the most relevant audio file. Audio scene recognition (ASR) involves identifying the location and surrounding where the audio was recorded. It is similar to visual scene recognition that involves identifying the environment of the image as a whole, with the only difference that here it is applied to audio data [1–3].

In this paper, we attempt to perform ASR with topic modelling. Topic modelling is a popular text mining technique that involves using the semantic structure of documents to group similar documents based on the high-level subject discussed. Assigning topic labels to documents helps for an efficient information retrieval by yielding more relevant search results. In recent times, researchers have applied topic modelling to audio data and achieved significant results. Topic modelling can be extended to ASR due to the presence of analogous counterparts between text documents and audio documents.

While an entire document can be split into words and then lemmatized, an audio document can be segmented at the right positions to derive the words and each frame can correspond to the lemmatization results. There are several advantages to using audio over video for classification tasks such as the ease of recording, lesser storage requirement, lesser pre-processing overhead and ease of streaming over networks. ASR has many useful applications [4]. ASR aids in the development of intelligent

J. Sangeetha · P. Umamaheswari · D. Rekha (✉)
Department of Computer Science and Engineering, Srinivasa Ramanujan Centre SASTRA University, Kumabakaonam 612001, India
e-mail: in_sona@src.sastra.edu

P. Singh et al. (eds.), *Machine Learning and Computational Intelligence Techniques for Data Engineering*, Lecture Notes in Electrical Engineering 998,
https://doi.org/10.1007/978-981-99-0047-3_9

agents by perceiving the surrounding environment for accurate information extraction, a concept that can be extended to home automation devices. ASR can also be used in the aqua-culture industry to accomplish classification algorithm with respect to the context based environment, and in order to evaluate the feed intake of prawns accuately. Acoustic controllers based on ASR have seen widespread usage in aquaculture industry to compute feed volumes for prawns and shrimp breeds.

The remaining part of the work is structured as given here. Section 2 contains the related work; Sect. 3 concisely gives two topic models: Latent Semantic Analysis (LSA) and Latent Dirichlet Allocation (LDA); Sect. 4 has focussed on the proposed framework; Sect. 5 illustrates the results of experimental analysis; Sect. 6 contains conclusion and the future enhancements.

2 Related Work

Audio scene recognition has been studied as a computational classification problem in the past. The setting of the recording could be any environment ranging from densely populated food markets and shopping malls to calm interior places like a family home. ASR can have many useful applications. It can be used in mobile devices which can make it smart [4]. It can also be useful in aquaculture industry [5], it makes the sound classification based on the context environment, that can help to calculate prawns consumption of feed more accurately; ASR is also used in smart homes [6] etc.

Leng et al. [7] devised a semi-supervised algorithm that focuses on the unlabelled samples within the margin band of SVM that was robust to sizes of both labelled and unlabelled training samples. Their proposed algorithm SSL_3C when applied to audio event classification was able to achieve significant classification accuracy post active learning. The samples had high confidence values and also being meaningful at the same time. This algorithm is suitable for several tasks as it significantly reduces the manual effort for labelling samples of large volumes.

Spoken document indexing is also another similar research area that is actively studied. The Speech Find project [8] team performed indexing on the NGSW repository vastly consisting of recordings from broadcasts and speeches of the twentieth century. The researchers used audio segmentation techniques combined with speech transcription to identify particular segments of the broadcast that were considered relevant.

ASR has also been widely performed on telephone speeches. Peskin et. al. [9] performed ASR on the Switchboard corpus and achieved 50–60% accuracy by using human transcripts for training.

Classifiers based on topic models are primarily used for text analysis applications, with Probabilistic Latent Semantic Analysis (PLSA) and LDA being the most prevalent [10–12]. Introduced by Hofmann, PLSA identifies topics of documents using a distribution of words and also it does not concern with the distribution of topics in a document. LDA addresses this concern by using Dirichlet prior for the document and word topics and eventually creating a Bayesian form of PLSA. Mesaros et al. [10] performed audio event detection with HMM model. PLSA was used for probabilities prior to the audio events which were then transformed to derive the transition probabilities. Hu et al. [11] improved the performance of LDA for audio retrieval by modelling with a Gaussian distribution. It uses a multivariate distribution for the topic and word distribution to alleviate the effects of vector quantization.

In this proposed work we adopted LSA and LDA to achieve ASR. As PLSA/LDA based ASR algorithms [12–14] have been compared with this proposed algorithm, it utilizes document event cooccurrence matrix, whereas in [12–14], document word cooccurrence matrix has been used for analyzing the topic. This method extracts the distribution of topics that would express the audio document in a better way, and then also we can attain better recognition results. Common audio events suppression and emphasizing unique topics are achieved by weighting the event distribution audio documents.

3 LSA and LDA

3.1 Latent Semantic Analysis (LSA)

LSA is a technique which uses vector-based representations for texts to map the text model using the terms. LSA is a statistical model which compares the similarity between texts based on their semantics. It is a technique used in information retrieval, analyzing relationships between documents and terms, identifying the hidden topics, etc., The LSA technique analyzes large corpus data and forms a document-term cooccurrence matrix to find the existence of the term in the document. It is a technique to find the hidden data in a document [15]. Every document and the terms are represented in the form of vectors with their corresponding elements related to the topics. The amount of match for a document or term is found using vector elements. The hidden similarities can be found by specifying the documents and terms in a common way.

3.2 Latent Dirichlet Allocation (LDA)

LDA is a generative probabilistic model for collections of discrete data. It is a Bayesian model which has three levels of hierarchy and each component of a collection will be modelled using the set of latent topics as a finite combination. Every term is derived from topics that are not observed directly. Every topic is modelled based on the set of probabilities as an infinite combination [16].

4 Framework of the Proposed Work

The proposed framework is shown in Fig. 3. First audio input vocabulary set is prepared, then the document-term cooccurrence matrix is generated to finally classify the audio.

4.1 Input Vocabulary Creation

Considering the audio vocabulary set as input, each frame is matched with a similar term in the vocabulary for training the model. Then the document-term cooccurrence matrix is counted which is represented as Z_{train} In the training set, the labels of the audio frames can be known previously to calculate the number of event term cooccurrence matrix X_{train} In the training dataset, if there are 'J' documents $\{d_1,d_2....d_J\}$and 'j' audio events $\{ae_1,ae_2,...ae_j\}$ and if the audio vocabulary set size is 'I', then the matrix I x J represents Z_{train} and the matrix I x j represents X_{train}. Then Y_{train} denotes the document event cooccurrence matrix j x J of a particular document d_h and for a particular event e_g. Y_{train} will take the form $[p_{aeh}{}^{dg}]\, j \times J$. *{p d g e h}* is the (*g,h*)th item of Y_{train}, which is the representation of the distribution of document *d g* on the event *e h*.

4.2 Event Term Cooccurrence Matrix

As many audio events occur simultaneously, the event term cooccurrence matrix X_{train} must be counted with care for various audio documents. We can annotate as many audio events but not more than 3 for a particular time interval. The audio frame containing multiple labels has been presented for all audio events with equal proportions to count the event term cooccurrence matrix in the statistics. For 'm' audio events, if we count the event term cooccurrence matrix, the result will be 1/m while.

Different annotators will produce different results for the same set of audio events for a given time interval. So we need at least three annotators to annotate the same set of audio events for a given time interval [17]. Finally, we retain the event labels

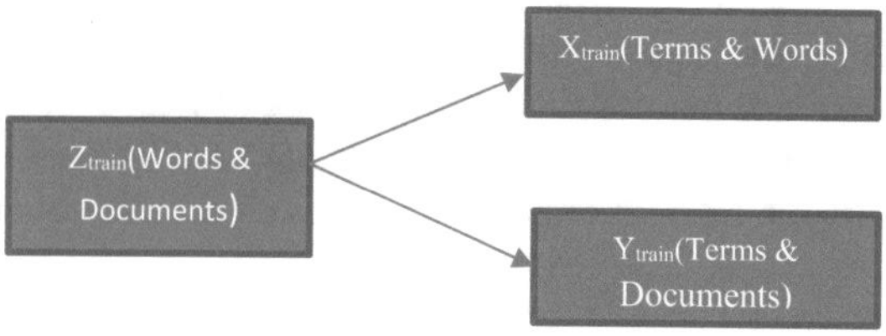

Fig. 1 Document event cooccurrence matrix (Training set)

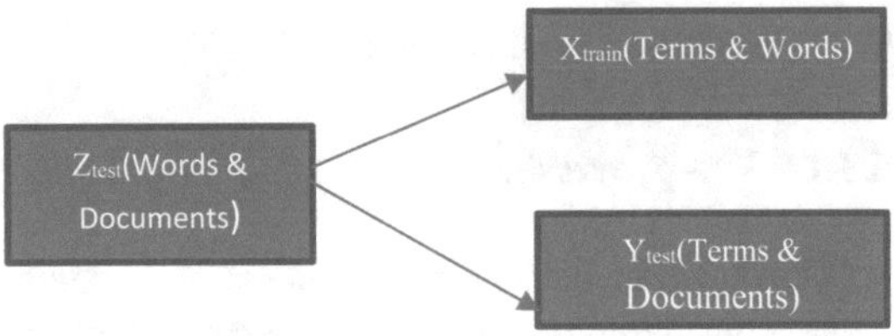

Fig. 2 Document event cooccurrence matrix (Test set)

annotated by more than one annotator and omit the rest. Here document-term cooccurrence matrix X_{test} of test case can be calculated by splitting the audio into terms and matching the frames with the terms. Having the event term cooccurrence matrix of test and training stages are the same, we derive the document event cooccurrence matrix Z_{train} of the test set which is similar to the training stage. The document event cooccurrence for the test and training sets are obtained through Latent Semantic Analysis matrix factorization. Instead of LSA matrix factorization, we can also obtain the Z_{train} by counting the number of occurrences.

Weighing the audio event's distribution is required for recognizing the influence of the events. Topic distribution along with its feature set is the input for the classifier. If the occurrence of the events reflects less topics, then they are less influential. But if the occurrence of the events reflects few topics, then they are more influential. Using entropy, we can find the influence of the events as mentioned in [18]. If there are t1 latent topics, then T will give the event topic distribution matrix [$p_{aeh}{}^{dg}$] t1 x j where $h = 1,2, \cdots, t\ 1, g = 1,2, \cdots, j$). The event distribution ae_h on the topic d_g is denoted by $p_{aeh}{}^{dg}$. So the event entropy can be computed by using the vector $E = E(ae_h)$, where $E(ae_h)$ denotes the value of entropy of the ae_hth event can be computed using the formula.

$$E(aeh) = -i = 1 paehd \log 2(paehd) \quad (1)$$

If the entropy value is too small the topic is very specific and if the entropy value is larger the audio event will be common to many topics. So, we choose audio events with smaller entropy values for classification. Using this entropy value, we calculate the coefficient to find the influence of an audio event [19]. Vector z represented as $z(ae_h)$ represents the coefficient of the event ae_h where the coefficient should be larger than or equal to 1. We can design it as.

$$z(aeh) = (ae) - |E(aeh) - \text{mean}(E)|/2\text{variance}(E) \quad (2)$$

$$z(aeh) \Leftarrow z(aeh)/\min(z) \quad (3)$$

The document event distribution in Y_{train} and Y_{test} can be found using the coefficient vector z. By reframing the formula for document event distribution, we get.

$paehdg \Leftarrow z(aeh).paehdg$ where $h = 1,2, \ldots,j$ and $g = 1,2, \ldots,J$.

4.3 Output Generation

4.3.1 Pre-processing

- The document-term matrix is taken as the input for topic models. The documents are considered as rows in the matrix and the terms as columns.
- The size of the corpus is equal to the number of rows and the vocabulary size is the number of columns.
- We should tokenize the document for representing the document with the frequency of the term like stem words removal, punctuation removal, number removal, stop word removal, case conversion and omission of low length terms.
- In a group of vocabulary, the index will map the exact document from where the exact term was found.

Here the distribution of the topics is taken as a feature set to achieve topic modelling. Y_{train} (Fig. 1) and Y_{test} (Fig. 2) are broken to find the distribution of topics for training and test documents of audio respectively. We can break Y_{train} as $Y1_{train}$ and $Y2_{train,}$ Y_{test} can be broken into $Y1_{train}$ and $Y2_{test}$ can be keeping $Y1_{train}$ fixed. $Y2_{train}$ is a L2 x J matrix if there are L2 latent topics and each of its column represents the training audio document's topic distribution. $Y2_{test}$ is a L2 x J test matrix if there are J test audio inputs and each of its column represents the test audio document's topic distribution. We consider this distribution of topics as a feature set for the audio documents to perform our classification model using SVM. We adopt a one–one multiclass classification technique in SVM to classify the audio scene which has been used in many applications [1, 20].

5 Experimental Results

The proposed algorithm is tested by two publicly available dataset IEEE AASP challenge and DEMAND (Diverse Environments Multi-channel Acoustic Noise Database) dataset [21]. There are 10 classes such as tube, busy street, office, park, Quiet Street, restaurant, open market, supermarket and bus. Each class consists of ten audio files which consists of 30 s long, sampled in 44.1 k Hz and stereo. Diverse Environments Multi-channel Acoustic Noise Database (DEMAND) dataset [22] offers various types of indoor and outdoor settings and eighteen audio scene classes are there, which includes kitchen, living, field, park, washing, river, hallway, office, cafeteria, restaurant, meeting, station, cafe, traffic, car, metro and bus. Each audio class includes 16 recordings related to 16 channels. For experiments, only the first channel recording is used and every recording is three hundred seconds long. Then it is sliced into 10 equal documents, with 30 s long each. As a summary, the dataset DEMAND contains 18 categories of audio scenes, each category has 10 audio files of 30 s long.

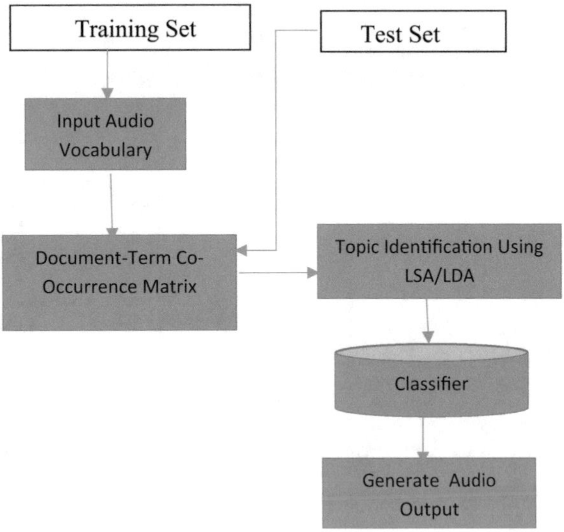

Fig. 3 Proposed framework

In this present work, audio documents have been partitioned into 30 ms-long frames spending 50% overlap of the hamming window; for every frame, 39 dimensional MFCCs features were obtained as the feature set; the distribution of topic is utilized for characterizing each audio document, which is given as the input for SVM, after carrying out topic analysis through LSA/LDA. One-to-one strategy is followed in SVM for multiclass type of classification and the kernel function has been taken as RBF (Radial Basis Function). The evaluation of algorithms have been done, in terms of classification accuracy,

$$\text{Accuracy} = \frac{\text{The number of correctly classified audio documents}}{\text{Thetotal number of audio documents in the test set}} * 100 \quad (4)$$

The following Table 1 shows the accuracy obtained for the LDA and LSA methods. In order to prove the proposed algorithm uses the prescribed format matrix for topic analysis is found to be more efficient than the existing algorithms which uses the conventional matrix to analyze the topic.

Based on the above result in Table 1, the algorithm trusts the prescribed matrix to carry out topic analysis. From the given analysis, we can conclude that, for a particular audio scene class, Many existing topic model work with ASR algorithms have approved SVM as a classification model [17–19]. So that in this work also, SVM is taken as a classification model to perform the classification and the results are tabled in Table 2.

Table 1 Classification performance document event (DE) cooccurrence matrix and document word (DW) cooccurrence matrix

Dataset	Topic model	Algorithm	Accuracy
AASP	LSA	Document event (DE) & LSA	45.6
		Document word (DW) & LSA	60.1
	LDA	Document event (DE) & LDA	46.9
		Document word (DW) & LDA	52.8
DEMAND	LSA	Document event (DE) & LSA	62.1
		Document word (DW) & LSA	81.3
	LDA	Document event (DE) & LDA	62.6
		Document word (DW) & LDA	76.5

Table 2 Performance of SVM on AASP and DEMAND

Data Set	Topic model	Accuracy (%)
AASP	LSA	61
	LDA	55
DEMAND	LSA	82
	LDA	77

6 Conclusion and Future Enhancement

In this proposed approach, new ASR algorithm which utilizes document event cooccurrence matrix for topic modelling instead of most widely used document word cooccurrence matrix. The adopted technique outperforms well than the existing matrix based topic modelling. To acquire the document event cooccurrence matrix in more efficient method, this proposed work uses a matrix factorization method. Even though this work found least results on AASP dataset, at least we have verified that using the existing matrix for analyzing the topic is much better to go with the proposed method matrix. As a future enhancement of our work, the deep learning models can be considered as a reference, and motivated in using the neural network to encompass in present system, by identifying the merits of topic models and neural networks, the recognition performance can be improved.

References

1. Leng Y, Sun C, Xu X et al (2016) Employing unlabelled data to improve the classification performance of SVM, and its application in audio event classification Knowl. Based Syst 98:117–129
2. Leng Y, Sun C, Cheng C et al (2015) Classification of overlapped audio events based on AT. PLSA, and the Combination of them Radio Engineering 24(2):593–603
3. Leng Y, Qi G, Xu X et al (2013) A BIC based initial training set selection algorithm for active learning and its application in audio detection. Radio Eng 22(2):638–649
4. Choi WH, Kim SI, Keum MS et al (2011) Acoustic and visual signal based context awareness system for mobile application. IEEE Trans Consum Electron 57(2):738–746
5. Smith DV, Shahriar MS (2013) A context aware sound classifier applied to prawn feed monitoring and energy disaggregation. Knowl Based Syst 52:21–31
6. Wang JC, Lee HP, Wang JF et al (2008) Robust environmental sound recognition for home automation. IEEE Trans Autom Sci Eng 5(1):25–31
7. Leng Y, Zhou N, Sun C et al (2017) Audio scene recognition based on audio events and topic model. Knowl-Based Syst 125:1–12. https://doi.org/10.1016/j.knosys.2017.04.001
8. Hansen JHL, Rongqing H, Bowen Z et al (2005) Speech Find advances in spoken document retrieval for a National Gallery of the Spoken Word. IEEE Trans Speech Audio Process 13:712–730
9. Peskin B et al (1996) Improvements in Switchboard recognition and topic identification. Proc ICASSP-96(I):303-306
10. Mesaros A, Heittola T, Diment A, Elizalde B, Shah A, Vincent E, Raj B, Virtanen T (2017) DCASE challenge setup: tasks, datasets and baseline system. In: DCASE 2017-workshop on detection and classification of acoustic scenes and events
11. Hu P, Liu W, Jiang W, Yang Z (2012) Latent topic med on Gaussian-LDA for audio retrieval. In: Chinese conference on pattern recognition, pp 556–563
12. Peng Y, Lu Z, Xiao J (2009) Semantic concept annotation based on audio PLSA model. In: Proceedings of the 17th ACM international conference on multimedia, pp 841–844
13. Lee L, Ellis DPW (2010) Audio-based semantic concept classification for consumer video. IEEE Trans Audio Speech Lang Process 18(6):1406–1416
14. Kim S, Sundaram S, Georgiou P et al (2009) Audio scene understanding using topic models. In: Proceedings of the neural information processing systems (NIPS) Workshop
15. Kherwa P, Bansal P (2017) Latent semantic analysis: an approach to understand semantic of text. In: 2017 international conference on current trends in computer, electrical, electronics and communication (CTCEEC) IEEE. 870–874 (2017)
16. Jelodar H, Wang Y, Yuan C, Feng X, Jiang X, Li Y, Zhao L (2019) Latent Dirichlet allocation (LDA) and topic modeling: models, applications, a survey. Multimedia Tools Appl 78(11):15169–15211
17. Wintrode J, Kulp S (2009) Techniques for rapid and robust topic identification of conversational telephone speech. In: Tenth annual conference of the international speech communication association
18. Leng Y, Zhou N, Sun C, Xu X, Yuan Q, Cheng C, Liu Y, Li D (2017) Audio scene recognition based on audio events and topic model. Knowl-Based Syst 125:1–12
19. Wang KC (2020) Robust audio content classification using hybrid-based SMD and entropy-based VAD. Entropy 22(2):183
20. Kim S, Sundaram S, Georgiou P, Narayanan S (2009) Audio scene understanding using topic models. In: Neural Information Processing System (NIPS) Workshop (Applications for Topic Models: Text and Beyond)
21. Stowell D, Giannoulis D, Benetos E, Lagrange M, Plumbley MD (2015) Detection and classification of acoustic scenes and events. IEEE Trans Multimedia 17(10):1733–1746
22. Thiemann J, Ito N, Vincent E (2013) The Diverse Environments Multi-channel Acoustic Noise Database (DEMAND): a database of multichannel environmental noise recordings. In: Proceedings of meetings on acoustics ICA2013. Acoustical Society of America, vol 19, no. 1, 035081

Diagnosis of Brain Tumor Using Light Weight Deep Learning Model with Fine Tuning Approach

Tejas Shelatkar and Urvashi Bansal

1 Introduction

A brain tumor is a cluster of irregular cells that form a group. Growth in this type of area may cause issues that be cancerous. The pressure inside the skull will rise as benign or malignant tumors get larger. This will harm the brain and may even result in death Pereira et al. [1]. This sort of tumor affects 5–10 people per 100,000 in India, and it's on the rise [12]. Brain and central nervous system tumors are also the second most common cancers in children, accounting for about 26% of childhood cancers. In the last decade, various advancements have been made in the field of computer-aided diagnosis of brain tumor. These approaches are always available to aid radiologists who are unsure about the type of tumor or wish to visually analyze it in greater detail. MRI (Magnetic Resonance Imaging) and CT-Scan (Computed Tomography) are two methods used by doctors for detecting tumor but MRI is preferred so researchers are concentrated on MRI. A major task of brain tumor diagnosis is segmentation. Researchers are focusing on this task using Deep Learning techniques [3]. In medical imaging, deep learning models have various advantages from the identification of important parts, pattern recognition in cell parts, feature extraction, and giving better results for the smaller dataset as well [3]. Transfer learning is a technique in deep learning where the parameters (weights and biases) of the network are copied from another network trained on a different dataset. It helps identify generalized features in our targeted dataset with help of features extracted from the trained dataset. The new network can now be trained by using the transferred parameters as initialization

T. Shelatkar (✉) · U. Bansal
Dr. B.R. Ambedkar National Institute of Technology, Jalandhar, India
e-mail: tejass.cs.20@nitj.ac.in

U. Bansal
e-mail: urvashi@nitj.ac.in

P. Singh et al. (eds.), *Machine Learning and Computational Intelligence Techniques for Data Engineering*, Lecture Notes in Electrical Engineering 998,
https://doi.org/10.1007/978-981-99-0047-3_10

(this is called fine-tuning), or new layers can be added on top of the network and only the new layers are trained on the dataset of interest.

Deep learning is a subset of machine learning. It is used to solve complex problems with large amounts of data using an artificial neural network. The artificial neural network is a network that mimics the functioning of the brain. The 'deep' in deep learning represents more than one layer network. Here each neuron represents a function and each connection has its weight. The network is trained using the adjustment of weights which is known as the backpropagation algorithm. Deep learning has revolutionized the computer vision field with increased accuracy on the complex data set. Image analysis employs a specific sort of network known as a convolutional network, which accepts photos as input and convolves them into a picture map using a kernel. This kernel contains weight that changes after training.

A frequent practice for deep learning models is to use pre-trained parameters on dataset. The new network can now be trained by using transferred parameters as initialization (this is called fine-tuning), or new layers can be added on top of the network and only the new layers are trained on the dataset of interest. Some advantages of transfer learning are it reduces the data collection process as well it benefits generalization. It decreases the training duration of a large dataset.

2 Motivation

The motivation behind this research is to build a feasible model in terms of time and computing power so that small healthcare systems will also benefit from the advancements in computer-aided brain tumor analysis. The model should be versatile enough so that it can deal with customized data and provide an acceptable result by using adequate time.

3 Literature Review

Various deep learning models have been employed for the diagnosis of brain tumor but very restricted research has been done by using object detection models. Some of the reviewed papers have been mentioned below.

Pereira and co-authors have used the modern deep learning model of the 3D Unet model which helps in grading tumor according to the severity of the tumor. It achieves up to 92% accuracy. It has considered two regions of interest first is the whole brain and another is the tumors region of interest [1].

Neelum et al. achieve great success in the analysis of the problem as they use pre-trained models DesNet and Inception-v3 model as which achieves 99% accuracy. Feature concatenation has helped a great deal in improving the model [4].

Mohammad et al. have applied various machine learning algorithms like decision tree, support vector machine, convolutional neural network, etc. as well as deep learning models, i.e., VGG 16, ResNET, Inception, etc. on the limited dataset of 2D images without using any image processing techniques. The most successful model was VGG19 which achieved 97.8% of F1 scope on top of the CNN framework. Some points stated by the author were that there is trade off between the time complexity and performance of the model. The ML method has lesser complexity and DL is better in performance. The requirement of a benchmark dataset was also stated by Majib et al. They have employed two methods FastAi and Yolov5 for the automation of tumor analysis. But Yolov5 gains only 85% accuracy as compared to the 95% of FastAI. Here they haven't employed any transfer learning technique to compensate for the smaller dataset [18].

A comprehensive study [7] is been provided on brain tumor analysis for small healthcare facilities. The author has done a survey that listed various challenges in the techniques. They have also proposed some advice for the betterment of techniques.

Al-masni et al. have used the YOLO model for bone detection. The YOLO method relieves a whooping 99% accuracy. So here we can see that the YOLO model can give much superior results in medical imaging [13].

Yale et al. [14] detected Melanoma skin disease using the YOLO network. The result was promising even though the test was conducted on a smaller dataset. The Dark Net Framework provided improved performance for the extraction of the feature. A better understanding of the working of YOLO is still needed.

Kang et al. [21] proposed a hybrid model of machine learning classifiers and deep features. The ensemble of various DL methods with a classifier like SVM, RBF, KNN, etc. The ensemble feature has helped the model for higher performance. But author suggested the model developed is not feasible for real-time medical diagnosis.

Muhammad et al. [18] have studied various deep learning and transfer learning techniques from 2015–2019. The author has identified challenges for the techniques to be deployed in the actual world. Apart from the higher accuracy, the researchers should also focus on other parameters while implementing models. Some concerns highlighted are the requirement of end-to-end deep learning models, enhancement in run time, reduced computational cost, adaptability, etc. The author also suggested integrating modern technologies like edge computing, fog and cloud computing, federated learning, GAN technique, and the Internet of Things.

As we have discussed various techniques are used in medical imaging and specifically on MRI images of brain tumor. Classification, segmentation, and detection algorithms were used but each one had its limitation. We can refer to Table 1 for a better understanding of the literature review.

4 Research Gap

Although classification methods take fewer resources, they are unable to pinpoint the specific site of a tumor. The segmentation methods which can detect exact locations take large amounts of resources. The existing models do not work efficiently on the comparatively smaller dataset for small healthcare facilities. Harder to imple-

Table 1 Literature review

Author and Year	Dataset	Objective	Technique	Limitations
Pereira et al. [1]	BRats 2017	Automatic tumor grading	3D Unet	Complex computation
Rehman et al. [2]	Brain tumor dataset by Cheng	To explore fine-tuning transfer learning model	AlexNet, GoogleNet, VGGNet	High time complexity
Ercan et al. [3]	Private dataset	Faster classification using R-CNN models	Faster-RCNN	Need improved performance for lesser data
Saba et al. [23]	BRats15-17	Optimize deep learning feature for classification	UNET + RNN + FULL CNN	Complex calculation
Neelum et al. [4]	Brain tumor dataset by Cheng	Feature extraction using concatenation approach	Inception and Desnet	2D dataset used
Montalbo [5]	Nanfang hospital dataset	Fine-tune Yolo model using low space and computation	Yolov4 using Transfer learning	Precise selection of tumor
Si-Yuan et al. [18]	ATLAS MRI dataset	To deploy pre-trained models for classification purposes	MobileNet	smaller dataset
Hammami et al. [17]	Visceral anatomy dataset	To develop a hybrid multi-organ detector	CycleGan and YOLO	Many outliers detected
Zuzana et al. [13]	CT based multiple bone dataset	Distinctive bone creation	YOLO	Improvement of accuracy
Jaeyong et al. [13]	Hybrid dataset	Classify MRI scans of brain tumor	DL method and ML classifier	Larger model size
Mohammad et al. [18]	Pathology institute dataset	To apply the hybrid model approach	VGG-SCNET	High-end processor required
Nadim et al. [7]	Brats18	To build fast deep learning models for brain tumor classification	Yolov5 and FastAI	Low accuracy
Futrega et al. [8]	Brats21	Experiment various Unet with architecture modification	Optimized Unet	Computational heavy

ment models by healthcare facilities with limited resources and custom data created. Human intervention is needed for feature extraction and preprocessing of the dataset.

5 Our Contribution

- To deploy a light weight model using the fine-tuning approach of pre-trained models.
- To create a model which can also be used on the smaller dataset by small healthcare facilities.

6 Characteristics improved using our Brain Tumor Analysis Model

6.1 Light Weight

Our model needs to consume less storage and computing resources. As we are keen on designing a model which can be used by smaller healthcare facilities. So model size must be smaller as well as it should be occupied lesser storage.

6.2 Reliability

The radiologist must beware of the false positive as they can't directly rely on the analysis as it may not be completely precise and the system should be only used by the proper radiologists as our system can't completely replace the doctors.

6.3 Time Efficiency

The system for brain tumor diagnosis must consume lesser time to be implemented in the real world. The time complexity must be feasible even without the availability of higher end systems at healthcare facilities.

7 Dataset

Various datasets are available for brain tumor analysis from 2D to 3D data. Since we are focusing on the MRI data set it includes high-grade glioma, low-grade glioma, etc. The images can be of 2D or 3D nature. The types of MRI are mostly of T1-weighted scans. Some datasets are (a) multigrade brain tumor data set, (b) brain tumor public data set, (c) cancer imaging archive, (d) brats, and (e) internet brain segmentation repository.

Brats 2020 is an updated version of the brats dataset. The Brats dataset has been used in organizing events from 2012 to up till now, they encourage participants to research their own collected dataset. The Brats 2017–2019 varies largely from all the previous versions. The Brats 20 is an upgraded version of this series. Figure 1 displays our selected dataset.

8 Deep Learning Based Brain Tumor Diagnosis Using Yolov5

8.1 Yolov5

Object detection technique accomplishes two objectives (a) Localization-object location in the image (b) Classification-identifies the object in the image. Yolo is an object detection model. Yolo means you only look once since it is the single stage detector. Since it is a single stage, it is very fast and due to its accuracy, it is a state of art detector. It has currently 5 versions with its initial launch in 2016 of yolov1 by Redmon. Yolo has many variants like tiny, small, multi-scale, and different backbone which has different feature extractors and backbone. It can train on different platforms Darknet, Pytorch, and Tensorflow. YOLOv5 is an object detection algorithm designed by Ultralytics which is well known for its AI study, combining lessons learned and best practices gleaned from millions of hours of work they are can develop Yolov5.

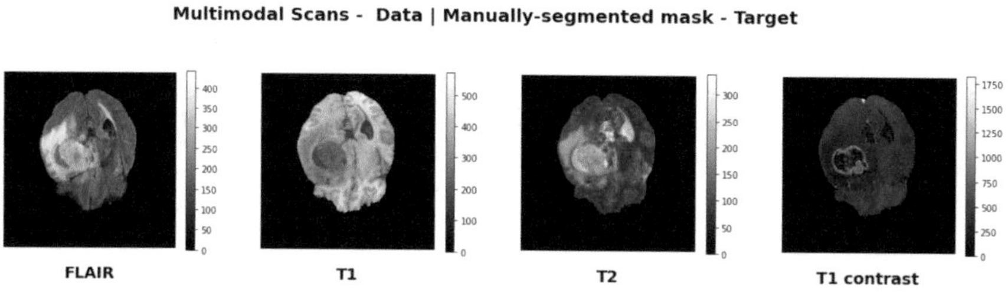

Fig. 1 Brats 2020 dataset

The three major architectural blocks of the YOLO family of models are the backbone, neck, and head. The backbone of YOLOv5 is CSPDarknet, which is used to extract features from photos made up of cross-stage partial networks. YOLOv5 Neck generates a feature pyramids network using PANet to do feature aggregation and passes it to Head for prediction. YOLOv5 Head has layers that provide object detection predictions from anchor boxes. Yolov5 is built on the Pytorch platform which is different from previous versions which are built on DarkNet. Due to this, there are various advantages like it has fewer dependencies and it doesn't need to be built from the source.

9 Proposed Model

As mentioned above we are going to use the state-of-the-art model Yolov5. The pre-trained weights are taken from COCO (Microsoft Common Objects in Context) dataset. Fine-tuning is done using these parameters. The model is trained using the BRat 2020 dataset. The model is fed with 3D scans of patients. Once the model is trained we input the test image to get information about the tumor. The new network can now be trained by using the transferred parameters as initialization (this is called fine-tuning), or new layers can be added on top of the network and only the new layers are trained on the dataset of interest. Some advantages of transfer learning are it reduces the data collection process as well as benefits generalization. It decreases the training duration of a large dataset.

Some preprocessing is needed before we train the model using the YOLO model, the area of the tumor must be marked by the box region. This can be done using the tool which creates a bounding box around the object of interest in an image. For Transfer learning we can use the NVIDIA transfer learning toolkit, we can feed the COCO dataset as it also supports the YOLO architecture. This fine-tunes our model and makes up an insufficient or unlabeled dataset. Afterward we can train our BRats dataset on our model. The environment used for development is Google Colab which gives 100 Gb storage, 12 GB Ram, and GPU support. The yolov5 authors have made available their training results on the COCO dataset to download and use their pre-trained parameters for our own model. For applying the yolov5 algorithm on our model we need a labeled dataset for training which is present in the brats dataset. Since we need to train it for better results on BRats dataset we will freeze some layers and add our own layer on top of the YOLO model for better results. Since we need a model which takes lesser space we will use the YOLOv5n model. As mentioned in the official repository YOLOv5 model provides us a mean average precision score of 72.4 with a speed of 3136 ms on the COCO dataset [25]. The main advantage of this model is smaller and easier to use in production and it is 88 percent smaller than the previous YOLO model [26]. This model is able to process images at 140 FPS. The pre-trained weights are taken from COCO (Microsoft Common Objects in Context) dataset. Fine-tuning is done using these parameters. The model is trained using the BRats 2020 dataset. Here specifically we are going to use the yolov5 nano model

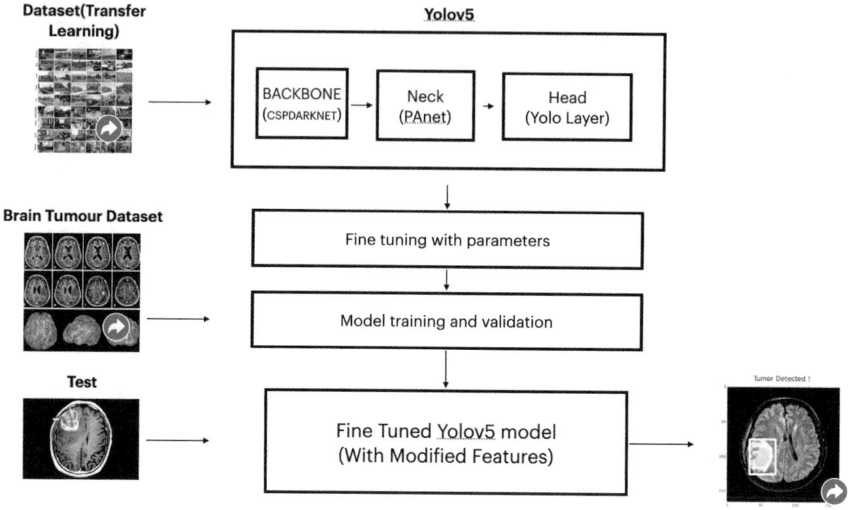

Fig. 2 Fine-tuned yolov5 model

since it has smaller architecture than the other models as our main priority is the size of the model. The YOLO model has a much lower 1.9 M params as compared to the other models. Our model needs a certain configuration to be able to perform on brain scans. Since the scanned data of Brats is complex, we perform various preprocessing on the data from resizing to masking. Since the image data is stored in nii format with different types of scans like FLAIR, T1, T2, it is important to process the dataset according to the familiarity of our model. The model is fed with scans of patients. For evaluating the results of our model we use the dice score jaccard score and map value but our main focus is on the speed of the model to increase the usability of the model. For training and testing the dataset is already partitioned for Brats. Our dataset contains almost 360 patient scans for training and 122 scans for patient scans for testing. The flow of our model is mentioned in Fig. 2. The yolov5 models provide with their yml file for our custom configuration so we can test the network according to our own provision. Since we have only 3 classes we will configure it into three. As well as many convolution layers must be given our parameters in the backbone or head of our model. Once the model is trained we can input the test image dataset on our models. The expected result of the model must be close to a dice score of 0.85 which compares to the segmentation models. This model takes up lesser storage and better speed in processing of brats dataset as compared to the previous models.

10 Conclusion

Various models and toolkits for brain tumor analysis have been developed in the past which has given us promising results but the viability of the model in terms of real-time application has been not considered. Here we present a deep learning-based method for brain tumor identification and classification using YOLOv5 in this research. These models are crucial in the development of a lightweight brain tumor detection system. A model like this with lesser computational requirements and relatively reduced storage will provide a feasible solution to be considered by various healthcare facilities.

References

1. Pereira S et al (2018) Automatic brain tumor grading from MRI data using convolutional neural networks and quality assessment. In: Understanding and interpreting machine learning in medical image computing applications. Springer, Cham, pp 106–114
2. Rehman A et al (2020) A deep learning-based framework for automatic brain tumors classification using transfer learning. Circuits Syst Signal Process 39(2): 757–775
3. Salçin Kerem (2019) Detection and classification of brain tumours from MRI images using faster R-CNN. Tehnički glasnik 13(4):337–342
4. Noreen N et al (2020) A deep learning model based on concatenation approach for the diagnosis of brain tumor. IEEE Access 8: 55135–55144
5. Montalbo FJP (2020) A computer-aided diagnosis of brain tumors using a fine-tuned YOLO-based model with transfer learning. KSII Trans Internet Inf Syst 14(12)
6. Dipu NM, Shohan SA, Salam KMA (2021) Deep learning based brain tumor detection and classification. In: 2021 international conference on intelligent technologies (CONIT). IEEE
7. Futrega M et al (2021) Optimized U-net for brain tumor segmentation. arXiv:2110.03352
8. Khan P et al (2021) Machine learning and deep learning approaches for brain disease diagnosis: principles and recent advances. IEEE Access 9:37622–37655
9. Khan P, Machine learning and deep learning approaches for brain disease diagnosis: principles and recent advances
10. Amin J et al (2021) Brain tumor detection and classification using machine learning: a comprehensive survey. Complex Intell Syst 1–23
11. https://www.ncbi.nlm.nih.gov/
12. Krawczyk Z, Starzyński j (2020) YOLO and morphing-based method for 3D individualised bone model creation. In: 2020 international joint conference on neural networks (IJCNN). IEEE
13. Al-masni MA et al (2017) Detection and classification of the breast abnormalities in digital mammograms via regional convolutional neural network. In: 2017 39th annual international conference of the IEEE engineering in medicine and biology society (EMBC). IEEE
14. Nie Y et al (2019) Automatic detection of melanoma with yolo deep convolutional neural networks. In: 2019 E-health and bioengineering conference (EHB). IEEE
15. Krawczyk Z, Starzyński J (2018) Bones detection in the pelvic area on the basis of YOLO neural network. In: 19th international conference computational problems of electrical engineering. IEEE
16. https://blog.roboflow.com/yolov5-v6-0-is-here/
17. Hammami M, Friboulet D, Kechichian R (2020) Cycle GAN-based data augmentation for multi-organ detection in CT images via Yolo. In: 2020 IEEE international conference on image processing (ICIP). IEEE

18. Majib MS et al (2021) VGG-SCNet: A VGG net-based deep learning framework for brain tumor detection on MRI images. IEEE Access 9:116942–116952
19. Muhammad K et al (2020) Deep learning for multigrade brain tumor classification in smart healthcare systems: a prospective survey. IEEE Trans Neural Netw Learn Syst 32(2): 507–522
20. Baid U et al (2021) The rsna-asnr-miccai brats 2021 benchmark on brain tumor segmentation and radiogenomic classification. arXiv:2107.02314
21. Kang J, Ullah Z, Gwak J (2021) MRI-based brain tumor classification using ensemble of deep features and machine learning classifiers. Sensors 21(6):2222
22. Lu S-Y, Wang S-H, Zhang Y-D (2020) A classification method for brain MRI via MobileNet and feedforward network with random weights. Pattern Recognit Lett 140:252–260
23. Saba T et al (2020) Brain tumor detection using fusion of hand crafted and deep learning features. Cogn Syst Res 59:221–230
24. Menze BH et al (2015) The multimodal brain tumor image segmentation benchmark (BRATS). IEEE Trans Med Imaging 34(10):1993–2024. https://doi.org/10.1109/TMI.2014.2377694
25. https://github.com/ultralytics/yolov5/
26. https://models.roboflow.com/object-detection/yolov5

Comparative Study of Loss Functions for Imbalanced Dataset of Online Reviews

Parth Vyas, Manish Sharma, Akhtar Rasool, and Aditya Dubey

1 Introduction

Please Google Play serves as an authentic application database or store for authorized devices running on the Android operating system. The application allows the users to look at different applications developed using the Android Software Development Kit (SDK) and download them. As the name itself indicates, the digital distribution service has been developed, released, and maintained through Google [1]. It is the largest app store globally, with over 82 billion app downloads and over 3.5 million published apps. The Google Play Store is one of the most widely used digital distribution services globally and has many apps and users. For this reason, there is a lot of data about app and user ratings. In Google Play shop Console, you may get a top-level view of various users' rankings on an application, your app's rankings, and precis facts approximately your app's rankings. An application can be ranked and evaluated on Google Play in the form of stars and reviews by the users. Users can rate the app only once, but these ratings and reviews can be updated at any time. The play store can also see the top reviews of certified users and their ratings [2]. These user ratings help many other users analyze your app's performance before using it. Different developers from different companies also take their suggestions for further product development seriously and help them improve their software.

Leaving an app rating is helpful to users and developers and the Google Play Store itself [3]. The goal of the Play Store as an app platform is to quickly display accurate and personalized results and maintain spam when searching for the app you need. Launch the app. This requires information about the performance of the app displayed through user ratings [4]. A 4.5-star rating app may be safer and more

P. Vyas · M. Sharma · A. Rasool · A. Dubey (✉)
Department of Computer Science and Engineering, Maulana Azad National Institute of Technology, Bhopal, India
e-mail: dubeyaditya65@gmail.com

P. Singh et al. (eds.), *Machine Learning and Computational Intelligence Techniques for Data Engineering*, Lecture Notes in Electrical Engineering 998,
https://doi.org/10.1007/978-981-99-0047-3_11

relevant than a 2-star app in the same genre. This information helps Google algorithms classify and download apps in the Play Store and provide high-quality results for a great experience [5]. The cheaper the app's ratings and reviews, the more people will download and use the Play Store services.

Natural language processing (NLP) has gained immense momentum in previous years, and this paper covers one such sub-topic of NLP: sentiment analysis. Sentiment analysis refers to the classification of sentiment present in a sentence, paragraph, or manuscript based on a trained dataset [6]. Sentiment analysis has been done through trivial machine learning algorithms such as k-nearest neighbors (KNN) or support vector machine (SVM) [7, 8]. However, for more optimization, for the search of this problem, the model selected for the sentiment analysis on Google Play reviews was the Bidirectional Encoder Representations from Transformers (BERT) model, a transfer learning model [9]. The BERT model is a pre-trained transformer-based model which tries to learn through the words and predicts the sentiment conveyed through the word in a sentence.

For this paper, selected models of BERT from Google BERT were implemented on textual data of the Google Play reviews dataset, which performed better than the deep neural networking models. This paper implemented the Google BERT model for sentiment analysis and loss function evaluation. After selecting the training model, the loss function to be evaluated was studied, namely the cross-entropy loss and focal loss. After testing the model with various loss functions, the f1 score was calculated with the Google Play reviews dataset, and the best out of the loss functions which could be used for sentiment analysis of an imbalanced dataset will be concluded [10].

2 Literature Review

With the increasing demand for balance and equality, there is also an increasing imbalance in the datasets on which NLP tasks are performed nowadays [6]. If a correct or optimized loss function is not used with these imbalance datasets, the result may appropriate the errors due to these loss functions. For this reason, many research papers have been studied extensively. At last, the conclusion was to compare the five loss functions and find which will be the best-optimized loss function for sentiment analysis of imbalanced datasets. This segment provides a literature review of the results achieved in this field.

For the comparison of loss functions first need was for an imbalanced dataset. Therefore, from the various datasets available, it was decided to construct the dataset on Google Play apps reviews manually and then modify the constructed dataset to create an imbalance [11]. The dataset was chosen as in the past and has been studied on deep learning sentiment analysis on Google Play reviews of customers in Chinese [12]. The paper proposes the various models of long short-term memory (LSTM), SVM, and Naïve Bayes approach for sentiment analysis [7, 13–15]. However, the dataset has to be prepared to compare the cross-entropy loss and the focal loss function. Focal loss is a modified loss function based on cross-entropy loss which is frequently used in imbalanced datasets. Thus, both the losses will be compared to check which loss will perform better for normal and both imbalanced datasets. Multimodal Sentiment Analysis of #MeToo Tweets using Focal Loss proposes the Roberta model of BERT, which is a robust BERT model which does not account for the bias of data while classification and for further reduction of errors due to misclassification of imbalance of dataset they have used the focal loss function [16].

After finalizing the dataset, the next topic of discussion is the model to be trained on this dataset. The research started with trivial machine learning models based on KNN and SVM [7, 8]. Sentiment Analysis Using SVM suggests the SVM model for sentiment analysis of pang corpus, which is a 4000-movie review dataset, and the Taboada corpus, which is a 400-website opinion dataset [7, 17, 18]. In sentiment Analysis of Law Enforcement Performance Using an SVM and KNN, the KNN model has been trained on the law enforcement of the trial of Jessica Kumala Wongso. The result of the paper shows that the SVM model is better than the KNN model [8]. But these machine learning algorithms like KNN and SVM are better only for a small dataset with few outliers; however, these algorithms cease to perform better for the dataset with such a large imbalance and large dataset. For training of larger dataset with high imbalance, the model was changed to the LSTM model.

The LSTM model is based on the recurrent neural network model units, which can train a large set of data and classify them efficiently [7, 15]. An LSTM model can efficiently deal with the exploding and vanishing gradient problem [19]. However, since the LSTM model has to be trained for classification, there are no pre-trained LSTM models. In the LSTM model, the model is trained sequentially from left to right and simultaneously from right to left, also in the case of the bi-directional LSTM model. Thus, LSTM model predicts the sentiment of a token based on its predecessor or successor and not based on the contextual meaning of the token. So, in search of a model which can avoid these problems, transfer learning model BERT was finally selected for data classification.

Bidirectional encoder representations from transformer abbreviated as BERT are a combination of encoder blocks of transformer which are at last connected to the classification layer for classification [20–22]. The BERT model is based on the transformers, which are mainly used for end-to-end speech translation by creating embeddings of sentences in one language and then using the same embeddings by the decoder to change them in a different language [20, 23]. These models are known as transfer learning because these models are pre-trained on a language such as BERT is trained on the Wikipedia English library, which is then just needed to be fine-tuned, and then the model is good to go for training and testing [21, 22]. Comparing BERT and the trivial machine learning algorithms in comparing BERT against traditional machine learning text classification, the BERT performed far better than the other algorithms in NLP tasks [24]. Similarly, with the comparison of the BERT model with the LSTM model in A Comparison of LSTM and BERT for Small Corpus, it was seen that the BERT model performed with better accuracy in the case of a small training dataset. In contrast, LSTM performed better when the training dataset was increased above 80 percent [25]. And also, Bidirectional LSTM is trained both from left-to-right to predict the next word, and right-to-left to predict the previous work.

But, in BERT, the model is made to learn from words in all positions, meaning the entire sentence and in this paper using Bidirectional LSTM made the model high overfit. At last BERT, the model was finalized for the model's training and checking the performance. In the case of the BERT model still, there is two most famous model, first is the Google BERT model, and the other is Facebook ai research BERT model "Roberta" [26]. In comparison, the Roberta model outperforms the BERT model by Google on the general language understanding evaluation benchmark because of its enhanced training methodology [27].

After finalizing the training model, the loss functions to be compared for imbalanced dataset evaluation were then studied through previous research. Each of the five loss functions has been described in further sections.

3 Loss Functions

3.1 Cross-Entropy Loss

Cross-Entropy loss is one of the most important cost functions. Cross-entropy is based on entropy and is a measure from the field of information theory that generally calculates the difference between two probability distributions [28, 30]. This is closely related to KL divergence, which calculates the relative entropy between two probability distributions, but it is different, while it can be thought of as cross-entropy, which calculates the total entropy between distributions. Cross-entropy is also associated with logistic loss and is often confused with log loss [29]. The two measures are taken from different sources, but both estimates calculate the same amount when used as a loss function in a classification model. They can be used interchangeably—used to optimize the classification model. When tackling machine learning or deep learning problems, use the loss/cost function to optimize your model during training. The goal is, in most cases, to minimize the loss function. The less loss, the better the model. The formula for cross-entropy loss for binary classes can be defined as

$$\mathrm{CE(binary)} = -(y\log(p) + (1 - y)\log(1 - p)) \tag{1}$$

where y is the binary indicator in 0 or 1, and p is the predicted probability of class being present in the instance or not. However, multiple classes for sentiment analysis have been used in this model. Thus, the formula for cross-entropy for multiple classes can be given as per Eq. (2).

$$\mathrm{CE} = -\sum_{c=1}^{M} y * \log(p) \tag{2}$$

where M refers to the total number of classes and the loss is calculated by summation of all the losses calculated for each class separately.

3.2 Focal Loss

To address the case of classification and object detection, a high imbalance focal loss was introduced [16, 30]. Starting with the cross-entropy loss to incur the high imbalance in any dataset, an adjusting factor is added to the cross-entropy loss, α for class 1 and $1-\alpha$ for class 0. Even since the α can differentiate between positive and negative examples, it is still unable to differentiate between easy and hard examples. The hard examples are related to the examples of the classification of the minority class. Thus, in the loss function, instead of α, a modulating factor is introduced in the cross-entropy loss to reshape the loss function to focus on hard negatives and down weight the easy examples. The modulating factor $(1 - \text{pt})^{\gamma}$ contains the tunable factor $\gamma \geq 0$, which changes to the standard cross-entropy loss when equated to zero. The focal loss equation is

$$\text{FL}(\text{pt}) = -(1 - \text{pt})^{\gamma}\log(\text{pt}) \tag{3}$$

In the above equation, pt is equal to if $y = 1$ and $\text{pt} = 1 - p$ in all other cases and y refers to the ground-truth value and the estimated probability for a class with $y = 1$, in Eq. (3), it can be derived that the focal loss has two main properties. The first property is that if the instance is misclassified, i.e., pt value is very low, the value of the complete loss remains unchanged as $1 - \text{pt}$ is approximately equal to 1. However, when the value of pt is very high or equal to one, the value of loss becomes very low, which in turn leads to giving more focus to misclassified instances.

4 Dataset

The dataset used in this manuscript is on the Google Play reviews dataset, which has been scrapped manually using the Google Play scrapper library based on NodeJS [11]. The data was scraped from Google Play based on the productivity category of Google Play. Various apps were picked up from the productivity category using the app on any website, and the info of the app was kept in a separate excel file which was then used for scrapping out the reviews of each app contained in the excel file. Finally, the data was scraped out, which contained the user's name; the user reviews the stars the user has given to the app, the user image, and other pieces of information that are not needed in the model training. The total number of user reviews was 15,746 reviews, out of which 50,674 reviews were of stars of 4 and above, 5042 reviews were of stars of 3, and 5030 reviews were for stars two and below. The model training was done using the reviews of the users and the stars which have been given to an app. Since the range of the stars given to an app was from 1 to 5, the range was to be normalized into three classes: negative, neutral, and positive. Therefore, the reviews containing stars from 1 to 2 were classified in the negative class. Reviews with three stars were classified in neutral class, and the reviews which contained stars from 4 and above were classified in the positive class. The text blob positive review can be seen in Fig. 1. Which shows the most relevant words related to the positive sentiment in a review. More is the relevancy of a word in review greater will be the size of the token. Since the comparison was done between the two-loss function on an imbalanced dataset, the data percentage for the classification was calculated, as shown in Table 1. As per Table 1, the dataset so formed was a balanced one. However, the imbalance was created to compare the two-loss functions for imbalance on the dataset. The number of neutral reviews decreased up to 20 percent for one iteration to form one dataset. Similarly, in the balanced dataset, neutral reviews were decreased up to 40 percent for the next iteration to create a second set of imbalanced datasets. The classification percentage for both datasets can be seen in the Table. Finally, these three datasets, namely, the dataset with balanced classifications, the dataset with 20 percent fewer neutral reviews, and finally the dataset with 40 percent fewer neutral datasets were used for training, and finally, a comparison of cross-entropy and focal loss was done on these datasets to see the difference in accuracy on using a weighted loss function on imbalanced datasets.

Fig. 1 Text blob for positive reviews

Table 1 Data percentage for different reviews in all the different datasets

Datasets	Positive review percentage	Negative review percentage	Neutral review percentage
Reviews without any imbalance	36	31	33
Reviews with 20 percent less neutral reviews	38	34	28
Reviews with 40 percent less neutral reviews	42	36	22

5 Methodology

The basic model structure is deployed as given in Fig. 2. Firstly, the Roberta model was trained on all the three datasets of Google Play reviews. After the model training, the model is tested with both the loss functions for accuracy and f1 score, calculated to compare the loss functions [16, 31]. The following steps are executed in the model for data processing and evaluation.

- Data Pre-Processing
  - Class Normalization—As per the previous section first the class normalization that is the stars in reviews will be normalized to positive, neutral, and negative classes.
  - Data Cleaning—In this phase, all characters of non-alphabet characters are removed. For example, Twitter hashtags like #Googleplayreview will be

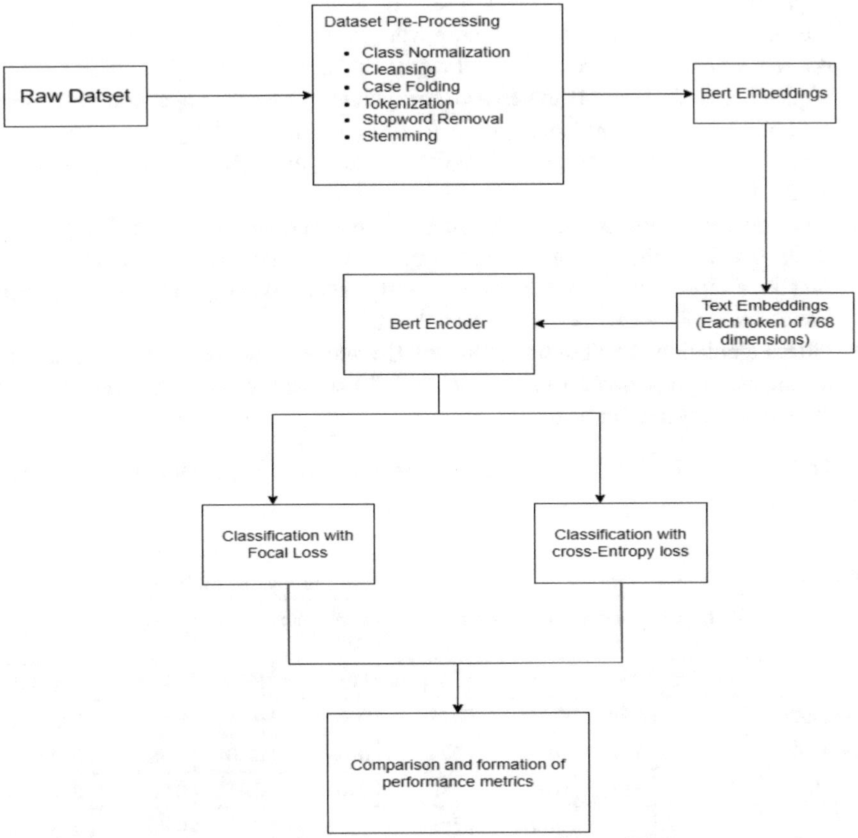

Fig. 2 Model architecture

removed as every review will be containing such hashtags, thus it will lead to errors in classification.
- Tokenization—In this step, a sentence is split into each word that composes it. In this manuscript, BERT tokenizer is used for the tokenization of reviews.
- Stop words Removal—All the irrelevant general tokens like of, and our which are generalized and present in each sentence are removed.
- Lemmatization—Complex words or different words having same root word are changed to the root word for greater accuracy and easy classification.

- The tokens present in the dataset were transformed into BERT embeddings of 768 dimensions which are converted by BERT model implicitly. Also, the advantage of BERT vectorizer over other vectorization methods like Word2Vec method is that BERT produces word representations as per the dynamic information of the words present around the token. After embeddings creation, the BERT training model was selected.
- There are two models of BERT available for BERT training one is "BERT uncased," and the other is "BERT case." In the case of the BERT uncased model, the input words are lowercased before workpiece training. Thus, the model does not remain case sensitive [21]. On the other hand, in the case of the BERT cased model, the model does not lowercase the input; thus, both the upper case and lower case of a particular word will be trained differently, thus making the process more time-consuming and complex. Therefore, in this paper, BERT uncased model is used.
- After training of the BERT model on first, the actual pre-processed Google Play reviews dataset, the testing data accuracy has been calculated with both the loss functions separately. Then the accuracy and f1 score has been calculated for each loss function for each column of the dataset.
- After calculating the f1-score score for Google Play reviews, the training and testing step is repeated for the dataset with 20 percent fewer neutral reviews and 40 percent less neutral dataset.

Lastly, Tables 2, 3, and 4 are plotted to ease the results and comparison, shown in the next section.

Table 2 Performance metrics for both focal loss and cross-entropy loss for balanced dataset

	Performance metrics (focal loss)				Performance metrics (cross-entropy loss)			
	Precision	Recall	F1-Score	Support	Precision	Recall	F1-Score	Support
Negative	0.80	0.73	0.76	245	0.88	0.84	0.86	245
Neutral	0.69	0.70	0.69	254	0.79	0.80	0.80	254
Positive	0.83	0.88	0.85	289	0.89	0.91	0.90	289
Accuracy			0.77	788			0.86	788

Table 3 Performance metrics for both focal loss and cross-entropy loss for 20 percent fewer neutral classes of the dataset

	Performance metrics (focal loss)				Performance metrics (cross-entropy loss)			
	Precision	Recall	F1-Score	Support	Precision	Recall	F1-Score	Support
Negative	0.77	0.81	0.79	245	0.87	0.84	0.86	245
Neutral	0.67	0.65	0.66	220	0.69	0.77	0.73	220
Positive	0.84	0.81	0.83	269	0.87	0.82	0.85	269
Accuracy			0.77	734			0.81	734

Table 4 Performance metrics for both focal loss and cross-entropy loss for 40 percent fewer neutral classes of the dataset

	Performance metrics (focal loss)				Performance metrics (cross-entropy loss)			
	Precision	Recall	F1-Score	Support	Precision	Recall	F1-Score	Support
Negative	0.82	0.87	0.84	243	0.91	0.91	0.91	243
Neutral	0.70	0.62	0.66	152	0.77	0.79	0.78	152
Positive	0.87	0.87	0.87	289	0.91	0.91	0.91	289
Accuracy			0.76	684			0.88	684

6 Training and Classification

The datasets trained on the BERT base uncased model is used for training and classification of the model. The datasets were split into train and test with 10 percent data for testing with a random seed. The value for gamma and alpha used in focal loss functions has been fixed at gamma = 2 and alpha value = 0.8. The epochs used for the training model in the case of BERT are fixed to three for all three datasets. Lastly, the f1 score was calculated for all the classes individually. Then the accuracy of the model was calculated using the f1 score itself, where the f1 score is defined as the harmonic mean of precision and recall of the evaluated model [10]. Further support and precision, and recall have been calculated for each of the classes, and then the overall support has been calculated for the model in each of the dataset cases [32, 33].

7 Results

In this paper, the model is trained at three epochs for three categories of data as follows:

- Google Play reviews dataset with balanced classes
- Google Play reviews dataset with 20 percent or less neutral classes
- Google Play reviews dataset with 40 percent or less neutral classes

Tabulated data for performance metrics for focal loss and cross-entropy loss for each class of positive, negative, and neutral and for the overall model has been tabulated for all three categories of the dataset in Tables 2, 3, and 4, respectively. The accuracy derived from training the dataset on BERT is between 0.75 and 0.78 for focal loss. On the other hand, for cross-entropy loss the accuracy is between 0.80 and 0.90. Low accuracy of training model is because the token embeddings created by BERT inbuilt tokenizer do not assign weights to the tokens as per their relevancy which causes errors in classification. The higher accuracy in case of cross-entropy loss vis-à-vis focal loss is attributed to the summation of different errors which are still present in the classification. As per the focal loss formula the hyperparameter γ leads to decrease in many errors which may be required for classification, thus cross-entropy loss performs better than focal loss. In the case of Table 2 of the dataset of balanced classes, it can be concluded that accuracy for cross-entropy loss is more than that in the case of focal loss. However, the focal loss has been a modified version of cross-entropy loss; still, the cross-entropy loss performs better in the balanced dataset. In Table 3, where the neutral classes have decreased by 20 percent to create a little imbalance, the cross-entropy loss has performed better than focal loss. A similar trend is visible in Table 4, where the cross-entropy loss has performed better than the focal loss for individual classes and the overall model. This confirms that although the focal loss has been a modified version of the cross-entropy loss, in the case of slightly imbalanced data still the cross-entropy loss outperforms the focal loss. Also, the focal loss is more focused on solving the imbalance problem in binary classification problems only. Focal loss is not suitable in the case of datasets where the classes are more than two classes.

8 Conclusion

As the reach of technology grows along with it grows the number of users on different apps to meet the benefits are providing and solving problems for them and making their life easier. As the users use other apps, they tend to give their reviews on the Google Play Store about how the app helped them or if they faced any problem with the app. Most developers take note of the reviews to fix any bugs on applications and try to improve the application more efficiently. Many times, reviews of different apps on the Google Play Store help other users use different apps. Good reviews on any application tend to grow faster. Similarly, other people's reviews can help you navigate your app better and reassure your downloads are safe and problem-free. And also, if the number of positive reviews is way more than the negative reviews, then negative reviews may get overshadowed, and the developer may not take note of the bugs. The loss function that showed better results is the cross-entropy loss function over the focal loss function. Focal loss doesn't differentiate on multiclass as cross-entropy loss is able to classify. Although focal loss is a modification of cross-entropy loss function, it is able to outperform only when the imbalance is high. In slight imbalanced data, the focal loss function ignores many loss values due to the modulating factor. In the future, more experiments will be conducted on different datasets to make conclusion that a particular loss function performs well with a particular model. Also, model can be further upgraded by comparing other loss functions for imbalanced data's most reliable loss function. Lastly, the upgradation on focal loss has to be done mathematically so that the loss function can perform well even in a multiclass dataset and slightly imbalanced dataset.

References

1. Malavolta I, Ruberto S, Soru T, Terragni V (2015) Hybrid mobile apps in the google play store: an exploratory investigation. In: 2nd ACM international conference on mobile software engineering and systems, pp. 56–59
2. Viennot N, Garcia E, Nieh J (2014) A measurement study of google play. ACM SIGMETRICS Perform Eval Rev 42(1), 221–233
3. McIlroy S, Shang W, Ali N, Hassan AE (2017) Is it worth responding to reviews? Studying the top free apps in Google Play. IEEE Softw 34(3):64–71
4. Shashank S, Naidu B (2020) Google play store apps—data analysis and ratings prediction. Int Res J Eng Technol (IRJET) 7:265–274
5. Arxiv A Longitudinal study of Google Play page, https://arxiv.org/abs/1802.02996, Accessed 21 Dec 2021
6. Patil HP, Atique M (2015) Sentiment analysis for social media: a survey. In: 2nd international conference on information science and security (ICISS), pp. 1–4
7. Zainuddin N, Selamat, A.: Sentiment analysis using support vector machine. In: International conference on computer, communications, and control technology (I4CT) 2014, pp. 333–337
8. Dubey A, Rasool A (2021) Efficient technique of microarray missing data imputation using clustering and weighted nearest neighbor. Sci Rep 11(1)
9. Li X, Wang X, Liu H (2021) Research on fine-tuning strategy of sentiment analysis model based on BERT. In: International conference on communications, information system and computer engineering (CISCE), pp. 798–802
10. Mohammadian S, Karsaz A, Roshan YM (2017) A comparative analysis of classification algorithms in diabetic retinopathy screening. In: 7th international conference on computer and knowledge engineering (ICCKE) 2017, pp. 84–89
11. Latif R, Talha Abdullah M, Aslam Shah SU, Farhan M, Ijaz F, Karim A (2019) Data scraping from Google Play Store and visualization of its content for analytics. In: 2nd international conference on computing, mathematics and engineering technologies (iCoMET) 2019, pp. 1–8
12. Day M, Lin Y (2017) Deep learning for sentiment analysis on Google Play consumer review. IEEE Int Conf Inf Reuse Integr (IRI) 2017:382–388
13. Abdul Khalid KA, Leong TJ, Mohamed K (2016) Review on thermionic energy converters. IEEE Trans Electron Devices 63(6):2231–2241
14. Regulin D, Aicher T, Vogel-Heuser B (2016) Improving transferability between different engineering stages in the development of automated material flow modules. IEEE Trans Autom Sci Eng 13(4):1422–1432
15. Li D, Qian J (2016) Text sentiment analysis based on long short-term memory. In: First IEEE international conference on computer communication and the internet (ICCCI) 2016, pp. 471–475 (2016)

16. Lin T, Goyal P, Girshick R, He K, Dollár P (2020) Focal loss for dense object detection. IEEE Trans Pattern Anal Mach Intell 42(2):318–327
17. Arxiv A Sentimental Education: Sentiment Analysis Using Subjectivity Summarization Based on Minimum Cuts, https://arxiv.org/abs/cs/0409058, Accessed 21 Dec 2021
18. Sfu Webpage Methods for Creating Semantic Orientation Dictionaries, https://www.sfu.ca/~mtaboada/docs/publications/Taboada_et_al_LREC_2006.pdf, Accessed 21 Dec 2021
19. Sudhir P, Suresh VD (2021) Comparative study of various approaches, applications and classifiers for sentiment analysis. Glob TransitS Proc 2(2):205–211
20. Gillioz A, Casas J, Mugellini E, Khaled OA (2020) Overview of the transformer-based models for NLP tasks. In: 15th conference on computer science and information systems (FedCSIS) 2020, pp. 179–183
21. Zhou Y, Li M (2020) Online course quality evaluation based on BERT. In: 2020 International conference on communications, information system and computer engineering (CISCE) 2020, pp. 255–258
22. Truong TL, Le HL, Le-Dang TP (2020) Sentiment analysis implementing BERT-based pre-trained language model for Vietnamese. In: 7th NAFOSTED conference on information and computer science (NICS) 2020, pp. 362–367 (2020)
23. Kano T, Sakti S, Nakamura S (2021) Transformer-based direct speech-to-speech translation with transcoder. IEEE spoken language technology workshop (SLT) 2021, pp. 958–965
24. Arxiv Comparing BERT against traditional machine learning text classification, https://arxiv.org/abs/2005.13012, Accessed 21 Dec 2021
25. Arxiv A Comparison of LSTM and BERT for Small Corpus, https://arxiv.org/abs/2009.05451, Accessed 21 Dec 2021
26. Arxiv BERT: Pre-training of Deep Bidirectional Transformers for Language Understanding, https://arxiv.org/abs/1810.04805, Accessed 21 Dec 2021
27. Naseer M, Asvial M, Sari RF (2021) An empirical comparison of BERT, RoBERTa, and Electra for fact verification. In: International conference on artificial intelligence in information and communication (ICAIIC) 2021, pp. 241–246
28. Ho Y, Wookey S (2020) The real-world-weight cross-entropy loss function: modeling the costs of mislabeling. IEEE Access 8:4806–4813
29. Zhou Y, Wang X, Zhang M, Zhu J, Zheng R, Wu Q (2019) MPCE: a maximum probability based cross entropy loss function for neural network classification. IEEE Access 7:146331–146341
30. Yessou H, Sumbul G, Demir B (2020) A Comparative study of deep learning loss functions for multi-label remote sensing image classification. IGARSSIEEE international geoscience and remote sensing symposium 2020, pp. 1349–1352
31. Liu L, Qi H (2017) Learning effective binary descriptors via cross entropy. In: IEEE winter conference on applications of computer vision (WACV) 2017, pp. 1251–1258 (2017)
32. Riquelme N, Von Lücken C, Baran B (2015) Performance metrics in multi-objective optimization. In: Latin American Computing Conference (CLEI) 2015, pp. 1–11
33. Dubey A, Rasool A (2020) Clustering-based hybrid approach for multivariate missing data imputation. Int J Adv Comput Sci Appl (IJACSA) 11(11):710–714

A Hybrid Approach for Missing Data Imputation in Gene Expression Dataset Using Extra Tree Regressor and a Genetic Algorithm

Amarjeet Yadav, Akhtar Rasool, Aditya Dubey, and Nilay Khare

1 Introduction

Missing data is a typical problem in data sets gathered from real-world applications [1]. Missing data imputation has received considerable interest from researchers as it widely affects the accuracy and efficiency of various machine learning models. Missing values typically occur due to manual data entry practices, device errors, operator failure, and inaccurate measurements [2]. A general approach to deal with missing values is calculating statistical data (like mean) for each column and substituting all missing values with the statistic, deleting rows with missing values, or replacing them with zeros. But a significant limitation of these methods was a decrease in efficiency due to incomplete and biased information [3]. If missing values are not handled appropriately, they can estimate wrong deductions about the data. This issue becomes more prominent in Gene expression data which often contain missing expression values. Microarray technology plays a significant role in current biomedical research [4]. It allows observation of the relative expression of thousands of genes under diverse practical states. Hence, it has been used widely in multiple analyses, including cancer diagnosis, the discovery of the active gene, and drug identification [5].

Microarray expression data often contain missing values for different reasons, such as scrapes on the slide, blotting issues, fabrication mistakes, etc. Microarray data may have 1–15% missing data that could impact up to 90–95% of genes. Hence, there is a need for precise algorithms to accurately impute the missing data in the dataset utilizing modern machine learning approaches. The imputation technique known as k-POD uses the K-Means approach to predict missing values [6]. This approach

A. Yadav · A. Rasool · A. Dubey (✉) · N. Khare
Department of Computer Science & Engineering, Maulana Azad National Institute of Technology, Bhopal 462003, India
e-mail: dubeyaditya65@gmail.com

P. Singh et al. (eds.), *Machine Learning and Computational Intelligence Techniques for Data Engineering*, Lecture Notes in Electrical Engineering 998,
https://doi.org/10.1007/978-981-99-0047-3_12

works even when external knowledge is unavailable, and there is a high percentage of missing data. Another method based on Fuzzy C-means clustering uses Support vector regression and genetic algorithm to optimize parameters [7]. The technique suggested in this paper uses both of these models as a baseline. This paper presents a hybrid method for solving the issue. The proposed technique applies a hybrid model that works on optimizing parameters for the K-Means clustering algorithm using an Extra tree regression and genetic algorithm. In this paper, the proposed model is implemented on the Mice Protein Expression Data Set and then its performance is compared with baseline models.

2 Literature Survey

Missing value, also known as missing data, is where some of the observations in a dataset are empty. Missing data is classified into three distinctive classes. These classes are missing completely at random (MCAR), missing at random (MAR), and missing not at random (MNAR) [2, 8]. These classes are crucial as missing data in the dataset generates issues, and the remedies to these concerns vary depending on which of the three types induces the situation. MCAR estimation presumes that missing data is irrelevant to any unobserved response, indicating any observation in the data set does not impact the chances of missing data. MCAR produces unbiased and reliable estimates, but there is still a loss of power due to inadequate design but not the absence of the data [2]. MAR means an organized association between the tendency of missing data and the experimental data, while not the missing data.

For instance, men are less likely to fill in depression surveys, but this is not associated with their level of depression after accounting for maleness. In this case, the missing and observed observations are no longer coming from the same distribution [2, 9]. MNAR describes an association between the propensity of an attribute entry to be missing and its actual value. For example, individuals with little schooling are missing out on education, and the unhealthiest people will probably drop out of school. MNAR is termed "non-ignorable" as it needs to be handled efficiently with the knowledge of missing data. It requires mechanisms to address such missing data issues using prior information about missing values [2, 9]. There must be some model for reasoning the missing data and possible values. MCAR and MAR are both viewed as "ignorable" because they do not require any knowledge about the missing data when dealing with it.

The researchers have proposed many methods to solve the accurate imputation of missing data. Depending on the type of knowledge employed in the techniques, the existent methodology can be classified into four distinct categories: (i) Global approach, (ii) Local approach, (iii) Hybrid approach, and (iv) Knowledge-assisted approach [10, 11]. Each of the approaches has distinct characteristics. Global methods use the information about data from global correlation [11, 12]. Two widely utilized global techniques are Singular Value Decomposition imputation (SVDimpute) and Bayesian Principal Component Analysis (BPCA) methods [13, 14]. These approaches are differentiated by their ability to retrieve missing data by capturing global correlation information. Nevertheless, they ignore the hidden local structure of data. SVDimpute delivers accurate results on time-series data sets with a minor error, but it has one disadvantage: it is not appropriate for non-time-series datasets. The BPCA approach has a tolerable computation error, suggesting that the bias raised by BPCA is less compared to prior methods [14]. However, the BPCA technique may not give accurate results if a gene has local similitude structures in a dataset.

Local approaches use the local information from the data. Some of the Local methods are K Nearest Neighbor Imputation (KNNimpute), Local Least Squares Imputation (LLSimpute), Gaussian Mixture Clustering imputation (GMCimpute), Collateral Missing Value Imputation (CMVE), Multiple Imputations by Chained Equations (MICE) and Classification and Regression Tree (CART), and Locally Auto-weighted Least Squares Method (LAW-LSimpute) [15–21]. When the number of samples is small, KNNimpute produces better results using local similarity, but it gives unsatisfactory outcomes on big data sets [15]. The performance of LLSimpute enhances when the number of closest neighbors represented by "k" becomes near to the number of samples [16]. Still, it degrades when "k" moves near to the number of instances. MICE-CART utilizes data imputation technology to comprehend complicated relationships with the tiniest accommodations [19, 22]. However, since CART-based and conventional MICE outcomes rely on inferior glitch representation, comparable validity is not ensured. GMCimpute is more effective because it can better utilize global association knowledge. But it has an issue with slower fitting [17]. The tailored nearest neighbor method executes satisfactorily with limited sample size and delivers superior accuracy than random forest techniques. In addition, in both time series and non-time series data sets, the CMVE method produces better results when the missing rate is higher [18]. Still, CMVE does not automatically determine the most optimistic number of terminating genes (k) from the data set. LAW-LSimpute optimizes convergence and reduces estimation errors, making it more reliable [20]. However, this method is not recommended if the missing rate is high.

The hybrid strategy appears to be derived by combining global and local data matrix correlations [23]. The hybrid approach may provide better imputation results than an individual technique. Lincmb, Hybrid Prediction Model with Missing Value Imputation (HPM-MI), KNN + Neural Network (NN), Genetic Algorithm (GA) + Support Hybrid approaches like SVR and Fuzzy C-means + SVR + GA come under hybrid systems [7, 23–25]. In terms of precision, selectivity, and sensitivity, HPM-MI outperforms other methods [23]. Case of imbalanced classification in a multi-class dataset creates a problem for the HPM-MI method. The GA + SVR model takes less time to compute, and the SVR clustering method produces a more realistic result [7]. This imputation technique has a problem with local minimization for some outlier data. Because of noise mitigation measures that improve computation accuracy, the KNN + GA strategy outperforms other NN-GA systems in terms of evaluation accuracy [7, 24]. However, some criteria must be chosen ahead of time, for instance, the sort of neural network to use and the proper parameters to use when training the model to fulfill performance standards [25]. In Knowledge-assisted approaches for imputation of missing value, domain knowledge from data is utilized. Fuzzy C-Means clustering (FCM) and Projection Onto Convex Sets (POCS) are some of the knowledge-assisted methods [1, 25]. FCM process missing value imputation using gene ontology annotation as external information. On the other hand, it becomes hard to extract and regulate prior knowledge. Furthermore, the computation time is increased.

3 About Genetic Algorithm, K-Means, and Extra Tree Regression

3.1 Genetic Algorithm

John Holland introduced a Genetic Algorithm in 1975. A genetic algorithm can be used to optimize both constrained and unconstrained problems using the process of natural selection [26]. The genetic algorithm constantly changes a population of individual solutions. The genetic algorithm randomly picks individuals from the current population who will be parents and utilizes them to produce descendants for the following generation. Over subsequent generations, the population grows toward an optimal solution. Figure 1 demonstrates the working of the Genetic algorithm.

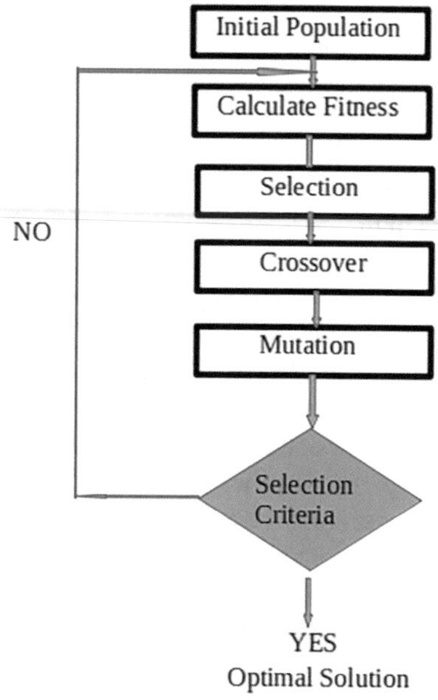

Fig. 1 Block diagram of genetic algorithm

3.2 K-Means Algorithm

K-means clustering is one of the prevalent unsupervised machine learning algorithms [27]. Unsupervised learning algorithms make deductions from datasets without pre-assigned training data labels. This method was proposed during the 1950s and 1960s. Researchers from diverse domains independently conceived proposals for K-Means. In 1956 Steinhaus was the first researcher to propose the algorithm. In 1967 MacQueen coined the term *K*-means. In the K-means algorithm, *k* number of centroids are first identified. Then each data point is allotted to the closest cluster while maintaining the centroids as small as feasible. K-Means minimize the following objective function:

$$J = \sum_{j=1}^{k} \sum_{i=1}^{n} \left\| x_i^j - c_j \right\|^2 \tag{1}$$

In Eq. (1) j indicates cluster, c_j is the centroid for cluster j, x_i represents case i, k represents number of clusters, n represents number of cases, and $|x_i - c_j|$ is distance functions.

In addition to K-Means clustering, each case x_i has a membership function representing a degree of belongingness to a particular cluster c_j. The membership function is described as

$$u_{ij} = \frac{1}{\sum_{k=1}^{c} \left(\frac{x_i - c_j}{x_i - c_k} \right)^{\frac{2}{m-1}}} \tag{2}$$

In Eq. (2), m is the weighting factor parameter, and its domain lies from one to infinity. c represents the number of clusters, whereas c_k represents the centroid for the kth cluster. Only the complete attributes are considered for revising the membership functions and centroids.

Missing value for any case x_i is calculated using the membership function and value of centroid for a cluster. Function used for missing value imputation is described as.

$$\text{Missing value} = \Sigma m_i * c_i \tag{3}$$

In Eq. (3), m_i is the estimated membership value for ith cluster, c_i represents centroid of a ith cluster and c is the number of clusters. Σ denotes summation of product of m_i and c_i.

3.3 Extra Tree Regression

Extra Trees is an ensemble technique in machine learning. This approach integrates the predictions from multiple decision trees trained on the dataset. The average predicted values from the decision trees are taken when estimating regression values in the Extra tree regression, while majority voting is performed for classification [28]. The algorithms like bagging and random forest generate individual decision trees by taking a bootstrap sample from the training dataset. In contrast, in the case of the Extra Tree regression, every decision tree algorithm is fitted on the complete training dataset.

4 About Dataset

For the implementation of the model, this paper uses the Mice Protein Expression Data Set from UCI Machine Learning Repository. The data set consists of the expression levels of 77 proteins/protein modifications. There are 38 control mice and 34 trisomic mice for 72 mice. This dataset contains eight categories of mice which are defined based on characteristics such as genotype, type of behavior, and treatment. The dataset contains 1080 rows and 82 attributes. These attributes are Mouse ID, Values of expression levels of 77 proteins, Genotype, Treatment type, and Behavior. Dataset is artificially renewed such that it has 1%, 5%, 10%, and 15% missing value ratios. All the irrelevant attributes such as MouseID, Genotype, Behavior, and Treatment are removed from the dataset. Next, 558 rows were selected from shuffled datasets for the experiment. For dimensionality reduction, the PCA (Principal Component Analysis) method was used to reduce the dimensions of the dataset to 20. To normalize the data values between 0 and 1, a MinMax scaler was used.

5 Proposed Model

This research proposes a method to evaluate missing values using K-means clustering optimized with an Extra Tree regression and a genetic algorithm. The novelty of the proposed approach is the application of an ensemble technique named Extra Tree regression for estimating accurate missing values. These accurate predictions with the genetic algorithm further help in the better optimization of K-Means parameters. Figure 2 represents the implementation of the proposed model. First, to implement the model on the dataset, missing values are created artificially. Then the dataset with missing values is divided into a complete dataset and an incomplete dataset. In the complete dataset, those rows are considered in which none of the attributes contains a missing value. In contrast, an incomplete dataset contains rows with attributes with one or more missing values.

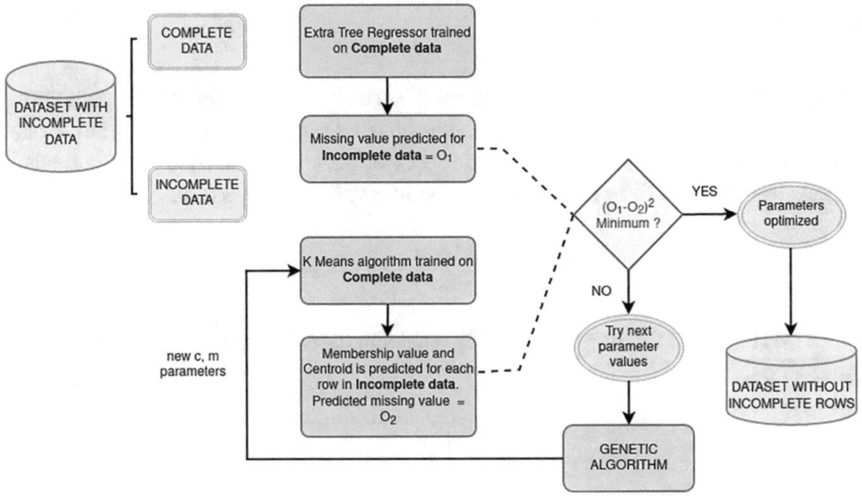

Fig. 2 Proposed model (KextraGa)

In the proposed approach, an Extra Tree regression and Genetic algorithm are used for the optimization of parameters of the K-Means algorithm. The Extra tree regression and K-Means model are trained on a complete row dataset to predict the output. Then, K-means is used to evaluate the missing data for the dataset with incomplete rows. K-means outcome is compared with the output vector received from the Extra Tree regression. The optimized value for c and m parameters is obtained by operating the genetic algorithm to minimize the difference between the Extra Tree regression and K-means output. The main objective is to reduce error function = $(X - Y)^2$, where X is the prediction output of the Extra Tree regression method and Y is the outcome of the prediction from the K-means model. Finally, Missing data are estimated using K-means with optimized parameters.

5.1 Experimental Implementation

The code for the presented model is written in Python version 3.4. The K-means clustering and Extra Tree regression are imported from the sklearn library. The number of clusters = 3 and membership operator value = 1.5 is fed in the K-Means algorithm. In the Extra tree regression, the number of decision trees = 100 is used as a parameter. The genetic algorithm uses 20 as population size, 40 as generations, 0.60 crossover fraction, and a mutation fraction of 0.03 as parameters.

6 Performance Analysis

The performance of the missing data imputation technique is estimated by calculating the mean absolute error (MAE), root mean squared error (RMSE), and relative classification accuracy (A) [29, 30]. MAE is an evaluation metric used with regression models. MAE takes the average of the absolute value of the errors.

$$\text{MAE} = \frac{1}{n}\sum\nolimits_{j=1}^{n}\left|\widehat{y}_j - y_j\right| \tag{4}$$

RMSE is one of the most commonly used standards for estimating the quality of predictions.

$$RMSE = \sqrt{\sum_{j=1}^{n}\frac{\left(\widehat{y}_j - y_j\right)^2}{n}} \tag{5}$$

In Eqs. (4) and (5), $\hat{y}_j$ represents predicted output and y_j represents actual output. "n" depicts total number of cases. The relative classification accuracy is given by.

$$A = \frac{c_t}{c} * 10 \tag{6}$$

In Eq. (6), c represents the number of all predictions, and c_t represents the number of accurate predictions within a specific tolerance. A 10% tolerance is used for comparative prediction, which estimates data as correct for values within a range of ±10% of the exact value.

7 Experimental Results

This section discusses the performance evaluation of the proposed model. Figures 3–4 shows box plots of the performance evaluation of the three different methods for the Mice Protein Expression Data Set, with 1%, 5%, 10%, and 15% missing values. In Box plots, the halfway mark on each box represents the median. The whiskers cover most of the data points except outliers. Outliers are plotted separately. Figure 3a compares three methods on the dataset with 1–15% missing data. Each box includes 4 results in the RMSE. The median RMSE values are 0.01466, 0.01781, and 0.68455. Figure 3b compares the MAE on the dataset with 1–15% missing values. The median MAE values are 0.10105, 0.10673, and 0.78131. Better performance is indicated from lower error. Figure 4 compares the accuracy of different models used for the experiment. This accuracy is estimated by computing the difference between the correct and predicted value using a 10% tolerance. Accuracy is calculated for three techniques executed on the dataset with 1–15% missing values. The median accuracy

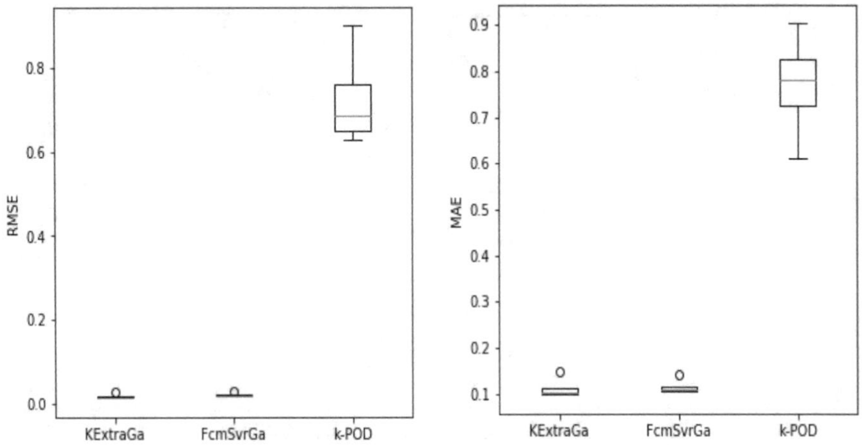

(a) RMSE (b) MAE

Fig. 3 Box plot for RMSE and MAE in three methods for 1–15% missing ratio

Fig. 4 Box plot for relative accuracy in three methods for 1–15% missing ratio

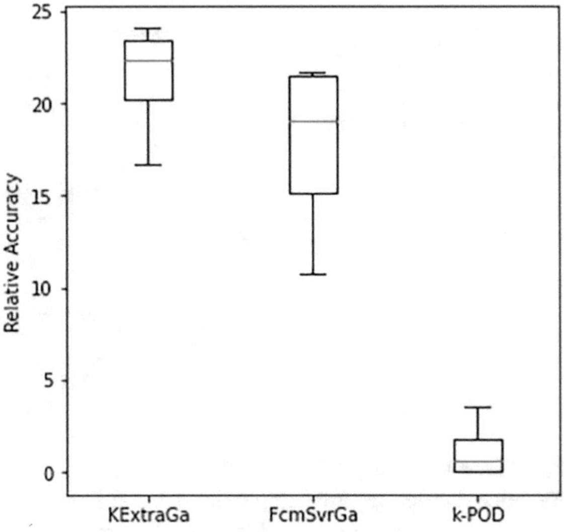

values are 22.32143, 19.04762, and 0.67. Better imputations are indicated from higher accuracy.

It is evident from the box plots that the proposed method gives the lowest RMSE and MAE error and the highest relative accuracy on the given dataset. Figure 5–6 represents a line graph of the performance evaluation of the three different methods against the missing ratios. Figure 5a illustrates that the hybrid K-Means and ExtraTree-based method has a lower RMSE error value compared to both methods for the mice dataset. Figure 5b indicates that the proposed hybrid K-Means and

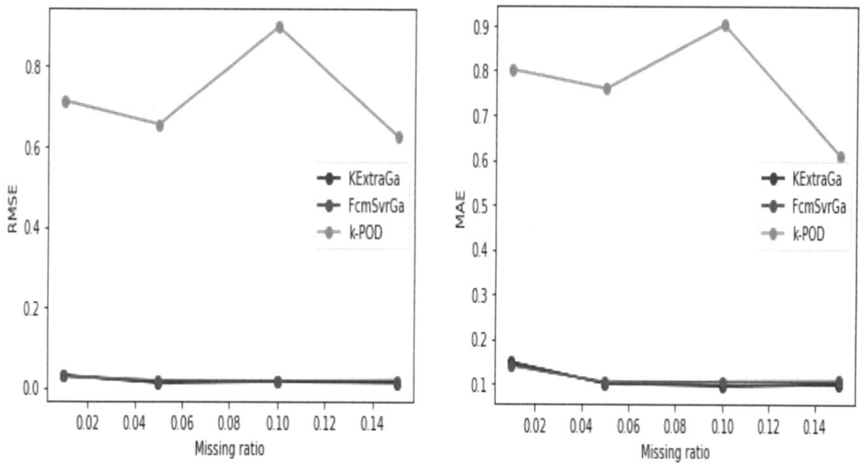

(a) RMSE (b) MAE

Fig. 5 RMSE and MAE comparison of different techniques for 1–15% missing ratio in the dataset

Fig. 6 Relative Accuracy comparison of different techniques for 1–15% missing ratio in the dataset

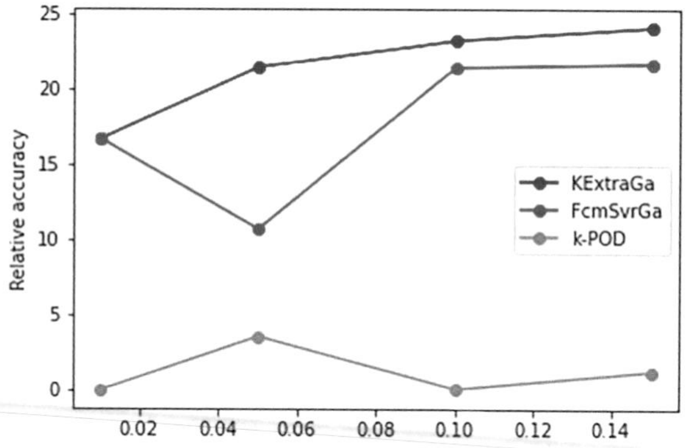

ExtraTree-based hybrid method has a lower MAE error value than both methods for the mice dataset. Figure 6 demonstrates that the accuracy of the evaluated and actual data with 10% tolerance is higher for the proposed method than the FcmSvrGa and k-POD method.

The graphs in Figs. 5–6 indicate that k-POD gives the highest error and lowest accuracy at different ratios [6]. The FcmSvrGa method gives a slightly lower error at a 1% missing ratio, but when compared to overall missing ratios KExtraGa method provides the lowest error than other baseline models. Furthermore, compared to other methods, the KExtraGa method gives better accuracy over each missing ratio. It is clearly illustrated from Figs. 3–6 that the proposed model KExtraGa performs better than the FcmSvrGa and k-POD method. The Extra Tree regression-based method achieves better relative accuracy than the FcmSvrGa and k-POD method. In addition, the proposed method also achieves overall less median RMSE and MAE error than both methods. There are some drawbacks to the proposed method. Training of Extra tree regression is a substantial issue. Although the training time of the proposed model is slightly better than FcmSvrGa, it still requires an overall high computation time.

8 Conclusion and Future Work

This paper proposes a hybrid method based on the K-Means clustering, which utilizes an Extra tree regression and genetic algorithm to optimize parameters to the K-Means algorithm. This model was applied to the Mice protein expression dataset and gave better performance than the other algorithms. In the proposed model, complete dataset rows were clustered based on similarity, and each data point is assigned a membership function for each cluster. Hence, this method yields more practical results as each missing value belongs to more than one cluster. The experimental results clearly illustrate that the KExtraGa model yields better accuracy (with 10% tolerance) and low RMSE and MAE error than the FCmSvrGa and k-POD algorithm. The limitation of the model proposed in this research paper has indicated a need for a fast algorithm. Hence, the main focus area for the future would be a reduction of the computation time of the proposed algorithm. Another future goal would be to implement the proposed model on a large dataset and enhance its accuracy [22].

References

1. Gan X, Liew AWC, Yan H (2006) Microarray missing data imputation based on a set theoretic framework and biological knowledge. Nucleic Acids Res 34(5):1608–1619
2. Pedersen AB, Mikkelsen EM, Cronin-Fenton D, Kristensen NR, Pham TM, Pedersen L, Petersen I (2017) Missing data and multiple imputation in clinical epidemiological research. Clin Epidemiol 9:157
3. Dubey A, Rasool A (2020) Time series missing value prediction: algorithms and applications. In: International Conference on Information, Communication and Computing Technology. Springer, pp. 21–36
4. Trevino V, Falciani F, Barrera- HA (2007) DNA microarrays: a powerful genomic tool for biomedical and clinical research. Mol Med 13(9):527–541
5. Chakravarthi BV, Nepal S, Varambally S (2016) Genomic and epigenomic alterations in cancer. Am J Pathol 186(7):1724–1735
6. Chi JT, Chi EC, Baraniuk RG (2016) k-pod: A method for k-means clustering of missing data. Am Stat 70(1):91–99
7. Aydilek IB, Arslan A (2013) A hybrid method for imputation of missing values using optimized fuzzy c-means with support vector regression and a genetic algorithm. Inf Sci 233:25–35
8. Dubey A, Rasool A (2020) Clustering-based hybrid approach for multivariate missing data imputation. Int J Adv Comput Sci Appl (IJACSA) 11(11):710–714
9. Gomer B (2019) Mcar, mar, and mnar values in the same dataset: a realistic evaluation of methods for handling missing data. Multivar Behav Res 54(1):153–153

10. Meng F, Cai C, Yan H (2013) A bicluster-based bayesian principal component analysis method for microarray missing value estimation. IEEE J Biomed Health Inform 18(3):863–871
11. Liew AWC, Law NF, Yan H (2011) Missing value imputation for gene expression data: computational techniques to recover missing data from available information. Brief Bioinform 12(5):498–513
12. Li H, Zhao C, Shao F, Li GZ, Wang X (2015) A hybrid imputation approach for microarray missing value estimation. BMC Genomics 16(S9), S1
13. Troyanskaya O, Cantor M, Sherlock G, Brown P, Hastie T, Tibshirani R, Botstein D, Altman RB (2001) Missing value estimation methods for DNA microarrays. Bioinformatics 17(6):520–525
14. Oba S, Sato Ma, Takemasa I, Monden M, Matsubara, Ki, Ishii S (2003) A Bayesian missing value estimation method for gene expression profile data. Bioinformatics 19(16), 2088–2096
15. Celton M, Malpertuy A, Lelandais G, De Brevern AG (2010) Comparative analysis of missing value imputation methods to improve clustering and interpretation of microarray experiments. BMC Genomics 11(1):1–16
16. Kim H, Golub GH, Park H (2005) Missing value estimation for DNA microarray gene expression data: local least squares imputation. Bioinformatics 21(2):187–198
17. Ouyang M, Welsh WJ, Georgopoulos P (2004) Gaussian mixture clustering and imputation of microarray data. Bioinformatics 20(6):917–923
18. Sehgal MSB, Gondal I, Dooley LS (2005) Collateral missing value imputation: a new robust missing value estimation algorithm for microarray data. Bioinformatics 21(10):2417–2423
19. Burgette LF, Reiter JP (2010) Multiple imputation for missing data via sequential regression trees. Am J Epidemiol 172(9):1070–1076
20. Yu Z, Li T, Horng SJ, Pan Y, Wang H, Jing Y (2016) An iterative locally auto-weighted least squares method for microarray missing value estimation. IEEE Trans Nanobiosci 16(1):21–33
21. Dubey A, Rasool A (2021) Efficient technique of microarray missing data imputation using clustering and weighted nearest neighbour. Sci Rep 11(1):24–29
22. Dubey A, Rasool A (2020) Local similarity-based approach for multivariate missing data imputation. Int J Adv Sci Technol 29(06):9208–9215
23. Purwar A, Singh SK (2015) Hybrid prediction model with missing value imputation for medical data. Expert Syst Appl 42(13):5621–5631
24. Aydilek IB, Arslan A (2012) A novel hybrid approach to estimating missing values in databases using k-nearest neighbors and neural networks. Int J Innov Comput, Inf Control 7(8):4705–4717
25. Tang J, Zhang G, Wang Y, Wang H, Liu F (2015) A hybrid approach to integrate fuzzy c-means based imputation method with genetic algorithm for missing traffic volume data estimation. Transp Res Part C: Emerg Technol 51:29–40
26. Marwala T, Chakraverty S (2006) Fault classification in structures with incomplete measured data using autoassociative neural networks and genetic algorithm. Curr Sci 542–548
27. Hans-Hermann B (2008) Origins and extensions of the k-means algorithm in cluster analysis. Electron J Hist Probab Stat 4(2)
28. Geurts P, Ernst D, Wehenkel L (2006) Extremely randomized trees. Mach Learn 63(1):3–42
29. Yadav A, Dubey A, Rasool A, Khare N (2021) Data mining based imputation techniques to handle missing values in gene expressed dataset. Int J Eng Trends Technol 69(9):242–250
30. Gond VK, Dubey A, Rasool A (2021) A survey of machine learning-based approaches for missing value imputation. In: Proceedings of the 3rd International Conference on Inventive Research in Computing Applications, ICIRCA 2021, pp. 841–846

A Clustering and TOPSIS-Based Developer Ranking Model for Decision-Making in Software Bug Triaging

Pavan Rathoriya, Rama Ranjan Panda, and Naresh Kumar Nagwani

1 Introduction

Meeting software development deadlines is a key challenge in software development. To meet deadlines, testing, and bug fixing activities should be managed in a prioritized and optimized manner. Bug triaging is the process of assigning newly reported bugs to the appropriate software developers. The person who handles the bug triage is called a trigger. The finding of expert developers includes the understanding of the developer's profile and domain in which the developers are comfortable fixing bugs. In recent years, many machine learning-based techniques have been proposed by researchers to automate the process of bug triaging. These machine Learning-based techniques analyzed the historical data and then discover the appropriate software developer for the newly reported bugs. The problem with these techniques is that the availability of the developers is not considered while assigning the newly reported bugs. A few developers may be heavily loaded with the assigned bugs, but most of the developers are free. So, consideration of the availability of developers should also be considered for better management of the triaging process. An effective bug triaging technique considers many attributes extracted from the software bug repositories. Most of the bug repositories maintain information about the developer's profile in terms of the developer's experience, bugs resolved and fixed, how many bugs are assigned to a developer, and so on. MADM (Multi-Attribute Decision Making) [1] techniques play a key role in solving and decision-making problems having multiple

P. Rathoriya (✉) · R. R. Panda · N. K. Nagwani
Department of Information & Technology, National Institute of Technology, Raipur, India
e-mail: pavan.mtech2020.it@nitrr.ac.in

R. R. Panda
e-mail: rrpanda.phd2018.cs@nitrr.ac.in

N. K. Nagwani
e-mail: nknagwani.cs@nitrr.ac.in

P. Singh et al. (eds.), *Machine Learning and Computational Intelligence Techniques for Data Engineering*, Lecture Notes in Electrical Engineering 998,
https://doi.org/10.1007/978-981-99-0047-3_13

attributes and generating the ranked list of solutions to such problems. TOPSIS [2] is one of the popular techniques under the MADM paradigm to solve problems. The main attributes for bug triaging include the consideration of the attributes, namely, the experience of developers in years (D), the number of assigned bugs (A), the number of newly assigned bugs (N), the number of fixed or resolved bugs (R), and the average resolving time (T). Software bugs are managed through online software bug repositories. For example, the Mozilla bugs are available online at 1 https://bugzilla.mozilla.org/users_profile?.user_id=X, where X is the id of the software bug.

In this paper, Sect. 2 presents motivation, and Sect. 3 presents some related work to bug triaging. In Sect. 4, the methodology is presented. Sect. 5 describes our model with an illustrative example. Sect. 6 covers some threats to validity, and Sect. 7 discusses the conclusion and future work.

2 Motivation

The problem with machine learning techniques mostly depends on the historical dataset for training and do not consider the availability of developers in bug triaging. For example, the machine learning algorithm can identify one developer as an expert for the newly reported bug, but at the same time, the developer might have been assigned numerous bugs at the same time as the developer is an expert developer. With the help of MCDM approaches, such a problem can be handled efficiently by considering the availability of the developer as one of the non-profit criteria (negative/lossy attribute or maximize/minimize attributes).

3 Related Work

Bug triaging is a decision-making problem in which a suitable developer who can solve the bug is identified. There have been a series of studies conducted by many researchers. They've gone through various methods. Several researchers used machine learning, deep learning, topic modeling, and MCDM methodologies, as discussed in the below paragraphs.

Different researchers used various machine learning algorithms in [3–10]. Malhotra et al. [6] used textual information for bug triaging using various machine learning algorithms on six open-source datasets: Mesos, Hadoop, Spark, Map Reduce, HDFS, and HBASE. Since textual information may contain a lot of unnecessary data, therefore result can be inconsistent, Shadhi et al. [9] have used Categorical fields of bag of word model (BOW) of bug report and combined both Categorical and textual data to show the results. Agrawal et al. [10] created the word2vec approach, which employs natural language processing (NLP) to construct a prediction model for determining the appropriate developer for bug triaging using various classification algorithms such as KNN, SVM, RF, and others. The challenges faced in the existing machine learning mechanism were that it is difficult to label bug reports with missing or insufficient label information, and most classification algorithms used in existing approaches are costly and inefficient with large datasets.

Deep learning techniques to automate the bug assignment process are another set of approaches that can be used with large datasets being researched by researchers [11–18]. By extracting features, Mani et al. [19] proposed the deep triage technique, which is based on the Deep Bidirectional Recurrent Neural Network with Attention (DBRN-A) model. Unsupervised learning is used to learn the semantic and syntactic characteristics of words. Similarly, Tuzun et al. [15] improved the accuracy of the method by using Gated Recurrent Unit (GRU) instead of Long-Short Term Memory (LSTM), combining different datasets to create the corpus, and changing the dense layer structure (simply doubling the number of nodes and increasing the layer) for better results. Guo et al. [17] proposed an automatic bug triaging technique based on CNN and developer activity, in which they first apply text preprocessing, then create the word2vec, and finally use CNN to predict developer activity. The problem associated with the deep learning approach is that, based on the description, a developer can be selected accurately, but availability and expertise can't be determined.

Several studies [19–23] have increased bug triaging accuracy by including additional information such as components, products, severity, and priority. Hadi et al. [19] have presented the Dependency-aware Bug Triaging Method (DABT). This considers both bug dependency and bug fixing time in bug triaging using Topic mode Linear Discriminant Analysis (LDA) and integer programming to determine the appropriate developer who can fix the bug in a given time slot. Iyad et al. [20] have proposed the graph-based feature augmentation approach, which uses graph-based neighborhood relationships among the terms of the bug report to classify the bug based on their priority using the RFSH [22] technique.

Some bugs must be tossed due to the inexperience of the developer. However, this may be decreased by selecting the most suitable developer. Another MCDM-based method is discussed in [24–28] for selecting the best developer. Goyal et al. [27] used the MCDM method, namely the Analytic Hierarchy Process (AHP) method, for bug triaging, in which the first newly reported bug term tokens are generated and developers are ranked based on various criteria with different weightages. Gupta et al. [28] used a fuzzy technique for order of preference by similarity to ideal solution (F-TOPSIS) with the Bacterial Foraging Optimization Algorithm (BFOA) and Bar Systems (BAR) to improve bug triaging results.

From the above-discussed method, it can be concluded that the existing methods do not consider the ranking of bugs or developers using metadata and multi-criteria decision-making for selecting the developer. And in reality, all the parameters/features are not equal, so the weight should be assigned explicitly for features and their prioritization. Other than these, in the existing method, developer availability is also not considered. Hence, these papers identify this gap and suggest a hybrid bug triaging mechanism.

4 Methodology

The proposed method explained in this section consists of the following steps (Fig. 1).–

1. Extract bugs from bug repositories
2. Preprocessing
3. Developer vocabulary and metadata generation
4. Matching of newly reported bugs with developer vocabulary
5. For filtered developers, extract metadata from 3
6. Apply AHP
7. Apply TOPSIS for ranking of developers.

In the first step, the bug data is collected from open sources like Kaggle or bugzilla. In the present paper, the dataset has 10,000 raw and 28 columns of attributes which contain information related to bugs, like the developer who fixed it, when the bug was triggered, when the bug was fixed, bud id, bug summary, etc. The dataset is taken from Kaggle.

In step two, preprocessing tasks are applied to the bug summary, for example, text lowercasing, stop word removal, tokenization, stemming, and lemmatization.

In the third step, developer metadata is extracted from the dataset. It consists of the following: developer name, total number of bugs assigned to each developer, number of bugs resolved by each developer, new bugs assigned to each developer, total experience of developer, average fixing time of developer to resolve all bugs. And developer vocabulary is also created by using bug developer names and bug summary.

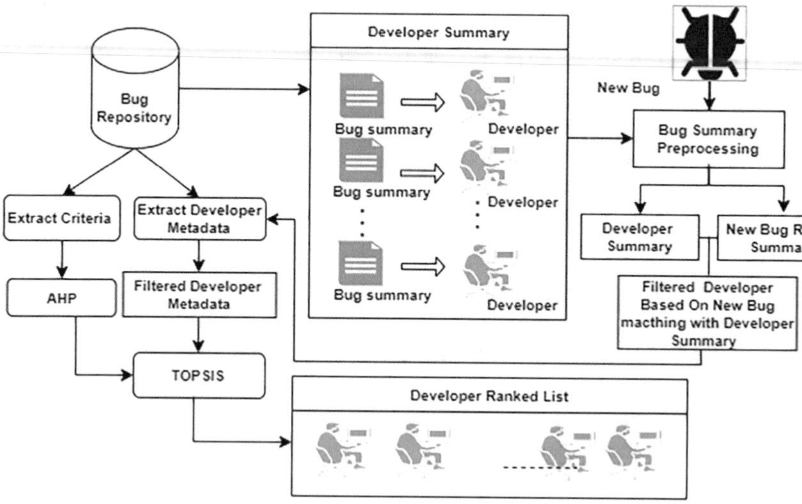

Fig. 1 Framework of overall proposed approach

In the fourth step, the newly reported preprocessed bug summary is matched with developer vocabulary using the cosine similarity [17] threshold filter. Based on similarity, developers are filtered from the developer vocabulary for further steps.

In the fifth step, developer metadata is extracted from step 3 only for the filtered developer from step 4.

In step six AHP method is applied to find the criteria weight. It has the following steps for bug triaging:

1. Problem definition: The problem is to identify the appropriate developer to fix the newly reported bug.
2. Parameter selection: Appropriate parameter (Criteria) are selected for finding their weight. It has the following criteria: name of developer (D), developer experience in year (E), total number of bugs assigned (A), newly assigned bugs (N), total bug fixed (R), and average fixing time (F).
3. Create a judgement matrix (A): An squared matrix named A order of $m \times m$ is created for pairwise comparison of all the criteria, and the element of the matrix the relative importance of criteria to the other criteria-

$$A_{m*m} = a_{ij} \tag{1}$$

where m is the number of criteria and a is the relative importance of criteria.

Their entry of an element into the matrix follows the following rules:

$$a_{ij} = 1/a_{ji} \tag{2}$$

$$a_{ii} = 1 \text{ for all } i \tag{3}$$

Here $i, j =$ 1,2.3……..m.

For relative importance, the following data will be used:

4. Then normalized the matrix A.
5. Then find the eigenvalue and eigenvector W^t.
6. Then a consistency check of weight will be performed. It has the following steps:

   i. Calculate λ_{max} by given Eq. (4):

$$\lambda_{max} = \frac{1}{n}\sum_{i=1}^{n}\frac{i_{\text{th}}\text{in}AW^t}{i_{\text{th}}\text{in}W^t} \tag{4}$$

   ii. Calculate the consistency index (CI) using Eq. (5):

$$\text{CI} = \frac{(\lambda_{max} - n)}{(n-1)} \tag{5}$$

Here n is the number of criteria.

iii. Calculate the consistency ration (CR) using Eq. (6):

$$\mathrm{CR} = \frac{\mathrm{CI}}{\mathrm{RI}} \tag{6}$$

Here RI is random index [24]. If the consistency ratio is less than 0.10, the weight is consistent and we can use the weight (W) for further measurement in next step. If not, repeat from step 3 of AHP and use the same step.

In step 7, the TOPSIS [27] model is applied for ranking the developer. The TOPSIS model has the following steps for developer ranking:

1. Make a performance matrix (*D*) for each of the selected developers with the order $m \times n$, where n is the number of criteria and m is the number of developers (alternatives),

   And the element of the matrix will be the respective value for the developer according to the criteria.
2. Normalize the Matrix Using the Following Equation

$$R_{ij} = a_{ij} / \sqrt{\sum_{k=1}^{m} a_{ij}^2} \tag{7}$$

3. Multiply the Normalized Matrix (R_{ij}) with the Calculated Weight from the Ahp Method.

$$V = (V_{ij})_{m \times n} = (W_i R_{ij})_{m \times n} \tag{8}$$

4. Determine the positive idea solution (best alternative) (A^*) and the anti-idea solution (A^-)(worst alternative) using the following equations.

$$A^* = \left[\left(\max_i V_{ij} | j \in J\right), \min_i V_{ij} | j \in J^{`}\right] = \{V_1^*, V_2^* I., V_n^*\} \tag{9}$$

$$A^- = \left[\left(\max_i V_{ij} | j \in J\right), \min_i V_{ij} | j \in J^{`}\right] = \{V_1^-, V_2^- ...., V_n^-\} \tag{10}$$

5. Find the Euclidean distance between the alternative and the best alternative called d^*, and similarly, from the worst alternative called d^-, using the following formula

$$d_i^* = \sqrt{\sum_{j=1}^{n} (v_{ij} - v_j^*)^2} \tag{11}$$

$$d_i^- = \sqrt{\sum_{j=1}^{n} (v_{ij} - v_j^i)^2} \quad (12)$$

6. Find the similarity to the worst condition (CC). It is also called the closeness ration. Using the following formula, the higher the closeness ration of an alternative, the higher the ranking

$$\mathrm{CC}_i = \frac{d_i^-}{d_i^* + d_i^-} \quad (13)$$

For bug triaging, the developer who has the highest closeness will be ranked first, and the lowest closeness developer will have the last rank for bug triaging.

5 Illustrative Example: A Case Study

In order to better explain the proposed model, an illustrative example is presented in this section. The data is taken from Kaggle repository, which consists of 10,000 bugs and 4,000 developers. Then one bug at a time is taken and preprocessing task is performed on bug summary and developer vocabulary is created. Then the newly reported bug summary similarity is checked with the developer vocabulary, and based on similarity top 5 developers are selected and its metadata (developer name (D), developer experience in year (E), total number of bugs assigned (A), newly assigned bugs (N), total bug fixed (R), and average fixing time (F)) is filtered on which TOPSIS method will be applied to rank them is shown in Table 1. For privacy reasons, the actual names of developers are not mentioned here. Now the AHP method will be applied for calculating the criteria weight. In the AHP method, first the goal is defined. Here the goal is to find the best appropriate developer for bug fixing.

In step 2, the criteria are selected. Here the criteria are the developer experience in a year (E), the total number of bugs assigned (A), newly assigned bugs (N), the total number of bugs fixed (R), and the average fixing time (F). In the next step, a pairwise judgement matrix is created. It is a 2D square matrix. By referring to Table 2 and Eqs. (2) and (3), you can see how element values are assigned based on the

Table 1 Relative importance of criteria

Importance value	Description
1	Identical importance
3	Reasonable importance
5	Strong importance
7	Very Strong importance
9	Extremely strong importance
2, 4, 6, and 8	Middle values

Table 2 Sample dataset for demonstrating TOPSIS-based model

Developer D	Experience (Years) E	Total bug assigned A	Newly assigned bugs N	Total fixed bugs R	Average fixing time (Days) F
D_1	10	40	3	35	7
D_2	5	15	0	12	9
D_3	3	25	7	16	8
D_4	5	2	0	1	3
D_5	7	20	4	4	4

Table 3 Pairwise comparison of criteria

	E	A	N	R	F
E	1.00	0.20	1.00	0.14	0.14
A	5.00	1.00	1.00	0.20	0.14
N	1.00	1.00	1.00	0.20	0.20
R	7.00	5.00	5.00	1.00	0.33
T	7.00	7.00	5.00	3.00	1.00

relative importance of criteria to other criteria. For example, in Table 4 criteria "total bug assigned" has strong importance, our "experience" hence assigns 5 and in vice versa case assigns 1/5, and criteria "total bug resolved" has very strong importance, our experience assigns 7 and in visa versa case 1/7 will be assigned. Similarly, other values can be filled by referring to Table 2, and diagonal value is always fixed to one. The resultant judgement matrix is given in table format in Table 3.

In next step, Table 4 will normalize and eigenvalue and eigenvector will be calculated, the transpose of eigenvector is the weight of criteria that is given in Table 4.

Then the calculated criteria weight consistency will be checked by following Eqs. (4), (5) and (6) and get the result shown below.

$$\lambda_{max} = 5.451$$

$$\text{Consistency Index (CI)} = 0.113$$

$$\text{Consistency Ration (CR)} = 0.10069$$

Table 4 Criteria weight

	E	A	N	R	T
Weight (W)	0.0497	0.102	0.069	0.295	0.482

Table 5 Weighted normalized matrix

Developer	E	A	N	R	T
D1	0.034	0.076	0.024	0.255	0.228
D2	0.017	0.029	0.000	0.087	0.293
D3	0.010	0.048	0.056	0.116	0.261
D4	0.017	0.004	0.000	0.007	0.098
D5	0.024	0.038	0.032	0.029	0.130

Since the CR $\simeq$ 0. 1, hence the weights are consistent and it can be used for further calculation.

Now in next Step TOPSIS method is applied, The TOPSIS method generates the first evolutionary matrix (*D*) of size $m*n$, where m is the number of alternatives (developers) and n is the number of criteria, and in our example, there are 5 developers and 5 criteria, so the evolution matrix will be 5^*5.In the next step, matrix D will be normalized by using Eq. (7), and then get the resultant matrix R, and, matrix R will be multiplied with weight W by using Eq. (8) and get the weighted normalized matrix shown in Table 5.

In the next step, the best alternative and worst alternative are calculated by using Eqs. 9 and 10 shown in Table 6.

Next, find the distance between the best substitute and the target substitute and also from the worst alternative using Eqs. (11) and (12). Then, using Eq. (13), get the closeness ratio that is shown in Table 7.

Generally, CC = 1 if the alternative has the best solution. Similarly, if CC = 0, then the alternative has the worst solution. Based on the closeness ration, D1 as the first rank, then D4 has the second rank, and respectively, D5, D3, and D2 have the third, fourth, and fifth ranks. Ranking Bar graph based on closeness ration is also shown in Fig. 2.

Table 6 Ideal solution (P*) and anti-ideal solution (P^-)

P*	0.034	0.076	0.000	0.255	0.098
P–	0.010	0.004	0.056	0.007	0.293

Table 7 Developer closeness ratio and rank

Developer	d^*	d^-	CC	Rank
D1	0.1011	0.3077	0.7526	1
D2	0.0996	0.0996	0.4975	5
D3	0.1445	0.1445	0.4983	4
D4	0.1325	0.2690	0.6700	2
D5	0.2950	0.2950	0.4984	3

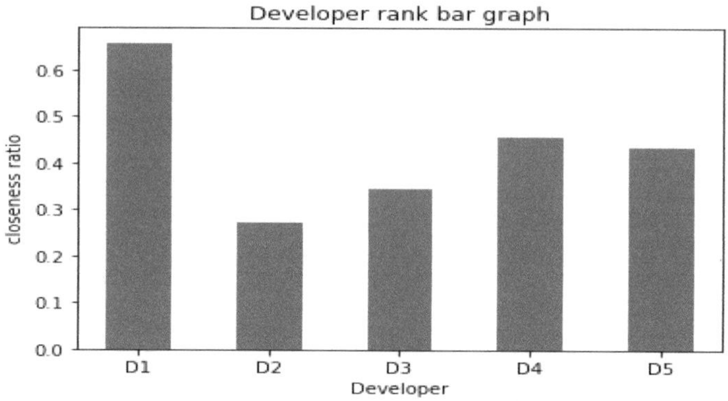

Fig. 2 Developer ranking bar graph

6 Threats to Validity

The suggested model poses a threat due to the use of the ahp approach to calculate the criterion weight. Because the judgement matrix is formed by humans, there may be conflict in the emotions of humans when assigning weight to criteria, and there may be a chance of obtaining distinct criteria weight vectors, which may affect the overall rank of a developer in bug triaging.

7 Conclusion and Future Scope

A new algorithm is proposed for bug triaging using hybridization of two MCDM algorithms respectively AHP for criteria weight calculation and TOPSIS for ranking of the developers with considering the availability of the developers. The future work can be applying other MCDM algorithms for the effective ranking of developers in bug triaging.

References

1. Yalcin AS, Kilic HS, Delen D (2022) The use of multi-criteria decision-making methods in business analytics: A comprehensive literature review. Technol Forecast Soc Chang 174, 121193
2. Mojaver M, et al. (2022) Comparative study on air gasification of plastic waste and conventional biomass based on coupling of AHP/TOPSIS multi-criteria decision analysis. Chemosphere 286, 131867
3. Sawarkar R, Nagwani NK, Kumar S (2019) Predicting available expert developer for newly reported bugs using machine learning algorithms. In: 2019 IEEE 5th International Conference

for Convergence in Technology (I2CT), pp. 1–4, doi: https://doi.org/10.1109/I2CT45611.2019.9033915
4. Chaitra BH, Swarnalatha KS (2022) Bug triaging: right developer recommendation for bug resolution using data mining technique. In: Emerging Research in Computing, Information, Communication and Applications. Springer, Singapore, pp. 609–618
5. Sun, Xiaobing, et al. "Experience report: investigating bug fixes in machine learning frameworks/libraries." *Frontiers of Computer Science* 15.6 (2021): 1-16.
6. Malhotra R, et al. (2021) A study on machine learning applied to software bug priority prediction. In: 2021 11th International Conference on Cloud Computing, Data Science & Engineering (Confluence). IEEE
7. Goyal A, Sardana (2019) Empirical analysis of ensemble machine learning techniques for bug triaging. In: 2019 Twelfth International Conference on Contemporary Computing (IC3). IEEE
8. Roy NKS, Rossi B (2017) Cost-sensitive strategies for data imbalance in bug severity classification: Experimental results. In: 2017 43rd Euromicro Conference on Software Engineering and Advanced Applications (SEAA). IEEE
9. Chowdhary MS, et al. (2020) Comparing machine-learning algorithms for anticipating the severity and non-severity of a surveyed bug. In: 2020 International Conference on Smart Technologies in Computing, Electrical and Electronics (ICSTCEE). IEEE
10. Agrawal R, Goyal R (2021) Developing bug severity prediction models using word2vec. Int J Cogn Comput Eng 2:104–115
11. Mani S, Sankaran A, Aralikatte R (2019) Deeptriage: Exploring the effectiveness of deep learning for bug triaging. In: Proceedings of the ACM India Joint International Conference on Data Science and Management of Data
12. Zhou C, Li B, Sun X (2020) Improving software bug-specific named entity recognition with deep neural network. J Syst Softw 165:110572
13. Liu Q, Washizaki H, Fukazawa Y (2021) Adversarial multi-task learning-based bug fixing time and severity prediction. In: 2021 IEEE 10th Global Conference on Consumer Electronics (GCCE), IEEE 2021
14. Zaidi SFA, Lee C-G (2021) One-class classification based bug triage system to assign a newly added developer. In: 2021 International Conference on Information Networking (ICOIN). IEEE, 2021
15. Tüzün E, Doğan E, Çetin A (2021) An automated bug triaging approach using deep learning: a replication study. Avrupa Bilim ve Teknoloji Dergisi 21:268–274
16. Mian TS (2021) Automation of bug-report allocation to developer using a deep learning algorithm. In: 2021 International Congress of Advanced Technology and Engineering (ICOTEN). IEEE, 2021
17. Guo S, et al. (2020) Developer activity motivated bug triaging: via convolutional neural network. Neural Process Lett 51(3), 2589–2606
18. Aung TWW, et al. (2022) Multi-triage: A multi-task learning framework for bug triage. J Syst Softw 184, 111133
19. Jahanshahi H, et al. (2021) DABT: A dependency-aware bug triaging method. J Syst Softw 2021, 221–230
20. Terdchanakul P, et al. (2017) Bug or not? Bug report classification using n-gram idf. In: 2017 IEEE international conference on software maintenance and evolution (ICSME). IEEE
21. Xi S.-Q., et al. (2019) Bug triaging based on tossing sequence modeling. J Comput Sci Technol 34(5), 942–956
22. Alazzam I, et al. (2020) Automatic bug triage in software systems using graph neighborhood relations for feature augmentation. In: IEEE Transactions on Computational Social Systems 7(5), 1288–1303
23. Nguyen U, et al. (2021) Analyzing bug reports by topic mining in software evolution. In: 2021 IEEE 45th Annual Computers, Software, and Applications Conference (COMPSAC). IEEE
24. Yadav V, et al. (2019) PyTOPS: A Python based tool for TOPSIS. SoftwareX 9, 217–222
25. James AT, et al. (2021) Selection of bus chassis for large fleet operators in India: An AHP-TOPSIS approach. Expert Syst Appl 186, 115760

26. Goyal A, Sardana N (2017) Optimizing bug report assignment using multi criteria decision making technique. Intell Decis Technol 11(3):307–320
27. Gupta C, Inácio PRM, Freire MM (2021) Improving software maintenance with improved bug triaging in open source cloud and non-cloud based bug tracking systems. J King Saud Univ-Comput Inf Sci
28. Goyal A, Sardana N (2021) Feature ranking and aggregation for bug triaging in open-source issue tracking systems. In: 2021 11th International Conference on Cloud Computing, Data Science & Engineering (Confluence). IEEE

GujAGra: An Acyclic Graph to Unify Semantic Knowledge, Antonyms, and Gujarati–English Translation of Input Text

Margi Patel and Brijendra Kumar Joshi

1 Introduction

One of the most challenging issues in NLP is recognizing the correct sense of each word that appears in input expressions. Words in natural languages can have many meanings, and several separate words frequently signify the same notion. WordNet can assist in overcoming such challenges. WordNet is an electronic lexical database that was created for English and has now been made available in various other languages [1]. Words in WordNet are grouped together based on their semantic similarity. It segregates words into synonym sets or synsets, which are sets of cognitively synonymous terms. A synset is a collection of words that share the same part of speech and may be used interchangeably in a particular situation. WordNet is widely regarded as a vital resource for scholars working in computational linguistics, text analysis, and a variety of other fields. A number of WordNet compilation initiatives have been undertaken and carried out in recent years under a common framework for lexical representation, and they are becoming more essential resources for a wide range of NLP applications such as a Machine Translation System (MTS).

The rest of the paper is organized as follows:

Next section gives a brief of Gujarati Language. Section 3 gives an overview of previous Relevant Work in this topic. Section 4 covers the description about each component of System Architecture for the software used to build WordNet graph with respect to Gujarati–English–Gujarati Language. Section 5 demonstrates the Proposed Algorithm for the same. Section 6 is about the Experiment and Results. Section 7 brings the work covered in this article to a Conclusion.

M. Patel (✉)
Indore Institute of Science and Technology, Indore, India
e-mail: margi.patel22@gmail.com

B. K. Joshi
Military College of Telecommunication Engineering, Mhow, India

P. Singh et al. (eds.), *Machine Learning and Computational Intelligence Techniques for Data Engineering*, Lecture Notes in Electrical Engineering 998,
https://doi.org/10.1007/978-981-99-0047-3_14

2 Gujarati Language

Gujarati is an Indo-Aryan language that is indigenous to the Indian state of Gujarat. Gujarati is now India's seventh most frequently spoken language in terms of native speakers. It is spoken by approximately 4.48% of the Indian population, totaling 46.09 million people [2]. It is the world's 26th most frequently spoken language in terms of native speakers, with about 55 million people speaking it [3, 4]. Initially, Gujarati writing was mostly used for commercial purposes, with literature Devanagari script used for literary. The poetry form of language is considerably older, and it has been enriched by the poetry of poets such as Narsinh Mehta [5]. Gujarati prose literature and journalism began in the nineteenth century. It is utilized in schools, the government, industry, and the media. The language is commonly spoken in expatriate Gujarati communities in the United Kingdom and the United States. Gujarati publications, journals, radio, and television shows are viewable in these communities.

3 Literature Review

Word Sense Disambiguation (WSD) is the task of identifying the correct sense of a word in a given context. WSD is an important intermediate step in many NLP tasks especially in Information extraction, Machine translation [N3]. Word sense ambiguity arises when a word has more than one sense. Words which have multiple meanings are called homonyms or polysemous words. The word mouse clearly has different senses. In the first sense it falls in the electronic category, the computer mouse that is used to move the cursor in computers and in the second sense it falls in animal category. The distinction might be clear to the humans but for a computer to recognize the difference it needs a knowledge base or needs to be trained. Various approaches have been proposed to achieve WSD: Knowledge-based methods rely on dictionaries, lexical databases, thesauri, or knowledge graphs as primary resources, and use algorithms such as lexical similarity measures or graph-based measures. Supervised methods, on the other hand make use of sense annotated corpora as training instances. These use machine learning techniques to learn a classifier from labeled training sets. Some of the common techniques used are decision lists, decision trees, Naive Bayes, neural networks, support vector machines (SVM).

Finally, unsupervised methods make use of only raw unannotated corpora and do not exploit any sense-tagged corpus to provide a sense choice for a word in context. These methods are context clustering, word clustering, and cooccurence graphs. Supervised methods are by far the most predominant as they generally offer the best results [N1]. Many works try to leverage this problem by creating new sense annotated corpora, either automatically, semi-automatically, or through crowdsourcing.

In this work, the idea is to solve this issue by taking advantage of the semantic relationships between senses included in WordNet, such as the hypernymy, the hyponymy, the meronymy, and the antonymy. The English WordNet was the first of its kind in this field to be developed. It was devised in 1985 and is still being worked on today at Princeton University's Cognitive Science Laboratory [6]. The success of English WordNet has inspired additional projects to create WordNets for other languages or to create multilingual WordNets. EuroWordNet is a semantic network system for European languages. The Dutch, Italian, Spanish, German, French, Czech, and Estonian languages are covered by the Euro WordNet project [7]. The BalkaNet WordNet project [8] was launched in 2004 with the goal of creating WordNets for Bulgarian, Greek, Romanian, Serbian, and Turkish languages. IIT, Bombay, created the Hindi WordNet in India. Hindi WordNet was later expanded to include Marathi WordNet. Assamese, Bengali, Bodo, Gujarati, Hindi, Kannada, Kashmiri, Konkani, Malayalam, Manipuri, Marathi, Nepali, Oriya, Punjabi, Sanskrit, Tamil, Telugu, and Urdu are among the main Indian languages represented in the Indo WordNet project [9]. These WordNets were generated using the expansion method, with Hindi WordNet serving as a kingpin and being partially connected to English WordNet.

4 Software Description

In this section, we describe the salient features of the architecture of the system. The Gujarati WordNet is implemented on Google Colaboratory platform. To automatically generate semantic networks from text, we need to provide some preliminary information to the algorithm so that additional unknown relation instances may be retrieved. We used Indo WordNet, which was developed utilizing the expansion strategy with Hindi WordNet as a pivot, for this purpose As a result, we manually created Gujarati antonyms for over 700 words as a tiny knowledge base.

4.1 Software Architecture

Initially, sentence in Gujarati Language is taken as input. A feature of text to speech is provided for those who are not aware about the pronunciation of the sentence that is given as input (Fig. 1).

Text Analysis Phase

The text analysis procedure then begins with the elimination of non-letter elements and punctuation marks from the sentence. This is followed by Tokenization of words. Each token is saved in a list. Like if input is જલ્દી ઠીક થઇ જાવ તેવી શુભકામના(jaldi thik thai jaav tevi shubhkamna), then output of tokenization phase will be ‚જલ્દી', 'ઠીક', 'થઇ', 'જાવ', 'તેવી', 'શુભકામના'..

Concept Extraction Phase

Then comes concept extraction phase. Here semantically related concepts for each token term are extracted from the IIT Synset, Gujarati Lexicon, or Bhagwad Go

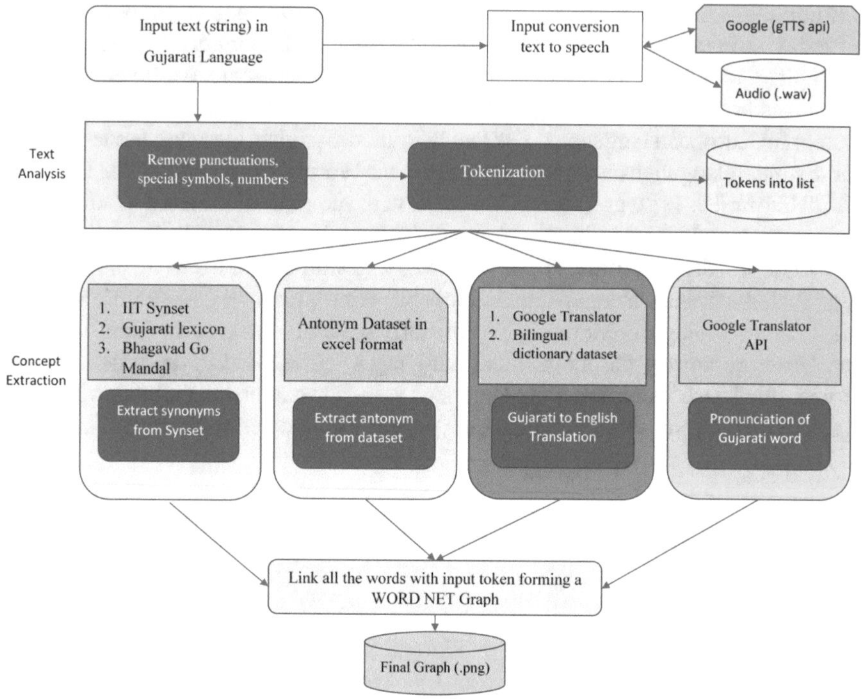

Fig. 1 Software architecture

Mandal which is used to create a collection of synonyms. Antonyms of each tokens are extracted from the Gujarati antonym knowledge base that was created manually of more than 700 words. English Translation of each token is searched either from google translator or Bilingual Dictionary Dataset. Google Translator API is used to fetch pronunciation of each token. Then, an acyclic graph is formed of token and its respective concept extracted in Concept Extraction phase.

5 Proposed Algorithm

This section describes a method for producing an acyclic graph, which is essentially a visualization tool for the WordNet lexical database. Through the proposed algorithm we wish to view the WordNet structure from the perspective of a specific word in the database using the suggested technique. Here we have focused on WordNet's main relation, the synonymy or SYNSET relation, antonym and the word's English translation.

This algorithm is based on what we will call a sense graph, which we formulate as follows. Nodes in the sense graph comprise the words wi in a vocabulary W together with the senses sij for those words. Labeled, undirected edges include word-sense edges ⟨wi, si,j⟩, which connect each word to all of its possible senses, and sense-sense edges sij, sij labeled with a meaning relationship r that holds between the two senses. WordNet is used to define their sense graph. Synsets in the WordNet ontology define the sense nodes, a word-sense edge exists between any word and every synset to which it belongs, and WordNet's synset-to-synset relations of synonymy, hypernymy, and hyponymy define the sense-sense edges. Figures 4 and 5 illustrate a fragment of a WordNet- based sense graph.

Key point to observe is that this graph can be based on any inventory of word-sense and sense-sense relationships. In particular, given a parallel corpus, we can follow the tradition of translation-as-sense-annotation: the senses of an Gujarati word type can be defined by different possible translations of that word in any other language. Operationalizing this observation is straightforward, given a word-aligned parallel corpus. If English word form ei is aligned with Gujarati word form gj, then ei(gj) is a sense of ei in the sense graph, and there is a word-sense edge ei, ei(gj). Edges signifying a meaning relation are drawn between sense nodes if those senses are defined by the same translation word. For instance, English senses Defeat and Necklace both arise via alignment હાર(Haar), so a sense-sense edge will be drawn between these sense nodes.

```
READ Input String
REPAT
  remove punctuation marks, stop words
UNTIL end_of_string
convert text to speech
   STORE audio_file
split string sentence to words
   FOR each word on the board
   SEARCH word's antonym from antonym dataset_list
   SEARCH word's synonyms from online_synset
   SEARCH word's translation & Pronunciation on google
   OBTAIN search results
   IF result is not null THEN
      COMPUTE results
      add result nodes with different color
       END IF
      generate word_net network graph
      STORE graph_image
      PRINT graph_image
   END FOR
```

6 Experiment and Result

For the experimental purpose more than 200 random sentences have been found from different Guajarati language e-books, e-newspapers, etc. A separate excel document (file contains 700+ words) named as 'Gujarati Opposite words.xlsx' keeping one word and its corresponding antonym in each row was created. Now for the generation of the word net graph, google colab is used as it's an online cloud service provided by google (standalone system having Jupiter Notebook can also be used).

Firstly, all the APIs are being installed using pip install command. Then importing required packages for processing of tokens like pywin, tensorflow tokenizer google translator, and netwrokx. Figure 2 displays the content of 'sheet 1' of excel file named 'Gujarati Opposite words.xlsx' using panda (pd) library. Then, the instance of 'Tokenzier' from 'keras' api is called for splitting the sentence into number of tokens as shown in Fig. 3.

Different color coding is used to represent different things. Like Light Blue is used to represent token in our input string, Yellow color is used to represent synonyms, Green color is used to represent opposite, Red color is used for English translation of the token, and Pink color is used to represent pronunciation of the token. Hence if no work has been done on a particular synset of the Gujarati WordNet, then acyclic graph will not contain yellow node. As in our example, synonym of હોય(hoy) is

```
import pandas as pd
df = pd.read_excel('/content/Gujarati Opposite Words.xlsx', sheet_name='Sheet1')
df
```

	word	antonym
0	કડવું	મીઠું
1	દૃશ્ય	અદૃશ્ય
2	ગામડિયું	શહેરી
3	જીવંત	મૃત
4	આઘાત	પ્રત્યાઘાત
...	...	...
765	અવનતિ	ઉન્નતિ
766	ઝાઝું	થોડું
767	પૂર્ણિમા	અમાવસ્યા
768	એકદેશીય	સર્વદેશીય
769	અખંડ	ખંડિત

770 rows × 2 columns

Fig. 2 Reading the excel file

```
[ ] import tensorflow as tf
    from tensorflow import keras
    from tensorflow.keras.preprocessing.text import Tokenizer
    sentences = ['પૂનમ ચંદ્ર આકર્ષક હોય છે.']
    tokenizer = Tokenizer()
    tokenizer.fit_on_texts(sentences)
    index_word = tokenizer.index_word
    print(len(index_word))
    for x in range (1,len(index_word)+1):
      print(index_word[x])

    5
    પૂનમ
    ચંદ્ર
    આકર્ષક
    હોય
    છે
```

Fig. 3 Reading string and generating tokens

not found so acyclic graph is not plotted for the same. Same way if antonym is not available in knowledge base then green node will be omitted and so on.

Thereafter, a custom function is created which uniquely read the token and calls different functions for obtaining respective value of synonyms, anonym, English translation, pronunciation, and then create an acyclic WordNet graph.

Finally, the result is being saved in different dot png files showing the acyclic WordNet graph for each token as shown in Fig. 4.

We have made acyclic graph to more than 200 sentences through our proposed system. In some of the cases, we faced challenges. One of which is:

when 'રામે પૃથ્વીલોક ત્યાગ કર્યું'(Ram e prativilok tyag kariyu) was given as input, then acyclic graph for the word 'પૃથ્વીલોક'(prathvilok) is shown in Fig. 5. Here, the linguistic resource that is used to extract synonyms of 'પૃથ્વીલોક'(Prathvilok) is Synset provided by IIT, ID 1427. The concept 'એ લોક જ્યાં આપણે બધાં પ્રાણીઓ રહીએ છીએ'means the place meant for all of us to live. But in Synset 'મૃત્યુલોક'(Mratyulok) is given as co-synonym of 'પૃથ્વીલોક'.

7 Conclusion

The application of a differential theory of lexical semantics was one of WordNet's core design concepts. WordNet representations are at the level of word meanings rather than individual words or word formations. In turn, a term's meaning is defined

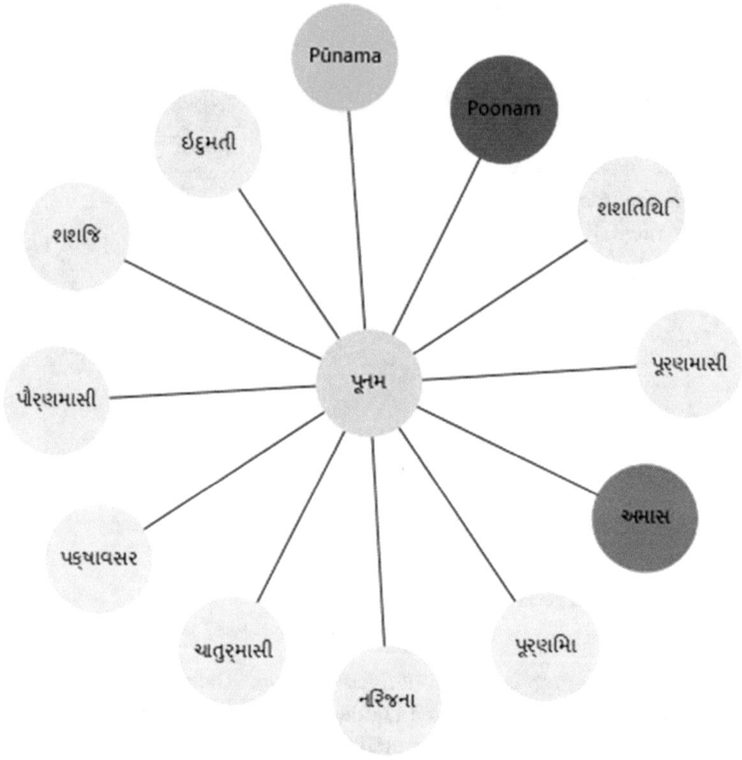

Fig. 4 Acyclic WordNet Graph for Word 'Poonam'

by simply listing the different word forms that might be used to describe it in a synonym set (synset). Through the proposed architecture, we extracted tokens from the inputted sentence. Synonyms, antonyms, pronunciation, and translation of these tokens are identified. Synonyms, antonyms, pronunciation, and translation of the tokens identified previously are then plotted to form an acyclic graph to give pictorial view. Different color coding is used to represent the tokens, its synonyms, its antonyms, its pronunciation, and its translation (Gujarati or English). We demonstrated the visualization of WordNet structure from the perspective of a specific word in this work. That is, we want to focus on a specific term and then survey the greater structure of WordNet from there. While we did not design our method with the intention of creating WordNets for languages other than Gujarati, we realize the possibility of using it in this fashion with other language combinations as well. Some changes must be made to the system's architecture, for example, in Concept Extraction phase, linguistic resources of other languages for providing needed synonyms have to be made available. But the overall design of displaying the information of the Gujarati WordNet can be easily applied in developing a WordNet for another language. We have presented an alternative means of deriving information about

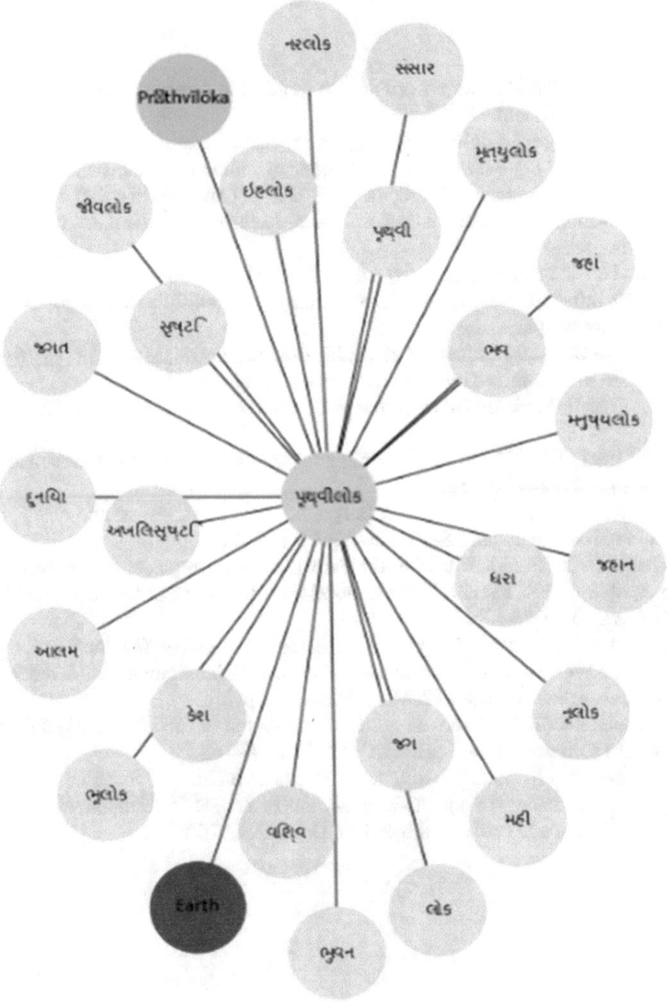

Fig. 5 Acyclic WordNet Graph for Word 'Prathvilok'

senses and sense relations to build sense-specific graphical space representations of words, making use of parallel text rather than a manually constructed ontology. Based on the graphs, it would be interesting to evaluate further refinements of the sense graph: alignment-based senses could be clustered.

References

1. Miller GA, Fellbaum C (2007) WordNet then and now. Lang Resour Eval 41(2), 209–214. http://www.jstor.org/stable/30200582
2. Scheduled Languages in descending order of speaker's strength - 2001". Census of India. https://en.wikipedia.org/wiki/List_of_languages_by_number_of_native_speakers_in_India
3. Mikael Parkvall, "Världens 100 största språk 2007" (The World's 100 Largest Languages in 2007), in National encyclopedia. https://en.wikipedia.org/wiki/List_of_languages_by_number_of_native_speakers
4. "Gujarati: The language spoken by more than 55 million people". The Straits Times. 2017–01–19. https://www.straitstimes.com/singapore/gujarati-the-language-spoken-by-more-than-55-million-people
5. Introduction to Gujarati wordnet (GCW12) IIT Bombay, Powai, Mumbai-400076 Maharashtra, India. http://www.cse.iitb.ac.in/~pb/papers/gwc 12-gujarati-in.pdf
6. Miller GA (1990) WordNet: An on-line lexical database. Int J Lexicogr 3(4):235–312. Special Issue.
7. Vossen P (1998) EuroWordNet: a multilingual database with lexical semantic networks. J Comput Linguist 25(4):628–630
8. Tufis D, Cristea D, Stamou S (2004) Balkanet: aims, methods, results and perspectives: a general overview. Romanian J Sci Technol Inf 7(1):9–43
9. Bhattacharya P (2010) IndoWordNet. In: lexical resources engineering conference, Malta
10. Narang A, Sharma RK, Kumar P (2013) Development of punjabi WordNet. CSIT 1:349–354. https://doi.org/10.1007/s40012-013-0034-0
11. Kanojia D, Patel K, Bhattacharyya P (2018) Indian language Wordnets and their linkages with princeton WordNet. In: Proceedings of the eleventh international conference on language resources and evaluation (LREC 2018), Miyazaki, Japan
12. Patel M, Joshi BK (2021) Issues in machine translation of indian languages for information retrieval. Int J Comput Sci Inf Secur (IJCSIS) 19(8), 59–62
13. Patel M, Joshi BK (2021) GEDset: automatic dataset builder for machine translation system with specific reference to Gujarati–English. In: Presented in 11th International Advanced Computing Conference held on 18th & 19th December, 2021

Attribute-Based Encryption Techniques: A Review Study on Secure Access to Cloud System

Ashutosh Kumar and Garima Verma

1 Introduction

Cloud computing is turning into the principal computing model in the future because of its benefits, for example, high asset use rate and saving the significant expense of execution. The existing algorithms for security issues in cloud computing are advanced versions of cryptography. Mainly cloud computing algorithms are concerned about data security and privacy-preservation of the user. Most solutions for privacy are based on encryption and data to be downloaded is encrypted and stored in the cloud. To implement the privacy protection of data owners and data users, the cryptographic data are shared and warehoused in cloud storage by applying Cyphertext privacy—Attribute-based encryption (CP-ABE). Algorithms like AES, DES, and so on are utilized for encoding the information before downloading it to the cloud.

The main three features of clouds define the easy allocation of resources, a platform for service management, and massive scalability to designate key design components of processing and clouds storage. A customer of cloud administrations might see an alternate arrangement of characteristics relying upon their remarkable requirements and point of view [1]:

- Location free asset pools—process and storage assets might be located anyplace that is the network available; asset pools empower reduction of the dangers of weak links and redundancy,

A. Kumar (✉) · G. Verma
School of Computing, DIT University, Dehradun, India
e-mail: ashugeu08@gmail.com

G. Verma
e-mail: garima.verma@dituniversity.edu.in

P. Singh et al. (eds.), *Machine Learning and Computational Intelligence Techniques for Data Engineering*, Lecture Notes in Electrical Engineering 998,
https://doi.org/10.1007/978-981-99-0047-3_15

- On-demand self-service—the capability to use, manage storage and allocation of storage, computing, and further business benefits voluntarily without relying upon support staff,
- Flexible costing—generally any cloud providers work on the "pay as you go" costing model,
- Network ubiquitous access—the capacity to work with cloud assets from any point with Internet access,
- Adaptable scalability—As the resource utilized by cloud users changes as on-demand or time to time so resource allocation in the cloud is done by the cloud itself as low-demand to peak demand.

Customary models of information security have regularly centered around network-driven and perimeter security, often with tools such as intrusion detection systems and firewalls. However, this methodology doesn't give adequate protection against APTs, special clients, or other guileful kinds of safety attacks [2]. The encryption execution should join a vigorous key administration solution for giving insistence that the keys are sufficiently secured. It's basic to review the whole encryption and key administration arrangement. Encryption works by cooperating with other focus data security advancements, gleaning increased security intelligence, to deliver an inclusive hybrid approach to the transaction with ensuring sensitive data transmission of the cloud [3].

In this way, any data-driven system should join encryption, key administration, minimal access controls, and security understanding to guarantee data in the cloud and give the basic level of safety [4]. By utilizing a hybrid approach that joins these fundamental parts, affiliations can additionally foster their security act more reasonably and successfully than by just worrying exclusively on ordinary association is driven security procedures [5].

A cloud computing architecture involves a front end and a back end. They partner with each other over an organization, generally the Internet. Any computer user is an example of the front end and the "cloud" section stands for the back end of the system (Fig. 1).

The front end of the cloud computing structure incorporates the client's gadgets (or it very well may be an organization association) and a couple of uses are needed for getting to the distributed computing system. All distributed computing systems don't give a comparable interface to customers. Web organizations like electronic mail programs use some current web programs like Firefox, Apple's Safari [3].

Various systems have some outstanding applications that give network admittance to their clients. The back end suggests some actual peripherals. In distributed computing, the back end is the cloud itself which may fuse distinctive processing machines, servers, and information stockpiling systems. Gatherings of these mists make an entire distributed computing framework. Theoretically, a distributed computing system can consolingly any kind of web application program, for instance, PC games to applications for data dealing with, diversion, and programming improvement. Typically, every application would have its steadfast server for administration.

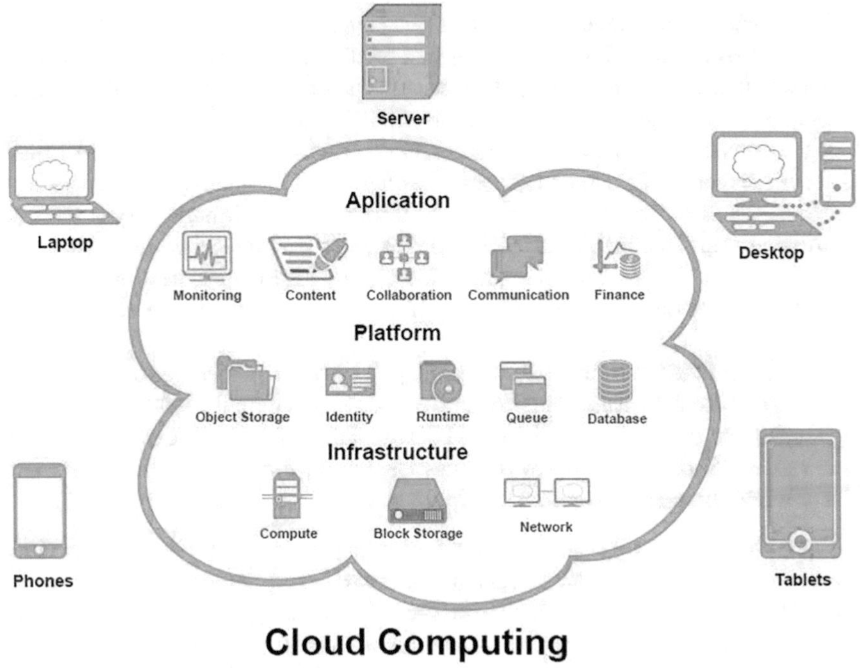

Fig. 1 Example of cloud computing systems [1]

The paper is divided into mainly five sections. Section one deals with an introduction to research work with an explanation of basic concepts in brief. The second section is about the background study of cloud computing security issues. The third section deals with the survey study of the existing studies which are useful as exploratory data for the research work, and to evaluate the review for new framework designing. This section also presents the tabular form of the survey studies. The fourth section describes the summary of the literature study done in Sect. 4 and also presents the research gap analysis. Last Sect. 5 contains concludes the paper.

2 Background of the Review Study

In the current conventional framework, there exist security issues for storing the information in the cloud. Cloud computing security incorporates different issues like data loss, authorization of cloud, multi-occupancy, inner threats, spillage, and so forth. It isn't difficult to carry out the safety efforts that fulfill the security needs of all the clients. It is because clients might have dissimilar security concerns relying on their motivation of utilizing the cloud services [5].

- **Data security management:** To certify any particular cloud service providers to hold for data storage, it must have verified security policy and life cycle. Analysis done to implement this concept recommends that CSPs use encryption strategies using keys to protect and securely transmit their data.
- **Data protection in the cloud:** Cloud service provider (CSP) has given a brilliant security layer for the owner and user. The client needs to guarantee that there is no deficiency of information or misuse of information for different clients who are utilizing a similar cloud. The CSPs should be equipped for receiving against digital assaults. Not all cloud suppliers have the capacity for data protection. Different techniques are being carried out to annihilate the security issues in cloud storage of data
- **Key management in cryptography:** Cryptography is a technique for covering data to conceal it from unapproved clients [6]. Communicated information is clouded and delivered in a ciphertext design that is inexplicable and unreadable to an unauthorized user. A key is utilized to change figure text to plain text. This key is kept hidden, and approved customers can move toward it [7] (Fig. 2).

Encryption is probably the most secure way of staying away from MitM assaults because regardless of whether the communicated information gets captured, the assailant would not be able to translate it. There exist data hypothetically secure plans that most likely can't be earned back the original investment with limitless figuring power—a model is a one-time cushion—yet these plans are harder to execute than the best hypothetically delicate however computationally secure components.

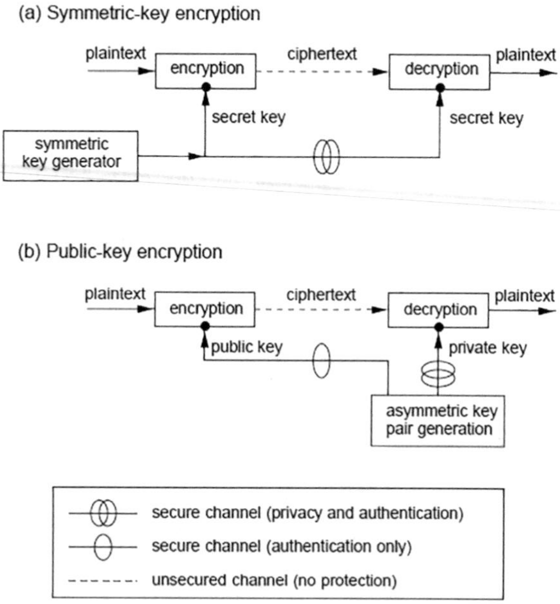

Fig. 2 Encryption process in cloud system [6]

- **Access controls:** The security concept in cloud system require the CSP to provide an access control policy so that the data owner can restrict end-user to access it from authenticated network connections and devices.
- **Long-term resiliency of the encryption system:** With most current cryptography, the capacity to maintain encoded data secret is put together not concerning the cryptographic calculation, which is generally known, yet on a number considered a key that should be utilized with the cryptographic algorithm to deliver an encoded result or to decode the encoded data. Decryption with the right key is basic. Decoding without the right key is undeniably challenging, and sometimes for all practical purposes.

3 Review Study

In 2018, Li Chunhua et al. [7] presented a privacy-preserving access control scheme named CP-ABE utilizing a multiple-cloud design. By working on the customary CP-ABE technology and presenting a proxy to steal the private key from the user, it needs to be certified that the user attribute set can be attained by any cloud, which successfully secures the protection of the client ascribes. Security analysis presents the effectiveness of the proposed scheme against man-in-the-middle attacks, user collusion, and replay attacks.

In 2018, Bramm et al. [8] developed a combined protocol named attribute management protocol for CP-ABE schemes grounded on the system called a BDABE-Blockchain Distributed Attribute-Based Encryption scheme. This development acknowledges storage, reversal of private attribute keys, and distributed issues-based adding a contract-driven structure, a blockchain. This upgraded both the security and effectiveness of key administration in distributed CP-ABE frameworks for the use of cloud data sharing.

In 2019, Wang et al. [9] projected a new hybrid secure cloud storage model with access control. This model is a grouping of CP-ABE and Ethereum blockchain.

In 2019, Sarmah [10] reviews the use of blockchain in distributed computing frameworks. First and foremost, the idea of blockchain is momentarily talked about with its benefits and disadvantages. Secondly, the idea of cloud computing is momentarily exhibited with blockchain technology. At last, earlier studies are explored and introduced in tabular form. It directs that the research gaps actually relate to the field of blockchain-dependent on cloud computing frameworks.

In 2020, Qin et al. [11] proposed a Blockchain-based Multi-authority Access Control conspire called BMAC for sharing information safely. Shamir's secret sharing plan and permission blockchain are acquainted with performance that each attribute is jointly controlled by multiple authorities to avoid the weak link of failure. Moreover, it took advantage of blockchain innovation to set up trust among numerous data owners which cause a reduction of computation and communication overhead on the user side.

Table 1 Tabular analysis of different Encryption-based cloud computing studies

Authorname/ year	Technique name	Methodology used	Results
Chunhua et al. [7]	CPABE	Improving the customary CP-ABE algorithm	Effectiveness against man-in-the-middle attacks, replay, and user collusion attack
Bramm et al. [8]	BDABE	For Ciphertext-policy, it developed a combined attribute management protocol	Enhanced both security and productivity of key generation in distributed CP-ABE structures
Wang et al. [9]	Ethereum blockchain technology	Combination of Ethereum blockchain and CP-ABE	Proposed a secure cloud storage framework
Sarmah [10]		A review study of the methods related to the uses of blockchain in cloud computing	The research gaps showed the developments in the blockchain method
Qin et al. [11]	BMAC	This is a mixture of secret information sharing plans and permission blockchain	Reduction of computation and communication overhead on the user side
Guo et al. [13]	O-R-CP-ABE	Assisting of blockchains in the IoMT ecosystem and cloud servers	In this proposition, the qualities of fine-grained access control are accomplishes

In Ref. [12], an analytical procedure is presented to review and compare the existing ABE schemes proposed. For KP-ABE and each sort of CP-ABE, the comparing access control circumstances are introduced and clarified by substantial examples.

In Ref. [13], the authors present proficiency in the online/offline revocable CP-ABE scheme with the guide of cloud servers and blockchains in the IoMT environment. This proposition accomplishes the qualities of user revocation, fine-grained access control, ciphertext verification fast encryption, and outsourced decryption.

From the existing studies examined above, we have acquired the inspiration to deal with digital data sharing with the help of attribute-based encryption and blockchain [14]. The tabular analysis of different attribute-based encryption and blockchain-based studies is presented in Table 1.

4 Review Summary

There are some points summarized after surveying the distinctive encryption-based cloud security strategies for late exploration improvements that are as per the following:

Computing is sorted by its utilization design. Cluster computing, distributed computing, and parallel computing are notable standards of these classifications. A cluster computing acts as a group of connected

systems that are firmly combined with rapid networks and work intently together. Whereas, distributed computing is an assortment of software or hardware frameworks that contain more than one task or stored data component yet show up as a single random process running under a tightly or loosely controlled system. In the distributed system, computers don't share a memory rather they pass messages nonconcurrent or simultaneously between them [15]. In addition, parallel processing is a type of calculation where a major process is divided into different minor or smaller processes so that these smaller processes can be simultaneously computed.

Cloud computing is a specific type of network, distributed, and utility computing and it takes a style of network registering where stable and virtualized assets are accessible as a service over the internet. Moreover, cloud computing technology gives numerous development elements, for example, on-demand, portal services, assets versatility, firewall applications, and so on. Nonetheless, these elements are affected by numerous security issues (security attacks and threats, key distribution, and cryptographic perspectives) due to an open environment related to cloud computing [16].

In distributed computing, the new emerging design models incorporate Grid registering, utility processing, and Cloud registering, which have empowered the use of wide variability of distributed computational assets as a unified resource. These new emerging design models of distributed computing, with the fast improvement of new systems services, are moving the whole computing standard in the direction of a new era of distributed computing.

With the quick progression of web insurgency, distributed storage has transformed into a critical game plan in our day-to-day routine. It has given different sorts of data stockpiling administrations for individuals and tries, making it attainable for customers to get to Internet resources and offer data at any place and whenever, it has conveyed mind-blowing comfort to our lives [17]. Such distributed storage systems have been incredibly useful and have procured expanding affirmation, in any case, as such kind of structures just depends upon a huge association with a single storage capacity to store and communicate data, in which the huge association is seen as a confided in an outsider, it certainly gets the failure point downside of depending on outsider administrations. Whether or not distributed storage systems are maintained for data availability, disseminated capacity administrations providers might in any case experience the ill effects of specific components of power Majeure led to the way that clients cannot permit their information [18]. Additionally, with the advancement of capacity innovation, the expense of capacity gadgets has become diminished. The cost of concentrated circulated stockpiling administrations comes generally from legitimate expenses, worker wages, server farm rentals, and so forth These costs are unaltered or gradually extended. The expense of the concentrated circulated stockpiling administrations will be higher [19].

In conventional distributed storage frameworks, assuming clients need to share their data that is stored in a third-party cloud server secretly, then a new model is required to make data accessible to only the person who is authenticated to do that. To fulfill this demand, a new technique named the attribute-based encryption mechanism (ABE) was developed. These algorithms are also enhanced by maintaining the confidentiality of the content and the user. Security is a significant prerequisite in distributed computing while we talk about data storage. There are several existing procedures used to carry out security in the cloud. Almost all ABE encryption

schemes require a key that is used to change cipher text to plain text. This key is kept private and just authorized clients can approach it [14].

Identified problem gaps

1. A longer period is needed for data access in most of the structures that caused the improvement of the consensus delay by misclassified data blocks.
2. In the majority of the cases, the proposed models have restricted effects over the proficiency of the information selection. The encryption queries on altered reports cause higher computational expenses.
3. In a review [20], the proposed strategy is neglected to offer types of assistance for the two servers and the clients. In any case, if a unified programming part is joined with the most common way of executing security highlights despises the managing fluctuated applications.
4. Real-time application and the services over shared organization are not prepared as expected. At times, the cloud infrastructure is interfered with during limitless no of nodes assessment.
5. They examined the recent blockchain approach by recognizing the potential threats. The framework upgraded the information intervention. The probability of software and hardware are as a rule compromised [21].

5 Conclusion

The paper's motive is to study the security issues in cloud computing systems and data storage capacity. The technologies related to cryptography must be assured to share data through the distributed environment. This paper gives an itemized depiction of different security models by utilizing different encryption techniques for cloud computing systems. Data losses are unavoidable in various certifiable complex framework structures. The paper presents a thorough review analysis of the different techniques of data security in cloud storage and various algorithms of encryption of the data.

References

1. Khalid A (2010) Cloud Computing: applying issues in Small Business. In: Intl. Conf. on Signal Acquisition and Process (ICSAP'10). pp. 278–281
2. KPMG (2010) From hype to future: KPMG's 2010 Cloud Computing survey. Available at: http://www.techrepublic.com/whitepapers/from-hype-to-futurekpmgs-2010-cloud-computing-survey/2384291
3. Rosado DG et al (2012) Security analysis in the migration to cloud environments. Futur Internet 4(2):469–487. https://doi.org/10.3390/fi4020469
4. Mather T et al (2009) Cloud security and privacy. O'Reilly Media Inc., Sebastopol, CA
5. Gartner Inc. Gartner identifies the Top 10 strategic technologies for 2011. Online. Available at: http://www.gartner.com/it/page.jsp?id=1454221
6. Bierer BE et al (2017) Data authorship as an incentive to data sharing. N Engl J Med 376(17):1684–1687. https://doi.org/10.1056/NEJMsb1616595
7. Chunhua L, Jinbiao H, Cheng L, Zhou K (2018) Achieving privacy-preserving CP-ABE access control with multi-cloud. In: IEEE Intl Conf on parallel & distributed processing with applications. p. 978-1-7281
8. Bramm G, et al. (2018) BDABE: Blockchain-based distributed attribute-based encryption. In: Proc. 15th international joint conf. on E-business and telecommunications (ICETE 2018), SECRYPT, vol 2, pp. 99–110
9. Wang S et al (2019) A secure cloud storage framework with access control based on Blockchain. IEEE Access 7:112713–112725. https://doi.org/10.1109/ACCESS.2019.2929205
10. Sarmah SS (2019) Application of Blockchain in cloud computing. Int. J. Innov. Technol. Explor. Eng. (IJITEE) ISSN: 2278-3075 8(12)
11. Qin X, et al. (2020) A Blockchain-based access control scheme with multiple attribute authorities for secure cloud data sharing. J. Syst. Archit. 112:2021, 101854, ISSN: 1383-7621
12. Zhang Y, et al. (2020) Attribute-Based Encryption for Cloud Computing Access Control: A Survey. ACM Computing Surveys, Article No. 83
13. Guo R, et al. (2021) O-R-CP-ABE: An efficient and revocable attribute-based encryption scheme in the cloud-assisted IoMT system. IEEE Internet Things J. 8(11):8949–8963. doi:https://doi.org/10.1109/JIOT.2021.3055541
14. Alharby M, et al. (2018) Blockchain-based smart contracts: a systematic mapping study of academic research 2018. In: Intl. Conf. on Cloud Comput., Big Data and Blockchain (ICCBB), vol 2018. IEEE, Fuzhou, pp. 1–6
15. Huh S, et al. (2017) Managing IoT devices using a block-chain platform. In: Proc. 2017 19th Intl. Conf. on Advanced Communication Technology (ICACT), Bongpyeong, Korea, February 19–22 2017
16. Armknecht F, et al. (2015) Ripple: Overview and outlook. In: Conti M, Schunter M, Askoxylakis I (eds) Trust and trustworthy computing. Springer International Publishing, Cham, Switzerland, 2015, pp. 163–180
17. Vasek M, Moore T (2015) There's no free lunch, even using Bitcoin: Tracking the popularity and profits of virtual currency scams. In: Lecture Notes in Computer Science, Proc. Intl. Conf. on Financial Cryptography and Data Security, San Juan, Puerto Rico. Springer, Berlin/Heidelberg, Germany, 44–61, Jan. 26–30 2015. doi:https://doi.org/10.1007/978-3-662-47854-7_4
18. Zhang J et al (2016) A secure system for pervasive social network-based healthcare. IEEE Access 4:9239–9250. https://doi.org/10.1109/ACCESS.2016.2645904
19. Singh S et al (2016) A survey on cloud computing security: Issues, threats, and solutions. J Netw Comput Appl 75:200–222. https://doi.org/10.1016/j.jnca.2016.09.002
20. Assad M, et al. (2007) Personis AD: Distributed, active, a scrutable model framework for context-aware services. In: Intl. Conf. on Pervasive Comput. Springer, Berlin; Heidelberg, pp. 55–72
21. Benet J (2015) IPFS-content addressed, Versioned. *File System (DRAFT 3)*. Available at: https://ipfs.io/ipfs/QmV9tSDx9UiPeWExXEeH6aoDvmihvx6jD5eLb4jbTaKGps, vol. P2p

22. Raghavendra S et al (2016) Index generation and secure multi-user access control over an encrypted cloud data. Procedia Comput Sci 89:293–300. https://doi.org/10.1016/j.procs.2016.06.062

Fall Detection and Elderly Monitoring System Using the CNN

Vijay Mohan Reddy Anakala, M. Rashmi, B. V. Natesha, and Ram Mohana Reddy Guddeti

1 Introduction

Fall detection had recently gained attention for its potential application in fall alarming system [1] and wearable fall injury prevention system. Falls are the leading cause of injury deaths among which the majority (over 80%) were people over 65 years of age. Among all the causes leading to falls, slipping was considered as the most frequent unforeseen triggering event. Foot slippage was found to contribute between 40 and 50% of fall-related injuries and 55% of the falls on the same level. Among all the fall intervention approaches, automatic fall event detection has attracted research attention recently [2], for its potential application in fall alarming system and fall impact prevention system. Nevertheless, the existing approaches have not satisfied the accuracy and robustness requirements of a good fall detection system [3]. Sensor data [4] is also used to deal with the classification problems such as fall detection for elderly monitoring. Nevertheless, existing fall detection research is facing three major problems. The first problem is concerned with detection performance, more specifically the balance between misdetection and false alarms. Due to the heterogeneity between subject motion features and ambiguity of within-subject activity characteristics, higher fall detection sensitivity is always found to be associated with higher false alarm rates [5]. Almost all of the current fall detection techniques are facing this issue to a varying degree. The second problem is that the target of detection is unclear. Different devices may actually detect the impact of a fall, the incapacity to rise/recover after a fall (post-fall impact), or the fall itself (the postural disturbance prior to the fall impact). From the perspective of preventing fall injuries directly using

V. M. Reddy Anakala (✉) · M. Rashmi · B. V. Natesha · R. M. Reddy Guddeti
National Institute of Technology Karnataka, Surathkal 575025, India
e-mail: anakalavijay51@gmail.com

R. M. Reddy Guddeti
e-mail: profgrmreddy@nitk.edu.in

P. Singh et al. (eds.), *Machine Learning and Computational Intelligence Techniques for Data Engineering*, Lecture Notes in Electrical Engineering 998,
https://doi.org/10.1007/978-981-99-0047-3_16

a wearable protection system, timely and accurate detection of a fall event prior to an impact is of utmost importance.

The aim of this paper is to design and implement a sensor-based fall detection system using movement-based sensor data by focusing on practical issues such as the user's convenience and power consumption. Accordingly, we presented a sensitivity and specificity-based high-performance fall detection system using threshold-based classification. The key contribution of this paper is an efficient deep learning model based on CNN for fall detection using movement-based sensor data of adults.

The remaining of the paper is organized as follows. Section 1 deals with the introduction. Section 2 focuses on related work, Sect. 3 presents the design and implementation details of the proposed system. The experimental setup, results, and discussion are presented in Sect. 4. Finally, Sect. 5 concludes the paper with important observations and the potential future directions.

2 Related Work

Various approaches have been proposed in the area of fall detection like the research done by Ozcan et al. [6] using a camera. The study uses the body movement for capturing the change in the orientation of the camera from which it can be concluded that the person has fallen or not. There are several commonly used sensors such as an accelerometer and a gyroscope for capturing movement-based sensor data. YanjunLi et al. [7] tried to use an accelerometer sensor. They conducted a small-scale experiment using a chipset named "Telos W" which is connected to the computer using a wireless connection that has determined its usage to give optimum performance in an indoor environment.

Fall detection methods based on thresholds are very common, due to the expected physical impact related to falls [8, 9]. In [8], different approaches for threshold setup on fall detection solutions using accelerometer-based method were evaluated. The tests were performed considering the best specificity for an ideal sensitivity (100%) in three different body places: waist, head, and wrist. Evaluating the solution with data acquired from two subjects who performed fall and Activities of Daily Life (ADLs). Pierleoni et al. [10] developed, a threshold-based method for fall detection using the combination of an accelerometer, gyroscope, and magnetometer. Placing the device at user's waist, the system was able to identify different characteristics of a fall event, including pre-fall analysis and aftermath position. The applied sensor fusion algorithm was based on the method of Madgwick et al. [9], a simplification of Kalman-filter approach. Galvao et al. [11] developed a multi-modal approach for fall detection, and a comparative study is done using various deep learning models.

The SisFall dataset [12] used in this experiment was captured using an embedded device consisting of a 1000 mA/h generic battery supply, an SD card, an analog accelerometer, Kinet's microcontroller, and an ITG3200 gyroscope. The device was attached to the waists of the participants. The data collected by these two sensors are plotted in terms of graphs of (angular velocity vs. time) and (acceleration versus time). The SisFall dataset is a measure of the movement-based sensor data of 8 real-world Adult Daily Living (ADL) activities, namely: fall, walk, jog, jump, up stair, down stair, stand to sit, and sit to stand. The dataset consists of 2706 ADL and 1798 falls, including data from 15 healthy independent elderly people.

The existing works clearly show that the fall detection system requires an effective means of pattern identification using the feature extraction mechanism, resulting in a more accurate feature representation of the input data. The obtained feature representation is then streamlined through a deep learning model that better understands the types of features to supervise the variances in the elderly daily activities by adopting a threshold-based classification strategy that maximizes the overall performance of feature extraction, classification, and computation speed.

3 Proposed Method

The overall architecture of the proposed Threshold-Based Fall Detection CNN (TB-FD-CNN) system is shown in Fig. 1. We employed SisFall sensor data that was preprocessed to generate a feature representation in an RGB bitmap image. The model with a single CNN channel is developed for fall detection trained with RGB images produced after preprocessing steps.

3.1 ADLs and Falls Comparison

Considering various daily activities, seven kinds of daily activities (i.e., walk, jog, go-upstairs, go-downstairs, jump, standup, sit-down) and fall are compared and analyzed. It can be observed that the bitmap of daily activities is different from that

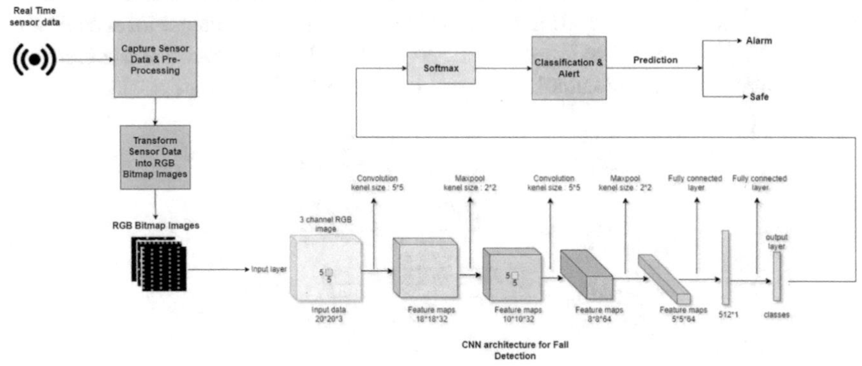

Fig. 1 Architecture of the proposed Threshold-Based Fall Detection CNN (TB-FD-CNN) system

of a fall, which makes it suitable for implementing a classification algorithm based on image recognition to identify falls from ADLs. CNN has excellent recognition accuracy for image recognition and detection, and LeNet [13] has been successful in character recognition when operating directly on 32 × 32-pixel images. Therefore, a CNN-based algorithm for fall detection according to the architecture of LeNet has been designed.

3.2 *The Visualization of the Bitmap Generation*

Human activity data captured in the SisFall dataset is divided into sliding window of 2 seconds each which contains 400 pieces of 3 axial accelerations and angular velocities, respectively [14]. This information can be summarized in a single RGB bitmap image. If the 3-axes of the human activity data are considered as the 3 channels of an RGB image, then the value of the XYZ axial data can be mapped with the values of the RGB channel data in an RGB image, respectively. Namely, each 3 axial data can be converted into an RGB pixel. The data cached into a single RGB image from 400 pieces of 3-axial data can be viewed as a bitmap with the size of 20 × 20 pixels. Fig. 2 illustrates the semantic way to map 3-trail accelerations and angular velocities into RGB bitmap image.

Since there is a mismatch between range of image pixel data which is from 0 to 255, and the ranges of accelerometer and gyroscope data are different, we need to normalize the data of acceleration and angular velocity to the range of 0–255 according to Eq. (1).

$$\text{result}_{\text{norm}} = \frac{255 \times (\text{sensor}_{\text{value}} + \text{sensor}_{\text{range}})}{2 \times \text{sensor}_{\text{range}}} \tag{1}$$

Sensor range refers to the acceleration or gyroscope's range value. The sensor value is the value that was measured. The result of the calculation is a normalized float value that is converted to an integer value. For instance, consider an acceleration dataset with X, Y, and Z-axis values of 5.947, –8.532, and 3.962. The accelerator's sensor range is 16 g. The calculated result is (174, 59, 159) using Eq. (1).

$$\text{result}_x = \frac{(5.947 + 16) \times 255}{32} = 174$$
$$\text{result}_y = \frac{(-8.532 + 16) \times 255}{32} = 59$$
$$\text{result}_z = \frac{(3.962 + 16) \times 255}{32} = 159$$

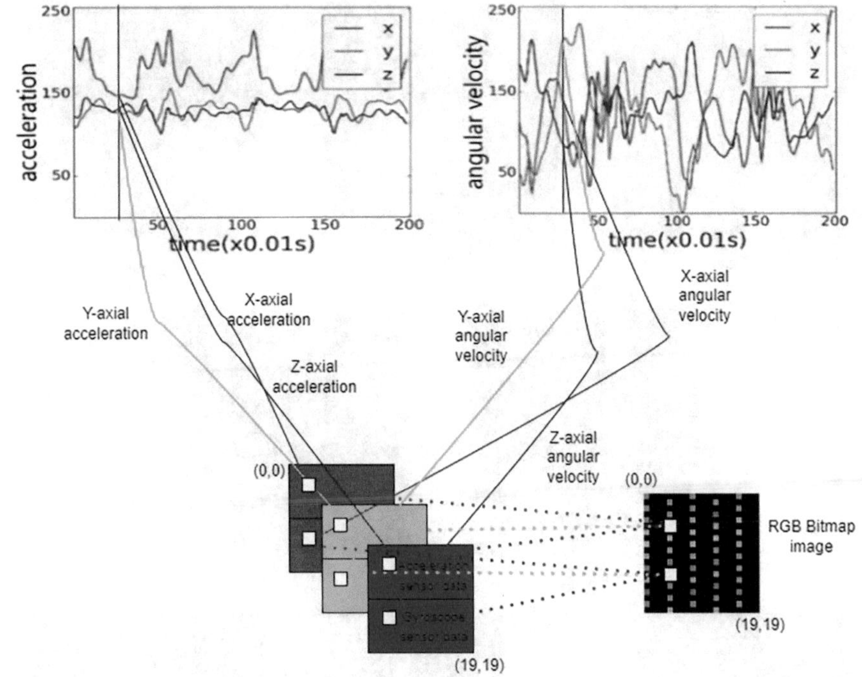

Fig. 2 Illustration of mapping 3 axial fall sensor, data into RGB bitmap

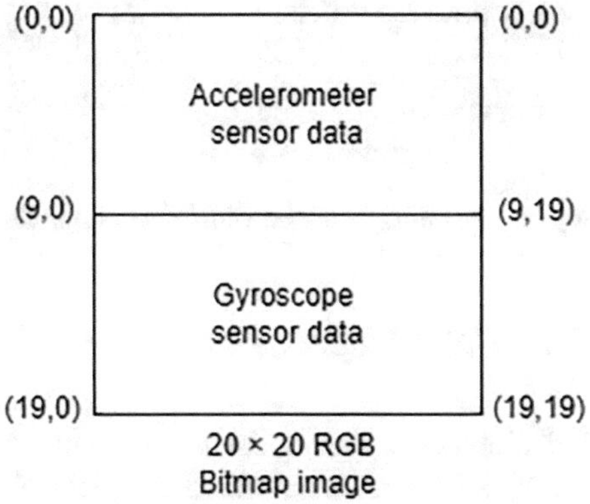

Fig. 3 An illustration of RGB bitmap transformation from SisFall sensor data. Top: accelerometer sensor data. Bottom: gyroscope sensor data

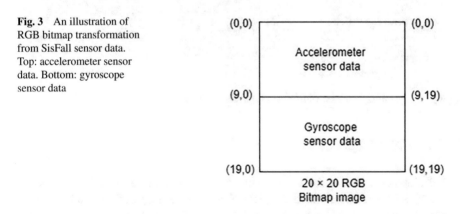

In Fig. 3 the first 200 data of the bitmap are 3 axial accelerations, and the latter 200 are 3 axial angular velocities. The data from (0, 0) to (9, 19) are 3 axial accelerations, and the data from (10, 10) to (19, 19) are 3 axial angular velocities (Fig. 4).

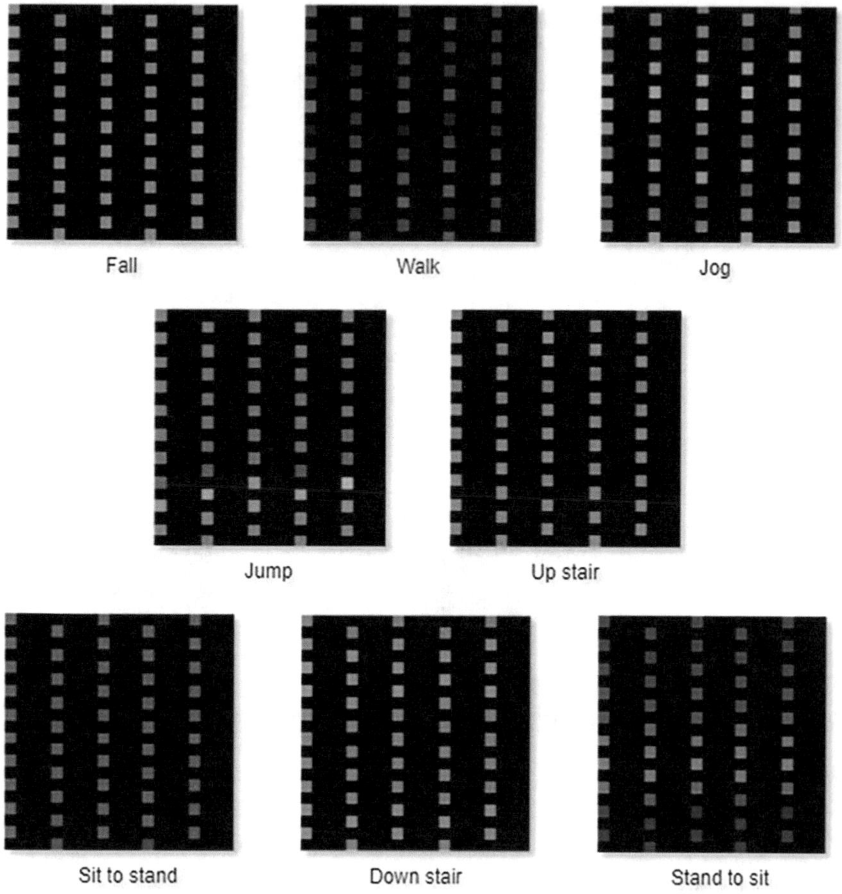

Fig. 4 The bitmap representation of actions in SisFall dataset [14]

3.3 CNN Model

Fall Detection: The model used to train the above preprocessed RGB bitmap images contains an input layer with 3 channels 20 × 20 RGB images, followed by the first convolutional layer with a kernel size of 5 × 5, which results in feature maps of size 18 × 18 × 32. Following the first convolution layer, a maxpool layer with a kernel size 2 × 2 produces in feature maps of size 10 × 10 × 32. Following that, a second convolutional layer with a kernel size of 5 × 5 produces feature maps of size 8 × 8 × 64. Following the second convolution layer, a maxpool layer with a kernel size 2 × 2 produces a feature map of size 5 × 5 × 64. Then, a fully connected layer produces a feature vector of size 512 × 1. Finally, a fully connected layer at the output layer is shown in Fig. 5.

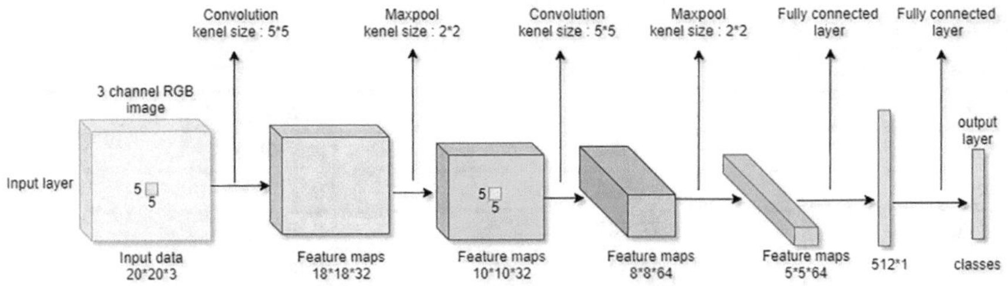

Fig. 5 CNN architecture for the proposed Fall Detection model using a single channel. The block in orange represents the dense layers, which is a result of convolutional and max-pooling operations and the red block represents the output layer. The output feature map size is shown on the top of each layer

Rectified Linear Unit (ReLU) activation function is followed after each convolutional operation in the CNN model. We used the Categorical Cross-Entropy loss function with Adam optimizer [15] to update the weights during the training process. However, the training experiments on our proposed model show better results when using a single fully connected convolutional layer after a pooling layer. The softmax function is used to generate a score for each class based on the computed trained weights.

Model Training: The proposed CNN model for the fall detection is trained with RGB bitmap images. The CNN model is trained with some predefined training parameters which are discussed in the later sections. The trained model is used to make the predictions to detect the various movement-based activities for elderly monitoring.

Training Parameters: We initialized the learning rate with 0.01, which generates a stable decrease in the loss function. The training process is affected by the learning rate, we used an adaptive learning rate which resulted in a stable improvement in the loss function. We initialized the model weights using a random function, a number of steps to 15000, and a batch step size to 64 for training the model. The training started with a gradual rise in the learning rate until it reached a peak stage at 2800 steps with a learning rate of 0.07 and it gradually decreased over the course of training and reached a learning rate of 0.58 and minimum loss of 0.2 at 15000 steps which ended the training process. The performance of the model and the duration of training depends on the input feature representation size and the dataset size.

Elderly Monitoring: The proposed CNN model for fall detection measures the required sensitivity and specificity for each specific class defining the threshold for classification. The threshold values of sensitivity and specificity are given in Table 1. The threshold values of sensitivity and specificity for each class are determined.

Table 1 Threshold values for classification

Class	Sensitivity	Specificity
Fall	1.00000	0.96475
Walk	0.97893	1.00000
Jog	0.96387	0.97232
Jump	0.95874	0.96745
Up stair	0.96236	0.96735
Down stair	0.97732	0.97345
Stand to sit	0.98754	0.97230
Sit to stand	0.97653	0.96632

4 Experimental Results and Analysis

4.1 Fall Detection

The fall is detected when sensitivity and specificity match the threshold values for classification, i.e., when they are equal to 1 and 0.96475, respectively. The sensor data considered comprises the measurement of angular velocity and acceleration of the ADLs and fall, which can be merged together to extract hidden features; by using such a data combination, we can perform fall detection. For faster response times of predictions, we can use a simple threshold-based classification algorithm for fall detection, which is the most popular and widely used since it is computationally less intensive than support vector machines and similar classification algorithms.

The fall detection capability, i.e., sensitivity, can be maximized by associating an appropriate Threshold (T) value for ADLs and falls. The threshold values are shown in Table 1 i.e.: sensitivity (SE), and specificity (SP) are computed using the Eqs. (2) and (3), respectively. The accuracy (AC) achieved during this experiment is 97.43%, and it is computed using Eq. (4). As shown in Table 2, CNN-based fall detection has outperformed most of the state-of-the-art methods.

$$SE = \frac{\text{True Positives}}{\text{True Positives} + \text{False Negatives}} \tag{2}$$

$$SP = \frac{\text{True Negatives}}{\text{True Negatives} + \text{False Positives}} \tag{3}$$

$$AC = \frac{SE + SP}{2} \tag{4}$$

Table 2 Comparison of our proposed system on fall detection

Method	Algorithm	Accuracy
K. Ozcan et al's method [6]	Modified HOG	86.66%
M. Kangas et al's protocol [8]	Threshold-based	96.67%
S. O. H. Madgwick et al's method [9]	MARG & IMU	90.37%
Shallow Siamese Network [16]	ReLU	93% ± 7%
Multi-sensor Fusion [17]	LiteFlowNet	95.23%
SVM [18]	ResNet	95.8 %
TB-FD-CNN (proposed)	CNN	**97.43%**

4.2 Computation Complexity

Preprocessing Time: The preprocessing delay comprises computing RGB bitmap images from the SisFall dataset. The RGB bitmap images require the transformation of a sequence of 400 pieces of 3 axial data captured by the sliding window as an RGB bitmap image of size 20 × 20 × 1. The overall computation time is widely affected by the feature representation size and the preprocessing delay. The prepossessing delay for RGB image transformation is 0.035 seconds. The computation time RGB bitmap image with 64 frames is 0.098 seconds. The results are obtained by using Intel Core i7-3770 @ 3.4 GHz, Graphics NVIDIA GeForce GTX 750 ti, 16 GB of RAM, and 64 bits operating system.

Training and Testing Time: The dataset size and complexity effects the overall training and the testing time, depending on the total number of frames available for training and testing. The Fall Detection model training time is less, since SisFall dataset size and type of data is sensor data. The sensor data for each type of elderly activity is converted into RGB bitmap images after preprocessing. The accuracy and loss of the proposed system are shown in Fig. 6. In this work 80% of the dataset

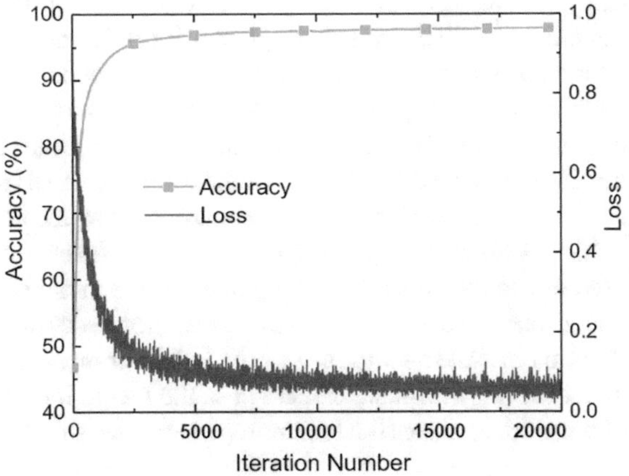

Fig. 6 Accuracy and loss of the proposed system

Table 3 Training and testing time of the dataset

Dataset	Iterations	Training			Testing		
		ADL	Falls	Training time (mins)	ADL	Falls	Testing time (mins)
SisFall	20158	2165	1439	23.87	541	359	3.59

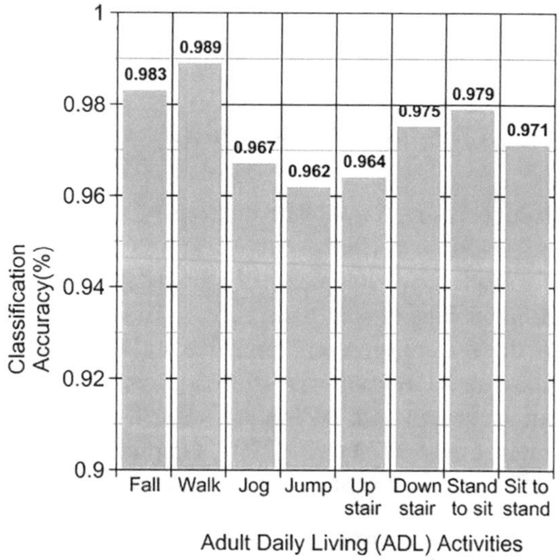

Fig. 7 Classification accuracy of Adult Daily Living (ADL) activities

is used for training and the remaining 20% of the dataset is used for testing. The training time, testing time, and train-test data split are shown in Table 3 (Fig. 7).

Discussion: The proposed fall detection system performs better than most of the state-of-the-art methods in a real-time environment. The computation and processing delay for fall detection using a CNN depends on the experimental setup used for the prediction. The system employed for the simulation uses the GPU architecture, which takes minimum delay for predictions, thus making it suitable for a real-time environment. The processing time depends on the memory and GPU used for RGB bitmap computation and the classification algorithm complexity. However, most of the CNN models differ in the number of layers used in a simple feed-forward neural network for processing. The Comparison in terms of computation time among the models is shown in Fig. 8. Additional delays can be avoided if the number of channels used was to be reduced, as using one CNN channel is less computationally intensive than using three channels. In this case, the processing time of the proposed method in terms of classification is relatively less compared to the existing state-of-the-art methods. As previously discussed in the methodology section, RGB transformation of sensor data makes the feature representation suitable for fall detection. The pattern

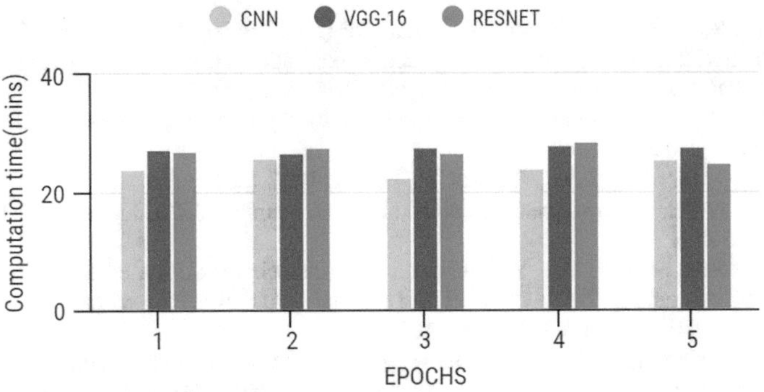

Fig. 8 Computation time comparison of various models

identification using the bitmap is computationally less intensive than using 3 axial data. The approach used is significantly better than most of the existing methods in computation time.

5 Conclusion

A sensor-based fall detection system is proposed for elderly monitoring. The system presented in this paper is easy to use, makes the interaction between a person and the system more natural, and is less expensive. The adopted classification strategy for elderly monitoring is used to calculate the accuracy from sensitivity and specificity, making it a feasible strategy for the minimum delay and faster computations in a real environment.

The proposed method outperforms most existing state-of-the-art methods, which were evaluated using SisFall movement-based sensor data of elderly activities in a real environment. Future work will mainly concern with: (i) Further improving the sensor data used for elderly monitoring with the help of moving wearable devices. (ii) Comparing the newly extracted features with the existing feature representation using the designed CNN model for testing the system's performance in a real environment.

References

1. Youngkong P, Panpanyatep W (2021) A novel double pressure sensors-based monitoring and alarming system for fall detection. In: 2021 second international symposium on instrumentation, control, artificial intelligence, and robotics (ICA-SYMP). IEEE, pp 1–5
2. Noury N, Rumeau P, Bourke A, ÓLaighin G, Lundy J (2008) A proposal for the classification and evaluation of fall detectors. Irbm 29(6):340–349
3. Mubashir M, Shao L, Seed L (2013) A survey on fall detection: principles and approaches. Neurocomputing 100:144–152

4. Gnanavel R, Anjana P, Nappinnai K, Sahari NP (2016) Smart home system using a wireless sensor network for elderly care. In: 2016 second international conference on science technology engineering and management (ICONSTEM). IEEE, pp 51–55
5. Nyan M, Tay FE, Murugasu E (2008) A wearable system for pre-impact fall detection. J Biomech 41(16):3475–3481
6. Ozcan K, Mahabalagiri AK, Casares M, Velipasalar S (2013) Automatic fall detection and activity classification by a wearable embedded smart camera. IEEE J Emerg Sel Top Circuits Syst 3(2):125–136
7. Li Y, Chen G, Shen Y, Zhu Y, Cheng Z (2012) Accelerometer-based fall detection sensor system for the elderly. In: 2012 IEEE 2nd international conference on cloud computing and intelligence systems, vol 3. IEEE, pp 1216–1220
8. Kangas M, Konttila A, Winblad I, Jamsa T (2007) Determination of simple thresholds for accelerometry-based parameters for fall detection. In: 2007 29th annual international conference of the IEEE engineering in medicine and biology society. IEEE, pp 1367–1370
9. Madgwick SO, Harrison AJ, Vaidyanathan R (2011) Estimation of imu and marg orientation using a gradient descent algorithm. In: 2011 IEEE international conference on rehabilitation robotics. IEEE, pp 1–7
10. Pierleoni P, Belli A, Palma L, Pellegrini M, Pernini L, Valenti S (2015) A high reliability wearable device for elderly fall detection. IEEE Sens J 15(8):4544–4553
11. Galvão YM, Ferreira J, Albuquerque VA, Barros P, Fernandes BJ (2021) A multimodal approach using deep learning for fall detection. Expert Syst Appl 168:114226
12. Sucerquia A, López JD, Vargas-Bonilla JF (2017) Sisfall: a fall and movement dataset. Sensors 17(1):198
13. LeCun Y, Bottou L, Bengio Y, Haffner P (1998) Gradient-based learning applied to document recognition. Proc IEEE 86(11):2278–2324
14. He J, Zhang Z, Wang X, Yang S (2019) A low power fall sensing technology based on fd-cnn. IEEE Sens J 19(13):5110–5118
15. Diederik K, Jimmy B et al (2014) Adam: a method for stochastic optimization, pp 273–297. arXiv:1412.6980
16. Bakshi S, Rajan S (2021) Few-shot fall detection using shallow siamese network. In: 2021 IEEE international symposium on medical measurements and applications (MeMeA). IEEE, pp 1–5
17. Lv X, Gao Z, Yuan C, Li M, Chen C (2020) Hybrid real-time fall detection system based on deep learning and multi-sensor fusion. In: 2020 6th international conference on big data and information analytics (BigDIA). IEEE, pp 386–391
18. Chen Y, Du R, Luo K, Xiao Y (2021) Fall detection system based on real-time pose estimation and svm. In: 2021 IEEE 2nd international conference on big data, artificial intelligence and internet of things engineering (ICBAIE). IEEE, pp 990–993

Precise Stratification of Gastritis Associated Risk Factors by Handling Outliers with Feature Selection in Multilayer Perceptron Model

Brindha Senthil Kumar, Lalhma Chhuani, Lalrin Jahau, Madhurjya Sarmah, Nachimuthu Senthil Kumar, Harvey Vanlalpeka, and Lal Hmingliana

1 Introduction

Outliers are unusual data points in the input data as they can lead to misinterpretation of data analysis and inferences, so to find a selective and systematic approach to handle the outliers that can yield better classification is a very significant process that helps to build efficient models for biological data [1]. Robust principal component analysis was one of the efficient methods to detect outliers. It had been used on RNA-seq data to detect outliers which involved data transformation in a subspace, construction of covariance matrix, computation of eigenvalues from the location and scatter matrix, followed by univariate analysis to find the outliers [2]. A compressed column wise robust principal component analysis method was used to detect outliers from hyperspectral images. The dimensionality reduction was done by Hadamard random projection technique, followed by outlier detection using sparse anomaly matrix as

B. S. Kumar · L. Hmingliana (✉)
Department of Computer Engineering, Mizoram University, Aizawl, Mizoram 796004, India
e-mail: lalhmingliana@mzu.edu.in

L. Chhuani · L. Jahau
Trinity Diagnostic Centre, Aizawl, Mizoram 796001, India

M. Sarmah
Department of Radio Imaging Technology, Regional Institute of Paramedical and Nursing Sciences, Zemabawk, Aizawl, Mizoram 796017, India

N. S. Kumar
Department of Biotechnology, Mizoram University, Aizawl, Mizoram 796004, India

H. Vanlalpeka
Deparment of Obstetrics and Gynecology, Zoram Medical College, Falkawn, Mizoram 796005, India

P. Singh et al. (eds.), *Machine Learning and Computational Intelligence Techniques for Data Engineering*, Lecture Notes in Electrical Engineering 998,
https://doi.org/10.1007/978-981-99-0047-3_17

the anomalies were sparse and they would not lie within the column subspace, any observation outside the columns subspace was considered as outliers [3].

Clustering algorithms, multiple circular regression model, and von Mises distribution models had used to detect outliers in circular biomedical data [4]. A combined approach of hierarchical clustering and robust principal component analysis were used to detect anomalies in gene expression matrices [5]. K-means clustering was used to identify the outliers in central nervous system disease, the clusters were generated by calculating the sum of squares between the datapoints, and had gained 10% accuracy than the multivariate outlier detection method [1]. Interquartile range (IQR) method is an efficient approach in detecting the outliers; generally, it tends to detect more numbers of observations as outliers. Though, IQR and MDist approaches had generated similar results, the MDist approach had identified a smaller number of observations as outliers. But both IQR and MDist methods produced a greater number of outliers when the contaminating share increased [6]. An optimized Isolation forest algorithm was proposed in detecting the outliers, it had an advantage in selecting the feature and locating the split point more accurately, the computation time taken to retrieve the best split point was less, and it required less trees to reach the convergence [7].

The present work was aimed to develop a model for sequential-selection of methods to detect and remove outliers and irrelevant information from gastritis data. A well-defined machine learning model was developed which gave accurate risk stratification accuracy to classify between presence/absence of *H. pylori*-associated gastritis. The above-two procedures were tested on the datasets: (i) outliers removed using three famous algorithms (Isolation forest, one-class SVM and Interquartile range method), (ii) outliers identified and replaced by median values (as the features are discrete), (iii) outliers replaced by median values + feature selection. Multilayer perceptron, AdaBoost, decision tree, logistic regression and Naive Bayes Bernoulli algorithms were devised on the above-two types of procedures and their performances were compared. The main focus of this study was to comprehensively analyze risk stratification accuracy based on the machine learning classifiers' performances on raw input gastritis data, outliers removed data, and outliers replaced by median values + feature extraction data. The present study had shown that classification accuracy had been greatly improved when the outliers were replaced by median values and non-informative features were eliminated by feature selection method.

2 Methods

2.1 Data Source

About 863 instances were classified into two classes: presence/absence of *H. pylori*-associated gastritis. There were 21 features which were collected from gastritis patients using a well-structured questionnaire. The features that are associated with

Table 1 Features and its data description

Features	Data description
Gender	Categorial
Age	Discrete
Alcohol	Categorial
Smokers	Categorial
Smokeless_Tobacco_Use	Categorial
Supari	Categorial
beetle_nut	Categorial
pan_zarda_pan	Categorial
Guthkha	Categorial
Sadha	Categorial
Khaini	Categorial
Tiranga	Categorial
Tuibur	Categorial
Nonveg	Categorial
RawUncookedfood	Categorial
Saum	Categorial
Salt_intake	Categorial
Pickles	Categorial
Water source	Categorial
Drinking water	Categorial
Sanitation	Categorial
H. pyloristatus	Categorial

this disease were considered for the analysis and their data descriptions were given in Table 1.

2.2 Data Pre-processing

Boxplot data distribution of the raw data is as in Fig. 1. Dataset1, dataset2, and dataset3 were created by removing outliers using Interquartile Range (IQR) method, one-class support vector machine (SVM) [8], and Isolation Forest method [9], respectively. Dataset4 was constructed by replacing outliers by median values [10]. Dataset5 was prepared by replacing outliers by median values + feature selection.

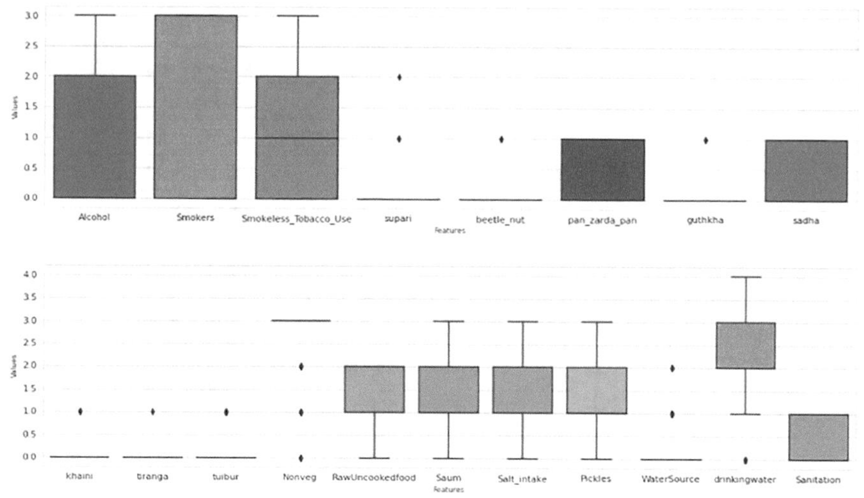

Fig. 1 Box plot data distribution of the gastritis features

2.3 Feature Selection

Lasso regression is used for feature selection, as the feature selection reduces data dimension and the model training time [11]. Lasso method is very efficient when data size is small with large number of features as in the present dataset.

$$\sum_{i=1}^{n}\left(y_i - \sum_{j} x_{ij}\beta_j\right)^2 + \lambda \sum_{j=1}^{q} |\beta_j| \quad (1)$$

where, λ is the shrinking parameter, when value of λ increases the number of features elimination increases in Eq. 1.

2.4 Learning Curves

Upon the number of iterations, the training loss and validation loss must have very less error gap to clearly conclude that data was not underfitting for the current models [12]. Learning curves were generated for seven classifiers (Multi-layer perceptron, decision tree, logistic regression, naïve bayes, random forest, support vector machine, and adaptive gradient boosting) to find the best fit for this dataset.

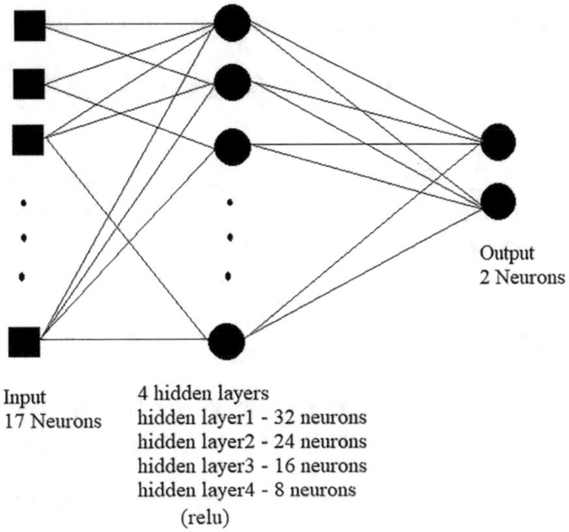

Fig. 2 Architecture of the proposed multilayer perceptron model (PMPM)

2.5 Data Modeling

2.5.1 Multilayer Perceptron

The input layer of the multilayer perceptron was fed with epidemiological features and followed by four sets of hidden layers. Relu is used as an activation function in the hidden layers which triggers the responsible neurons of the output layer (Fig. 2).

Relu activation function

$$\begin{aligned} f(x) = 0 \text{ when } x &< 0, \\ x \text{ when } x &\geq 0 \end{aligned}$$

weights Computation

$$f\left(b + \sum_{i=1}^{n} x_i w_i\right) \quad (2)$$

where, f is activation function, x is input, w is weight, and b is the bias in Eq. 2.

2.5.2 Adaptive Gradient Boosting

Decision Trees (DT) were used as weak learners in the Adaptive gradient boosting algorithm, and the number of estimators was set to 100 to generate the model. The model alters the weight each time when data points were misclassified, thus error rate was minimized [13]. Error was calculated using the formula given:

$$E_j = \Pr_i D_j\left[h_j(x_i) \neq y_i\right] = \sum_D h_j(x_i) \neq y_i D_j(i) \quad (3)$$

where, E_j is Error rate, $\Pr_i \sim D_j$ probability of random sample i from the dataset D_j, h_j is the hypothesis of the weak learner, x_i is the independent variables, y_i is the target variable and j is the iteration number in Eq. 3.

2.5.3 Decision Tree

In this ID3 algorithm: (i) original set S as the root node, (ii) on every iteration the entropy was calculated for the attribute from set S, (iii) smallest entropy value attribute was selected to proceed future, (iv) S was split based on the attribute selected in step iii, (v) the splitting of the tree continuous till all the attributes were chosen from set S. In this work, the maximum depth of the tree was set to 3, and entropy was calculated before splitting the nodes [14].

$$\text{Entropy}(S) = -p_+ \log_2 p_+ - p_- \log_2 p_- \quad (4)$$

where, p_+ is presence of *H. pylori* gastritis class, p_- is absence of *H. pylori* gastritis class in Eq. 4.

2.5.4 Logistic Regression

Logistic Regression (LR) finds the probability of an outcome for a given set of inputs [15]. A score between 0 and 1 for a candidate answer with attribute values $x_1, x_2, x_3, \ldots, x_n$ was calculated using the following logistic function.

$$f(x) = \frac{1}{1 + e^{-(\beta_0 + \sum_{i=1}^{n} \beta_i x_i)}} \quad (5)$$

where, β_0 is the intercept; $\beta_1, \ldots \beta_n$ are weights of the attributes in Eq. 5.

2.6 *Naive Bayes Bernoulli*

Naive Bayes Bernoulli (NBB) classifier is based on the Bayes theorem which has an assumption that attributes are conditionally independent. It is used for discrete datasets and greatly reduces the computation time.

2.7 Data Package

Pre-processing phase, data models, data visualization, and analyses were done in Python version 3 Jupyter Notebook using scikit-learn packages. Dataset was divided into 70% for training and 30% for testing. Figure 3 shows the flow chart of data preprocessing, developing machine learning models, and comparing the risk stratification accuracy between the models.

3 Results and Discussion

The dataset used in the present study comprised of 21 features which includes diet, lifestyle, and environmental factors. A total of 863 patients' records were collected for this study, out of which 370 were males and 493 were females. There are 475 patient records positive for *H. pylori*-associated gastritis and 388 records negative for *H. pylori*-associated gastritis. This paper presents an effective technique to detect and remove outliers and irrelevant information from gastritis data and has developed a well-defined machine learning model which gives accurate risk stratification accuracy to classify between presence/absence of *H. pylori*-associated gastritis. Two types of protocols have been utilized in this work to handle datasets with outliers: (i) outliers were removed using three famous algorithms (Isolation forest, one-class SVM, and Interquartile range method), (ii) outliers were replaced by median values, and applied feature selection. A set of five well-defined classifiers (Proposed Multilayer perceptron, AdaBoost, decision tree, logistic regression, and Naive Bayes Bernoulli) were applied on the above-two types of protocols and their

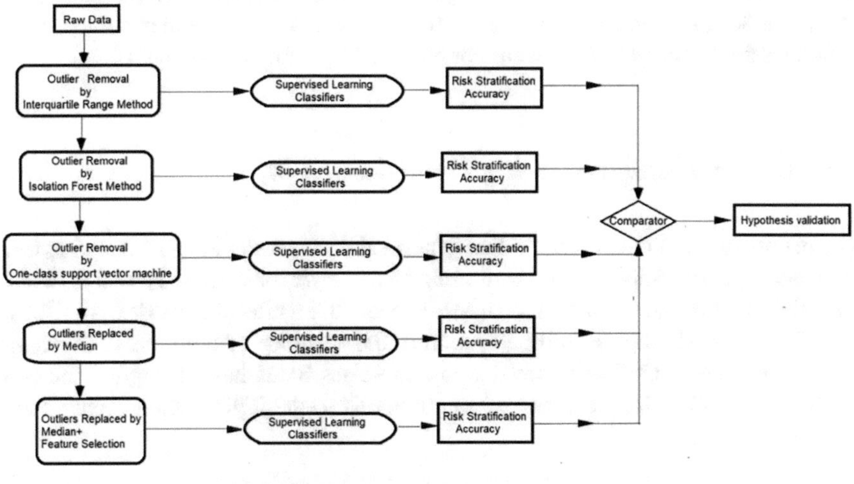

Fig. 3 Flow chart for data preparation and data modeling

performances have been compared. The main idea behind this study was to comprehensively analyze the risk stratification accuracy of PMPM, AdaBoost, DT, LR, and NBB classifiers, respectively with raw input gastritis data, outliers removed, outliers replaced by median values, outliers replaced with median values + feature extraction. The present study had shown the classification accuracy greatly improved when outliers replaced with median values + feature extraction (Tables 2 and 3, Fig. 4) [10].

3.1 Original Dataset

The raw dataset contains approximately 10% outliers. To show the accuracy gets boosted when outliers are handled with appropriate techniques, all 21 features of the dataset were utilized to develop five classifiers: PMPM, AdaBoost, LR, DT, and NBB. The fine-tuned PMPM has high accuracy of 74%, when compared to other classifiers: NBB, AdaBoost, LR, and DT, of 66%, 65%, 64%, and 61% respectively (Table 2 and Fig. 4).

3.2 Outliers Removed Using Interquartile Range Method

IQR method has identified and removed 8% of the dataset as outliers, resulting in 431 positive *H. pylori*-associated gastritis and 360 negative *H. pylori*-associated gastritis records. The resultant dataset was subjected to develop five classifiers. PMPM has an accuracy of 76% while other models AdaBoost, LR, DT, and NBB accuracies were 63%, 63%, 65%, and 65%, respectively. Based on the risk stratification accuracies of five models, it is well evident that PMPM had a mild increase in accuracy, while other classifiers performances were not remarkable (Table 2 and Fig. 4).

3.3 Outliers Removed Using One-Class SVM

The results of the one-class SVM had generated 418 records for positive *H. pylori*-associated gastritis and 358 records for negative *H. pylori*-associated gastritis, which was 10% of the dataset. PMPM has an accuracy of 78% while other models AdaBoost, LR, DT, and NBB showed mild elevation in the accuracy of 67%, 68%, 68%, and 66%, respectively. Outlier removal using one-class SVM has a marginal increase in the risk stratification accuracy when compared to the IQR method (Table 2 and Fig. 4).

Table 2 Comparisons of accuracies of all classifiers on raw, outliers removed, outliers replaced by median, and outlier replaced by median values + feature selection data

Dataset	Types of data-preprocessing	Classifiers	Accuracy in %
Raw data	Original	PMPM	72
		AdaBoost	65
		Decision tree	61
		Logistic regression	64
		Naive Bayes Bernoulli	66
Dataset1	Outliers removed by IQR method	PMPM	76
		AdaBoost	63
		Decision tree	65
		Logistic regression	63
		Naive Bayes Bernoulli	65
Dataset2	Outliers removed by one-class SVM	PMPM	78
		AdaBoost	67
		Decision tree	68
		Logistic regression	68
		Naive Bayes Bernoulli	66
Dataset3	Outliers removed by Isolation forest method	PMPM	79
		AdaBoost	68
		Decision tree	69
		Logistic regression	69
		Naive Bayes Bernoulli	67
Dataset4	Outliers replaced by median	PMPM	84
		AdaBoost	75
		Decision tree	71
		Logistic regression	70
		Naive Bayes Bernoulli	71
Dataset5	Outliers replaced by median and feature selection	PMPM	92
		AdaBoost	70
		Decision tree	73
		Logistic regression	71
		Naive Bayes Bernoulli	70

Table 3 Data models and their hyperparameters

Data models	Hyperparameters
PMPM	optimizer = 'lbfgs', learning rate = 0.001, momentum = 0.99
AdaBoost	n_estimators = 100
Decision tree	criterion = 'entropy', max_depth = 3

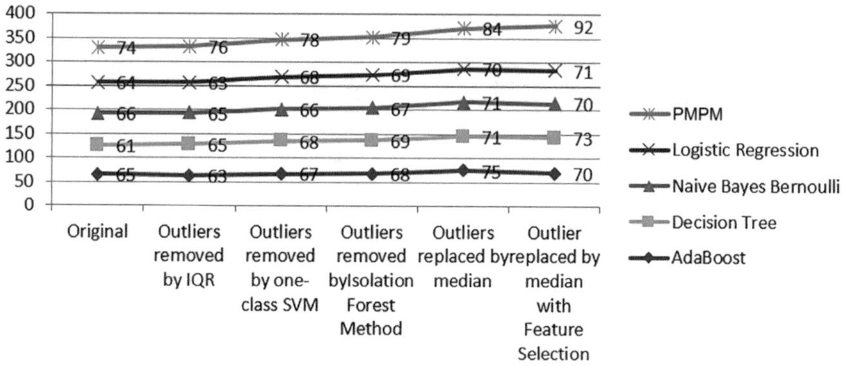

Fig. 4 Graphical representation of accuracies based on the classifiers

3.4 Outlier Removed Using Isolation Forest

Isolation forest has eliminated 7% of the data as outliers, it has resulted in 416 records positive for H. pylori-associated gastritis and 360 records negative for *H. pylori*-associated gastritis. By observing the five sets of accuracies acquired, PMPM has the highest accuracy among them of 79% whereas AdaBoost, LR, DT, and NBB accuracies were 68%, 69%, 69%, and 67%, respectively (Table 2 and Fig. 4).

3.5 Outliers Replaced by Median

From the above three methods used to identify the outliers, IQR method showed high percentage of outliers 10%. But still, model performances on IQR outlier deleted dataset did not perform well. From the literature, its well-understood that replacing outliers with median values have increased the stratification accuracy [16]. So, median values were replaced with detected outliers by the IQR method. It was shown that PMPM has an accuracy of 84% which was a 5% increase after replacing the missing values with median. Subsequently, other models AdaBoost, LR, DT, and NBB had also performed better on this dataset with accuracies of 75%, 70%, 71%, and 71%, respectively (Table 2 and Fig. 4). Overall the risk stratification accuracies of all five models were above 70%.

3.6 Outliers Replaced by Median Values + Feature Selection

Features of importance of Lasso regression method were shown in Table 4. Five machine learning algorithms were tested on the above ten selected features to find risk stratification accuracy. PMPM model had given highest risk stratification accuracy

Table 4 Feature selected from lasso regression

S. no	Features
1	Smoking
2	pan_zarda_pan
3	Sadha
4	Tuibur (aqueous tobacco extract)
5	Raw food
6	Saum (fermented pork fat)
7	Salt intake
8	Water source
9	Drinking water
10	Sanitation

of 92%, which was 20% more in increase of the accuracy from the raw data, whereas, other models AdaBoost, LR, DT, and NBB had a satisfactory performance of 70%, 71%, 73%, and 70%, respectively (Table 2 and Fig. 4).

4 Benchmarking Machine Learning Systems

Logistic Regression, AdaBoost, Bernoulli Naïve Bayes, Decision tree, and Proposed Multilayer Perceptron Models (PMPM) were utilized to classify gastritis dataset and pre-processed using different techniques. From Fig. 4, it is very clear that the PMPM had performed consistently well, and the risk stratification accuracy gradually increases from 74 to 92% on raw and pre-processed data (replacing median with outlier and feature selection), respectively. Multilayer perceptron classifier was able to generate highly accurate models not only in the present work but in many sub-domains of biology, in the classification of genus and species [17], and identifying chronic kidney disease patients [18].

Methods IQR, one-class SVM, and Isolation forest had improved the accuracy from 76 to 79%, but the accuracy was further improved to 84% when the outliers were replaced by median values (Fig. 4). Isolation forest feature section and classification methods had produced an accuracy of 92.26% by replacing missing values by group median and outliers by median values [16]. C4.5 classifier had produced 89.5% accuracy on diabetes dataset where outliers were detected by IQR [19]. Other machine learning algorithms such as Logistic regression, AdaBoost, NBB, and Decision tree had shown gradual improvement in the performance as their accuracies in every pre-processing method. Nevertheless, from the present study, it is a clear indication that the machine learning models that were trained with the selective data preprocessing strategies will definitely improve the classification accuracy.

5 Risk Factors for Gastritis-Associated *H. Pylori*

From the Tables 2 and 4, Fig. 4 results, the diet and lifestyle features: smoking, pan with zarda, sadha, tuibur (aqueous tobacco extract), raw food, saum (fermented pork fat), salt intake, water source, drinking water, and sanitation were significant cause for *H-pylori* associated gastritis in Mizo population. These features were well-studied in other populations and they were significant factors to cause gastric cancer because gastritis leads to gastric cancer. Many literatures had reported that smoking, salty food intake, and fermented, pickled or smoked food were causes for *H. pylori* infection. The feature selection of this study showed smoking, saum (fermented pork fat), and excess salt intake as few among the features for causing *H. pylori*-associated gastritis [20]. The food preservation methods by using salt and smoking were found to be a source of *H. pylori* infection associated gastric ulcer as it is known that in Mizoram, people preserve their foods with smoking and salt [21]. This study's results showed sanitation was also one of the factors for causing *H. pylori* infection as this infection spreads faster in crowded living conditions. Salads from raw vegetables and fruits were also found to have resistant and virulent strains of *H. pylori* [22]. Studies had shown raw cabbage and lettuce cultured were found to have positive for *H. pylori*, this is very evident in this present study as in Mizoram raw cabbage is extensively used in form of pickles and to make spicy salads. Polluted water sources were also causing *H. pylori*, in this paper water sources were from stagnated pounds and lakes and these were major sources of *H. pylori* infection [23].

6 Conclusion

A suitable machine learning algorithm was developed and its performance was tested on all different types of pre-processed datasets. The proposed multilayer perceptron model had produced highest accuracy among the other chosen classifiers (logistic regression, decision tree, naïve bayes and AdaBoost). Outlier replaced by median values + feature selection dataset had produced highest accuracy of 92% by the PMPM when the original data had an accuracy of 72% for the same classifier.

Acknowledgements The Authors thank the DBT-Uexcel project (BT/551/NE/U-Excel/2014), Mizoram University sponsored by the Department of Biotechnology (DBT), New Delhi, Govt. of India. The authors thank Subhojith Mukherjee and David K. Zorinsanga who helped in data collection.

Author Contribution BSK, LJ, MS, NSK, HV, and LH planned the work. LC, LJ, MS, and HV helped in sampling. LC did the sampling. BSK, and LH carried out the data analysis; LH and NSK supervised the work; BSK and NSK wrote the manuscript. All authors contributed to the final editing.

Ethical Approval The ethical committee of Civil Hospital Aizawl, Mizoram (B.12018/1 /13-CH(A)/ IEC/ 36) as well as the Mizoram University ethical committee approved the work.

Conflict of Interest Statement The authors declare that they have no conflict of interest.

References

1. Qiu Y, Cheng X, Hou W, Ching W (2015) On classification of biological data using outlier detection. In: 12th ınternational symposium on operations research and its applications in engineering technology and management, pp 1–7. https://doi.org/10.1049/cp.2015.0617
2. Chen X, Zhang B, Wang T, Bonni A, Zhao G (2020) Robust principal component analysis for accurate outlier sample detection in RNA-Seq data. BMC Bioinform 21:269. https://doi.org/10.1186/s12859-020-03608-0
3. Sun W, Yang G, Li J, Zhang D (2018) Hyperspectral anomaly detection using compressed columnwise robust principal component analysis. In: IEEE international symposium on geoscience and remote sensing, pp 6372–6375. https://doi.org/10.1109/IGARSS.2018.8518817
4. Satari SZ, Khalif KMNK (2020) Review on outliers identification methods for univariate circular biological data. Adv Sci Tech Eng Sys J 5:95–103. https://doi.org/10.25046/aj050212
5. Selicato L, Esposito F, Gargano G et al (2021) A new ensemble method for detecting anomalies in gene expression matrices. Mathematics 9:882. https://doi.org/10.3390/math9080882
6. Domański PD (2020) Study on statistical outlier detection and labelling. Int J of Auto and Comput 17:788–811. https://doi.org/10.1007/s11633-020-1243-2
7. Zhen L, Liu X, Jin M, Gao H (2018) An optimized computational framework for ısolation forest. Math Probl Eng 1–13. https://doi.org/10.1155/2018/2318763
8. Fujita H, Matsukawa T, Suzuki E (2020) Detecting outliers with one-class selective transfer machine. Knowl Inf Syst 62:1781–1818. https://doi.org/10.1007/s10115-019-01407-5
9. Liu FT, Ting KM, Zhou ZH (2012) Isolation-based anomaly detection. ACM Trans Knowl Disco from Data 6:1–39. https://doi.org/10.1145/2133360.2133363
10. Kwak SK, Kim JH (2017) Statistical data preparation: management of missing values and outliers. Korean J Anesthesiol 70:407–411. https://doi.org/10.4097/kjae.2017.70.4.407
11. Tibshirani R (1996) Regression shrinkage and selection via the lasso. J Royal Stat Soci Ser B (Methodol) 58:267–288
12. Zhang C, Vinyals O, Munos R, Bengio S (2018) A study on overfitting in deep reinforcement learning. arXiv:1804.06893
13. Rahman S, Irfan M, Raza M, Ghori KM, Yaqoob S, Awais M (2020) Performance analysis of boosting classifiers in recognizing activities of daily living. Int J Environ Res Public Health 17:1082. https://doi.org/10.3390/ijerph17031082
14. Mantovani RG, Horváth T, Cerri R, Junior SB, Vanschoren J, Carvalho ACPLF (2018) An empirical study on hyperparameter tuning of decision trees. arXiv:1812.02207
15. Thomas WE, David OM (2017) Exploratory study. Research methods for cyber security. In: Thomas WE, David OM (eds) Syngress, pp 95–130. https://doi.org/10.1016/B978-0-12-805349-2.00004-2
16. Maniruzzaman M, Rahman MJ, Al-MehediHasan M, Suri SH, Abedin MM, El-Baz A, Suri JS (2018) Accurate diabetes risk stratification using machine learning: role of missing value and outliers. J Med Syst 42:92. https://doi.org/10.1007/s10916-018-0940-7
17. Sumsion GR, Bradshaw MS, Hill KT, Pinto LDG, Piccolo SR (2019) Remote sensing tree classification with a multilayer perceptron. PeerJ 7:e6101. https://doi.org/10.7717/peerj.6101
18. Sharifi A, Alizadeh K (2020) A novel classification method based on multilayer perceptron-artificial neural network technique for diagnosis of chronic kidney disease. Ann Mil Health Sci Res 18:e101585. https://doi.org/10.5812/amh.101585
19. Nnamoko N, Korkontzelos I (2020) Efficient treatment of outliers and class imbalance for diabetes prediction. Artific Intellig in Med 104:101815. https://doi.org/10.1016/j.artmed.2020.101815

20. Assaad S, Chaaban R, Tannous F, Costanian C (2018) Dietary habits and Helicobacter pylori infection: a cross sectional study at a Lebanese hospital. BMC Gastroenterol 18:48. https://doi.org/10.1186/s12876-018-0775-1
21. Muzaheed (2020) Helicobacter pylori Oncogenicity: mechanism, prevention, and risk factors. Sci World J 1–10. https://doi.org/10.1155/2020/3018326
22. Yahaghi E, Khamesipour F, Mashayekhi F et al (2014) Helicobacter pylori in vegetables and salads: genotyping and antimicrobial resistance properties. BioMed Res Int 2014:757941. https://doi.org/10.1155/2014/757941
23. Ranjbar R, Khamesipour F, Jonaidi-Jafari N, Rahimi E (2016) Helicobacter pylori in bottled mineral water: genotyping and antimicrobial resistance properties. BMC Microbiol 16:40. https://doi.org/10.1186/s12866-016-0647-1

Portfolio Selection Using Golden Eagle Optimizer in Bombay Stock Exchange

Faraz Hasan, Faisal Ahmad, Mohammad Imran, Mohammad Shahid, and Mohd. Shamim Ansari

1 Introduction

The portfolio selection problem (PSP) is one of the challenging issues in the field of economic and financial management. The main target of the portfolio selection is to find the best combination of stocks for investors that maximize the returns with minimum risk. There are various trade off exists between risks and returns. Therefore how to decide the best portfolio in which assets to invest the available capital is subjective to the decision maker. Various types of returns are related to various risk levels and there was not a standard portfolio available which fulfills the needs of all investors. Optimization process is a tool to increase the profit for investors and help them in the decision-making situations with their investment goals [1, 2].

The first approach proposed by Markowitz [1] for portfolio selection problem was mean-variance model. The main aim of this model was to provide the maximum return with minimum risk of the unconstrained portfolio. This approach can be formulated in terms of quadratic programming (QP). Markowitz's theory cannot optimize both objectives (risk and return) with the constraints such as cardinality,

F. Hasan
Department of Computer Science and Engineering, Koneru Lakshmaiah Education Foundation, Guntur, India
e-mail: faraz.hasan@live.in

F. Ahmad
Workday Inc., Pleasanton, USA

M. Imran
Department of Computer Science, Aligarh Muslim University, Aligarh, India

M. Shahid (✉) · Mohd. S. Ansari
Department of Commerce, Aligarh Muslim University, Aligarh, India
e-mail: mdshahid.cm@amu.ac.in

P. Singh et al. (eds.), *Machine Learning and Computational Intelligence Techniques for Data Engineering*, Lecture Notes in Electrical Engineering 998,
https://doi.org/10.1007/978-981-99-0047-3_18

transaction cost or lot effectively. To find out the optimal solution, several types of work has been done with MV portfolio model using contemporary techniques. There are significant number of work reported in literature for MV model but the important constraints namely cardinality, budget, and lower/upper bound constraints are ignored when the money is allocated by investor among various stocks. However, when the model incorporates more constraints, optimization becomes more complex and classic and deterministic approaches fail to produce satisfactory results. In such circumstances, metaheuristic approaches such as evolutionary, swarm intelligence and nature inspired approaches are advised to obtain better results [2, 3].

In this paper, a novel portfolio optimization model using golden eagle optimizer (GEO) [4] has been proposed with the aim of optimizing return and risk. GEO is a nature inspired approach that mimics the haunting behavior of the golden eagle. GEO uses two processes Viz. cruise and attack for exploration and exploitation of solution space respectively. To conduct the performance evaluation, an experimental study has been conducted with performance comparison on execution time, optimal solutions at efficient frontier obtained by the proposed GEO, ABC and IWO on S&P BSE dataset (30 stocks) of Indian stock exchange.

The organization of this paper is as: Sect. 2 presents the related work of the field. The mathematical formulation of the problem is represented in Sect. 3. Section 4, depicts the solution model of GEO problem. In Sects. 5 and 6, the simulation results and conclusion have been discussed respectively.

2 Related Work

For portfolio optimization problem, asset allocation is one of the most important issues in financial management. Over a decade, many approaches have been developed such as evolutionary approaches (EA) [5–8], swarm intelligence approaches (SIA) [9–17] and nature inspired approaches (NIA) [18–20] which provide better solutions.

Bili Chen [5] proposed an approach which is a multi-objective evolutionary framework supporting non-dominated sorting with local search to obtain the solution. Hidayat et al. [6] suggested a genetic algorithm based model that computes the portfolio risk using absolute standard deviation. Jalota [7] depicts an approach in fuzzy environment to investigate the impact of various sets of lower and upper bounds on assets. Shahid et al. [8] presents an evolutionary computation based model namely stochastic fractal search, modeling growth behavior of nature to solve risk budgeted portfolio problem, maximizing the shape ratio.

The two extended set-based algorithms with substantial gains were proposed by Erwin in [9]. Zaheer et al. [10] recommended a hybrid particle swarm optimization (HPSO) metaheuristic algorithm from financial toolbox in MATLAB. For analysis, data is taken from Shanghai Stock Exchange (SSE). Cura et al. [11] reported a heuristic approach using artificial bee colony for mean-variance portfolio optimization problem with cardinality-constrained providing the better solution. Sahala et al.

[12] proposed an approach in which improved Quick Artificial Bee Colony (iqABC) method with Cardinality-constrained mean-variance (CCMV) model and reported better results on Sharpe ratio and return values.

Further, Stumberger et al. [13] suggested a genetic algorithm with inspired elements and hybridized artificial bee colony algorithm to establish a better balance between diversification and intensification to solve portfolio problem. Abolmaali et al. [14] demonstrates an approach to construct constrained portfolio by using Ant Colony optimization Algorithm. Kalayci [15] designed a hybrid-integrated mechanism with critical components using artificial bee colony optimization, continuous ant colony optimization, and genetic algorithms to solve portfolio selection problem with cardinality constraints. Suthiwong et al. [16] have presented ABC algorithm and applied Sigmoid-based Discrete Continuous model to solve the stock selection problem for optimizing both diversity, investment return and model robustness. Rezani et al. [17] have presented a cluster-based ACO algorithm to solve the portfolio optimization problem by using iterative k-means algorithm to optimize the sharpe ratio.

A nature inspired optimization approach based on squirrel search algorithm (SSA) is reported in [18] by exploiting the gliding mechanism of small mammals to travel long distances. Shahid et al. [19] presented an invasive weed optimization (IWO) based solution approach for risk budgeted constrained portfolio problem. Here, sharp ratio of the constructed portfolio is optimized on BSE 30 dataset. Gradient-based optimization (GBO) method is used to design an unconstrained portfolio in [20]. Sefiane et al. [21] demonstrates a Cuckoo Optimization Algorithm (COA) which provides better results in comparison to ant colony algorithm (ACO) and genetic algorithm (GA).

3 The Problem Statement

In stock exchange, a portfolio (P) with N number of stocks has been formulated i.e., $P = \{\mathrm{st}_1, \mathrm{st}_{2,}, \ldots \mathrm{st}_N\}$ using their corresponding weights $\{W_1, W_{2,}, \ldots W_N\}$. The predicted returns of the stocks are considered as $\{R_1, R_{2,}, \ldots R_N\}$. Then, portfolio risk and return can be estimated as

$$\mathrm{Risk}_P = \sum_i^N \sum_j^N W_i * W_j * \mathrm{CV}_{\mathrm{ij}} \tag{1}$$

$$\mathrm{Return}_P = \sum_1^N R_i * W_i \tag{2}$$

where W_i and W_j are the weights for st_i and st_j respectively. $\mathrm{CV}_{\mathrm{ij}}$ is called the covariance of the portfolio returns. Here the main aim of the problem is to obtain the optimal values for return and risk. Therefore, problem statement can be taken as weighted sum of risk and return and written as

$$\mathrm{Min}(Z) = \mu * \mathrm{Risk}_P - (1 - \mu) * \mathrm{Return}_P \tag{3}$$

(i) $\sum_{i=1}^{N} w_i = 1$
(ii) $w_i \geq 0$
(iii) $a \leq w_i \leq b$

Here (i) represent the budget constraint. It restricts the method to explore weights having sum equal to one (100%). (ii) constraint restricts the short sell. Next, (iii) express the boundary constraint, which imposes lower and upper bounds for asset weights in the portfolio. In the above problem, the constraints are represented as linear with convex feasible region and a repair method is used to handle these constraints. Whenever lower or upper bounds are violated, then, respective weights are replaced by the lower or upper bound values respectively. To maintain budget constraint (sum equal to 1), normalization approach is used such as each stock weight is divided by sum of the total weights of the portfolio.

4 Proposed Strategy

In this section, all the details about proposed Golden Eagle Optimizer [4] have been discussed. The haunting behavior of Golden Eagle has been formulated using metaheuristic to solve typical portfolio selection problem. There are characteristics of the golden eagle that are spiral motion, prey selection, cruising and attacking. The processes e.g., cruise and attack are used for exploring and exploiting search space for the problem. These unique features help them to continuously monitor the targeted prey and to find out a proper angle for attack. When the pray is found out by golden eagle, it memorizes the exact location and continues to encircle it. It downs slowly with altitude and concurrently comes closer by making the hypothetical circle smaller and smaller around the targeted prey. If could not find a better location, it circled around the prey continuously. If eagle gets another alternative, it flies around new prey in a circle and ignores the previous one. But the final attack is in a straight line.

4.1 Attack (Exploitation)

The attacking process can be expressed with the help of a vector, which denotes the current and last positions of the eagle. The attack vector can be computed through Eq. (4).

$$\overrightarrow{A_i} = \overrightarrow{X_f^*} - \overrightarrow{X_i} \tag{4}$$

where $\overrightarrow{A_i}$ denotes as attack vector (eagle), $\overrightarrow{X_f^*}$ denotes the best prey location visited by f (eagle), and $\overrightarrow{X_i}$ denotes the current location of ith eagle.

4.2 Cruise (Exploration)

To compute the cruise vector, first of all compute the tangent equation of hyperplane, which is expressed by Eq. (5).

$$h_1x_1 + h_2x_2 + \cdots + h_nx_n = d, \sum_{j=1}^{n} h_nx_n = d \tag{5}$$

where $\overrightarrow{H_i} = [h_1, h_2, h_3 \ldots\ldots h_n]$ is denoted as normal vector and $\overrightarrow{X_i} = [x_1, x_2, x_3 \ldots\ldots x_n]$ are denoted by the variables vector. For iteration i the cruise vector of golden eagle *i* is represented by Eq. (6) as follows

$$\sum_{j=1}^{n} a_jx_j = \sum_{j=1}^{n} a_j^* x_j^* \tag{6}$$

where $\overrightarrow{A_i} = [a_1, a_2, a_3 \ldots\ldots a_n]$ is denoted as attack vector, $\overrightarrow{X_i} = [x_1, x_2, x_3 \ldots\ldots x_n]$ are denoted as variables vector, and $X^* = x_1^*, x_2^*, x_3^* \ldots\ldots x_n^*$ are denoted as the locations of the selected prey. For golden eagle, to find a point *C* randomly on the hyperplane, the following stepwise approach is given below.

Step 1 First we select one variable randomly out of n and fixed it.
Step 2 For all the variables assign the random values except the *k*-th variable as this one has been fixed in previous stage.
Step 3 The value of *k*-th variable obtained by Eq. (7) as

$$C_k = \frac{d - \sum_{jj \neq k} a_j}{a_k} \tag{7}$$

where C_k *is* denoted as *k*-th element *C*, a_j is denoted as *j*-th element of the attack vector $\overrightarrow{A_i}$, and *d* is denoted in Eq. (5). Equation (8) depicted the general form *C* (destination point) on the hyperplane.

$$\overrightarrow{C_i} = \left(c_1 = \text{random}, c_2 = \text{random}, \ldots, c_k = \frac{d - \sum_{jj \neq k} a_j}{a_k}, \ldots, c_n = \text{random} \right) \tag{8}$$

The displacement of eagles is the summation of attack and vector: it is denoted by Eq. (9).

$$\Delta x_i = \overrightarrow{r_1} p_a^t \frac{\overrightarrow{A_i}}{\left|\overrightarrow{A_i}\right|} + \overrightarrow{r_2} p_c^t \frac{\overrightarrow{C_i}}{\left|\overrightarrow{C_i}\right|} \tag{9}$$

where p_a^t is denoted as attack coefficient in tth iteration and p_c^t *is* denoted as cruise coefficient in tth iteration. Cruise and attack vectors are computed by using Eq. (10).

$$\left|\overrightarrow{A_i}\right| = \sqrt{\sum_{j=1}^{n} a_j^2}, \left|\overrightarrow{C_i}\right| = \sqrt{\sum_{j=1}^{n} c_j^2} \tag{10}$$

In $(t+1)$th iteration the position of the golden eagles can be computed by Eq. (11)

$$x^{t+1} = x^t + \Delta x_i^t \tag{11}$$

The intermediate values can be computed with the help of linear transition and expressed by Eq. (12).

$$\begin{cases} p_a = p_a^0 + \frac{t}{T}\left|p_a^T - p_a^0\right| \\ p_c = p_c^0 + \frac{t}{T}\left|p_c^T - p_c^0\right| \end{cases} \tag{12}$$

where T denotes maximum iterations, t denotes the current iteration, p_a^0 and p_a^T denoted as the initial and final values of propensity to attack (p_a), respectively. Similarly, p_c^0 and p_c^T are as the initial and final propensity to cruise (p_c), respectively. After mathematical modeling of golden eagle optimization problem, the algorithm is given below:

Algorithm 1: GEO algorithm based solution approach

GEO () Input: Initialization of the golden eagles population
1. Compute fitness function
2. Initialize p_a, p_c and population memory
3. **for** each iteration t
4. **Update**
5. // p_a *and* p_c *through* (Eq. 12)
6. **for** each golden eagle *i*
7. Select a prey randomly
8. Compute attack vector // $\vec{A}$ through (Eq. 4)
9. **if** length of attack vector≠0

(Compute cruise vector $\vec{C}$ *) // Using (Eqs. 5 to 8)*

10. Compute step vector Δx

//using Eqs. 9 to 11

11. **Update** *// positions through (Eq.11)*
12. Compute fitness function

// for new positions generated

13. Replace old positions with new ones

// As per fitness function

14. **end if**
15. **end for**
16. **end for**
17. **Output:** Optimal solution

Table 1 Control parameters for ABC and proposed GEO algorithms

Algorithms	Parameters specifications
Common parameters	Size of initial population = 500, Iteration number = 200, No. of μ = 200, Runs = 10
GEO	Attack propensity = [6.5–2.5], Cruise propensity = [0.0002-0]
ABC	Limit = 3, Onlookers = 50%, Employed = 50%, Scouts = 1
IWO	Smin = 0; Smax = 50; Exponent = 2; sigma_initial = 0.9; sigma_final = 0.001

5 Experimental Results

To verify the experimental study, the dataset of S&P Bombay stock exchange (BSE) have been used on MATLAB with configuration (Intel processor i7 (R) and 16 GB RAM). This dataset carries 30 stocks monthly for the financial year from 1st April 2010 to 31st March 2020. For performance comparison, the results of GEO approach, ABC and IWO have been compared for same objective and environment. The codes for the algorithms GEO, IWO and ABC are available at matlabcentral/fileexchange/84430-golden-eagle-optimizer-toolbox, https://abc.erciyes.edu.tr/, and www.yarpiz.com respectively. The parameter setting of the experiments for the comparative analysis are listed in Table 1 as follows.

The constraints imposed such as fully invested constraints and boundary constraint, are satisfied in the portfolio produced by the proposed GEO, IWO and ABC based solution methods. Here, experiments were conducted for twenty different runs to avoid fluctuations in the results. For both the solution approaches i.e., GEO, ABC and IWO, the best values, of fitness, risk and return have been presented for various μ in Table 2. It is clear from Table 2 and Fig. 1a that the proposed GEO is performing better than ABC and IWO on account of the optimal solutions of the fitness on efficient frontiers produced by 200 values of μ, $0 \leq \mu \leq 1$. Table 2 shows the fitness, risk and return only for equally interfaced 11 values of μ. In Fig. 1b, the execution time behavior of the algorithms namely ABC, IWO and GEO has been shown. These figures represent the performance of GEO, which is much better than ABC and IWO when compared regarding achieved objective value and execution time.

6 Conclusion

In this work, a new metaheuristic algorithm based solution approach namely Golden Eagle Optimizer (GEO), is suggested to solve the portfolio optimization problem. The main objective of using GEO was to find out the optimal fitness value, which is the weighted sum of risk and return. This algorithm initially starts with a random population and mimics the hunting method of golden eagles to achieve the optimum

Table 2 Comparative results of objective values between GEO, ABC and IWO

S. no	μ	Algorithms	Z	Return	Risk
1	0	GEO	−0.02254	0.0225	0.0238
		ABC	−0.02270	0.0227	0.0238
		IWO	−0.02026	0.0203	0.0094
2	0.1005	GEO	−0.01806	0.0217	0.0144
		ABC	−0.01805	0.0221	0.0183
		IWO	−0.01743	0.0205	0.0100
3	0.2010	GEO	−0.01505	0.0203	0.0059
		ABC	−0.01499	0.0205	0.0068
		IWO	−0.01462	0.0200	0.0067
4	0.3015	GEO	−0.01260	0.0198	0.0041
		ABC	−0.01256	0.0198	0.0043
		IWO	−0.01220	0.0191	0.0038
5	0.4020	GEO	−0.01034	0.0193	0.0030
		ABC	−0.01030	0.0189	0.0025
		IWO	−0.00998	0.0186	0.0028
6	0.5025	GEO	−0.00811	0.0188	0.0024
		ABC	−0.00814	0.0189	0.0025
		IWO	−0.00788	0.0184	0.0025
7	0.6030	GEO	−0.00600	0.0185	0.0023
		ABC	−0.00595	0.0185	0.0023
		IWO	−0.00590	0.0185	0.0024
8	0.7035	GEO	−0.00394	0.0184	0.0022
		ABC	−0.00392	0.0183	0.0021
		IWO	−0.00384	0.0179	0.0021
9	0.8040	GEO	−0.00187	0.0180	0.0021
		ABC	−0.00185	0.0176	0.0020
		IWO	−0.00185	0.0177	0.0020
10	0.9045	GEO	0.00003	0.0148	0.0016
		ABC	0.00004	0.0146	0.0016
		IWO	0.00003	0.0153	0.0016
11	1.0000	GEO	0.00124	0.0070	0.0012
		ABC	0.00124	0.0077	0.0012
		IWO	0.00124	0.0073	0.0012

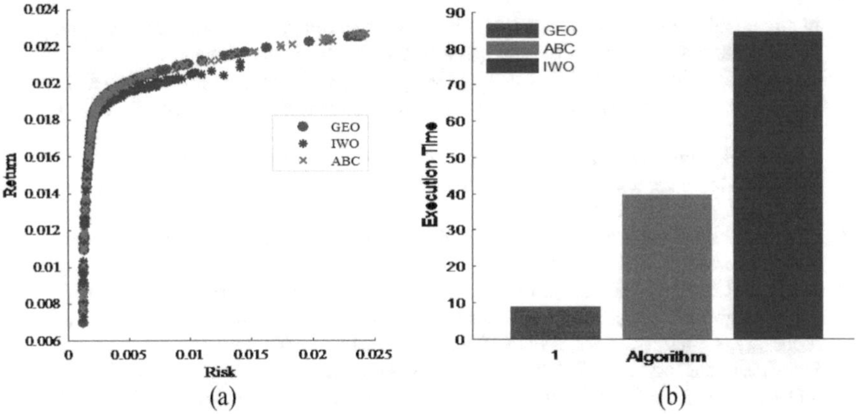

Fig. 1 **a** Efficient frontiers obtained by GEO, ABC and IWO. **b** Execution time of GEO, ABC and IWO

solution. To conduct the performance evaluation, an experimental evaluation has been conducted with a comparative study of proposed GEO based solution results with the results of ABC on S&P BSE dataset (30 stocks). Study shows the better performance of proposed GEO based solutions among ABC and IWO on account of execution time, and obtained optimal solutions on efficient frontiers.

This work can be extended for multi-objective version of GEO to obtain tradeoff solutions on pareto front along with some more complex constraints imposed on to be constructed portfolio to investors for effective decision making.

References

1. Markowitz HM (1952) Portfolio selection. J Financ 7(1):77–91
2. Deng GF, Lin WT, Lo CC (2012) Markowitz-based portfolio selection with cardinality constraints using improved particle swarm optimization. Expert Syst Appl 39(4):4558–4566
3. Tollo DG, Roli A (2008) Metaheuristics for the portfolio selection problem. Int J Opera Res 5(1):13–35
4. Mohammadi-Balani A, Nayeri MD, Azar A, Taghizadeh-Yazdi M (2021) Golden eagle optimizer: a nature-inspired metaheuristic algorithm. Comput Ind Eng 152:107050
5. Chen B, Lin Y, Zeng W, Xu H, Zhang D (2017) The mean-variance cardinality constrained portfolio optimization problem using a local search-based multi-objective evolutionary algorithm. Appl Intell 47(2):505–525
6. Hidayat Y, Lesmana E, Putra AS, Napitupulu H, Supian S (2018) Portfolio optimization by using linear programing models based on genetic algorithm. In: IOP conference series: materials science and engineering 2018. vol 300, no 1. IOP Publishing, pp 012001
7. Jalota H, Thakur M (2018) Genetic algorithm designed for solving portfolio optimization problems subjected to cardinality constraint. Int J Syst Assur Eng Manag 9(1):294–305
8. Shahid M, Ansari MS, Shamim M, Ashraf Z (2022) A stochastic fractal search based approach to solve portfolio selection problem. In: Gunjan VK, Zurada JM (eds) Proceedings of the 2nd

international conference on recent trends in machine learning, IoT, smart cities and applications 2021. Lecture notes in networks and systems, vol 237. Springer, Singapore
9. Erwin K, Engelbrecht A (2020) Improved set-based particle swarm optimization for portfolio optimization. In: IEEE symposium series on computational intelligence (SSCI) 2020. IEEE, pp 1573–1580
10. Zaheer KB, AbdAziz MIB, Kashif AN, Raza SMM (2018) Two stage portfolio selection and optimization model with the hybrid particle swarm optimization. MATEMATIKA: Malaysian J Ind Appl Math 125–141
11. Cura T (2021) A rapidly converging artificial bee colony algorithm for portfolio optimization. Knowl-Based Syst 233:107505
12. Sahala AP, Hertono GF, Handari BD (2020) Implementation of improved quick artificial bee colony algorithm on portfolio optimization problems with constraints. In: AIP conference proceedings 2020, vol 2242, no 1. AIP Publishing LLC, pp 030008
13. Ray J, Bhattacharyya S, Singh NB (2019) Conditional value-at-risk-based portfolio optimization: an ant colony optimization approach. In: Metaheuristic approaches to portfolio optimization 2019. IGI Global, pp 82–108
14. Abolmaali S, Roodposhti FR (2018) Portfolio optimization using ant colony method a case study on Tehran stock exchange. J Account 8(1)
15. Kalayci CB, Polat O, Akbay MA (2020) An efficient hybrid metaheuristic algorithm for cardinality constrained portfolio optimization. Swarm Evolut Comput 54
16. Suthiwong D, Sodanil M, Quirchmayr G (2019) An Improved quick artificial bee colony algorithm for portfolio selection. Int J Comput Intell Appl 18(01):1950007
17. Rezani MA, Hertono GF, Handari BD (2020) Implementation of iterative k-means and ant colony optimization (ACO) in portfolio optimization problem. In: AIP conference proceedings 2020, vol 2242, no 1. AIP Publishing LLC, p 030022
18. Jain M, Singh V, Rani A (2019) A novel nature-inspired algorithm for optimization: squirrel search algorithm. Swarm Evol Comput 44:148–175
19. Shahid M, Ansari MS, Shamim M, Ashraf Z (2022) A risk-budgeted portfolio selection strategy using invasive weed optimization. In: Tiwari R, Mishra A, Yadav N, Pavone M (eds) Proceedings of international conference on computational intelligence 2021. Algorithms for intelligent systems. Springer, Singapore, pp 363–371
20. Shahid M, Ashraf Z, Shamim M, Ansari MS (2022) A novel portfolio selection strategy using gradient-based optimizer. In: Saraswat M, Roy S, Chowdhury C, Gandomi AH (eds) Proceedings of international conference on data science and applications 2021. Lecture notes in networks and systems, vol 287. Springer, Singapore, p 287
21. Sefiane S, Bourouba H (2017) A cuckoo optimization algorithm for solving financial portfolio problem. Int J Bank Risk Insur 5(2):47

Hybrid Moth Search and Dragonfly Algorithm for Energy-Efficient 5G Networks

Shriganesh Yadav, Sameer Nanivadekar, and B. M. Vyas

1 Introduction

Mobile network technologies are still growing in terms of technology. The 5th generation (5G) mobile network was first deployed in the year 2020 [13]. The 5G network satisfies the needs for quality of experience (QoE) and quality of service (QoS) [5]. The network performance is tested with help of network QoS. The good QoS plays a significant role in the comfort of 5G network users. 5G networks are typically segregated as enhanced mobile broadband (eMBB). Massive machine-type communication (mMTC) and ultra-reliable low latency communication (URLLC) [2]. Also, the 5G mobile networks meet the requirements for ever-growing data traffic, produced by the rising count of cellular devices [7]. The main features of 5G networks include higher data rate, low latency, and large bandwidth [1]. The server selection in communication networks depends on the measurement of QoS [11]. In 5G networks device to device communication (D2D) gave the ability to reduce power utilization, improve spectrum efficiency and eventually enhance network capacity [9]. On the other hand, D2D in 5G networks introduced a new technical challenge, which includes mode selection, resource allocation, and power allotment. We propose an optimal power allotment model, which directly guarantees the energy efficiency of 5G networks while assuring the QoS. We propose a hybrid Moth Search and Dragonfly Algorithm that is the combined advantages of both Moth Search Algorithm (MSA) [12] and Dragonfly Algorithm (DA) [3]. Figure 1 shows the 5G network architecture. A Base Station (BS) is the central communication module that connects a huge network. This BS is surrounded by several sub-networks such as pico-net, Femto-net, and

S. Yadav (✉) · B. M. Vyas
Pacific Academy of Higher Education and Research University, Udaipur 313003, India
e-mail: yadavshri11@gmail.com

S. Nanivadekar
A.P. Shah Institute of Technology, Thane 400615, India

P. Singh et al. (eds.), *Machine Learning and Computational Intelligence Techniques for Data Engineering*, Lecture Notes in Electrical Engineering 998,
https://doi.org/10.1007/978-981-99-0047-3_19

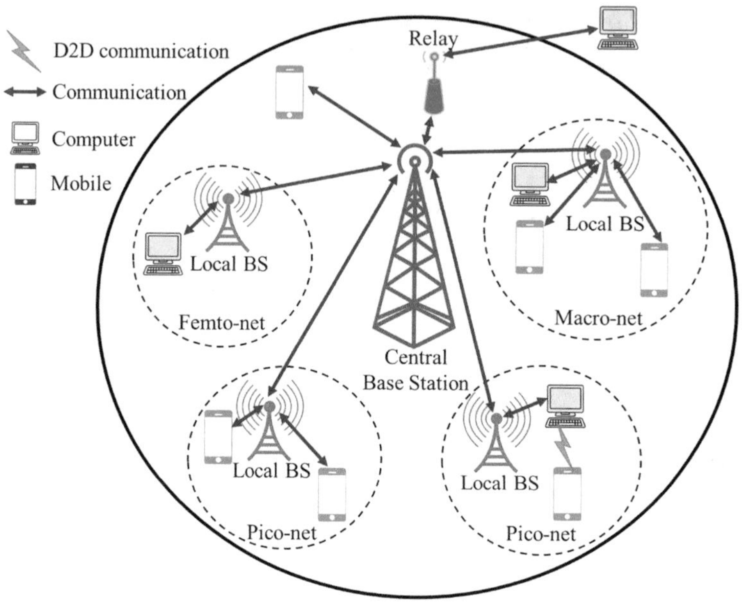

Fig. 1 5G network architecture. A Base Station (BS) is the central communication module that connects a large network. Several sub-networks, such as pico-net, Femto-net, and macro-net, each containing their own local base station are formed around this BS

macro-net. Each sub-network has a local BS. These sub-networks differ in size and the amount of device communication links that their local BS can allocate. Devices in these sub-networks communicate with one another via the local BS. Two devices connect directly with each other in the pico-net on the bottom right side; this is known as Device-to-Device (D2D) communication. This D2D communication then communicates with the local BS. Mobile devices that are not inside the range of these sub-networks can communicate with other devices via the main BS.

2 Literature Review

Khoza et al. [4] propose an algorithm to reduce traffic congestion in ad-hoc vehicular networks. They utilized the hybrid ant colony algorithm, which is a combination of particle swarm and colony optimization algorithms, to select the best route to the vehicles while maintaining the Quality of Service (QoS). Rathore et al. [8] developed an optimization algorithm with whale and grey wolf optimization algorithm. They used this to improve clustering in Wireless Sensor Networks (WSN). They demonstrated an improvement of 67% in delay and an improvement of 55.78% in packet delivery ratio. The consumption of energy was improved by 88.56% and the life of the network increased by 59.81%. Tan et al. [10] enhanced the combination of Genetic Algorithm (GA) and Particle Swarm Optimization (PSO) algorithms. They did this to improve the efficiency of Device-to-Device (D2D) communication. It took 200 generations for the GA to converge and for the GA and PSO algorithm, the system capacity was 2 and 0.6 devices better than the simple particle swarm optimization algorithm. Maddikunta et al. [6] developed a hybrid method that uses the whale and moth-flame optimization algorithm. They aimed to improve the load of the clusters in the network and thus make the network more energy efficient. Nature-inspired algorithms have been frequently used for improving energy efficiency in communication networks. Their low-complexity and efficiency make them a great option to be applied to 5G networks.

3 Methodology

The peak and average transmit power are denoted by and respectively. The packets were separated from the frame at the data-link layer and also at the physical layer into the bit-streams. The channel power gain was set to constant for a given time frame with a predetermined length. The frame time was set to be less than fading coherence duration. The probability density function (PDF) for Nakagami-n channel distribution is indicated by is formulated by $Pr(\gamma) = \frac{\gamma^{n-1}}{\tau(n)}(\frac{n}{\bar{\gamma}})^n exp(-\frac{n}{\bar{\gamma}}\gamma), \gamma \geq 0$, where stand for the Gamma function, correspond to the fading constraint of Nakagami-distribution, point out the instant channel SNR, and $\bar{\gamma}$ point out average SNR at receiver (Fig. 2).

3.1 Delay-Bounded QoS Provisioning

Depending on LDP the queue length process q(t) gets converged in distribution to arbitrary parameter q (∞) as shown in Eq. 1, where q_{th} refers to bound of queue length and Θ refers to QoS exponent.

$$\lim_{q_{\text{th}} \to \infty} \left(\frac{\log \left(Pr \left\{ q\left(\infty\right) > q_{\text{th}} \right\} \right)}{q_{\text{th}}} \right) = \Theta \tag{1}$$

A large Θ symbolizes a speedy rate of decay that indicates a higher QoS necessity, whereas, a small Θ indicates a slower rate of decay, which indicates a lower QoS necessity

Fig. 2 Flowchart for proposed MS-DA algorithm

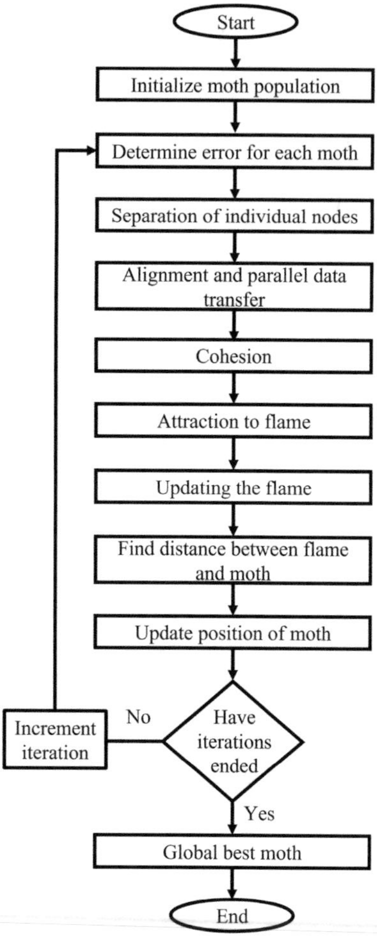

3.2 EPE Under QoS Provisioning

The power utilization $P_o(\Theta)$ is modeled under statistical QoS provisioning given by $P_o(\Theta) = \alpha E_\gamma\{P_t(\Theta, \gamma)\} + P_c = \alpha\overline{p_t}(\Theta) + P_c$, where α refers to mean transmission power utilization coefficient, E_γ refers to the expectation regarding instant CSI γ, $P_t(\Theta, \gamma)$ refers to the instant power allotment related to specified γ and QoS exponent Θ .

Accordingly, alpha ϵ $[1, \Theta)$ refers to the reciprocal of power amplifier efficiency that lies among $(0, 1]$, P_c refers to circuit power utilization.

The EPE, indicated by $\varepsilon\ (\Theta)$ is the attained effectual capacity for each unit power that is computed as by $\varepsilon\ (\Theta) = \Delta\frac{c_\Theta}{p_\Theta}$, wherein, $c_\Theta \& p_\Theta$ refers to effective capacity and power utilization respectively.

By optimal allocation of power, the energy efficiency can be improved, which ensures higher QoS.

3.3 Optimal Power Allocation Via MS-DA Model

Objective Function The major intention of the present research work is to improve the Energy Efficiency of 5G network systems, thereby ensuring higher QoS. The path gain vector of parallel singular valued channels is indicated as $\lambda = (\lambda_1, \lambda_2, \lambda_3, ...\lambda_M)^t$, wherein λ_m $(1 \leq m \leq M)$ correspond to gain of mth channel path and t represent the transpose. Assume $\gamma_m = \lambda_m^2$ $(1 \leq m \leq M)$ as mth channel power gain. The NSI for MIMO is portrayed as $\widetilde{v} = \Delta(\Theta, \lambda)$ & indicated by $P_m(\widetilde{v})$ $(1 \leq m \leq M)$ for m^{th} channel.

The instant rate for MIMO system, indicated by $V_n(\widetilde{v})$, is formulated by $V_n(\widetilde{v}) = L_f C \sum_{m=1}^{M} \log(1 + P_m(\widetilde{v})\gamma_m)$ and $V_n(\widetilde{v}) = L_f C \sum_{m=1}^{M} \log\left(1 + P_m(\widetilde{v})\lambda_m^2\right)$

The objective function of the developed work is shown in Eq. 2. Figure 3 shows the solution encoding of the presented model. The fitness function for the proposed MS-DA optimization algorithm has been defined using the objective function. This algorithm works to reduce this value to the lowest in order to optimize the network.

$$O = \frac{\arg\min}{P(\widetilde{v})} \left\{ E_{\lambda_1} .. E_{\lambda_m} \left[\prod_{m=1}^{M} \left[1 + P_m(\widetilde{v})\lambda_m^2\right]^{-\beta} \right] \right\} \tag{2}$$

Proposed MS-DA Algorithm Although the conventional DA model offers exact estimation; it can only resolve continuous problems. Therefore, to overcome the disadvantages of conventional DA, the theory of MSA is amalgamated with it to initiate a novel model. Hybridized optimizations are found to be capable of certain search issues. DA consists of 2 stages: "(i) Exploration and (ii) Exploitation" which are formulated as follows: The modeling for separation is computed as revealed by $Y_i = -\sum_{i=1}^{U}(S - S_i)$. Here,$S_l$ indicates the lth nearer individual position, signifies the position of the present individual, and reveals the nearby individual's count.

Likewise, the alignment formula is modeled as specified by $R_i = \frac{\sum_{i=1}^{U} Q_l}{U}$, where Q_l symbolizes the velocity of lth nearby individual. Moreover, the cohesion formula is specified by $O_i = \frac{\sum_{i=1}^{U_b} S_l}{U_b} - S$, where S_l symbolizes the position of lth nearer individual. As per the conventional model, if random integer 'ra' is greater than 0.5, the food update occurs as in $F_i = S^+ - S$, else, attraction to food is assigned as zero. However, in the proposed work, if random integer 'ra' is greater than 0.5, the food update takes place as in $F_i = S^+ - S$, where S^+ point out food source position and refer to the present position of the individual. Else if random integer 'ra' is lesser than 0.5, the food update takes place based on the position update of the MSA model by $S(it + 1) = S(it) + \eta l(s)$, in which $l(s)$ refers to levy distribution and η refers to scaling factor.

Distraction to enemy is shown in $E_i = S^- + S$, in which the enemy position is designated by S^-. For updating the dragonfly's position, two vectors such as step (ΔS) and position (S) are computed as specified below.

The step vector is formulated by $\Delta S\,(it+1) = (qY_i + aR_i + cO_i + fF_i + bEn_i) + h\Delta S\,(it)$, where, point out the separation weight, Y_i denote the separation of ith individual, O_i denote the ith individual cohesion, c point out cohesion weight, R_i and F_i signifies the alignment and food resources of ith individual, a refers to the alignment weight, f correspond to food factor, b symbolize enemy factor, h points out the inertia weight, En_i refers to enemy's position of ith individual and it signifies iteration counter. Following the assessment of the step vector, the position is calculated as per $S\,(it+1) = S\,(it) + \Delta S\,(it+1)$, where refer to the present iteration.

To enhance the stochastic performance of dragonflies, it is necessary to flutter in the exploration space in the absence of the nearest solutions. In such conditions, the positions of dragonflies are updated as by $S\,(it+1) = S\,(it) + Levy\,(z) \times S\,(it)$, here z indicates the dimension of the position vectors. The Levy flight is computed as $Levy\,(x) = 0.01 \times \frac{r_1 \times \delta}{|r_2|^{\frac{1}{\eta}}}$, in which η denote a steady value and r_1, r_2 indicate arbitrary integers. Furthermore, δ is given in Eq. 3, wherein $\Gamma\,(x) = (x-1)$.

$$\delta = \left[\frac{\Gamma\,(1+\eta) \times \sin\left(\frac{\pi\eta}{2}\right)}{\Gamma\left(\frac{1+\eta}{2}\right) \times \eta \times 2^{\left(\frac{\eta-1}{2}\right)}}\right]^{\frac{1}{\beta}} \tag{3}$$

Algorithm 1 reveals the pseudo-code of the presented MS-DA model and Fig. 2 shows the flowchart for the algorithm.

```
Data: Initialization
Initialize step values i = 1,2,3,...,n
while The end condition is not satisfied do
    Compute the objective values of all dragonflies
    if ra > 0.5 then
        Update attraction to food as shown in Eq. (12)
    else
        Update attraction to food based on position update of MSA as shown in Eq. (13)
        Update food source and enemy
        Update h, q, a, c, f, and b
        Y, R, O, F, and E as per Eqs. (9)–(14)
        Update the neighboring radius
    end
    if neighbor > 1 then
        Update velocity and position based on Eqs. (15) and (16)
    else
        Update position as per Eq. (17)
    end
end
```

Algorithm 1: MS-DA algorithm

4 Results and Discussions

Figure 3 shows a comparison between different algorithms such as artificial bee colony (ABC), Moth search algorithm (MSA), Dragonfly algorithm (DA), and proposed hybrid moth search and dragonfly algorithm (MS-DA). Where theta is the QoS exponent that is proportional to the length of the queue waiting for service. It was observed that MSA is the fastest convergence. It was observed that with respect to theta the convergence value of cost function remains constant. The MSA has the fastest convergence and all the algorithms except proposed get converged with less than 10 iterations. The proposed MS-DA algorithm does not have the fastest convergence as the moth-flame optimization algorithm has been combined with the dragonfly algorithm. This slows the convergence process of the moths.

Figure 4 shows energy efficiency variation concerning SNR variation. To establish the capability of MIMO 3 combination of antenna were tested which are 1 antenna (Fig. 4a), 2-antenna (Fig. 4b), 3-antenna (Fig. 4c) in all six algorithms were tested under identical conditions. These algorithms are LAG indicated in dark blue, GRAD indicated in red, ABC indicated in orange, DA indicated in violet, MSA indicated in green, and proposed (MSA-DA) indicated in light blue. Figure 4a shows how the energy efficiency varied when we change SNR –5 to –25 in steps of –5. It was observed LAG algorithm outperformed the proposed algorithm in the case of SNR of –10 and –15 dB for all other cases with a single antenna the proposed system is at least 3 times better than all other methods. MSA algorithm has the least efficiency through the comparison the values of GRAD and DA algorithm do not change with SNR, on the other hand, ABC algorithm decreases with SNR. Figure 4b shows energy efficiency with 2 antennae in all cases the proposed algorithm is 3 times better with respect to others methods in comparison. Also, an energy efficiency increase was observed from 8 to 10 as SNR varied from –5 to –25 dB. Figure 4c shows the proposed algorithm has constant efficiency above 14. The GRAD system was the Second best for this 3-antenna configuration. ABC algorithm performance decrease with SNR. DA, MSA, LAG efficiency remains unchanged. We have varied values of θ and computed efficiency in bit per joules as shown in Fig. 4. Efficiency versus θ variation performance with 1 antenna system is shown in Fig. 4d. As the θ decreases –2 to –0.5 MS -DA algorithm efficiency reduces for all the cases DA, ADC, GRAD, and LAG Algorithm the θ doesn't have any effect at all. MSA and MS-DA algorithm talks significantly at a value of –1.5. Figure 4e shows performance with 2-antenna system. As the θ decreases –2 to –0.5 value with DS-MA peak around 1.5. Overall efficiency value compares to all the other algorithms is at least three times more. All the algorithms except LAG and MS-DA are θ invariant. Figure 4f shows performance with 3-antenna system. It is observed that the value of efficiency is independent of θ when the 3-antenna system is used. Only ABC algorithm value decreases θ the DS-MA efficiency is at least 4 times than other algorithms. The overall efficiency with 3-antenna systems is 60% more than 1 and 2 antenna systems.

Figure 5 shows fitness versus θ variation plot for 1, 2, and 3 antenna systems. Figure 5a shows the performance of fitness versus SNR with 1 antenna system.

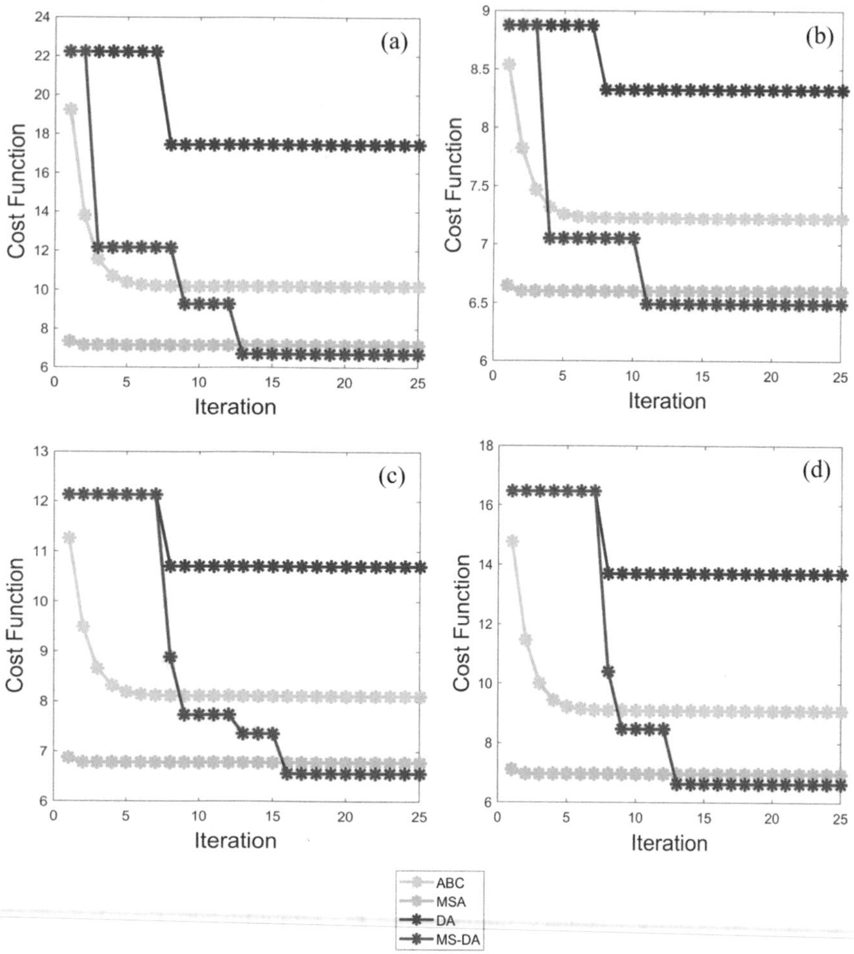

Fig. 3 Cost function convergence plot attended for different values of θ **a** for the value of $\theta = -2$ rad. The MS-DA is the proposed algorithm that took the longest to converge with a minimum cost function of 6 after 13 iterations. **b** for the value of $\theta = -0.5$ rad. The MS-DA algorithm could reach a cost function of 6.5. For the value of $\theta = -1$ rad **c** and –1.5 rad **d** steep slope of cost function was observed from the 6th iteration up till the 13th iteration where the cost function was reduced to a minimum level of 6.8 and which was lowest compared to others

MSA algorithm performance was found out to be at least 3 fold better than the other algorithms. The DS-MA algorithm can achieve a fitness value of 10 for all the SNR values ranging from –25dB to –15dB. Figure 5b shows the performance of fitness function with 2 antenna systems. As the SNR increases from –25 dB to –5dB the overall antenna system fitness function decreases. The maximum fitness value of 85 can be achieved with the MSA algorithm at –25dB. The DS-MA can reach the fitness value above 6, for SNR ranging from –10 dB to –25 dB. Figure 5c depicts

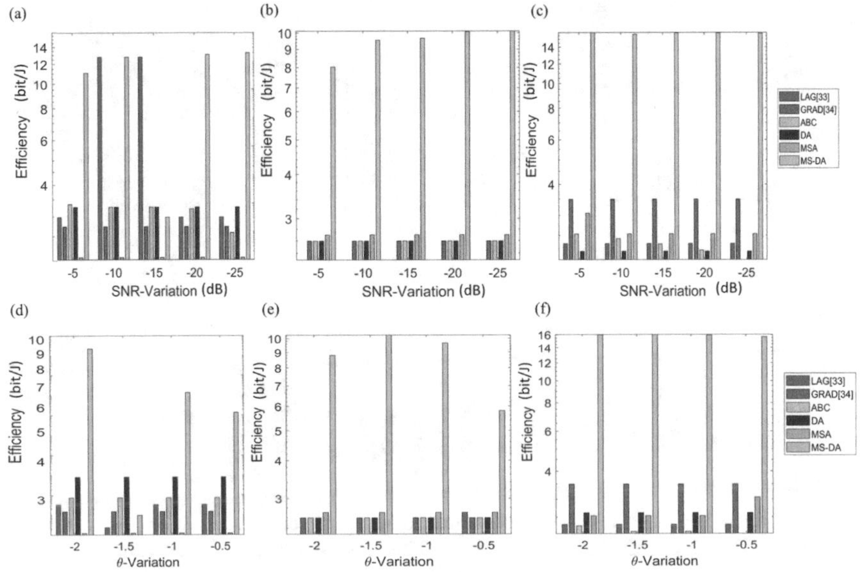

Fig. 4 SNR vs efficiency for different antenna counts. **a** The efficiency of the proposed algorithm was found to be better when SNR was –5, –10, –20, –25. **b** Throughout the SNR range, the proposed algorithm performed at least 3 times better than existing methods with two antenna systems. **c** SNR versus efficiency graph for three antenna count. **d** Efficiency versus θ variation with 1 antenna system. **e** Efficiency versus θ variation with 2-antenna system. **f** Efficiency versus θ variation with 3-antenna system

the performance with 3-antenna system. As the SNR increases from –25 to –5 dB, the overall fitness value decreases from 30 to 8. The GRAD algorithm performs well with 3 antenna systems compared to any other algorithms mentioned. For the DS-MA algorithm, the value is above 7 for all the values of SNR below –10 dB till –25 dB. Figure 5d shows the performance of fitness versus θ with 1 antenna system. As θ decreases to –2° to –0.5° the value of fitness for MS-DA algorithm decreases from 10^8 to 10^3. For all the other algorithms the fitness value is less than 10^2 and doesn't vary with θ variation. Figure 5e shows the performance of fitness with respect to θ for 2 antenna systems. The overall fitness value reduces from 10^8 to 10^3. The DS-MA only shows the fitness of 10 at θ of –0.5°. Figure 5f shows the performance with 3 antenna systems. Here the overall value fitness takes the value of 10^2 for most of the algorithms are inversely proportional to the increasing value to θ (–2 to –0.5). The DS-MA value reached the max value of 8 at θ= –0.5°.

Since our algorithm is a combination of the moth-flame and dragonfly algorithms, it combines the qualities of both to give better results. The bandwidth consumed by the network is as seen in Fig. 6a. The consumption of network bandwidth was the least for the proposed MS-DA algorithm. Figure 6b shows the graph of fitness versus iterations. It can be seen that out of the three algorithms proposed hybrid MS-DA

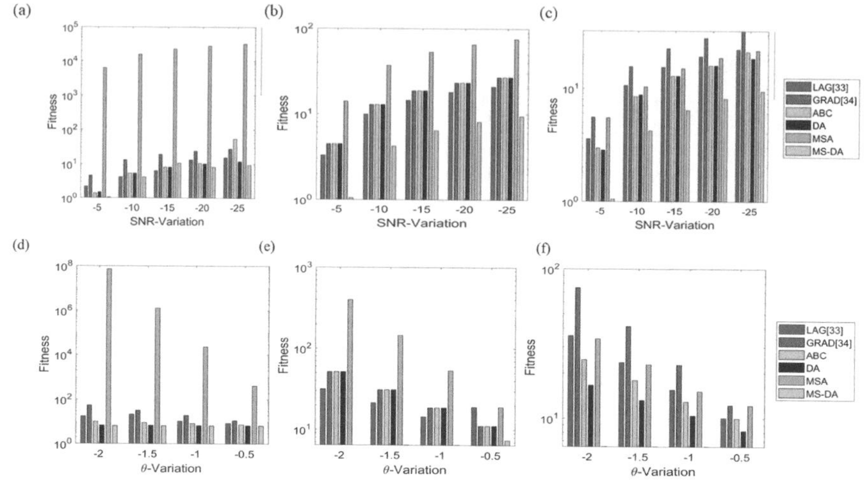

Fig. 5 **a** Fitness versus SNR variation performance with 1 antenna system. **b** Fitness versus SNR variation performance with 2-antenna system. **c** Fitness versus SNR variation performance with 3-antenna system. **d** Fitness versus θ variation with 1 antenna system. **e** Fitness versus θ variation with 2-antenna system. **f** Fitness versus θ variation with 3-antenna systems

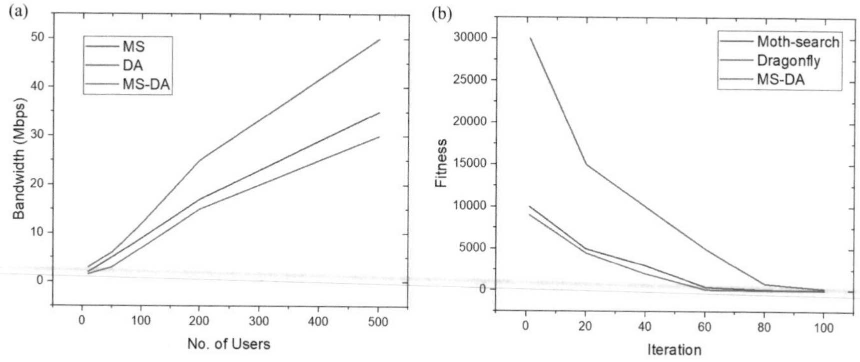

Fig. 6 **a** The consumption of network bandwidth for the MS, DA, and MA-DA algorithms. **b** Fitness versus number of iterations

algorithm showed the best curve for the fitness function. In comparison to the MA and DA algorithms, the MS-DA showed better results.

As a scope for the future, this algorithm can be combined with another type of nature-inspired algorithm to improve the efficiency even further. Also, as this algorithm perform well on 5G networks, in the future it can also be applied in 6G communication networks. Further, in our current work, we have used our technology for the transmission of data and not audio. Because the audio transmission is slower, we are solely concerned with data transmission. However, the proposed algorithm can be evaluated in the future with audio transmission as the primary focus.

5 Conclusions

We have developed an optimization-based power allocation scheme that increased the EPE and ensured QoS over MIMO-oriented 5G networks. In addition, the statistical QoS-driven green power allocation system was examined for increasing the EPE. Predominantly, this research intended to make an optimal power allocation model, for which the MS-DA model was introduced in this work. The advantage of the presented MS-DA scheme was proved over the existing models in terms of varied measures like fitness and efficiency. Particularly, the presented MS-DA model at SNR = –25 dB achieved an efficiency value of around 16 bits/Joul and at least 4 times better than DA, MSA, ABC, LAG, and GRAD models reported in the literature. MS-DA performs well even with a single antenna. The proposed MS-DA method can achieve a fitness of 30 with one antenna only. Also, MS-DA has achieved the least cost function in the longer run proving to be one of the most efficient algorithms for optimized power allocation.

Compliance with Ethical Standards

Conflicts of Interest

Authors S. Yadav, S. Nanivadekar, and B. Vyas declare that they have no conflict of interest.

Involvement of Human Participant and Animals

This article does not contain any studies with animals or Humans performed by any of the authors. All the necessary permissions were obtained from the Institute Ethical Committee and concerned authorities.

Information About Informed Consent

Informed consent was not required as there were no participant

Funding Information

No funding was involved in the present work.

Author Contributions Conceptualization was done by S. Yadav (SY), S. Nani-Vadekar (SN), and B. Vyas (BV). All the simulations were performed by SY. Manuscript writing—original draft preparation SY and SN. Review and editing were SY and SN. Visualization work carried out by SY.

Acknowledgements Authors would like to thank colleagues from Pacific Academy of Higher Education and Research University.

References

1. Abd EL-Latif AA, Abd-El-Atty B, Venegas-Andraca SE, Mazurczyk W (2019) Efficient quantum-based security protocols for information sharing and data protection in 5g networks. Futur Gener Comput Syst 100:893–906
2. Condoluci M, Mahmoodi T (2018) Softwarization and virtualization in 5g mobile networks: Benefits, trends and challenges. Comput Netw 146:65–84
3. Jafari M, Chaleshtari MHB (2017) Using dragonfly algorithm for optimization of orthotropic infinite plates with a quasi-triangular cut-out. Eur J Mech-A/Solids 66:1–14

4. Khoza E, Tu C, Owolawi PA (2020) Decreasing traffic congestion in vanets using an improved hybrid ant colony optimization algorithm. J Commun 15(9):676–686
5. Li W, Wang J, Yang G, Zuo Y, Shao Q, Li S (2018) Energy efficiency maximization oriented resource allocation in 5g ultra-dense network: Centralized and distributed algorithms. Comput Commun 130:10–19
6. Maddikunta PKR, Gadekallu TR, Kaluri R, Srivastava G, Parizi RM, Khan MS (2020) Green communication in iot networks using a hybrid optimization algorithm. Comput Commun 159:97–107
7. Monge MAS, González AH, Fernández BL, Vidal DM, García GR, Vidal JM (2019) Traffic-flow analysis for source-side ddos recognition on 5g environments. J Netw Comput Appl 136:114–131
8. Rathore RS, Sangwan S, Prakash S, Adhikari K, Kharel R, Cao Y (2020) Hybrid wgwo: whale grey wolf optimization-based novel energy-efficient clustering for eh-wsns. EURASIP J Wirel Commun Netw 2020(1):1–28
9. Ricart-Sanchez R, Malagon P, Salva-Garcia P, Perez EC, Wang Q, Calero JMA (2018) Towards an fpga-accelerated programmable data path for edge-to-core communications in 5g networks. J Netw Comput Appl 124:80–93
10. Tan TH, Chen BA, Huang YF (2018) Performance of resource allocation in device-to-device communication systems based on evolutionally optimization algorithms. Appl Sci 8(8):1271
11. Thomas R, Rangachar M (2018) Hybrid optimization based dbn for face recognition using low-resolution images. Multimed Res 1(1):33–43
12. Wang GG (2018) Moth search algorithm: a bio-inspired metaheuristic algorithm for global optimization problems. Memetic Comput 10(2):151–164
13. Yang S, Yin D, Song X, Dong X, Manogaran G, Mastorakis G, Mavromoustakis CX, Batalla JM (2019) Security situation assessment for massive mimo systems for 5g communications. Futur Gener Comput Syst 98:25–34

Automatic Cataract Detection Using Ensemble Model

Ashish Shetty, Rajeshwar Patil, Yogeshwar Patil, Yatharth Kale, and Sanjeev Sharma

1 Introduction

A cataract is a cloudy area in the eye that causes visual loss [1]. Fading colors, hazy or double vision, halos surrounding light, difficulty with bright lights, and difficulty seeing at night are all symptoms of this condition. Cataract can often develop in one or both eyes. Cataracts cause half of all cases of blindness and 33% of visual impairment worldwide [2].

Detection of this disease takes a long time to give back the result of the test. There is a need for a Computer-Aided Diagnosis system that can automate the process. Artificial intelligence, machine learning, and deep learning are the cutting-edge technologies that are used to solve great challenges in the world, encompassing the medical field also.

Deep learning is a subset of machine learning. Deep Learning is based on a neural network with three or more layers. Deep Learning algorithms can determine which features are most important for the prediction and predict according to it. Deep learning models consist of multiple layers of interconnected nodes according

A. Shetty (✉) · R. Patil · Y. Patil · Y. Kale · S. Sharma
Indian Institute of Information Technology, Pune, India
e-mail: ashishshetty@cse.iiitp.ac.in

R. Patil
e-mail: rajeshwarpatil19@cse.iiitp.ac.in

Y. Patil
e-mail: yogeshwarpatil19@cse.iiitp.ac.in

Y. Kale
e-mail: yatharthkale19@cse.iiitp.ac.in

S. Sharma
e-mail: sanjeevsharma@iiitp.ac.in

P. Singh et al. (eds.), *Machine Learning and Computational Intelligence Techniques for Data Engineering*, Lecture Notes in Electrical Engineering 998,
https://doi.org/10.1007/978-981-99-0047-3_20

to the output of previous layers, the weights are adjusted to optimize the prediction or categorization.

It is important to develop a system that is cheaper and easily available for everyone. Deep learning provides fast and automatic solutions to the detection of diseases once trained with different ocular images. Various models have produced accurate results on medical data which shows the fact that these models are capable of learning patterns from medical data and thus capable of producing a promising result in the medical field.

We present an ensemble deep learning model for categorizing ocular images into two categories: cataract and non-cataract. Three independent CNN models are combined using a stack ensemble to get the final model. Multiple evaluation criteria, such as the AUC-ROC curve, precision, and recall, are used to assess the models.

The rest of the paper is organized as follows: Sect. 2 discusses the Literature Survey. Section 3 discusses the Materials and Methods used. Section 4 discusses the experiments and results. Section 5 concludes the paper with future trends.

2 Literature Survey

A learning ensemble approach is presented as a way of improving diagnostic accuracy in this article [16]. The three independent feature sets extracted from each fundus image are wavelet, sketch, and texture-based. For each feature set, two base learning models are constructed, namely, support vector machine and back propagation neural network. Lastly, we examine ensemble methods of majority voting and stacking to combine the multiple base learning models for final fundus image classification.

Yang et al. [17] in this paper, A neural network classifier is proposed to be used for automatic cataract detection by analyzing retinal images. An enhanced Top-bottom hat transformation is proposed in the pre-processing stage to improve the contrast between the foreground and the object, and a trilateral filter is utilized to reduce picture noise. The luminance and texture messages of the image are retrieved as classification features based on the study of the preprocessed image. The classifier is built using a two-layer backpropagation (BP) neural network. Patients' cataracts are classed as normal, moderate, medium, or severe depending on the degree of clarity of the retinal image.

Dong et al. [5] this paper proposes a method for obtaining features from retinal images. Firstly, the maximum entropy method is used to preprocess the fundus images. To identify even more distinct features of fundus images automatically, a deep learning network that is based on Caffe is used. Last, several representative classification algorithms are used to identify automatically extracted features. Compared to features extracted by deep learning and wavelet features extracted from retinal vascular, SVM (support vector machines) and Softmax are used for cataract classification. Finally, cataract images are classified into normal, slight, medium, or severe four-class with an accuracy of 84.7% for SVM classifier and 90.82% for softmax classifier.

Zhang et al. [18] proposed a method based on a deep convolutional neural network which consists of eight layers, first five layers are convolutional layers and the last three layers are fully-connected layers, and output softmax which produces a distribution over four classes, namely, non-cataractous, mild, moderate and severe for this model achieved an accuracy of 86.69%. For cataract detection in which the model classifies images into two classes, cataract and non-cataract models achieved the best accuracy of 93.52%.

In his paper [6], the author proposes an automated method for automatically grading nuclear cataract severity from slit-lamp images. Firstly local filters are acquired through the clustering of images. Then the learned filters are fed to a convolutional neural network which is followed by a recursive neural network that helps in extracting higher-order features. Support vector regression is applied to these features to classify cataract grades.

An automatic computer-aided method is presented in the paper [12] to detect normal, mild, moderate, and severe cataracts from fundus images. Automated cataract classification is performed using the pretrained convolutional neural network (CNN). The AlexNet model is used as a pretrained model. Using the pretrained CNN model, features are extracted and then applied to a support vector machine (SVM) classifier accuracy of the model is 92.91%.

3 Materials and Methods

3.1 Methodology

As shown in Fig. 1 the steps used to detect and classify cataract images using deep learning begins by looking for existing models and the way to implement them. The next step is data collection and applying the needed pre-processing to improve and enhance the images. Different data augmentation techniques are applied. Design a predictive deep learning model and train it on the collected images. The results of the trained model are evaluated and then tested on the testing data (images) to find the performance of the model.

The images from the datasets are resized to 224 × 224. Rescaling is applied to the images by transforming every pixel value from the range [0–255] to [0–1]. Data augmentation is applied to the dataset, namely, horizontal-flip and rotation-range to increase the dataset size and also make the trained model more robust to real-life data (Fig. 1).

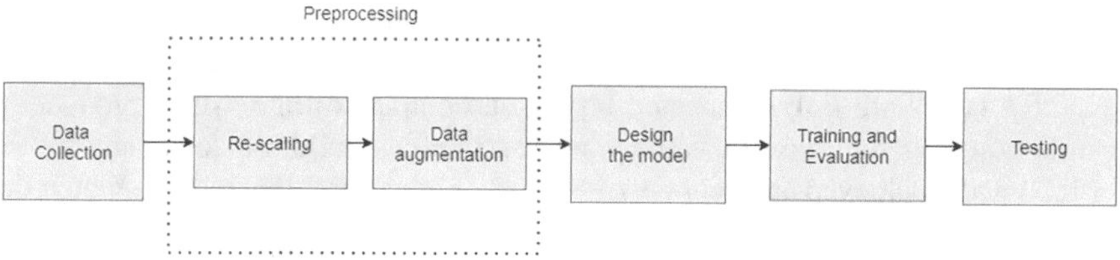

Fig. 1 Flow graph

3.2 Dataset

Ocular Disease Intelligent Recognition (ODIR) [10] is a structured ophthalmic database of 5,000 patients with age, color fundus photographs from left and right eyes, and doctors' diagnostic keywords from doctors. Annotations were labeled by trained human readers with quality control management. They classify the patient into eight labels including Normal (N), Diabetes (D), Glaucoma (G), Cataract (C), Age-related Macular Degeneration (A), Hypertension (H), Pathological Myopia (M), Other diseases/abnormalities (O) (Fig. 2).

Dataset [9] has 100 cataract images and 300 normal images which is present on kaggle (Fig. 3)

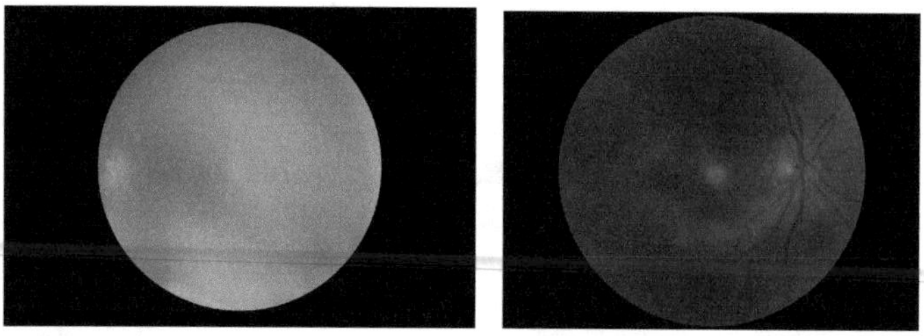

(a) Cataract image **(b)** Non-cataract image

Fig. 2 Images from first dataset [10]

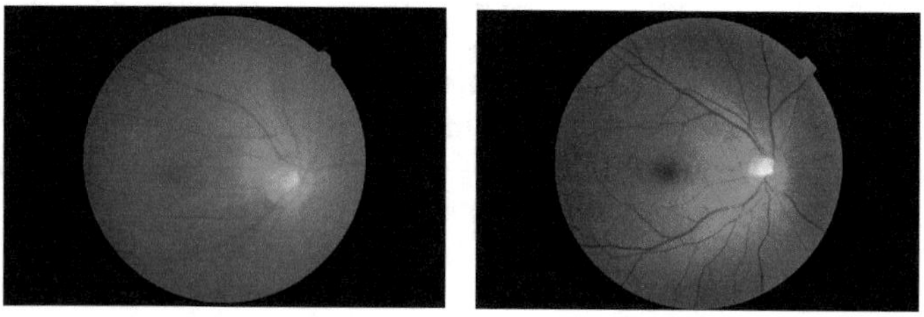

(a) Cataract image **(b)** Non-cataract image

Fig. 3 Images from second dataset [9]

3.3 Proposed Designed

Xception

The model is based on the depthwise separable convolution layers model. Xception achieved 79% top one accuracy and 94.5% top-five accuracy on the ImageNet dataset which has over 15 million labeled high-resolution images belonging to roughly 22,000 categories. The Xception architecture has 36 convolutional layers forming the feature extraction base of the network [4] (Fig. 4).

DenseNet201

This model consists of convolutional neural network that is 201 layers deep. DenseNet201 achieved 77.3% top one accuracy and 93.6% top-five accuracy on the ImageNet dataset which has over 15 million labeled high-resolution images belonging to roughly 22,000 categories. The model has 20,242,984 parameters [8] (Fig. 5).

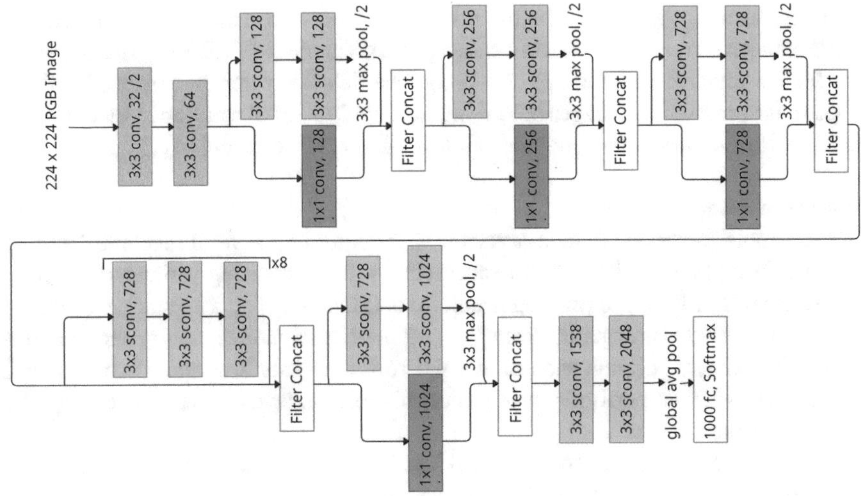

Fig. 4 Xception model architecture [13]

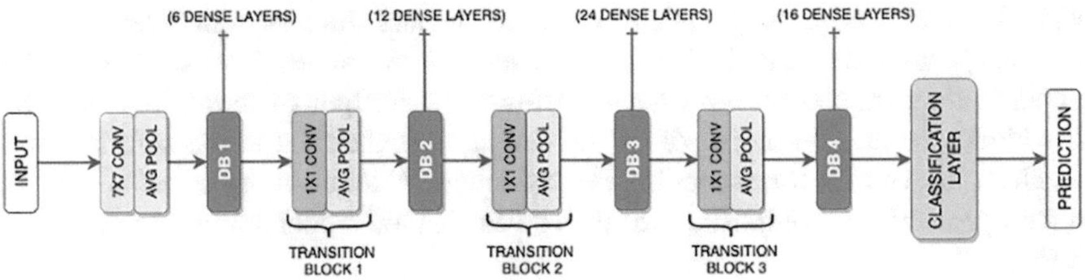

Fig. 5 Densenet model architecture

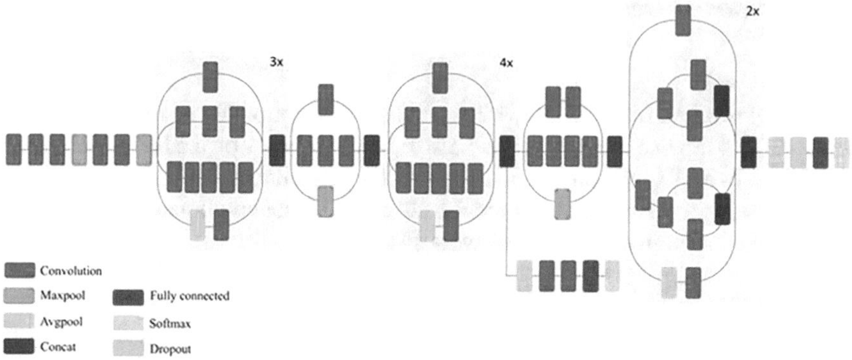

Fig. 6 InceptionV3 model architecture [14]

InceptionV3

It is a convolutional neural network designed to reduce computing costs without reducing accuracy and to make the architecture easy to extend or adapt without sacrificing performance or efficiency. InceptionV3 achieved 77.9% top one accuracy and 93.7% top-five accuracy on the ImageNet dataset which has over 15 million labeled high-resolution images belonging to roughly 22,000 categories.InceptionV3 has 23,851,784 parameters and has 159 layer deep architecture [14] (Fig. 6).

Ensemble model

The ensemble method is a meta-algorithm for combining several machine-learning models. Ensembles can be used for several tasks like decreasing variance(Bagging), bias (Boosting) [3] and improving predictions (Stacking) [7]. The stacking method is used to combine information from several predictive models and generate a new model. Stacking highlights each model where it performs best and discredits each base model where it performs poorly. For this reason, stacking is used to improve the model's prediction.

The ensemble model is built by stacking three convolutional models trained on the cataract dataset. The first model is trained using transfer learning on the Xception model and some custom layers, the second model is trained using transfer learning on the InceptionV3 model and some custom layers, and the last model is trained using transfer learning on the Densenet201 model and some custom layers.

Finally, the three individual models are stacked ensembles. This stack ensemble model is then trained on the cataract dataset. The output of these models is fed to a hidden layer. The output of the hidden layer is then fed to the softmax layer which has two nodes corresponding to the 2 labels. After training, the label with the highest probability is output as a result. The ensembled model architecture is shown in Fig. 7.

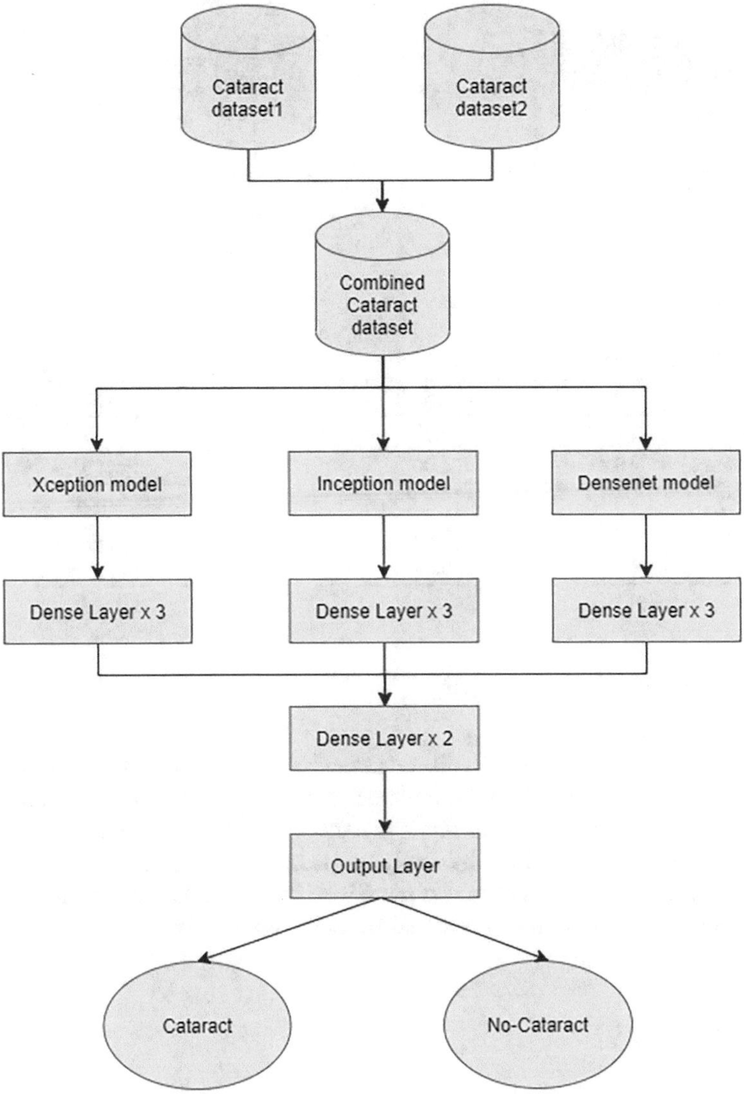

Fig. 7 Ensemble model architecture

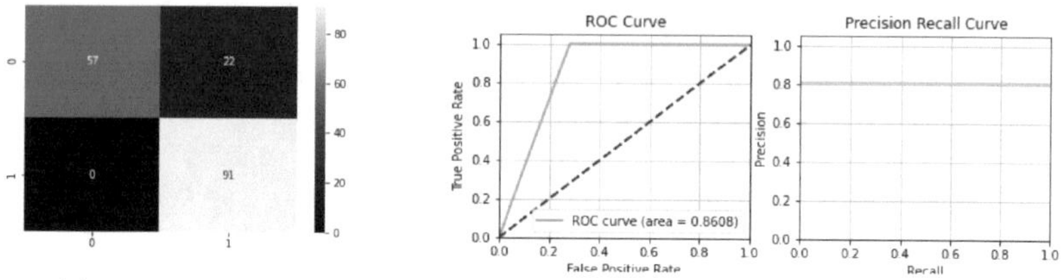

(a) Confusion matrix (b) AUC-ROC curve

Fig. 8 Evaluation metrics for first model

Table 1 Classification report

	Precision	Recall	f1-score	Support
Cataract	1.00	0.72	0.84	79
No_cataract	0.81	1.00	0.89	91
Accuracy			0.87	170
Macro avg	0.90	0.86	0.87	170
Weighted avg	0.90	0.87	0.87	170

4 Experiments and Results

Firstly individual models were built for detecting cataract disease in binary format using pretrained models. Then these models are combined using a stack ensemble. The class with maximum probability is predicted by the final ensemble model.

Three models were trained by applying transfer learning to Xception, InceptionV3, and Densenet201 pretrained models (These models are pretrained on the Imagenet dataset). Custom layers are added to the models for relevant results.

4.1 First model

The top layers of the pretrained Xception model is first removed and the parameters of the model are frozen. Then three dense layers and a softmax layer is added to the network. The entire model is then trained on the cataract dataset. The cataract dataset is built by combining two datasets [9, 10] in order to have more images for training and for robustness so that the model can work well in real-world scenarios. The confusion matrix and AUC-ROC [11] curve for the model is shown in Fig. 8 respectively. The Classification Report is presented in Table 1.

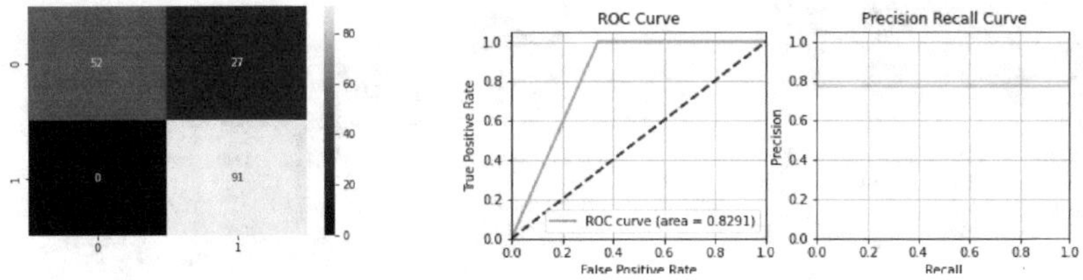

(a) Confusion matrix (b) AUC-ROC curve

Fig. 9 Evaluation metrics for second model

Table 2 Classification report

	Precision	Recall	f1-score	Support
Cataract	1.00	0.66	0.79	79
No_cataract	0.77	1.00	0.87	91
Accuracy			0.84	170
Macro avg	0.89	0.83	0.83	170
Weighted avg	0.88	0.84	0.84	170

4.2 Second Model

The top layers of the pretrained InceptionV3 model are first removed and the parameters of the model are frozen. Then three dense layers and a softmax layer are added to the network. The entire model is then trained on the cataract dataset. The cataract dataset is built by combining two datasets [9, 10] in order to have more images for training and for robustness so that the model can work well in real-world scenarios. The confusion matrix and AUC-ROC [11] curve for the model is shown in Fig. 9. The classification report is presented in Table 2.

4.3 Third Model

The top layers of the pretrained Densenet201 model are first removed and the parameters of the model are frozen. Then three dense layers and a softmax layer are added to the network. The entire model is then trained on the cataract dataset. The cataract dataset is built by combining two datasets [9, 10] in order to have more images for training and for robustness so that the model can work well in real-world scenarios. The confusion matrix and AUC-ROC [11] curve for the model are shown in Fig. 10 respectively. The classification report is presented in Table 3.

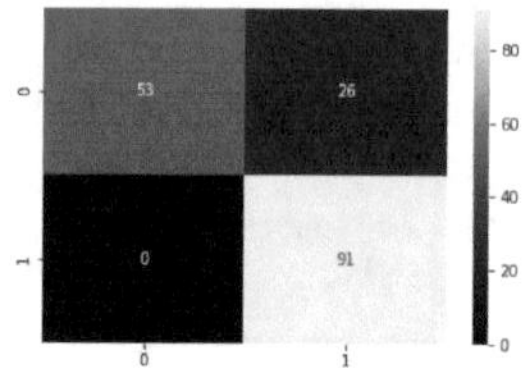

(a) Confusion matrix

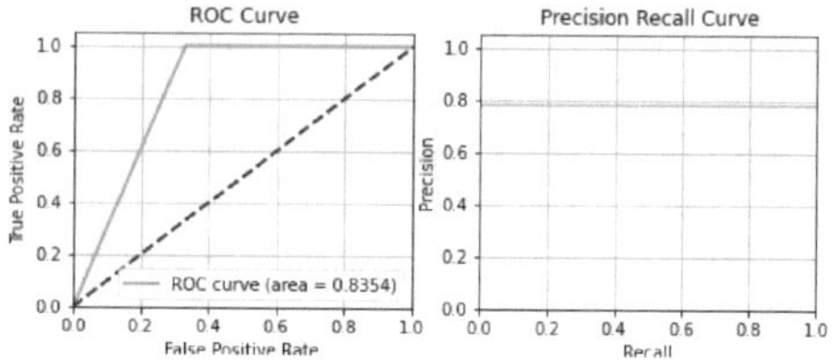

(b) AUC-ROC curve

Fig. 10 Evaluation metrics for third model

Table 3 Classification report

	Precision	Recall	f1-score	Support
Cataract	1.00	0.67	0.80	79
No_cataract	0.78	1.00	0.88	91
Accuracy			0.85	170
Macro avg	0.89	0.84	0.84	170
Weighted avg	0.88	0.85	0.84	170

4.4 *Ensemble Model*

We load the three trained models, remove their last softmax layer, and freeze their weights (i.e., set them as non-trainable). The three models are then stacked to produce a stacked ensemble. Essentially, the output layers of all the models are connected to a hidden layer, which is then connected to a softmax layer, which contains two nodes representing the 2 labels and outputs the label with the highest probability (score).

The stacked ensemble's input is in the form ([Xtrain, Xtrain, Xtrain], Ytrain) and ([Xtest, Xtest, Xtest], Ytest) to compensate input for all models and a single labels array to check the model's performance. This ensemble is trained with a learning rate of 0.0001 and adam optimizer. An accuracy of 91.18% is achieved on the validation data. The confusion matrix, AUC-ROC curve, and classification report of stack ensemble on testing it on validation data are shown in Fig. 11 and Table 4, respectively. Ensemble models are beneficial in that whatever is learned by individual models contributes to the ensemble model, so if a model misses some information and the others pick it up, or vice-versa, it will increase the performance of the whole ensemble model.

Ensemble models are beneficial in that whatever is learned by individual models contributes to the ensemble model, so if a model misses some information and the others pick it up, or vice-versa, it will increase the performance of the whole ensemble model.

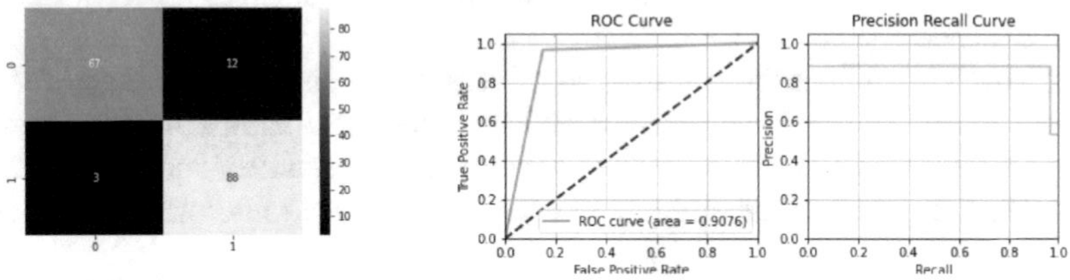

(a) Confusion matrix **(b)** AUC-ROC curve

Fig. 11 Evaluation metrics for ensemble model

Table 4 Classification Report

	Precision	Recall	f1-score	Support
C	0.96	0.85	0.90	79
N	0.88	0.97	0.92	91
Accuracy			0.91	170
Macro avg	0.92	0.91	0.91	170
Weighted avg	0.92	0.91	0.91	170

5 Comparative Study

The previous study has used one dataset for detecting cataract disease. The dataset contains less number of images (from the same source) for some of the classes which can lead to overfitting and result in a less robust model in the real world.

We propose a model which is trained on two datasets belonging to different sources which makes our model more robust. Since the quality of images can differ from different imaging conditions in the real world, by using multiple datasets the model can address this issue. Our model achieves an accuracy of 91.18% for detecting Cataracts in ocular images.

6 Conclusion and Future Scope

The paper presents a deep learning approach for the automatic detection of cataract (an eye disease). Three models were built by applying transfer learning on Xception, InceptionV3, and Densenet201 models. These models were then combined using stack ensembling.

Xception, InceptionV3, and Densenet201 models were used by applying transfer learning to them. These models are trained individually and then combined by using stacked ensembling. Additionally, 2 dense layers are added for improving combined accuracy. The ensembled model is then trained on the combined cataract dataset and achieved an accuracy of 91.18%.

The above model can be extended to classify more diseases. Accuracy for the classification of diseases can be improved by extensive training. More ensemble [15] learning techniques can be used for improving the accuracy and robustness of the model. More images for different classes can be added to normalize the model more.

References

1. Cataracts. https://www.nei.nih.gov/learn-about-eye-health/eye-conditions-and-diseases/cataracts
2. Vision impairment and blindness. https://www.who.int/en/news-room/fact-sheets/detail/blindness-and-visual-impairment
3. Bühlmann P (2012) Bagging, boosting and ensemble methods. In: Handbook of computational statistics. Springer, Berlin, pp 985–1022
4. Chollet F (2017) Xception: deep learning with depthwise separable convolutions. In: Proceedings of the IEEE conference on computer vision and pattern recognition, pp 1251–1258
5. Dong Y, Zhang Q, Qiao Z, Yang, J-J (2017) Classification of cataract fundus image based on deep learning. In: *2017 IEEE international conference on imaging systems and techniques (IST)*. IEEE, pp 1–5
6. Gao X, Lin S, Wong TY (2015) Automatic feature learning to grade nuclear cataracts based on deep learning. IEEE Trans Biomed Eng 62(11):2693–2701 <error l="305" c="Invalid command: paragraph not started." />
7. Güneş F, Wolfinger R, Tan P-Y (2017) Stacked ensemble models for improved prediction accuracy. In: Proc Static Anal Symp, pp 1–19
8. Huang G, Liu Z, Van Der Maaten L, Weinberger KQ (2017) Densely connected convolutional networks. In: Proceedings of the IEEE conference on computer vision and pattern recognition, pp 4700–4708
9. jr2ngb (2019) Cataract dataset. https://www.kaggle.com/jr2ngb/cataractdataset
10. Larxel (2020) Ocular disease recognition. https://www.kaggle.com/andrewmvd/ocular-disease-recognition-odir5k
11. Sarang N (2018) Understanding auc-roc curve. Towards Data Sci 26:220–227 Science 26:220–227
12. Pratap T, Kokil P (2019) Computer-aided diagnosis of cataract using deep transfer learning. Biomed Signal Process Control 53:101533
13. Srinivasan K, Garg L, Datta D, Alaboudi AA, Jhanjhi NZ, Agarwal R, Thomas AG (2021) Performance comparison of deep cnn models for detecting driver's distraction. CMC-Comput Mater Continua 68(3):4109–4124
14. Szegedy C, Liu W, Jia Y, Sermanet P, Reed S, Anguelov D, Erhan D, Vanhoucke V, Rabinovich A (2015) Going deeper with convolutions. In: Proceedings of the IEEE conference on computer vision and pattern recognition, pp 1–9
15. Wolpert DH (1992) Stacked generalization. Neural Netw 5(2):241–259
16. Yang J-J, Li J, Shen R, Zeng Y, He J, Bi J, Li Y, Zhang Q, Peng L, Wang Q (2016) Exploiting ensemble learning for automatic cataract detection and grading. Comput Methods Programs Biomed 124:45–57

17. Yang M, Yang J-J, Zhang Q, Niu Y, Li J (2013) Classification of retinal image for automatic cataract detection. In: 2013 IEEE 15th international conference on e-health networking, applications and services (Healthcom 2013). IEEE, pp 674–679
18. Zhang L, Li J, Han H, Liu B, Yang J, Wang Q et al (2017) Automatic cataract detection and grading using deep convolutional neural network. In: 2017 IEEE 14th international conference on networking, sensing and control (ICNSC). IEEE, pp 60–65

Nepali Voice-Based Gender Classification Using MFCC and GMM

Krishna Dev Adhikari Danuwar, Kushal Badal, Simanta Karki, Sirish Titaju, and Swostika Shrestha

1 Introduction

Natural Language Processing (NLP) refers to the evolving set of computer and AI-based technology that allow computers to learn, understand and produce content in human languages [1]. The technology works closely with speech/voice recognition and text recognition engines. Automatic gender classification in today's time plays a significant role in numerous ways in many domains. Most of these voice detection systems detect voice by reading word sequencing. Majority of these voice detection systems detect voice by reading word sequencing. This research work involves a voice classification based on wave frequency of the voice of different people. This model can automatically identify the gender using Nepali voice. Being a low resource language, the works in NLP involving Nepali language is already limited [2]. This work is also a motivation to researchers working in the field of Nepali NLP.

The Mel frequency cepstral coefficients (MFCCs) of the audio signal are a small set of features that concisely describe the overall shape of a cepstral envelope. MFCCs are commonly used as features in speech recognition systems such as the systems which can automatically recognize numbers spoken into a telephone and also used in music information retrieval applications such as genre classification, audio similarity measures, etc. [3].

A Gaussian mixture model is a probabilistic model that assumes all the data points are generated from a mixture of a finite number of Gaussian distributions with unknown parameters. It is a universally used model for generative unsupervised learning or clustering. It is also called Expectation-Maximization Clustering or EM Clustering and is based on the optimization strategy. Gaussian Mixture models are

K. D. A. Danuwar (✉) · K. Badal · S. Karki · S. Titaju · S. Shrestha
Khwopa Engineering College, Libali-08, Bhaktapur, Nepal
e-mail: 74krishnadev@gmail.com

P. Singh et al. (eds.), *Machine Learning and Computational Intelligence Techniques for Data Engineering*, Lecture Notes in Electrical Engineering 998,
https://doi.org/10.1007/978-981-99-0047-3_21

used for representing Normally Distributed subpopulations within an overall population. The advantage of Mixture models is that they do not require which subpopulation a data point belongs to. It allows the model to learn the subpopulations automatically. This constitutes a form of unsupervised learning [4].

2 Literature Review

A number of research works have been done to identify gender from a voice. So, classification of gender using speech is not a new thing in the field of machine learning. A new system using Bootstrapping on the identification of speech was introduced in which the model detects gender from voice. This system used different machine learning algorithms such as Neural Network, k Nearest Neighbors (KNN), Logistic Regression, Naive Bayes, Decision Trees and SVM Classifiers which shows more than 90% performance [5]. Another system is a combination of neural networks which is content based multimedia indexing segments and piece wise GMM and every segment duration being one second. This showed 90% accuracy for different channels and languages [6]. Likewise in 2019, Gender Classification Through Voice and Performance Analysis by using Machine Learning Algorithms uses different machine learning algorithms such as KNN, SVM, Naïve Bayes, Random Forest, and Decision Tree for gender classification [7].

Another model was found which gave 90% accurate output and had used multimedia indexing of voices channel in 2007 for voice classification [8]. An SVM is used on discriminative weight training to detect gender. This algorithm consists of a finest weighted Mel frequency Cepstral Coefficient (MFCC) which uses Minimum Classification Error as a basis and results in a gender decision rule [9]. Since the decision space is less in problems involving gender classification, SVM models perform well in such problems involving smaller decision spaces [10]. Another SVM introduced in 2008, generates nearly 100% accurate results [11]. A model which used GMM for 2 stage classifiers for better accurate output and less complexity with more than 95% accurate result [12].

Similarly, in 2015, Speaker Identification Using GMM with MFCC was also tested against the specified objectives of the proposed system with an accuracy of 87.5% [13]. In 2019, a research was done to identify gender using Bengali voice which used three different algorithms for their comparative study. In this method, the Gradient Boosting algorithm gave an accuracy of 99.13% and by the Random Forest method, the accuracy was 98.25% likewise by the Logistic Regression method the accuracy was 91.62% [14]. More likely, it seems that the gender classification research is done in different languages. In the Nepali Language, there is no previous research done on Gender Classification by using voice.

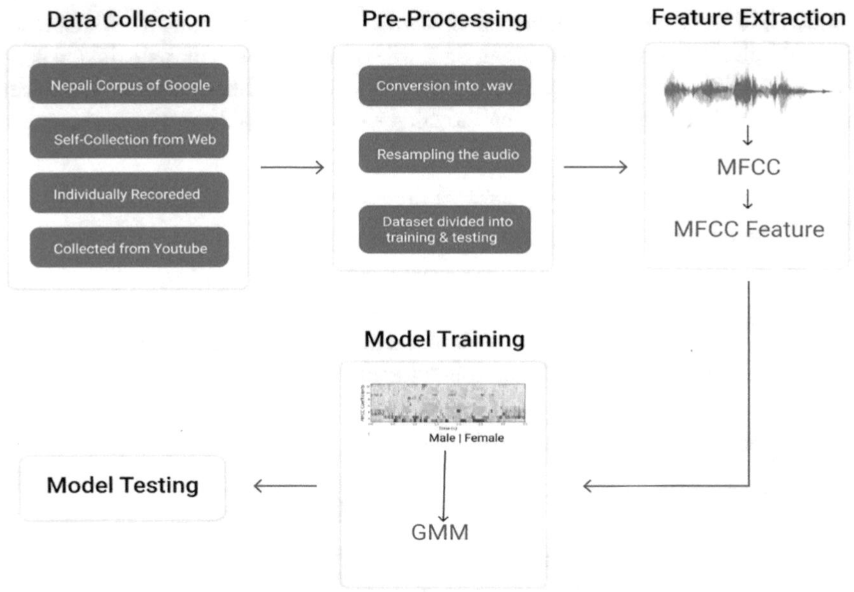

Fig. 1 Pipeline for gender identification

3 Methodology

3.1 Data Collection

Data collection was done considering many methods, we collected different voice samples from a website we developed for data collection, clipping Nepali YouTube videos, recording voice samples from smartphones, and Nepali female voice corpus of Google [15]. Most of the age of those speakers are 15–55. Voice data are edited by the software named WavePad Sound Editor. Most of the collected audio data were in 128kbps. Most of the voice data are 4–10 s long. We collected 10,000 (male voice = 4900 and female voice = 5100) data samples from nearly 500 people. The voice data collected was first divided into female and male categories where the female is 51% of total data and male is 49% of total data (Fig. 1).

3.2 Data Processing

Data preprocessing is the important step that helps us to eliminate noise, split it into appropriate training and test sets and make the dataset ready for the algorithm [16, 17]. Initially, the silence present in the data was trimmed. An audio segment can have 6 consecutive silent frames. Then this processed audio signal was sampled at

16000 Hz. All the collected data were converted to .wav file format. The dataset was divided into training data and testing data where 12.24% of total male data was test set and 11.76% of total female data was test set. Since the dataset is almost balanced, no oversampling or under-sampling methods were performed.

3.3 Feature Extraction

Here, we have used MFCC for feature extraction. It is one of the most effective and popular processes of feature extraction of the human voice (Figs. 2 and 3).

To find MFCCs we followed the following steps as shown in Fig. 4.

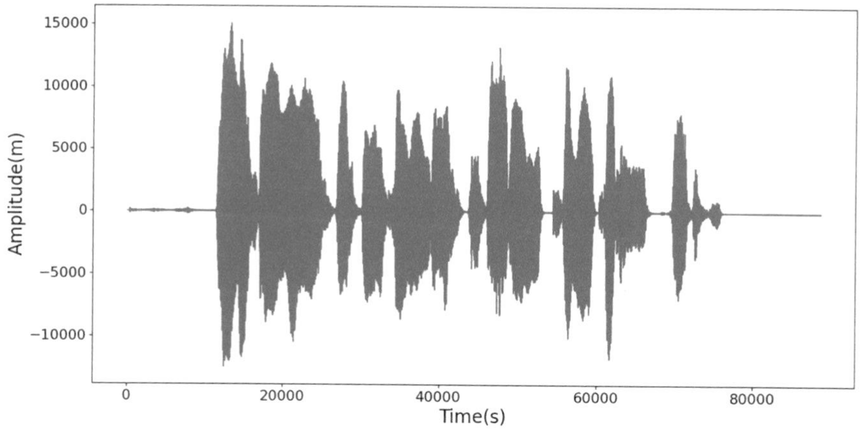

Fig. 2 The waveform of a male voice

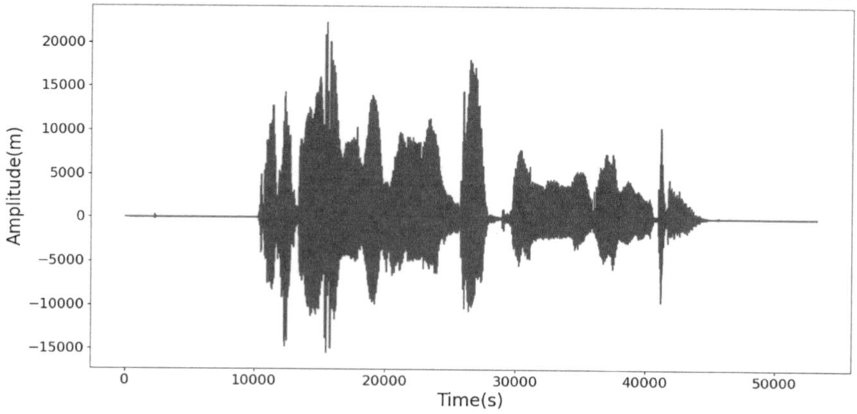

Fig. 3 The waveform of a female voice

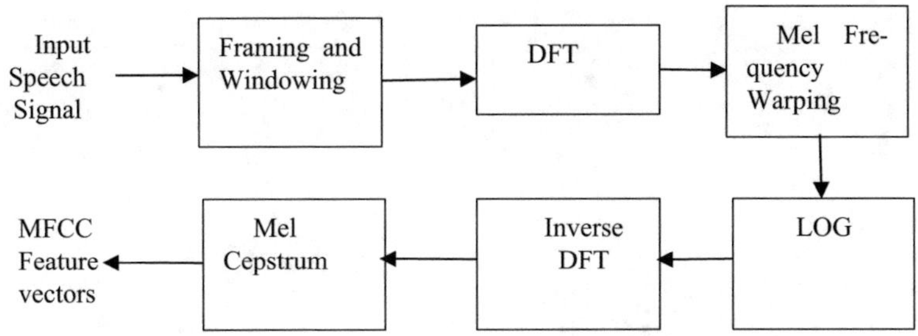

Fig. 4 Steps of feature extraction

To detect and understand the pitches of those voices in a linear manner we have to use mel scale [18]. Frequency scale can be converted to mel scale using following formula:

$$M(f) = 1125\ln(1 + \frac{f}{700}) \tag{1}$$

$$M^{-1}(m) = 700\left(e^{\left(\frac{m}{1125}-1\right)}\right) \tag{2}$$

The steps of MFCC feature extraction are given as follows:

- The standard size for framing audio signal is 25 ms but the range between 20 and 40 ms is considered good.
- This step mainly focuses on DFT of the frames. There are 16 MFCC features generated using GMM models.

$$S_i(K) = \sum_{N}^{n=1} S_i(n)h(n)e^{-i2\pi Kn/N} \tag{3}$$

$$f(x) = \frac{1}{\sigma\sqrt{2\pi}}e^{-\frac{1}{2}(\frac{x-\mu}{\sigma})^2} \tag{4}$$

- This step mainly focuses on Mel spaced filter-bank which is a preliminary collection of 20–40 filters. These filters are used as the periodogram power spectral in preceding step.
- Log filter-bank energies were computed for every energy from preceding step.
- Lastly, cepstral coefficients were calculated by transforming filter-bank energies into discrete cosine [14].

We extracted 16 features of MFCC finally (Figs. 5 and 6).

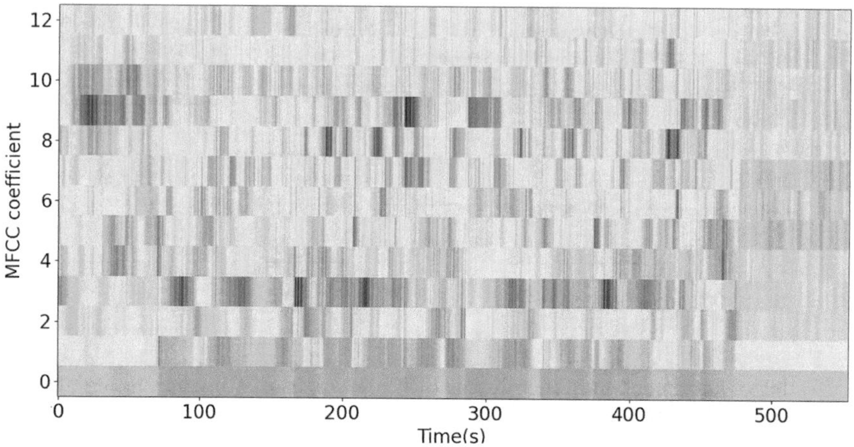

Fig. 5 Male MFCC

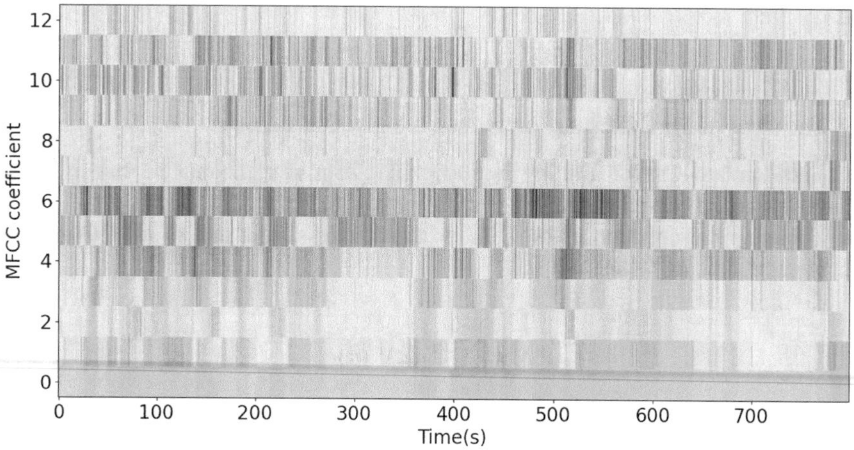

Fig. 6 Female MFCC

3.4 Model Training

We have used Gaussian Mixture Model for the classification of a male and female voice. Gaussian Mixture Model is a probabilistic model and uses the soft clustering approach for distributing the points in different clusters. Gaussian Mixture Model is based upon the Gaussian Distributions (or the Normal Distribution). It has a bell-like curve with data points symmetrically distributed around the mean value. In one dimensional space, the probability density function of Gaussian distribution is given by:

$$f\left(x|\mu,\sigma^2\right) = \frac{1}{\sqrt{2\pi\sigma^2}}e^{-\frac{(x-\mu)^2}{2\sigma^2}} \tag{5}$$

The pdf for 3D space is given as:

$$f\left(x|\mu,\sum\right) = \frac{1}{\sqrt{2\pi|\sum|}}e^{[-\frac{1}{2}(x-\mu)^t\sum^{-1}(x-\mu)]} \tag{6}$$

where x is the input vector, is the 2D mean vector, and is the 2×2 covariance matrix. The covariance matrix will define the shape of the curve. For d-dimension, we can generalize the same. Thus, this multivariate Gaussian model would have x and as vectors of length d and would be a $d * d$ covariance matrix. Hence, for a dataset with d features, we would have a mixture of k Gaussian distributions (where k is equivalent to the number of clusters), each having a certain mean vector and variance matrix. The mean and variance value for each Gaussian is assigned using a technique called Expectation-Maximization (EM) [7, 19]. We have to understand this technique before we dive deeper into the working of the Gaussian Mixture Model. We used 2 different covariance types first one is tied and the second one is diagonal for training our data (Fig. 7).

Covariance type tied means they have the same shape, but the shape may be anything. Covariance type diagonal means the contour axes are oriented along the coordinate axes, but otherwise, the eccentricities may vary between components.

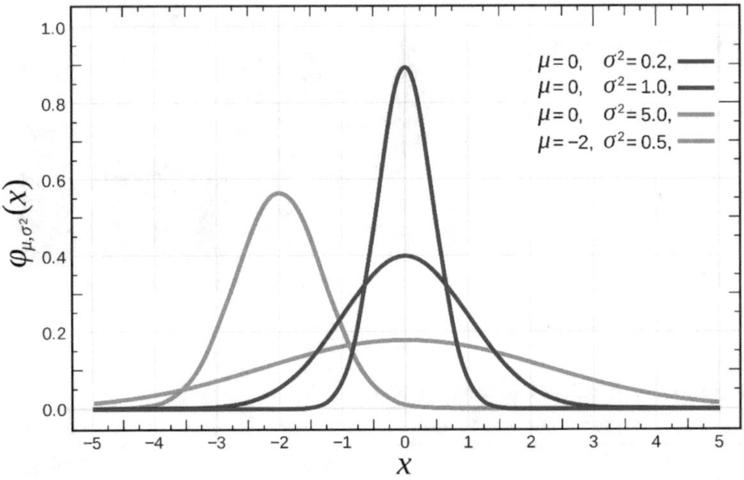

Fig. 7 Gaussian distribution

4 Experiments and Results

First, we split the data into the training set and testing set. We used the Gaussian Mixture model to train our data in two different ways. The first one considers covariance type as tied and another one considers covariance type as diagonal. For analyzing our result we used a confusion matrix.

Confusion matrix of covariance type Tied and Diagonal are plotted (Figs. 8 and 9).

After completing the training of the model, we tested it. We test the model with 600 male and 600 female voices for covariance type Tied and Diagonal. We calculate the performance of both trained models as given (Tables 1 and 2).

Fig. 8 Confusion matrix for covariance type diagonal

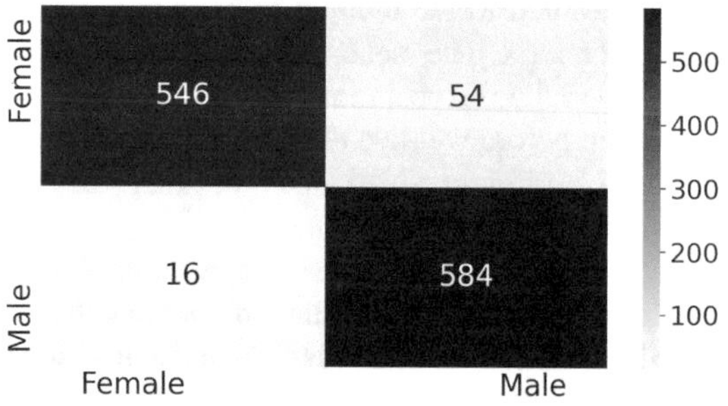

Fig. 9 Confusion matrix for covariance type tied

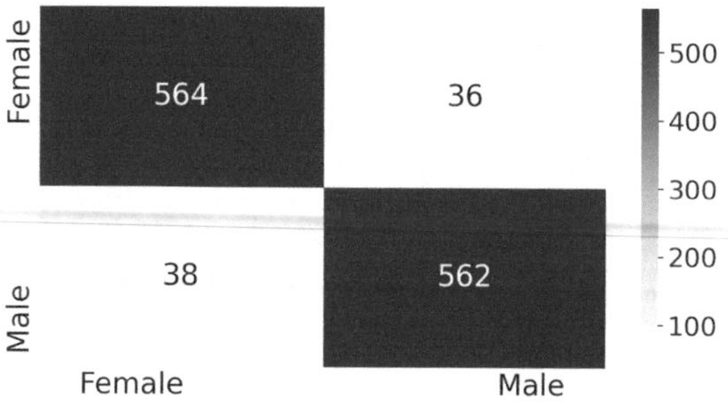

Table 1 Performance with covariance type of tied

	Precision	Recall	F1-score	Support	Accuracy
Male	0.936	0.940	0.938	600	0.938
Female	0.940	0.940	0.940	600	0.938
Micro Avg	0.938	0.938	0.938	600	0.938
Macro Avg	0.938	0.94	0.938	600	0.938
Weighted Avg	0.938	0.94	0.938	600	0.938

Table 2 Performance with covariance type of Diag's

	Precision	Recall	F1-score	Support	Accuracy
Male	0.92	0.97	0.94	600	0.941
Female	0.97	0.91	0.94	600	0.941
Micro Avg	0.94	0.94	0.94	600	0.940
Micro Avg	0.94	0.94	0.94	600	0.940
Weighted Avg	0.94	0.94	0.94	600	0.940

Therefore, GMM with covariance type Diagonal gave better accuracy which is 94.16%. So, it will be the best fit for us.

5 Conclusion

In this paper, we detected the gender based on Nepali voice. Here, we worked with nearly 500 speakers' voices and trained the voice using GMM, and detected the gender. We had 10,000 data samples from which we train 8800 and found the training accuracy of covariance type Tide is 91.65% and the covariance type diagonal is 95.4%. Similarly, the testing accuracy from covariance tied is 93.8% and the covariance type diagonal is 94.16%. We found that the accuracy of the covariance type diagonal is more accurate than that of the covariance type tide. Out of 600 each tested male and female voices, we found that in type tide the error ratio of male and female voices are nearly equal whereas in type diagonal error ratio is different as the female error is too much more than the male error. This paper focuses only on the adult Nepali voice, in near future, we would extend it to children and old-aged people.

References

1. Thapa S, Adhikari S, Naseem U, Singh P, Bharathy G, Prasad M (2020) Detecting Alzheimer's disease by exploiting linguistic information from Nepali transcript. In: International conference on neural information processing. Springer, Cham, pp 176–184
2. Adhikari S, Thapa S, Naseem U, Singh P, Huo H, Bharathy G, Prasad M (2022) Exploiting linguistic information from Nepali transcripts for early detection of Alzheimer's disease using natural language processing and machine learning techniques. Int J Hum Comput Stud 160:102761
3. Badhon SS, Rahaman MH, Rupon FR (2019) A machine learning approach to automating bengali voice based gender classification, pp 55–61
4. Mahboob T, Khanum M (2015) Speaker identification using gmm with mfcc, p 126 (2015)
5. Hasan MR, Jamil M (2004) Speaker identification using mel frequency cepstral coefficients, pp 565–568
6. Reynolds (2009) Gaussian mixture models, pp 659–663
7. Tzanetakis G (2005) Audio-based gender identification using bootstrapping, pp 432–433

8. Chen H (2005) Voice-based gender identification in multimedia applications, pp 179–198
9. Prasad B (2019) Gender classification through voice and performance analysis by using machine learning algorithms, pp 1–11
10. Thapa S, Adhikari S, Ghimire A, Aditya A (2020) Feature selection based twin-support vector machine for the diagnosis of Parkinson's disease. In: 2020 IEEE 8th R10 humanitarian technology conference (R10-HTC). IEEE, pp 1–6
11. Chang S-I (2009) Discriminative weight training-based optimally weighted mfcc for gender identification, pp 1374–1379
12. Lee K-H, Kang S-I (2008) A support vector machine-based gender identification using speech signal, pp 3326–3329
13. Hu Y, Wu D (2012) Pitch-based gender identification with two-stage classification, pp 211–225
14. Sodimana K, Pipatsrisawat K (2018) A step-by-step process for building TTS voices using open source data and framework for Bangla, Javanese, Khmer, Nepali, Sinhala, and Sundanese, pp 66–70
15. Dr. Kavitha R, Nachammai N (2014) Speech based voice recognition system for natural language processing, pp 5301–5305
16. Ghimire A, Jha AK, Thapa S, Mishra S, Jha AM (2021) Machine learning approach based on hybrid features for detection of phishing URLs. In: 2021 11th international conference on cloud computing, data science & engineering (Confluence). IEEE, pp 954–959
17. Thapa S, Singh P, Jain DK, Bharill N, Gupta A, Prasad M (2020) Data-driven approach based on feature selection technique for early diagnosis of Alzheimer's disease. In: 2020 international joint conference on neural networks (IJCNN). IEEE, pp 1–8
18. Singh A, Build better and accurate clusters with Gaussian mixture models. In: Analytics Vidhya. https://www.analyticsvidhya.com/blog/2019/10/gaussian-mixture-models-clustering/
19. Normal Distribution. In: Wikipedia. https://en.wikipedia.org/wiki/Normal_distribution

Analysis of Convolutional Neural Network Architectures for the Classification of Lung and Colon Cancer

Ankit Kumar Titoriya and Maheshwari Prasad Singh

1 Introduction

Cancer causes the highest number of death worldwide. According to the article of the World Health Organization (WHO) [1], the highest number of newly recorded cancer cases was breast cancer (2.26 million cases) in 2020. Lung cancer (2.21 million cases) and colon cancer (1.93 million cases) also occurred most frequently. Most cancer patients died due to lung cancer (1.80 million deaths) in 2020. Following that 0.935 million patients died from colon cancer. Both these cancers are fatal for the worldwide population. Early-stage diagnosis of cancer may be helpful for the patient. For this, microscopic analysis of biopsy or histopathology is the most prominent way for diagnosis. This method requires specialized and experienced pathologists. Physicians use it to check the type and grade of cancer. It is highly costly and time-consuming. It requires in-depth studies like gland segmentation and mitosis detection etc. Still, pathologists analyze the histopathological images themselves for each patient. Therefore, the probability of making a mistake and getting a report late to a patient is high. Currently, some problems occur in the diagnosis of cancer through histopathology. First, the number of pathologists needed in health centers is still not enough worldwide. Second, whatever pathologists are available, those people are not that experienced. Third, histopathology is so complex that the slight carelessness of the pathologist can lead to a wrong decision. That is why artificial intelligence can contribute to the decision-making process without error. The recent

A. K. Titoriya (✉) · M. Prasad Singh
Department of Computer Science and Engineering, National Institute of Technology Patna, Patna, India
e-mail: ankitt.ph21.cs@nitp.ac.in

M. Prasad Singh
e-mail: mps@nitp.ac.in

P. Singh et al. (eds.), *Machine Learning and Computational Intelligence Techniques for Data Engineering*, Lecture Notes in Electrical Engineering 998,
https://doi.org/10.1007/978-981-99-0047-3_22

advancement in artificial intelligence has developed an interest for researchers to create a reliable system for diagnosis.

Deep learning is a part of artificial intelligence, whose architectures are used extensively in image classification. Many medical image classification studies related to deep learning architectures have been experimented so far. Recent research makes one thing clear. For such complex problems, it may be beneficial to build a Computer-Aided Diagnosis (CAD) system that will assist pathologists in diagnosis. The main challenges of this type of system are how to interpret the complexity of histopathology images. Researchers have been doing continuous exploration on the automated imaging of cancer for many years. It is still challenging due to the complexity of cancer images. This paper observes the accuracies of the 19 different existing CNN architectures on the LC25000 dataset namely, (i) SqueezeNet (ii) GoogleNet (iii) ResNet18 (iv) NasNet Mobile (v) ShuffleNet (vi) AlexNet (vii) VGG19 (viii) VGG16 (ix) DarkNet19 (x) MobileNetV2 (xi) DenseNet201 (xii) DarkNet53 (xiii) InceptionV3 (xiv) Xception (xv) NasNet Large (xvi) ResNet101 (xvii) Efficient-Netb0 (xviii) ResNet50 (xix) Inception-ResNetV2. This study also compares the result with the previously published research articles on this dataset.

This paper starts with a literature review survey in Sect. 2. An overview of the proposed work follows in Sect. 3. Section 4 presents the experimental setup. Section 5 discusses the results of the experiments. Section 6 concludes the paper.

2 Related Works

For a long time, researchers have been putting continuous effort into artificial intelligence and medical imaging. In [2], Lee Lusted was the first one to identify the possibilities of computers in medical diagnostics in 1955. In [3], Lodwick et al. computerized chest X-rays for the first time eight years later to build CAD systems and used them to detect lung cancer. Lung cancer detection using chest radiographic images was one of the most explored CAD applications in the 1970s and 1980s. On the other hand, the development of Deep Learning approaches has completely transformed this area. In all sorts of cancer diagnoses, researchers are applying both deep learning and non-deep learning-based learning approach.

In lung and colon cancer, Andrew A. Borkowski et al. publish a dataset of histology images named LC25000 in [1]. For making the dataset, the authors use the facilities of James A. Haley Veterans' Hospital, Tampa, Florida, USA. In [4], Nishio et al. propose a CAD system for classification. In this, the authors use two different datasets (1) the Private Dataset (94 images) and (2) the LC25000 Dataset (25,000 images). This paper also uses classical feature extraction and machine learning algorithms for classification. This work only uses lung cancer images. The study achieves accuracy between 70.83% and 99.33%.

In [5], Das et al. suggest a cancer detection system to detect the cancerous area in medical images. The study uses Brain Tumor Detection, BreCHAD, SNAM, and LC25000 datasets. This paper applies segmentation-based classification for the datasets and achieves 100% accuracy for lung cancer images. The study only uses the lung cancer images from the LC25000 dataset. In [6], Masud et al. propose a classification method. This paper uses two-dimensional Discrete Fourier transform, single-level discrete two-dimensional wavelet transform, and unsharp masking for feature extraction. For classification, the paper uses a CNN and achieves an accuracy of 96.33%. In [7], Wang et al. publish a python package that uses CNN and SVM for classification. The study uses four histopathology datasets and achieves an accuracy of 94%. In [8], Toğaçar proposes an approach to classify with a feature extraction method. The study uses Manta-Ray Forging and Equilibrium optimization algorithms to optimize the feature extracted through Darknet-19 Architecture. The paper achieves an accuracy of 99.69% on the LC25000 dataset. In [9], Ali et al. use a multi-input dual-stream capsule network with and without preprocessing. The study achieves an accuracy of 99.33%. The study employs transfer learning on the dataset.

In [10], Garg et al. use CNN features and SVM to classify. The paper uses this approach on only colon cancer images. This study works only on binary classification. In [11], Phankokkruad employs ensemble learning using three CNN architectures. The study uses VGG16, ResNet50V2, and DenseNet 201 for the ensemble model. The final proposed model achieves an accuracy of 91%. In [12], Lin et al. propose a Pyramidal Deep-Board Learning method to improve the classification accuracy of a CNN architecture. The study uses ShuffLeNetV2, EfficientNetb0 and ResNet50 architectures. This method achieves the highest accuracy of 96.489% with ResNet50. In [13], Mohalder et al. use classical feature extraction methods like LBP and Hog filter. The study employs various classifiers like XGBoost, Random Forest, K-Nearest Neighbor, Decision Trees, Linear Discriminant Analysis, Support Vector Machine, and Logistic Regression. The study achieves an accuracy of 99% through XGBoost. From this survey, we have learned that very few experiments have been done on LC25000 so far. Therefore, we have decided to implement feature extraction methods with CNN.

In [14], Fan et al. propose a transfer learning model for image classification. The study also uses SVM based classification model and Softmax based classification model to classify the lung and colon cancer dataset. Transfer learning modal achieves an accuracy of 99.44%. In [15], Adu et al. propose a dual horizontal squash capsule network for classification. The study modifies the traditional CapsNet. This method achieves an accuracy of 99.23%.

3 Proposed Work

The proposed work uses the feature extraction method using various pre-trained CNN models. Figure 1 shows the working of the proposed methodology. This method has these steps namely, (i) Image acquisition and preprocessing (ii) Feature extraction (iii) Classification.

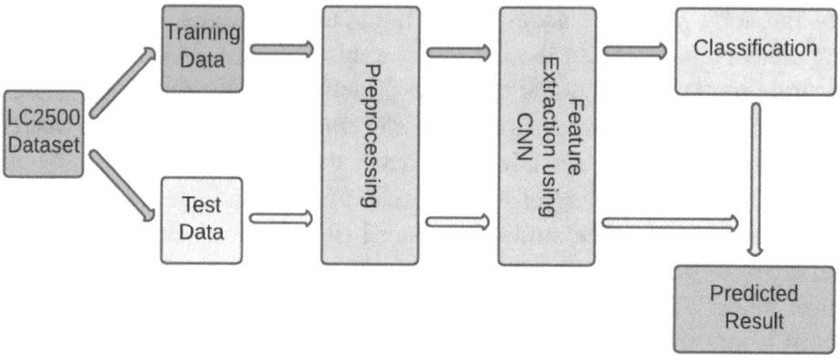

Fig. 1 Work flow of generalized model for feature extraction and classification

3.1 Image Acquisition and Preprocessing

The proposed work takes raw images and then preprocesses them according to the CNN network. In general, medical labs use various methods for the acquisition part. Similarly, in histopathology, pathologists capture all the pictures with the help of a camera. After this, labs send these images to the doctors with their assessments. These images may be treated as raw images. The labs assemble several raw images and pathologists' assessments to form a dataset. The LCP25000 and BrakeHis are good examples of it that are publically available to researchers. In these datasets, all the images usually have the same size or are in some coded form. CNN and all classical feature extraction methods use images either in PNG or in JPEG format. In which we can know the value of each pixel. Therefore, the approach first converts the images from coded form to PNG or JPEG format.

The proposed work uses only CNN for feature extraction that works better than classical feature extraction methods. This paper uses different input sizes (227 × 227, 224 × 224, 256 × 256, 299 × 299, 331 × 331) of images for different CNN architectures. Therefore, there is a need to resize the images according to all the architectures. There are many other methods for preprocessing like augmentation, translation, and filters. The paper uses image resizing. The proposed work divides the dataset into two parts for supervised learning. This paper uses the dataset in the ratio of 70:30 for training and testing respectively.

3.2 Feature Extraction

CNN is a neural network architecture that processes multi-dimensional data such as images and time-series data. It learns the features from the input data. Based on it, existing CNN architectures classify the data and update the weights under supervised learning. It is called CNN because of the convolution operator, which solves the most complex problems. It consists of many pooling and convolutional layers and extracts features in the network at each layer. There are many filters in the convolutional layer, which perform convolution operations on the images. The pooling layer mainly reduces the dimensionality of the previous layer. There are two ways to use CNN. These ways are feature extraction and transfer learning.

In feature extraction, the proposed work passes the images through the CNN and extracts the features from the last layer. After this, any classifier algorithm such as Support Vector Machine (SVM) can classify the data. In transfer learning, the proposed work uses the complete architecture to train it on training data. The network also updates the weight and biases by using validation data. Research reveals that transfer learning works better than the feature extraction method on a large dataset. However, for a small and skewed dataset, feature extraction works better. This work uses feature extraction due to the skewness of LCP2500.

3.3 Classification

For classification, the proposed work uses multi-class SVM on the outcome of feature extraction. The objective of multi-class SVM is to predict a hyperplane in n-dimensional space that classifies the data points to their classes. This hyperplane must be at a maximum distance from all data points. The data points with the shortest distance to the hyperplane are support vectors. This study also predicted the test accuracy using the test data feature shown in Fig. 1. This work uses a one-versus-one (OVO) approach. Due to the limitations of SVM, it uses n(n-1)/2 binary SVM to classify n classes. OVO breaks down the multi-class problem into multiple binary classification problems. Figure 2 shows the example of multi-class SVM for three classes.

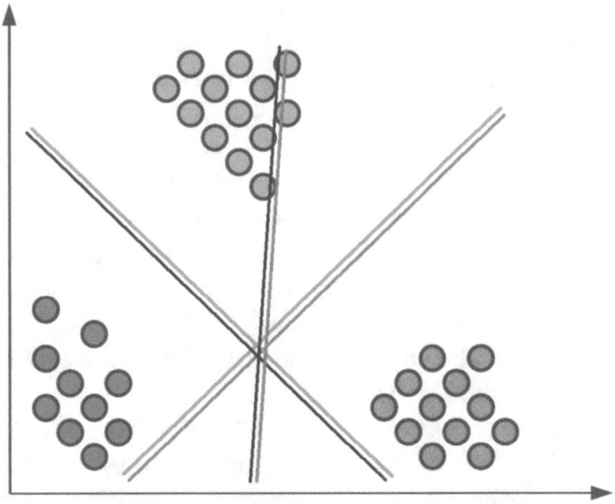

Fig. 2 Example of multi-class SVM

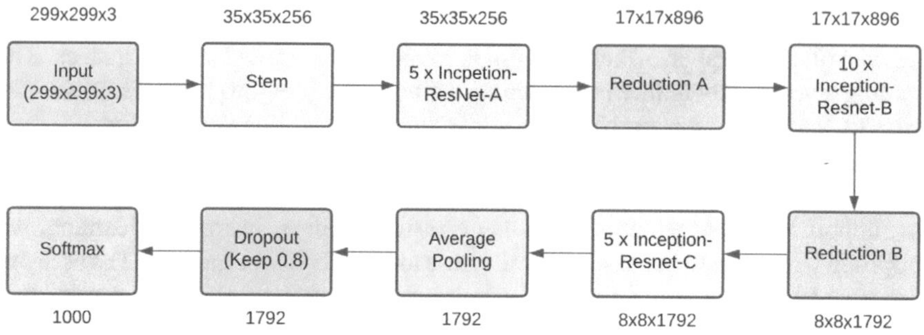

Fig. 3 Block diagram of Inception-ResNet V2

3.4 Inception-ResNet V2

Inception-ResNet V2 is a hybrid inception version of CNN with improved recognition performance. It can train more than a million images from a dataset with the help of 164 layers and 55.9 million parameters. The network's input size is 299 × 299. It is a combination of the Inception model and Residual connections. This network learns very accurate feature representations for classification. Due to the residual connection, the network has fewer degradation problems. Figure 3 shows the architecture of Inception-Resnet V2.

4 Experimental Setup

The proposed work uses the LCP25000 dataset, which contains 25,000 images of size 768 × 768 pixels. The dataset has five classes with 5000 JPEG images for each. Further, authors [16] augment the 1250 images to create a total no. of 25,000 images. The authors [16] employ various augmentation techniques for images. The original size of the 1250 images is 1024 × 768. This dataset is Health Insurance Portability and Accountability Act (HIPAA) compliant and validated. Colon_image_sets and lung_image_sets are two subfolders of the lung-colon image set. The colon image sets subdirectory has two secondary nested folders namely (i) Colon_aca (Colon Adenocarcinomas) (ii) Colon_n (Colon Benign). The lung_image_sets subfolder has three secondary subfolders namely (i) lung_aca(lung adenocarcinomas) (ii) lung_scc (lung squamous cell carcinomas) (iii) lung_n (lung benign). Each class contains 5000 images. Table 1 shows the distribution of the dataset. Figure 4 shows the sample images of the LCP25000 Dataset.

During preprocessing every CNN requires a different input size as per the requirement of different CNN architectures. The proposed work uses MATLAB's augmentedImageDatastore library to resize the training and testing data. This work does not apply any other data augmentation technique.

Table 1 LCP25000 dataset structure

Classes	No. of images
Colon Adenocarcinomas	5000
Colon Benign	5000
Lung Adenocarcinomas	5000
Lung Squamous Cell Carcinomas	5000
Lung Benign	5000
Total	25,000

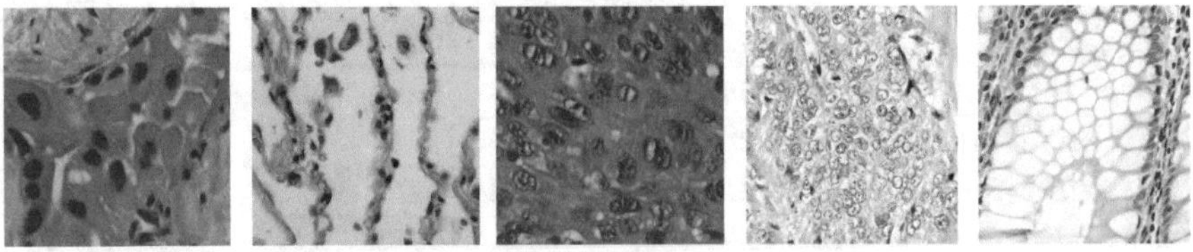

Lung Adenocarcinomas | Benign lung tissues | Lung squamous cell carcinomas | Colon Adenocarcinomas | Benign colonic tissues

Fig. 4 Sample images of LCP25000 dataset

During feature extraction and classification, the proposed work uses the dataset for five classes namely. (I) Lung Adenocarcinomas (II) Benign lung tissues (III) Lung squamous cell carcinomas (IV) Colon Adenocarcinomas (V) Benign colonic tissues. The proposed work extracts the features through CNN using MATLAB's Deep Learning toolbox. For classification, this study applies multi-class classification using the fitcecoc library of MATLAB's Statistics and Machine Learning Toolbox.

5 Experimented Results

This study determines the results based on the five forms of cancer. Table 2 shows the accuracies of 19 different CNN networks on the LC25000 dataset. This experiment uses fivefold validation to avoid overfitting. The approach experiments five times with random images in training and testing data. As can be seen in Table 2, Inception-Resnet V2 performs well in comparison to other networks on this dataset. It achieves an accuracy of 99% in Fold 5, which is the highest among all experiments. Figure 5 shows the confusion matrix of it.

Table 3 shows the performance measure of the confusion matrix of Inception-ResNet V2 as per the class. This study calculates TPR (Sensitivity or True Positive Rate), TNR (Specificity or True Negative Rate), PPV (Precision or Positive Predictive Value), NPV (Negative Predictive Value), FPR (False Positive Rate), FDR (False Discovery Rate), FNR (Miss Rate or False Negative Rate), ACC (Accuracy), F1

Table 2 Accuracies of various networks implemented on LC25000

Network	Input Size	Fold 1	Fold 2	Fold 3	Fold 4	Fold 5	Avg
SqueezeNet [17]	227 × 227	97.54	97.05	97.69	97.44	97.11	97.366
GoogleNet [18]	224 × 224	97.4	97.32	97.12	97.67	97.51	97.404
ResNet18 [19]	224 × 224	98.25	98	98.04	98.01	97.99	98.058
NasNet Mobile [20]	224 × 224	98.14	98.27	98.43	98.05	98.51	98.28
ShuffleNet [21]	224 × 224	98.17	98.6	98.31	98.65	98.45	98.436
AlexNet [22]	227 × 227	98.56	98.44	98.45	98.61	98.51	98.514
VGG19 [23]	224 × 224	98.93	98.51	98.61	99.12	98.71	98.776
VGG16 [23]	224 × 224	98.8	98.76	95.69	98.76	98.89	98.78
Darknet19	256 × 256	98.8	98.93	98.95	98.71	98.73	98.824
MobileNetV2 [24]	224 × 224	99.12	98.16	99.24	99.24	99.2	99.192
DenseNet201 [25]	224 × 224	99.4	99.43	99.39	99.37	99.28	99.374
DarkNet53	256 × 256	99.51	99.41	99.43	99.6	99.4	99.47
InceptionV3 [26]	299 × 299	99.6	99.37	99.51	99.43	99.49	99.48
Xception [27]	299 × 299	99.44	99.48	99.61	99.56	99.43	99.504
NasNet Large [20]	331 × 331	99.6	99.67	99.55	99.43	99.59	99.568
ResNet101 [19]	224 × 224	99.63	99.65	99.61	99.55	99.56	99.6
EfficientNetb0 [28]	224 × 224	99.57	99.77	99.57	99.56	99.57	99.608
ResNet50 [19]	224 × 224	99.72	99.67	99.67	99.57	99.63	99.652
Inception-ResNetV2 [29]	299 × 299	99.72	99.71	99.65	99.75	99.79	99.724

(F1-score) and MCC (Matthews Correlation Coefficient) of the confusion matrix. Table 4 shows the comparison study with the previously published article.

6 Conclusion

Cancer is one of the deadly diseases, which is increasing continuously. The mortality rate is decreasing due to this disease. Lung and colon cancer is also the most common cancer in the whole world. According to doctors, the survival of the patient in these two cancers becomes impossible without proper diagnosis. This work presents an approach to detecting lung and colon cancer using artificial intelligence. The proposed work applies the same method with all the other existing CNN architectures. This experiment concludes. The networks that work well with ImageNet may not necessarily do equally well with other datasets. Inception-Resnet V2 has the highest accuracy achieved by the feature extraction method on the LC25000 dataset. In the future, this study may be beneficial for the researchers working on this dataset.

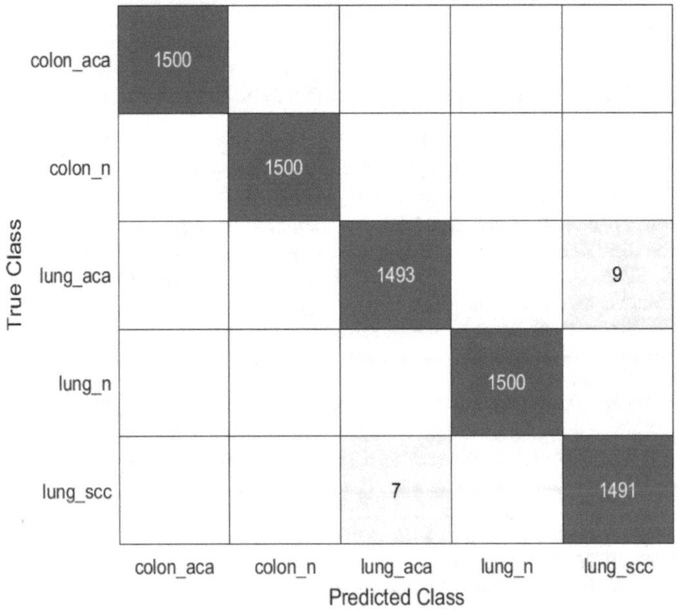

Fig. 5 Confusion matrix of Inception-ResNet V2

Table 3 Measures for the confusion matrix of Inception-ResNet V2 as per the class

Class	TPR	TNR	PPV	NPV	FPR	FDR	FNR	ACC	F1	MCC
colon_aca	1	1	1	1	0	0	0	1	1	1
colon_n	1	1	1	1	0	0	0	1	1	1
lung_aca	0.994	0.998	0.995	0.998	0.001	0.004	0.006	0.997	0.994	0.993
lung_n	1	1	1	1	0	0	0	1	1	1
Lung_scc	0.995	0.998	0.994	0.998	0.001	0.006	0.004	0.997	0.994	0.993

Table 4 Study comparison on same dataset

Article	Accuracy (%)
Nishio et al. [4]	99.43
Masud et al. [6]	96.33
Wang et al. [7]	94
Toğaçar [8]	99.69
Ali et al. [9]	99.33
Phankokkruad et al. [11]	91
Lin et al. [12]	96.489
Mohalder et al. [13]	99
Fan et al. [14]	99.44
Adu et al. [15]	99.23
Proposed work	99.72

References

1. Cancer (2021) World Health Organization, WHO, 21 Sept. www.who.int/news-room/fact-sheets/detail/cancer
2. Lusted LB (1955) Medical electronics. N Engl J Med 252(14):580–585
3. Lodwick GS, Keats TE, Dorst JP (1963) The coding of roentgen images for computer analysis as applied to lung cancer. Radiology 81(2):185–200
4. Nishio M, Nishio M, Jimbo N, Nakane K (2021) Homology-based image processing for automatic classification of histopathological images of lung tissue. Cancers 13(6):1192
5. Das UK, Sikder J, Salma U, Anwar AS (2021) Intelligent cancer detection system. In: 2021 international conference on intelligent technologies (CONIT). IEEE, pp 1–6
6. Masud M, Sikder N, Nahid AA, Bairagi AK, AlZain MA (2021) A machine learning approach to diagnosing lung and colon cancer using a deep learning-based classification framework. Sensors 21(3):748
7. Wang Y, Yang L, Webb GI, Ge Z, Song J (2021) OCTID: a one-class learning-based Python package for tumor image detection. Bioinformatics 37(21):3986–3988
8. Toğaçar M (2021) Disease type detection in lung and colon cancer images using the complement approach of inefficient sets. Comput Biol Med 137:104827
9. Ali M, Ali R (2021) Multi-input dual-stream capsule network for improved lung and colon cancer classification. Diagnostics 11(8):1485
10. Garg S, Garg S (2020) Prediction of lung and colon cancer through analysis of histopathological images by utilizing pre-trained CNN models with visualization of class activation and saliency maps. In: 2020 3rd artificial intelligence and cloud computing conference, pp 38–45
11. Phankokkruad M (2021) Ensemble transfer learning for lung cancer detection. In: 2021 4th international conference on data science and information technology, pp 438–442
12. Lin J, Han G, Pan X, Chen H, Li D, Jia X, Han C (2021) PDBL: improving histopathological tissue classification with plug-and-play pyramidal deep-broad learning. arXiv preprint. arXiv:2111.03063
13. Mohalder RD, Talukder KH (2021) Deep learning based colorectal cancer (CRC) tumors prediction. In: 2021 12th international conference on computing communication and networking technologies (ICCCNT). IEEE, pp 01–06
14. Fan J, Lee J, Lee Y (2021) A transfer learning architecture based on a support vector machine for histopathology image classification. Appl Sci 11(14):6380
15. Adu K, Yu Y, Cai J, Owusu-Agyemang K, Twumasi BA, Wang X (2021) DHS-CapsNet: dual horizontal squash capsule networks for lung and colon cancer classification from whole slide histopathological images. Int J Imag Syst Technol 31(4):2075–2092
16. Borkowski AA, Bui MM, Thomas LB, Wilson CP, DeLand LA, Mastorides SM (2019) Lung and colon cancer histopathological image dataset (lc25000). arXiv preprint arXiv:1912.12142
17. Iandola FN, Han S, Moskewicz MW, Ashraf K, Dally WJ, Keutzer K (2016) SqueezeNet: AlexNet-level accuracy with 50x fewer parameters and< 0.5 MB model size. arXiv preprint arXiv:1602.07360
18. Szegedy C, Liu W, Jia Y, Sermanet P, Reed S, Anguelov D, Rabinovich A (2015) Going deeper with convolutions. In Proceedings of the IEEE conference on computer vision and pattern recognition, pp 1–9
19. He K, Zhang X, Ren S, Sun J (2016) Deep residual learning for image recognition. In: Proceedings of the IEEE conference on computer vision and pattern recognition, pp 770–778
20. Zoph B, Vasudevan V, Shlens J, Le QV (2018) Learning transferable architectures for scalable image recognition. In: Proceedings of the IEEE conference on computer vision and pattern recognition, pp 8697–8710
21. Zhang X, Zhou X, Lin M, Sun J (2018) Shufflenet: an extremely efficient convolutional neural network for mobile devices. In: Proceedings of the IEEE conference on computer vision and pattern recognition, pp 6848–6856
22. Krizhevsky A, Sutskever I, Hinton GE (2012) Imagenet classification with deep convolutional neural networks. In: Advances in neural information processing systems, 25

23. Simonyan K, Zisserman A (2014) Very deep convolutional networks for large-scale image recognition. arXiv preprint arXiv:1409.1556
24. Sandler M, Howard A, Zhu M, Zhmoginov A, Chen LC (2018) Mobilenetv2: Inverted residuals and linear bottlenecks. In: Proceedings of the IEEE conference on computer vision and pattern recognition, pp 4510–4520
25. Huang G, Liu Z, Van Der Maaten L, Weinberger KQ (2017) Densely connected convolutional networks. In: Proceedings of the IEEE conference on computer vision and pattern recognition, pp 4700–4708
26. Szegedy C, Vanhoucke V, Ioffe S, Shlens J, Wojna Z (2016) Rethinking the inception architecture for computer vision. In: Proceedings of the IEEE conference on computer vision and pattern recognition, pp 2818–2826
27. Chollet F (2017) Xception: deep learning with depthwise separable convolutions. In: Proceedings of the IEEE conference on computer vision and pattern recognition, pp 1251–1258
28. Tan M, Le Q (2019) Efficientnet: rethinking model scaling for convolutional neural networks. In International conference on machine learning. PMLR, pp 6105–6114
29. Szegedy C, Ioffe S, Vanhoucke V, Alemi AA (2017) Inception-v4, inception-resnet and the impact of residual connections on learning. In: Thirty-first AAAI conference on artificial intelligence

Wireless String: Machine Learning-Based Estimation of Distance Between Two Bluetooth Devices

Mritunjay Saha, Hibu Talyang, and Ningrinla Marchang

1 Introduction

Wireless technology has spawned the stupendous growth in the usage of mobile devices. Moreover, applications that leverage the all-important mobility features are on the rise. Several such applications make use of the location of the device with respect to some reference location. One important application is object tracking. For instance, one could connect her mobile phone and smartwatch when she leaves her home. Then, in the event of the distance between the mobile phone and the watch is more than some pre-defined threshold (say 2 m), an alarm may be set off. This will help in the prevention of misplacing or losing one's mobile phone.

Several positioning systems exist. One such is the well-known worldwide satellite-based Global Positioning System (GPS). However, GPS is generally considered to be unsuitable for indoor positioning due to the reason that a GPS receiver usually requires line-of-sight visibility of the satellite.

Research has been on for positioning using Bluetooth [1] technology. Bluetooth is a standard that is designed to provide low power, short-range wireless connection between mobile devices such as mobile phones, tablets, laptops, cars, display monitors, etc. Its radio coverage ranges from 10 m to 100 m depending on its type. It can be used to connect a master device with up to 7 slave devices in a network called a piconet. The two most attractive features of Bluetooth behind its success are its low power requirement and the ability for Bluetooth devices to automatically connect with each other when they come within radio range of each other [2].

Out of several methods for location positioning using Bluetooth, RSSI-based methods are common [2–9]. These methods use RSSI for estimating the distance

M. Saha · H. Talyang · N. Marchang (✉)
North Eastern Regional Institute of Science and Technology, Nirjuli, Itanagar 791109, Arunachal Pradesh, India
e-mail: nm@nerist.ac.in

P. Singh et al. (eds.), *Machine Learning and Computational Intelligence Techniques for Data Engineering*, Lecture Notes in Electrical Engineering 998,
https://doi.org/10.1007/978-981-99-0047-3_23

between two Bluetooth devices. However, estimation accuracy depends on the ability to measure the received power level precisely [8].

Motivated by the above research findings, we present a study that attempts to estimate the distance between two Bluetooth devices with the help of machine learning, viz., regression. First, we introduce a hybrid method in which GPS coordinates are obtained and transmitted from the slave to the master device. The master device then estimates the distance between them using the coordinates. Normally, methods such as Haversine formula are used to calculate the distance between two GPS coordinates. However, sufficient accuracy can be achieved only if the coordinates are miles apart. Hence, we adopt another method, viz., regression for distance estimation. We generate a dataset of the coordinates and the actual distance measurements between the two devices (explained in detail in later sections). Then, we apply several regression algorithms for estimating the distance between any two positions. Second, we apply the same regression algorithms on an existing dataset [11] which consists of RSSI values w.r.t. two Bluetooth devices and the actual distance between them from IEEE dataport. The summary of the contributions is as follows:

1. We propose a hybrid positioning system based on both GPS and Bluetooth and employ it to generate a dataset.
2. We explore and confirm the viability of using Machine Learning, viz., regression for estimating the distance between two Bluetooth devices with the help of the generated dataset.
3. We also present a comparison of regression results between the generated dataset and an existing dataset [11].

The rest of the paper is organized as follows. Section 2 represents related works, which is followed by the development of the ML (Machine Learning)-based approach in Sect. 3. The simulation results are analyzed in Sect. 4 and conclusions and future directions are presented in Sect. 5.

2 Related Works

Some research efforts have been expended on location positioning or distance estimation w.r.t. Bluetooth devices. Two relevant recent applications developed to fight the COVID-19 pandemic are Aarogya Setu [12] and 1point5 [13]. The Aarogya Setu app which was developed by the National Informatics Centre (NIC) under the Government of India tracks an individual's interaction with a COVID-19 positive suspect with the help of a social graph. Bluetooth is used to monitor the the proximity of a mobile device to another mobile device. It alerts even if a person with a mobile device unknowingly comes near the device of someone who has tested positive. 1point5 [13] from the United Nations Technology Innovation Labs scans nearby mobile devices and alerts by vibrating a person's device when another device enters a perimeter of 1.5 m around the device. It uses Bluetooth RSSI signals and allows the user to adjust the distance between 1.5 m and 2.5 m.

A distance estimation scheme based on RSSI is presented in [3]. The RSSI values are filtered first using a median filter for removing outliers. Then, the processed values are converted to distance values using a function. Finally, noise reduction is performed using a Kalman filter. A Euclidean distance correction algorithm is proposed in [4] for indoor positioning system. Preliminary work on statistically estimating the distance between two devices from time series of RSSI readings is presented in [5]. However, only the RSSI distribution has been presented. A patent on a method and apparatus for measuring the distance between two Bluetooth devices is given in [6]. First, a Bluetooth device transmits distance measurement radio waves to another device. Then, based on the intensity of the distance measurement radio waves received as a reply from the slave device, the distance between the devices is calculated. In a similar work, the master device sends a signal to the slave device [2]. On receiving the return signal from the slave device, the master calculates the distance between them by determining the delay between the first signal and the return signal.

An indoor position service for Bluetooth Ad hoc Networks is given in [7], in which a model that describes the relationship between RSSI and distance between the two Bluetooth devices is given, Similarly, positioning based on RSSI values which are used to estimate distance according to a simple propagation model is given in [8]. In [9], triangulation methods are used along with RSSI values for positioning. However, the downside of using triangulation is the requirement of fixed reference points. The general assumption in all these works is that distance is the only factor that affects signal strength. However, in reality, RSSI values can be affected by a wide variety of factors such as attenuation, obstruction, etc. Moreover, the accuracy of the distance estimation depends on the precision with which devices are able to measure the power level.

It can be thus concluded that distance estimation between two Bluetooth devices is not a trivial task. Therefore, this study takes a novel approach in that it learns from past experiences to estimate the distance between two Bluetooth devices.

3 Distance Estimation Between Bluetooth Devices as a Regression Problem

This section describes the use of machine learning (viz., regression) for estimating the distance between two Bluetooth devices. First, we discuss how the dataset is built. Second, we discuss seven regression algorithms, viz., Linear Regression (LR), Ridge, LASSO (Least Absolute Shrinkage and Selection Operator), Elastic Net, Decision Tree (DT), Random Forest (RF), and K-Nearest Neighbors (KNN) which we apply on the created dataset and an existing dataset. The proposed approach is best described by the diagram in Fig. 1.

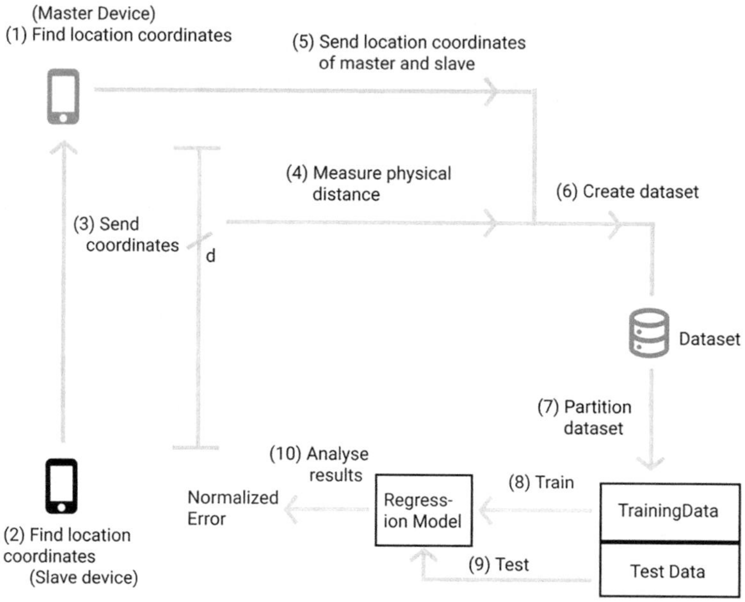

Fig. 1 Distance estimation between bluetooth devices using regression

3.1 Generating the Dataset

The Bluetooth devices are connected to transfer data between them. In this case, we transfer the location coordinates of the devices with the help of Bluetooth. An android application has been developed using Java, which uses location coordinates of Google Maps [10]. Initially, a connection is set up between the master and the slave devices. Then, the slave sends its coordinates to the master. The master receives the coordinates of slave. At the same time, it also gets its own coordinates. Moreover, the measurement accuracy is also logged.

Thus, an example (or a data point) in the dataset is denoted by a vector $x = (l_m^a, l_m^o, a_m, l_s^a, l_s^o, a_s, d)$, where l_m^a, l_m^o and a_m denote the latitude, longitude, and accuracy measurements of the master device respectively. Similarly, l_s^a, l_s^o, and a_s denote the latitude, longitude, and accuracy measurements of the slave device respectively. Moreover, *d* denotes the actual physical distance between the devices in meters (measured physically). physically.

The datasets are subdivided into six types based on the six scenarios under which the experiments were conducted: Hand-to-Hand (HH), Hand-to-Pocket (HP), Hand-to-Backpack (HB), Backpack-to-Backpack (BB), Pocket-to-Backpack (PB), and Pocket-to-Pocket (PP). These scenarios are chosen so that they are the same as the ones used in the existing dataset [11] which we use for comparison. For instance, the scenario, HH means that both the mobile devices are held in the hand. The

GPS coordinates are fetched at an interval of 5 secs. Moreover, experiments were conducted indoors as well as outdoors.

3.2 Regression

Regression falls under the ambit of supervised machine learning. It is a special case of classification. Classification algorithms learn the features of examples in the input dataset, which are already labeled or classified into classes. Then, they use this learning to classify new examples. In regression, the classes (or labels) are not categorical values but continuous values (known as target values). In other words, regression algorithms learn from the examples given to them and use this learning to predict (or estimate) the target value of a new example. The training of the regression model, which may be computationally intensive can be done offline. The representative (both linear and non-linear) regression algorithms that are used are described in the following subsections.

Linear Regression (LR) Linear Regression learns up a linear function that captures the relationship between the dependent variable (target value) and the independent variables (input features). It is known to be the simplest form of regression. The dependent variable is continuous in nature. This algorithm works well when there is a linear relation between independent and dependent variables and the examples are independent.

Least Absolute Shrinkage and Selection Operator (LASSO) LASSO regression is a variant of LR suitable for data that exhibit high multicollinearity (high correlation of features with each other). It makes use of L1 regularization technique in the objective function. Regularization is used to overcome the over-fitting problem. Over-fitting is the phenomenon of a model performing well on training data while performing poorly on test data (new examples).

Ridge Ridge regression is similar to LASSO regression in that it is well-suited for data with high multicollinearity. However, the difference is that Ridge uses L2 regularization whereas LASSO uses L1 regularization. Ridge regression is known to more computationally efficient than LASSO regression.

Elastic Net Regression Elastic Net regression is a hybridization of Ridge and LASSO in that it uses a combination of both L1 and L2 regularization. Like both Ridge and LASSO, it is assumes data with high multicollinearity.

Decision Tree (DT) Decision tree uses a tree structure for building the regression model. It subdivides the examples in the dataset into subsets based on the values of the input features. This process is repeated for the subsets in a recursive manner such that an associated decision tree is ultimately created. The final result is a tree with decision nodes and leaf nodes. Based on the values of the input features of a new example, the tree is traversed and the target value was generated.

Random Forest (RF) The DT algorithm is known to cause over-fitting. To overcome this, output of multiple Decision Trees is merged to generate the final output in the Random Forest algorithm. The decision trees are randomly generated. Hence, the name 'Random' Forest. It is known to be computationally efficient. It can be considered as an example of an ensemble method in that simpler models are used to build a more complicated model while exploiting the advantage of the simpler models.

K-Nearest Neighbors (KNN) KNN regression involves first determining the K examples (neighbors) which are nearest to the candidate example for which we wish to predict the target value, where K is some pre-defined parameter. Then, an aggregate function (e.g., average) is applied to the target values of neighbors to estimate the target value of the candidate example. This is known to be a simple method. However, it is also generally known to behave poorly when the data distribution is sparse.

4 Performance Evaluation

This section presents the simulation results. We consider the six types (viz., HH, HP, HB, BB, PB, PP) of dataset generated (discussed in Sect. 3.1), which consists of about 650 examples each both for indoor and outdoor settings. The target (distance between the devices) values range from 1 m to 10 m. The mobile devices used were of the specifications: realme3i (GPS/A-GPS/Gnolass, Bluetooth 4.2) and realme1.

Simulations were conducted on each type of dataset separately and then also on the combined dataset. We also carry out experiments on an existing dataset [11] for comparison. Unless otherwise stated, the training data size and the testing data size percentages are 80 and 20% respectively, which are randomly chosen from the dataset. Each point in the plots is an average result of 10 random runs. For our simulation, we use the *Python3.7* programming language with the *scikit-learn* package. The section is divided into two parts: (i) results obtained for each dataset type, and (ii) results obtained for combined dataset (consolidation of all types). The performance metric used is the normalized mean absolute error (NMAE), which is given by

$$NMAE = \frac{\sum_{i=1}^{n} |d_i - \hat{d}_i|}{\sum_{i=1}^{n} d_i} \tag{1}$$

where, d_i is the actual distance (target value) and $\hat{d}_i$ is the predicted target value of the ith test example respectively. Here, n is the number of test examples.

4.1 Comparison Using Separate Datasets

In this subsection, we illustrate how the different regression algorithms perform on the various types of dataset. Figure 2 shows the results for dataset type, HH for KNN (refer Fig. 2a) and other algorithms such as LR, DT, and so on (refer Fig. 2b). Two separate graphs are used to represent the results as KNN has several variants based on its type and the value of its parameter, K. The label 'RSS-HH' denotes the plot generated using the existing dataset [11] under the scenario Hand-to-Hand (HH). Similarly, the labels 'GPS-IN-HH' and 'GPS-OUT-HH' denotes the plot using the dataset generated in this study (refer Sect. 3.1) under the scenario, HH for indoors and outdoors respectively. Similar notations are followed in later graphs. The label 'KNN-2-uni' denotes that KNN algorithm is used where $K = 2$ and the distance metric used is 'uniform'. By 'uniform', it means that each of the K nearest neighbors are given equal weightage while taking the average of their target values for prediction. On similar lines, the label 'KNN-2-dist' denotes the use of KNN algorithm in which $K = 2$ and the distance metric is 'distance', meaning that weight given to the K nearest neighbors while averaging is inversely proportional to their distance from the candidate example for which the target value is being predicted.

From the figure (refer Fig. 2), we observe that regressors, KNN, DT, and RF give an NMAE less than 0.1 (in most cases less than 0.05) for the datasets, GPS-IN and GPS-OUT whereas there is no regressor that gives an NMAE of less than 0.1 for the existing dataset, RSS [11]. We also notice that algorithms based on a linear function such as LR, LASSO, Ridge, and Elastic Net perform poorly for all the datasets. This shows that the relationship between the input features and target value cannot be captured with the help of a linear function.

Figures 3, 4, 5, 6 and 7 illustrate the results for the rest of the types of dataset. Out of all the regressors, KNN, DT, and RF are the only three algorithms out of the seven which largely give NMAE less than 0.1 for the generated datasets, viz., GPS-IN and GPS-OUT. Moreover, out of these three, DT and RF show consistently good performance (NMAE less than 0.02) for all scenarios whereas KNN sometimes give NMAE greater than 0.1 (for instance, Fig. 5). However, for all scenarios, no

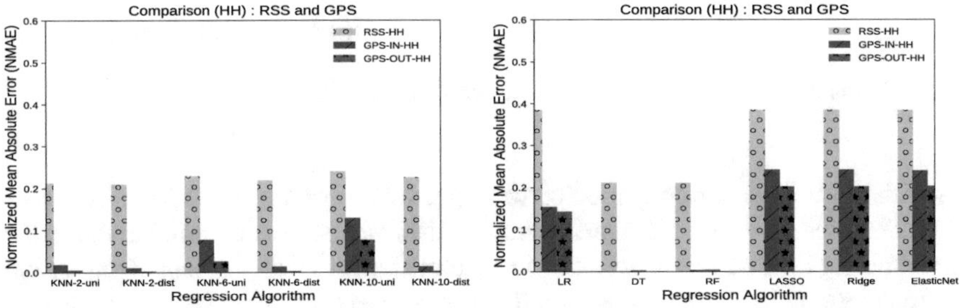

(a) KNN (b) LR, DT, RF, etc.

Fig. 2 Normalized mean absolute error [Hand-to-Hand]

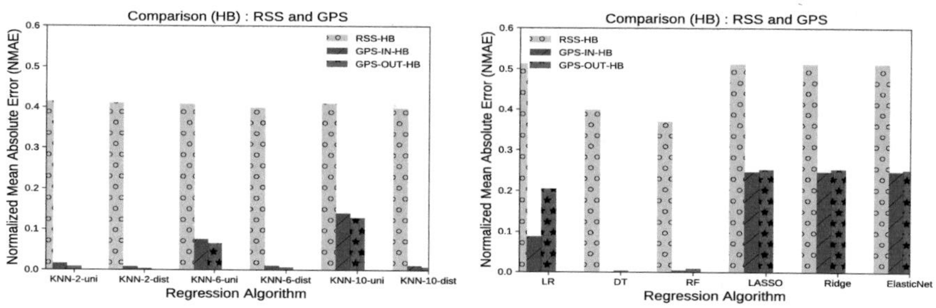

(a) KNN (b) LR, DT, RF, etc.

Fig. 3 Normalized mean absolute error [Hand-to-Backpack]

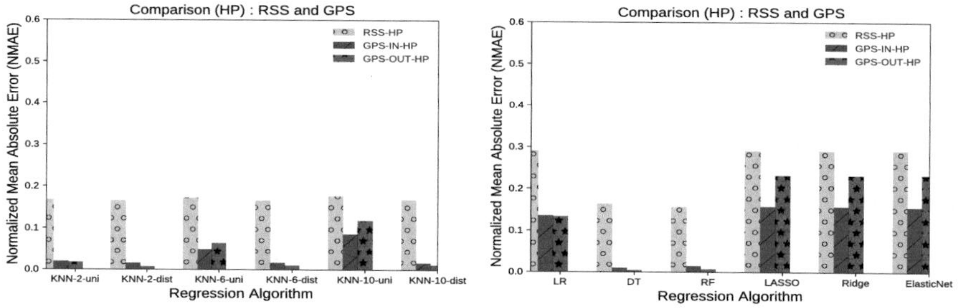

(a) KNN (b) LR, DT, RF, etc.

Fig. 4 Normalized mean absolute error [Hand-to-Pocket]

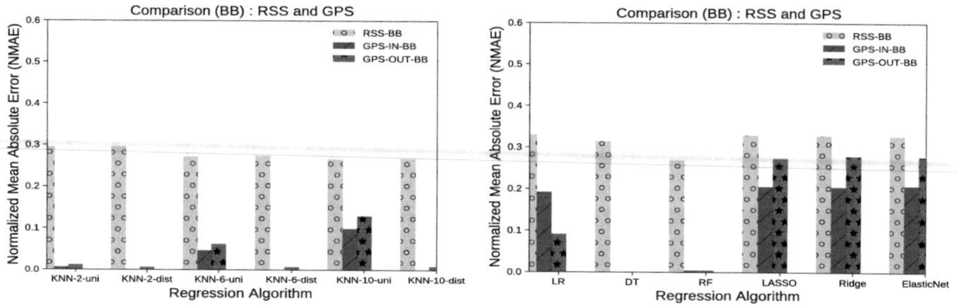

(a) KNN (b) LR, DT, RF, etc.

Fig. 5 Normalized mean absolute error [Backpack-to-Backpack]

algorithm is able to correctly predict the target values for the existing dataset (RSS) [11]. Therefore, we conclude that it is more reliable to use GPS coordinates than RSS in estimating the distance between two Bluetooth devices with the help of regression.

To determine the effect, if any, on the test size percentage on the results, we run the simulation. We find that there is no visible effect of the test size percentage on the performance of the regressors under all scenarios. Since the results are similar for

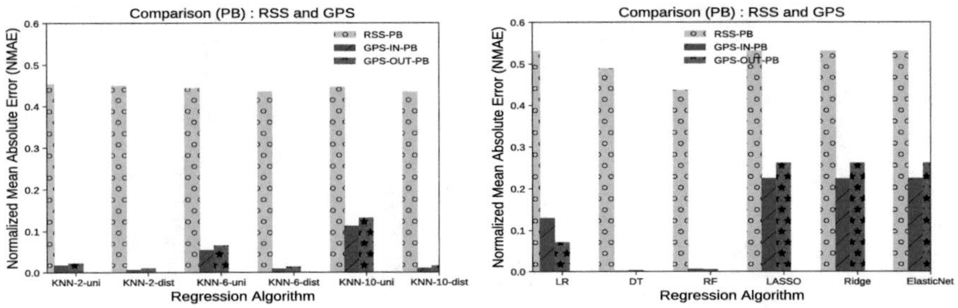

(a) KNN (b) LR, DT, RF, etc.

Fig. 6 Normalized mean absolute error [Pocket-to-Backpack]

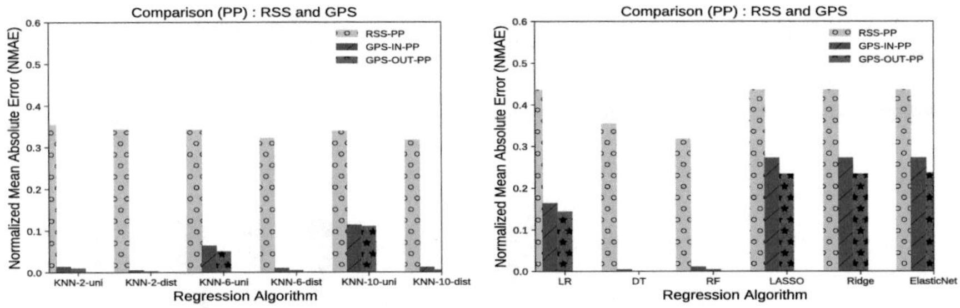

(a) KNN (b) LR, DT, RF, etc.

Fig. 7 Normalized mean absolute error [Pocket-to-Pocket]

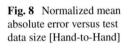

Fig. 8 Normalized mean absolute error versus test data size [Hand-to-Hand]

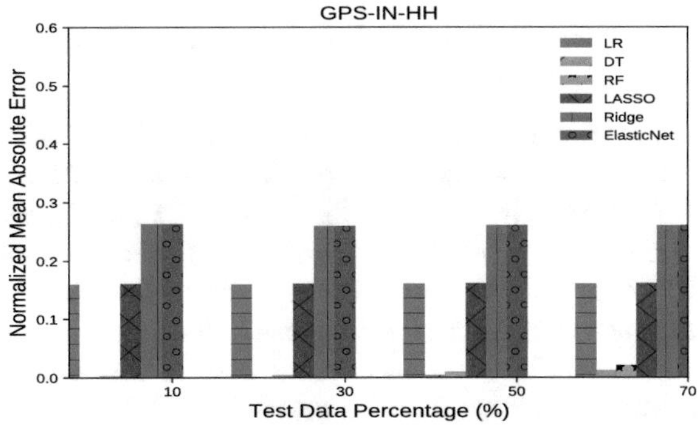

all scenarios, we show only for one, HH (refer Fig. 8). From the figure, we observe that there is a slight increase in NMAE as the test size percentage increases for both DT and RF. Even then, it is well within 0.1.

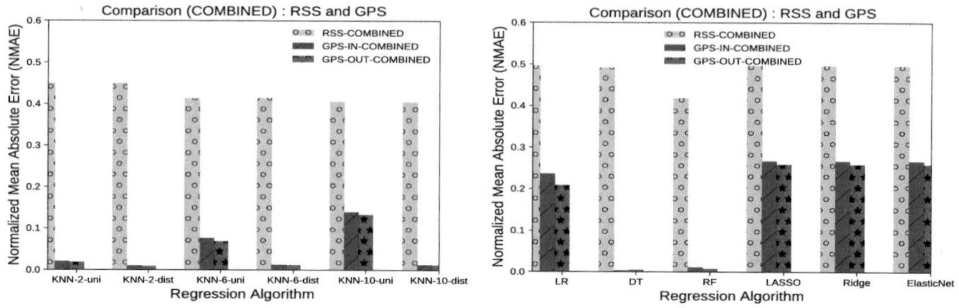

(a) KNN (b) LR, DT, RF, etc.

Fig. 9 Normalized mean absolute error [Combined]

Fig. 10 Normalized mean absolute error versus test data size [Combined]

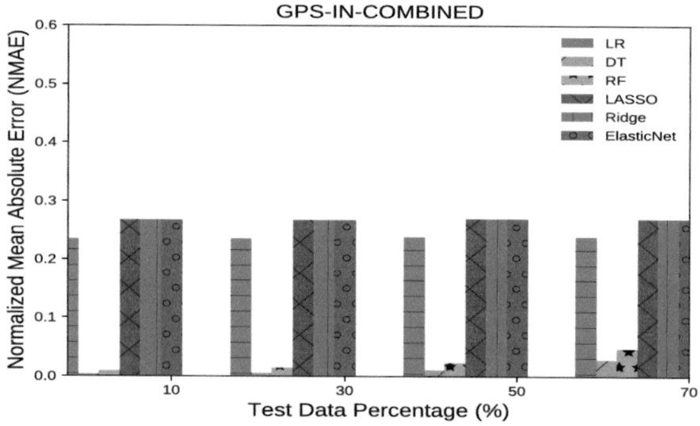

4.2 Comparison Using Combined Dataset

Over-fitting may occur when the number of examples in the dataset is small. Hence, we combine the six separate subsets of data into a single dataset, which we call the 'COMBINED' dataset for GPS-IN, GPS-OUT, and RSS. Figure 9 illustrates the performance of the algorithms. We find similar results as in the previous case. DT and RF outperform the rest of the algorithms, showing very good performance (NMAE even less than 0.01). Next in line is KNN largely gives an NMAE of 0.1, even though for the 'uniform' distance metric case, it is sometimes greater than 0.1. This also confirms the basic *intuition* that the weighting factor that depends on the distance of the neighbors in KNN does indeed give better performance (refer Fig. 9a). As seen in the previous case, no regressor is able to predict correctly for the existing dataset [11]. We conclude this from the graphs in Fig. 9, in which the best NMAE any regressor is able to give for the dataset, RSS [11] is about 0.4. This means that if the actual distance is 10m, the estimated distance could be around 14m or 6m, which is pretty off the mark.

The effect of test data size percentage on the performance in the case of combined dataset is shown in Fig. 10. Only result for the GPS-IN environment is shown as it is similar to that of other environments. As seen in the previous case (refer Fig. 8), we observe that there there is a slight increase in the NMAE as the test size percentage increases for both DT and RF. The other regressors perform poorly for all percentages.

5 Conclusions

In conclusion, this study is an attempt to estimate the distance between two Bluetooth devices with the help of machine learning, viz., regression. For this, instead of using RSSI as is commonly done, we use the GPS coordinates of the devices to build a dataset. On applying seven representative regression algorithms (both linear and non-linear) on the dataset, we find that two non-linear regressors successfully estimate the distance with very good precision. Another non-linear regressor, KNN performs well when the weighting factor is inversely proportional to the distance of the neighbor. The rest of them, which are linear regressors perform very poorly, which confirms that the relationship between the input features and the target feature cannot be captured by a linear function.

Additionally, the same algorithms were applied on an existing dataset [11] which consists of RSSI values between two Bluetooth devices. We observe that none of them are able to build a regression model successfully which brings us to the conclusion that the use of GPS coordinates rather than RSSI helps better in estimating the distance between two Bluetooth devices. However, the downside of using GPS coordinates is that a connection needs to be established between the devices and data (GPS coordinates) need to be transmitted which would consume more time and energy than in the case of using RSSI values. Additionally, the dataset that is built is in no way comprehensive. It would be interesting to use a collection of various mobile devices and build the dataset under uncontrolled environments, which could be taken up in the future.

Acknowledgements Ths work is partially supported by DST-SERB, Government of India under grant EEQ/2017/000083

References

1. Muller N (2001) Bluetooth demystified. McGraw-Hill, New York
2. Kalayjian NR (2008) Location discovery using bluetooth. US patent 2008/0125040 AI, date: May 29
3. Huang J, Chai S, Yang N, Liu L (2017) A novel distance estimation algorithm for bluetooth devices using RSSI. In: 2nd international conference on control, automation, and artificial intelligence (CAAI 2017)
4. Wang Y et al (2016) Indoor positioning system using Euclidean distance correction algorithm with bluetooth low energy beacon. In: IEEE international conference on internet of things and applications, pp 243–247

5. Naya F, Noma H, Ohmura R, Kogure K (2005) Bluetooth-based indoor proximity sensing for nursing context awareness. In: Ninth IEEE international symposium on wearable computers (ISWC05)
6. Jung J-H (2008) Method and apparatus for measuring distance between bluetooth terminals. US patent 2008/0036647 AI, date: Feb. 14
7. Thapa K (2003) An indoor positioning service for bluetooth Ad Hoc networks. In: MICS 2003, Duluth, MN, USA
8. Kotanen A, Hnnikinen M, Leppkoski H, Hminen T (2003) Experiments on local positioning with bluetooth. In: International conference on information technology: computers and communications, pp 297–303
9. Wang Y, Yang X, Zhao Y, Liu Y, Cuthbert L (2013) Bluetooth positioning using RSSI and triangulation methods. In: IEEE conference on consumer communications and networking (CCNC'13), pp 837–842. https://doi.org/10.1109/CCNC.2013.6488558
10. Google Maps. https://www.google.com/maps
11. Ng PC, Spachos P (2020) RSS_HumamHuman. https://github.com/pc-ng/rss/_HumanHuman, https://doi.org/10.21227/rd1e-6k71
12. Aarogya Setu. https://www.mygov.in/aarogya-setu-app
13. 1point5. https://github.com/UNTILabs/1point5

Function Characterization of Unknown Protein Sequences Using One Hot Encoding and Convolutional Neural Network Based Model

Saurabh Agrawal, Dilip Singh Sisodia, and Naresh Kumar Nagwani

1 Introduction

Functional characterization of recently evolved protein sequences is crucial to understand the various molecular and cellular processes [1], mutants of bacteria and virus [2], and drug repositioning [3]. An exponential number of new protein sequences are grown with the advancements in omics projects and mutations of the bacterial proteins [4]. PSL is important for the function prediction of UPS [5], though function of protein sequences relies on the localization of the cell in which it exists [6]. In-Vivo techniques are reliable for function identification of protein sequences; however they are lab oriented, time taking, and costly for large number of protein sequence samples [7]. In-Silico approaches based on deep learning techniques handle the time and cost constraints and became increasingly important for the functional characterization of UPS [8].

Exploration of useful and harmful characteristics of the UPS is significant for biological studies and applications. Compatible representation of the protein sequence [9] is requisite for the In-Silico based deep learning model [10]. Selection of optimum activation and optimization function for prediction of protein sequence is another challenge for the development of the deep learning based functional characterization model [11]. OCNN model is proposed in the present work for function

S. Agrawal (✉) · D. S. Sisodia · N. K. Nagwani
Department of Computer Science and Engineering, National Institute of Technology, GE Road, Raipur, Chhattisgarh 492010, India
e-mail: sagrawal.phd2018.cs@nitrr.ac.in

D. S. Sisodia
e-mail: dssisodia.cs@nitrr.ac.in

N. K. Nagwani
e-mail: nknagwani.cs@nitrr.ac.in

P. Singh et al. (eds.), *Machine Learning and Computational Intelligence Techniques for Data Engineering*, Lecture Notes in Electrical Engineering 998,
https://doi.org/10.1007/978-981-99-0047-3_24

characterization of the UPS in context with G+ bacterial protein, which can also handle undermentioned gaps and challenges.

- Representation of variable-length protein sequence residues in compatible format and length.
- Identification of optimum combination of activation and optimization function.

The main motivation behind the present study is to identify the pathogenic and nonpathogenic characteristics of the newly evolved UPS in context with the G+ bacterial proteins. In this paper OHE and CNN-based OCNN model is proposed for functional characterization of UPS with reference to the PSL. Key contributions of the present work are given as follows.

- Transformation of protein sequence residues in compatible format using One-Hot encoding.
- Standardization and normalization of the protein Sequence length through capping and padding.
- Prediction of the protein sequence functions with optimum combination of the activation and optimization function.

2 Related Work

Huge numbers of the protein sequences are not annotated yet, simultaneously novel protein sequences have been evolved and added to the proteomics archives. In-Silico approaches based on deep learning can offer the computerized method for handling complex and big data rapidly [12]. Deep learning models namely CNN, Long Short Term Memory (LSTM), and hybrid of CNN-LSTM are widely utilized for the function prediction and subcellular localization of the protein sequences. Emerging human pathogen identification technique was defined by the Shanmugham and Pan [13]. Audagnotto and Peraro [14] proposed an In-Silico prediction tool for the analysis of post transitional modifications and its effects in the structure and dynamics of protein sequence. Mondal et al. developed the Subtractive genome analysis model through an In-Silico approach to identify potential drug targets [15]. With the advancement in genome sequencing techniques and mutation of the protein sequences a huge amount of protein sequences have been evolved [9]. Novel UPS plays a crucial role for biological investigations and its applications [16], if their suitable subcellular localization has been identified [17].

Transformation of the protein sequence residues in numeric format is prerequisite for the CNN based functional characterization model. Various encoding and feature extraction techniques are available for the conversion of protein sequences into the suitable numeric format. Agrawal et al. [18] proposed a deep and shallow feature based protein function characterization model using One-Hot encoding. Elabd et al. proposed various amino acid encoding methods for the deep learning applications [19]. One hot encoding based deep-multi-modal protein function classification model was developed by Giri et al. [20]. Choong and Lee [21] defined the evolution of convolution neural network model for DNA sequence prediction using one hot encoding. Different deep learning approaches namely CNN and LSTM have been utilized for the subcellular localizations [22, 23] and function characterization [24, 25] of the known and unknown protein sequence. CNN has been compatible for the feature extraction form the sequential data, that's why CNN based deep learning methods have been utilized for functional characterization of protein sequences [26]. A CNN based motif detector model was proposed by Zhou et al. [27]. Kulmanov et al. proposed a function categorization model for protein, where the model features of protein sequence residues are learned through CNN [24].

3 Methodology

Methods and techniques applied towards the development of OHE and CNN based protein function characterization model are illustrated in Fig. 1. 473 protein sequence samples of the benchmark G+ dataset with four different subcellular localizations and 50 independent protein sequence samples of the UPS dataset exclusive of localizations are used for the experimentation. As the initial stage of processing protein sequence residues are encoded in digital form using OHE, the length of protein sequences has been standardized and normalized through capping and padding. Encoded and preprocessed sequence samples are convoluted in the hidden layer of the OCNN model using ReLU, TanH, and Sigmoid activation functions. Next, convoluted features of the known protein sequence samples have been used for PSL prediction through the Adam and SGD functions in the optimization layer. Performance of the OCNN model has been validated with fivefold cross validation using accuracy, precision, recall, and f1-score. The validated OCNN model has been further utilized for function characterization of UPS in context with G+ bacterial proteins.

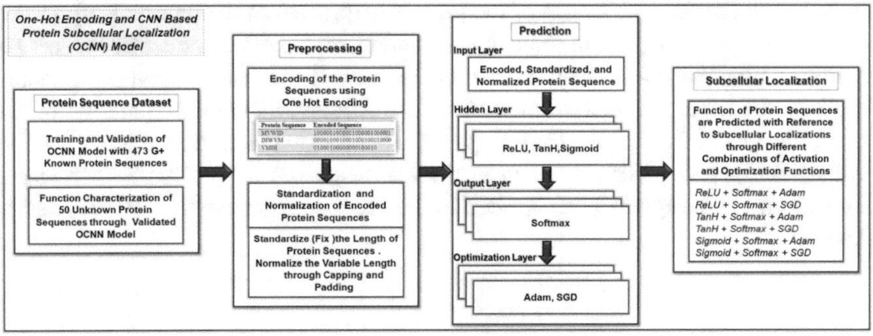

Fig. 1 OHE and CNN based protein subcellular localization model

3.1 Protein Dataset

A Gram-Positive dataset had been developed by Chou and Shen [28]. G+ dataset consists 523 protein sequence samples with different four PSLs such as C1:.Cell Membrane, C2:.Cell Wall, C3:.Cytoplasm, and C4:.Extracellular. However, protein sequence's functions rely on localization of the cell. G+ dataset is accessible through the web link[1] as on date January 8, 2022.

In this paper, protein sequence samples of the G+ dataset have been split as known and unknown samples. Approximately 90% samples from each class of the G+ dataset with respective PSL have been taken for the known protein sequence dataset. Rests of the 10% samples are utilized for the UPS dataset where PSL has not been considered.

3.2 Preprocessing

As essential preprocessing protein sequence residues are first transformed in digital format using OHE. Next, protein sequence length has been standardized and normalized through capping and padding.

Protein sequences have been made up by different combinations of twenty different amino acid residues. Therefore a single residue has been encoded with a twenty bit code through OHE as depicted in Table 1, although OHE signify the residues according to the total number of different types of the residues.

Table 2 demonstrates the two different protein sequences with seven and four residues after and before encoding. Fixed length input is the essential requirement for the CNN based protein function characterization model, therefore capping and padding is employed in this work for the standardization and normalization of the protein sequence length. Generally the length of the sequence has been defined according to the average length of the protein sequences in the dataset [29]. In this work the length of the protein sequence is standardized as 400 residues per sequence. Now normalization of the sequences which are smaller or larger than the standard length is required. Sequences are normalized through capping or padding of the bits in the encoded sequence [30]. Suppose we have standardized the length as five residues as given in Table 2. Then in the first sequence "MSGEVLS" capping is required and padding has been applied in the second sequence "MISP" as given in Table 3.

[1] http://www.csbio.sjtu.edu.cn/bioinf/Gpos-multi/Data.htm.

Table 1 Encoding of protein sequence residues using one-hot encoding

Protein sequence residues	One hot encoding
A	10000000000000000000
C	01000000000000000000
D	00100000000000000000
E	00010000000000000000
F	00001000000000000000
G	00000100000000000000
H	00000010000000000000
I	00000001000000000000
K	00000000100000000000
L	00000000010000000000
M	00000000001000000000
N	00000000000100000000
P	00000000000010000000
Q	00000000000001000000
R	00000000000000100000
S	00000000000000010000
T	00000000000000001000
V	00000000000000000100
W	00000000000000000010
Y	00000000000000000001

Table 2 Encoded protein sequence residues using one-hot encoding

Sequence	Encoded sequence
MSGEVLS	00000000001000000000000000000000000100000000010000000000000000001000 0000000000000000000000000000001000000000001000000000000000000000000 010000
MISP	0000000000100000000000000001000000000000000000000000000100000000000 0000010000000

Table 3 Standardized and normalized encoded protein sequence with five residues

Sequence	Encoded sequence with standardization and normalization
MSGEVLS	0000000000100000000000000000000000010000000001000000000000000001000000 000000000000000000000000000100
MISPX	0000000000100000000000000001000000000000000000000000000100000000000000 001000000000000000000000000000

3.3 Prediction Using Convolutional Neural Network

CNN has been competent for evaluating the spatial information included in the encoded protein sequence [31]. CNN can exploit the abstract-level representation of the complete sequence by integrating the features of the residue level [32]. In this work the CNN model has been established with four fundamental layers namely input, convolution, subsampling, and dense-output for functional characterization of protein sequences.

In the proposed model feature maps of the *i*th layer have been signified through $\boldsymbol{L_i}$ which is given in Eq. 1. Where $\boldsymbol{W_i}$ signifies the weight-matrix, offset-vector is represented by $\boldsymbol{b_i}$ and h (.) symbolizes the activation function.

$$\mathbf{L_i} = \mathrm{h}(\mathrm{L_{i-1}[CONV]W_i + b_i}) \tag{1}$$

Convoluted feature map has been sampled in the subsampling layer according to the defined rule as formulated in Eq. 2.

$$\mathbf{L_i} = \mathrm{subsampling}(\mathrm{L_{i-1}}) \tag{2}$$

CNN can extract the features from the dense layer as well as probability distribution $\mathcal{F}$. Next, a multilayer data transformation has been employed for mapping of the input sequence as convoluted deep features as given in Eq. 3.

$$\mathcal{F}(\mathbf{i}) = \mathrm{Map}(\mathrm{C} = \mathrm{c_i}|\mathrm{L_0};\ (\mathrm{W, b})) \tag{3}$$

where the *i*th label class has been represented through $\mathbf{c_i}$, the input sequence has been signified through $\boldsymbol{L_0}$, and $\mathcal{F}$ denotes the feature-expression. However, loss function $\boldsymbol{F(W, b)}$ is minimized through convolution. Simultaneously final loss function $\boldsymbol{E(W, b)}$ is controlled by norm, and intensity of over-fitting is adjusted by the parameter $\boldsymbol{\epsilon}$.

$$\mathbf{E(W, b)} = \mathrm{F(W, b)} + \frac{\mathrm{e}}{2}\mathrm{W^T W} \tag{4}$$

Generally Gradient descent technique can be utilized with CNN for the updates in the network parameter $(\boldsymbol{W, b})$ and optimization in diverse layers, although backpropagation has been controlled through learning rate λ.

$$\mathbf{W_i} = \mathrm{W_i} - \lambda \frac{\partial \mathrm{E(W, b)}}{\partial \mathrm{W_i}} \tag{5}$$

$$\mathbf{b_i} = \mathrm{b_i} - \lambda \frac{\partial \mathrm{E(W, b)}}{\partial \mathrm{b_i}} \tag{6}$$

In this work three activation functions such as ReLU, TanH, and Sigmoid are utilized for the convolution of encoded protein sequence samples, while function has been predicted using Adam and SGD optimization functions.

3.4 Performance Measures

OCNN model's performance has been indexed through accuracy, precision, recall, and f1-score [33] which can be calculated through Eqs. 7 to 10.

$$\text{Accuracy} = \frac{1}{n}\sum_{i=1}^{n}\frac{TP_i + TN_i}{TP_i + TN_i + FP_i + FN_i} \tag{7}$$

$$\text{Precision} = \frac{1}{n}\sum_{i=1}^{n}\frac{TP_i}{TP_i + FP_i} \tag{8}$$

$$\text{Recall} = \frac{1}{n}\sum_{i=1}^{n}\frac{TP_i}{TP_i + FN_i} \tag{9}$$

$$\text{f1} - \text{score} = \frac{1}{n}\sum_{i=1}^{n}\frac{2TP_i}{2TP_i + FP_i + FN_i} \tag{10}$$

Total number of the accurately predicted PSLs of the ***i***th class is signified through $\boldsymbol{TP_i}$ and $\boldsymbol{TN_i}$ although incorrectly predicated localizations have been represented by $\boldsymbol{FP_i}$ and $\boldsymbol{FN_i}$, where $\boldsymbol{n}$ is the total number of classes. Percentage of the correctly predicted samples has been calculated as accuracy. Precision measures the exactness and recall measures the completeness of the OCNN model. F-measure is the evaluation of the accuracy, computed through precision and recall.

4 Results and Discussion

An OHE and CNN based PSL model is proposed in the present work for function characterization of the UPS in context with G+ protein. The OCNN model is developed using Keras library and implemented through Python 3.7.10. The obtained results have been discussed in the subsequent sections.

4.1 Results

A known G+ protein sequence dataset is used for training and validation of the OCNN model. Where, G+.dataset contains 473 protein sequence samples with four subcellular localizations. Six different combinations of the three activation and two optimization functions are employed in the proposed model. Classification report of the OCNN model with the known protein sequence dataset is depicted in Table 4.

Combination of the *Sigmoid-Softmax-Adam* outperforms the other combination of the activation and optimization function. However OHE converts the protein sequence residues in the real and sparse data, therefore the Adam optimization function performs higher than SGD with all activation functions.

The validated OCNN model is further utilized for function characterization of the UPS. In this work the UPS dataset is comprised of 50 independent protein sequences. Subcellular localization of the independent protein sequence samples have been not considered for the experimentation, while localizations are externally utilized to verify the performance of the OCNN model. As illustrated in Table 5 a combination of the *ReLU-Softmax-Adam* outperforms the other combinations for functional characterization of UPS.

Table 4 Classification report of the known G+ protein sequences with OCNN model

Hidden layer	Output layer	Optimization layer	Accuracy	Precision	Recall	F1-Score
ReLU	Softmax	Adam	91.48	92.27	91.38	92.42
ReLU	Softmax	SGD	90.43	89.29	89.86	90.91
TanH	Softmax	Adam	89.23	90.15	90.26	90.48
TanH	Softmax	SGD	87.54	86.72	87.44	87.69
Sigmoid	Softmax	Adam	**92.94**	**93.46**	**93.21**	**92.76**
Sigmoid	Softmax	SGD	83.05	84.25	83.57	84.26

Table 5 Classification report of the unknown protein sequences

Hidden layer	Output layer	Optimization layer	Accuracy	Precision	Recall	F1-Score
ReLU	Softmax	Adam	**64.83**	**65.35**	**64.39**	**64.87**
ReLU	Softmax	SGD	58.01	57.96	59.22	57.42
TanH	Softmax	Adam	57.32	56.38	56.47	57.32
TanH	Softmax	SGD	56.96	56.76	55.32	55.49
Sigmoid	Softmax	Adam	63.56	62.23	63.72	63.23
Sigmoid	Softmax	SGD	53.89	53.52	54.94	53.23

4.2 Discussion

Diverse variants of the bacterial protein have been exponentially evolving due to the pathogen-host interaction and mutations of bacteria, while novel variant could reveals the pathogenic as well as nonpathogenic characteristics. Therefore, PSL prediction is imperative to understand the useful and harmful functions of the UPS. OCNN model is proposed for the fast accurate function characterization of the UPS in context with the G+ bacterial protein using OHE and CNN. In the present paper OHE is employed for the transformation of protein sequence residues in the digital form, however, OHE preserves the spatial information. Encoded sequences are standardized and normalized to fix the length of the protein sequences, while consistent length and type of input is requisite for the CNN based prediction model. CNN exploits the abstract level and hidden information of the protein sequences through the different combinations of the activation and optimization functions. In this study two aspects are accountable for the improvement in processing speed and performance of the OCNN model: (a) Transformation of the protein sequences in numeric form with perseverance of spatial information. (b) Prediction of encoded protein sequence through CNN using different combinations of the activation and optimization functions.

5 Conclusion

Functional characterization of protein sequence is crucial for different biotechnological applications such as study of cellular and molecular process, bacteria and virus mutation, treatment plan, and repositioning of drug. In the present study PSL based computational framework is proposed to identify the function of UPS using OHE and CNN. G+ bacterial benchmark with known and UPS samples have been used for the experimentation. The OCNN model achieves 92.94% accuracy through the combination of Sigmoid, Softmax, and Adam functions with known G+ protein sequence samples. Further, the validated model has been utilized for the PSL prediction of UPS, although the function of the protein sequence relies on the localization in which it resides. The OCNN model attains 64.83% accuracy with the independent protein sequence samples of the UPS dataset through the combination of ReLU, Softmax, and Adam functions. The proposed model has been also compatible for the functional prediction of UPS in context with other bacterial and virus protein sequence datasets. Future scope of the present work is to develop a multisite PSL prediction model.

References

1. Lei X, Zhao J, Fujita H, Zhang A (2018) Predicting essential proteins based on RNA-Seq, subcellular localization and GO annotation datasets. Knowl-Based Syst 151:136–148. https://

doi.org/10.1016/j.knosys.2018.03.027
2. Guo H, Liu B, Cai D, Lu T (2018) Predicting protein–protein interaction sites using modified support vector machine. Int J Mach Learn Cybern 9:393–398. https://doi.org/10.1007/s13042-015-0450-6
3. Sureyya Rifaioglu A, Doğan T, Jesus Martin M, Cetin-Atalay R, Atalay V (2019) DEEPred: automated protein function prediction with multi-task feed-forward deep neural networks. Sci Rep 9:1–16.https://doi.org/10.1038/s41598-019-43708-3
4. Zhang J, Yang JR (2015) Determinants of the rate of protein sequence evolution. Nat Rev Genet 16:409–420. https://doi.org/10.1038/nrg3950
5. Tahir M, Khan A (2016) Protein subcellular localization of fluorescence microscopy images: employing new statistical and Texton based image features and SVM based ensemble classification. Inf Sci 345:65–80. https://doi.org/10.1016/j.ins.2016.01.064
6. Wan S, Mak MW (2018) Predicting subcellular localization of multi-location proteins by improving support vector machines with an adaptive-decision scheme. Int J Mach Learn Cybern 9:399–411. https://doi.org/10.1007/s13042-015-0460-4
7. Ranjan A, Fahad MS, Fernandez-Baca D, Deepak A, Tripathi S (2019) Deep robust framework for protein function prediction using variable-length protein sequences. IEEE/ACM Trans Comput Biol Bioinf 1–1. https://doi.org/10.1109/tcbb.2019.2911609
8. Almagro Armenteros JJ, Sønderby CK, Sønderby SK, Nielsen H, Winther O (2017) DeepLoc: prediction of protein subcellular localization using deep learning. Bioinformatics (Oxford, England) 33:3387–3395.https://doi.org/10.1093/bioinformatics/btx431
9. Agrawal S, Sisodia DS, Nagwani NK (2021) Augmented sequence features and subcellular localization for functional characterization of unknown protein sequences. Med Biol Eng Comput 2297–2310. https://doi.org/10.1007/s11517-021-02436-5
10. Shi Q, Chen W, Huang S, Wang Y, Xue Z (2019) Deep learning for mining protein data. Brief Bioinform 1–25. https://doi.org/10.1093/bib/bbz156
11. Wang Y, Li Y, Song Y, Rong X (2020) The influence of the activation function in a convolution neural network model of facial expression recognition. Appl Sci (Switzerland) 10. https://doi.org/10.3390/app10051897
12. Vassallo K, Garg L, Prakash V, Ramesh K (2019) Contemporary technologies and methods for cross-platform application development. J Comput Theor Nanosci 16:3854–3859. https://doi.org/10.1166/jctn.2019.8261
13. Shanmugham B, Pan A (2013) Identification and characterization of potential therapeutic candidates in emerging human pathogen mycobacterium abscessus: a novel hierarchical In Silico approach. PLoS ONE 8. https://doi.org/10.1371/journal.pone.0059126
14. Audagnotto M, Dal Peraro M (2017) Protein post-translational modifications: In silico prediction tools and molecular modeling. Comput Struct Biotechnol J 15:307–319. https://doi.org/10.1016/j.csbj.2017.03.004
15. Mondal SI, Ferdous S, Jewel NA, Akter A, Mahmud Z, Islam MM, Afrin T, Karim N (2015) Identification of potential drug targets by subtractive genome analysis of Escherichia coli O157:H7: an in silico approach. Adv Appl Bioinform Chem 8:49–63. https://doi.org/10.2147/AABC.S88522
16. Weimer A, Kohlstedt M, Volke DC, Nikel PI, Wittmann C (2020) Industrial biotechnology of Pseudomonas putida: advances and prospects. Appl Microbiol Biotechnol 104:7745–7766. https://doi.org/10.1007/s00253-020-10811-9
17. Zhang T, Ding Y, Chou KC (2006) Prediction of protein subcellular location using hydrophobic patterns of amino acid sequence. Comput Biol Chem 30:367–371. https://doi.org/10.1016/j.compbiolchem.2006.08.003
18. Agrawal S, Sisodia DS, Nagwani NK (2021) Long short term memory based functional characterization model for unknown protein sequences using ensemble of shallow and deep features. Neural Comput Appl 4. https://doi.org/10.1007/s00521-021-06674-4
19. Elabd H, Bromberg Y, Hoarfrost A, Lenz T, Franke A, Wendorff M (2020) Amino acid encoding for deep learning applications. BMC Bioinform 21:1–14. https://doi.org/10.1186/s12859-020-03546-x

20. Giri SJ, Dutta P, Halani P, Saha S (2021) MultiPredGO: deep multi-modal protein function prediction by amalgamating protein structure, sequence, and interaction information. IEEE J Biomed Health Inform 25:1832–1838. https://doi.org/10.1109/JBHI.2020.3022806
21. Choong ACH, Lee NK (2017) Evaluation of convolutionary neural networks modeling of DNA sequences using ordinal versus one-hot encoding method. In: 1st international conference on computer and drone applications: ethical integration of computer and drone technology for humanity sustainability, IConDA 2017. 2018 Jan, pp 60–65. https://doi.org/10.1109/ICONDA.2017.8270400.
22. Sønderby SK, Sønderby CK, Nielsen H, Winther O (2015) Convolutional LSTM networks for subcellular localization of proteins. Lect Notes Comput Sci (including subseries Lecture Notes in Artificial Intelligence and Lecture Notes in Bioinformatics) 9199:68–80. https://doi.org/10.1007/978-3-319-21233-3_6
23. Wei L, Ding Y, Su R, Tang J, Zou Q (2018) Prediction of human protein subcellular localization using deep learning. J Parall Distrib Comput 117:212–217. https://doi.org/10.1016/j.jpdc.2017.08.009
24. Kulmanov M, Khan MA, Hoehndorf R (2018) DeepGO: predicting protein functions from sequence and interactions using a deep ontology-aware classifier. Bioinformatics 34:660–668. https://doi.org/10.1093/bioinformatics/btx624
25. Gao R, Wang M, Zhou J, Fu Y, Liang M, Guo D, Nie J (2019) Prediction of enzyme function based on three parallel deep CNN and amino acid mutation. Int J Mol Sci 20. https://doi.org/10.3390/ijms20112845
26. Kulmanov M, Hoehndorf R, Cowen L (2020) DeepGOPlus: improved protein function prediction from sequence. Bioinformatics 36:422–429. https://doi.org/10.1093/bioinformatics/btz595
27. Zhou J, Lu Q, Xu R, Gui L, Wang H (2017) CNNsite: Prediction of DNA-binding residues in proteins using Convolutional Neural Network with sequence features. In: Proceedings—2016 IEEE international conference on bioinformatics and biomedicine, BIBM 2016, pp 78–85. https://doi.org/10.1109/BIBM.2016.7822496
28. Shen H-B, Chou K-C (2009) Gpos-mPLoc: a top-down approach to improve the quality of predicting subcellular localization of gram-positive bacterial proteins. Protein Pept Lett 16:1478–1484. https://doi.org/10.2174/092986609789839322
29. Lipman DJ, Souvorov A, Koonin EV, Panchenko AR, Tatusova TA (2002) The relationship of protein conservation and sequence length. BMC Evol Biol 2:1–10. https://doi.org/10.1186/1471-2148-2-20
30. Sercu T, Goel V (2016) Advances in very deep convolutional neural networks for LVCSR. In: Proceedings of the annual conference of the international speech communication association, INTERSPEECH. 08–12-September-2016, pp 3429–3433. https://doi.org/10.21437/Interspeech.2016-1033
31. Wang L, Wang HF, Liu SR, Yan X, Song KJ (2019) Predicting protein-protein interactions from matrix-based protein sequence using convolution neural network and feature-selective rotation forest. Sci Rep 9:1–12. https://doi.org/10.1038/s41598-019-46369-4
32. Zhou S, Chen Q, Wang X (2013) Active deep learning method for semi-supervised sentiment classification. Neurocomputing 120:536–546. https://doi.org/10.1016/j.neucom.2013.04.017
33. Sharma R, Dehzangi A, Lyons J, Paliwal K, Tsunoda T, Sharma A (2015) Predict gram-positive and gram-negative subcellular localization via incorporating evolutionary information and physicochemical features Into Chou's General PseAAC. IEEE Trans Nanobiosci 14:915–926. https://doi.org/10.1109/TNB.2015.2500186

Prediction of Dementia Using Whale Optimization Algorithm Based Convolutional Neural Network

Rajalakshmi Shenbaga Moorthy, Rajakumar Arul, K. Kannan, and Raja Kothandaraman

1 Introduction

The term Dementia which gives rise to difficulty in thinking, memory loss, slowly degrading the mental ability is a severe cognitive disorder which needs to be detected in advance. In other words, Dementia is a neurogenerative disorder in brain which mostly occurred due to Alzheimer's Disease (AD) [1]. Dementia leads to misfunctioning of the brain which paves way for lack of recognition, acknowledging, thinking and behavioral skills of the individuals. The individuals affected with Dementia struggles to control emotions and also forgot everything. Symptoms of Dementia varies from person to person. Once occurred it is not possible to cure but there exist several mechanisms to predict the occurrence of Dementia. Various forms of Dementia include Alzheimer's Disease, Vascular Dementia, Lewy body Dementia, and Parkinson's Disease. The various stages of Dementia include mild dementia, moderate dementia and severe dementia which is shown in Fig. 1.

Detecting or predicting the Dementia is very essential as the disease is progressive and irreversible. Also, the person with Dementia starts with mild stage where the person suffers with occasional forgetfulness slowly progressing to moderate stage where the person requires assistance for doing daily activities and further severe progression to last stage in which person may lose their physical ability also. Thus,

R. Shenbaga Moorthy (✉)
Sri Ramachandra Institute of Higher Education and Research, Chennai, Tamil Nadu, India
e-mail: srajiresearch@gmail.com

R. Arul
Vellore Institute of Technology (VIT) Chennai Campus, Chennai, India

K. Kannan
Sree Vidyanikethan Engineering College, Tirupathi, Andra Pradesh, India

R. Kothandaraman
SRM Institute of Science and Technology, Ramapuram, Chennai, India

P. Singh et al. (eds.), *Machine Learning and Computational Intelligence Techniques for Data Engineering*, Lecture Notes in Electrical Engineering 998,
https://doi.org/10.1007/978-981-99-0047-3_25

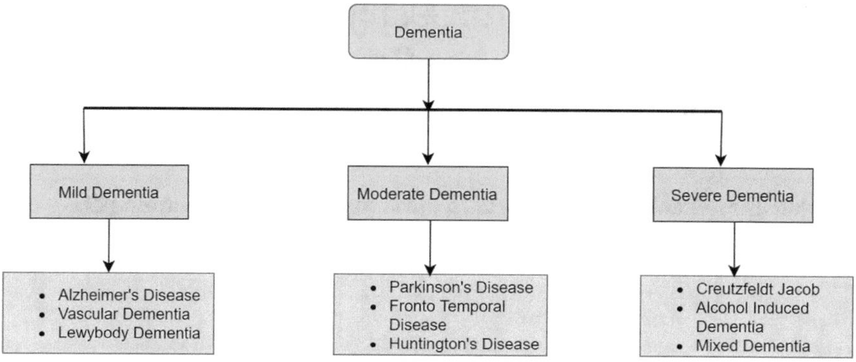

Fig. 1 Various stages and forms of Dementia

prediction of occurrence of Dementia is essential as the symptoms also vary from person to person. In order to give better prediction, deep learning algorithms may be employed to work with image datasets [2].

The datasets for the process are taken from the real world Kaggle repository. The datasets are trained over the Whale Optimization Algorithm (WOA) based Convolution Neural Network (CNN) with the desire to find whether the person is affected with mild, moderate, or very moderate Dementia or not affected with Dementia. CNN has been widely used to handle medical datasets [3]. Various variations of CNN include LeNet—5, ALexNet, VGG-16, Inception—V1, Inception—V3, ResNet-50 [4]. The hyperparameters in CNN are essentially numeric values, thus finding that the optimal value is really a NP_Hard problem and thus requires any approximation algorithms or metaheuristic algorithms as a solution. The hyperparameters like dropout rate and batch size play a crucial role in the convergence of CNN with better accuracy. Since accuracy is a prime concern in medical field and optimization algorithms intends to find optimal value of the hyperparameters of CNN, the proposed WOA based CNN achieves good accuracy with minimum loss. Recently, researchers intend to fine tune the parameters of the machine learning algorithm for enhancing the performance [19].

The major contributions of the paper are given as:

- Learning the hyperparameters of CNN with WOA
- The proposed WOA-CNN is compared with CNN in terms of accuracy and loss.

The remaining paper is organized as follows: Sect. 1 briefly gives the introduction about Dementia and motivation behind the problem statement. Section 2 gives various existing mechanisms available to predict Dementia. Section 3 gives the proposed system where the whale optimization algorithm has been used to tune the hyperparameters of Convolutional Neural Network. Section 4 gives the experimental results comparing the proposed work with other existing works. Finally, Sect. 5 concludes the work with Future scope.

2 Related Work

Automatically detecting the major form of dementia known as Alzheimer's disease has been diagnosed using the voice recordings as the input to additive logistic regression with an accuracy of 86.5%. [5]. Automatic speech recognition has been done through two pipelines of feature extraction techniques namely manual pipeline and fully automated pipeline [6]. Normal conversational speech has been used for detecting Dementia using unsupervised voice activity detection and speaker diarization. The designed methodology is obtained with an average recall of about 0.645 [7]. Neurological screening campaign had been conducted in Southern Italy for 5 days which included neurological and neuropsychological examinations. Mild form of Dementia had been detected for 39% of screened patients [8]. Transfer learning approach has been used to classify Alzheimer's disease. The idea behind transfer learning approach is that the model has been pre-trained with large sized datasets and tested using ADEsSS datasets [9]. Alzheimer's disease has been diagnosed using Alzheimer's disease Neuroimaging initiative (ADNI) and Open access series of imaging studies (OASIS) datasets. The machine learning algorithms logistic regression and Support Vector Machine obtained 99.43% and 99.10% for the ADNI dataset. Similarly, logistic regression and random forest obtained 84.33% and 83.92% accuracy for the OASIS dataset [10].

Magnetic Resonance Imaging (MRI) image was given as input which is preprocessed using morphological operations. Convolutional Shape local binary texture (CSLBT) was used to extract features and Linear Discriminant Analysis (LDA) was used to reduce features [11]. Transcriptions of spontaneous speech for detecting Dementia were done through the Single Layer Convolutional Neural Network [12]. Elastic Net Regularization (ELR) in feature selection has been introduced in deep learning architecture with the aid to increase accuracy and reduce the processing time. The features extracted using CNN were fed to Principal Component Analysis (PCA) and ELR for performing the next level feature selection. The selected features were fed to Extreme Machine Learning (EML) for better classification of images [13]. WOA had been applied to find an optimal value for filter and weights in the fully connected layer for optimal recognition of texture [14]. WOA based CNN had been effectively used to predict breast cancer. WOA was utilized to optimize the parameters of CNN. The designed WOA CNN had been processed for the BreakHis histopathological dataset [15]. Metaheuristic based Optimized CNN classifier had been used to detect the existence of skin cancer. WOA had been used to optimize the weights and biases in CNN. The designed model had been evaluated on the Dermis Digital Database Dermquest Database [16]. Canonical Particle Swarm Optimization (C-PSO) based Convolutional Neural Network had been designed to find the hyperparameters of CNN. The designed CPSO-CNN outperformed well than CNN with a random hyperparameter and also it reduced the computational cost [17].

From the study, it has been observed that when a metaheuristic algorithm is integrated with CNN, the performance of the CNN had been improved.

3 Proposed WOA Based CNN

The CNN architecture, taken into account for processing grey scale images of Dementia include AlexNet. AlexNet has 8 layers in which 5 layers are convolutional and 3 layers are Fully Connected. Maxpooling layer is not taken into account as it does not have the parameters involved. The hyperparameters that play key role in the convergence of AlexNet are filter size, values for filters, weights in the fully connected layer, dropout rate and batch size. In this paper, the dropout rate and batch size are taken into account and the optimal values are found by using the Whale Optimization Algorithm. Since the dropout rate has a serious effect with overfitting, the optimal value for the dropout rate prevents the model from overfitting. Batch size not only affects the computational time of the CNN but also the accuracy. Thus, in this paper mini batch size is considered as another hyperparameter that is found using WOA.

The original input dimension of ALexNet is 227 * 227 * 1. At each layer the width and height of the output image size is determined using Eqs. (1) and (2) respectively.

$$w_j \leftarrow \frac{w_i - F + 2P}{S} \tag{1}$$

$$h_j \leftarrow \frac{H_i - F + 2P}{S} \tag{2}$$

where w_j and h_j are the width and height of the jth layer respectively. F represents the dimensions of the filter size, P represents padding, S represents stride size. Finally, w_i and h_i are width and height of the previous ith layer. Figure 2 represents the proposed integrated WOA based AlexNet architecture.

Whale Optimization Algorithm is a nature inspired metaheuristic algorithm, which is inspired by the hunting behavior of whales [20]. The population is represented by whales. Each whale is represented by two dimensions viz, mini batch size and dropout rate. WOA algorithm intends to balance between exploration and

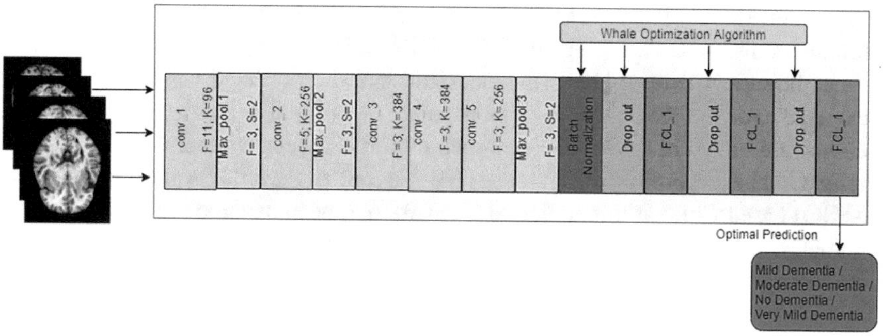

Fig. 2 Hyperparameter tuning using whale optimization Algorithm in AlexNet

exploitation in order to find the global optimal solution. Each whale is represented as W_i. Each whale is represented with two dimensions viz, mini batch size and drop out rate. At each iteration, each whale tends to move towards the best solution which is represented in Eq. (3)

$$\mathrm{W_i^t} \leftarrow \mathrm{W_*^{t-1}} - \mathrm{A} * \left|\mathrm{C} * \mathrm{W_*^{t-1}} - \mathrm{W_i^{t-1}}\right| \tag{3}$$

where t represents the current iteration number. W_*^{t-1} represents the position of the best whale at the previous iteration $t-1$. W_i^{t-1} represents the position of the ith whale at previous iteration. The coefficients A and C are used to assign the weights for the position of the best whale and the distance between the position of the best whale and position at the iteration $t-1$. The computation of A and C are shown in Eqs. (4) and (5) respectively.

$$\mathrm{A} \leftarrow 2 * \mathrm{a} * \mathrm{r} - \mathrm{a} \tag{4}$$

$$C \leftarrow 2 * r \tag{5}$$

The variable $\boldsymbol{r}$ is randomly initialized between 0 and 1. The variable a is initialized to 2 and is decreased to 0 as the iteration proceeds as shown in Eq. (6).

$$a = 2 - t * \frac{2}{num\_iter} \tag{6}$$

During the exploitation phase, either the whales update their position based on the position of the best agent or they tend to update it spirally. The choice between best agent and spiral movement is chosen based on the probability of a variable to be less than 0.5 or not. The updating of the whale's position spirally is given in Eq. (7).

$$\mathrm{W_i^t} \leftarrow \mathrm{W_*^{t-1}} + \left|\mathrm{W_i^{t-1}} - \mathrm{W_*^{t-1}}\right| * \mathrm{e^{bl}} * \cos(2\pi l) \tag{7}$$

During the exploration phase, the whales tend to update the position based on the random agent which is shown in Eq. (8).

$$\mathrm{W_i^t} \leftarrow \mathrm{W_*^{t-1}} - \mathrm{A} * \left|\mathrm{C} * \mathrm{W_{rand}^{t-1}} - \mathrm{W_i^{t-1}}\right| \tag{8}$$

The choice between exploitation and exploration is made by using the variable A. If the value of A is less than 1, then exploitation takes place else exploration takes place. The fitness function taken into account for evaluating the whale is accuracy and it is shown in Eq. (9).

$$MaxFit(W_i) \leftarrow Accuracy \tag{9}$$

The working of WOA for finding the hyperparameters mini batch size and dropout rate is shown in Algorithm 1.

Algorithm: Hyperparameter Learning using WOA

Input: $Num\_Whales = 20$; $num\_iter = 100$; $Dataset$

Output: Mini batch size and dropout rate

```
Initialize W ← {W_1, W_2, ..., W_Num_Whales}
Initialize t ← 1
while t ≤ num_iter
    For each whale W_i
        Compute Fitness using Equation (9)
    End For
    Choose the Best Whale W_*^t ← {W_i | Max (Fit(W_i))}
    For each whale W_i
        Compute a using Equation (6)
        Compute A using Equation (4)
        Compute C using Equation (5)
        Generate probability pr ∈ (0,1)
        If pr < 0.5 then
            If A < 1 then
                Compute position of the whale using Equation (3)
            Else
                Compute position of the whale using Equation (8)
            End If
        Else
            Compute position of the whale using Equation (7)
        End If
    End For
End while
Return W_*^t
```

4 Experimental Results

The experimentation is carried out on Dementia dataset taken from Kaggle repository where the dataset is viewed as a multi class problem. Total number of images in the dataset is 6400, out of which, 2240 instances belong to very mild dementia, 896 instances belong to mild dementia, 64 instances belong to moderate dementia and 3200 instances belongs to non-dementia. The training and test instances are divided in the range of 70:30 i.e., 70% of instances are viewed as training instances and 30% instances are viewed as test instances. Table 1 represents the details of the Dementia dataset taken from Kaggle [18].

4.1 Comparison of Accuracy for Various Values of Dropout Rate and Mini Batch Size

The accuracy of CNN AlexNet is compared for various values of the dropout rate. Figure 3 represents the accuracy obtained for various values of the dropout rate from 0.1 to 1.0. From the figure, it is observed that when the value of the dropout rate is 0.8, the accuracy of WOA-CNN is 98.987%. Also, when the dropout rate is 0.9, the accuracy is decreased by 1.99%. Also, the accuracy of WOA-CNN for the dropout rate = 1.0 is 97.011 which is nearly the same as for the dropout rate = 0.9. From the dropout rate 0.7, to 0.8, the accuracy is improved by 1.58%.

Also, the accuracy is measured for mini batch size 100, 200 and 400. It is observed that when the batch size is set as 100, the WOA-CNN achieves maximum accuracy of 98.987%. Table 2 represents the accuracy obtained for various values of Mini batch size.

Table 1 Details of dataset

Class	Number of instances	Training set	Test set
No dementia	2560	1792	768
Mild dementia	717	502	215
Very mild dementia	1792	1254	538
Moderate dementia	52	36	16

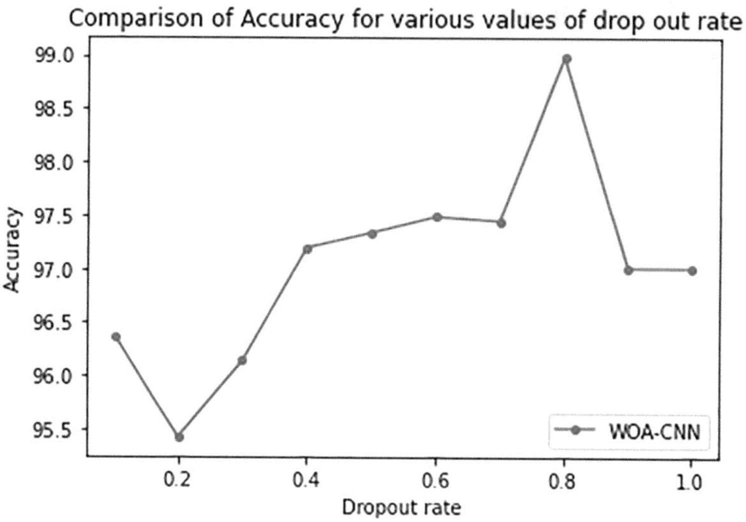

Fig. 3 Comparison of accuracy for various values of dropout rate

Table 2 Accuracy obtained for various values of mini batch size

Mini batch size	Accuracy
100	98.987
200	97.43
400	97.321

4.2 *Comparison of Accuracy*

Having found the dropout rate and mini batch size as 0.8 and 100 by WOA-CNN, the accuracy is measured for each epoch. It is evident from Fig. 4, that accuracy of WOA-CNN is higher than Conventional CNN. In other words, the WOA-CNN achieves, 3.46% greater accuracy than CNN. The reason for the increase in accuracy is because the drop-out rate and mini batch size are determined using the whale optimization algorithm for maximizing accuracy. As WOA algorithm, tends to balance the exploration and exploitation, it finds the optimal value of dropout rate and minibatch size which really prevents the model from overfitting thereby maximizing accuracy.

4.3 *Comparison of Loss*

Next level of comparison is measured for the loss value across epoch for dropout rate = 0.8 and mini batch size = 100. The graphical representation of Loss value of the proposed WOA-CNN and CNN is shown in Fig. 5. At the epoch 15, there

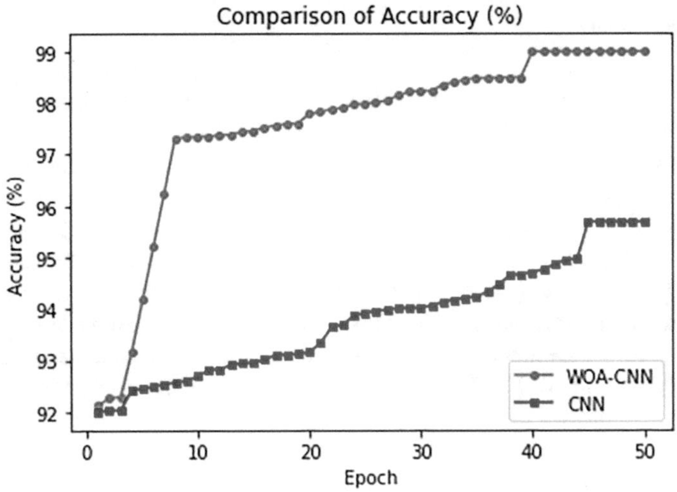

Fig. 4 Comparison of accuracy for various epoch at dropout rate = 0.8 and Mini Batch Size = 100

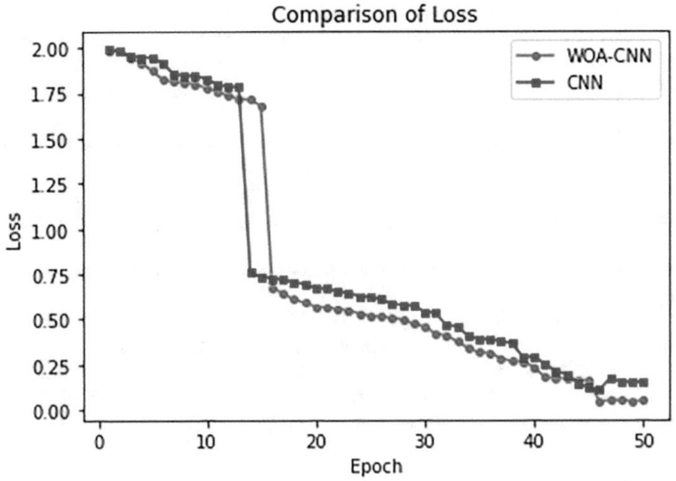

Fig. 5 Comparison of loss for various Epoch dropout rate = 0.8 and Mini Batch Size = 100

is a decrease in the loss value from 1.679 to 0.673 in WOA-CNN. And, then there is gradual decrease in the loss value. Also, the loss of the proposed WOA-CNN is 2.16% less than CNN.

5 Conclusion

Dementia, a serious neuro disorder affects the daily activities of humans, which is non curable. The essential of predicting Dementia in advance is studied as it is irreversible. Whale Optimization algorithm based Convolutional Neural Network had been experimented to predict Dementia in advance. The reason behind the integration of Whale Optimization algorithm with CNN is to prevent the model from overfitting which really degrades the performance of the model. The WOA intends to find the optimal values of the hyperparameters mini batch size and dropout rate thereby maximizing the performance of the AlexNet architecture. The experimentation was carried out on the Dementia dataset taken from Kaggle and the proposed WOA-CNN achieves loss of 0.047 which is 2.16% less than the conventional AlexNet architecture.

References

1. Mirheidari B, Blackburn D, Walker T, Reuber M, Christensen H (2019) Dementia detection using automatic analysis of conversations. Comput Speech Lang 1(53):65–79
2. Pouyanfar S, Sadiq S, Yan Y, Tian H, Tao Y, Reyes MP, Shyu ML, Chen SC, Iyengar SS (2018) A survey on deep learning: algorithms, techniques, and applications. ACM Comput Surv (CSUR). 51(5):1–36
3. Yadav SS, Jadhav SM (2019) Deep convolutional neural network based medical image classification for disease diagnosis. J Big Data 6(1):1–8
4. Aloysius N, Geetha M (2017) A review on deep convolutional neural networks. In: 2017 International conference on communication and signal processing (ICCSP), 6 April, pp 0588–0592. IEEE
5. Luz S, de la Fuente S, Albert P (2018) A method for analysis of patient speech in dialogue for dementia detection. arXiv e-prints. arXiv-1811
6. Weiner J, Engelbart M, Schultz T (2017) Manual and automatic transcriptions in dementia detection from speech. In: Interspeech 2017 Aug, pp 3117–3121
7. Weiner J, Angrick M, Umesh S, Schultz T (2018) Investigating the effect of audio duration on dementia detection using acoustic features. In: Interspeech 2018, pp 2324–2328
8. De Cola MC, Triglia G, Camera M, Corallo F, Di Cara M, Bramanti P, Lo BV (2020) Effect of neurological screening on early dementia detection in southern Italy. J Int Med Res 48(10):0300060520949763
9. Zhu Y, Liang X, Batsis JA, Roth RM (2021) Exploring deep transfer learning techniques for Alzheimer's dementia detection. Front Comput Sci, May; 3
10. Alroobaea R, Mechti S, Haoues M, Rubaiee S, Ahmed A, Andejany M, Bragazzi NL, Sharma DK, Kolla BP, Sengan S. Alzheimer's disease early detection using machine learning techniques.
11. Ambili AV, Kumar AS, El Emary IM (2021) CNN approach for dementia detection using convolutional SLBT feature extraction method. In: Computational vision and bio-inspired computing 2021. Springer, Singapore, pp 341–352
12. Meghanani A, Anoop CS, Ramakrishnan AG (2021) Recognition of Alzheimer's dementia from the transcriptions of spontaneous speech using fasttext and CNN models. Front Comput Sci 3:624558
13. Shrestha K, Alsadoon OH, Alsadoon A, Rashid TA, Ali RS, Prasad PW, Jerew OD (2021) A novel solution of an elastic net regularisation for dementia knowledge discovery using deep learning. J Exp Theor Artif Intell 5:1–23

14. Dixit U, Mishra A, Shukla A, Tiwari R (2019) Texture classification using convolutional neural network optimized with whale optimization algorithm. SN Appl Sci 1(6):1–1
15. Rana P, Gupta PK, Sharma V (2021) A novel deep learning-based whale optimization algorithm for prediction of breast cancer. Braz Arch Biol Technol 7:64
16. Zhang L, Gao HJ, Zhang J, Badami B (2019) Optimization of the convolutional neural networks for automatic detection of skin cancer. Open Med 15(1):27–37
17. Wang Y, Zhang H, Zhang G (2019) cPSO-CNN: an efficient PSO-based algorithm for fine-tuning hyper-parameters of convolutional neural networks. Swarm Evol Comput 1(49):114–123
18. Dataset. https://www.kaggle.com/tourist55/alzheimers-dataset-4-class-of-images
19. Moorthy RS, Pabitha P (2021) Prediction of Parkinson's disease using improved radial basis function neural network. CMC-Comput Mater Continua 68(3):3101–3119
20. Mirjalili S, Lewis A (2016) The whale optimization algorithm. Adv Eng Softw 1(95):51–67

Goodput Improvement with Low–Latency in Data Center Network

M. P. Ramkumar, G. S. R. Emil Selvan, M. Mahalakshmi, and R. Jeyarohini

1 Introduction

The Data Center is a facility for the housing of computational and capacity frameworks interconnected through a similar network called Data Center Network [1, 2]. In recent years, the global data center business is being extended quickly to enhance TCP [3–5]. The Transmission Control Protocol is not efficient for transferring more number of data and packets in the data center network [3, 6]. In the data center network, TCP has been associated with issues like throughput collapse, packet delay, heavy congestion, and increment of flow completion time [7]. The data center TCP flow characteristics are long queuing delay in switches (i.e.) the main reason for increment in flow completion time in the Data Center Network. Heavy congestion is made from data traffic in flow [2, 8]. Repeated packet loss and queue accuracy in packets are the causes of congestion [1, 9]. These are some of the issues affecting the DCN process. So, there is a need for modifying the existing transport protocol, for an efficient transfer of data in DCN.

M. P. Ramkumar · G. S. R. Emil Selvan · M. Mahalakshmi (✉)
Department of Computer Science and Engineering, Thiagarajar College of Engineering, Madurai, Tamilnadu, India
e-mail: maha93427@gmail.com

M. P. Ramkumar
e-mail: ramkumar@tce.edu

G. S. R. Emil Selvan
e-mail: emil@tce.edu

R. Jeyarohini
Department of Electronics and Communication Engineering, PSNA College of Engineering and Technology, Dindigul, Tamilnadu, India
e-mail: rjreee2008@gmail.com

P. Singh et al. (eds.), *Machine Learning and Computational Intelligence Techniques for Data Engineering*, Lecture Notes in Electrical Engineering 998,
https://doi.org/10.1007/978-981-99-0047-3_26

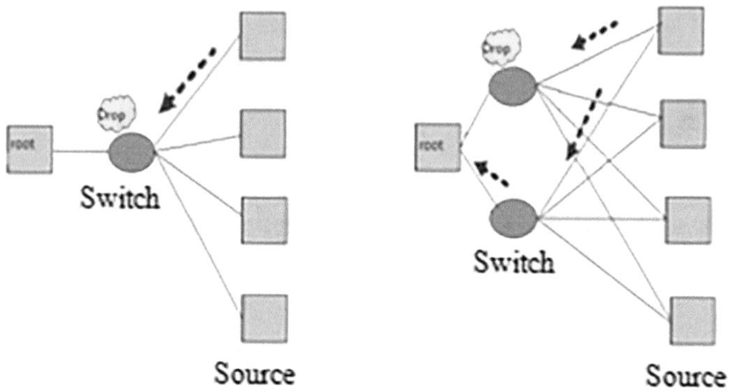

Fig.1 Single path versus multipath

Recently, the new data center architecture, such as the dual homed network topology has been introduced to offer more advanced cumulative bandwidth by taking the benefits of multiple paths [1, 10]. Multipath Transmission Control Protocol (MPTCP) is one of the proposed approaches for TCP protocols. The main advantage of MPTCP includes better aggregate throughput, data weight adjustment, more paths for data transferring, and reduces the number of not handled and unoccupied links. Figure 1 shows the scenario of single and multiple path packets sending. In the single path, the packet gets dropped but in multipath, the packet traverses through another sub-flow, so no congestion occurs in multipath networking and it can efficiently transfer packets from the source to destination.

In this paper, the proposed methodology is used to improve the performance of MPTCP in the Data Center Network called Enhanced Multipath Transmission Control Protocol. The Enhanced Multipath Transmission Control Protocol is efficient for small size packet flows and also large size packet flows. EMPTCP effectively expands the nontraffic paths and successfully completes the data transmission. The main goal of the enhanced protocol is to decrease the completion time of data flows and queuing delay. The implementation of the proposed methodology (EMPTCP) is done using network simulator and their performance is evaluated. Our experiment results show that EMPTCP considerably gives the greater performance than normal TCP and MPTCP.

2 Related Work

This section discusses some related work of the TCP improvement of data center and resolves the TCP related problems in DCN. The main problem of Transmission protocol in DCN is TCP incast, Timeouts, Latency, and Queue buildup [1, 11]. Some existing approaches are trying to resolve the TCP related problems. They are A^2DTCP, L^2DCT, DCTCP, D^2TCP and ICTCP.

DCTCP (Data Center Transmission Control Protocol) is a first modified approach for TCP. This protocol is effective for TCP incast problem [2]. In order to generate, feedback in a network using the Explicit Congestion Notification (ECN) method to end hosts, DCTCP is used [2, 12]. In the perspect of data transmission, DCTCP gives low latency and high burst tolearance [13]. The main problem of DCTCP is TCP Incast problem, where it sends more than 35 nodes in one aggregator so the performance will be failed.

ICTCP (Incast Congestion Control for Transmission Control Protocol) is another TCP modified protocol [14]. The primary objective of using ICTCP is to reduce the packet loss [14–16]. This protocol effectively handles the TCP incast problem in DCN. The main problem in ICTCP is that the sender and receiver assume that it uses the same switches for data transfer so that the space used for the buffer is unknown. Whereas at the sender side, no changes required and at the receiver side it is modified [11].

D^2TCP (Dead-Line Aware Data center Transmission Control Protocol) is one of the data center transport [17]. It handles the deadline aware of data transmission and packet transaction without delay. For allocating bandwidth, this protocol uses a direct and distributed approach. D^2TCP also avoid the Queue buildup, TCP Incast and High tolerance [12, 17]. The main issue of D^2TCP is increase in flow completion time (latency) and delay packet sending (TCP timeout).

L^2DCT (Low Latency Data Center Transport) is a kind of Transmission Control Protocol. Minimizing the latency is the main objective of the protocol [12]. L^2DCT's main methodology is additive increase mechanism. This method is used to find the previously sent amount of data [18]. This protocol reduces the completion time for short flows [11, 18]. L^2DCT protocol is slightly modified from DCTCP protocol [19]. The problem of L^2DCT is for any urgent flow, and it provides no superior precedence. Also it provides no deadline awareness for the data transmission.

A^2DTCP (Adaptive Acceleration Data Center Transmission Control Protocol) is one of the recent modification of the Transmission control protocol [12]. This protocol satisfies the latency and timeout problems. Deadline awareness is the major concern of the protocol. Many flows are to meet their deadline without delay. The issue of A^2DTCP is it can't be used for ECN mechanism deployment as it has no congestion detecting scheme.

3 Enhanced Multipath Transmission Control Protocol

This section describes on how to evaluate the MPTCP performance in Data Center Network and then design the new enhanced Multipath Transmission Control Protocol.

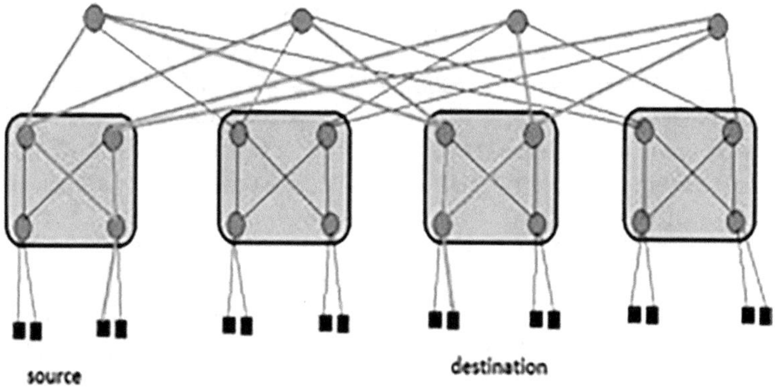

Fig.2 MPTCP in dual-homed fat tree topology

3.1 Multipath Transmission Control Protocol (MPTCP)

MPTCP is a replacement of TCP that can efficiently improve the throughput and performance. MPTCP have multiple paths so that it can transfer data efficiently compared with other TCP modified protocols [1]. During data transmission, when one path switch was failed, the packet automatically transfers to another path for a destination. In this protocol, each packet has the same amount of time for transferring data. Equal weighted congestion control algorithm is used for MPTCP protocol [1, 20, 21]. This algorithm controls the congestion or every path when sending a packet. MPTCP is effective for dual homed topology (fat tree, B-cube). MPTCP is capable of solving incast collapse, time delay problem and goodput improvement [1, 22]. Figure 2 shows multipath data transmission from source to destination in dual-homed fat tree network topology.

3.2 Packet Sprinkle

Packet Sprinkling is an effective method for small weighed packet transfer. This Sprinkling method is processed from the single path using a single congestion window. In this packet sprinkling method, data volume and transmission time will be fixed. Only smaller weighted packets are transferred by the packet sprinkling method. For example, 1 MB is the fixed value for packet sprinkling, less than 1 MB weighted data can only be sent from the source to destination. It is effective for small data flow completion time. This method is used for load balancing on small and large volume of data packets and it also eliminates the traffic from the core layer. Figure 3 shows that the packet sending from source to destination is in single sub-flow for the packet sprinkling method.

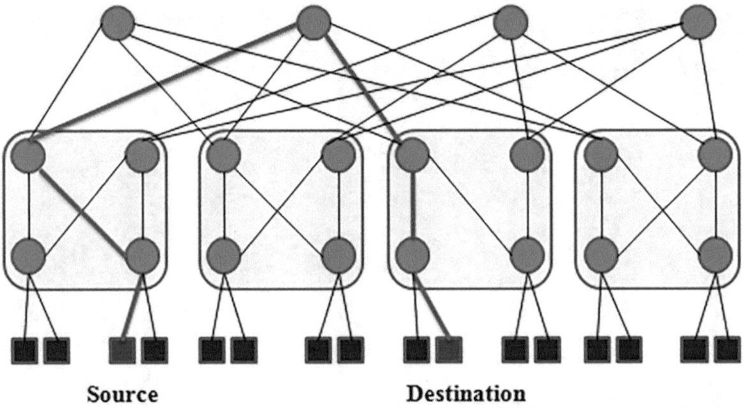

Fig. 3 Packet sprinkling from source to destination

4 Design of Proposed Protocol

In this section, the overall performance of EMPTCP is described. Enhanced MPTCP is a combined process for MPTCP and Packet sprinkling. At First, the packet sprinkling method is implemented and then the MPTCP method is processed. Packet sprinkling method is used only for small volume data packets (<300 KB), and this method will be deactivated after transmission of small volume data packets. The EMPTCP process is explained by the following steps.

Step 1: Data packets sending from source.

Step 2: At first, small volume data packets (<300 KB) will be sent from source to destination in the single path with a single congestion window with the help of the packet sprinkling method.

Step 3: Then it shuts down the packet sprinkling process.

Step 4: Second, a large volume of the data packet is sent within fixed amount of time from source to destination using multipath with the help of MPTCP.

Step 5: Finally all packets are received from the destination within the predefined time.

4.1 Architecture

Figure 4 shows the architecture of the fat tree network. The Dual-homed network topology is used for this Enhanced MPTCP. The fat tree network topology is selected for this protocol implementation. Fat tree topology is efficient for the Data Center Network. This fat tree topology is designed for k-ary fat tree multi-hop topology. In this method, the k value is 4. To find the number of host in the topology it is calculated by this formula $K^3/4$ that means 16 hosts are used. Core switch in the

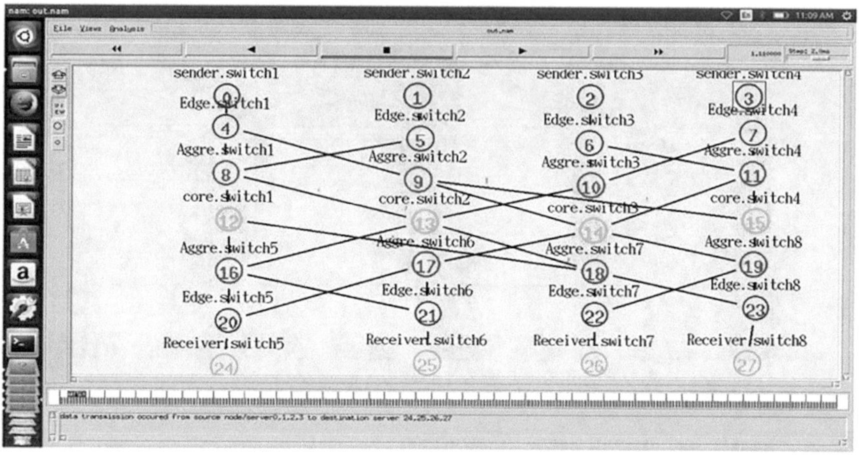

Fig. 4 Packet sending for Fat tree network topology

topology is calculated by this formula $K^2/4$ that means 4 are used. The number of edge and aggregate switches for this topology is calculated from K/2 that means 2. Each edge and aggregate switch has a K port that means 4 ports.

5 Implementation

Implementation of EMPTCP using Network Simulator (NS-2.35) and EMPTCP protocol is compared with MPTCP. Many-to-one communication pattern is used for this topology. The dual homed fat tree network topology is designed for 4 core switches, 8 aggregate switches, 8 edge switches and 8 host switches [23, 24]. In this topology, 28 nodes are running a permutation traffic matrix.

In the packet sprinkling method 300 KB packets are sent from the source to destination. Then 3000 KB packets are sent from the source to destination using MPTCP [25]. After the sending of packets trace file graphs are generated. Figure 5 shows the trace file graph for loss rate at the core layer switch.

Figure 6 shows the trace file graph for time delay acquired at the receiver. Figure 7 shows the number of packet losses in the core layer switch. This loss rate is calculated as the difference between a number of packets sending at source and number of packets leaving from the core layer switch.

Figure 8 shows the trace file graph of a number of packets received at the sink. The receiver gets packets from the sender [26]. Packet receiving are recorded in the trace file. Packet delivery ratio is the ratio between the number of packets received and sent from the destination and the source respectively and it is recorded in the trace file. Figure 9 indicates the packet delivery ratio through trace file graph.

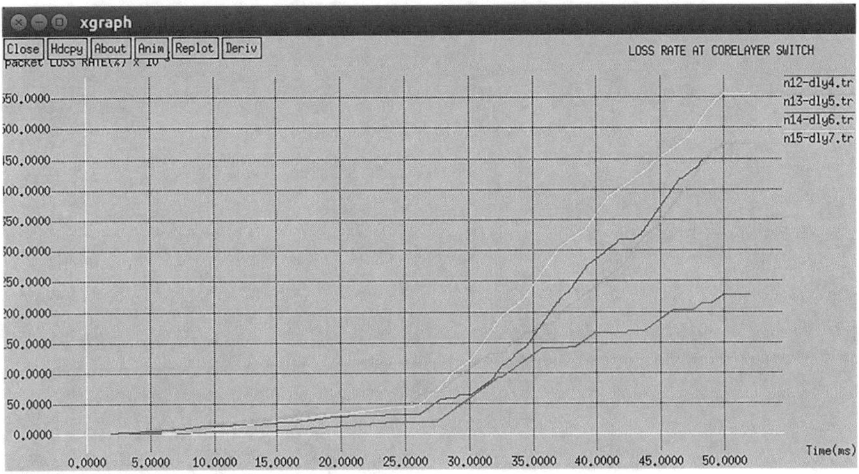

Fig. 5 Representation of core layer loss rate in trace file graph

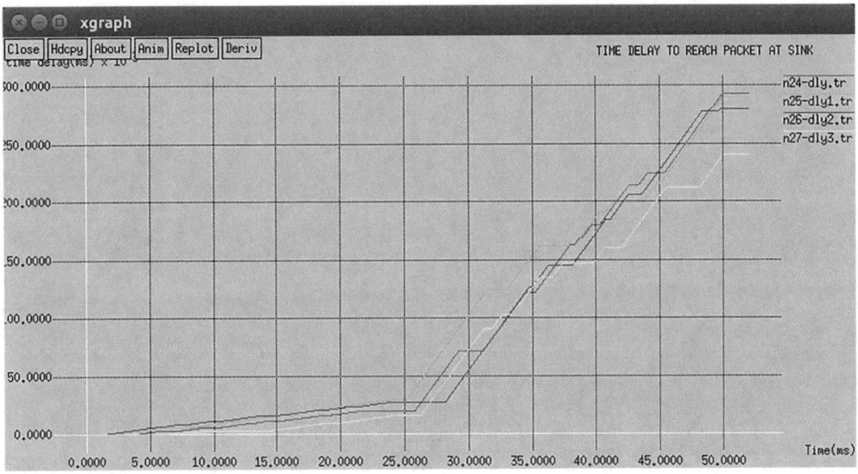

Fig. 6 Representation of time delay to reach packet at receiver in trace file graph

6 Performance Analysis

This part illustrates the performance analysis of the MPTCT and EMPTCP Data Center Network protocols. The protocols are evaluated by considering various factors like Goodput, Packet loss rate and Finishing time. Figure 10 shows the Goodput comparison between MPTCP and EMPTCP. Goodput is computed by considering the file size that is transmitted and the time taken for transmission.

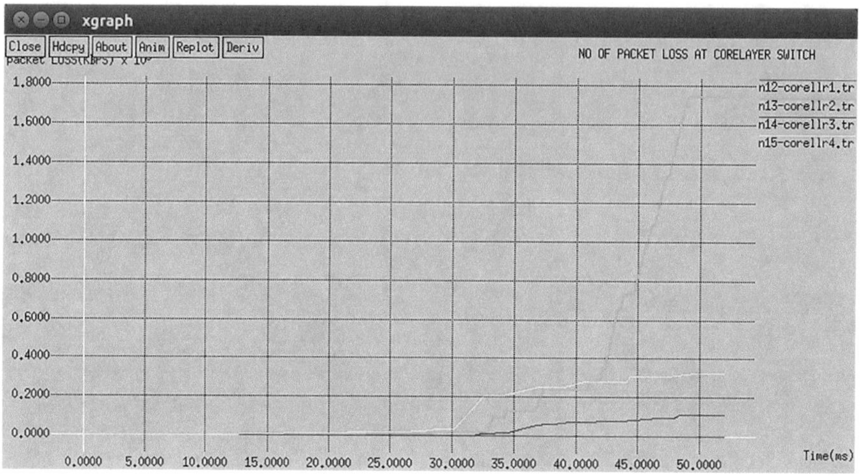

Fig. 7 Trace file graph for number of packet loss at core layer switch

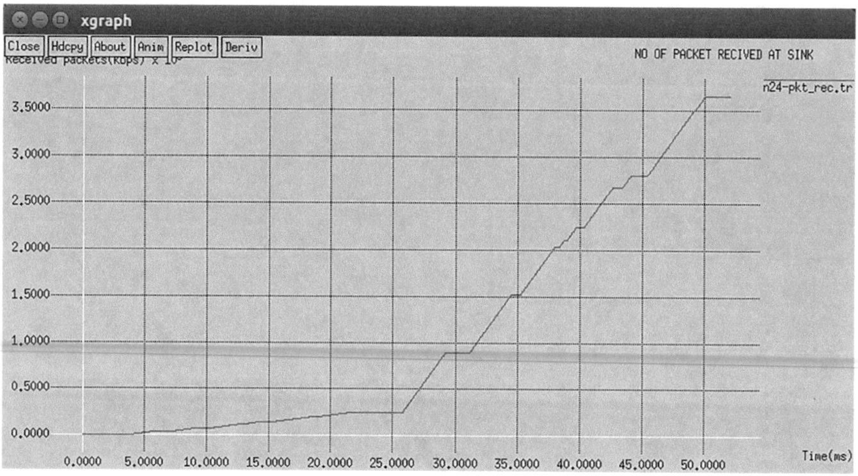

Fig. 8 Trace file graph for Number of packets received at sink

Table 1 shows the mean goodput comparison for MPTCP and EMPTCP based on hotspot degree and goodput value. Goodput is calculated by the ratio of a number of packets delivered to the total delivery time. EMPTCP have high goodput value than MPTCP.

Figure 11 shows the comparison graph of core switch loss rate at MPTCP and EMPTCP. The packet gets dropped at the core switch is called as core loss rate. The core layer loss rate of EMPTCP is lesser than MPTCP.

Table 2 show the mean comparison for core layer loss rate at MPTCP and EMPTCP. Packets are lost after a core layer is called Core Layer Rate. The packet

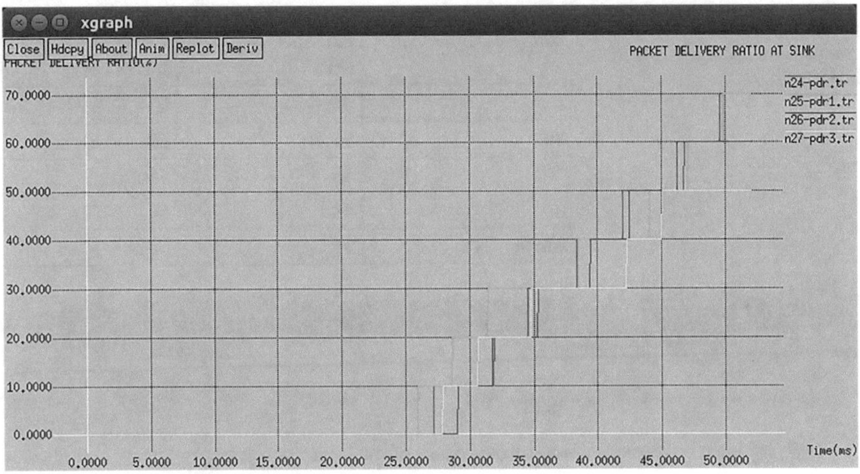

Fig. 9 Trace file graph for packet delivery ratio

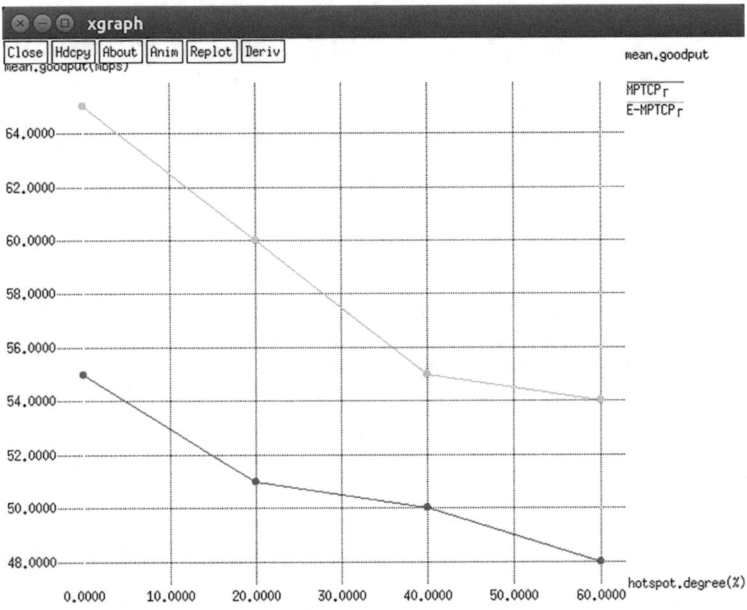

Fig. 10 Goodput comparison between MPTCP and EMPTCP

loss rate is calculated by the ratio of a number of lost packets to the number of received packets.

Table 1 Goodput comparison for MPTCP and EMPTCP

Hotspot degree (%)	Goodput (Mbps) MPTCP	Goodput (Mbps) EMPTCP
5	55	65
20	51	60
40	50	55
60	48	54
Mean	51	58.5

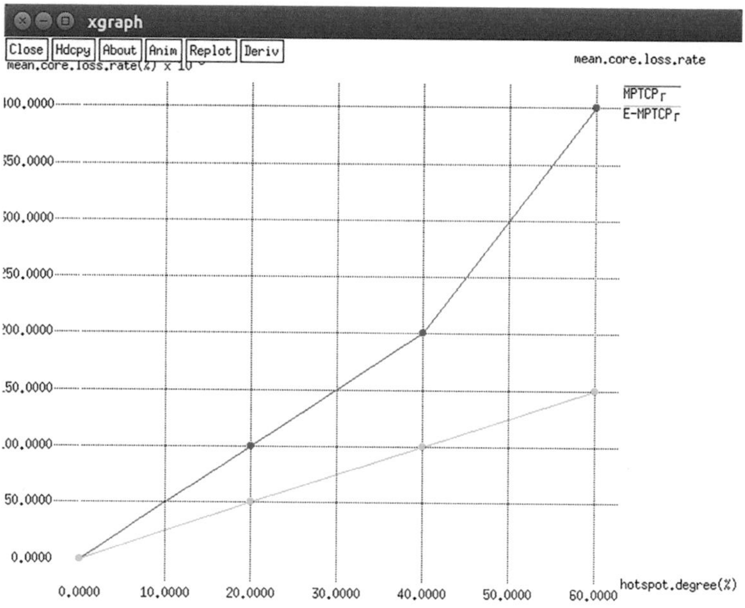

Fig. 11 MPTCP and EMPTCP comparison: core loss rate

Table 2 Core layer loss rate comparison for MPTCP and EMPTCP

Hotspot degree (%)	Packet loss rate (%) MPTCP	Packet loss rate (%) EMPTCP
5	0	0
20	0.1	0.05
40	0.2	0.1
60	0.4	0.15
Mean	0.175	0.075

Figure 12 shows the Finishing time comparison for MPTCP and EMPTCP.EMPTCP finishing time is faster than MPTCP finishing time. So EMPTCP is efficient for data transfer.

Table 3 shows the mean finishing time comparison between MPTCP and EMPTCP. This MPTCP and EMPTCP protocols' Finishing time are compared with fixed amount of hotspot degree. Table 4 shows the overall Comparison of Finish Time, Goodput, and Core Layer Loss Rate for MPTCP and EMPTCP Based on Mean Value.

During our performance analysis, Enhanced Multipath TCP is efficient for transferring data packets in Data Center Network. EMPTCP have achieved high goodput, low packet loss rate, and low latency.

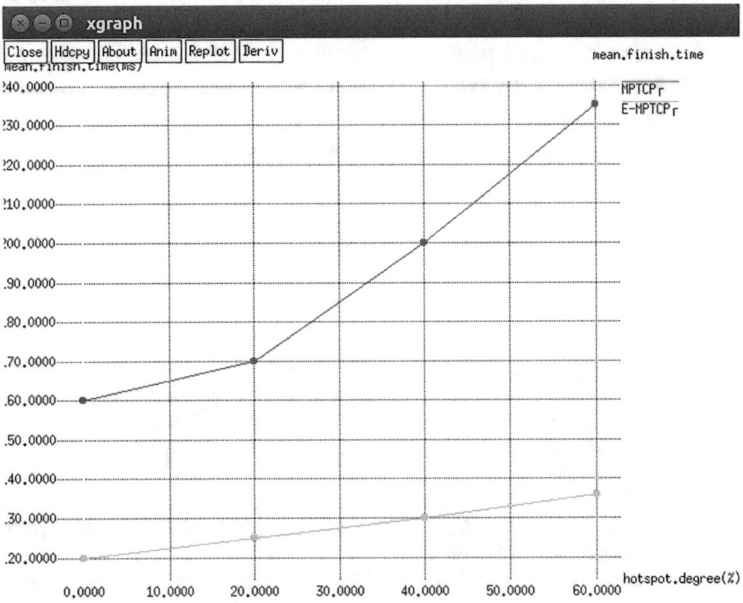

Fig. 12 Finishing time comparison for MPTCP and EMPTCP

Table 3 Finishing Time Comparison for MPTCP and EMPTCP

Hotspot degree (%)	Time (ms) MPTCP	Time (ms) EMPTCP
5	0	0
20	0.1	0.05
40	0.2	0.1
60	0.4	0.15
Mean	0.175	0.075

Table 4 Mean comparison for MPTCP and EMPTCP

Transport protocol	Finishing time (ms)	Goodpout (Mbps)	Core layer loss rate (%)
MPTCP	191	51	0.175
EMPTCP	127.75	58.5	0.075

7 Conclusion and Future Work

In this paper the proposed protocol EMPTCP (i.e.) modified mechanism of MPTCP effectively transmits data using multiple paths. EMPTCP achieves high goodput for a large volume data packets and low time delay for a small volume data packet. Packet sprinkling method is additionally added for the MPTCP mechanism. MPTCP is effective for a large volume of data packets whereas Packet sprinkling is efficient for a small volume data packet [27]. Enhanced MPTCP reduces the traffic for data flows and improves the throughput. Small volume data packets are sent to the packet sprinkling method so less time is taken for packet transferring. EMPTCP has better performance than the data center network protocols such as TCP, DCTCP, and MPTCP.

References

1. Yao J, Pang S, Rodrigues JJ, Lv Z, Wang S (2021) Performance evaluation of MPTCP incast based on queuing network. In: IEEE transactions on green communications and networking
2. Yan S, Wang X, Zheng X, Xia Y, Liu D, Deng W (2021) ACC: automatic ECN tuning for high-speed datacenter networks. In: Proceedings of the 2021 ACM SIGCOMM 2021 conference, August, pp 384–397
3. Zhang Q, Ng KK, Kazer C, Yan S, Sedoc J, Liu V (2021). MimicNet: fast performance estimates for data center networks with machine learning. In: Proceedings of the 2021 ACM SIGCOMM 2021 conference, August, pp 287–304
4. Casoni M, Grazia CA, Klapez M, Patriciello N (2017) How to avoid TCP congestion without dropping packets: an effective AQM called PINK. Comput Commun 103:49–60
5. Pei C, Zhao Y, Chen G, Meng Y, Liu Y, Su Y, ... Pei D (2017) How much are your neighbors interfering with your WiFi delay?. In: 2017 26th international conference on computer communication and networks (ICCCN), July. IEEE, pp 1–9
6. Alipio M, Tiglao NM, Bokhari F, Khalid S (2019) TCP incast solutions in data center networks: a classification and survey. J Netw Comput Appl 146:102421
7. Lu Y, Ma X, Xu Z (2021) FAMD: a flow-aware marking and delay-based TCP algorithm for datacenter networks. J Netw Comput Appl 174:102912
8. Xu Y, Shukla S, Guo Z, Liu S, Tam ASW, Xi K, Chao HJ (2019) RAPID: Avoiding TCP incast throughput collapse in public clouds with intelligent packet discarding. IEEE J Sel Areas Commun 37(8):1911–1923
9. Tsiknas KG, Aidinidis PI, Zoiros KE (2021) On the fairness of DCTCP and CUBIC in cloud data center networks. In: 2021 10th international conference on modern circuits and systems technologies (MOCAST), July. IEEE, pp 1–4
10. Mirzaeinnia AC, Mirzaeinia M, Rezgui A (2020) Latency and throughput optimization in modern networks: a comprehensive survey. arXiv preprint. arXiv:2009.03715

11. Bai W, Hu S, Chen K, Tan K, Xiong Y (2020) One more config is enough: saving (DC) TCP for high-speed extremely shallow-buffered datacenters. IEEE/ACM Trans Netw 29(2):489–502
12. Mukerjee MK, Canel C, Wang W, Kim D, Seshan S, Snoeren AC (2020) Adapting {TCP} for reconfigurable datacenter networks. In: 17th {USENIX} symposium on networked systems design and implementation ({NSDI} 20), pp 651–666
13. Zhang T, Wang J, Huang J, Chen J, Pan Y, Min G (2017) Tuning the aggressive TCP behavior for highly concurrent HTTP connections in intra-datacenter. IEEE/ACM Trans Netw 25(6):3808–3822
14. Kumar VA, Das D, IEEE SM (2021) Data sequence signal manipulation in multipath TCP (MPTCP): the vulnerability, attack and its detection. Comput Secur 103:102180
15. Noormohammadpour M, Raghavendra CS (2017) Datacenter traffic control: understanding techniques and tradeoffs. IEEE Commun Surv Tutor 20(2):1492–1525
16. Ramkumar MP, Narayanan B, Selvan GSR, Ragapriya M (2017) Single disk recovery and load balancing using parity de-clustering. J Comput Theor Nanosci 14(1):545–550. ISSN 15461955
17. Zhang X, Zhu Q (2021) Statistical tail-latency bounded QoS provisioning for parallel and distributed data centers. In: 2021 IEEE 41st international conference on distributed computing systems (ICDCS), July. IEEE, pp 762–772
18. Nguyen TA, Min D, Choi E, Tran TD (2019) Reliability and availability evaluation for cloud data center networks using hierarchical models. IEEE Access 7:9273–9313
19. Dittmann L (2016) Data center network. 34350—broadband networks. Technical University of Denmark
20. Ramkumar MP, Balaji N, Rajeswari G (2014) Recovery of disk failure in RAID-5 using disk replacement algorithm. Int J Innov Res Sci, Eng Technol 3(3):2319–8753. ISSN (Online). ISSN (Print): 2347–6710, 2358–2362
21. Zhu L, Wu J, Jiang G, Chen L, Lam SK (2018) Efficient hybrid multicast approach in wireless data center network. Futur Gener Comput Syst 83:27–36
22. Gulhane M, Gupta SR (2014) Data center transmission control protocol an efficient packet transport for the commoditized data center. Int J Comput Sci Eng Technol (IJCSET) 4:114–120
23. Modi T, Swain P (2019) FlowDCN: flow scheduling in software defined data center networks. In: 2019 IEEE international conference on electrical, computer and communication technologies (ICECCT), February. IEEE, pp 1–5
24. Srijha,V, Ramkumar MP (2018) Access time optimization in data replication. In: 2018 2nd international conference on trends in electronics and informatics (ICOEI), May. IEEE, pp 1161–1165
25. Chefrour D (2021) One-way delay measurement from traditional networks to SDN: a survey. ACM Comput Surv (CSUR) 54(7):1–35
26. Lin D, Wang Q, Min W, Xu J, Zhang Z (2020) A survey on energy-efficient strategies in static wireless sensor networks. ACM Trans Sens Netw (TOSN) 17(1):1–48
27. Ramkumar MP, Balaji N, Vinitha Sri S (2013) Stripe and disk level data redistribution during scaling in disk arrays using RAID 5. In: PSG—ACM conference on intelligent computing, 26–27 April, Vol 1, Article No. 74, pp 469–474

Empirical Study of Image Captioning Models Using Various Deep Learning Encoders

Gaurav and Pratistha Mathur

1 Introduction

In the realm of computer vision and natural language processing, image captioning is an extremely important research topic. It does not only detect the important objects from an image but also determines the relationship between the various objects so that a meaningful and relative caption can be generated. Only detecting the image objects is not the sole purpose of image captioning. Representing these image objects along with their properties into a fined sentence is the main purpose of image captioning. Image captioning can be helpful for visually impaired people [1]. There are more fields where image captioning can be used like biomedicine, the military, education etc.

Initially, Template based methods were implemented in which a fixed size template was used for fill up with image objects and their properties. The retrieval based approach was used in which some images were used to match with the query image and a caption was generated with the help of captions of these images. There were some limitations in these approaches of missing out important objects. Recent approaches for the images captioning uses simple encoder and decoder based methods [2, 3]. Encoder takes the image as input and extracts the important features and the use of decoder is to convert the features into an appropriate caption [4, 5]. In this paper, we have compared the performance of image captioning models with various image encoders such as Visual Geometry Group (VGG), Residual Networks (ResNet), InceptionV3 while using the Gated Recurrent Unit (GRU) as the decoder for the purpose of text generation.

Gaurav (✉) · P. Mathur
Manipal University Jaipur, Jaipur, Rajasthan, India
e-mail: gauravsingla31@gmail.com

P. Mathur
e-mail: pratistha.mathur@jaipur.manipal.edu

P. Singh et al. (eds.), *Machine Learning and Computational Intelligence Techniques for Data Engineering*, Lecture Notes in Electrical Engineering 998,
https://doi.org/10.1007/978-981-99-0047-3_27

These various encoders are different based on their architectures. VGG16 and VGG19 takes input of image size 224 × 224. Then image passes through various convolution layers with kernel size of 3 × 3. The fully connected layers come after the convolutional layers. In VGG16 and VGG19, there are 13 convolutional layers and 16 convolutional layers respectively. InceptionV3 encoder takes image of size 299 × 299 as input and uses kernels with different sizes such as 1 × 1, 3 × 3 and 5 × 5. It uses the concept of batch normalization. The network's last layer is a 1 × 1 × 1000 linear layer. ResNet also takes an image of size 299 × 299 as the input. In ResNet, the filter expansion layer comes after the inception block and batch normalization is used on top of traditional layers only.

For image captioning, there are a variety of datasets accessible. Most popular are Flickr8K dataset, Flickr30K dataset and MSCOCO dataset [6]. Flickr8k contains 8,092 images and each image is represented using five captions. Flickr30k dataset contains 30,000 images with each image having five captions. MSCOCO contains around 3,28,000 images on various objects. In this paper, the image captioning model with various encoders and GRU as decoder has been implemented using the Flickr8k dataset. The results are compared on the basis of BLEU score using Flickr8K Dataset. BLEU score metrics is the most popular among all evaluation metrics. BLEU score refers to the level of match between the generated caption and referenced caption. BLEU score comes in between 0 and 1. Score '0' means there is no match at all between the referenced caption and generated caption. BLEU score as 1 means the generated caption is exactly same as the referenced caption [7].

2 Related Works

2.1 Past Work

There have been a variety of techniques for image captioning. Initially, Template based methods were implemented in which a fixed size template was used for fill up with image objects and their properties. This method was good enough to generate image captions but there was some limitation like fixed size of the generated caption. Later on, a Retrieval based approach was used in which some images were matched with the query image and the final caption was used to generate using the captions of these related images. Still, there were some limitations of missing out important objects existing in the query image. Recently approaches are based on using encoder and decoder [2, 3]. Encoder takes the image as input and extracts the important features and the use of decoder is to convert the features into an appropriate caption [4, 5]. There are many more encoder decoder based models that have been designed for the task of image captioning. Attention mechanisms have been used to generate more semantical and syntactical accurate captions. Image based attention mechanisms and textual based attention mechanisms have been used to incorporate more accuracy and relativeness into the captions.

A model emphasises multi-task learning helps to increase model generality and performance [8]. The Deep Hierarchical Encoder Decoder Network, which uses deep networks' representation power to efficiently combine vision and language semantics at a high level in the production of captions, was introduced for image captioning [9]. One method employs adaptive and dense net attention mechanisms [10]. A model is presented for a domain-specific image caption generator that uses a semantic ontology to deliver natural language descriptions for a given domain by combining the object and attribute information with attention processes to construct captions [11]. There are some models that have focused only upon the attention mechanism for better caption generation. To provide more accurate captions, a two-part attention model was developed, combining an attention model at the word level with another attention model at the sentence level [12].

2.2 *Datasets*

Various datasets are available for image captioning. Flickr8k is the dataset which contains 8,000 images. Each image in the dataset contains five captions. Flickr30k is the dataset which contains 30,000 images and each image is represented using five captions. MSCOCO is the dataset which contains 3,28,000 images related to various objects[13]. In this paper, the dataset that has been used is Flickr8k. A brief summary of Flickr8k dataset is shown in the Table 1.

3 Image Captioning

Image captioning is the process of generating a meaningful caption for the given image and it has been a very interesting research topic in recent times. There have been many approaches for image captioning. Deep learning-based image captioning methods can be seen in the Fig. 1. The recent approach for the image captioning is to use the encoding process to extract the features of an image and decoding process to use the features and generate an appropriate caption. Initially, Template based methods were implemented in which a fixed size template was used for fill up with

Table 1 Summary of Flicrk8k dataset

	Dataset size	Training data size	Testing data size	Development data size
Flickr8K Dataset	8,092 Images	6,000 Images	1,000 Images	1,000 Images
	Number of captions	Total number of captions	Number of unique words	Max length of any caption
	5 per Image	40,000	8763	40 Words

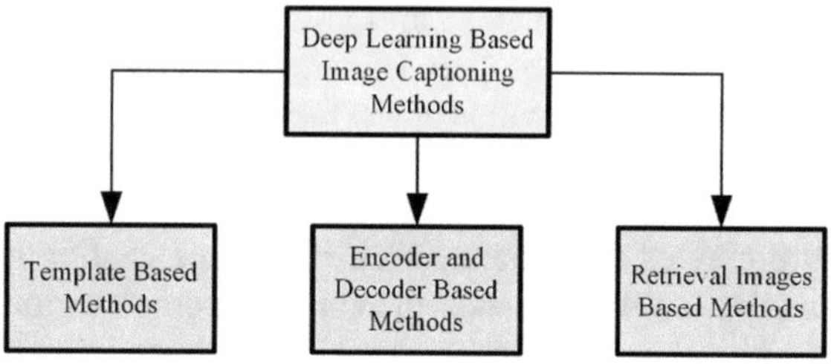

Fig. 1 Deep learning-based image captioning methods

image objects and their properties. This method was good enough to generate image captions but there were some limitations like fixed size of the generated caption. Later on, Retrieval based approach was used in which some images were matched with the query image and the final caption is used to generate using the captions of these related images. Still, there were some limitations of missing out important objects that are existing in the query image. Later on, the main focus given onto the encoder and decoder based methodology [2, 3].

Among the deep learning-based methods for image captioning, encoder and decoder-based approach is highly used because of its performance. The most common encoder used is the Convolutional Neural Network (CNN). It extracts the features of an image and either uses it for any specific task or transfers it to any other network for further use. The most common decoder used is Long Short-Term Memory (LSTM) that takes the features of the image from CNN and then produces the caption by generating one word at each processing of the network. Here in this paper, we have used the Gated Recurrent Unit (GRU) as a decoder. A simple encode and decoder-based framework can be seen in the Fig. 2.

In the next section, we have represented the architecture of various convolutional neural networks that we have used as the encoder for the implementation of image captioning process. VGG16, VGG19, Resnet, InceptionV3 are the already trained networks that are used as encoders for image classification, object detection and other image related works. Here, we have used these networks as encoders and tested each and every encoder one by one using GRU as the decoder.

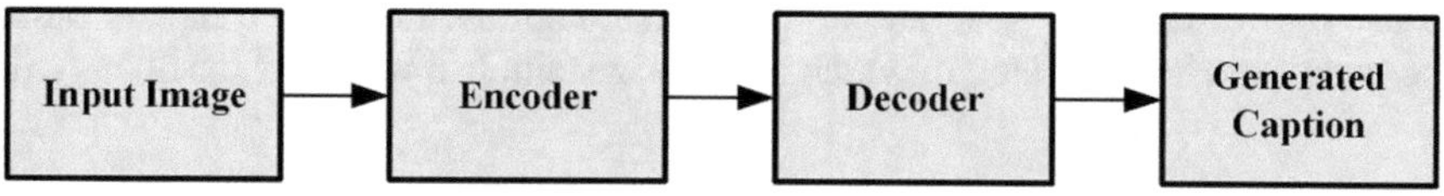

Fig. 2 A simple architecture of encoder and decoder-based image captioning framework

3.1 Encoders

Image captioning is achieved using the process of encoding and decoding. Encoders are used to obtain important features of the image and decoders are used to interpret those features and generate the appropriate caption for the image. In this section, we will discuss predefined encoders that are used in image captioning.

VGG16 and VGG19. VGG16 is an encoder that takes the image as input of size 224 × 224 and it is passed through 13 convolutional layers and 3 fully connected layers. VGG19 is similar to VGG16 but the former has 16 convolutional layers [14]. The architecture of VGG16 is shown in Fig. 3. VGG19 also has almost the same architecture as VGG16. The difference between VGG16 and VGG19 is that there are three layers of Conv 3 × 3 (256 channels) and Conv 3 × 3 (512 channels) in VGG16 while there are four layers Conv 3 × 3 (256 channels) and Conv 3 × 3 (512 channels) in VGG19.

InceptionV3. InceptionV3 encoder takes input image of size 299 × 299 and applies different kernel sizes e.g., 1 × 1, 3 × 3, 5 × 5. It uses the concept of batch normalization. The final layer of the neural network is the Linear layer with size 1 × 1x1000. Its architecture is shown in Fig. 4.

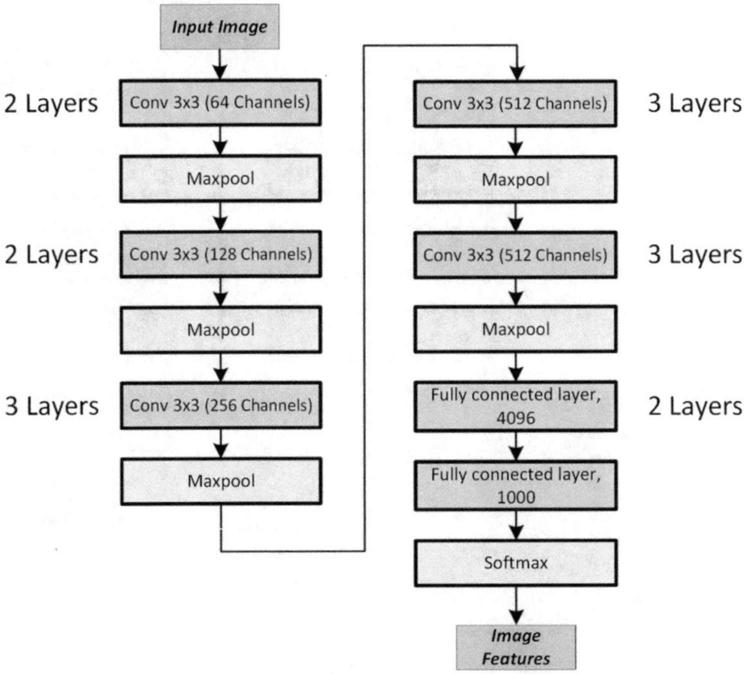

Fig. 3 VGG16 architecture

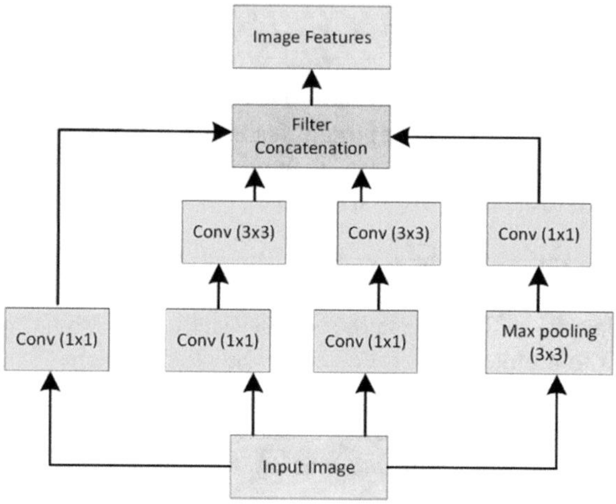

Fig. 4 InceptionV3 architecture

ResNet. ResNet takes the input image of size 299 × 299. The filter expansion layer comes after the inception block. Its architecture is shown in Fig. 5.

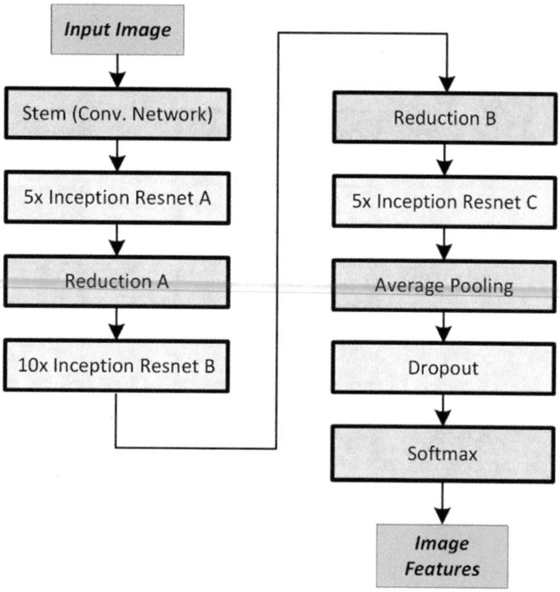

Fig. 5 ResNet architecture

3.2 Gated Recurrent Unit (Decoder)

GRU is a decoder that has been used to generate the caption using the features received from the encoder. Here, we have seen different encoders like VGG16, VGG19, InceptionV3, ResNet. GRU works similar to LSTM but it has more advantages over to LSTM. LSTM was invented in 1995–97 while GRU was invented in 2014. GRU is more efficient computation wise and exposes complete memory and hidden layers as compared to LSTM. Different models have been tested taking all these encoders one by one along with GRU as a decoder. GRU uses two different gates such as Reset gate and Update gate. These gates are used to decide that either to retain or forget the information. We can see the architecture of the Gated Recurrent Unit in Fig. 6.

The following equations show the use of update cell, reset cell and generation of new hidden state using previous hidden state and other parameters:

$$u_t = \sigma(W^u i_t + V^u h_{t-1}) \tag{1}$$

$$r_t = \sigma(W^r i_t + V^r h_{t-1}) \tag{2}$$

$$h_t^{'} = \tanh(W i_t + r_t.V h_{t-1}) \tag{3}$$

$$h_t = u_t.h_{t-1} + (1 - u_t).h_t^{'} \tag{4}$$

Here u_t is the update cell or gate that is used to determine which information from the pass is to retain for the future use. It takes the input vector multiplied by its own weight vector. The previous hidden state is multiplied by its own weight and then sigmoid activation function is applied onto the summation of both the results as shown in Eq. (1). Next gate or cell r_t is used to decide which information to forget.

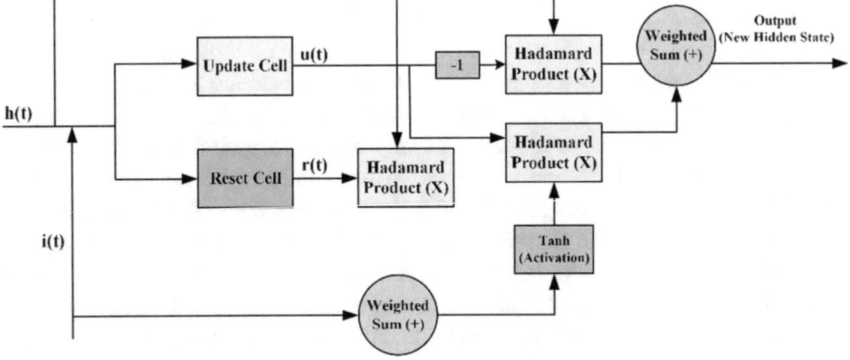

Fig. 6 GRU architecture

On processing, it is almost similar to the update gate. It takes the input vector and multiplies it by its own weight vector. The previous hidden state is multiplied by its own weight and then sigmoid activation function is applied onto the summation of both the results as shown in Eq. (2).

At time step t, once the update gate and reset gate value is obtained, we calculate the memory content. Current input vector is multiplied by weight W. Previous hidden state value is multiplied by weight V and resultant vector is multiplied by vector r_t. Then summation of both the results is obtained and tanh activation function is applied to obtain the memory content as shown in Eq. (3).

Final h_t is calculated that contains the information for the network and is applied as one of the inputs for the next time step. It is calculated by multiplying the previous hidden state and updating the cell state. Memory content is multiplied by negation of the update cell state. Summation of these two vector multiplications provide us the new hidden state or the output of the network at current time step t as shown in Eq. (4).

In the Fig. 6, the symbol (+) represents the weighted sum of the two matrices and (X) represents the Hadamard product. The Hadamard product is the product of two matrices in a similar fashion to matrix addition where the corresponding elements are multiplied with each other to form the elements of the newly generated matrix. The example of Hadamard matrix can be seen in fig. [15].

GRU is a recurrent neural network and it repeats the same process again and again until some specific or mentioned information is received. In our process of image captioning, GRU will generate one word at one time step. It will use the generated output and new input vector to produce the next word. This process will be repeated till the final caption is generated as per the model is trained.

4 Experiments

Various frameworks have been implemented using Flickr8k dataset. In each of the model, different encoders like VGG16, VGG19, InceptionV3, ResNet has been used and GRU is the common decoder for all the models. Flickr8k is the dataset that contains 6,000 images for training and 1,000 images for testing and 1,000 for validation respectively. During training images features are extracted and image attention helps to identify the important features of the image. The global features are passed only once and the local features are passed at every time step during caption generation. GRU decoder receives global features of the image and the local features of the image and produces the caption. The important training parameters are Learning rate, Optimizer etc. Learning rate is a parameter that is used to adjust the weights used in the network w.r.t. loss gradient. Learning rate is taken as 0.01. Keeping the low value means that there will be slow change in existing weights to reach at a value where loss is minimum. It is taken as 0.01, so that there is no missing of any local minima where loss is minimum [16]. Formula for adjusting weights w.r.t. loss gradient is given in Eq. (5).

Optimizer is RMSProp. This optimizer is like the gradient descent algorithm along with a momentum. Bilingual Evaluation Understudy (BLEU) is evaluated for the generated caption [8].

$$\text{new}_{\text{weights}} = \text{old}_{\text{weights}} - \text{learning rate} * \text{gradient} \tag{5}$$

4.1 Result Analysis

In this paper, the various image captioning models using VGG16, VGG19, InceptionV3, Resnet as encoders and GRU as a decoder has been implemented. Results of these different combinations of models is presented using BLEU scores. There are some more evaluation metrics e.g. METEOR (Metric for Evaluation for Translation with Explicit Ordering), CIDEr (Consensus based Image Description Evaluation) etc. that shows the performance of image captioning models [17]. BLEU score metrics is the most popular among all evaluation metrics. BLEU score refers the level of match between the generated caption and referenced caption. BLEU score comes in between 0 and 1. When score is 0, it means there is no match at all between referenced caption and generated caption. BLEU score as 1 means the generated caption is exactly same as referenced caption. There are four different types of BLEU scores referred as BLEU-1, BLEU-2, BLEU-3 and BLEU-4. Referenced captions are a list of captions which is matched with the generated caption. In BLEU-1, one word (one-gram) is matched in between the referenced caption and generated caption. Number of words matched decides the BLEU score. To calculate BLEU-2 score, two words are matched at a time. To calculate the BLEU-3 score, three words are matched at a time. To calculate BLEU-4 score, four words are matched at a time. BLEU score is highly useful to check the quality of generated caption. In Table 2, we have shown BLEU-1, BLEU-2, BLEU-3, BLEU-4 scores for all different combinations of encoders and decoders taken into consideration. VGG16 as the encoder and GRU as the decoder provided the scores as 0.6938, 0.5897, 0.4923 and 0.3992 respectively. VGG19 as the encoder and GRU as the decoder provided the scores as 0.7109, 0.5998, 0.4985 and 0.4056 respectively. InceptionV3 as the encoder and GRU as the decoder provided the scores as 0.7302, 0.6098, 0.4912 and 0.4098 respectively. ResNet as the encoder and GRU as the decoder provided the scores as 0.7389, 0.6185, 0.5081 and 0.4192 respectively.

As per the results available, it can be observed that the best BLEU scores is obtained when ResNet has been taken as an encoder into the image captioning model. The graphical representation of the result analysis is shown in Fig. 7. In the graphical representation, the analysis has been shown using bar graph as a comparative analysis of various models.

In Fig. 8, a sample output is shown using an image from Flickr8k dataset with the generated caption. The caption has been generated using a model with Resnet as the encoder, GRU as the decoder.

Table 2 BLEU-1 to BLEU-4 scores based on various encoders

	BLEU-1	BLEU-2	BLEU-3	BLEU-4
VGG16	0.6938	0.5897	0.4923	0.3992
VGG19	0.7109	0.5998	0.4985	0.4056
InceptionV3	0.7302	0.6098	0.4912	0.4098
ResNet	0.7389	0.6185	0.5081	0.4192

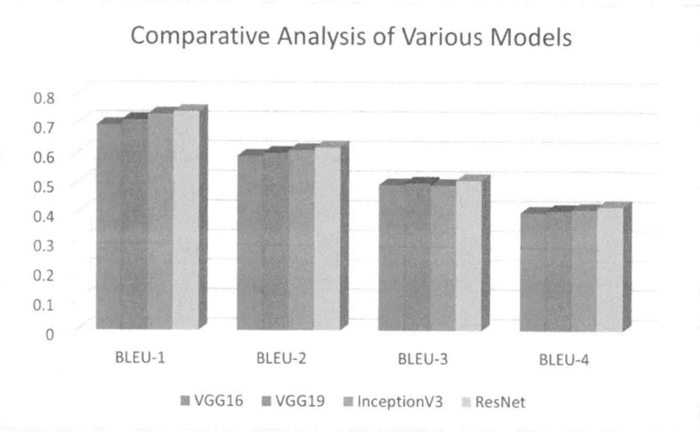

Fig. 7 Comparative analysis of various models based on different encoders

A man in a hat stands next to a skier with a blue hat, presenting photographs.

Fig. 8 A sample image from Flickr8k dataset with generated caption

5 Conclusion

In this paper, image captioning has been implemented using different encoders such as VGG16, VGG19, InceptionV3, ResNet. GRU has been taken as a common decoder for all the models. Flickr8K is the dataset that has been used for the training of the model. BLEU score is the evaluation metric that represents the level of matching between the referenced captions and generated captions. As per the result analysis, it can be clearly seen that the model with ResNet as an encoder and GRU as a decoder has provided the best results among all the models. We have taken the combination of a simple encoder and decoder. In the future, our work will be towards the implementation of image captioning models using some attention mechanism for better results in the field of image captioning.

References

1. Pal A, Kar S, Taneja A, Jadoun VK (2020) Image captioning and comparison of different encoders. J Phys Conf Ser 1478. https://doi.org/10.1088/1742-6596/1478/1/012004
2. Yang Z, Liu Q (2020) ATT-BM-SOM: a framework of effectively choosing image information and optimizing syntax for image captioning. IEEE Access 8:50565–50573. https://doi.org/10.1109/ACCESS.2020.2980578
3. Wang H, Wang H, Xu K (2020) Evolutionary recurrent neural network for image captioning. Neurocomputing 401:249–256. https://doi.org/10.1016/j.neucom.2020.03.087
4. Lu X, Wang B, Zheng X, Li X (2017) Sensing image caption generation. IEEE Trans Geosci Remote Sens 56:1–13
5. Wang B, Zheng X, Qu B, Lu X, Member S (2020) Remote sensing image captioning. IEEE J Sel Top Appl Earth Obs Remote Sens 13:256–270
6. Liu M, Li L, Hu H, Guan W, Tian J (2020) Image caption generation with dual attention mechanism. Inf Process Manag 57:102178. https://doi.org/10.1016/j.ipm.2019.102178
7. Gaurav, Mathur P (2011) A survey on various deep learning models for automatic image captioning. J Phys Conf Ser 1950. https://doi.org/10.1088/1742-6596/1950/1/012045
8. Wang C, Yang H, Meinel C (2018) Image captioning with deep bidirectional LSTMs and multi-task learning. ACM Trans Multimed Comput Commun Appl 14 (2018). https://doi.org/10.1145/3115432
9. Xiao X, Wang L, Ding K, Xiang S, Pan C (2019) Deep hierarchical encoder-decoder network for image captioning. IEEE Trans Multimed 21:2942–2956. https://doi.org/10.1109/TMM.2019.2915033
10. Deng Z, Jiang Z, Lan R, Huang W, Luo X (2020) Image captioning using DenseNet network and adaptive attention. Signal Process Image Commun. 85:115836. https://doi.org/10.1016/j.image.2020.115836
11. Han SH, Choi HJ (2020) Domain-specific image caption generator with semantic ontology. In: Proceedings—2020 IEEE international conference on big data and smart computing (BigComp), pp 526–530. https://doi.org/10.1109/BigComp48618.2020.00-12
12. Wei H, Li Z, Zhang C, Ma H (2020) The synergy of double attention: combine sentence-level and word-level attention for image captioning. Comput Vis Image Underst 201:103068. https://doi.org/10.1016/j.cviu.2020.103068
13. Kalra S, Leekha A (2020) Survey of convolutional neural networks for image captioning. J Inf Optim Sci 41:239–260. https://doi.org/10.1080/02522667.2020.1715602

14. Yu N, Hu X, Song B, Yang J, Zhang J (2019) Topic-Oriented image captioning based on order-embedding. IEEE Trans Image Process 28:2743–2754. https://doi.org/10.1109/TIP.2018.2889922
15. Part 14 : Dot and Hadamard Product | by Avnish | Linear Algebra | Medium, https://medium.com/linear-algebra/part-14-dot-and-hadamard-product-b7e0723b9133. Last Accessed 07 Feb 2022
16. Understanding learning rates and how it improves performance in deep learning | by Hafidz Zulkifli | Towards Data Science. https://towardsdatascience.com/understanding-learning-rates-and-how-it-improves-performance-in-deep-learning-d0d4059c1c10. Last Accessed 08 Feb 2022
17. Zakir Hossain MD, Sohel F, Shiratuddin MF, Laga H (2019) A comprehensive survey of deep learning for image captioning. ACM Comput Surv 51. https://doi.org/10.1145/3295748.

SMOTE Variants for Data Balancing in Intrusion Detection System Using Machine Learning

S. Sams Aafiya Banu, B. Gopika, E. Esakki Rajan, M. P. Ramkumar, M. Mahalakshmi, and G. S. R. Emil Selvan

1 Introduction

In many countries, cyber-attacks have become a serious threat due to the vulnerability of critical infrastructure. Different components such as sensors and actuators are used in data acquisition (SCADA) and supervisory control systems to monitor critical infrastructures, including pipelines and refineries. The introduction of viruses and malware is one of the most notable types of attacks that can lead to serious consequences. Such types of attacks can cause irreparable damage to the infrastructure. This prompts to research efficient ways to safeguard control systems digitally from attacks that are emerging without warning from cyberspace.

As a result, it has become necessary to improve the control-system security. To predict cyber-attacks, many active research studies include different machine learning methods. So, to prevent further harm and to identify diverse attacks, Intrusion detection systems (IDSs) are invented. If the total number of the class of events of interest is less than the class of normal events, then the situation is known as Class imbalance. Class imbalance in multiclass classification is a major issue as large classes tend to overwhelm standard classifiers and small ones are ignored. As the model predicts the value of the majority classes for all predictions, accuracy is not helpful with the samples that are imbalanced. Thus Precision, Recall, False Positive Rate (FPR),

S. Sams Aafiya Banu (✉) · B. Gopika · E. Esakki Rajan · M. P. Ramkumar · M. Mahalakshmi · G. S. R. Emil Selvan
Department of Computer Science and Engineering, Thiagarajar College of Engineering, Madurai, India
e-mail: samsaafiya@gmail.com

M. P. Ramkumar
e-mail: ramkumar@tce.edu

G. S. R. Emil Selvan
e-mail: emil@tce.edu

P. Singh et al. (eds.), *Machine Learning and Computational Intelligence Techniques for Data Engineering*, Lecture Notes in Electrical Engineering 998,
https://doi.org/10.1007/978-981-99-0047-3_28

False Negative Rate (FNR) and False Alarm Rate (FAR) are considered to compare the imbalanced set of samples. Class imbalance is handled by a sampling-based approach. This approach modifies the imbalanced samples between the minority class and the majority class. Synthetic Minority Oversampling Technique (SMOTE) variants are compared with different evaluation metrics to identify the best out of them for this dataset.

2 Literature Survey

Machine learning, a subset of artificial intelligence, learns from the historical data to derive a solution based on prediction, classification, or regression towards a problem [1].

An IDS model based on the Boruta feature selection approach and a Random Forest Classifier using the NSL-KDD dataset was developed. The evaluation was carried out in terms of accuracy, recall, and True negative rate where they all provide a value of 0.99. The running time of the entire model has increased to an average of 2400 s while working on finding the optimal parametric values [2].

An IDS using Naive Bayes as the embedded feature extraction approach with the Support Vector Machine (SVM) as the classifier using four different datasets UNSW-NB15, CICIDS2017, NSL-KDD, and Kyoto 2006+ was modeled. The SVM classifier was used to classify the normal and attack class using the transformed dataset. The NB-SVM model has proved the accuracy of 93.75%, 98.92%, 99.35%, and 98.58% respectively on the four datasets. Though the accuracy works well, the False alarm rate of 7.33 in the UNSWNB15 dataset seems to be much higher, which is heavily due to the an imbalanced dataset [3].

A hybrid ensemble approach of combining J48 Decision Tree (DT) and Support Vector Machine (SVM) on the KDD CUP '99 dataset was proposed. The feature selection approach of Particle Swarm Optimization (PSO) had been adopted. The model showed an accuracy of 99.1%, 99.2%, and 99.1% respectively, a detection rate of 99.6% for all three splits and False Alarm Rate of 1.0%, 0.9%, 0.9% respectively. The author adopted a different split ratio which adds no information or noticeable change in metric [4].

The importance of dimensionality reduction of dataset features on the performance of a classification model was described using the NSL-KDD dataset. Random Forest had been adopted to compare the performance of the model with and without feature selection and proved an accuracy of 99.63% and 99.34% respectively. The only drawback is that a single classifier has been used for evaluation [5].

An ensemble approach for the anomaly-based IDS was proposed using the CSECI-CIDS2018. The main objective of the research had been to identify the best filter-based feature selection approach between chi-square and spearman's rank coefficient. The spearman's rank coefficient had proven better results in terms of accuracy 0.985, precision 0.980, recall 0.986, and f1 score 0.983 when adopted with logistic regression classifier. However, the model holds a limitation of misclassification rate of 0.935 concerning the infiltration attack class [6].

3 Proposed Methodology

This paper focuses on finding the best approach among the various resampling techniques of handling the imbalance data problem in multiclass classification. To attain these goals the following steps are applied (see Fig. 1).

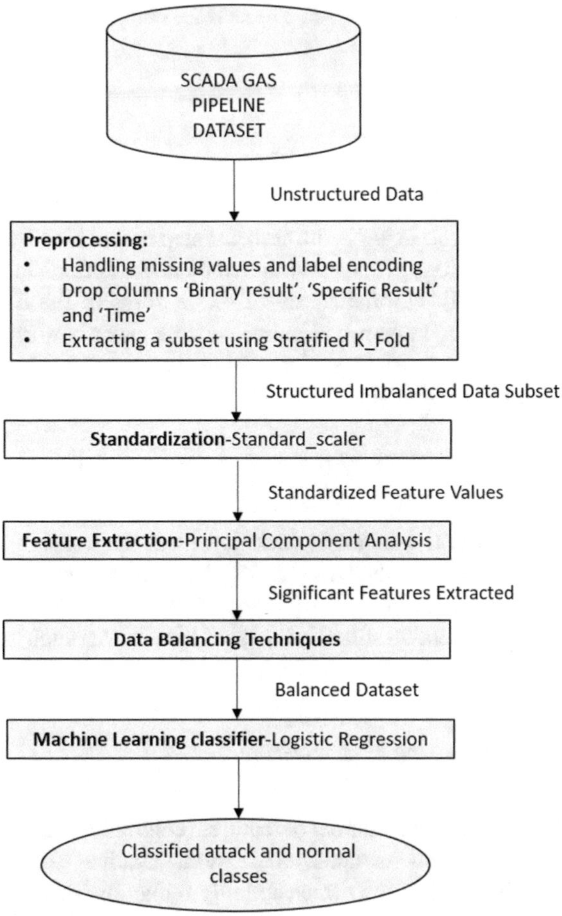

Fig. 1 Overview of proposed methodology

3.1 Dataset Description

In this study, the dataset used emerged at the Mississippi State University [7–10]. Sensors, actuators, communication networks and supervisory control are contained in this testbed. MTU (Master Terminal Unit) and HMI (Human Machine Interface) are contained in the supervisory control. The SCADA system is operated by HMI, which provides an interface. The Man-in-the-middle which is a cyber-attack approach has been applied to incorporate fabrication, interruption, threat type, interception, and modification [11]. The types of attack are Normal, Naive Malicious Response Injection, Complex Malicious Response Command, Malicious State command Injection, Malicious Parameter Command Injection, Malicious Function Code Injection, Denial of Service Interruption, Reconnaissance [12]. CSV and ARFF file formats are available for this dataset. The dataset comprises 274,627 instances with 20 columns [13].

3.2 Data Preprocessing

Data preprocessing is the eminent step in machine learning [14]. Raw input is transformed into a proper and understandable format in this step by handling missing values [15]. Normalization of the features of the dataset or range of independent variables is known as Feature scaling or data normalization [16]. To rescale the features, a standard scalar method is used. The formula for scaling the values is given in the Eq. (1).

$$Z = \frac{x - \mu}{\sigma} \tag{1}$$

where σ is the standard deviation from the average and μ is the average which is contained by all the features [17]. Instead of processing the entire dataset, a subset of the dataset has been considered for evaluation. For this purpose, stratified k-fold has been adopted. Stratified k-fold splits the dataset such that the class ratio is maintained.

3.3 Feature Extraction

Reduction of the number of features (columns) and retaining maximum information is known as Dimensionality Reduction. Principal Component Analysis (PCA) is used, which is one of the dimensionality reduction techniques. PCA extracts the features as principal components, which is a linear combination of features. These principal components cumulatively provide the maximum information of the original dataset.

3.4 Data Balancing Techniques

Most of the datasets in the real world, the extremely imbalanced data's distribution causes imbalance data problem. If the number of instances of every class is about to be equal, then many machine learning models will work better [18, 19]. The majority class dominates the minority class in imbalance data problems. Thus, the performance is not reliable as the majority classes are inclined by the classifiers [20]. The data which are imbalanced are dealt by various strategies. The efficient method to solve the data that are imbalanced is the sampling based approach. In general, this approach is categorized as Under-Sampling [21], Over-Sampling [22] and Hybrid-Sampling [23]. The adopted resampling approaches are as follows:

Over-Sampling Methods. The minority class weight is increased by making new samples which are of minority class in over sampling.

Synthetic Minority Oversampling Technique (SMOTE). In this method, the number of samples which is of minority class is increased by producing new instances. Feature space has class samples for every target class and their neighbors that are nearer are taken by this algorithm. Then, the newly created samples combine the features of the target case with the features of its neighbors. The existing minority samples are not replicated by the new instances [24, 25].

SMOTE Support Vector Machine (SSVM). The newly created minority samples are generated along with directions towards their nearest neighbors from the existing minority class instances in this method The SVM model is used by the newly created minority samples near borderlines that help to set boundaries between classes which is focused by SSVM [26].

Hybrid Methods

SMOTE Edited Nearest Neighbor (SENN). SMOTE, which is an over sampling model and Edited Nearest Neighbor (ENN), which is an under sampling model are combined that enhances the result. Root sample is successively selected from the minority samples for the merging of the new sample by SMOTE. ENN is employed for the elimination of noise samples. This well-known method is called SENN [27].

SMOTE Tomek (STOMEK). SMOTE, which is an over sampling model, is connected to Tomek links, which is an under sampling model that enhances the result. STOMEK focuses on cleaning intersecting points for every class that is disseminated in the sample space. STOMEK is a common hybrid method [27].

3.5 Machine Learning

This paper uses Logistic Regression. The probability of a target variable is predicted by Logistic Regression. It is a supervised learning classification algorithm. Implementation, Interpretation and efficiency to train is easier with Logistic Regression. To support multiclass classification, logistic regression is parameterized with the "One-Vs-Rest" approach.

4 Experimental Implementation and Evaluation

4.1 Evaluation Metrics

Accuracy. A metric to evaluate the overall performance of the model, evaluates the ratio of correctly classified instances to the overall total number of instances in the dataset as mentioned in Eq. (2).

$$\text{Accuracy} = \frac{TP + \text{TN}}{\text{TP} + \text{FP} + \text{TN} + \text{FN}} \tag{2}$$

where TP = True Positive, TN = True Negative, FP = False Positive, FN = False Negative.

Precision. Precision evaluates the ratio of correctly classified instances of the attack class to the overall predicted instances of attack class as mentioned in Eq. (3) [28, 29]

$$\text{Precision} = \frac{TP}{\text{TP} + \text{FP}} \tag{3}$$

Recall. Recall evaluates the ratio of correctly classified instances of the attack class to overall instances of the attack class in the entire dataset under evaluation. It is also known as Sensitivity or True Positive Rate as mentioned in Eq. (4).

$$\text{Recall} = \frac{TP}{\text{TP} + \text{FN}} \tag{4}$$

False Positive Rate (FPR). False Positive Rate evaluates the ratio of wrongly classified instances of the attack class to the overall instances of the normal class, as mentioned in Eq. (5).

$$\text{False Positive Rate} = \frac{FP}{\text{TN} + \text{FP}} \tag{5}$$

False Negative Rate (FNR). False Negative Rate evaluates the ratio of wrongly classified instances of the normal class to the overall instances of the attack class as mentioned in Eq. (6).

$$\text{False Negative Rate} = \frac{FN}{\text{TP} + \text{FN}} \tag{6}$$

False Alarm Rate (FAR). It is the mean of FPR and FNR. This depicts the rate of misclassification by the classifier as mentioned in Eq. (7).

$$\text{False Alarm Rate} = \frac{FPR + FNR}{2} \tag{7}$$

F-measure. F-measure is a statistical technique that examines the accuracy of the AI model built by evaluating the weighted mean of precision and recall as mentioned in Eq. (8).

$$\text{F} - \text{measure} = 2 * \frac{\text{Precision}\cdot\text{Recall}}{\text{Precision} + \text{Recall}} \tag{8}$$

4.2 Performance Evaluation on Data Preprocessing

The multiclass classification is performed by encoding all 7 types of attack classes as 1 to 7 and the normal class as 0. Stratified kfold is used to get a subset of the dataset with the same class ratio of the original dataset. It is difficult to process the whole dataset and it is time-consuming so a subset of data is analyzed by stratified kfold. nsplits = 2 is used and four files are created in the first iteration. One file is taken and used in the second iteration. In the second iteration, the dataset is reduced and this subset is taken for further processing.

It is observed from Table 1 that the stratified k-fold produces a subset of the dataset, with the same class distribution of the original dataset. For example, in the original dataset, class 0 has 214,580 instances and with a class ratio of 78.13%. In the extracted dataset class 0 is reduced to 53,821 instances with a class ratio maintained as 78.13%. To rescale the features of the extracted dataset, a standard scalar is used.

4.3 Performance Evaluation on Feature Extraction

It is observed that from PCA (see Fig. 2), the first 10 principal components depict 96 to 97% of the information. Thus the significant linear combinations of features

Table 1 Original and extracted dataset

Class	Original dataset #Instances	Extracted dataset #Instances	% Instances
0-Normal	214,580	53,821	78.13
1-NMRI	7753	1939	2.82
2-CMRI	13,035	3259	4.75
3-MSCI	7900	1975	2.88
4-MPCI	20,412	5103	7.43
5-MFCI	4898	1224	1.78
6-DoS	2176	544	0.79
7-Recon	3874	968	1.41

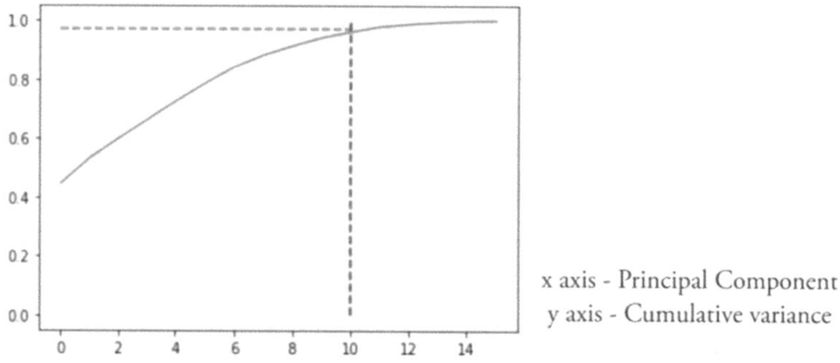

Fig. 2 Principal *component analysis*

have been extracted and the number of features is reduced from 17 to 10 using PCA. Further processing is done with these reduced features.

4.4 Performance Evaluation Without Data Balancing Technique

The performances of logistic regression applied to the original imbalanced dataset with the extracted features are evaluated. One of the performance metrics, confusion matrix is represented in Table 2.

While evaluating without data balancing techniques, all attack class instances from 1 to 7 have been wrongly classified as normal class and all normal class instances have been perfectly classified as normal which leads to increased accuracy of 78%. Thus it clearly depicts accuracy that is not a right metric while evaluating a highly imbalanced dataset.

Table 2 Confusion matrix without data balancing

Class	0	1	2	3	4	5	6	7
0	53,821	0	0	0	0	0	0	0
1	1853	0	0	0	0	0	0	0
2	3199	0	0	0	0	0	0	0
3	1983	0	0	0	0	0	0	0
4	5159	0	0	0	0	0	0	0
5	1175	0	0	0	0	0	0	0
6	526	0	0	0	0	0	0	0
7	941	0	0	0	0	0	0	0

Table 3 Precision, Recall, F1, FPR, FNR and FAR Score for each class

Class	Precision	Recall	F1-Score	FPR	FNR	FAR
0	0.78	1.00	0.88	1.00	0.00	0.5
1	0.00	0.00	0.00	0.00	1.00	0.5
2	0.00	0.00	0.00	0.00	1.00	0.5
3	0.00	0.00	0.00	0.00	1.00	0.5
4	0.00	0.00	0.00	0.00	1.00	0.5
5	0.00	0.00	0.00	0.00	1.00	0.5
6	0.00	0.00	0.00	0.00	1.00	0.5
7	0.00	0.00	0.00	0.00	1.00	0.5

It is observed in Table 3 that, with respect to class 0 (majority class), since all instances have been correctly predicted, the Logistic regression achieves high precision, recall and F1-score. Whereas with respect to the classes 1 to 7, due to heavy class imbalance it leads to zero precision, recall and F1-score. FPR and FNR must be nearly equal to zero. FAR must be more less than 0.5. Without data balancing technique FPR, FNR, FAR are not in their respective range, because all instances of classes 1 to 7 are falsely predicted as class 0. This leads to an increased range in FPR of class 0 and FNR of class 1 to 7. Therefore data balancing techniques are needed.

4.5 Performance Evaluation on Different Data Balancing Techniques

To overcome the disadvantages caused by without using any data balancing technique and to improvise FPR, FNR and FAR, data balancing techniques are implemented. The results after applying data balancing techniques are described (see Fig. 3).

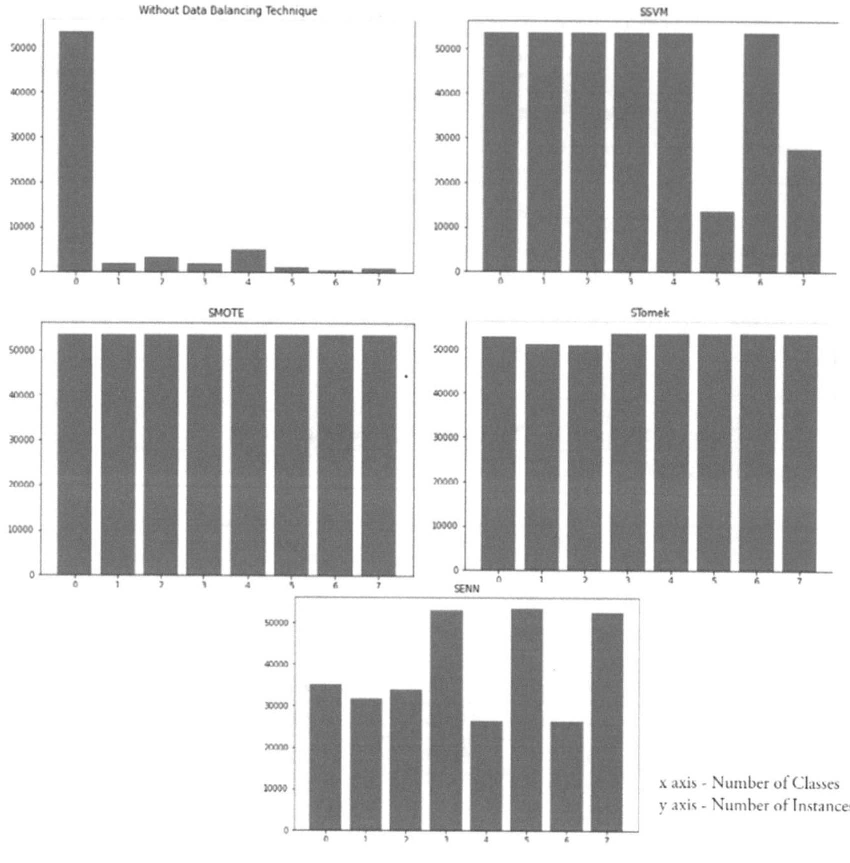

Fig. 3 Comparison of resampling methods

Without data balancing approach, the classes are imbalanced as out of 68,657 records, 78% of the data is held by class 0. By using data balancing techniques, the numbers of instances in minority classes have been increased at various levels by various techniques. In case of SSVM and SENN, it overcomes with the major drawback say over fitting. Whereas SMOTE and STOMEK blindly increases all the minority class instances equal to the majority class 0 i.e. 53,821. This purely may lead to over fitting of data. The major difference between SSVM and SENN is, SSVM deals with the other drawback of overlapping classes which SENN does not.

Logistic Regression is implemented on balanced data generated by different data balancing techniques like SSVM, SMOTE, STomek, SENN.

From Table 4, it is observed that in this dataset, the Logistic Regression classifier with SSVM has achieved 59% accuracy, 62% precision, 58% recall and 59% F1-score which is acceptable and better than other resampling methods.

From the results in Table 5, SMOTE and STOMEK results in higher values of FNR. SENN results in higher value of False Negative rate and False Alarm Rate.

Table 4 Performance metrics based on logistic regression on various resampling methods

Logistic regression	SSVM	SMOTE	STomek	SENN
Accuracy	**0.595**	0.428	0.430	0.205
Precision	**0.623**	0.546	0.536	0.189
Recall	**0.588**	0.428	0.429	0.195
F1-Score	**0.590**	0.357	0.360	0.136

Thus SSVM outperforms other resampling methods with higher Precision, Recall, F1-score, Accuracy and lower FPR, FNR and FAR, nearly equal to zero and less than 0.5.

5 Conclusion and Future Work

Cyber attack detection and security are imperative to the SCADA system. Such Intrusion detection systems are implemented using machine learning. The Class imbalance problem is one of the main objectives in those systems. Large number of normal samples and very few attack samples are contained in the SCADA dataset. To eliminate the class imbalance problem in multiclass classification, the dataset is preprocessed and the features are reduced before applying resampling techniques. Machine learning classifiers are used to evaluate the performance of resampling methods. The best resampling technique among the various data handling methods that is imbalanced, such as SMOTE, SSVM, STOMEK and SENN is shown in this study. Showing the importance of the borderline instances of the minority class, SSVM outperforms other SMOTE variants by achieving the highest precision and recall score with 62% and 58% respectively and FPR, FNR and FAR with a range below 0.5 and near to zero.

The major drawback observed is running time of the ML model. This can be resolved by using cloud computing resources such as apache-spark, Hadoop etc. which will be a promising scope for future work. In case of adopting DL models as the future work, it will further reduce the hindrance of separate feature engineering steps in ML.

Table 5 Performance metrics for various resampling methods on different classes

Class	FPR				FNR				FAR			
	SSVM	SMOTE	STomek	SENN	SSVM	SMOTE	STomek	SENN	SSVM	SMOTE	STomek	SENN
0	0.024	0.111	0.110	0.189	0.671	0.850	0.854	0.332	0.347	0.481	0.482	0.260
1	0.093	0.107	0.106	0	0.488	0.353	0.341	1	0.290	0.230	0.224	0.5
2	0.092	0.0001	0.0001	0.005	0.497	0.974	0.970	1	0.295	0.487	0.485	0.502
3	0.082	0.002	0.003	0.090	0.237	0.985	0.979	0.976	0.160	0.493	0.491	0.533
4	0.073	0.369	0.364	0.003	0.379	0.245	0.252	0.918	0.226	0.307	0.308	0.461
5	0.006	0.015	0.016	0.110	0.519	0	0	1	0.263	0.007	0.008	0.555
6	0.05	0.001	0.002	0.052	0.368	0.990	0.988	0.845	0.209	0.495	0.495	0.448
7	0.042	0.046	0.046	0.483	0.131	0.178	0.179	0.364	0.087	0.112	0.113	0.423

References

1. Abid Salih A, Abdulazeez AM (2021) Evaluation of classification algorithms for intrusion detection system: a review. J Soft Comput Data Mining 2(1):31–40
2. Iman AN, Ahmad T (2020) Improving intrusion detection system by estimating parameters of random forest in Boruta. In: 2020 international conference on smart technology and applications (ICoSTA). IEEE, pp 1–6
3. Gu J, Lu S (2021) An effective intrusion detection approach using SVM with naïve Bayes feature embedding. Comput Secur 103:102158
4. Kumari A, Mehta AK (2020) A hybrid intrusion detection system based on decision tree and support vector machine. In: 2020 IEEE 5th international conference on computing communication and automation (ICCCA). IEEE, pp 396–400
5. Sah G, Banerjee S (2020) Feature reduction and classifications techniques for intrusion detection system. In: 2020 international conference on communication and signal processing (ICCSP). IEEE, pp 1543–1547
6. Fitni QRS, Ramli K (2020) Implementation of ensemble learning and feature selection for performance improvements in anomaly-based intrusion detection systems. In: 2020 IEEE international conference on industry 4.0, artificial intelligence, and communications technology (IAICT). IEEE, pp. 118–124
7. Srijha V, Ramkumar MP (2018) Access time optimization in data replication. In: 2018 2nd international conference on trends in electronics and informatics (ICOEI). IEEE), pp 1161–1165
8. Morris T, Srivastava A, Reaves B, Gao W, Pavurapu K, Reddi R (2011) A control system test bed to validate critical infrastructure protection concepts. Int J Crit Infrastruct Prot 4(2):88–103
9. Morris TH, Thornton Z, Turnipseed I (2015) Industrial control system simulation and data logging for intrusion detection system research
10. Turnipseed IP (2015) A new SCADA dataset for intrusion detection research (Order No. 1596111). Available from ProQuest Dissertations & Theses Global. (1710049699)
11. Ullah I, Mahmoud QH (2017) A hybrid model for anomaly-based intrusion detection in SCADA networks. In: 2017 IEEE international conference on big data (Big Data). IEEE, pp 2160–2167
12. Choubineh A, Wood DA, Choubineh Z (2020) Applying separately cost-sensitive learning and Fisher's discriminant analysis to address the class imbalance problem: a case study involving a virtual gas pipeline SCADA system. Int J Crit Infrastruct Prot 29:100357
13. Nazir S, Patel S, Patel D (2021) Autoencoder based anomaly detection for scada networks. Int J Artif Intell Mach Learn (IJAIML) 11(2):83–99
14. Meena SM, Ramkumar MP, Asmitha RE, GSR ES (2020) Text summarization using text frequency ranking sentence prediction. In: 2020 4th international conference on computer, communication and signal processing (ICCCSP). IEEE, pp 1–5
15. Kotsiantis SB, Kanellopoulos D, Pintelas PE (2006) Data preprocessing for supervised leaning. Int J Comput Sci 1(2):111117
16. Ghorbani R, Ghousi R (2020) Comparing different resampling methods in predicting Students' performance using machine learning techniques. IEEE
17. Pal KK, Su KS (2016) Preprocessing for image classification by convolutional neural networks. In: Proceedings of IEEE international conference recent trends electronic, information and communication technology (RTEICT), pp 17781781
18. Ramkumar MP, Balaji N, Vinitha Sri S (2013) Stripe and disk level data redistribution during scaling in disk arrays using RAID 5. In: PSG—ACM conference on intelligent computing, 26–27 April, vol 1, Article No 74, pp 469–474
19. Longadge R, Dongre S (2013) Class imbalance problem in data mining review. arXiv:1305.1707
20. Kotsiantis S, Kanellopoulos D, Pintelas P (2006) Handling imbalanced datasets: a review. GESTS Int Trans Comput Sci Eng 30(1):2536
21. Liu XY, Wu J, Zhou ZH (2009) Exploratory undersampling for class imbalance learning. IEEE Trans Syst Man, Cybern B, Cybern 39(2):539550

22. Yap BW, Rani KA, Rahman HAA, Fong S, Khairudin Z, Abdullah NN (2013) An application of oversampling, undersampling, bagging and boosting in handling imbalanced datasets. In: Proceedings of 1st international conference on advanced data information engineering (DaEng). Springer, Singapore, pp 1322
23. Galar M, Fernandez A, Barrenechea E, Bustince H, Herrera F (2012) A review on ensembles for the class imbalance problem: Bagging-, boosting-, and hybrid-based approaches. IEEE Trans Syst, Man, Cybern C, Appl Rev 42(4):463484
24. Ramkumar MP, Balaji N, Rajeswari G (2014) Recovery of disk failure in RAID-5 using disk replacement algorithm. Int J Innov Res Sci, Eng Technol 3(3). ISSN, 2358–2362
25. Chawla NV, Bowyer KW, Hall LO, Kegelmeyer WP (2002) SMOTE: synthetic minority over-sampling technique. J Artif Intell Res 16:321357
26. Tang Y, Zhang YQ, Chawla NV, Krasser S (2009) SVMs modeling for highly imbalanced classification. IEEE Trans Syst, Man, Cybern B. Cybern 39(1):281288
27. Batista GE, Prati RC, Monard MC (2004) A study of the behavior of several methods for balancing machine learning training data. ACM SIGKDD Explor Newslett 6(1):2029
28. Ramkumar MP, Balaji Narayanan, Selvan GSR, Ragapriya M (2017) Single disk recovery and load balancing using parity de-clustering. J Comput Theor Nanosci 14(1):545–550. ISSN 15461955
29. Alzahrani AO, Alenazi MJ (2021) Designing a network intrusion detection system based on machine learning for software defined networks. Future Internet 13(5):111

Grey Wolf Based Portfolio Optimization Model Optimizing Sharpe Ratio in Bombay Stock Exchange

Mohammad Imran, Faraz Hasan, Faisal Ahmad, Mohammad Shahid, and Shafiqul Abidin

1 Introduction

The portfolio optimization problem is a well-known problem in the financial management. A portfolio includes various financial securities such as bonds and stocks owned by an organization or by an individual. How to select the best combination of securities is the most important and challenging topic in the domain under consideration. Individual people and organizations prefer to invest in portfolios rather than a single asset as it contributes to the maximum return on the possible reduced risk. Therefore, this problem can be considered as the construction of a portfolio (combination of the securities) that optimizes the portfolio risk and expected return [1].

In 1952, Markowitz is the pioneer of modern portfolio theory, proposed the mean–variance model for portfolio optimization which is known as Markowitz model for portfolio selection [2]. This model initially considers only the fully investing constraint and other constraints such as boundary constraint, cardinality constraint, etc., are ignored as these constraints make the problem very complex to solve. Therefore, the Markowitz model and some other conventional models could not deal this problem with large numbers of securities and constraints efficiently in uncertain

M. Imran · S. Abidin
Department of Computer Science, Aligarh Muslim University, Aligarh, India

F. Hasan
Department of Computer Science and Engineering, Koneru Lakshmaiah Education Foundation, Guntur, India
e-mail: faraz.hasan@live.in

M. Shahid (✉)
Department of Commerce, Aligarh Muslim University, Aligarh, India
e-mail: mdshahid.cm@amu.ac.in

F. Ahmad
Workday Inc, Pleasanton, USA

P. Singh et al. (eds.), *Machine Learning and Computational Intelligence Techniques for Data Engineering*, Lecture Notes in Electrical Engineering 998,
https://doi.org/10.1007/978-981-99-0047-3_29

environment of the share markets. So, the meta-heuristic approaches have been applied to solve the portfolio selection problem for obtaining satisfactory results. Meta-heuristics optimization approaches are categorized such as evolutionary algorithms (EA), Swarm Intelligence algorithms, nature inspired algorithms, bio-inspired algorithms, human based meta-heuristic algorithms and many more [2, 3].

In this paper, a novel portfolio selection model using the grey wolf optimization (GWO) approach has been proposed. GWO [4] is a swarm intelligence approach inspired by the haunting process of the grey wolf. This natural haunting behavior of the grey wolf is modeled for solving the portfolio optimization problem. The major contributions are listed as follows:

- GWO based portfolio selection model to optimize the Sharpe ratio of the portfolio has presented.
- The solution model is implemented by using MATLAB R2018a.
- To conduct the performance evaluation, an experimental study has been conducted with a comparative study of the proposed GWO model with Genetic Algorithm (GA).
- For analysis, dataset (30 stocks) of the S&P BSE Sensex of Indian stock exchange is used.
- Study reveals the better performance of the proposed strategy than GA on account of convergence rate, execution time and obtained optimal value of the objective.

Organization of the paper is as: Sect. 2 presents the related literature of the domain. The problem formulation is depicted in Sect. 3. In Sect. 4, the solution model of GWO has been explained. Simulation results and conclusion are given in Sects. 5 and 6 respectively.

2 Related Work

The portfolio selection problem is an NP hard problem from the field of the financial management. As of today, evolutionary algorithms (EA) [5–8], swarm intelligence algorithms (SIA) [9–16], nature inspired algorithms (NIA) [17–19], and gradient based [20], neighborhood search [21] have become attractive approaches for the researcher to develop a model to solve the portfolio selection problem. Chang et al. [5] proposed Genetic Algorithm (GA), Tabu Search (TS) and Simulated Annealing (SA) for unconstrained and constrained portfolio selection problem [5]. In [6], the authors suggested various measures of risk for portfolio optimization problems using GA. This algorithm collects three different types of risk such as semi-variance, mean absolute deviation and variance with skewness based upon the MV model. In this paper [7], the author applied Stochastic Fractal Search (SFS) for the portfolio selection problem optimizing Sharpe ratio and results are tested on the S&P BSE 30 stocks dataset. In [8], the risk budgeted model optimizing Sharpe ratio has been presented and compared with the Particle Swarm Optimization (PSO) based solution method. In 2009, the fast Particle Swarm Optimization (PSO) based solution method has been reported in [9] and another work is reported by using PSO for the constrained portfolio selection problem [10]. Further, a two stage hybrid version for portfolio optimization is presented in [11]. Deng et al. [12] presented an Ant Colony Optimization (ACO) based solution for the mean–variance portfolio problem. Tuba et al. made a hybrid model of ABC and firefly for mean–variance portfolio selection with cardinality constraint [13]. Kalayci et al. recommend a solution using the artificial bee colony approach to solve the cardinality constraints portfolio problem using infeasibility tolerance procedure [14]. Artificial bee colony based model for fuzzy portfolio selection has been proposed by Gao et al. [15]. In 2021, Cura developed a fast converging ABC based solution algorithm for cardinality constrained portfolio [16].

Amir et al. developed a portfolio selection model with a multi-objective approach using the invasive weed method transforming into a single-objective programming model with fuzzy normalization [17]. Shahid et al. also applied invasive weed optimization for risk budgeted portfolio selection optimizing Sharpe ratio [18]. A squirrel search based meta-heuristic approached is applied on the constrained portfolio problem and reported the superior performance among the considered peers [19]. In [20], Sharpe ratio is maximized with the mean variance model by using a gradient based optimizer. Akbay et al. [21] proposed a parallel variable neighborhood search algorithm which decides combinations of assets in portfolio, and the cardinality constrained portfolio selection problem is solved by this method's the results states that proposed algorithm is efficient than others.

3 The Problem Formulation

From stock exchange, a portfolio (P) having N number of stocks has been considered by i.e. $P = \{st_1, st_{2,}, \ldots st_N\}$ with their corresponding weights $\{W_1, W_{2,}, \ldots W_N\}$. The respective expected returns of the stocks are as $\{R_1, R_{2,}, \ldots R_N\}$. Portfolio risk can be expressed as

$$\mathrm{Risk_P} = \sqrt{\sum_i^N \sum_j^N W_i * W_j * CV_{ij}} \tag{1}$$

where W_i and W_j are the weights of st_i and st_j respectively. CV_{ij} is the covariance's of the returns.

Here, Sharpe Ratio (SR_P) is considered as the objective function as it optimizes both the parameters (return and risk). In case of equity based securities, risk free return is zero. Therefore, the sharpe ratio of the portfolio constructed and problem statement can be formulated as

$$\mathrm{Max}(SR_P) = \max\left(\frac{\sum_1^N R_i * W_i}{\mathrm{Risk_P}}\right) \tag{2}$$

subject to the constraints

(i) $\sum_{i=1}^{N} w_i = 1$
(ii) $w_i \geq 0$
(iii) $a \leq w_i \leq b$

Here, (i) represents the budget constraint and (ii) constraint restricts the short sell. Further, (iii) is the floor/ceiling constraint that imposes a lower and upper bounds for assets weights in the portfolio. A repair method is used to handle these constraints. Whenever, lower or upper bounds are violated, then, respective weights are replaced by the lower or upper bound value respectively. To maintain budget constraint (sum equal to 1), a normalization approach is used such that each stock weight is divided by the sum of the total weights of the portfolio.

4 Proposed Strategy

In this subsection, the proposed grey wolf [4] based solution method has been explained in detail. In this meta-heuristic, the haunting behavior of the grey wolf has been modeled for solving the complex portfolio selection problem. The various processes including the social hierarchy, tracking, encircling, and attacking prey are discussed.

In order to mathematically model the social hierarchy of wolves when designing GWO, we consider the fittest solution as the alpha (α). Consequently, the second, third and fourth best solutions are named beta (β), delta (δ) and omega (ω) respectively. The rest of the candidate solutions are assumed to be gamma (γ). In the GWO algorithm the hunting (optimization) is guided by α, β, δ and ω. The γ wolves follow these four wolves. Here, during the hunt, grey wolves encircle the prey and mathematically modeled this encircling behavior by using Eqs. (3) and (4)

$$\vec{D} = \left| \vec{C} \cdot \vec{X}_p(g) - \vec{X}(g) \right| \tag{3}$$

$$\vec{X}(g+1) = \vec{X}_p(g) - \vec{A} \cdot \vec{D} \tag{4}$$

where g represents the current generations, $\vec{A}$ and $\vec{C}$ are coefficient vectors, $\vec{X}$ is the position of the prey, and $\vec{X}$ is the position of a grey wolf. Now, $\vec{A}$ and $\vec{C}$ are computed as:

$$\vec{A} = 2\vec{a} \cdot \vec{r_1} - \vec{a} \tag{5}$$

$$\vec{C} = 2 \cdot \vec{r_2} \tag{6}$$

where $\overrightarrow{a}$ is a control parameter and it is linearly decreased from 2 to 0 over the number of generations and r1, r2 in [0,1]. Grey wolves identify the position of the prey, encircle them under the guidance of the alpha. Occasionally, in the hunting, the beta and delta may participate. In simulation of the hunting behavior, it is assumed that the alpha (best individual solution) beta, delta and omega have better idea about the position of the prey. Therefore, the first three best solutions are saved and the remaining search agents (gamma) update their positions in accordance to the location of the leaders. The formulas are written for this as follows

$$\overrightarrow{D_\alpha} = \left|\overrightarrow{C_1} \cdot \overrightarrow{X_\alpha} - \vec{X}\right|, \overrightarrow{D_\beta} = \left|\overrightarrow{C_2} \cdot \overrightarrow{X_\beta} - \vec{X}\right|,$$
$$\overrightarrow{D_\delta} = \left|\overrightarrow{C_2} \cdot \overrightarrow{X_\alpha} - \vec{X}\right|, \overrightarrow{D_\omega} = \left|\overrightarrow{C_4} \cdot \overrightarrow{X_\omega} - \vec{X}\right| \tag{7}$$

$$\overrightarrow{X_1} = \overrightarrow{X_\alpha} - \overrightarrow{A_1} \cdot \left(\overrightarrow{D_\alpha}\right), \overrightarrow{X_2} = \overrightarrow{X_\beta} - \overrightarrow{A_2} \cdot \left(\overrightarrow{D_\beta}\right),$$
$$\overrightarrow{X_3} = \overrightarrow{X_\delta} - \overrightarrow{A_3} \cdot \left(\overrightarrow{D_\delta}\right), \overrightarrow{X_4} = \overrightarrow{X_\omega} - \overrightarrow{A_4} \cdot \left(\overrightarrow{D_\omega}\right) \tag{8}$$

$$\vec{X}(g+1) = \frac{\overrightarrow{X_1} + \overrightarrow{X_2} + \overrightarrow{X_3} + \overrightarrow{X_4}}{4} \tag{9}$$

In other words, alpha, beta, delta and omega estimate the location of the prey, and other wolves update their locations randomly around encircled prey. After this, grey wolves attack on the prey as it stops moving. Mathematically it is modeled by decreasing the value of $\overrightarrow{a}$ which decreases the range of $\overrightarrow{A}$ as $\overrightarrow{A}$ is a random value ∈ [−2a, 2a]. It is evident that the encircling mechanism suggested shows exploration. Grey wolves usually search as per the location of the alpha, beta, delta and omega. For searching the prey, they diverge from each other and converge again for attacking the prey. This divergence is modeled by vector $\overrightarrow{A}$ assigning the random values greater than 1 or less than −1. The component favoring exploration is $\overrightarrow{C} \in [0, 2]$. This assists GWO to maintain a more random behavior during optimization for avoiding a local optima not only during the initial generations but also in the final generations. And, a is decreased from 2 to 0 for exploration and exploitation, respectively. Finally, the GWO is stopped as per the end criterion. The algorithmic template of GWO is given by Fig. 1.

5 Experimental Results

In this section, an experimental study is presented for the performance evaluation of the suggested model by using the dataset of S&P BSE Sensex of Indian stock exchange using MATLAB running on the processor Intel i7(R) with 16 GB RAM. This dataset contains monthly returns of 30 stocks for the financial year 1st April 2010 to 31st March 2020. For performance comparison, the results of the GWO based

```
GWO ()
1.  Input: Initialization of the grey wolf population, a, A and C
2.  Repair the population
3.  Fitness evaluation
    // as per objective i.e. Sharpe ratio
4.  Find four best leaders from the population
5.  While stopping condition Do
6.  For each point in population
7.  Update
    //Create the different point by Eq. (10)
8.  End For
9.  Update
    // a, A, C
10. Repair the updated individuals
11. Calculated the fitness function and obtain best individual
12. g=g+1
13. End While
14. Output: Find global optimal solution
```

Fig. 1 GWO algorithm based solution approach

solution approach have been compared with the results produced by the genetic algorithm (GA) for the same objective and environment. The MATLAB codes of GWO and GA are available at: seyedalimirjalili.com, www.yarpiz.com.

The parameter settings of the experiments for GWO and GA algorithms for comparative analysis are listed in Table 1 as follows:

The constraints imposed such as fully invested constraints and boundary constraint, in the problem formulation are satisfied in the optimal portfolio produced by the proposed GWO and GA based solution methods. Here, experiments are conducted for twenty different runs to avoid fluctuations in the results. For both the solution approaches i.e., GWO and GA, the average value, and standard deviation along with the best, worst value have been presented in Table 2. It is clear from Table 2 that the proposed GWO is performing better than GA on account of the objective parameter (Sharpe ratio).

Further, convergence behavior of the algorithms has been presented in the convergence curves depicting the best fitness value of the Sharpe Ratio with respect to the number of generations and execution times of GA and GWO as shown in the Figs. 2

Table 1 Control parameters for GA and proposed GWO algorithms

Algorithms	Parameters specifications
Common parameters	Size of initial population = 100, Iteration number = 100
GWO	a = [2-0], Number of leaders = 4
GA	Single-point crossover probability = 0.7, Polynomial mutation probability = 0.3

Table 2 Comparative results of Sharpe ratio between GWO and GA

Sharpe ratio					
S.No	GA	Proposed (GWO)	S.No	GA	Proposed (GWO)
1	0.39912956	0.400282839	11	0.39952824	0.400294665
2	0.39872635	0.400153986	12	0.39929673	0.397424552
3	0.39915631	0.400199646	13	0.39775699	0.400256577
4	0.39846927	0.400481524	14	0.39739663	0.400521316
5	0.39898049	0.40029957	15	0.39850596	0.399895534
6	0.39922763	0.397231324	16	0.39861202	0.398609442
7	0.39906307	0.400173561	17	0.3994055	0.400363452
8	0.39666674	0.400311904	18	0.39955047	0.400457248
9	0.39831109	0.400440098	19	0.39890134	0.400545344
10	0.39932357	0.400169961	20	0.39693495	0.397453189
Best				0.39955047	**0.400545344**
Worst				0.39666674	**0.397231324**
Average				0.39864715	**0.399778287**

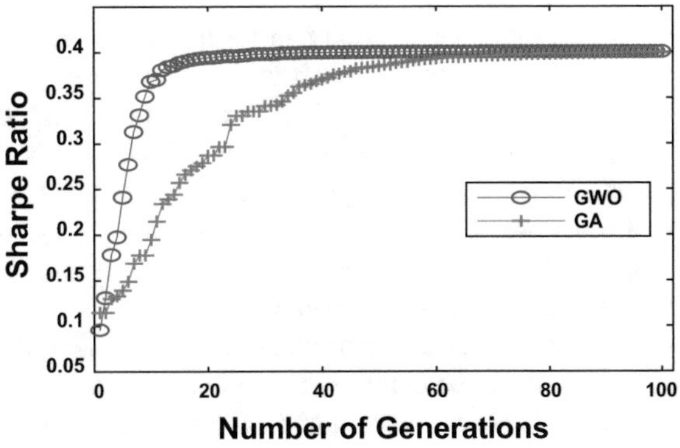

Fig. 2 Convergence curves of the GA and GWO

and 3 respectively. The performance of the GWO is far better than GA for convergence rate, and an optimal fitness value is achieved apart from the execution time.

6 Conclusion

In this paper, an attempt has been made by using the grey wolf optimizer algorithm for giving the solution to the portfolio selection problems. The main objective of using GWO was to improve the Sharpe's ratio of the constructed portfolio. An approach with parameter free penalties was used to control the various constraints

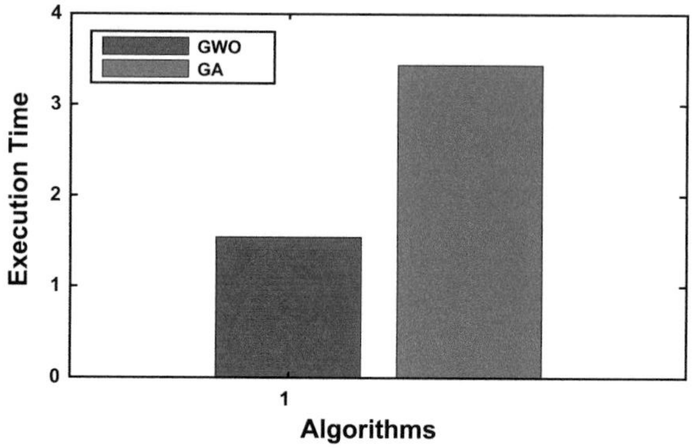

Fig. 3 Execution times of GA and GWO

of the constructed portfolio. GWO algorithm is applied to obtain Share's ratio to an optimum level. The results of the study show that the proposed GWO algorithm is having an edge over GA algorithm in terms of giving solution to the portfolio selection problem.

This work can be extended to improve or overcome the limitations of the proposed work. Some of them have been listed as expected future directions:

- GWO solution approach can be further improved for more satisfactory results with some additional constraints such as cardinality, transaction cost etc.
- The performance of GWO can be tested by comparing with other evolutionary/swarm optimization models having the same objective using a more real dataset to find a better place in the literature.
- The work can be further extended by obtaining the Pareto Front describing the trade-off between the risk and return using multi-objective optimization with a non-dominant sorting approach.
- The work can be tested on a standard benchmark on various performance measures such as mean percentage error, median percentage error, variance of return error, mean of return error. Statistical analysis can be also carried out for statistical significance.

References

1. Markowitz HM (1952) Portfolio selection. J Financ 7(1):77–91
2. Markowitz H (1991) Portfolio selection: efficient diversification of investments. Massachusetts, Cambridge
3. Di Tollo G, Roli A (2008) Metaheuristics for the portfolio selection problem. Int J Oper Res 5(1):13–35
4. Mirjalili S, Mirjalili SM, Lewis A (2014) Grey wolf optimizer. Adv Eng Softw 69:46–61
5. Chang TJ, Meade N, Beasley JE, Sharaiha YM (2000) Heuristics for cardinality constrained portfolio optimisation. Comput Oper Res 27(13):1271–1302

6. Chang TJ, Yang SC, Chang KJ (2009) Portfolio optimization problems in different risk measures using genetic algorithm. Expert Syst Appl 36(7):10529–10537
7. Shahid M, Ashraf Z, Shamim M, Ansari MS (2022) An evolutionary optimization algorithm based solution approach for portfolio selection problem. IAES International Journal of Artificial Intelligence, 11(3), 843.
8. Shahid M, Ansari MS, Shamim M, Ashraf Z (2022) A stochastic fractal search based approach to solve portfolio selection problem. In: Gunjan VK, Zurada JM (eds) Proceedings of the 2nd international conference on recent trends in machine learning, IoT, smart cities and applications 2021. Lecture Notes in Networks and Systems, vol 237. Springer, Singapore
9. Wang W, Wang H, Wu Z, Dai H (2009) A simple and fast particle swarm optimization and its application on portfolio selection. In: 2009 international workshop on intelligent systems and applications. IEEE, pp 1–4
10. Zhu H, Wang Y, Wang K, Chen Y (2011) Particle swarm optimization (PSO) for the constrained portfolio optimization problem. Expert Syst Appl 38(8):10161–10169
11. Zaheer KB, Abd Aziz MIB, Kashif AN, Raza SMM (2018) Two stage portfolio selection and optimization model with the hybrid particle swarm optimization. MATEMATIKA: Malaysian J Ind Appl Math 125–141
12. Deng GF, Lin WT (2010) Ant colony optimization for Markowitz mean-variance portfolio model. In: International conference on swarm, evolutionary, and memetic computing. Springer, Berlin, Heidelberg, pp 238–245
13. Tuba M, Bacanin N (2014) Artificial bee colony algorithm hybridized with firefly algorithm for cardinality constrained mean-variance portfolio selection problem. Appl Math Inf Sci 8(6):2831
14. Kalayci CB, Ertenlice O, Akyer H, Aygoren H (2017) An artificial bee colony algorithm with feasibility enforcement and infeasibility toleration procedures for cardinality constrained portfolio optimization. Expert Syst Appl 85:61–75
15. Gao W, Sheng H, Wang J, Wang S (2018) Artificial bee colony algorithm based on novel mechanism for fuzzy portfolio selection. IEEE Trans Fuzzy Syst 27(5):966–978
16. Cura T (2021) A rapidly converging artificial bee colony algorithm for portfolio optimization. Knowl-Based Syst 233:107505
17. Rezaei Pouya A, Solimanpur M, Jahangoshai Rezaee M (2016) Solving multi-objective portfolio optimization problem using invasive weed optimization. Swarm Evolut Comput 28:42–57
18. Shahid M, Ansari MS, Shamim M, Ashraf Z (2022) A risk-budgeted portfolio selection strategy using invasive weed optimization. In: Tiwari R, Mishra A, Yadav N, Pavone M (eds) Proceedings of international conference on computational intelligence 2020. Algorithms for intelligent systems. Springer, Singapore
19. Dhaini M, Mansour N (2021) Squirrel search algorithm for portfolio optimization. Expert Syst Appl 178:114968
20. Shahid M, Ashraf Z, Shamim M, Ansari MS (2022) A novel portfolio selection strategy using gradient-based optimizer. In: Saraswat M, Roy S, Chowdhury C, Gandomi AH (eds) Proceedings of international conference on data science and applications 2021. Lecture Notes in Networks and Systems, vol 287. Springer, Singapore
21. Akbay MA, Kalayci CB, Polat O (2020) A parallel variable neighborhood search algorithm with quadratic programming for cardinality constrained portfolio optimization. Knowl-Based Syst 198:105944

Fission Fusion Behavior-Based Rao Algorithm (FFBBRA): Applications Over Constrained Design Problems in Engineering

Saurabh Pawar and Mitul Kumar Ahirwal

1 Introduction

There are various problems in the world related to the engineering domain and these computational problems have required efficient ways to solve them One of the ways to solve this problem is the use of intelligent decision-making techniques. Evolutionary algorithms are generally referred to as population-based algorithms [1]. A nature-inspired technique plays an important role in the computing environment. In various areas of science, swarm intelligence (SI) finds its utility in solving various design problems along with real-world problems. SI-based algorithms are based on the behavioral aspect of the animals, insects. The author had classified the SI-based algorithms into two categories namely Animal-based and insect-based. Animal-based SI algorithms include Wolf Based algorithm, Monkey Algorithms, and their variants. In insect-based algorithms, examples include ant colony optimization (ACO), butterfly algorithm, bee-inspired algorithms, etc. [2]. Genetic algorithms are also a branch of evolutionary algorithms. Many of the algorithm processes are random and this type of technology allows to set the level of randomization and control, to get optima [3]. Author Yang had explained the challenges and the open problems in this nature-inspired optimization algorithm [4].

Author Rao had given one algorithm 'Jaya' which is a powerful algorithm that does not require any particular algorithm-specific parameter [5]. Later the same author had given three simple parameters-less equations for solving optimization problems called Rao Algorithms. These algorithms also don't require any algorithm-specific parameter control, and these are metaphor-less algorithms [6]. Researchers used

S. Pawar (✉) · M. K. Ahirwal
Maulana Azad National Institute of Technology, Bhopal, Madhya Pradesh, India
e-mail: saurabhpawar1396@gmail.com

P. Singh et al. (eds.), *Machine Learning and Computational Intelligence Techniques for Data Engineering*, Lecture Notes in Electrical Engineering 998,
https://doi.org/10.1007/978-981-99-0047-3_30

this algorithm and solved various optimization problems including real-world problems e.g., multi-objective optimization for solving selected thermodynamic cycles, optimal reactive power dispatch problem, optimal weight design problem of the spur gear train, etc. [7–9]. This algorithm is also used to solve mechanical design problems as well [10]. Various variants to the original algorithms are also developed by researchers like the self-adaptive multi-population optimization algorithm [11].

Monkey algorithms are SI-based algorithms that are inspired by the behaviors of monkeys. But there are several variants of it such as monkey search, Spider monkey optimization, etc., Hybrid monkey search has been proposed by various researchers [12]. Spider Monkey Optimization (SMO) uses fission–fusion social behavior in which fission is nothing but the task of splitting into groups while foraging the food whenever the scarcity of food occurs. In this case, they divide the group known as the parent group into several small groups and unite in the evening. The process of merging is called a fusion [13, 14]. Identification of the plant leaf disease using a variant called exponential spider monkey optimization, and discrete spider monkey optimization for traveling salesperson problems are some of the applications with variants of the spider monkey optimization algorithm [15, 16]. SMO includes parallelism to get the optimal solution because after fission each group individually works concerning the local leaders, so this idea is used to solve the problems using the Rao algorithm, and the new fission–fusion behavior-based Rao algorithm has been proposed which is given in this paper.

The rest of the paper is arranged as follows: Sect. 2 gives an idea of the original Rao algorithm and SMO background, Sect. 3 focuses on the proposed methodology. The experimental setup with result analysis have been discussed in Sect. 5 and the conclusion along with the future work are given in Sect. 6.

2 Background

Rao Algorithm is a metaphor less algorithm, which does not use any parameter in initialization at the starting of the algorithm. In addition to this, any control parameters are also not required so that algorithm has shown great popularity. The author had tested this algorithm over various benchmark functions [6]. Following are the simplified questions (1), (2), (3) where Xnew is a new candidate which is calculated using Xold, Xbest, and Xworst. Here Xbest is the best candidate and Xworst is the worst candidate and Xrandom is a random candidate taken from all candidates after initializing the population. Using the function evaluation value, a maximum number of iterations has been calculated. Fitness is calculated using the objective function for the entire population. The best and worst solutions are noted down and using this best solution, worst solution, and the equations given below, the new candidate values are calculated, and the comparison is made between the old and new values and accordingly the best value is selected in every iteration.

$$X_{new} = X_{old} + r(X_{best} - X_{worst}) \tag{1}$$

$$\begin{aligned} Xnew &= Xold + r_1(Xbest - Xworst) \\ &\quad + r_2(|Xoldor\ Xrandom| - |Xrandomor\ Xold|) \end{aligned} \tag{2}$$

$$X_{new} = X_{old} + r_1(X_{best} - |X_{worst}|) + r_2(|X_{old} \text{or } X_{random}| - (X_{random} \text{or } X_{old})) \tag{3}$$

In SMO, the social behavior of spider monkeys is taken into consideration while calculating the optima [13]. Generally, this monkey lives in a group that is mostly leaded by the oldest female monkey. When the food scarcity situation occurs, this leader divides the entire group into small groups and this process is called Fission. Later, the subdivided groups go in a different direction for the food, and they are led by one leader. Unite group is also called a parent group and its leader is called a global leader while leaders of subgroups are called local leaders. These subgroups again merge in the evening and this merging process is termed as the fusion process. There are six phases of this SMO algorithm which includes the leader phase, learning phase, and decision phase for both the local as well as the global leader.

3 Proposed Methodology

In the proposed Fission Fusion behavior-based Rao algorithm, the original version of the Rao algorithm is modified using the Fission Fusion behaviors of Spider monkey optimization. The main difference is in the update equations of the original Rao algorithm. The best candidate is always taken as global best and the worst candidate is always considered as the local worst. It means, for finding optima we move towards the global best and move away from the local worst solution, and this can also help to avoid local optima of various subpopulations. Following are the Eqs. (4), (5), (6) used to find a new candidate in FFBBRA. In the equation, 'r' is the random number from 0 to 1 which is used to add some percentage of change in the original values of candidate 'X'.

$$X' = X + r1\left(X_{Global\_best} - X_{Local\_worst}\right) \tag{4}$$

$$\begin{aligned} X' &= X + r1(XGlobal\_best - XLocal\_worst) \\ &\quad + r2(|X \text{ or } Xrandom| - |Xrandomor\ X|) \end{aligned} \tag{5}$$

$$X' = X + r1(XGlobal\_best - |XLocal\_worst|) + r2(|X \text{ or } Xrandom| - (Xrandomor\ X)) \quad (6)$$

The following algorithm gives an idea about the implementation of FFBBRA. At the initial stage the number of variables, iteration maximum count is initialized along with the upper and lower bound as per function requirement. Then initial solution set is calculated, and the global best solution is detected. Then the population is divided into the subpopulation, and local worst solutions are searched, and using both the global best and the local worst solution it is updated. Now there is the last step of comparison of a good solution. As randomness is key in solution finding, its a need to always go towards the best solution. So, the best among the new and old solutions is checked and continued with the best for further iteration. And after completion of iterations, the best value is returned as an optimum value received by this algorithm. Here the subpopulation and the individual calculation of the subpopulations play a vital role for parallelism so, considering future modifications based on parallel implementation of this FFBBRA, here in the algorithm, the population can be divided based on the number of processors. The performance of the proposed methodology is tested over some benchmark functions, and it is observed that the performance is improved by this modified approach.

The complexity of the proposed methodology mainly depends on factors like fitness function, Initialization, and solution update process. Fitness function is always defined based on the problem. If we consider population size as N then for initialization we require $O(N)$ time. Considering iteration number as T, problem dimensions as D, and the number of processors or population divide criteria as P which define the number of subpopulations after dividing the entire population, then the complexity of the updating process can be given as $O(N) + O(T(N/P * D + 1))$ where 1 indicates some constant number of operations required while updating. As we are only dividing the populations into chunks and merging them, the computational complexity is approximately the same as the earlier approach.

Algorithm 1: FFBBRA Framework

```
BEGIN
Initialize the number of candidates or solutions – 'n',
the number of variables or dimensions- 'd', lower and
upper boundaries of variables 'LB' and 'UB' respectively,
function evaluation-FE, termination criteria -
'Max_iter'. number of processors 'p' and iteration
counter iter = 0;
Generate the candidate's initial solution and find its
fitness value (F1);
While iter < Max_iter:
    iter = iter + 1;
    For i = 1 → n
          Identify the best solution based on fitness
          value and find global best solution.
    End for
    For d = 1 → p
          Find local worst in the subpopulation d;
          Calculate X' using equations (4) or (5) or (6);
          Check boundaries and apply bound values in case
          violated;
    End for
    For i = 1 → n
          Evaluate F2 Fitness value for updated solution;
          If F2 is better than F1
                  F1 = F2;
          End if
    End for
End while
Return the final solution.
END
```

4 Experimental Setup

The experiment is performed over 7 unconstrained functions and the 3 constrained benchmark design problems. Codes are written in python hence, the testing of the algorithm is performed using open-source Google Colab. Performance is compared by reimplementing the Rao algorithm along with FFBBRA. While experimenting, the population is divided into 4 subpopulations. In the future, this count can be decided based on the processors. Table 1 shows the list of unconstrained benchmark functions used for the experiment [17, 18]. In the table, Column 'C' is characteristics, and abbreviations used inside it are 'U'- unimodal, 'M' for Multimodal, 'S' for separable, and 'N' for Non-separable with unconstrained problems, some Engineering design

Table 1 Details of unconstrained benchmark functions

No	Function	Formula	Dim	Search range	C	Optimum solution
FU_1	Bird function	$\sin(x)\,(e^{(1-\cos(y)2)}) + \cos(y)\,(e^{(1-\sin(x)2)}) + (x-y)^2$	2	$[-2\pi, 2\pi]$	NM	−106.7645
FU_2	SCHAFFER function	$0.5 + \sin^2((x^2 + y^2)^2) - 0.5/(1 + 0.001(x^2 + y^2))^2$	2	[−100, 100]	NM	0
FU_3	Six-hump camel function	$4x^2 - 2.1x^4 + (x^6)/3 + xy - 4x^2 + 4y^4$	2	[−5, 5]	NM	−1.0316
FU_4	DROP WAVE function	$\cos(12\,(x^2 + y^2))/0.5\,(x^2 + y^2) + 2$	2	[−5.12, 5.12]	UN	−1
FU_5	McCormick function	$\sin(x + y) + (x-y)^2 - 1.5x + 2.5y + 1$	2	[−1.5,3] [−3,3]	NM	−1.9133
FU_6	ADJIMAN function	$\cos(x).\sin(y) - \frac{x}{y2+1}$	2	[−1, 2] [−1,1]	NM	−2.0281
FU_7	Power Sum function	$\sum_{i=1}^{d} \lvert x_i \rvert^{\,i+1}$	10	[−1,1]	US	0

problems are also solved. Following are the design problems along with the models used.

4.1 *Cantilever Beam Problem*

The Cantilever Beam problem has an objective to minimize the weight. It has generally 5 elements. Those hollow elements are in cross-section. One end is rigidly supported and at one end the vertical load is applied where the beam is free. Side length is a design parameter. The model of the problem is constructed as follows [19].

Design variables = 5 [a, b, c, d, e]

Objective to min: $f(x) = 0.0624\,(a + b + c + d + e)$

Subject to: $\frac{61}{a3} + \frac{37}{b3} + \frac{19}{c3} + \frac{7}{d3} + \frac{1}{e3} <= 1$

Variable range: $0.01 <= a, b, c, d, e <= 100$.

4.2 Three Bar Truss Design Problem

In the area of civil engineering, there is a structural optimization problem called three bar truss design problem. It has 2 design variables. The objective is to minimize the weight subject to buckling constraints, stress, deflection. This problem has a difficult constrained search space so referred by various researchers while testing. The model is given as follows—[19].

Objective Function: min $z = ((2\sqrt{2}\,a) + b)\,L$

Subject to: $(\sqrt{2}\,a + b)\,P - \sigma\,(\sqrt{2}\,a^2 + 2ab) <= 0$

$Pb - \sigma\,(\sqrt{2}\,a^2 + 2ab) <= 0$

$P - \sigma\,(\sqrt{2}\,b + a) <= 0$

Variable range: $0 <= a, b <= 1$ where, $\sigma = 2$ KN/cm^2, $L = 100$ cm, and $P = 2$ KN/cm^2

4.3 Pressure Vessel Problem

Another problem that is tested using FFRRBA is the pressure vessel (PV) design problem. It is targeted to find the minimum cost of the cylindrical pressure vessel which comprises the cost for welding, forming, and material. The thickness of the shell (T_s), head thickness (T_h), shell radius (R), cylinder length (L) are the design variables. Formation of the problem can be expressed as given below [19]:

Let's consider $[T_s, T_h, R, L] = [a, b, c, d]$

Objective to minimize, $z = 0.6224acd + 1.7781bc^2 + 3.1661a^2d + 19.84b^2c$

Subject to: $g1 = (0.0193\,c) - a <= 0$

$g2 = (0.00954\,c) - b <= 0$

$g3 = 1{,}296{,}000 - \frac{4}{3}\,\pi\,c^3 - \pi dc^2 <= 0$

$g4 = d - 240 <= 0$

Variable range: $0 <= a, b <= 99$ and $10 <= c, d <= 200$.

5 Result Discussion

For checking the proposed methodology on the platform of google collab, FFBBRA tested on 7 unconstrained functions [FU_1 - FU_7] as shown in Table 2 and three constrained design problems as shown in Table 3 by considering 50,000 function evaluations and the code is run over 10 independent runs taking the population size as 20 for each benchmarking function. After running the script for Rao algorithms and the respective FFBBRA algorithms, results were obtained in the form of the best, worst, mean, standard deviation, and mean function evaluation. These results are compared to check the performance of this proposed methodology. The new approach also works well for benchmark functions whose optima is not at origin

Table 2 Optimum solutions of unconstrained benchmark functions are listed in Table 1

		Rao 1	Rao 2	Rao 3	FFBB-RA 1	FFBB-RA 2	FFBB-RA 3
FU_1	Best	−106.7877	−106.78773	−106.7877	−106.787	−106.787	−106.7877
	Worst	−106.749	−106.787733	−106.7463	**−106.787**	−106.7877	−87.52028
	Mean	−106.78131	−106.787733	−106.7772	**−106.78773**	−106.7877	−102.92904
	S.D	0.011430	6.3552e−15	0.01229	**4.49e−15**	**0.0**	7.7043
	MFE	12,994.0	9434.0	18,600.0	**9912.0**	15,228.0	19,626.0
FU_2	Best	0.0	0.0	0.0	0.0	0.0	0.0
	Worst	0.0	0.0	0.0	0.0	0.0	0.0
	Mean	0.0	0.0	0.0	0.0	0.0	0.0
	S.D	0.0	0.0	0.0	0.0	0.0	0.0
	MFE	5530	1044	5074	**2446**	3994	**1984**
FU_3	Best	−1.031628	−1.0316284	−1.03162845	−1.0316284	−1.031628	−1.031628
	Worst	−1.031606	−1.0316284	−1.031624	**−1.0316284**	−1.0316284	−1.03162
	Mean	−1.031624	−1.03162845	−1.03162	−1.03162845	−1.031628	−1.03162
	S.D	6.6823e−06	0.0	1.26026e−06	**1.857e−16**	0.0	**3.741e−07**
	MFE	11,984	4446	33,538	**4360**	14,844	**32,730**
FU_4	Best	−1.0	−1.0	−1.0	−1.0	−1.0	−1.0
	Worst	−1.0	−1.0	−0.99990	−0.93624	−1.0	**−1.0**
	Mean	−1.0	−1.0	−1.0	−0.987249	−1.0	**−1.0**
	S.D	0.0	0.0	2.724055e−05	0.02550	0.0	**0.0**
	MFE	19,626	1914	18,764	**12,714**	8112	**3418**
FU_5	Best	−1.91322	−1.9132229	−1.91321	−1.91322	−1.91322295	−1.9132217

(continued)

Table 2 (continued)

		Rao 1	Rao 2	Rao 3	FFBB-RA 1	FFBB-RA 2	FFBB-RA 3
	Worst	−1.91322	1.2283696	−1.91301	−1.91322	**−1.91322295**	−1.91318
	Mean	−1.91322	−1.599063689	−1.91315618	−1.91322	−1.9132229	−1.9132094
	S.D	2.2204e−16	0.9424777960	6.1195e−05	2.2204e−16	2.2204e−16	**1.1665e−05**
	MFE	2538	1706	28,304	**1500**	**2694**	**16,776**
FU_6	Best	−2.0218067	−2.02180678	−2.02180678	−2.02180678	−2.02180678	−2.02180678
	Worst	−2.0218067	−2.02180678	−2.02180678	−2.02180678	−2.02180678	−2.02180678
	Mean	−2.0218067	−2.02180678	−2.02180678	−2.02180678	**−2.02180678**	−2.02180678
	S.D	0.0	0.0	0.0	0.0	0.0	0.0
	MFE	1476	482	1086	**1458**	1280.0	**976**
FU_7	Best	0.0	0.0	0.0	0.0	0.0	0.0
	Worst	0.0	0.0	0.0	0.0	0.0	0.0
	Mean	0.0	0.0	0.0	0.0	0.0	0.0
	S.D	0.0	0.0	0.0	0.0	0.0	0.0
	MFE	49,884	49,196	49,948	49,976.0	49,972.0	**33,408**

Table 3 Optimum solutions of constrained design problems

		Rao 1	Rao 2	Rao 3	FFBBRA 1	FFBBRA 2	FFBBRA 3	Jaya
FC_1	Best	1.3399569	1.33995636	1.33995664	**1.3399563**	1.3399566	**1.3399564**	1.3399567
	Worst	1.33995	1.33995636	1.3399599	1.33995	1.3399606	**1.3399567**	1.3399583
	Mean	1.339958	1.339956363	1.339957	**1.339956**	1.339957591	**1.339956**	1.3399575
	S.D	1.71273e−06	2.0542e−09	9.5681e−07	**1.80e−07**	1.08205e−06	**1.0757e−07**	4.592e−07
	MFE	39,474.0	43,462.0	43,228.0	42,578.0	39,670.0	**39,620.0**	43,936.0
FC_2	Best	263.895848	263.8958433	263.895865	**263.895844**	263.8958454	**263.89584**	263.895844
	Worst	263.8960244	282.8427124	263.89601	**263.895855**	263.895937	**263.895922**	263.8960626
	Mean	263.89591	265.790530	263.89591	**263.89584**	263.8958752	**263.89585**	263.895889
	S.D	4.81023e−05	5.6840607	4.1967e−05	**2.555e−06**	2.9929e−05	**2.1516e−05**	6.041e−05
	MFE	39,376.0	38,588.0	32,142.0	**36,538.0**	40,358.0	33,640.0	39,642.0
FC_3	Best	5520.137400	5517.05479	5530.65873	5520.77230	5521.5088	**5517.15651**	5517.058832
	Worst	5736.097825	155,426.617	42,745.4199	5765.80807	**42,745.4199**	**5766.1381**	5767.02004
	Mean	5619.34396	57,742.53280	9319.20774	5635.45065	**13,059.18314**	**5643.31177**	5608.391867
	S.D	84.64444	53,496.76498	11,142.22	**82.319330**	**14,843.26264**	**85.805781**	80.8083
	MFE	28,150.0	10,676.0	33,100.0	36,514.0	30,366.0	39,922.0	34,212.0

as like earlier algorithms. In the case of performance, FFBBRA 1 and FFBBRA 3 show better results than the earlier approaches. FFBBRA 2 is not having that much improvement because of the randomness used in the equations. Functions Tested in Table 1 are mostly multimodal and non-separable functions. So, the new approach is showing better results in this type of function. Using FFBBRA1, the mean function evaluation value is improved in all the cases of unconstrained problems. In the six hump camel function, the worst solution is improved by using FFBBRA1, so that it automatically improves the standard deviation value. The same thing is also observed in the Drop Wave function when solved using FFBBRA3. As FFBBRA splits the population and moves towards the global best solution and moves away from the local worst solution, this approach will improve the performance because of independent updating candidates. This results in avoiding the local optima and improves the worst solutions as well in many cases.

FFBBRA-1, FFBBRA-2, and the FFBBRA-3 are independent to each other. FFBBRA-3 and FFBBRA-1 had shown good performance than the earlier approach so, preference will be given to them. There are various problems including optimal reactive power dispatch [7], which were solved only by using one of the Rao algorithms. Because of simplicity and parameterless approach in all three forms, any one of them can be picked up to solve a real-time optimization problem.

Statistical Analysis

A quantitative decision can be taken about the problem using the process of statistical testing where the evaluated data set is compared hypothetically. In this paper, statistical testing on the design problems is applied to all algorithms using the non-parametric test called the 'Wilcoxon signed-rank test' [20]. This test mainly depends on the order of the observations of various samples. Table 4 gives a comparative idea of the best value obtained in Rao 1, Rao 3, FFBBRA 1, FFBBRA 3, Multi-Verse Optimizer from Mirjalili [19], Cuckoo search [19], and Jaya Algorithm. Using the results based on comparative analysis from Table 4, the rank summary is calculated and given in Table 5. The result shows that FFBBR-3 received the lowest rank in all other algorithms. FFBBRA-1 is in the second position. It shows that these algorithms prove superior performance among other algorithms. Here superior performance doesn't mean that the algorithm is better than other algorithms given in the table which will also lead to violation of the 'no free lunch theorem' [18]. Here this performance is showing that it is better than other algorithms considered in this work only.

Graphical Analysis

Convergence graphs plotted against the unconstrained functions are shown in Figs. 1 and 2. For convergence graph the Jaya algorithm, Rao algorithm, and FFBBRA Algorithm were compared. The proposed methodology also tested on some engineering design problems like the three bar trust problem. This problem shows improved performance in the case of FFBBRA1 and FFBBRA3. When the cantilever beam design problem is solved using FFBBRA1, convergence it is faster as compared to the Jaya and Rao algorithm as shown in Fig. 3. Similarly, for the Pressure vessel design problem, FFBBRA3 had improved the performance as shown in Fig. 4.

Table 4 Comparison of results of all design problems

Algo	Rao 1	Rao 3	FFBBRA- 1	FFBBRA- 3	MVO [19]	JAYA	CS [19]
FC_1	1.3399569	1.3399566	**1.3399563**	**1.3399564**	1.3399595	1.3399567	1.33999
FC_2	263.895848	263.89586	**263.895844**	**263.89584**	263.895849	263.895844	263.9716
FC_3	5520.13740	5530.6587	5520.77230	**5517.15651**	6060.8066	**5517.058832**	N.A

Table 5 Rank summary statistical result

Algorithm	Rao 1	Rao 3	href	FFBB-RA3	MVO [19]	JAYA	CS [19]
FC_1	5	3	1	2	6	4	7
FC_2	4	6	2	1	5	3	7
FC_3	3	5	4	2	6	1	7
Total:	12	14	**7**	**5**	17	8	21

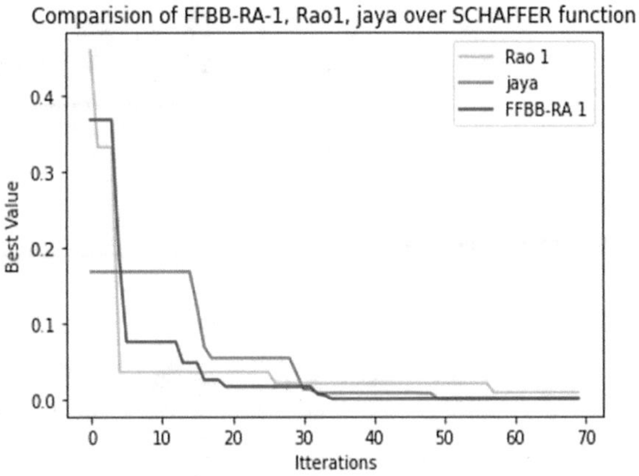

Fig. 1 Convergence graphs of UF_2

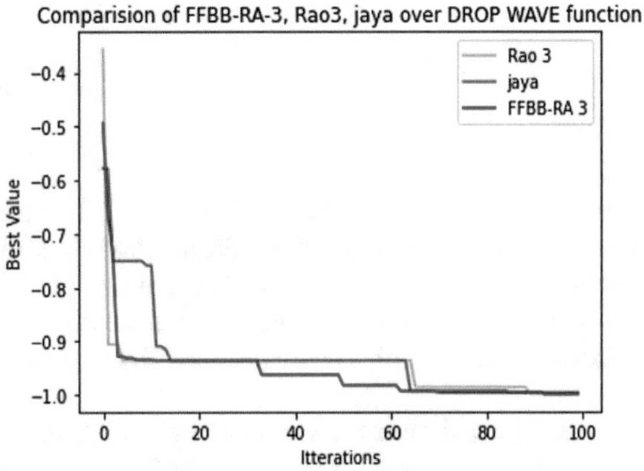

Fig. 2 Convergence graphs of the UF_4

Results obtained from unconstrained and constrained problems investigated the exploitation and exploration abilities of the proposed FFBBRA. To check exploitation ability, 7 unconstrained benchmark functions are considered in which FU_4 and FU_7 are unimodal while the remaining are multimodal functions. FU_1-FU-3 are low dimension multimodal functions used to investigate the exploration ability. While updating the candidate, it moves towards the global best candidate and moves away

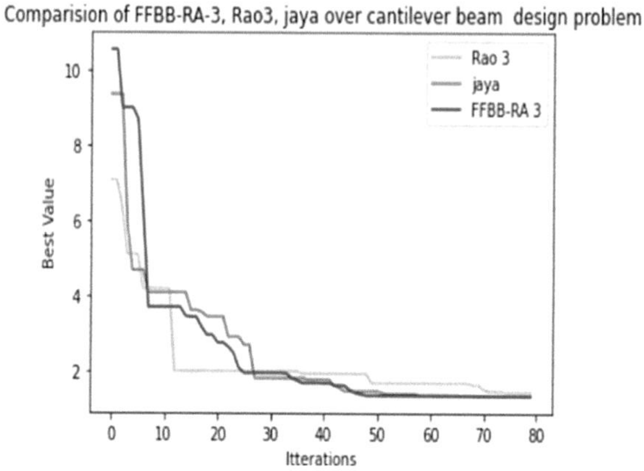

Fig. 3 Convergence graphs of Cantilever beam

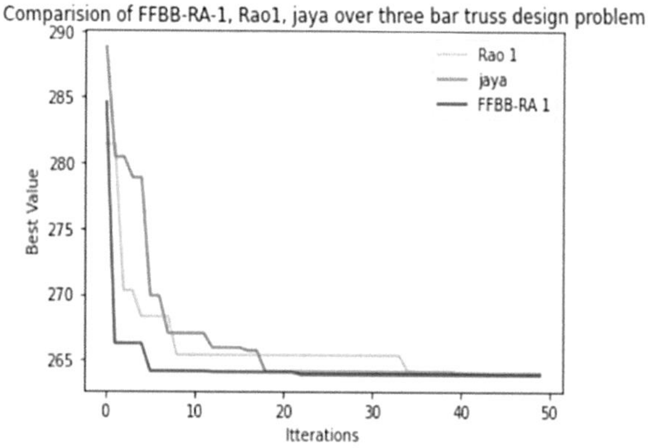

Fig. 4 Convergence graphs of three bar truss

from the local worst solution. The result of the convergence curve and optimum values of the design problems indicates that FFBBRA generally outperforms the mentioned competitors. It proves that the combination of the Rao algorithm with SMO in which dividing populations into subpopulations and updating concerning the local worst and global best can effectively control the exploration and exploitation balance.

FFBBRA had even though shown good performance, still, we believe there is the scope of improvement in case of FFBBRA-2 compared to the Rao-2 algorithm because of randomness and the modulus operator present in the equation. For unconstrained problems, the proposed methodology is only tested with existing Rao algorithms for comparison. In further research, comparison can be done with various other metaheuristic algorithms as well. The fact that can't be ignored is FFBBRA is more competitive enough with other algorithms. As this can be assumed from the 'no free lunch' theorem [18] that in the future, performance can be checked using parallelism on real-time problems with comparison to various metaphor-less algorithms.

6 Conclusion and Future Scope

This paper proposed a novel methodology of the Fission–Fusion social behavior-based Rao algorithm (FFBBRA), and the Rao algorithm equations using the social behavior of spider monkeys. The new approach finds optima by moving towards the global best and moving away from the local worst solutions of the subpopulation. In this work, FFBBRA was tested over multimodal unconstrained benchmark functions along with engineering design problems for validating the efficiency of the methodology. Statistical testing was performed to show the statistical significance over real-life industrial applications like a PV design problem, Three bar truss design problem, and Cantilever beam are used to check and verify its performance.

The result given in this paper is just a primary analysis. In the future, this algorithm can be implemented in a parallel way similar to other algorithms like parallel PSO [21]. As of now, for testing purposes, the population is divided manually into m subpopulations, but the same can be done based on the number of CPUs for effective utilization of the available resources. High-performance scientific computing can play a vital role to reduce the time required for the execution using various libraries like OpenMP to speed up the performance and the efficiency of the existing algorithm.

References

1. Mohammadi F, Hadi M (2019) Evolutionary computation, optimization and learning algorithms for data science. In: Optimization, learning and control for interdependent complex networks. Springer
2. Chakraborty A, Kar A (2017) Swarm intelligence: a review of algorithms. In: Nature-inspired computing and optimization vol 10. Springer, pp 475–494
3. Jenna C (2014) An introduction to genetic algorithms. Sr Proj 1:40
4. Yang X-S (2020) Nature-inspired optimization algorithms: challenges and open problems. Elsevier J Comput Sci 101–104
5. Rao R (2016) Jaya: a simple and new optimization algorithm for solving constrained and unconstrained optimization problems. Int J Ind Eng Comput 7:19–34
6. Rao R (2020) Rao algorithms: three metaphor-less simple algorithms for solving optimization problems. Int J Ind Eng Comput 11:107–130
7. Hassan M, Kamel S, El-Dabah M, Khurshaid T, Dominguez-Garcia T (2021) Optimal reactive power dispatch with time-varying demand and renewable energy uncertainty using Rao-3 algorithm. IEEE Access 9:23264–23283
8. Rao R, Pawar R (2020) Optimal weight design of a spur gear train using Rao algorithms. In: ICSISCET 2019, vol 13. Springer Nature Switzerland AG, pp 351–362
9. Rao R, Keesari H (2020) Rao algorithm for multi-objective optimization of selected thermodynamics cycles. Springer-Verlag London Ltd., part of Springer Nature Journal (2020)
10. Rao R, Pawar R (2020) Constrained design optimization of selected mechanical system components using Rao algorithms. Appl Soft Comput 89:106–141
11. Rao R, Pawar R (2020) Self-adaptive multi-population Rao algorithms for engineering design optimization. Appl Artif Intell 34(3):187–250
12. Vasundhara R, Sathya S (2017) Monkey behaviour based algorithms—A survey. Int J Intell Syst Appl (IJISA) 9(12):67–86

13. Sharma H, Bansal J (2019) Spider monkey optimization for algorithm. Chapter in Studies in Computational Intelligence
14. Agrawal V, Rastogi R, Tiwari D (2018) Spider monkey optimization algorithm. Int J Syst Assur Eng Manag 9:929–941
15. Kumar S, Sharma B, Sharma V, Sharma H, Bansal J (2018) Plant leaf disease identification using exponential spider monkey optimization. Sustain Comput Inf Syst
16. Akhand M, Safial I, Shahriyar S, Siddique N, Adeli H (2020) Discrete spider monkey optimization for traveling salesman problem. Appl Soft Comput 86
17. Momin J, Xin-She Y (2013) A literature survey of benchmark functions for global optimization problems. Int J Math Model Numer Optimiz 4(2):150–194
18. Xin-She-Yang, Xin-She-He, Qin-Wei F (2020) Mathematical Framework for algorithm analysis. Chapter 7, Nature inspired computation and SI. Elsevier, pp 89–108
19. Mirjalili S, Mirjalili S, Hatamlou A (2016) Multi-verse optimizer: a nature-inspired algorithm for global optimization. Neural Comput Appl 27:495–513
20. Wilcoxon F (1945) Individual comparisons by ranking methods. Biomaterials 6:80–83
21. Lalwani S, Sharma H, Satapathy S, Deep K, Bansal J (2019) A survey on parallel particle swarm optimization algorithms. Arab J Sci Eng 44:2899–2923

A Novel Model for the Identification and Classification of Thyroid Nodules Using Deep Neural Network

Rajshree Srivastava and Pardeep Kumar

1 Introduction

The thyroid nodules are the most common nodular tumor in the adult population. The early diagnosis of this tumor is essential. There are many imaging modalities for screening thyroid nodules, but ultrasonography (USG) is widely used as it is cost-effective and real-time [1]. A thyroid nodule can be defined as a lump of nodes present in the thyroid region of the neck. These thyroid nodules can be benign or malignant. In most cases, these nodules, through USG examination are found to be benign. Some of the characteristics of benign nodules are regular shape, while malignant nodules have irregular shapes, hypo-echogenicity, etc. The person with a high risk of malignant nodules is recommended for surgery, while medicines and regular follow-ups are suggested for benign nodules. The traditional diagnostic method based on doctors' expert knowledge has one of the limitations that they are heavily dependent on the person's knowledge and experience. Thus, sometimes a double screening scheme system has been applied in the hospitals by employing an additional expert, which results in time-consuming and extra expenditure [2]. In medical research, a computer-aided diagnosis system is developed to diagnose the disease, which has proved successful in many applications like lung cancer, brain tumor, breast cancer, etc. [3]. Thyroid nodule formation mostly occurs when there is an excess of thyroxine hormone in the body [4]. For thyroid nodule diagnosis cases, an automated system is helpful to differentiate benign and malignant nodules. Thyroid imaging reporting and data system (TI-RADS) have assigned some scores based on the characteristics of the thyroid USG images. These scores help us to pre-classify the USG images into benign/malignant cases [5]. The traditional methods mainly focus on designing or

R. Srivastava (✉) · P. Kumar
Department of Computer Science and Engineering, Jaypee University of Information Technology, Wakhnaghat, Himachal Pradesh, India
e-mail: rajshree.srivastava27@gmail.com

P. Singh et al. (eds.), *Machine Learning and Computational Intelligence Techniques for Data Engineering*, Lecture Notes in Electrical Engineering 998,
https://doi.org/10.1007/978-981-99-0047-3_31

selecting better hand-crafted features [6]. Traditional machine learning (ML) techniques like support vector machine (SVM), neural network, decision tree (DT) [7], etc., fail to give good results with the raw data. While in the case of deep learning (DL), it automatically finds the features and classifies the models. It transforms the original data into a higher level through simple non-linear models [8]. This paper proposes a novel model for identifying and classifying thyroid nodules using a deep neural network. This paper is organized in the following manner: Sect. 2 covers the related work, Sect. 3 focuses on the proposed work, Sect. 4 discusses the experimental work and result analysis and Sect. 5 covers the conclusions.

2 Related Work

In recent years, many ML and DL technologies have been used to identify and classify thyroid nodules. Nugroho et al. [9] proposed a CAD system that used a fusion of features extraction methods, namely gray level co-occurrence matrix (GLCM), histogram, and gray level run length matrix (GLRLM) method. The proposed model has achieved 89.74% accuracy, 88.89% sensitivity, and 91.67% specificity using multi-layer perceptron (MLP) as the classifier. One of the limitations is the smaller dataset, i.e., 39 USG thyroid images collected from the local hospital and other feature extraction techniques and classifiers can be explored. Ko et al. [10] proposed a deep convolution neural network (DCNN) for thyroid nodules diagnosis. They have used the single image as a representative. The model has achieved 88% accuracy, specificity 82% and sensitivity 91%. Some of the limitations are it takes high computation time; the size of the dataset is less and failed to address the problem of noise. Xie et al. [11] proposed a novel hybrid model that used DL and handcrafted features (LBP) for feature extraction from the images and achieved 85% accuracy. They have used 623 thyroid USG images collected from Shanghai Tenth People's hospital for the experiment. Authors have designed a model that combines the high-level features extracted from CNN and maps its corresponding LBP feature to improve the accuracy of the model. If it failed to address the problem of noise different ML/DL classifiers can be explored. Sun et al. [12] proposed the hybrid method to classify thyroid nodules where they have combined DL-based technique and statistical features together. A total of 104 statistical features were extracted, and principal component analysis and t-test is used for the feature reduction. Once the best features were obtained different classifiers like logistic regression, naïve Baeys and support vector machine (SVM) are used for the classification along with the VGG-16 model. The model has achieved 86.5% accuracy. The work can be extended on the hybrid DL model. Zhu et al. [13] proposed an automatic VGG-19 deep convolutional neural network (DCNN) based model to classify thyroid and breast lesions. They have collected a dataset from the Ethics committee of Shanghai Hospital. Failure to address the problem of noise. Their model has achieved 86.5% accuracy, specificity 87.7%, and sensitivity 86.7%. Further work can be extended to the fusion of handcrafted features.

3 Proposed Work

This section covers the proposed methodology adopted for this work. The first phase is data collection, the second phase is pre-processing, where RGB to Gray scale conversion, image resizing, cropping, noise removal and ROI extraction are performed to maintain uniformity. In the third phase, features are extracted using an intensity-based and GLCM based methods. In the last phase, DNN is used to classify thyroid nodules. Figure 1 shows the workflow of the proposed model.

3.1 Data Collection Phase

In this work, public and collected datasets are considered. The public dataset is an open database DDTI (Digital Database of thyroid Imaging) [14], having 188 malignant and 107 benign i.e., total of 295 thyroid USG images. The second dataset was collected from Kriti Scanning Center, Prayagraj, U.P., India, duly approved by NABH [15], having 226 malignant and 428 benign i.e., a total of 654 thyroid USG images. The duration for the dataset collected is from July 2020 to March 2021.

3.2 Pre-processing Phase

Pre-processing is an important part for the development of any model [14]. In our work, initially, the images were 560 × 360 pixels in RGB format, it's been resized to 256 × 256 pixels for the uniformity using Eq. 1:

$$B = imresize(A, scale) \tag{1}$$

where A: input image

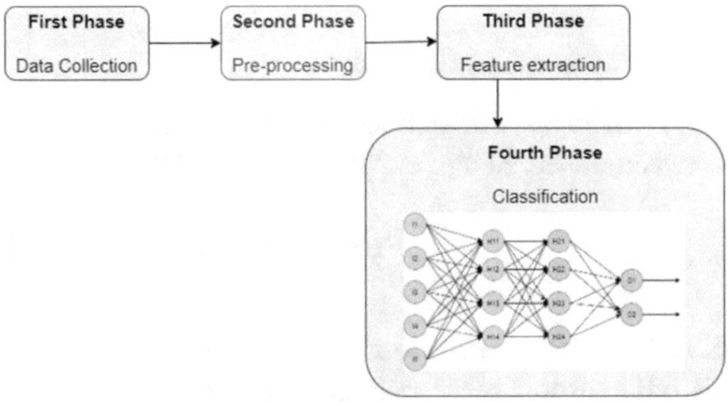

Fig. 1 Workflow of the proposed model

RGB to Grayscale conversion is performed using Eq. 2:

$$I = rgb2gray(RGB) \tag{2}$$

ROI extraction is performed by the radiologists, gaussian blur function is used to remove noise using Eq. 3:

$$h = fspecial(type) \tag{3}$$

where fspecial: returns h as a correlation kernel, h: creates a 2D filter

Further, we have labelled benign thyroid USG images as '1' and malignant as '0'.

3.3 Feature Extraction Phase

The intensity and GLCM methods are used for extraction of features [16]. Intensity is defined as the average pixel values, variance, asymmetry, or standard deviation of the whole input image. The GLCM method is a texture feature extraction method that describes the relationship between the neighboring pixels [17]. After that, the extracted features are normalized to [0, 1] using Eq. 4:

$$normalized\ z_i = \frac{(x_i - \min(x))}{\max(x) - \min(x)} \tag{4}$$

where x_i: ith normalized term; max(x): maximum value; min(x): minimum value

The features which are extracted using intensity and GLCM methods are described below:

3.3.1 Mean: It is defined as the ratio of all pixel values to the total number of pixels in an image [18]. It can be computed using Eq. 5:

$$Mean = \sum\nolimits_{i,j=1}^{n} \boldsymbol{P}(i, j) \tag{5}$$

where P: co-occurrence matrix

3.3.2 Variance: It is defined as moments in probability describing, i.e., the distance between observed and expected values [19]. It can be computed using Eq. 6:

$$Variance = \Sigma_{i,j=1}^{n}(i - j)^2 \boldsymbol{P}(i, j) \tag{6}$$

3.3.3 Standard Deviation: It is dispersion or variation of the brightness of the pixels [20]. Equation 7 represents the equation to solve the standard deviation.

$$Standard\ Deviation = \sqrt{\sum\nolimits_{i,j=1}^{n} \boldsymbol{P}(i,j)(i-\mu)^2} \tag{7}$$

where μ: mean

3.3.4 Skewness: It is a measure of symmetry and can be computed using Eq. 8:

$$Skewness = \sum\nolimits_{i,j=1}^{n} (i-\mu)^3 \boldsymbol{P}(i,j) \tag{8}$$

3.3.5 Energy: It is a parameter to measure the similarity of an image [21]. It is defined using Eq. 9:

$$Energy = \sum\nolimits_{i,j=1}^{n} \boldsymbol{P}(i,j)^2 \tag{9}$$

3.3.6 Contrast: It measures the intensity of a pixel and its neighbor over the image [22]. It can be calculated using Eq. 10:

$$Contrast = \sum\nolimits_{i,j=1}^{n} \boldsymbol{P}(i,j)(i-j)^2 \tag{10}$$

3.3.7 Correlation: It measures the spatial dependencies between the pixels and be computed using Eq. 11:

$$Correlation = \sum\nolimits_{i,j=1}^{n-1} \boldsymbol{P}(i,j)\frac{(i-\mu)(j-\mu)^2}{\sigma} \tag{11}$$

where σ: standard deviation

3.3.8 Homogeneity: It is a measure of local homogeneity of an image [23]. It can be measured using Eq. 12:

$$Homogeneity = \sum\nolimits_{i,j=1}^{n} \frac{\boldsymbol{P}(i,j)}{1+(i-j)^2} \tag{12}$$

3.4 Classification Phase

A simple neural network (NN) consists of (i) an input layer, (ii) hidden layer and (iii) output layer. All these layers have nodes that are connected with some weights. A DNN is a neural network that has several layers [24]. The most common DNN used is the Feed-Forward Dense Neural Network (FF-DNN) [25]. It is widely used for classification and prediction problems [26]. DNN can be computed using Eq. 13

Table 1 Parameter settings of DNN

Parameters	Values
Optimizer	Adam
No. of epochs	124
Batch size	20
No. of neurons	8
Activation	Sigmoid
Loss function	MSE
Random state	50
Stop criteria	End of epoch

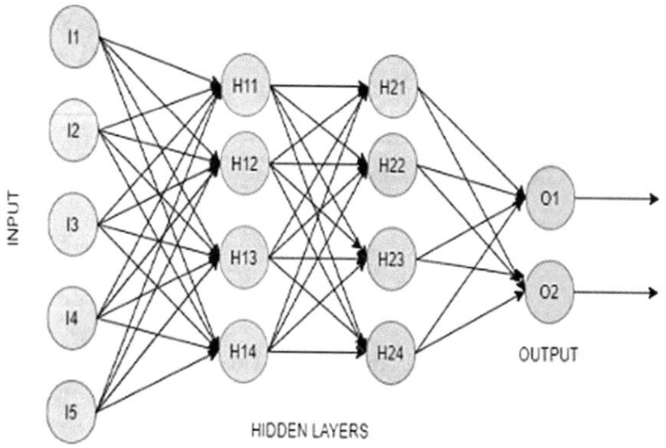

Fig. 2 Architecture of DNN

and its output using Eq. 14.

$$z = x1 * w1 + x2 * w2 + \cdots + xn * wn + b * 1 \tag{13}$$

$$\hat{y} = aout = sigmoid(z) \tag{14}$$

input: x1, x2 … xn, bias: b, weights: w1, w2, wn, sigmoid (z): activation function.

In DNN, it has more than one hidden layer. In our work, 2 hidden layers are used. The parameter settings of DNN are given in Table 1. Figure 2 shows the architecture of DNN.

3.5 Proposed Algorithm

Initially, the datasets were collected and various pre-processing steps like resizing the image datasets using Eq. 1 is performed. The conversion of RGB to greyscale is done using Eq. 2, extraction of ROI, removing noise using gaussian blur function, labeling of datasets, benign as ‘1’ and malignant as ‘0’ was performed. After that, GLCM and intensity-based features are extracted using Eqs. 5–12. To normalize the extracted features to [0,1], min–max function is applied using Eq. 4 and classification is performed using Eqs. 13 and 14. Various parameters of DNN are initialized like epoch size, the number of hidden layers, neurons, batch size, etc. The training and testing ratio is set and the model is trained to evaluate its performance. If the model performance is not maximum, then parameters of DNN are updated and trained. Figure 3 shows the complete process of the proposed model.

Algorithm 1 *Identification and Classification of Thyroid Nodules Using DNN*

Input: Image Dataset
Output: Prediction
Step 1: Input the dataset.
Step 2: Perform pre-processing step, resize the image dataset, convert RGB to grey scale, extract ROI, remove noise using non-median filter, label dataset benign as ‘1’ and malignant as ‘0’.
Step 3: Extract features using intensity and GLCM based feature extraction techniques.
Step 4: Normalize the extracted features to [0,1].

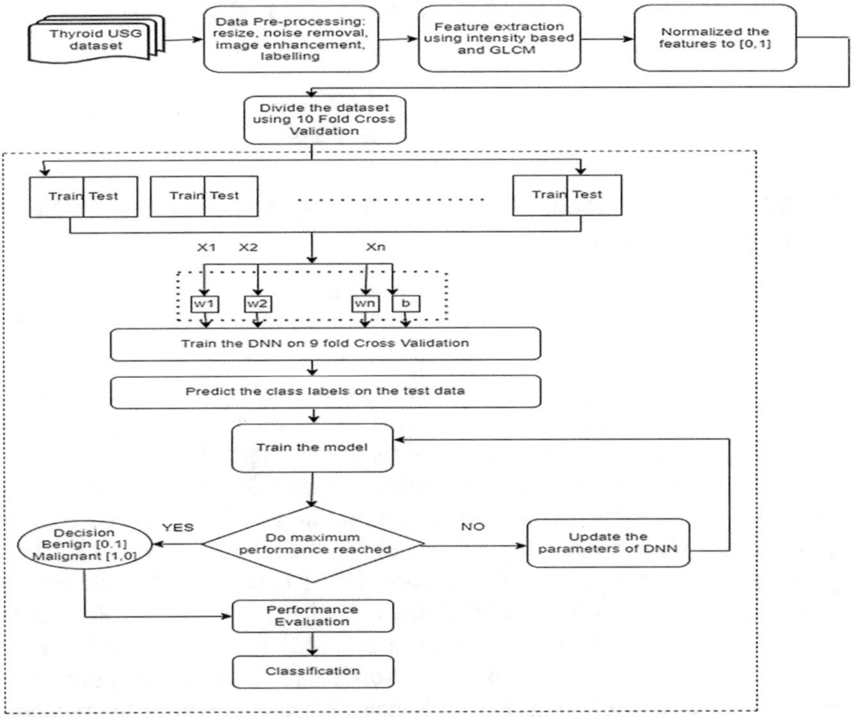

Fig. 3 Complete process of the proposed model

Step 5: Initialize the parameters of DNN like epoch size, number of hidden layers, neurons, batch size etc.
Step 6: Train the model
Step 7: Compute the performance
Step 8: Prediction

4 Experimental Work and Result Analysis

The code was executed on MATLAB 2016B, intel i5, 8th generation. To validate the performance of the model 10 cross validation and 50–50% hold out method is used for dataset-1 and dataset-2. The performance of the model is evaluated using three parameters which are discussed below:

4.1 Accuracy: It is defined as the percentage of correctly classified instances. It can be computed using Eq. 15:

$$Accuracy = \frac{TP + TN}{TP + TN + FP + FN} \tag{15}$$

4.2 Specificity: It is a measure of how well a test can identify true negative. It can be computed using Eq. 16:

$$Specificity = \frac{TN}{TN + FP} \tag{16}$$

4.3 Sensitivity: It is a measure of how well a test can identify true positives. It can be computed using Eq. 17:

$$Sensitivity = \frac{TP}{TP + FN} \tag{17}$$

For a better understanding, we have re-named public datasets as dataset-1 and the collected dataset as dataset-2. Table 2 compares the proposed model and state-of-the-art models based on accuracy, sensitivity and specificity. Figure 4 shows performance comparison of the study based on 10 cross validation and 50–50% hold out method for dataset-1 and dataset-2. It can be concluded from the figure that there is 2% to 4% improvement in the performance evaluation of the model using 10 cross validation. Figure 5 shows the comparison of the proposed model and state-of-the art models based on accuracy. It can be concluded from the figure that the proposed model has achieved an accuracy of 90.9% on dataset-1 and an accuracy of 92.85% on dataset-2 which is higher than the other state-of-the-art models for thyroid nodules identification and classification. Figure 6 shows the comparison of the proposed model and state-of-the-art models based on sensitivity and specificity. The proposed

Table 2 Comparison of state-of-the art models and proposed model based on accuracy, sensitivity and specificity

Models Ref. ID	Accuracy (%)	Sensitivity (%)	Specificity (%)
Nugroho et al. [9]	89.74	88.89	91.67
Ko et al. [10]	88	91	82
Xie et al. [11]	85	–	–
Sun et al. [12]	86.5	–	–
Zhu et al. [13]	86.5	86.7	87.7
This study (50–50% hold out method)			
Dataset-1	87.56	88.11	86.95
Dataset-2	89.14	90.81	88.15
Proposed Model (10 cross-validation)			
Dataset-1	90.90	91.75	89.87
Dataset-2	92.85	93.68	91.78

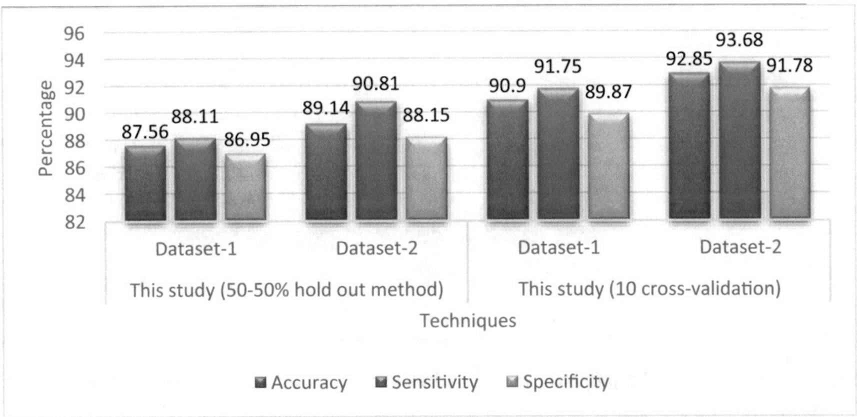

Fig. 4 Performance comparison of the study based on 10 cross-validation and 50–50% hold out method on dataset-1 and dataset-2

model has achieved sensitivity of 91.75% and specificity of 89.87% on dataset-1 and sensitivity of 93.68% and specificity of 91.78% on dataset-2.

5 Conclusion

This paper presents a novel model for the classification of thyroid nodules using a deep neural network. The proposed model is evaluated on public and collected datasets. A total of eight features, namely mean, variance, standard deviation, skewness, contrast, correlation, energy, and homogeneity are extracted using intensity and

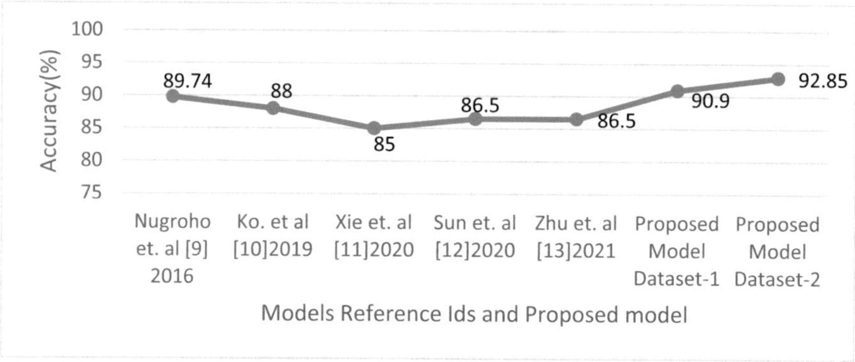

Fig. 5 Comparison of the proposed model and state-of-the art models on accuracy

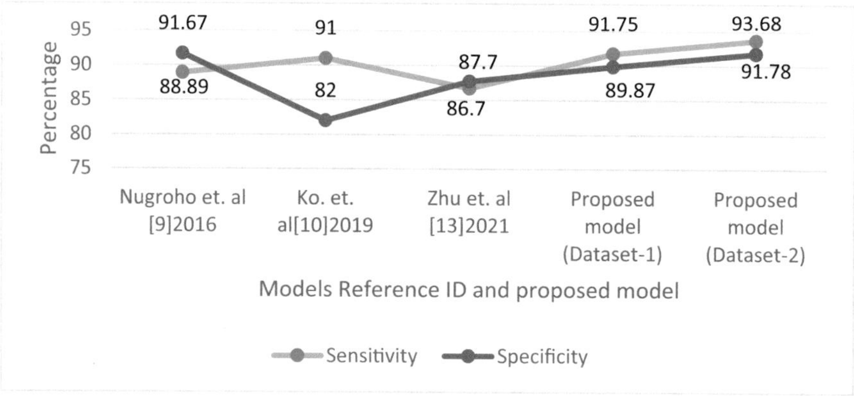

Fig. 6 Comparison of the proposed model and state-of-the art models based on sensitivity and specificity

GLCM methods. In the end, DNN is used to classify the thyroid nodules. Experiments are conducted on public and collected datasets with 50–50% hold out method and tenfold cross validation to validate the performance of the proposed model. It is inferred from the results that the proposed model performs better with an accuracy of 90.9%, sensitivity of 91.75% and specificity of 89.97% on the public dataset and an accuracy of 92.85%, sensitivity of 93.68% and specificity of 91.72% on the collected dataset using tenfold cross validation. It is evident that the proposed model is competitive to other state-of-the-art models for classifying thyroid nodules. In the future, to achieve better outcomes, researchers can incorporate the fusion of deep learning, segmentation, and boundary detection techniques to classify thyroid nodules.

References

1. Meiburger KM, Acharya UR, Molinari F (2018) Automated localization and segmentation techniques for B-mode ultrasound images: a review. Comput Biol Med 92:210–235
2. Jung NY, Kang BJ, Kim HS, Cha ES, Lee JH, Park CS, ... Choi JJ (2014) Who could benefit the most from using a computer-aided detection system in full-field digital mammography? World J Surg Oncol 12(1):1–9
3. Cheng CH, Liu WX (2018) Identifying degenerative brain disease using rough set classifier based on wavelet packet method. J Clin Med 7(6):124
4. Srivastava R, Kumar P (2021) BL_SMOTE ensemble method for prediction of thyroid disease on imbalanced classification problem. In: Proceedings of second international conference on computing, communications, and cyber-security. Springer, Singapore, pp 731–741
5. Pedraza L, Vargas C, Narváez F, Durán O, Muñoz E, Romero E (2015) An open access thyroid ultrasound image database. In: 10th international symposium on medical information processing and analysis, January, vol 9287. International Society for Optics and Photonics, p 92870W
6. Nguyen DT, Pham TD, Batchuluun G, Yoon HS, Park KR (2019) Artificial intelligence-based thyroid nodule classification using information from spatial and frequency domains. J Clin Med 8(11):1976
7. Abiodun OI, Jantan A, Omolara AE, Dada KV, Mohamed NA, Arshad H (2018) State-of-the-art in artificial neural network applications: a survey. Heliyon 4(11):e00938
8. Ma L, Ma C, Liu Y, Wang X (2019) Thyroid diagnosis from SPECT images using convolutional neural network with optimization. Comput Intell Neurosci
9. Nugroho HA, Rahmawaty M, Triyani Y, Ardiyanto I (2016) Texture analysis for classification of thyroid ultrasound images. In: 2016 international electronics symposium (IES), September. IEEE, pp 476–480
10. Ko SY, Lee JH, Yoon JH, Na H, Hong E, Han K, Kwak JY (2019) Deep convolutional neural network for the diagnosis of thyroid nodules on ultrasound. Head Neck 41(4):885–891
11. Xie J, Guo L, Zhao C, Li X, Luo Y, Jianwei L (2020). A hybrid deep learning and handcrafted features based approach for thyroid nodule classification in ultrasound images. In: Journal of physics: conference series, vol 1693, No. 1, December. IOP Publishing, p 012160
12. Sun H, Yu F, Xu H (2020) Discriminating the nature of thyroid nodules using the hybrid method. Math Probl Eng
13. Zhu YC, AlZoubi A, Jassim S, Jiang Q, Zhang Y, Wang YB, ... Hongbo DU (2021) A generic deep learning framework to classify thyroid and breast lesions in ultrasound images. Ultrasonics 110:106300
14. Srivastava R, Kuma P (2021) A hybrid model for the identification and classification of thyroid nodules in medical ultrasound images. Int J Modell, Identif Contr (IJMIC). [In Press]
15. https://www.nabh.co/frmViewCGHSRecommend.aspx?Type=Diagnostic%20Centre&cityID=94
16. Das HS, Roy P (2019) A deep dive into deep learning techniques for solving spoken language identification problems. In: Intelligent speech signal processing. Academic Press, pp 81–100
17. Singh GAP, Gupta PK (2019) Performance analysis of various machine learning-based approaches for detection and classification of lung cancer in humans. Neural Comput Appl 31(10):6863–6877
18. Monika MK, Vignesh NA, Kumari CU, Kumar MNVSS, Lydia EL (2020) Skin cancer detection and classification using machine learning. Mater Today: Proc 33:4266–4270
19. Murugan A, Nair SAH, Kumar KS (2019) Detection of skin cancer using SVM, random forest and kNN classifiers. J Med Syst 43(8):1–9
20. Chunmei X, Mei H, Yan Z, Haiying W (2020) Diagnostic method of liver cirrhosis based on MR image texture feature extraction and classification algorithm. J Med Syst 44:1–8
21. Francis SV, Sasikala M, Saranya S (2014) Detection of breast abnormality from thermograms using curvelet transform based feature extraction. J Med Syst 38(4):1–9

22. Suganthi M, Madheswaran M (2012) An improved medical decision support system to identify the breast cancer using mammogram. J Med Syst 36(1):79–91
23. Karimah FU, Harjoko A (2017) Classification of batik kain besurek image using speed up robust features (SURF) and gray level co-occurrence matrix (GLCM). In International conference on soft computing in data science, November. Springer, Singapore, pp 81–91
24. Nanglia P, Kumar S, Mahajan AN, Singh P, Rathee D (2021) A hybrid algorithm for lung cancer classification using SVM and neural networks. ICT Express 7(3):335–341
25. Liu W, Wang Z, Liu X, Zeng N, Liu Y, Alsaadi FE (2017) A survey of deep neural network architectures and their applications. Neurocomputing 234:11–26
26. Wen T, Xie G, Cao Y, Cai B (2021) A DNN-based channel model for network planning in train control systems. IEEE Trans Intell Transp Syst (Early access) 1–8

Food Recipe and Nutritional Information Generator

Ayush Mishra, Ayush Gupta, Arvind Sahu, Amit Kumar, and Pragya Dwivedi

1 Introduction

Having a healthy diet throughout the life course is important for an individual for preventing poor nutrition in all its forms along with a range of non-communicable diseases and conditions. However, an increase in the production of processed foods, growing urbanization, etc., has had major impacts on our lives which have led to changes in our lifestyles and dietary patterns. Individuals are now taking processed food which is high in fat and also having high salt/sodium, and in addition, many people do not consume healthy foods such as fruits, vegetables, and whole grains. As predicted by the World Health Organization, the number of overweight adults has reached an alarming level. More importantly, obesity also causes many types of diseases such as gastrointestinal diseases, chronic diseases, etc. Individual preferences, beliefs, cultural traditions, geographical and environmental aspects along with their busy lifestyles make them neglect appropriate behaviors. As Food enthusiasts ourselves, we wanted to build a solution for all other fellow foodies to get holistic information about their food based on factors like ingredients, cuisine, method to prepare, etc.

India is a blend of hundreds of cultures and civilizations. With 28 states and different cultures, it is a storehouse of a plethora of cuisines. The method of cooking and ingredients in a recipe varies from place to place. Hence, it is all the more important for us to have appropriate knowledge of our dish, especially with the indigents as well the nutritional value. Since the spread of COVID-19, people have been sitting in their homes for the last 2 years. Everybody wants to keep themselves

A. Mishra (✉) · A. Gupta · A. Sahu · A. Kumar · P. Dwivedi
Computer Science and Engineering, Motilal Nehru National Institute Of Technology, Prayagaraj 211004, India
e-mail: ayushgsu2018@gmail.com

P. Dwivedi
e-mail: pragyadwi86@mnnit.ac.in

P. Singh et al. (eds.), *Machine Learning and Computational Intelligence Techniques for Data Engineering*, Lecture Notes in Electrical Engineering 998,
https://doi.org/10.1007/978-981-99-0047-3_32

healthy, people are becoming more health conscious, but there are not many food websites, especially for Indian food to provide them with a complete guide for their food. However, we can use the data collected by web scraping over several food websites to further study deep neural networks and bridge the gap which is present in the absence of sufficient data. Thus, we have developed a nutritional information generator which helps people to choose healthy food in terms of their nutritional value.

2 Related Work

There have been a few works in the past on food image identification and calorie estimation, which we found quite helpful while developing our application and they guided us in the right direction that helped us in gaining a better understanding of what came before us. They include but are not limited to:

- Indian Food Image Classification with Transfer Learning [1].
- Deep learning and machine vision for food processing [2].
- Deep Convolutional GAN-Based Food Recognition Using Partially Labeled Data [3].
- Highly Accurate Food/Non-Food Image Classification Based on a Deep CNN [4].
- Deep Learning-Based Food Image Recognition for Computer-Aided Dietary Assessment [5].
- Deep Learning-Based Food Calorie Estimation Method in Dietary Assessment [6].

Bossard et al. tried to tackle the food recognition issue, they introduced a method that used random forests to mine discriminative parts. A little later, Min et al. proposed a multi-attribute theme-based modeling approach that included different types of attributes such as cuisine style, flavors, course types, ingredient types, etc. [7]. Chen et al. attempted to predict ingredient, cutting, and cooking attributes from food images by finding out that food cutting style and the ingredients used for cooking play a significant role in food's appearances [8]. Few recent works explore spatial layout [9]. Bossard et al. in 2014 used a Food-101 food images dataset consisting of 101000 images equally divided among 101 categories to set a baseline accuracy of 50.8% [10]. Salvador et al., in 2017, introduced the Recipe1M+ dataset, of over a million cooking recipes and 13 million food images [11]. Currently, this is the largest publicly available collection of food-related data.

3 Dataset

Since fewer experiments have been done on Indian foods, the available datasets on Indian foods contain either less number of images or less number of food items; hence, they cannot be directly used for this project. So, we have used a combination of Indian foods present in the Food-101 dataset and another Indian Food Dataset that we have prepared by writing a web crawler tool with alternating proxies to download around 20,000 images from those websites along with all of their ingredients, preparation, and other related information. The current dataset has around 24,000 images for the training set, 3000 images for the validation set, and another 3000 images for the test set to follow the 80/10/10 split. Moreover, we have used the image augmentation techniques such as flipping, cropping, and mirroring for all of the training images. All images used for training had a resolution of $400 \times 400 \times 3$ and with normalization by dividing the RGB values by 255.

4 Methodology

We started with building a food identification system by experimenting with models such as inceptionv3, ResNet-50, ResNet-101, and DenseNet-121. Once the food has been classified the second phase model estimates the calorific value of the food item from the image based on Mask R-CNN.

Finally, we connected both the parts to predict recipes corresponding to the image along with their nutritional information (Fig. 1).

4.1 Food Image Classification

To build our food image classification system we first started with a baseline neural network model consisting of four convolutional layers and one fully connected layer. Different data augmentation techniques like the padding, the cropping, and the horizontal flipping are mostly used to train deep neural networks. However, most approaches used in training neural networks only use very elementary types of data augmentation approaches. While neural network architectures have been analyzed in great depth, less focus has been put into discovering and inventing new and strong types of the data augmentation policies that capture data invariances. As a result of the task of classification of food images in this project, we have applied the random transforms methods like zoom, flip, warp, rotate, lighting, and contrast randomly. Random transform techniques will help in increasing the variety of image samples and prevents overfitting. Once the image has been classified then we look at its ingredients to classify them either as vegetarian or non-vegetarian. For this project, we experimented with six different models. We have chosen these models specifi-

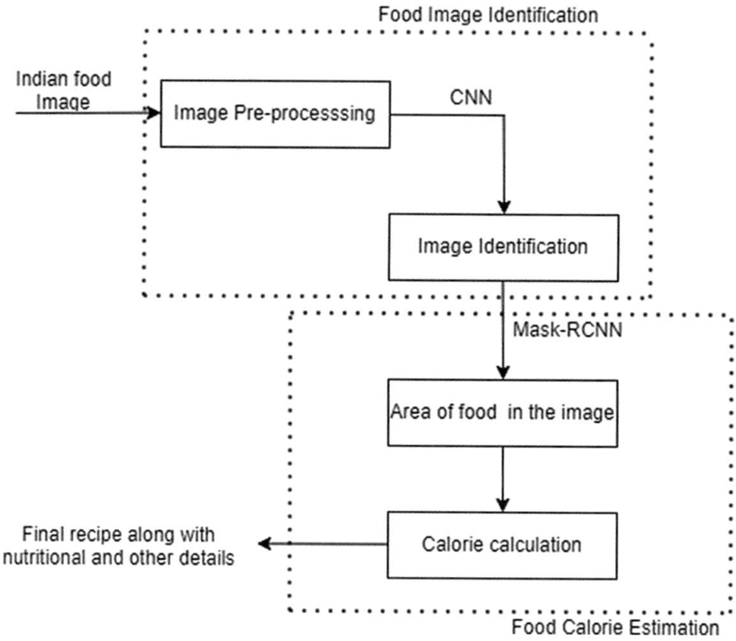

Fig. 1 Flowchart

cally because of their capability to handle all the difficulties and issues that arise in training neural networks with a large number of hidden layers. We have used two baseline models and other models including ResNet-50, ResNet-101, Inceptionv3 CNN models, and their variations. Very deep neural networks suffer from vanishing and exploding gradient problems so it becomes difficult to train the model. To overcome this problem, we use ResNets that make use of skip connections in which the activation from one layer is taken and is added to another layer even much deeper in neural network which allows us to train large neural networks even with layers greater than 100. ResNets are built by stacking some residual blocks. Each residual block adds a shortcut/skip connection before the second activation. These networks can go deeper without hurting performance.

$$a^l = g(z^l + w^{l-2,l} a^{l-2}) \tag{1}$$

a^l the activations (outputs) of neurons in layer l,
g is the activation function for layer l,
$w^{l-1,l}$ is the weight matrix for neurons between layer l–1 and l, and
$z^l = w^{l-1,l} a^{l-1} + b^l$.

DenseNets simplify the connectivity pattern between the neural network layers introduced in other architectures. We connect all layers with each other. DenseNets solve the same problem as the ResNets.

$$a^l = g\left(z^l + \sum_{k=2}^{K} w^{l-k,l} a^{l-k}\right) \quad (2)$$

a^l the activations (outputs) of neurons in layer l,
g is the activation function for layer l,
$w^{l-1,l}$ is the weight matrix for neurons between layer l-1 and l, and
$z^l = w^{l-1,l} a^{l-1} + b^l$.

When designing a layer for a ConvNet, we might have to pick if we want a 1 × 3 filter, or 3 × 3, or 5 × 5, or a pooling layer. And so the basic idea is that instead of we need to pick one of these filter sizes or pooling layers we want and commit to that, we can just select them all and concatenate all and this is called the inception module. It lets the network learn whatever parameters it wants to use, whatever the combinations of these filter sizes it wants. These network architectures are more complicated, but they also work remarkably well. Inception networks help to gain high-performance and efficient usage of computing resources.

In transfer learning we start with patterns learned from solving a related task and use that learned knowledge on a new task. All of the transfer learning models we have used for developing this application, are pretrained on the ImageNet dataset, consisting of 1000 different classes.

4.2 Food Calorie Estimation

Once the food has been classified, we have built a model based on Mask R-CNN that will calculate the food calorie based on the food area. Mask R-CNN is a state-of-the-art image segmentation method developed by Kaiming He [12]. Instance segmentation is the method by which we are able to separate different instances of objects in an image/video. For example, if in a food image, there are two types of chapatis, then they will be segmented using two different masks. We have used a 25 cm diameter plate as a reference object for the estimation of the size of food detected in images.

Mask R-CNN takes an image as an input and gives masks of the identified items, and bounding boxes. The Masks in R-CNN are the binary-coded monophonic matrices of the dimensions of the input image that denotes the boundaries of the identified object. We needed the real sizes of food items, which is not always possible through camera images alone. So, we have taken a referencing approach in which we reference the food objects to the size of some identified object to extract the actual dimensions of the food present in that specific image. The pixels-per-inch-square is calculated using the actual size of the plate in real life.

$$(\text{pixels\_per\_sq\_cm}) = (\text{plate's\_pixel\_ar})/(\text{actual\_plate\_ar}) \quad (3)$$

$$(\text{real\_food\_ar}) = (\text{detected\_food\_pixel\_ar})/(\text{pixels\_per\_sq\_cm}) \quad (4)$$

Once we have got the food area and plate area in pixels, and since we have the radius of the plate, we can easily determine the food area and hence, an approximate estimate of calories contained in the plate using our data of calories per sq cm of that specific food item.

5 Evaluation/Results

5.1 Food Image Identification

Experiments with Different Baseline Models We started with different pretrained models in order to see which one worked the best for our dataset. After experimenting with different architectures such as DenseNets, ResNets, InceptionV3, and VGG16, we came to the conclusion that ResNets were more suitable for this problem, as it brought out the best performance with the lowest mean squared error loss value.

Experiments with Adding More Convolutional Layers We also experimented by adding a few more convolutional layers along with the pooling layers on top of the baseline model, since deeper networks have a greater chance of success. Hence, by adding a few more layers to the baseline models, we have improved the performance marginally.

Experiments with Different Learning Rates, Dropout layer, and Batch size Since learning rates, batch size, and dropout layers considerably affect the performance of the models, we tried varying them to analyze the resultant outcome (Figs. 2, 3, 4 and 5).

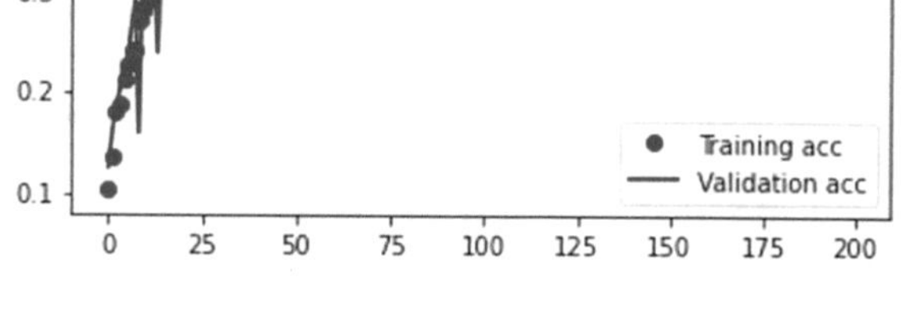

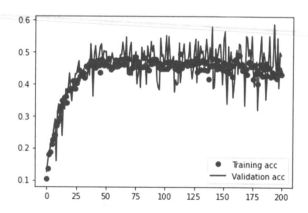

Fig. 2 5 Convolutional and 2 Dense Layer + Dropout

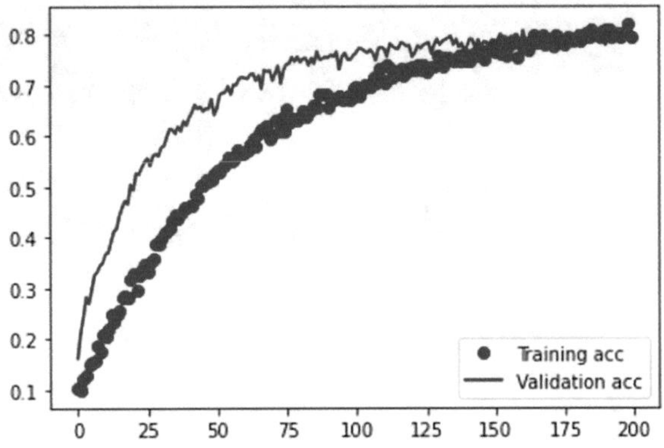

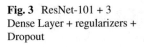

Fig. 3 ResNet-101 + 3 Dense Layer + regularizers + Dropout

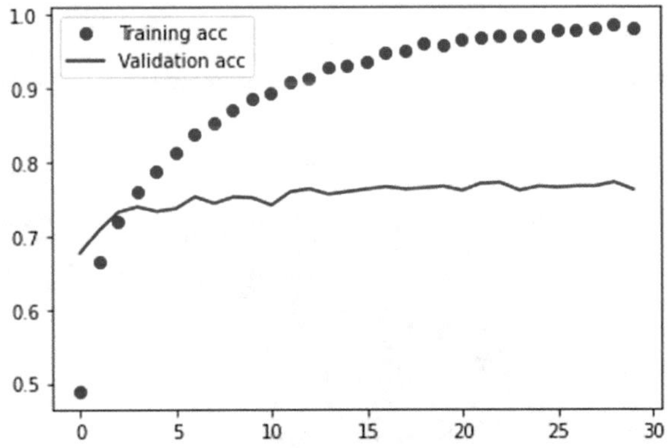

Fig. 4 InceptionV3 and 1 Dense Layer + Dropout

5.2 *Calorie Estimation*

The below table shows the predicted and actual calorific content of a few of the food items.

Food items	Calorific value (in Cal)		
	Predicted	Actual	Error
Samosa (5)	208.42	260	51.58
Chicken Curry	208.42	260	51.58
Aloo Parantha (1)	246.20	300	53.80
Dhokla (5)	321.93	400	78.07
Uttpam (1)	186.47	210	23.53
Wheat Paratha (1)	195.32	180	15.32
Vada Pao (1)	175.41	195	19.59
Bhel Puri	265.73	290	24.27
Khichdi	231.89	215	16.89
Gajar Halwa	386.22	353	33.22

Values within parenthesis indicate the quantity of that item in the image.

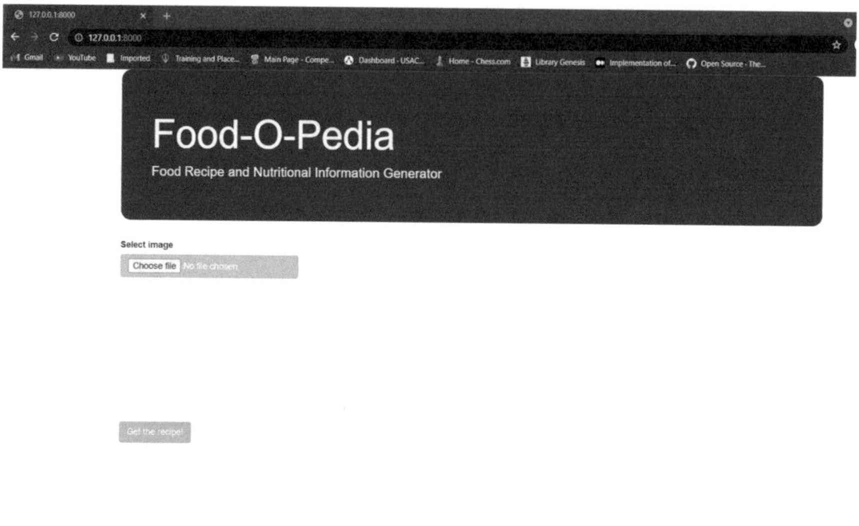

Fig. 5 Home Page of our Application

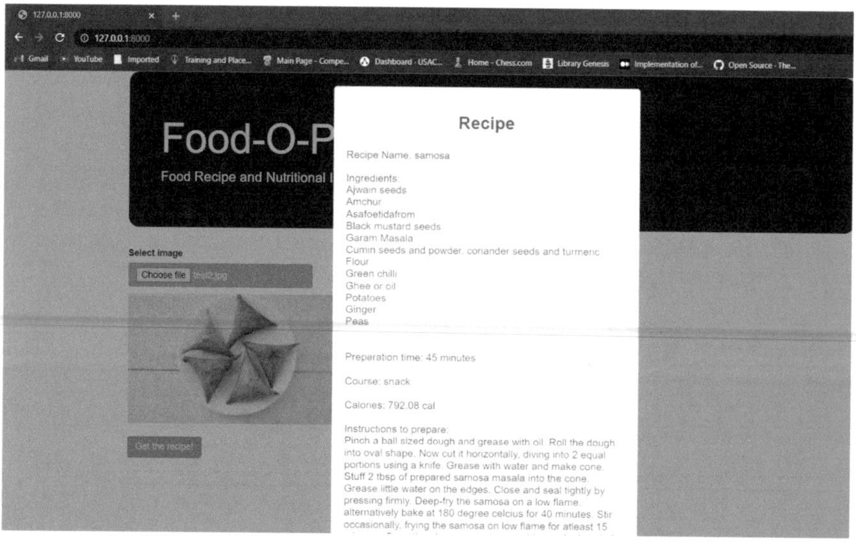

Fig. 6 Samosa

5.3 *Final Output*

Below are some of the images and outputted recipes for running our application (Figs. 6 and 7).

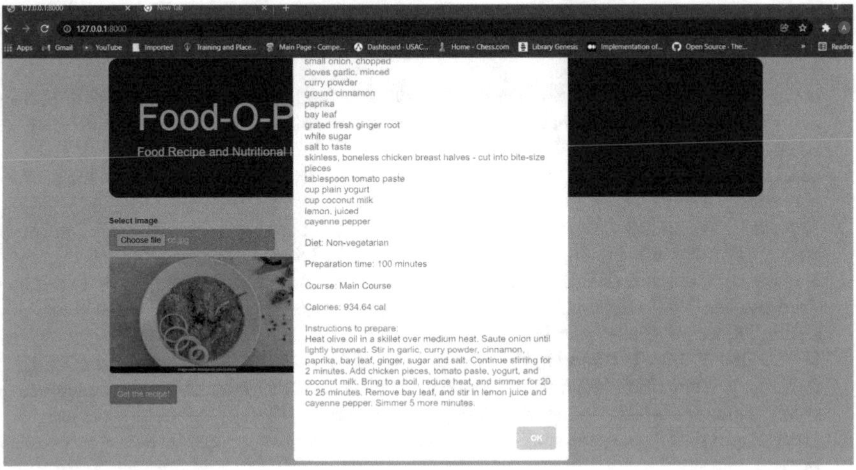

Fig. 7 Chicken curry

6 Conclusion

We tested different model architectures for food classification tasks against our custom Indian food dataset. The best-performing model was pretrained ResNet with some unfrozen layers along with some dropout layers. We also tried tuning approaches such as unfreezing the top layers of pretrained models, methods of data augmentation, image preprocessing, dropout layers, and learning rate/optimizer hyperparameters. From our work, we concluded that Mask R-CNN can be used for food calorie estimation since it can compute a mask for every food object in the image.

The limitation of our application is that it predicts calorific values based on the area of the food item. Hence, the image of the food has to be taken from the top angle so that we get the top view of the food item. Also, different orientations of the food items can result in different predicted calorific values. However, if these conditions are correctly met, our application can successfully predict a good approximate calorific value as is shown in the table. In the future, we can implement a model that takes orientation also into consideration.

The use of computer vision in dietary intake is an emerging field of computer science. Our application has demonstrated the identification of food from food images using image processing and rough calorie estimation with decent accuracy. As it is a rapidly emerging field the systems have to adapt to the pace of improvement. One of the most sought improvements is the addition of automatic calorie estimation for all kinds of Indian food. Another improvement can be the advent of a complete dietary management system based on the techniques which have been proposed above that can aid in the selection of food types and nutrient cycles.

References

1. Rajayogi JR, Manjunath G, Shobha G (2019) Indian food image classification with transfer learning. In: 2019 4th international conference on computational systems and information technology for sustainable solution (CSITSS), vol 4. IEEE, pp 1–4
2. Zhu L, Spachos P, Pensini E, Plataniotis KN (2021) Deep learning and machine vision for food processing: a survey. Curr Res Food Sci 4:233–249
3. Mandal B, Puhan NB, Verma A (2018) Deep convolutional generative adversarial network-based food recognition using partially labeled data. IEEE Sens Lett 3(2):1–4
4. Kagaya H, Aizawa K (2015) Highly accurate food/non-food image classification based on a deep convolutional neural network. In: International conference on image analysis and processing. Springer, Cham, pp 350–357
5. Liu C, Cao Y, Luo Y, Chen G, Vokkarane V, Ma Y (2016) Deepfood: Deep learning-based food image recognition for computer-aided dietary assessment. In: International conference on smart homes and health telematics. Springer, Cham, pp 37–48
6. Liang Y, Li J (2017) Deep learning-based food calorie estimation method in dietary assessment. arXiv:1706.04062
7. Min W, Jiang S, Wang S, Sang J, Mei S (2017) A delicious recipe analysis for exploring multi-modal recipes with various attributes. In: Proceedings of the 25th ACM international conference on Multimedia, pp 402–410.
8. Chen J-J, Ngo C-W, Chua T-S (2017) Cross-modal recipe retrieval with rich food attributes. In: Proceedings of the 25th ACM international conference on multimedia, pp 1771–1779
9. He H, Kong F, Tan J (2015) DietCam: multiview food recognition using a multikernel SVM. IEEE J Biomed Health Inform 20(3):848–855
10. Bossard L, Guillaumin M, Van Gool L (2014) Food-101-mining discriminative components with random forests. In: European conference on computer vision. Springer, Cham, pp 446–461
11. Salvador A, Hynes N, Aytar Y, Marin J, Ofli F, Weber I, Torralba A (2017) Learning cross-modal embeddings for cooking recipes and food images. In: Proceedings of the IEEE conference on computer vision and pattern recognition, pp 3020–3028
12. He K, Gkioxari G, Dolláir P, Girshick R (2017) Mask r-cnn. In: Proceedings of the IEEE international conference on computer vision, pp 2961–2969

Can Machine Learning Algorithms Improve Dairy Management?

Rita Roy and Ajay Kumar Badhan

1 Introduction

Economic strain requires expanding the proficiency in the dairy creation, which had shown up with the high-yield of the dairy cows, enormous heard and the solid development towards accessible lodging frameworks like this [1]. Working on animal government assistance on the homestead can intensify benefits. It can decrease costs connected with medical care and helpless yields and work on the manageability and productivity of dairying [2].

Horticulture creation information is broadly accessible, yet they are not utilized to illuminate the creation of essential assignments [3]. Until now, we can gauge their latent capacity, and in this manner, using this information is testing. In this manner, measuring their potential information the board has an opportunity to further develop their business [4]. In human medication, there is a capability of the machine learning (ML) algorithms that have perceived that utilization of the methods has further developed diagnostics in various sicknesses like coronary illness and diabetes. So forth, for example, irregular woodland, AI models can hold downright information and are unfeeling toward missing qualities [5]. Besides, they can break down enormous datasets, which are regularly hard to assess with customary factual models. This features ML procedures' possibilities for dairy cultivating [6].

Dairy farming is a type of agriculture that involves the long-term production of milk, which is then processed (either on the farm or at a dairy plant, both of which can

R. Roy (✉)
Department of Computer Science and Engineering, GITAM Institute of Technology, GITAM (Deemed to be University), Visakhapatnam, Andhra Pradesh, India
e-mail: ritaroy1311@gmail.com

A. K. Badhan
Department of Computer Science, Lovely Professional University, Phagwara, Punjab, India
e-mail: ajay.27337@lpu.co.in

P. Singh et al. (eds.), *Machine Learning and Computational Intelligence Techniques for Data Engineering*, Lecture Notes in Electrical Engineering 998,
https://doi.org/10.1007/978-981-99-0047-3_33

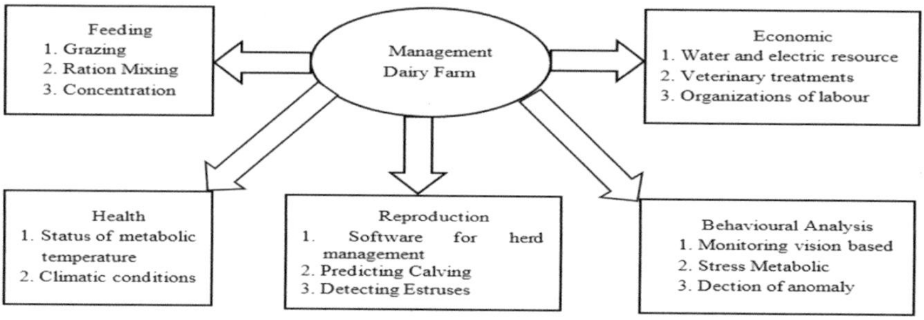

Fig. 1 Structure frame of dairy management

be referred to as a dairy) for the eventual sale of a dairy product. The dairy industry is a constantly changing industry [7] (Fig. 1).

ML Techniques among different strategies present a way to evaluate these datasets, progressively opening up on many ranches and farms. AI is a subfield of man-made reasoning [8]. These ML techniques come under three different categories: unsupervised, supervised, and semi-supervised learning. Supervised learning is used to classify the data by predefined class labels and target values. At present, with special packages of python, statistical approaches have made it easier to analyze the data concerning AI [9]. Traditional research or statistical techniques provide very biased results among various methods presented by earlier researchers.

The current paper focused on providing the best data classification, which is integrated from the intelligent wearable device of the cow. The data is taken from an online source [10]. Those sensors provide the information of all its behavior, etc., collected and classified using various ML techniques like SVM, bagging and boosting techniques, K-nearest neighbor, and hybridization techniques that can provide the best prediction for future data. This paper offers the comparison result among all the classifications as mentioned earlier methods.

2 Literature Review

Authors in [11] designed a scoring system that will choose an ML algorithm to identify the risk of ketosis in the cow. Authors [4] discuss how to increase the cow's milk yield by using robotic milking. Authors in [12] presented a review on ML techniques in the agriculture sectors discussing the challenges and opportunities. Authors in [13] proposed a lightweight channel-wise model for determining the pig postures. Authors in [3] discussed the role of ML techniques in the area of animal farming and highlighted the challenges as well as future trends. Authors in [6] developed a model of measuring the com milk production performance using computer vision and 3D images of the cow. Authors in [14] presented a review of

ML applications in the area of dairy farming. Authors in [15] used image processing techniques to diagnose the pet animals' infections and diseases.

In [16] he applications of ML for the management of dairy farming and presented the challenges and future opportunities are discussed. In [17] ML techniques like random forest modelling for the milk yield in dairy management are used. In [7] ML techniques to measure the health of the grazing cows and measured the accuracy of the algorithm used. It is necessary to develop and validate a model for identifying the behaviour and posture of the dairy cows using ML techniques. In [18] the research focuses on developing predictive models using machine learning techniques to identify factors that influence farmers' decisions, forecast farmers' demands for breeding services, and predict farmers' decisions.

In [19] the objective is to create and test an innovative procedure for analyzing multi-variable time-series generated by the Automatic Milking System (AMS), focusing on herd segmentation, to aid dairy livestock farm management. This study aims to create and test a cluster-graph model using AMS data to automatically group cows based on production and behavioral characteristics. Authors in [20] used 3D sensors for predicting calving time in dairy cattle. It focused on the ML techniques for monitoring the behaviour of dairy calves using the machine vision-based method. Authors in [21] explored surveying similar examples in the field. This study explores the potential of using machine learning in dairy cattle breeding and considers the possible selection of a dataset for training the model to find a quantitative description of the critical parameters in breeding tasks.

In [22] ML techniques for predicting metabolic status in dairy cows, determining whether individual cows' metabolic status can be clustered based on plasma values in early lactation, and testing machine learning algorithms to predict metabolic status using on-farm cow data. Authors in [23] investigated the use of big data to add value to farm decision-making and the factors and processes that influence farmer engagement with and use of big data, such as institutional arrangements. Authors in [9] presented a predictive model suitable for routine field applications that can be an effective strategy for improving dairy herd lameness status. More data about lameness is needed to improve the model's performance. Authors in [24] describe the first step in creating a deep learning-based computer vision system that can recognize individual cows in real-time, detect their positions, actions, and movements, and record time history outputs for each animal.

In [8] decision-based instruments can utilize information from the executive's administrations and examination to take advantage of information streams from ranches and other financial well-being and rural sources. An application programming connection point is a dynamic instrument that associates information investigation apparatuses to primary cow, group, and financial information. This connection point permits the client to associate with an assortment of dairy applications without ultimately uncovering the fundamental framework model's intricacies or understanding the impacts of different choices on the ideal results.

3 Methodologies

3.1 General Outlooks and Findings

Horticultural farmlands can benefit significantly from information models. In any case, the accessibility and execution of models, sensors, and Internet of Things devices were evaluated.

3.2 Prediction Models for Water and Electricity Consumption

ML calculations could work on water and power utilization forecast on field-based dairy ranches or farming. Compared to previous studies' multiple linear regression models, the continuous expectation error for water (support vector machine) was increased by 54% and for power utilization (artificial neural network) by 23%. This provided a tool for dairy farmers and policymakers to break down natural variables in field-based dairy farming [16]. The recently evolved help vector machine could anticipate prairie-based dairy ranches' power utilization with an overall mistake of 10.4% at the homestead level and 5.0% across all farms remembered for the review; the creators likewise introduced a functional methodology of diminishing energy requests by 4% when groundwater was utilized to precool the milk.

3.3 Body Condition Scoring

Screening their body condition score is one method for detecting cows' physiological condition (BCS). Instructors utilize the standard BCS scoring to determine the singular cows and the gatherings' prosperity status. The BCS (on a size of 1–5 or 1–10) mirrors the fat reserve funds of cows and can consequently exhibit the requirement for changes in chief consideration or age [8]. Machine vision has been utilized to eliminate BCS through two-layered (2-D), three-layered (three dimensional), and warm imaging. In any case, the exhibition of such calculations can be improved [25]. The pleasant nature of named information is significant to accomplish dependable ML expectations or characterizations.

3.4 Behavior Classification Based on Sensor

Their dataset included information from cows seen in standing estrus; further research is needed to determine the meaning of action bunches during estrus [26]. The development made the groups investigate the potential for estrus acknowledgement and effectively-recognized estrus in 90% of the cows, while 10% of estrus events were missing and 17% were bogus positive. The makers asserted that their discoveries were better than past investigations. It utilized a sporadic forest area classifier to recognize lying, standing, walking, and mounting conduct in bulls on the field from accelerometer information [27]. It found high connections for lying and reasonably high connections for standing, strolling, and mounting behaviour when contrasted with camera insights.

3.5 Grouping the Feeding of Cows

Cluster evaluation can work on the examination of creature explicit information. Group diagram models were effectively utilized on time-series information of cows drained in programmed draining frameworks to classify crowd qualities and characterize the cow's dependency on five unique boundaries (number of days by day draining methodology, equality, ordinary day by day action, draining routineness, and cow body weight). To investigate conduct and creation highlights, k-cluster bunching models were carried out for each of these boundaries [28]. This data can be handled to bunch the creatures into a person, taking care of gatherings naturally. AI strategies were applied to enormous datasets, effectively inferring nourishing groups, distinguishing cows in danger of clinical mastitis, just as constantly foreseeing the beginning of clinical mastitis, with generally significant degrees of responsiveness and particularity. Knowing and carrying out this data can help work on the soundness of the individual cow and the whole group's wellness [29].

3.6 Grazing

It was focused on that the field-based dairy frameworks have financial benefits, as the immediate field use emphatically diminishes creation costs. The detailed and altogether higher pay upholds this speculation in touching frameworks through a significant decrease of work costs. Brushing frameworks are further helpful because of calving [30]. Also, top energy requests are better synchronized with leading grass development. This methodology has been executed to manage time-related feed consumption, which can be valuable for assessing field admission rates.

Table 1 shows the behavior and posture of the cows. Here in this paper, we considered sternal recumbence right and sternal recumbence left, standing and still standing posture of the cow and provided their definitions. Behaviors of the cow are resting, ruminating is also considered, and feeding of the cow is examined using the intelligent wearable device which will come under IoT.

Table 2 shows the posture and behavior total observation and hours of observation. These observations can classify the posture and behavior using various ML techniques.

Table 1 Behavior and posture of cows

	Behavior or posture	Definition
Posture	Stand	Stand position of the cow with four legs
	Sternal recumbence Right	Cows are positioned on the sternum, with their hind legs to the left
	Sternal recumbence Left	Cows are positioned on the sternum, with their hind legs on the right
	Still, Stand	Cows actively consume feed from the manger, including chewing it in the manger space
Behavior	Resting	Cows are not doing any moments
	Ruminating	The moment of chewing starts with chew regurgitated bolus and is completed with bolus swallowed back. Either this can be done in a standing or lying position
Feeding	Eating	Feed actively at the manger, including chewing it in the manger space

Table 2 Time spent in each posture and behavior

Posture	Total observation	Hours of observation
Sternal recumbence Right (SRR)	67,772	3.77
Sternal recumbence Left (SRL)	115,734	6.44
Stand (S)	273,223	15.3
Behaviour		
Feeding	84,207	4.69
Moving	84,401	4.70
Resting	141,056	7.80
Ruminating	53,755	2.9
Still stand	93,326	5.28

Hybridization algorithm:

Step 1: Load the important libraries.
Step 2: Predicting classes from the data set using hyperplanes, identified as one class and another −1.
Step 3: Weight has applied, inputting the weighted base hyper plan model and identity wrong classified d data point.
Step 4: Use the loss function

$$c(x, y, f(x)) = \begin{cases} 0 & \\ 1 - y * f(x) \end{cases} if\ yxf[x] \geq 1 \tag{1}$$

Optimized weights are calculated gradients

$$\frac{\partial}{\partial_{w_k}}[[w^2]] = 2w_k \tag{2}$$

$$\frac{\partial}{\partial w_k}(1 - y_i < x_i, w >) = \{0, \quad - y_i x_{ik}\, if\ y_i < x_i, w \gg 1 \tag{3}$$

Step 5: Increasing the weight of the wrongly classified data points.
Go to steps 6
Else Steps 3
Step 6: End

4 Results and Discussion

The hybridization techniques model adopts marginal separations between different classes and can handle extensive dimensional data compared to other techniques. This technique can increase memory management, and Adaboost is less inclined to overfitting as the information boundaries are not mutually advanced [31]. The exactness of weak classifiers can be improved by utilizing Adaboost. These days, Adaboost arranges text and pictures rather than paired characterization issues. This advantage made the hybridization more accurate than other algorithms like vanilla SVM, Adabosting, KNN, and CART. We used an animal data inventory dataset from Dallas open data, and comparative analysis is provided below.

Table 3 shows the various evaluations of the classification techniques like SVM, adabost, KNN, CART and hybridization. Considering all the evaluation parameters, Cohen kappa, Precision, Recall, and accuracy, we can observe that Hybrization techniques provide the highest values.

Table 4 shows the complete study of all ML techniques with all posture of the cow-like SRL, SRR and S, and their comparison results evaluated in terms of precision, recall, accuracy and Cohen kappa. The results show that hybridization techniques are providing the highest compared with other ML algorithms.

Figure 2 represents the graphical representation of posture comparison with various evaluation parameters. This shows that all ML algorithms better classify this label data, but the Hybrization technique offers the highest accuracy.

Table 3 Prediction of posture in different models

Model	Cohen Kappa	Precision	Recall	Accuracy
SVM	0.9	0.8	0.79	0.9
Adabosting	0.85	0.85	0.8	0.93
KNN	0.82	0.79	0.81	0.91
CART	0.82	0.81	0.84	0.92
Hybridization	0.95	0.9	0.88	0.96

Table 4 Cohen Kappa, precision, recall and accuracy of posture

Behaviour	Model	Cohen Kappa	Precision	Recall	Accuracy
SRL	SVM	0.9	0.88	0.83	0.89
SRR	SVM	0.87	0.86	0.82	0.88
S	SVM	0.86	0.87	0.81	0.81
SRL	Adabosting	0.95	0.89	0.85	0.9
SRR	Adabosting	0.93	0.87	0.79	0.93
S	Adabosting	0.94	0.86	0.80	0.92
SRL	KNN	0.90	0.85	0.82	0.89
SRR	KNN	0.81	0.86	0.83	0.88
S	KNN	0.85	0.87	0.84	0.91
SRL	CART	0.88	0.87	0.83	0.9
SRR	CART	0.86	0.84	0.84	0.89
S	CART	0.87	0.86	0.82	0.87
SRL	Hybridization	0.96	0.86	0.82	0.93
SRR	Hybridization	0.97	0.89	0.83	0.94
S	Hybridization	0.96	0.87	0.84	0.92

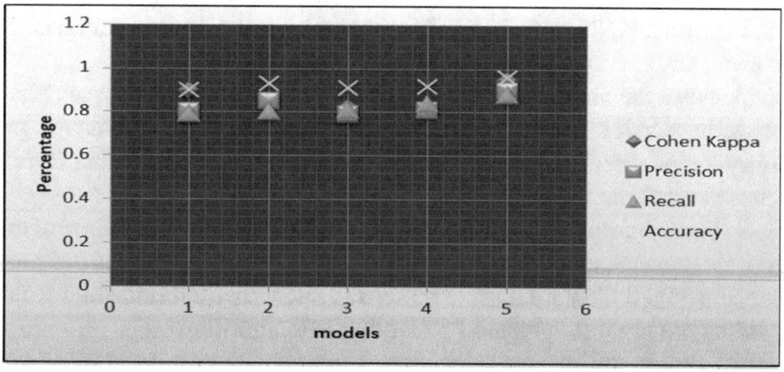

Fig. 2 Comparison of models for posture

Figure 3 represents the graphical representation of all the cow's posture on-farm, evaluated by parameters Cohen kappa, precision, recall, and accuracy. Table 5 provided data about various behaviors of the cows and their evaluation values of ML classification algorithms.

Figure 4 shows the graphical representation of the behavior of the cow. Figure 5 shows the visual representation of the cow's behaviour measured using parameters like Cohen kappa, precision, recall, and accuracy.

Previous research like [17] shows that dairy management and ML techniques such as random forest modelling increase milk yield. In [13] the previous study proposed a

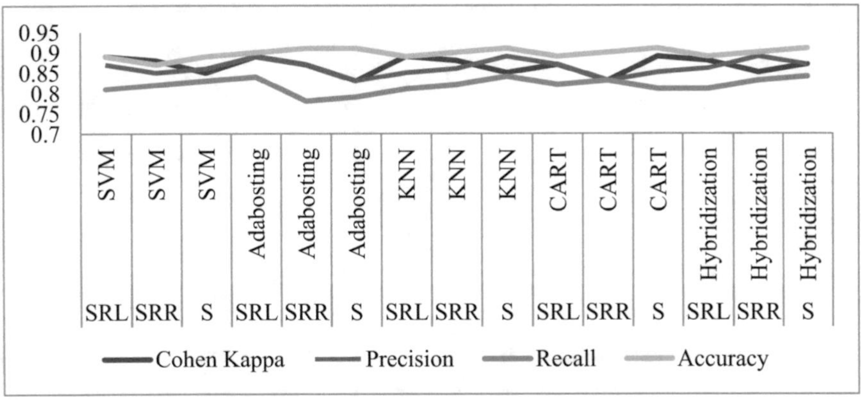

Fig. 3 Comparison of different models with each posture

Table 5 Prediction of behavior in different models

Model	Cohen Kappa	Precision	Recall	Accuracy
SVM	0.9	0.81	0.81	0.91
Adabosting	0.91	0.84	0.83	0.94
KNN	0.85	0.79	0.83	0.92
CART	0.88	0.83	0.85	0.93
Hybridization	0.98	0.95	0.90	0.97

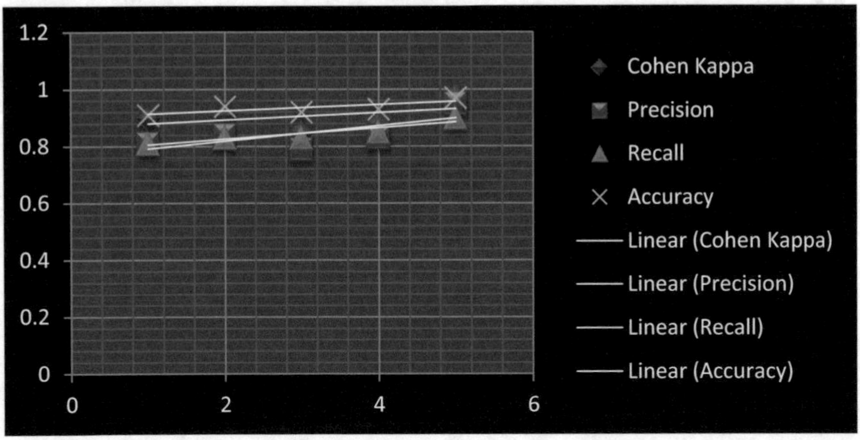

Fig. 4 Comparison of models for behavior

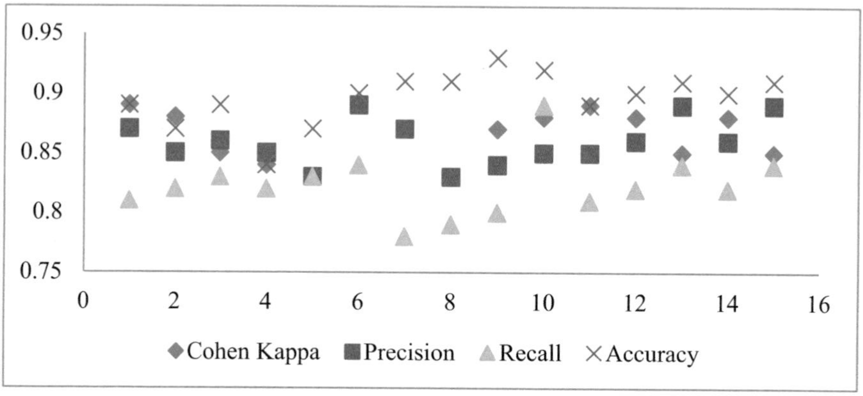

Fig. 5 Cohen Kappa, precision, recall and accuracy of behavior

lightweight channel-wise model for pig posture determination. Like [14], the earlier studies presented an overview of ML applications in dairy farming. Earlier, it was highlighted that [6] created a methodology for assessing com milk supply performance using machine learning and 3D cattle images. The focus is to generate and check an inventive method for analyses of multivariable time series produced by the Automatic Milking System (AMS), focusing on herd fragmentation, to aid milk livestock farm management [19]. This research aims to research and develop a cluster-graph model based on AMS data to instantly group cows based on manufacturing and behavioral characteristics. Authors [20] predicted calving time in milk production using 3D detectors. It concentrated on machine learning for observing dairy cattle behaviour using a machine vision-based method. Previously [21] explored surveying comparable examples in the field. This study examines the potential of using machine learning in dairy cow breeding and the possibility of selecting a dataset for training a model to find a quantitative description of the crucial components.

5 Conclusion

The investigated writing demonstrates that a wide range of parameters influences the performance and health of dairy cows. These should be closely monitored to improve the efficiency of dairy farms. As a result, various information sources should be linked. These frequent bombs are caused by the limited accessibility of public datasets, business sensors that do not provide a standardized information framework and seller lock-in. Moreover, we reason that those numerous analysts have perceived the capability of ML, and it is currently an ideal opportunity to begin carrying out these fantastic assets in multidisciplinary joint efforts between dairy and information researchers to understand their possible effect. Future research can be done using other ML techniques.

References

1. Morota G, Ventura RV, Silva FF, Koyama M, Fernando SC (2018) Big data analytics and precision animal agriculture symposium: machine learning and data mining advance big data analysis in precision animal agriculture. J Anim Sci
2. Nayeri S, Sargolzaei M, Tulpan D (2019) A review of traditional and machine learning methods applied to animal breeding. Anim Health Res Rev
3. Neethirajan S (2020) The role of sensors, big data and machine learning in modern animal farming. Sens Bio-Sens Res
4. Gálik R, Lüttmerding G, Boďo Š, Knížková I, Kunc P (2021) Impact of heat stress on selected parameters of robotic milking. Animals
5. Mukherjee S, Chittipaka V (2021) Analysing the adoption of intelligent agent technology in food supply chain management: empirical evidence. FIIB Bus Rev
6. Shorten PR (2021) Computer vision and weigh scale-based prediction of milk yield and udder traits for individual cows. Comput Electron Agric
7. Contla Hernández B, Lopez-Villalobos N, Vignes M (2021) Identifying health status in grazing dairy cows from milk mid-infrared spectroscopy by using machine learning methods. Animals
8. Ferris MC, Christensen A, Wangen SR (2020) Symposium review: dairy brain—Informing decisions on dairy farms using data analytics. J Dairy Sci
9. Warner D, Vasseur E, Lefebvre DM, Lacroix R (2020) A machine learning based decision aid for lameness in dairy herds using farm-based records. Comput Electron Agric
10. Datasets. https://www.kaggle.com/datasets?search=animal
11. Satoła A, Bauer EA (2021) Predicting subclinical ketosis in dairy cows using machine learning techniques. Animals
12. Benos L, Tagarakis AC, Dolias G, Berruto R, Kateris D, Bochtis D (2021) Machine learning in agriculture: a comprehensive updated review. Sensors
13. Luo Y, Zeng Z, Lu H, Lv E (2021) Posture detection of individual pigs based on lightweight convolution neural networks and efficient channel-wise attention. Sensors
14. Shine P, Murphy MD (2022) Over 20 years of machine learning applications on dairy farms: a comprehensive mapping study. Sensors
15. Chugh A, Makkar P, Aggarwal S, Sharma S, Singh YK (2020) Approach of image processing in diagnosis and medication of fungal infections in pet animals. Int J Innov Res Comput Sci Technol (IJIRCST)
16. Cockburn M (2020) Application and prospective discussion of machine learning for the management of dairy farms. Animals
17. Bovo M, Agrusti M, Benni S, Torreggiani D, Tassinari P (2021) Random forest modelling of milk yield of dairy cows under heat stress conditions. Animals
18. Balasso P, Marchesini G, Ughelini N, Serva L, Andrighetto I (2021) Machine learning to detect posture and behavior in dairy cows: information from an accelerometer on the animal's left flank. Animals
19. Mwanga G, Lockwood S, Mujibi DF, Yonah Z, Chagunda MG (2020) Machine learning models for predicting the use of different animal breeding services in smallholder dairy farms in Sub-Saharan Africa. Trop Anim Health Prod
20. Soysal Y, Ayhan Z, Eştürk O, Arıkan MF (2009) Intermittent microwave–convective drying of red pepper: drying kinetics, physical (colour and texture) and sensory quality. Biosyst Eng
21. Fadul M, Bogdahn C, Alsaaod M, Hüsler J, Starke A, Steiner A, Hirsbrunner G (2017) Prediction of calving time in dairy cattle. Anim Reprod Sci
22. Sedighi T, Varga L (2021) Evaluating the Bovine tuberculosis eradication mechanism and its risk factors in England's cattle farms. Int J Environ Res Public Health
23. Tarjan L, Šenk I, Pracner D, Rajković D, Štrbac L (2021) Possibilities for applying machine learning in dairy cattle breeding. In: 2021 20th international symposium INFOTEH-JAHORINA (INFOTEH)

24. Tassinari P, Bovo M, Benni S, Franzoni S, Poggi M, Mammi LM, Mattoccia S, Di Stefano L, Bonora F, Barbaresi A, Santolini E (2021) A computer vision approach based on deep learning for the detection of dairy cows in free stall barn. Comput Electron Agric
25. Dev RD, Badhan Ak, Roy R (2020) A study of artificial emotional intelligence for human—Robot interaction. J Crit Rev
26. Garcia R, Aguilar J, Toro M, Pinto A, Rodriguez P (2020) A systematic literature review on the use of machine learning in precision livestock farming. Comput Electron Agric
27. Roy R, Dev DR, Prasad VS (2020) Socially intelligent robots: evolution of human-computer interaction. J Crit Rev
28. Mukherjee S, Baral MM, Venkataiah C, Pal SK, Nagariya R (2021) Service robots are an option for contactless services due to the COVID-19 pandemic in the hotels. Decision
29. Cioffi R, Travaglioni M, Piscitelli G, Petrillo A, De Felice F (2020) Artificial intelligence and machine learning applications in smart production: progress, trends, and directions. Sustainability
30. Nosratabadi S, Ardabili S, Lakner Z, Mako C, Mosavi A (2021) Prediction of food production using machine learning algorithms of multilayer perceptron and ANFIS. Agriculture
31. Roy R, Giduturi A (2019) Survey on pre-processing web log files in web usage mining. Int J Adv Sci Technol

Flood Severity Assessment Using DistilBERT and NER

S. N. Gokul Raj, P. Chitra, A. K. Silesh, and R. Lingeshwaran

1 Introduction

Natural disasters are inevitable and cause a severe impact on people's lives, property and belongings. Flood is the most occurring natural disaster worldwide and causes a huge collapse in the economy of the country. Unlike other disasters, it is observed that social media is more likely to be used on a larger scale as a tool for seeking help during floods [1]. During floods, posts related to affected individuals, injured people, missing people, infrastructure and utility damage, volunteering or donation and other relevant posts are observed to be shared. With the advancement of technology, the severity of floods can be monitored and if utilized effectively, its impact can also be reduced. One such way of effective utilization is identifying flood related posts from social media platforms and analyzing them. Twitter is one of the most commonly used social media platforms in which millions of people share millions of tweets publicly. Ever since the emergence of Twitter, it has not only gained popularity among the people but also their trust. It has 206 million active users as per the second quarter of 2021 [2]. Twitter has proved to be one of the efficient platforms in providing useful information about the disaster during the occurrence of the disaster [3]. This research uses Twitter as a source of data because it has a lot of information available. During the recent floods in Chennai, people posted a lot of tweets related to floods and those tweets form the basis for this research. It shall be noted that this research focuses on post-disaster analysis but it can also be applied in the fields of event detection or early warning, depending on the characteristics of the disaster. The approach in this paper outperforms the current procedure in disaster management such as remote sensing-based methods as they face a major disadvantage of temporal lags of approximately 48–72 h in providing the information related to disaster [4].

S. N. Gokul Raj · P. Chitra · A. K. Silesh (✉) · R. Lingeshwaran
Department of CSBS, Thiagarajar College of Engineering, Madurai, Tamil Nadu, India
e-mail: sileshak12092003@gmail.com

P. Singh et al. (eds.), *Machine Learning and Computational Intelligence Techniques for Data Engineering*, Lecture Notes in Electrical Engineering 998,
https://doi.org/10.1007/978-981-99-0047-3_34

Deep learning and Natural Language Processing are becoming revolutionary in computing huge data and arriving at meaningful solutions. One such NLP model, DistilBERT is outracing other traditional machine learning models in terms of performance [5]. DistilBERT has a wide range of applications in language modelling which makes it a better model for text classification. The spatial analysis provides a better understanding of the disaster zone. Named Entity Recognition is a Natural Language Processing technique that can scan the entire text automatically and figure out some basic entities in a text and categorize them into predefined classes. In this paper, NER [6] is proposed to identify the location mentioned in the text (tweets) for performing spatial analysis. Temporal analysis is also performed by computing a time graph. Using this spatiotemporal analysis, the place and time with the highest frequency of damage due to floods can be identified. The contribution of the proposed work is to figure out the locations and time of high frequency damage due to flood and assist the volunteers to take immediate actions for the speedy recovery of the people.

The rest of the paper is organized as follows: In Sect. 2, previous works related to disaster management are discussed in detail. The proposed methodology is explained in Sect. 3. The results obtained are discussed in Sect. 4. In Sect. 5, the final conclusion and future enhancements are discussed.

2 Related Works

Social media data from various platforms can be used for several purposes like natural disaster planning and risk mitigation according to the work proposed in 2019 [7]. The challenging part in the classification of the tweets is that most of the previous works were keyword-based approaches. In 2013, authors [8] developed a burst detector for earthquakes in Australia and New Zealand. The authors used the keywords "earthquake" and "#eqnz" to identify messages related to earthquakes. The ideology behind this approach is that the frequency of these keywords increases if there is an earthquake and thus it can be detected. They used a keyword-based approach which has clear limits in expanding to other events or languages or collecting all available information. In 2015, [9] used a similar keyword filtering in the process of extraction of disaster related Tweets. Disastro, a real time Twitter-based disaster response system was proposed in 2019 which aims to classify tweets into 'donation and 'rescue' using machine learning algorithms [10]. In 2020, the authors [11] identified disaster related tweets using Natural Language Processing techniques and CART model using decision tree classification algorithm.

Most of the existing related studies were based on geotags from the posts for identifying the location of the post. A method was developed by [12] using geotagged tweets for detecting disaster related events. The research [13] classified highly relevant tokens used in geotagged tweets during disasters. The main drawback of geotag based approach is that it is generally lesser than 1% of the total tweets [14]. This is because most of the users do not prefer to turn on their geolocation. In 2018, the authors [4] proposed an NLP model, cascading Latent Dirichlet Allocation for classification of tweets which has the main disadvantage of identification of subtopics which is limited by the recognition rate of a topic in the first iteration. In consequence, Tweets, which are not present in a topic in the first step, cannot be used for further cascading.

In 2022, the authors [15] proposed a deep learning-based assessment of flood severity using social media data. The authors classified tweets and images taken during the flood and assessed the severity of the disaster. They used geo-referenced tweets for their assessment, which is the main drawback of the proposed approach. Also, the temporal factors are not taken into consideration for the assessment. Due to the lack of temporal factors, there might be a chance of considering the outdated tweets for assessment.

Most of the approaches discussed in this section show two major shortcomings. Most of them were keyword-based approaches and almost all the approaches involved geotag based location detection. The spatial and temporal features were not taken into consideration in the previous works for the analysis of results. The approach in this paper tries to overcome all these shortcomings and proposes a methodology that combines the classification of tweets, spatial and temporal analysis for measuring the severity of the disaster. Also, the proposed approach involves the identification of location even without geotags added in the tweet.

3 Proposed Methodology

The proposed methodology in this paper consists of four main stages. In the first stage, the tweets are collected from Twitter using Twitter API [16]. In the second stage, the collected tweets are preprocessed and tokenized according to the requirements of the classification model. In the third stage, the preprocessed tweets are classified into "flood related' and "non flood related" tweets using DistilBERT model and the flood related tweets are further preprocessed to satisfy the requirements of Named Entity Recognition. In the final stage, NER is implemented to find the location mentioned in the tweet for spatial analysis and a time graph is plotted for temporal analysis.

Figure 1b depicts the overall process of the proposed methodology. The tweets collected from Twitter API are preprocessed, tokenized and sent to the training step of the DistilBERT model. Then the model is fine-tuned with the preprocessed tweets so that it can recognize flood related tweets effectively. Flood related tweets are segregated and further preprocessed for implementing NER [6] to identify the location in the tweets and spatiotemporal analysis is performed.

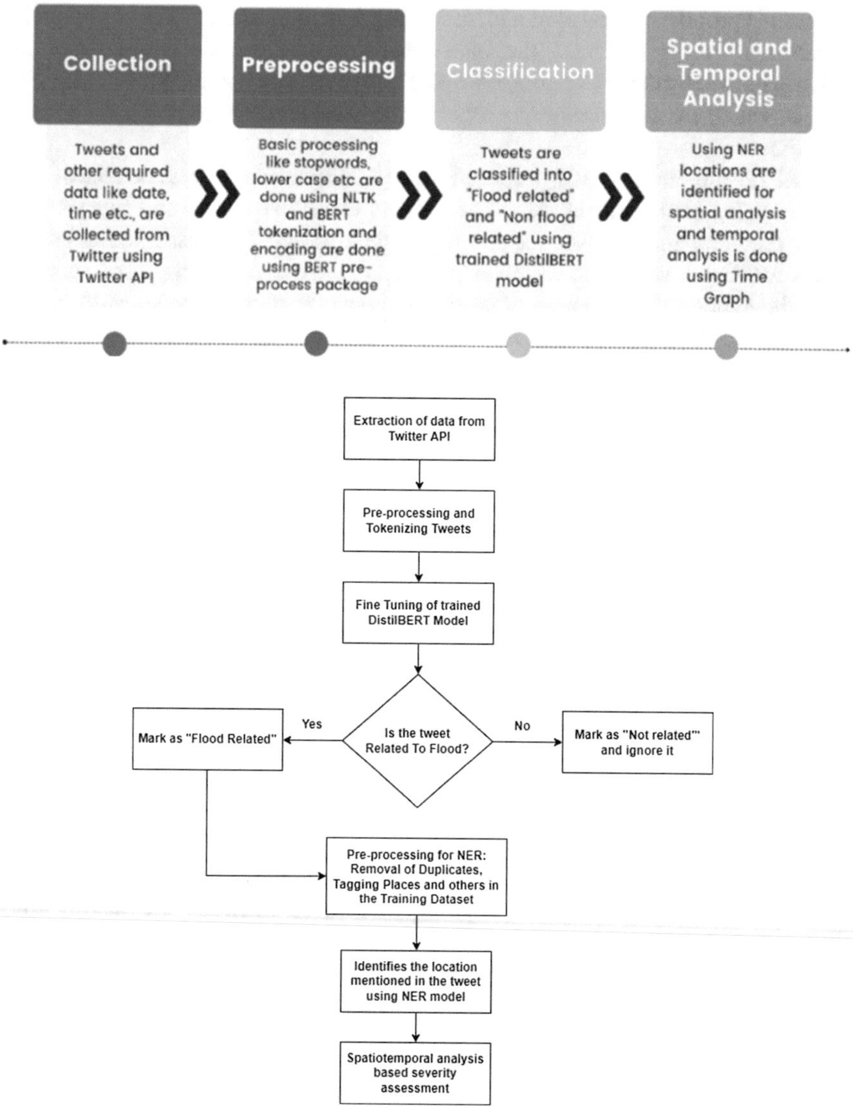

Fig. 1 **a** Stages involved in the proposed methodology. **b** Flow diagram of the proposed methodology

3.1 Extraction of Tweets

The dataset used for flood severity assessment includes tweets comprising the hashtags #HeavyRains, #ChennaiFloods collected using Twitter API. The language of the extracted tweets is restricted to English. The dataset comprises various features

Table 1 Overview of hashtags used

Hashtags	Tweet count	Unique tweets
#HeavyRains	759	313
#ChennaiFloods	801	435

Table 2 Preprocessing tasks

Subtask	Description
Removal of URLs, emojis	Hyperlinks starting with http, https, etc., along with emojis, emoticons are removed
Removal of whitespaces and punctuations	Punctuations like (,), (;), (:), etc., and whitespaces including newline characters are removed
Removal of symbols and mentions	Symbols such as @, #, $, %, etc., are removed and @mentions are replaced with AT_USER
Removal of stop words	Stop words like 'you', 'is', 'the', etc., are removed
DistilBERT preprocess and encoding	Preprocessed texts are tokenized, padded and then encoded
Splitting of dataset	The entire dataset is divided into two parts: 80% for train and 20% for test

like tweeted time, text content of the tweet, etc., First, 1560 tweets were extracted and among them, 748 unique tweets were selected for further analysis (Table 1).

3.2 *Preprocessing for DistilBERT*

Removal of noise from the tweets is important before fine-tuning the DistilBERT model as noise reduces the performance of the model. Noise is unwanted content in the text like hyperlinks, emojis, etc., Python RegEx, NLTK library, DistilBERT preprocessor and encoder are used to preprocess the tweets i.e., remove noise from data and encode them for fine-tuning the model. Table 2 depicts the entire preprocessing task.

3.3 *Classification of Texts*

Though neural network algorithms like CNN, RNN, LSTM and logistic regression can also be used for the classification of texts, BERT outperforms all these algorithms in terms of accuracy and speed. Bidirectional Encoder Representations from Transformers (BERT) is initiated by Google in the year 2018 along with the introduction of transformers [17]. BERT is basically a Transformer Encoder Stack that generates contextualized embeddings. It looks at the context of the statement and generates the meaningful number representation for a given word. BERT generates embeddings for the entire sentence. It also has larger feedforward neural networks of 768 hidden units and 12 attention heads. BERT was trained with 2500 million words from Wikipedia and 800 million words from different books [17]. And it was trained using Masked Language Model (MLM) and Next Sentence Prediction (NSP).

DistilBERT [5] is a distilled form of BERT which is a smaller, faster and lighter Transformer model based on the BERT architecture. DistilBERT has 40% fewer parameters than BERT and 60% faster than it [5]. Both BERT and DistilBERT do not require labelled data for fine-tuning. Fine-tuned DistilBERT model is used for the classification of text in the proposed work. The DistilBERT model is trained to classify whether the text is related to flood or not.

Model training. The DistilBERT model is fine-tuned with tokenized text from the preprocessing step. A functional model is created using three DistilBERT layers and neural network layers. A dropout layer is included and eventually, the output layer will classify whether the given text is flood related or not. The model is trained with the 'Adam' optimizer along with the 'Binary Cross—Entropy' loss function. The model is trained with 100 epochs and metrics like Binary Accuracy, Precision and Recall are computed for measuring the efficiency of the model. The results are discussed in Sect. 4.

3.4 Preprocessing and Implementation of NER

Tweets related to flood extracted from the previous step are further preprocessed for implementing NER. The preprocessing involves adding corresponding location tags mentioned in the tweet. Out of 748 tweets, 166 tweets are classified as flood related by DistilBERT. After adding location tags to those 166 tweets, the final dataset contains 305 tweets for further training. The preprocessed text is encoded with LabelEncoder from the sklearn library. The dataset is split into 80% for training and 20% for testing. Named Entity Recognition (NER) is a Natural Language Processing algorithm in which the named entities are predefined categories chosen according to the use case such as names of people, organizations, places, etc. NER assigns a class to each token (usually a single word) in a sequence [6]. Therefore, NER is also referred to as token classification. NER is used for location identification in this paper.

NER is trained with the processed dataset using NERModel and NERArgs packages imported from simple transformers library by HuggingFace. The model is trained with 10 epochs along with the learning rate of $1e^{-4}$. Performance metrics like Evaluation loss, F1 score, Precision, Accuracy and recall are computed. These metrics are explained in Sect. 4.

3.5 Spatiotemporal Modelling

The location mentioned in the tweets is obtained from NER and the most frequently occurring locations are taken for spatial analysis. The time when the tweets are posted is obtained from the Twitter API and analyzed for computing the peak time of tweets to know the exact time of the occurrence of the flood. The final results obtained are analyzed in Sect. 4.

4 Results and Discussions

4.1 Performance Metrics

The various performance metrics [18] used to evaluate the models are.

a. **Accuracy:** Ratio of correctly predicted data points to the total number of data points.

$$\text{Accuracy} = \frac{(tp + tn)}{(tp + tn + fp + fn)} \tag{1}$$

b. **Precision:** Ratio of the number of true positives (tp) to the number of true positives (tp) plus the number of false positives (fp).

$$\text{Precision} = \frac{(tp)}{(tp + fp)} \tag{2}$$

c. **Recall:** Ratio of the number of true positives (tp) to the number of true positives (tp) plus the number of false negatives (fn).

$$\text{Recall} = \frac{(tp)}{(tp + fn)} \tag{3}$$

d. **F1 score:** Harmonic mean of precision and recall.

$$\text{F1score} = \frac{(2 * Precision * Recall)}{(Precision + Recall)} \tag{4}$$

4.2 Text Classification

The performance of the DistilBERT model is compared with other models such as SVM, ANN, CNN, Bi-LSTM and BERT. Table 3 shows the performance metrics

Table 3 Performance comparison of various models

Model	Accuracy		Precision		Recall	
	Train	Test	Train	Test	Train	Test
SVM	0.83	0.82	0.81	0.80	0.80	0.82
ANN	0.82	0.81	0.84	0.83	0.81	0.84
CNN	0.87	0.86	0.88	0.88	0.87	0.85
Bi-LSTM	0.89	0.89	0.88	0.88	0.88	0.87
BERT	0.92	0.92	0.93	0.92	0.94	0.92
DistilBERT	0.99	0.99	0.99	0.98	0.98	0.98

of other text classification models implemented for comparison purposes. From Table 3, it can be inferred that DistilBERT has the highest accuracy of 99% for text classification. BERT and DistilBERT are exclusively Natural Language Processing models based on transformers which makes it perform better than traditional machine learning models like SVM and deep learning models like ANN, CNN and Bi-LSTM. For training machine learning and deep learning models, Word2Vec embeddings of text data are used. DistilBERT retains 50% of BERT's layers. Furthermore, pooler and token-type embeddings, present in the architecture of BERT, are also removed in DistilBERT [19]. These are the reasons for achieving increased accuracy for DistilBERT compared to BERT, which makes DistilBERT a more suitable model for the proposed technology.

Figure 2a shows the accuracy graph and Fig. 2b shows the loss graph of the DistilBERT text classification model. It is inferred that accuracy gradually increases at each epoch and reaches a maximum of 99% accuracy. The loss gradually decreases at each epoch and reaches a minimum of less than 0.04. Figure 3 shows the confusion matrix of the DistilBERT model. Out of 160 tested tweets, 7 are misclassified which is relatively good compared to other models.

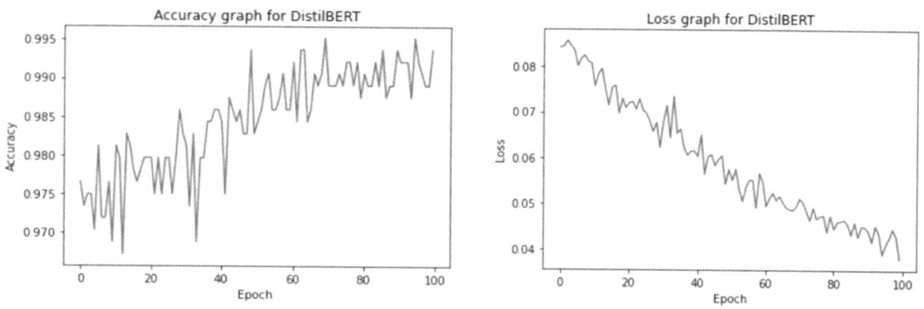

Fig. 2 **a** Accuracy graph for DistilBERT. **b** Loss graph for DistilBERT

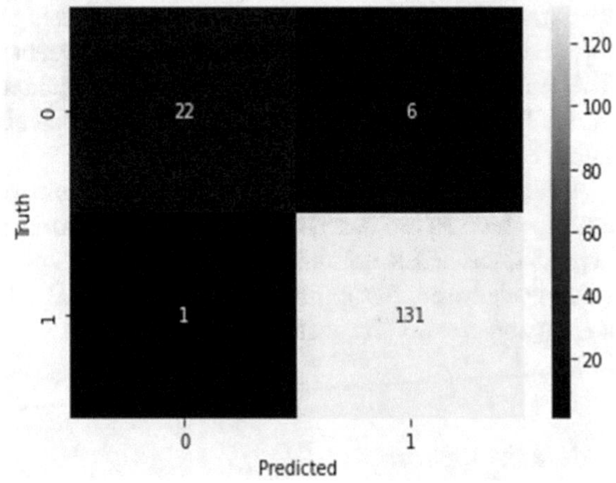

Fig. 3 Confusion matrix of DistilBERT model

4.3 Spatiotemporal Analysis

Most frequently occurring locations are obtained using NER. The main advantage of using NER is that it does not require geotags to identify the location. Figure 4a represents the frequency of the locations mentioned in the tweets. From Fig. 4a, it is evident that the impact of the flood is more severe in places like Nungambakkam and MRC Nagar than in places like Anna Nagar and VR Mall. The places which are mentioned in less than one or no tweets are placed in the category of others. Test scores of NER for various performance metrics are shown in Table 4.

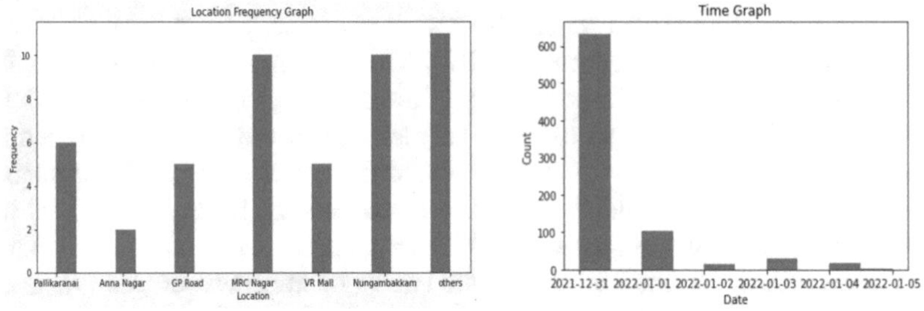

Fig. 4 **a** Location frequency graph. **b** Time graph

Table 4 Performance metrics of NER

Performance metric	Test score
Evaluation loss	0.70
Precision	0.87
Recall	0.84
F1 score	0.86

Figure 4b shows the temporal graph of tweets volume. From this graph, it is inferred that most of the rain and flood related tweets were posted on 31st Dec 2021. This shows that there was heavy rain on that particular day. It can also be noted that on 1st Jan 2022, there is a considerable impact created by the previous day due to flood.

Both DistilBERT and NER improve the results of the proposed technology as both the models are based on deep neural networks and are capable of computing huge datasets. NER eliminates the requirements of geotag which makes it a more suitable algorithm for finding the location. Thus, DistilBERT and NER appear to be a great combination for disaster management.

4.4 Discussions

This paper highlights the use of social media data for analyzing the time and location when the impact of the flood is severe and taking necessary recovery steps according to the impact. It also has certain advantages over other existing methods which include classification of texts using a non-keyword-based approach and identifying the location without geotags. Instead of collecting data using sensors, the collection of tweets avoids temporal lags and allows us to respond immediately to the situation. Most of the existing methods use geotag based approach for spatial analysis which has a major drawback of specifying that the geolocation of tweets is not available unless the user wishes to share the location publicly. Therefore, this paper proposes DistilBERT for classification and NER for location identification. This paper also utilizes spatial and temporal information for post disaster management.

NER primarily uses the location specified in the tweet. Consider a scenario where a user has geolocation turned on but did not mention the location explicitly in the tweet. Such tweets will be ignored by NER. This is a rare case and also has very less impact on the accuracy. If the severity of the disaster is very high in a particular location where the active Twitter users are very less due to network problems, it might be a challenging situation for applying the proposed methodology. The removal of fake tweets can also be difficult if the size of the dataset is very large. It is hard to guarantee that a tweet or retweet describes the real damage status that a person experienced during the flood because some tweets mention two or more locations, and the impact might not reflect the exact damage level regarding each location.

5 Conclusion and Future Works

Social media is a huge public platform that provides us with a large amount of data. Real time data can be extracted from social media like Twitter and can be used for the analysis of several events. The approach proposed in this paper is one such way of identifying the location and time of a well-known natural disaster, flood, where its impact is high. The tweets taken from Twitter are classified using a finely tuned DistilBERT. The DistilBERT model is fine-tuned with an epoch value of 100 to classify the tweets as related to flood or not. The performance of the DistilBERT is compared with other baseline models like SVM, ANN, CNN, Bi-LSTM and BERT. After analyzing the accuracies of various models, it is inferred that DistilBERT performs well. Also, while performing DistilBERT the accuracy rate is maximized up to 99%. After classifying, the tweets related to flood are utilized for the identification of the location and time of the severity of disaster with the help of the NER algorithm and temporal analysis. In the future, the location identification method can be improved with better semantic analysis. Further, different kinds of data other than text such as images, videos, etc., can be used for flood severity assessment.

References

1. Hashim KF, Ishak SH, Ahmad M (2015) A study on social media application as a tool to share information during flood disaster. ARPN J Eng Appl Sci 10(3):959–967
2. Statista webpage, https://www.statista.com/statistics/242606/number-of-active-twitter-users-in-selected-countries/, last accessed 07 Jan 2022
3. Pourebrahim N, Sultana S, Edwards J, Gochanour A, Mohanty S (2019) Understanding communication dynamics on Twitter during natural disasters: a case study of Hurricane Sandy. Int J Disaster Risk Reduct 37:101176
4. Resch B, Usländer F, Havas C (2018) Combining machine-learning topic models and spatiotemporal analysis of social media data for disaster footprint and damage assessment. Cartogr Geogr Inf Sci 45(4):362–376
5. Sanh V, Debut L, Chaumond J, Wolf T (2019) DistilBERT, a distilled version of BERT: smaller, faster, cheaper and lighter. arXiv preprint arXiv:1910.01108
6. Mohit B (2014) Named entity recognition. In: Natural language processing of semitic languages. Springer, Berlin, Heidelberg, pp. 221–245
7. Niles MT, Emery BF, Reagan AJ, Dodds PS, Danforth CM (2019) Social media usage patterns during natural hazards. PLoS ONE 14(2):e0210484
8. Robinson B, Power R, Cameron M (2013) A sensitive twitter earthquake detector. In: Proceedings of the 22nd international conference on world wide web, pp. 999–1002
9. De Albuquerque JP, Herfort B, Brenning A, Zipf A (2015) A geographic approach for combining social media and authoritative data towards identifying useful information for disaster management. Int J Geogr Inf Sci 29(4):667–689
10. Abirami S, Chitra P (2019) Real time twitter based disaster response system for indian scenarios. In: 2019 26th International Conference on High Performance Computing, Data and Analytics Workshop (HiPCW). IEEE, pp. 82–86
11. Goswami S, Raychaudhuri D (2020) Identification of disaster-related tweets using natural language processing: International Conference on Recent Trends in Artificial Intelligence, IOT, Smart Cities & Applications (ICAISC-2020). IOT, Smart Cities & Applications (ICAISC-2020), 26 May 2020
12. Zhang C, Zhou G, Yuan Q, Zhuang H, Zheng Y, Kaplan L, ... Han J (2016) Geoburst: Real-time local event detection in geo-tagged tweet streams. In: Proceedings of the 39th International ACM SIGIR conference on Research and Development in Information Retrieval, pp. 513–522
13. Hodas NO, Ver Steeg G, Harrison J, Chikkagoudar S, Bell E, Corley, CD (2015) Disentangling the lexicons of disaster response in twitter. In: Proceedings of the 24th International Conference on World Wide Web, pp. 1201–1204

14. Malik M, Lamba H, Nakos C, Pfeffer J (2015) Population bias in geotagged tweets. In: proceedings of the international AAAI conference on web and social media, vol. 9, No. 4, pp. 18–27
15. Kanth AK, Chitra P, Sowmya GG (2022) Deep learning-based assessment of flood severity using social media streams. Stochastic Environmental Research and Risk Assessment, 1–21
16. Twitter Developer Page, https://developer.twitter.com/en/docs, last accessed 01 May 2022
17. Devlin J, Chang MW, Lee K, Toutanova K (2018). Bert: Pre-training of deep bidirectional transformers for language understanding. ArXiv preprint arXiv:1810.04805
18. Hossin M, Sulaiman MN (2015) A review on evaluation metrics for data classification evaluations. Int J Data Min & Knowl Manag Process 5(2):1
19. Basu P, Roy TS, Naidu R, Muftuoglu Z, Singh S, Mireshghallah F (2021) Benchmarking differential privacy and federated learning for bert models. ArXiv preprint arXiv:2106.13973

Heart Disease Detection and Classification using Machine Learning Models

Saroj Kumar Chandra, Ram Narayan Shukla, and Ashok Bhansali

1 Introduction

The heart is one of the most important organs in the human body [1]. It circulates blood to other parts of the body. The blood contains food, oxygen, water, minerals, and many more substances. If its flow is disrupted, then it may lead to serious health issues including death. The heart is part of the cardiovascular system [2]. The cardiovascular system transports nutrients and oxygen-rich blood to the whole body. It also carries deoxygenated blood back to the lungs. Its major parts include the heart, blood vessels, and blood. The abnormal cardiovascular system leads to serious health complications such as coronary artery disease, heart attack, high blood pressure, and stroke. Cardiovascular diseases occur due to physical inactivity, using tobacco products, a poor diet, and obesity [3]. It has been reported approximately 17.9 million people lose their lives every year due to heart attack [4]. Heart Attack occurs due to atherosclerosis which restricts blood flow to a wide area of the heart [5]. Restriction of blood leads to damage to the heart muscle due to which the functioning of the heart is completely stopped and may cause death. The primary factors which mostly contribute to the arrival of atherosclerosis are not known till now. However, scientists have given attention to free radical damage to cholesterol which circulates low-density lipoproteins [6]. Also, they have acknowledged seven dietary factors including two promoters and five protective features for developing coronary heart disease. Myocardial infarction, the most serious complication of coronary heart disease, represents an amalgamation of having two distinct effects of dietary factors [5]. In the report, it has been observed that lowering serum homocysteine-containing folic acid reduces the risk of cardiovascular diseases. Medicines are used as a preventive measure for cardiovascular disease. However, their usage is limited to single risk factors which do not cover a large population.

S. K. Chandra (✉) · R. N. Shukla · A. Bhansali
Computer Science and Engineering, OP Jindal University, Raigarh, India
e-mail: saroj.chandra@opju.ac.in

P. Singh et al. (eds.), *Machine Learning and Computational Intelligence Techniques for Data Engineering*, Lecture Notes in Electrical Engineering 998,
https://doi.org/10.1007/978-981-99-0047-3_35

Heart attack symptoms include tightfistedness or affliction majorly in the chest, neck, back, and arms, tiredness, dizziness, abnormal heartbeat, and consternation. Risk factors cover both changeable and unchangeable. The changeable factors include age, sex, and family background. The changeable factors include smoking, high cholesterol, high blood pressure, fatness, deficiency of proper diet as well as exercise, and a huge amount of stress. Arterial reclamation to medication, ECG, and bypass surgery is the most used treatment method in case of heart attacks [7, 8].

In the literature, many machine learning models have been used to detect and classify heart diseases such as support vector machine, logistic regression, decision tree, Naive Bayes classifier, K-means, and K-nearest neighbor and random forest [9–14]. Archana et al. and Devansh et al. have used machine learning models for the prediction of a heart attack. In the manuscript, the authors have used the KNN algorithm on the dataset with different features in the dataset. Also, they have performed the quantitative analysis with support vector machine, decision tree, and linear regression with confusion matrix [9, 10]. Hidayet et al. have used an SVM classifier with a feature selection process [11]. Asha et al. used Naïve Bayes classifier in the prediction model [12]. R. Chitra et al. have used an unsupervised fuzzy C-means classifier in the prediction model with 13 features [13]. Farman et al. proposed a deep learning and feature fusion model in heart attack disease prediction. In the model, they have used information gain and conditional probability to select key features from the dataset [14]. In the present work, a rigorous comparison study has been performed to show the effectiveness of the models. Precision, Recall, and F1-score have been used for comparative study. Also, the AROC curve has been used to visually analyze the performance of the models.

The present work is organized as follows. The flow chart of the proposed model is presented in Sect. 2. Section 3 discusses the results obtained by the proposed model.

2 Proposed Methodology and Algorithm Design

Flowchart of the present work is shown in Fig. 1. The work has been divided mainly into dataset loading, data preprocessing, dividing dataset into training and testing, and then applying data to machine learning models and evaluations. A brief description of each of the steps is presented in the following section.

Data Set: Dataset taken into consideration contains 299 (105 women and 194 men patients in the age range between 40 and 95 years old) records with heart disease. It has been collected by Faisalabad Institute of Cardiology and at the Allied Hospital in Faisalabad (Punjab, Pakistan) in 2015 [15]. The dataset includes 13 key features which lead to a heart attack on the account of imbalance. The features include age, anemia, high blood pressure, creatinine phosphokinase, diabetes, ejection fraction, sex, platelets serum creatinine, serus sodium, time, and death event. These features with their normal values have been taken into account to train the model. The key features with their normal range are tabulated in Table 1. The dataset includes clinical, body, and lifestyle information. Some features are binary: anemia, high blood pres-

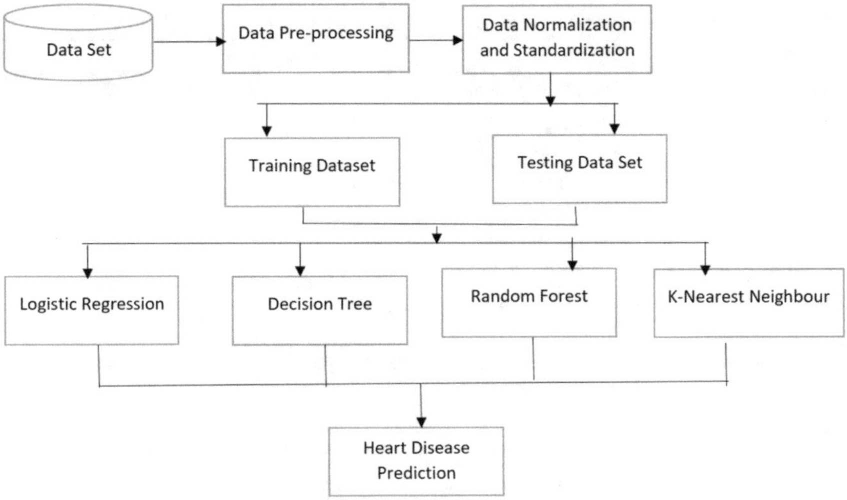

Fig. 1 Proposed flowchart for heart disease detection and classification

sure, diabetes, sex, and smoking (Table 1). The patients with 36% lower hematocrit levels have been considered amaemianic. The dataset does not contain the definition of high blood pressure. In the features, creatinine phosphokinase (CPK) is an enzyme level in blood. It flows into the blood when the tissue gets damaged. A high level of CPK indicates heart failure or injury. The ejection fraction is the percentage of blood the left ventricle pumps out with each contraction. Serum creatinine indicates waste product produced by creatine upon muscle breakdown. It plays an important role in checking kidney functionalities. Its high level in the body of the patient indicates renal dysfunction or kidney failure [16]. Sodium is a key measure to check muscles and nerve functioning. It is a routine blood examination to check sodium levels in the blood. Its abnormal range leads to heart failure [17]. The death event feature indicates whether the patient died or survived before the end of the follow-up period, which was 130 days on average.

Pre-Processing and Normalization and Standardization: In this step, missing value problem, data standardization, normalization, and outliers analysis are being done. In the present work, the dataset used does not have missing values and hence need not be handled. The models have been evaluated with and without normalization and data standardization. Outliers are present in the dataset used, it has been handled with the quintile method in python.

Dividing the Dataset and Model Selection: The 80% data in the dataset has been used to train the model. The remaining 20% data has been used to test the model. In the random state, 42 has been chosen to select the same data values and same result in the subsequent simulation of the models. After dividing the dataset into training and testing, many machine learning models such as logistics regression, K-nearest

Table 1 Meanings, measurement units, and intervals of each feature of the dataset

Feature	Explanation	Measurement	Range
Age	Age of the patient	Years	[40, …, 95]
Anaemia	Decrease of red blood cells or hemoglobin	Boolean	0, 1
High blood pressure	If a patient has hypertension	Boolean	0, 1
Creatinine phosphokinase (CPK)	Level of the CPK enzyme in the blood	mcg/L	[23, …, 7861]
Diabetes	If the patient has diabetes	Boolean	0, 1
Ejection fraction	Percentage of blood leaving the heart at each contraction	Percentage	[14, …, 80]
Sex	Woman or man	Binary	0, 1
Platelets	Platelets in the blood	kiloplatelets/mL	[25.01, …, 850.00]
Serum creatinine	Level of creatinine in the blood	mg/dL	[0.50, …, 9.40]
Serum sodium	Level of sodium in the blood	mEq/L	[114, …, 148]
Smoking	If the patient smokes	Boolean	0, 1
Time	Follow-up period	Days	[4,…,285]
(Target) death event	If the patient died during the follow-up period	Boolean	0, 1

neighbor, random forest, and support vector machine have been applied in heart attack disease prediction.

Machine Learning Model Evaluation: The machine learning models used for heart attack disease prediction have been quantitatively evaluated with precision, recall, and F1-Score. AROC AROC (Area under Receiver Operating Characteristic curve) has been used to visually analyze the performance of the machine learning models used.

3 Results and Discussion

All the experimental work has been performed on Python programming on Windows 10. The hardware used in simulation work is tabulated in Table 2. The dataset used for the experimental purpose has 13 features. Correlation analysis has been done on these features. Its values are presented in Fig. 2. Correlation measures how one feature is related to another. Its value lies between 0 and 1. It gives the degree of association between the compared features. If the value of one feature increases/decreases leads

Table 2 Hardware configuration

Hardware	Capacity
CPU clock speed	1.60 Ghz
RAM	8 GB
L1 cache Memory	256 KB
L2 cache memory	1 MB
L3 cache memory	6 MB

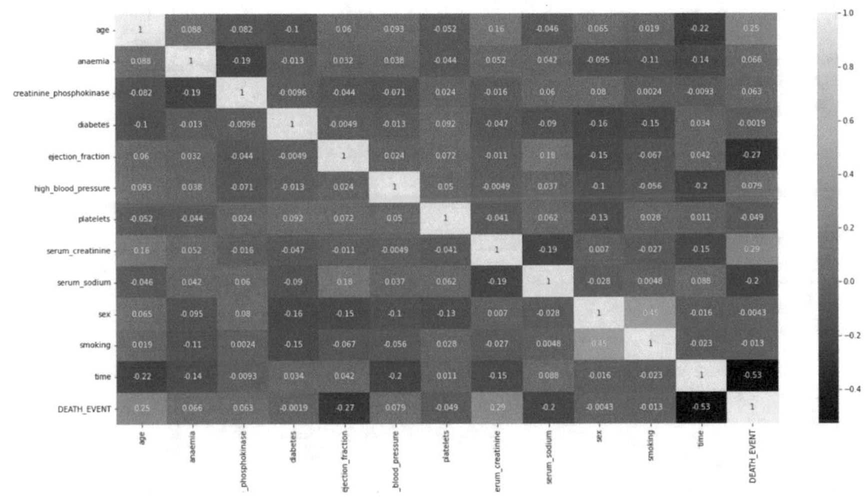

Fig. 2 Correlation measures among different features in the dataset

to an increase/decrease in another variable, those features are said to be positively correlated to each other. If one feature increase leads to a decrease in another variable, then they are said to be negatively correlated. If an increase/decrease in one feature does not affect another feature, then they are said to be uncorrelated to each other. The presence of correlated features leads to complexity and hence should be removed. From Fig. 2, it can be analyzed that all the features taken into consideration have a negligence correlation. One feature named time has been dropped because its presence does not contribute to heart disease prediction. The remaining 12 features have been taken into account for heart disease prediction.

Precision, Recall, and F1-score have been considered for quantitative evaluation of the proposed model [18]. Three different scenarios have been considered for evaluation of the presented model such as without using normalization and standardization, with normalization and standardization, and with normalization and standardization plus model tuning. The results obtained have been tabulated in Tables 3, 4, and 5. The precision measures the ratio between true positives and total positive calculated positives. Its value lies between 0 and 1. Recall measures the ratio between true

Table 3 Quantitative analysis without Normalization and Standardization

Methods	Class	Precision	Recall	F1 Score
Logistic regression [10]	0	0.65	0.97	0.78
	1	0.88	0.28	0.42
K-nearest neighbor [9]	0	0.58	0.83	0.68
	1	0.40	0.16	0.23
Random forest [14]	0	0.69	0.89	0.78
	1	0.73	0.44	0.55
Support vector machine [11]	0	0.67	0.86	0.70
	1	0.67	0.40	0.50

Table 4 Quantitative analysis with Normalization and Standardization

Methods	Class	Precision	Recall	F1 Score
Logistic regression	0	0.67	0.96	0.75
	1	0.67	0.40	0.50
K-nearest neighbor	0	0.67	0.94	0.79
	1	0.82	0.36	0.50
Random forest	0	0.70	0.86	0.77
	1	0.71	0.48	0.57
Support vector machine	0	0.67	0.86	0.75
	1	0.67	0.40	0.50

positives and total actual positives in the dataset. Its value lies between 0 and 1. The values closer to 1 gives higher performance, while values closer to 0 represent lower performance.

It can be analyzed from Table 3 that logistic regression gives a higher performance in heart attack disease prediction as compared to other models such as KNN, random forest, and support vector machine without using normalization and standardization techniques. Also, it can be easily analyzed from Table 4 that KNN gives higher performance measurement after applying the normalization and standardization technique. Model tuning is one of the optimization techniques to get the best parameters for machine learning models. The results obtained after applying model tuning in the KNN, random forest, and support vector machine are tabulated in Table 5. It can be easily analyzed from Table 5 that KNN performs better than other models used for comparative study. Sometimes, both precision and recall play an important

Table 5 Quantitative analysis with Normalization, Standardization, and Model Tuning

Methods	Class	Precision	Recall	F1 Score
Logistic regression	0	0.67	0.96	0.75
	1	0.67	0.40	0.50
K-nearest neighbor	0	0.59	1.00	0.74
	1	1.00	0.04	0.08
Random forest	0	0.59	0.94	0.73
	1	0.56	0.08	0.14
Support vector machine	0	0.67	0.86	0.75
	1	0.67	0.40	0.50

role in model selection. In such cases, F1 scores are calculated which is the harmonic mean of precision and recall. Its value lies between 0 and 1. The values closer to 1 represent a better model as compared to models which have values near 0. It can be analyzed from Tables 3, 4, and 5 that on an average logistic regression gives higher F1-score without and with using normalization and standardization.

AROC (Area under Receiver Operating Characteristic curve) has been used for visual inspection of the presented models [19]. By visual inspection of Fig. 3, it is analyzed that the proposed model achieves higher performance. It can be analyzed from Fig. 3 that logistic regression and random forest cover almost the same area but logistic regression performs better as discussed in the comparative study without using normalization and standardization. Also, It can be analyzed from Fig. 4 that KNN covers a higher area and hence has higher performance in heart attack disease prediction as compared to other models taken into consideration with normalization and standardization techniques. Also, It can be analyzed from Fig. 5 that KNN covers a higher area and hence has higher performance in heart attack disease predication as compared to other models taken into consideration with normalization and standardization and model tuning technique.

After rigorous comparative study with different models with different scenarios, it has been found that logistic regression performs better than other models considered for comparative study without using normalization and standardization. It has been also observed that the KNN gives higher performance measurement after applying the normalization and standardization techniques. Model tuning is one of the optimization techniques to get the best parameters for machine learning models. It has been found that logistic regression gives higher performance in heart disease detection and classification with model tuning techniques. Overall, It is found that on an average logistic regression gives higher performance ratio.

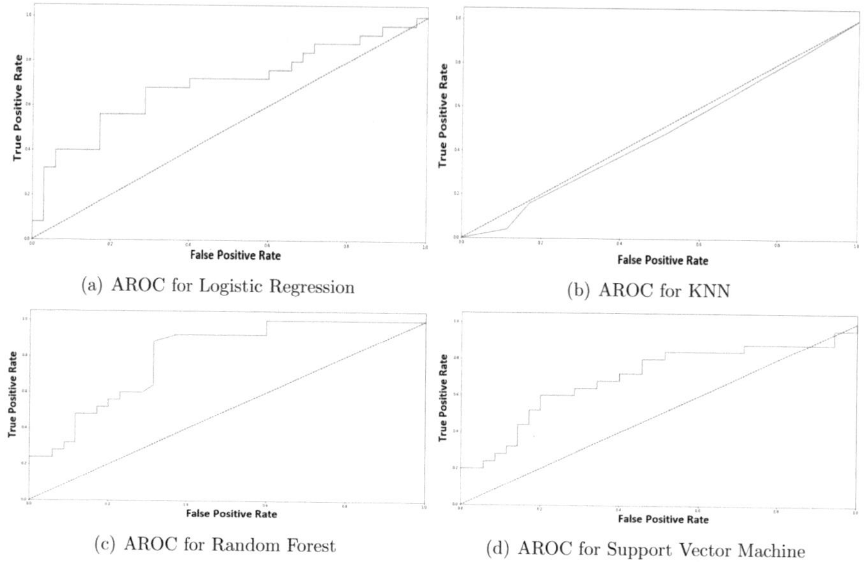

(a) AROC for Logistic Regression

(b) AROC for KNN

(c) AROC for Random Forest

(d) AROC for Support Vector Machine

Fig. 3 AROC for various machine learning models without using Normalization and Standardization

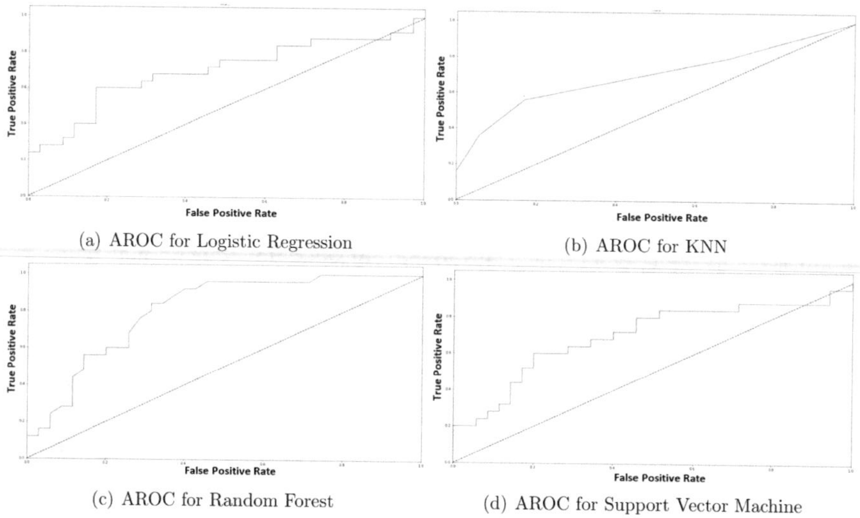

(a) AROC for Logistic Regression

(b) AROC for KNN

(c) AROC for Random Forest

(d) AROC for Support Vector Machine

Fig. 4 AROC for various machine learning models with Normalization and Standardization

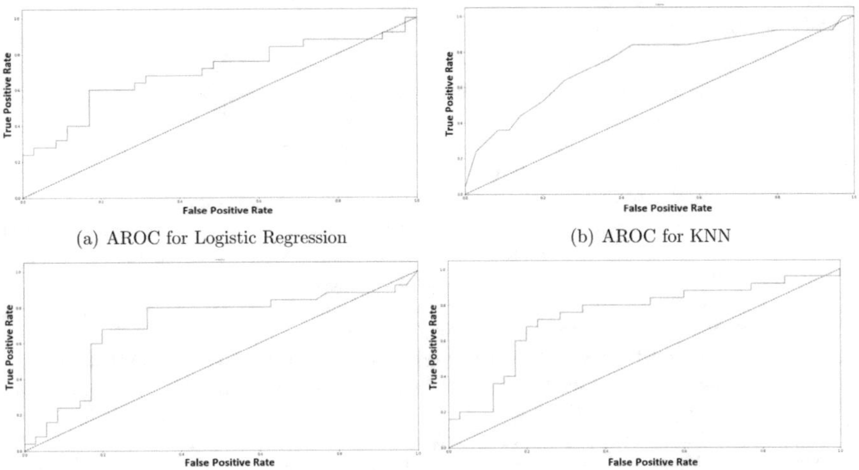

(a) AROC for Logistic Regression

(b) AROC for KNN

(c) AROC for Random Forest

(d) AROC for Support Vector Machine

Fig. 5 AROC for various machine learning models with Normalization and Standardization and Model Tuning

4 Conclusion

Machine learning models have been used for the detection and classification of heart disease. Logistic regression, K-nearest neighbor, random forest, and support vector machine have been used in the proposed model for heart disease detection and classification. The models have been evaluated with precision, recall, and F1-score. AROC (Area under Receiver Operating Characteristic curve) has also been used for evaluation of the current work. Three cases have been considered in the present work for evaluations such as data with and without normalization and standardization. Also, the models have been analyzed with model tuning with normalization and standardization. It has been found that logistic regression performs better than other models considered for comparative study without using normalization and standardization. It has been also observed that the KNN gives higher performance measurement after applying the normalization and standardization techniques. Model tuning is one of the optimization techniques to get the best parameters for machine learning models. It has been found that logistic regression gives higher performance in heart disease detection and classification with model tuning techniques. It has been found after analyzing all the different scenarios that an average logistic regression gives higher performance ratio. In the future, more datasets can be used to analyze the performance of the presented work.

References

1. TEoE Britannica, Heart. https://www.britannica.com/science/heart
2. Medical News Today. https://www.medicalnewstoday.com/articles/151444
3. Heart attack. https://www.webmd.com/heart-disease/guide/heart-disease-heart-attacks
4. Cardiovascular diseases (CVDs). https://www.who.int/news-room/fact-sheets/detail/
5. Heart attack. https://www.mayoclinic.org/diseases-conditions/heart-attack
6. Ellulu MS, Ismail P, Khazáai H, Rahmat A, Abed Y, Ali F (2016) Atherosclerotic cardiovascular disease: a review of initiators and protective factors. Inflammopharmacology 24:1–10. http://dx.doi.org/10.1007/s10787-015-0255-y, https://doi.org/10.1007/s10787-015-0255-y
7. Yasue HE, Mizuno Y (2019) Coronary artery spasm—clinical features, pathogenesis and treatment. Proc Jpn Acad Ser B Phys Biol Sci 95(2):53–66
8. Fass R, Achem S (2011) Noncardiac chest pain: epidemiology, natural course and pathogenesis, J Neurogastroenterol Motil 17:110–123. http://dx.doi.org/10.5056/jnm.2011.17.2.110, https://doi.org/10.5056/jnm.2011.17.2.110
9. Alim MA, Habib S, Farooq Y, Rafay A (2020) Robust heart disease prediction: a novel approach based on significant feature and ensemble learning model. In: 2020 3rd international conference on computing, mathematics and engineering technologies (iCoMET), pp 1–5. http://dx.doi.org/10.1109/iCoMET48670.2020.9074135, https://doi.org/10.1109/iCoMET48670.2020.9074135
10. Shah D, Patel SB, Bharti SK (2020) Heart disease prediction using machine learning techniques. SN Comput Sci 1:345
11. Takci H (2018) Improvement of heart attack prediction by the feature selection methods. Turk J Electr Eng Comput Sci 26:1–10. http://dx.doi.org/10.3906/elk-1611-235, https://doi.org/10.3906/elk-1611-235
12. Rajkumar A, Reena M, Diagnosis of heart disease using datamining algorithm. Global J Comput Sci Technol 10
13. Jegan C (2013) Heart attack prediction system using fuzzy c means classifier. IOSR J Comput Eng 14:23–31. http://dx.doi.org/10.9790/0661-1422331, https://doi.org/10.9790/0661-1422331
14. Ali F, El-Sappagh S, Islam SMR, Kwak D, Ali A, Imran M, Kwak KS (2020) A smart healthcare monitoring system for heart disease prediction based on ensemble deep learning and feature fusion. Inf Fusion 63:208–222
15. Ahmad T, Munir A, Bhatti SH, Aftab M, Raza MA (2017) Survival analysis of heart failure patients: a case study. PLOS ONE 12(7):1–8. http://dx.doi.org/10.1371/journal.pone.0181001, https://doi.org/10.1371/journal.pone.0181001, https://doi.org/10.1371/journal.pone.0181001
16. Carissa Stephens R, What is a creatinine blood test? https://www.healthline.com/health/creatinine-blood
17. U. of Illinois, What is a sodium blood test? https://www.healthline.com/health/sodium-blood
18. Chandra SK, Bajpai MK, Fractional mesh-free linear diffusion method for image enhancement and segmentation for automatic tumor classification. Biomed Signal Process Control 58. http://dx.doi.org/10.1016/j.bspc.2019.101841, https://doi.org/10.1016/j.bspc.2019.101841
19. Chandra SK, Bajpai MK (2019) Two-sided implicit euler based superdiffusive model for benign tumor segmentation. In: 2019 IEEE region 10 symposium (TENSYMP), pp 12–17. http://dx.doi.org/10.1109/TENSYMP46218.2019.8971188, https://doi.org/10.1109/TENSYMP46218.2019.8971188

Recognizing Indian Classical Dance Forms Using Transfer Learning

M. R. Reshma, B. Kannan, V. P. Jagathyraj, and M. K. Sabu

1 Introduction

India is primarily a cultural destination in which its living culture and performing arts promote cultural tourism in the country's tourism market. The Indian tourism slogan Incredible India itself gives the intellect. The word diversity is proportionate to India, which abides by the country's rich heritage. The perfect blend of languages, cultures, religions, arts, customs, and traditions mirrored throughout the length and breadth of the nation. These are the key features that attract international tourism and magnify domestic tourism. For most art forms, the plot or storyline was the religious and philosophical thoughts in the social context at any moment in history. These have high significance as they give a glimpse of India's bewilderingly diverse cultural life. The prominent and popular styles in this art diverge from region to region. At times they merge and mingle inseparably.

We live in a digital age, where everything trends to computerize. Performing arts, an integral part of human civilization, was imported into this digitization. The Indian Classical Dance or "Shastriya Nirta" encloses different performing arts related to Indian culture. The heritage preservation of Indian classical dance forms holds a pivotal role as it will give the forthcoming generation an insight into the amalgamation of Indian heritage. The performing and living art forms in India are awe-inspiring to tourists, and they are fascinated by gathering knowledge about them. We must uphold the authenticity of the Indian classical dance forms. These open up new challenges in Computer Science to conserve, analyze and categorize different classical dance forms. The main applications of such study will be in computer vision and informa-

M. R. Reshma (✉) · B. Kannan · M. K. Sabu
Department of Computer Applications, CUSAT, Kochi, Kerala, India
e-mail: reshmamr@cusat.ac.in

V. P. Jagathyraj
School of Management Studies, CUSAT, Kochi, Kerala, India

P. Singh et al. (eds.), *Machine Learning and Computational Intelligence Techniques for Data Engineering*, Lecture Notes in Electrical Engineering 998,
https://doi.org/10.1007/978-981-99-0047-3_36

(a) Bharatanatyam (b) Kathakali (c) Mohiniyattam

Fig. 1 Indian classical dance forms

tion retrieval. Artificial Intelligence intelligence and Deep deep Learning learning advancements provide tools and techniques to address these challenges.

However, In almost all studies, the focus was on analyzing the pose or gesture of a particular dance form. Furthermore, there are no publicly available datasets to test Indian dance forms. This paper proposes a framework capable of recognizing Indian dance forms. The framework relies on deep learning methods to classify Indian Classical Dance forms. The three categories of Indian Classical Dance contemplated in this research are Bharatanatyam (Tamil Nadu), Kathakali (Kerala), and Mohiniyattam (Kerala) (Fig. 1).

The earlier works on Indian classical dance form recognition are less in number. Maximum research done in this domain focused on a single dance form [7, 18] or in the analysis of their poses or gestures [2, 13]. In [13], they designed a gesture recognition algorithm using the skeleton of a human body generated using a Kinect sensor for discriminating between five different emotions of the Indian classical dance and achieved an accuracy of 86.8%. A model to recognize different hand gestures in Bharatanatyam, using Artificial artificial Neural neural Network network proposed [18], classifying the images based on the positioning of the fingers. A framework that classifies the foot postures called stanas in Indian classical dance using machine learning models and achieved an accuracy of 85.95% using Naive Bayes [17]. A trained Deep Stana Classifier, [16] stanas from the video frames are identified, and they obtained an accuracy of 91% in recognizing the stanas. A glove-based and vision-based gesture recognition method was used according to dance forms and employed different classifiers [5]. Mallik et al. [10] classified Bharatnatyam and Odissi and used Scalescale-Invariant invariant Feature feature Transform transform to recognize Indian dance forms attained 92.78% for recognition. In [9], they used CNN to classify the poses/mudras of Bharatanatyam alone and achieved an accuracy of 89.92%.

Kishore et al. [8] used the Adaboost AdaBoost classifier to classify Bharatanatyam poses, also using a segmentation model to extract and recognize human movements in the video sequence, and achieved an accuracy of 90.89%. Ankita et al. applied Deep deep Spatiospatio-temporal descriptors to identify Indian Classical Dance forms in videos [3] and achieved an accuracy of 75.83%. A pose descriptor based on the histogram of oriented optical flow is used in [15] to represent each frame of a dance video and classified three forms of dance using a support vector machine model achieved an accuracy of 86.67%.The 28 Bharatanatyam single-hand mudras using an Artificial artificial Neural neural Network network and attained 93.64 percent accuracy [2]. The HOG features are used with a support vector machine to classify the dance images [7]. Single- Hand hand Gestures gestures of Bharatanatyam Mudra classification using Transfer transfer Learning learning [12] using Transfer learning and yields an accuracy of 94.56%. In [14], a spatiotemporal feature model for classifying Indian Classical Dance. In [11], they used Deep deep Learning learning and Convolution convolution Neural neural Network network to classify and identify Indian Classical Dance forms and attained an accuracy of 78.88% using the resnetResNet34 architecture. While reviewing the literature, most of the earlier researchers concentrated on a particular Indian classical dance form. A few pieces of research conveyed out in recognizing the Indian classical dance form. Maximum of the works done in Indian classical dance focused on mudras/poses and gesture recognition. Generally, these come under the class of Human action recognition, which is one of the extensive interests in the computer vision community.

The catalog of later sections in the paper isare as follows: In the Methodology Section, there are subsections like datasets, feature extraction, classification, etc. This section succeeds with the implementation, experimental results and analysis, and finally, the conclusion and future work as the last section.

2 Methodology

In the proposed work, we are creating a framework capable of recognizing the classical dance forms of India. As part of this research, we are restricting the number of dance forms. Due to the lack of a public dataset for Indian Classical Dance Forms, a collection of handcrafted images from the Internet is used. Bharatanatyam, Kathakali, and Mohiniyattam are the three dance forms for the dance form recognition framework. A dance form recognition framework must identify the dance form and classify it into the correct class. A pipeline of the framework is given in Fig. 2 below.

It includes the following steps: the images from the dataset are fed to the transfer learning model to extract the features. These features applied to the classifier give the final classification result. The pre-trained models were applied to tackle enormous problems in different domains like Computer Vision and Natural Language Processing. Here, we are using pre-trained transfer learning models for feature extraction. We are using a Deep deep Neural neural Network network for the classification task in the proposed pipeline.

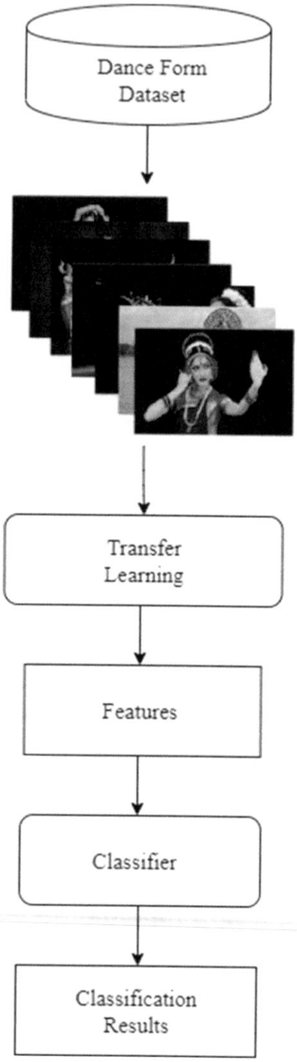

Fig. 2 The proposed transfer learning pipeline

2.1 Dataset

The dataset for this study is comprised of digital images of Indian classical dance forms collected from the internet. Our dataset comprises 340 Handcrafted images of varying sizes that belong to three dance forms. The dataset contains three classes Bharatanatyam, Kathakali, and Mohiniyattam. The images in the dataset are distributed evenly among the three classes.

2.2 Feature Extraction

One of the imperative tasks in the proposed framework is feature extraction. Features extraction is the key for improving the success of classification problems. The most distinctive and informative set of features is weighted to recognize the images individually. Manual feature extraction is incomplete, and designing and validating those takes a long time In the proposed framework, for feature extraction, we are using the Deep deep Learning learning technique called Transfer transfer learning.

In transfer learning the knowledge, a model obtained by training on a large dataset such as ImageNet [4] can apply to an application with a smaller dataset. It helps to remove the prerequisite of a larger dataset. Also, it reduces the training period compared with the one when developed from scratch.

Pre-Trained Neural Networks: The pre-trained models used here are Xception, VGG16, ResNet50, InceptionV3, MobileNetV2, and DenseNet121. After training on the ImageNet dataset, these models are now used to learn features from the Indian Classical Dance dataset.

2.3 Classification

Lastly, the proposed framework involves classification. It is the process of organizing a set of data into classes, and here the data is images. The algorithm used for this task is known as a classifier. They observe the data using the known features appropriately and categorize them into groups. In this study, we have a multi-class classification problem in which three classes or outcomes are there. In our example, Bharatanatyam, Kathakali, and Mohiniyattam are the three classes. We are using the deep learning classifier called Deep Neural Network (DNN). In comparability of the results obtained by our proposed method, we are using other classifiers called CNN and an ensemble classifier.

Deep Neural Network (DNN): Deep neural networks (DNN) excel in the tasks of image recognition. Neural networks are computing systems inspired by human brain structure to recognize patterns. A neural network becomes deep when the number of hidden layers is more than in traditional neural networks. The deep networks may contain hundreds of them when traditional networks have up to three hidden layers. The proposed DNN model designed here consists of a stack of layers. Figure 2 shows the detailed architecture of the proposed DNN model. The output of the first layer becomes the input to the next succeeding layer, and it continues likewise to the following layers. The top layers are dense layers with the activation function ReLU (Rectified Linear Unit). As this is a multi-class classification problem, the last layer activation function used is Softmax, the loss function is categorical cross-entropy, and the optimizer used is Adam.

Convolution Neural Network (CNN): The Convolution Neural Network (CNN) has enhanced performance in image classification tasks [1]. CNN accepts inputs and assigns importance to the notable features in the images, and it will be able to differentiate the images based on learned features. Its architecture is a stack of distinct layers, which transform input into output through differentiable functions.

Ensemble Classifier: An ensemble voting classifier is a model which combines more than one classifier result to obtain a final prediction [6]. It trains on classifiers and predicts an output/class based on the result with the highest probability class as the final output. The ensemble classifier builds here with three Machine machine learning models, namely Logistic logistic Regressionregression, Gaussian Naive Bayes, Random random Forest forest Classifierclassifier, and a voting classifier (Fig. 3).

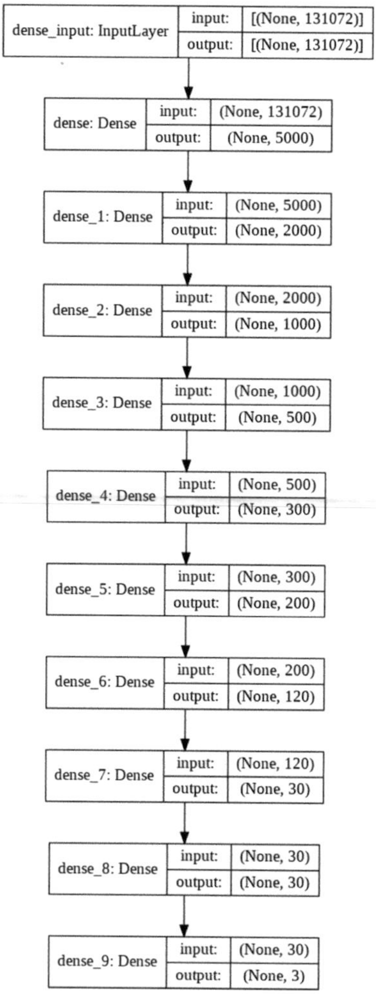

Fig. 3 The proposed sequential DNN model

3 Implementation

The Indian classical dance form recognition system must extract unique features from the different dance form images and classify them into the appropriate class. The proposed pipeline for transfer learning, is shown in Fig. 2.

A few changes in traditional transfer learning create this novel method. The inclusive architecture of the proposed model incorporates two parts the feature extractor and a classifier. The images from the dataset are taken one at a time and divided into four quarters. These quarters are loaded to the pre-trained architecture for feature extraction. The results obtained were appended to create a feature set. It completes the feature extraction part of the architecture. For classification, we are using a Deep Neural Network (DNN). The evaluation metrics used are accuracy and confusion matrix. In testing, we adopt the state-of-the-art Convolution Neural Network (CNN) and an Ensemble ensemble Classifier classifier for comparison studies.

4 Results and Analysis

Dataset consists of images of Indian classical dance forms, varying in size, collected from the Internet. The main aim of our framework is to classify the input image correctly into its classes. Multi-class classification began on the dataset with proposed transfer learning and DNN. This section deals with the results achieved by the proposed framework and both CNN and Ensemble classifier. The achieved results of proposed architectures compared with the CNN and Ensemble ensemble classifier.

In the proposed framework, pre-trained models are applied to the dataset for feature extraction, and then it is classified by using a Deep Neural Network (DNN). Among the different pre-trained models—the transfer learning model resnetResNet50 with Deep Neural Network (DNN) obtain 97.05% accuracy. Considering the results obtained by other pre-train models with DNN, the proposed architecture better classifies Indian Classical Dance forms. A summary of the accuracy of all the pre-trained models with DNN is in Table 1. The confusion matrix help to visualize the neural network decisions. The decision of heatmaps analyzed from the x-axis,

Table 1 Comparison of the accuracy of pre-trained models + DNN

Pre-train models	Accuracy
ResNet 50	97.05
VGG16	95.58
Xception	88.23
MobileNetV2	89.7
DenseNet121	95.58
InceptionV3	86.76

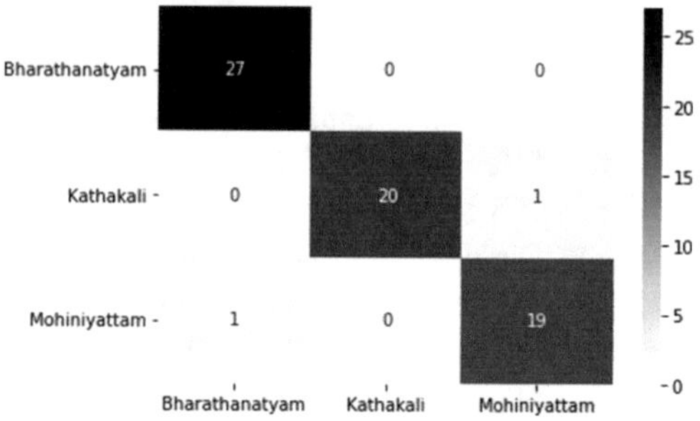

Fig. 4 Confusion matrix of ResNet50+DNN model

Table 2 Comparison of the accuracy of Traditional CNN in five-folds

Runs	Accuracy
1st	88.23
2nd	91.17
3rd	82.35
4th	91.17
5th	89.99
Mean ± SD	88.58±3.68

we have the predicted labels, and on the y-axis, we have the correct results. Here, we have a three-class problem and actual classification values depicted as the diagonal. The values in the boxes are counts of each class. Figure 4 shows the result obtained with the combination of ResNet50 with DNN.

For comparing comparison, the results obtained by the proposed framework state-of-the-art Convolution Neural Network (CNN) and ensemble classifier used. Table 2 shows the classification accuracy using CNN with 5five-fold cross-validation. The average accuracy result achieved using the CNN is 88.58±3.68 (Mean ± SD).The confusion matrix to visualize the result obtained in the five runs of Traditional traditional CNN in Fig 5 (Table 2).

We designed an ensemble classifier using Machine machine Learning learning classification models, which include Logistic logistic Regressionregression, Gaussian Naive Bayes, Random random Forest forest Classifierclassifier, and Voting voting Classifierclassifier. The classification results using the proposed ensemble classifier with individual machine learning models are in Table 3. The voting classifier obtained an accuracy of 84 ± 0.04, but among the models used in the ensemble classifier, Logistic logistic Regression regression attained a better result of 95 ± 0.00.

The confusion matrix in Fig. 6 show the result obtained with the combination of ResNet50 with Ensemble Classifiers.

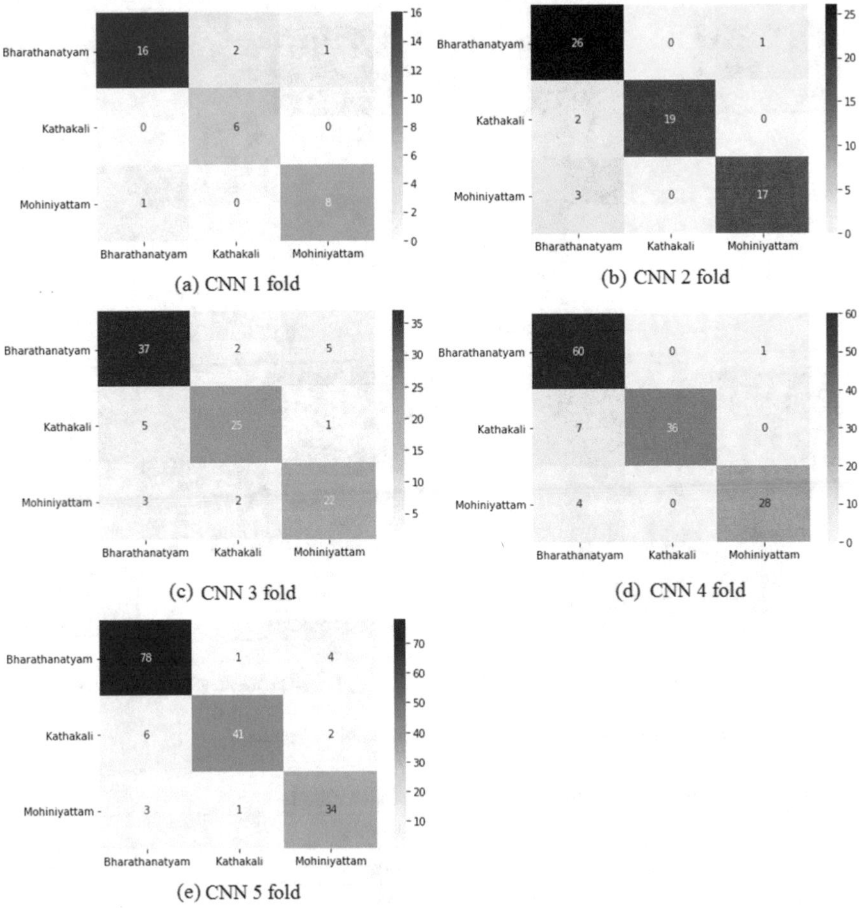

Fig. 5 **a**, **b**, **c**, **d**, **e** Confusion matrix of Traditional CNN

Table 3 Comparison of the accuracy obtained with ResNet 50 + Ensemble Classifier

Classifiers	Accuracy
Logistic regression	95 ± 0.00
Random forest classifier	75 ± 0.02
Gaussian Naive Bayes	83 ± 0.65
Voting classifier	84 ± 0.04

Table 4 compiles the classification accuracy obtained using the proposed model resnetResNet50 with DNN, CNN, and Ensemble ensemble Classifierclassifier. Our result shows that resnetResNet50 is more accurate for feature extraction when compared to the other pre-trained models. It improves the performance in the classification of Indian classical dance forms compared with other models.

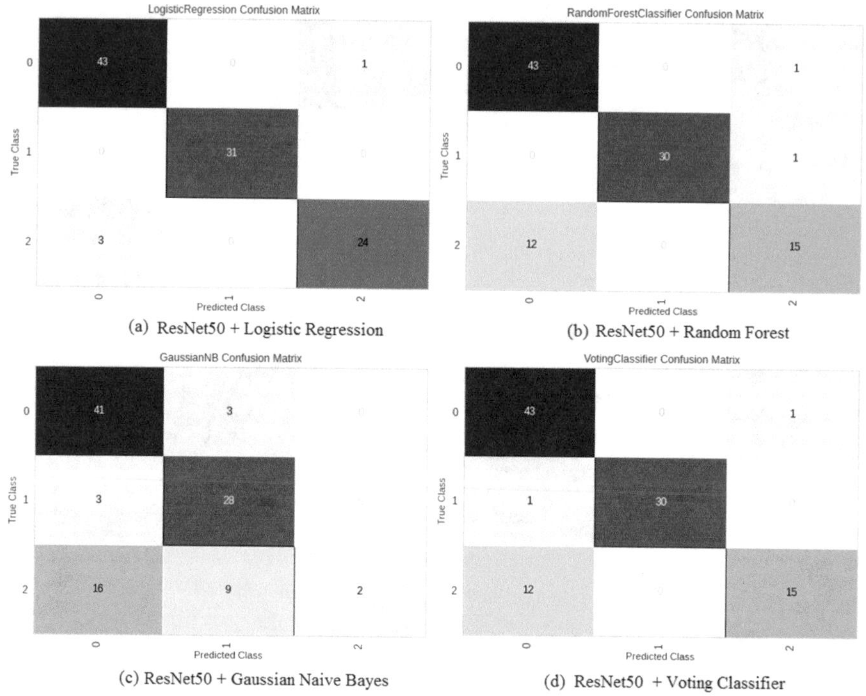

(a) ResNet50 + Logistic Regression

(b) ResNet50 + Random Forest

(c) ResNet50 + Gaussian Naive Bayes

(d) ResNet50 + Voting Classifier

Fig. 6 **a**, **b**, **c**, **d** Confusion matrix of ResNet50 + Ensemble Classifiers

Table 4 Comparison of the performance accuracy of all methods

Classifiers	Accuracy
ResNet50 + DNN	97.05
ResNet50 + CNN	88.58
ResNet50 + Ensemble classifier	84.04

5 Conclusion

We proposed a framework capable of recognizing Indian Dance forms. The core components in this framework consist of Deep deep Learning learning using transfer learning and Deep Neural Networks (DNN). The proposed model can overcome one of the significant classification problems using the Convolution Neural Network (CNN) model: the lesser training data. The pre-trained transfer learning model is used to handle this issue. The proposed method is evaluated in terms of accuracy and confusion matrix. It gained commendable accuracy in the classification. The method showed outperforming results when compared toagainst the results using a CNN and an Ensemble ensemble Classifierclassifier.

References

1. Saad A, Abed MT, Saad A-Z (2017) Understanding of a convolutional neural network. In: 2017 international conference on engineering and technology (ICET). IEEE, pp 1–6
2. Anami BS, Bhandage VA (2019) A comparative study of suitability of certain featu es in classification of Bharatanatyam mudra images using artificial neural network. Neural Process Lett 50(1):741–769
3. Bisht A et al (2017) Indian dance form recognition from videos. In: 2017 13th international conference on signal-image technology and internet-based systems (SITIS). IEEE, pp 123–128
4. Jia D (2009) Imagenet: a large-scale hierarchical image database. In: 2009 IEEE conference on computer vision and pattern recognition. IEEE, pp 248–255
5. Devi M, Saharia S, Bhattacharyya DK (2015) Dance gesture recognition: a survey. Int J Comput Appl 122(5)
6. Dietterich TG (2000) Ensemble methods in machine learning. In: International workshop on multiple classifier systems. Springer, Berlin, pp 1–15
7. Gautam S, Joshi G, Garg N (2017) Classification of Indian classical dance steps using HOG features. Int J Adv Res Sci Eng (IJARSE) 6(08)
8. Kishore PVV et al (2018) Indian classical dance action identification and classification with convolutional neural networks. Adv Multimed
9. Kumar KVV et al (2018) Indian classical dance action identification using adaboost multiclass classifier on multifeature fusion. In: 2018 conference on signal processing and communication engineering systems (SPACES). IEEE, pp 167–170
10. Mallik A, Chaudhury S, Ghosh H (2011) Nrityakosha: preserving the intangible heritage of indian classical dance. J Comput Cult Herit (JOCCH) 4(3):1–25
11. Naik AD, Supriya M (2020) Classification of indian classical dance images using convolution neural network. In: 2020 international conference on communication and signal processing (ICCSP). IEEE, pp 1245–1249
12. Parameshwaran AP et al (2019) Transfer learning for classifying single hand gestures on comprehensive Bharatanatyam Mudra dataset. In: Proceedings of the IEEE/CVF conference on computer vision and pattern recognition workshops
13. Saha S et al (2013) Gesture recognition from indian classical dance using kinect sensor. In: 2013 fifth international conference on computational intelligence, communication systems and networks. IEEE, pp 3–8
14. Samanta S, Chanda B (2014) Indian classical dance classification on manifold using jensen-bregman logdet divergence. In: 2014 22nd international conference on pattern recognition. IEEE, pp 4507–4512
15. Soumitra S, Pulak P, Bhabatosh C (2012) Indian classical dance classification by learning dance pose bases. In: 2012 IEEE workshop on the applications of computer vision (WACV). IEEE, pp 265–270
16. Shailesh S, Judy MV (2020) Automatic annotation of dance videos based on foot postures. Indian J Comput Sci Eng 5166. Engg Journals Publications-ISSN 976
17. Shailesh S, Judy MV (2020) Computational framework with novel features for classification of foot postures in Indian classical dance. Intell Decis Technol 14(1):119–132
18. Soumya CV, Ahmed M (2017) Artificial neural network based identification and classification of images of Bharatanatya gestures. In: 2017 international conference on innovative mechanisms for industry applications (ICIMIA). IEEE, pp 162–166

Improved Robust Droop Control Design Using Artificial Neural Network for Islanded Mode Microgrid

Shraddha Gajbhiye and Navita Khatri

1 Introduction

Today, dispersed generation (DG) and inexhaustible energy resources, i.e. the force of the wind, sun, and tide, are connected to the social network through power inverters. Before coupling to the social network [1–3], they usually form micro-networks. Many inverter area units must inevitably be managed in parallel for dynamic and/or profitable applications, due to the appropriateness of power electronic devices with high currents [4]. Another reason is that the inverters managed in parallel offer the system redundancy and high reliability that large customers demand. For parallel connected inverters, the disadvantage is the ability to share the load between them. For this, an important approach is droop management [5–7], which is generally used in regular power generation systems [8]. And the benefit is that no external conference mechanism is required between the inverters [9, 10], which allows a significant distribution of linear and / or nonlinear masses, [11–14]. In some cases, an external conversation means that the area unit is still accepted to share the load [15] and is rebuilding the voltage and frequency of the microgrid [1, 16]. On the other hand, the management of microgrids is more complex in an associated isolated way than in a network.

connected way. When it is connected to the network, the frequency and voltage of the microgrid are managed by the social network. Oppositely, the management of frequency and voltage and even the current is regulated and isolated by a microgrid. The degree of the power imbalance between DG units, frequency deviation, and unbalanced load voltages lead to the presence of circulating current. Therefore,

S. Gajbhiye (✉) · N. Khatri
Electrical and Electronics Engineering Department, SVVV, Indore, India
e-mail: shraddhagajbhiye@gmail.com

N. Khatri
e-mail: navitakhatri@svvv.edu.in

P. Singh et al. (eds.), *Machine Learning and Computational Intelligence Techniques for Data Engineering*, Lecture Notes in Electrical Engineering 998,
https://doi.org/10.1007/978-981-99-0047-3_37

this is the absolute requirement for an efficient controller to manage the voltage and frequency balance. The controller also offers a tolerable distribution of force between DG units in isolated operations [17]. The basic objective of an isolated solution is to maintain the distribution of power among numerous DGs. When the operation of the microgrid switch switches in isolation, the frequency, and the potential difference, E are maintained as when many inverters are operating in parallel and sharing the load. These inverters regulate the amplitude of frequency and voltage and have a completely different functioning mode than an alternative converter [18]. The controller also provides the essential run power. Yet, the problem of power quality [19] communal by the presence of transient circulating current with electrical resistance [7] can lead to system imbalance and is therefore not safe for inverters [20] due to similar output voltages. In an isolated microgrid, the loads of different DG units must be distributed correctly. Several control methods have been proposed for single-phase voltage source inverters (VSIs) [1]. The best methods used are the techniques of fall control, predictive control, inactive control, and cascade linear control [5–7]. Predictive control and static control occupy a wide range of controls that, in the meantime, have been dedicated to inverter control [7, 20–22]. To fuel current protests such as uneven power distribution, control design complexity, slow response, and nonlinearity, the artificial neural system is used to control the power distribution, voltage, and frequency of the inverter. Voltage source in a microgrid.

This research aims to expand the robust virtual impedance droop control technology with an artificial neural network [21], which copes with PR control for voltage and current loops and the implementation of current regulators in secondary loops for a single-phase microgrid for an LV network. It will have a base frequency of 50 $\pm$ 1 Hz for inverters connected in parallel, taking into account linear loads. Simulation studies are performed to validate and determine the potential of the proposed controller in comparison with the conventional robust static controller.

2 Droop Control Approach

The system accuracy and stability in microgrid are managed alone by voltage control when more than one micro source are associated. This voltage control lowers the voltage and reactive power vacillation.

The control strategy called droop control is when differing DGs are connected to the microgrid in electrical networks which are complex, the distribution of power between them is made properly. Droop control also allows plug-in features very well to the complex power system [4].

In power dispersion, the work of droop control is that it controls the reactive power based on voltage control and it controls the active power based on frequency control.

By managing the active and reactive power of the system, the frequency and voltage can be manipulated.

An inverter can be designed as a remark voltage source with an output impedance.

Fig. 1 Model of inverter

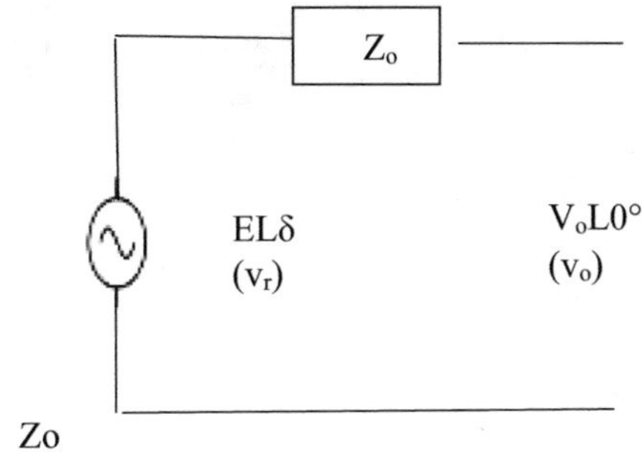

As shown in Fig. 1, the reactive power and real power delivered to the terminal via the output impedance Zo are

$$P = \left(\frac{EV_0}{Z_0}\cos\delta - \frac{V_0^2}{Z_0}\right)\cos\theta + \frac{EV_0}{Z_0}\sin\delta\sin\theta \quad (1)$$

$$Q = \left(\frac{EV_0}{Z_0}\cos\delta - \frac{V_0^2}{Z_0}\right)\sin\theta - \frac{EV_0}{Z_0}\sin\delta\cos\theta \quad (2)$$

where δ is the phase difference between the supply and the load, often called the power angle.

For an inductive impedance, θ = 90°. Then

$$P = \frac{EV_0}{Z_0}\sin\delta \quad \text{and} \quad Q = \frac{EV_0}{Z_0}\cos\delta - \frac{V_0^2}{Z_0}$$

when δ is small,

$$P = \frac{EV_0}{Z_0}\delta \quad \text{and} \quad Q = \frac{E - V_0}{Z_0}V_0$$

and, roughly,

$P \sim \delta$ and $Q \sim -V_0$

Hence, the droop control strategy takes the form

$$E_i = E^* - n_i Q_i \quad (3)$$

$$w_i = w^* - m_i P_i \quad (4)$$

where E^* is the rated RMS voltage of the inverter and ω^* is the rated frequency.

For a resistive impedance, θ = 0°, then

$$P = \frac{EV_0}{Z_0}\cos\delta - \frac{EV_0^2}{Z_0} \quad \text{and} \quad Q = -\frac{EV_0}{Z_0}\delta,$$

when δ is small,

$$P = \frac{EV_0}{Z_0}V_0 \quad \text{and} \quad Q = -\frac{EV_0}{Z_0}\delta$$

and, roughly,

P ~ V_0 and Q ~ −δ.

Hence, the droop control strategy takes the form

$$E_i = E^* + n_i p_i \tag{5}$$

$$w_i = w^* + m_i q_i \tag{6}$$

In relation to the power ratings of the inverter, the load is shared, and the droop coefficients of the inverters are chosen to be in converse extent to their power ratings [14], i.e. mi and ni charge to be chosen to fulfil

$$n_1 S^* = n_2 S^* \tag{7}$$

$$m_1 S^* = m_2 S^* \tag{8}$$

As shown in the above equations, the droop control approach is not able to correctly share both reactive and active power at the same time, when numerical errors, noises, and disturbances exist.

To avoid the above problems, virtual impedance robust droop control with ANN has emerged.

3 Robust Droop Controller

Using Eq. (5), it can be rewritten as

$$\Delta E_i = E_i - E^* = -n_i p_i \tag{9}$$

And the voltage Ei can be equipped by assimilating ΔEi, i.e.

$$E_i = \int_0^t \Delta E_i dt \tag{10}$$

As $\Delta Ei = 0$ (so the covert power is shipped off the grid without mistake), as proposed in [15, 17, 19] for the grid-connected approach. But, it does not work for the isolated mode, Ei can't be zero and on the grounds that the real power Pi is controlled by the load. This is the reason various controllers must be utilized for the standalone way and the grid-associated way, individually. At the point when the activity way changes, the regulator additionally should be changed. Additionally, when the load increments, the load voltage V0 drops, and this is additionally dropped because of the droop control. The modesty of the coefficient ni implies the more modesty of the voltage drop. Nonetheless, the coefficient ni requires a drop $E^* - Vo$ and should be feedback with a particular goal in mind to get a quick reaction. To ensure that the voltage stays inside a specific required reach, the load voltage is to the fundamental standards of the control hypothesis. It tends to be added to ΔEi through an intensifier Ke.

The chances of estimated frequency to stay inside a specific required reach in the.

parallel associated inverters generally are very unmanageable and will direct in contributing to an error value.

4 Proposed Control Algorithm

An approach was carried out to change the virtual droop control using an artificial neural network for frequency control as a primary controller, and in a secondary controller, as shown in Fig. 2, the current controller is used to reduce the effect of the impedance imbalance and the response time has been improved. To use a current controller, the inverter current is compared to the grid current, to produce an error signal. This error signal is used in the PR controller to produce a switching sequence.

The artificial neural network algorithm based on droop control was used to reduce obstacles in conventional droop control. Electricity is extracted from the social network. Artificial Neural Networks are a collection of neurons, simple and immensely interconnected processing units. They work together in a friendly way to solve problems. They are of the nonlinear type, with outstanding properties such as fault tolerance and self-learning ability, which make them powerful and perfectly suited for smooth handling. Artificial neural networks can resolve the connection between information and output factors even when there are no clear or effectively calculable correlations if enough information is available for training. Artificial

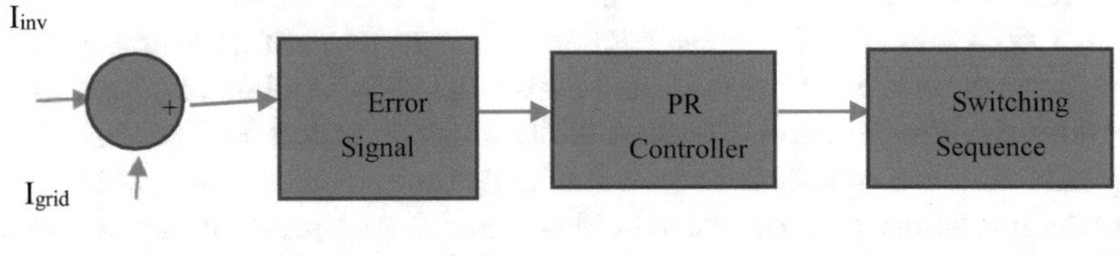

Fig. 2 Block diagram of current controller

neural network strategies are moderately easy to implement and do not require prior information on the framework model. They have amazing example recognition skills and are adaptable to the model, making them attractive for dealing with real-world problems. When applying artificial neural methods to a particular topic, there are informational, modelling, and preparation questions that must be considered. The artificial neural network controller must be trained using the robust droop control Eqs. (1) and (2) to achieve optimal performance. In a closed-loop control environment, the artificial neural system has the nature of a repeating network, as the output of the control-loop equations serves as feedback to the system at the next time step. To train the repeating network of Eqs. (5) and (6), we need to calculate the desired output frequencies and voltages of two inverters using the decay coefficient for different real and reactive powers. When the inputs are received, the calculations begin and the calculated results are sent back to the input of the artificial neural network in the subsequent step. The Levenberg–Marquardt (LM) algorithm is used to train the network and the closest feedforward neural network (FFNN) is obtained. Thereafter, the ANN Δw weight setting can be collected for a training epoch, and the system weights are updated to

$$w_{\text{updated}} = w + \Delta w$$

When the process meets a benchmark, it can be put to a halt. The network weights are adjusted repeatedly to optimize the functions.

5 Results and Discussion

The structure shown in Fig. 3, and with the proposed method, has been verified in the 2015 version of MATLAB Simulink, consisting of two numbers of single-phase inverters. The primary loop regulator actually consists of virtual static impedance control with an artificial neural system, and the current regulator is used as a secondary regulator to achieve proper power distribution for the linear load having an impedance magnitude of $Z = 0.25$ and impedance angle of $\theta = 46.1°$. Table 1 shows various parameter values and Table 2 shows the load change plan for the microgrid system.

The essential findings are shown to indicate the performance of the proposed ANS-based robust droop control technique.

The related curves from the analysis with the proposed controller are shown in Figs. 4, 5, 6(b), and 7(b). The same parameters are used to test with the conventional robust droop controller and the relevant curves are shown in Figs. 6(a) and 7(a).

Besides attaining equitable power distribution, the proposed ANS-based robust droop management approach keeps the islanded microgrid voltage and frequency within the limitations as can be seen in Figs. 4 and 5. As a result, it can be concluded that the suggested method effectively controls the voltage and frequency of the islanded microgrid and that it operates well in all scenarios.

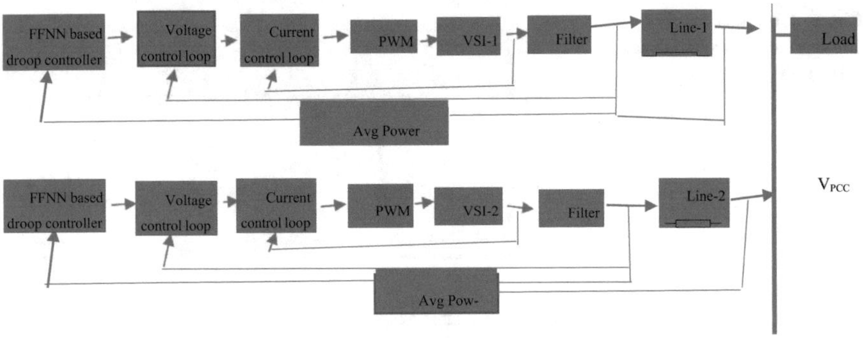

Fig. 3 Microgrid network control structure

Table 1 Parameter list

Description	Value for inverter-1	Value for inverter-2
DC bus voltage	363 V	367 V
Nominal bus frequency	50 Hz	50 Hz
Output volt-age (rms)	219.5 V	221.0 V
Filter Inductance	1.36e−3 H	1.29e−3 H
Filter Capacitance	11e−6 F	11e−6 F
Virtual Line impedance, DG1	0.19 + j535 μΩ	0.19 + j535 μΩ
Amplifier (Ke)	10	10
Frequency droop cooefficient (m)	3×10^{-5}	3×10^{-5}
Amplitude droop coefficient (n)	8×10^{-5}	8×10^{-5}

Table 2 Load change plan for the microgrid system

Time (in sec)	P demand (kw)	Q demand (kw)
0–0.15	9.5	2.4
0.15–0.35	15.5	3.1
0.35–0.55	11.5	2.4
0.55–0.75	14.5	3.1
0.75–0.95	8	2.4
0.95–1.15	4	1.5
1.15–1.35	13	1.9
1.35–1.5	9.5	1.7

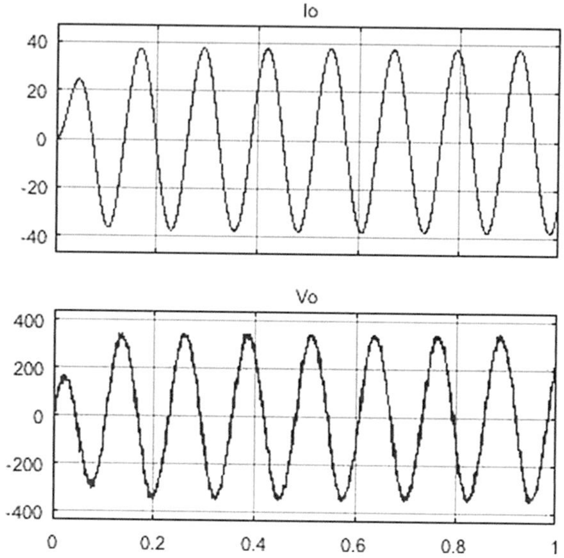

Fig. 4 Output current and voltage using the proposed controller

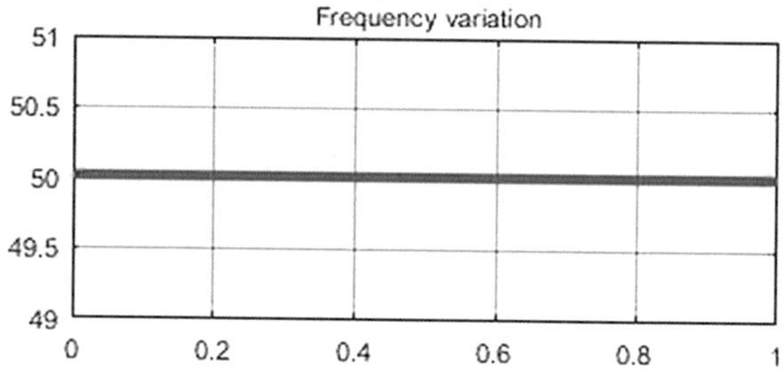

Fig. 5 Output frequencies of two inverters using the proposed method

Figure 6a, b shows the real power output of DG1 and DG2 utilizing conventional droop control and ANS-based droop control, respectively. Although the sharing accuracy of real power and reactive power of robust droop control using ANN are the same as the conventional droop control, the performance of the proposed method can be seen clearly and is better in the proposed method to convey information.

The results are compared with the existent literature [23] and [21]; to authorize the overall performance of the proposed ANN-based robust droop control technique, the results are compared. The line and load data for this persistence are taken from the existing literature and are shown in Tables 1 and 2. The line impedance is of a complex type.

The achievement index is the steady state current sharing error between the two DGs and that indicates the accuracy of the energy allocation of the proposed method, and the relevant graphs are shown in Figs. 8 and 9.

When both the DGs are given the load demand, it can be acclaimed that the steady state current sharing error is around 0.12A. The results are shown in Table 3. As the

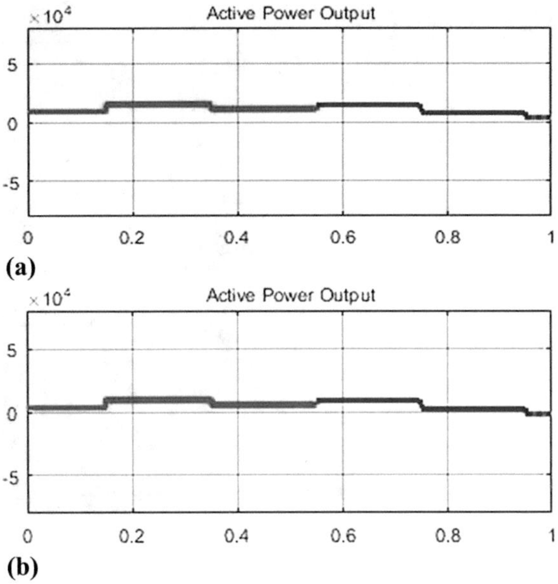

Fig. 6 **a** Real power output using conventional droop control. **b** Real power output using the proposed controller

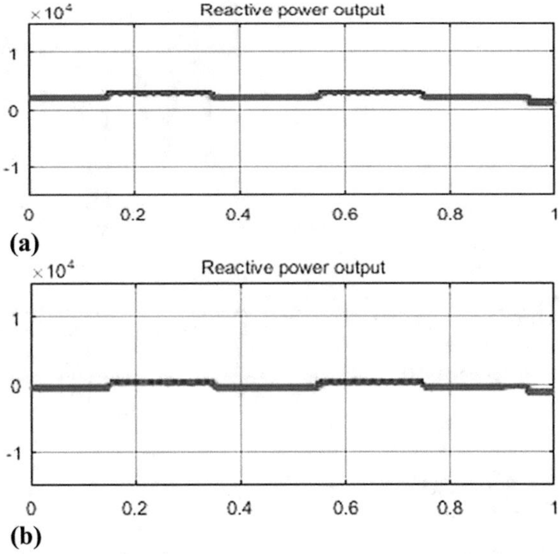

Fig. 7 **a** Reactive power output using conventional droop control. **b** Reactive power output using the proposed controller

line impedance is of complex nature, the better power sharing to minimize current sharing error between the DGs is got using the proposed ANN-based robust droop control method compared to the prevailing methods.

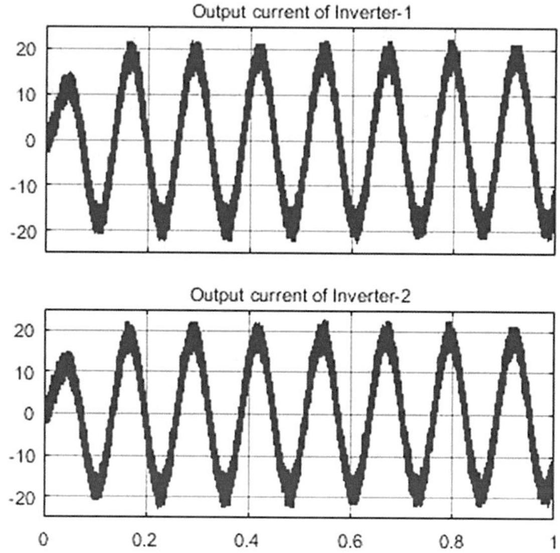

Fig. 8 Output current of DG1 and DG2

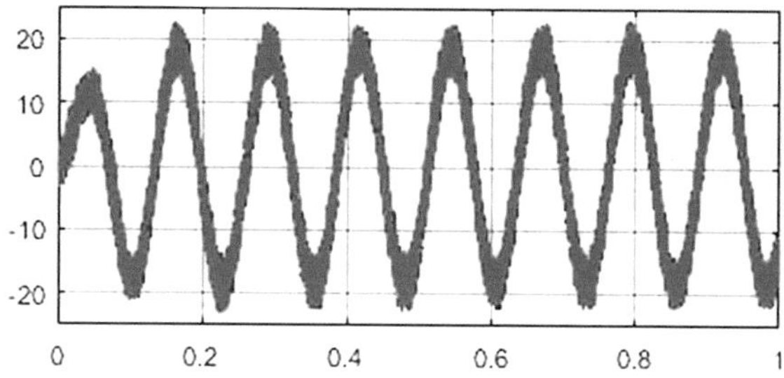

Fig. 9 Current Sharing by two inverters

Table 3 Comparison of steady state current sharing error between DG1 and DG2

Method proposed in [23]	Method proposed in [21]	Proposed ANN-based method
0.8	0.3	0.12

6 Conclusion

The islanded microgrid uses virtual impedance robust droop control with the artificial neural system as the primary controller, and the current controller is used as a secondary, tand he management and operation are examined and simulated. The ANS is trained using data sets collected from two single-phase DGs. To overcome the disadvantage of the power-sharing issue, frequency, and stabilization problem in the generalized droop control technique, this method is recommended as the essential control. To check and approve the work of the proposed method, it is tested on two DG systems with a linear load. From the frequency and power-sharing graph, it very well may be achieved that the proposed strategy in primary and secondary control defeats the disadvantage of generalized robust droop control, stabilizes the frequencies within the range, and appropriately divides the actual amount of reactive and active power between the DG units taking into account the output impedance of inverters and complex line impedance.

References

1. Guerrero JM, Berbel N, Matas J, De Vicuña LG, and Miret J (2006) Decentralized control for parallel operation of distributed generation inverters in microgrids using resistive output impedance. *IECON Proc. (Industrial Electron. Conf.*, pp. 5149–5154, doi: https://doi.org/10.1109/IECON.2006.347859
2. Barklund E, Pogaku N, Prodanovic M, Hernandez-Aramburo C, Green TC (2008) Energy management in autonomous microgrid using stability-constrained droop control of inverters. IEEE Trans Power Electron 23(5):2346–2352. https://doi.org/10.1109/TPEL.2008.2001910
3. Guerrero JM, Vásquez JC, Matas J, Castilla M, García de Vicuna L (2009) Control strategy for flexible microgrid based on parallel line-interactive UPS systems. IEEE Trans Ind Electron 56(3):726–736. https://doi.org/10.1109/TIE.2008.2009274
4. Gajbhiye S, Khatri N (2020) New approach for modeling of robust droop control for equal load sharing between parallel operated inverters. 7(4), 11–27, 2020
5. Iyer SV, Belur MN, Chandorkar MC (2010) A generalized computational method to determine stability of a multi-inverter microgrid. IEEE Trans Power Electron 25(9):2420–2432. https://doi.org/10.1109/TPEL.2010.2048720
6. Majumder R, Chaudhuri B, Ghosh A, Majumder R, Ledwich G, Zare F (May2010) Improvement of stability and load sharing in an autonomous microgrid using supplementary droop control loop. IEEE Trans Power Syst 25(2):796–808. https://doi.org/10.1109/TPWRS.2009.2032049
7. De Brabandere K, Bolsens B, Van den Keybus J, Woyte A, Driesen J, Belmans R (Jul.2007) A voltage and frequency droop control method for parallel inverters. IEEE Trans Power Electron 22(4):1107–1115. https://doi.org/10.1109/TPEL.2007.900456
8. Chen CL, Wang Y, Lai JS (2009) Design of parallel inverters for smooth mode transfer microgrid applications. *Conf. Proc. – IEEE Appl. Power Electron. Conf. Expo. - APEC* pp. 1288–1294. doi: https://doi.org/10.1109/APEC.2009.4802830
9. Díaz G, González-Morán C, Gómez-Aleixandre J, Diez A (Feb.2010) Scheduling of droop coefficients for frequency and voltage regulation in isolated microgrids. IEEE Trans Power Syst 25(1):489–496. https://doi.org/10.1109/TPWRS.2009.2030425
10. Sadabadi MS, Shafiee Q, Karimi A (May2017) Plug-and-play voltage stabilization in inverter-interfaced microgrids via a robust control strategy. IEEE Trans Control Syst Technol 25(3):781–791. https://doi.org/10.1109/TCST.2016.2583378
11. Miveh MR, Rahmat MF, Ghadimi AA, Mustafa MW (Feb.2016) Control techniques for three-phase four-leg voltage source inverters in autonomous microgrids: a review. Renew Sustain Energy Rev 54:1592–1610. https://doi.org/10.1016/J.RSER.2015.10.079
12. Vijayakumari A, Devarajan AT, Devarajan N (Jun.2015) Decoupled control of grid connected inverter with dynamic online grid impedance measurements for micro grid applications. Int J Electr Power Energy Syst 68:1–14. https://doi.org/10.1016/J.IJEPES.2014.12.015
13. Roslan MA, Ahmad MS, Isa MAM, Rahman NHA (2017) Circulating current in parallel connected inverter system. PECON 2016 - 2016 IEEE 6th Int. Conf. Power Energy, Conf. Proceeding, pp. 172–177. doi: https://doi.org/10.1109/PECON.2016.7951554

14. Monica P, Kowsalya M, Tejaswi PC (2017) Load sharing control of parallel operated single phase inverters. Energy Proc 117:600–606. https://doi.org/10.1016/j.egypro.2017.05.156
15. Sreekumar, Khadkikar V (2015) Nonlinear load sharing in low voltage microgrid using negative virtual harmonic impedance. BT - IECON 2015 - 41st Annual Conference of the IEEE Industrial Electronics Society, Yokohama, Japan, November 9–12, 3353–3358. doi: https://doi.org/10.1109/IECON.2015.7392617
16. Zhong QC, Weiss G (Apr.2011) Synchronverters: Inverters that mimic synchronous generators. IEEE Trans Ind Electron 58(4):1259–1267. https://doi.org/10.1109/TIE.2010.2048839
17. Zhong QC (2013) Robust droop controller for accurate proportional load sharing among inverters operated in parallel. IEEE Trans Ind Electron 60(4):1281–1290. https://doi.org/10.1109/TIE.2011.2146221
18. Guerrero JM, García de Vicuña L, Matas J, Castilla M, Miret J (Aug.2005) Output impedance design of parallel-connected UPS inverters with wireless load-sharing control. IEEE Trans Ind Electron 52(4):1126–1135. https://doi.org/10.1109/TIE.2005.851634
19. Guerrero JM, De Vicuña LG, Miret J, Matas J, Cruz J (2004) Output impedance performance for parallel operation of UPS inverters using wireless and average current- sharing controllers. *PESC Rec. - IEEE Annu. Power Electron. Spec. Conf.*, vol. 4, pp. 2482–2488, 2004, doi: https://doi.org/10.1109/PESC.2004.1355219
20. Guerrero JM, Hang L, Uceda J (2008) Control of distributed uninterruptible power supply systems. IEEE Trans Ind Electron 55(8):2845–2859. https://doi.org/10.1109/TIE.2008.924173
21. Vigneysh T, Kumarappan N (Sep.2016) Artificial Neural Network Based Droop-Control Technique for Accurate Power Sharing in an Islanded Microgrid. Int. J. Comput. Intell. Syst. 9(5):827–838. https://doi.org/10.1080/18756891.2016.1237183
22. Patel UN, Gondalia D, Patel HH (2015) Modified droop control scheme for load sharing amongst inverters in a micro grid. Adv. Energy Res. 3(2):81–95. https://doi.org/10.12989/eri.2015.3.2.081
23. Yao W, Chen M, Matas J, Guerrero JM, Qian ZM (Feb.2011) Design and analysis of the droop control method for parallel inverters considering the impact of the complex impedance on the power sharing. IEEE Trans Ind Electron 58(2):576–588. https://doi.org/10.1109/TIE.2010.2046001

AI-Driven Prediction and Data Analytics for Neurological Disorders—A Case Study of Multiple Sclerosis

Natasha Vergis, Sanskriti Shrivastava, L. N. B. Srinivas, and Kayalvizhi Jayavel

1 Introduction

Multiple sclerosis (MS) is a long-term autoimmune illness in which the immune system incorrectly attacks the central nervous system, which includes the brain and spinal cord. The neurological system is frequently destroyed in MS, including the myelin coating, nerve fibers, and even the cells that generate myelin. If the afflictions are not serious, they disappear in a short period of time, but if they are, they can lead to lasting alterations in the spinal cord [1]. Sclerosis is the name given to these irreversible alterations, and multiple sclerosis is the name given to the condition since the lesions appear in various and diverse places in the body. The immune system of the body reacts unnaturally as a result of this condition, producing inflammation and damage to bodily components. Fatigue, trouble walking, stiffness, vision issues, depression, weakness, dizziness, emotional changes, cognitive changes, and other symptoms are common in MS patients. MS hasn't been linked to any recognized causes [2].

N. Vergis (✉) · S. Shrivastava · L. N. B. Srinivas · K. Jayavel
Department of Networking and Communications, School of Computing, College of Engineering and Technology, SRM Institute of Science and Technology, SRM Nagar, Kattankulathur, Chennai, Tamilnadu 603203, India
e-mail: nv3668@srmist.edu.in

S. Shrivastava
e-mail: rk6850@srmist.edu.in

L. N. B. Srinivas
e-mail: srinival@srmist.edu.in

K. Jayavel
e-mail: kayalvij@srmist.edu.in

P. Singh et al. (eds.), *Machine Learning and Computational Intelligence Techniques for Data Engineering*, Lecture Notes in Electrical Engineering 998,
https://doi.org/10.1007/978-981-99-0047-3_38

MS is divided into four types: clinically isolated syndrome (CIS), relapsing–remitting MS (RRMS), primary progressive MS (PPMS), and secondary progressive MS [1, 2].

Multiple sclerosis symptoms are variable and unpredictable. Early MS identification is critical because it allows patients and their doctors to seek treatment and make plans for the future [1]. With no known cause for MS and an estimated 2.8 million cases globally, researchers and medical experts are focusing on effective therapies as well as techniques that can assist in anticipating the severity and particular symptoms of patients. It is one of the most common neurological conditions and is most identified in young people between the ages of 20 and 45, with females having a higher prevalence. As a result, it has a major societal impact in terms of prejudice and stigma. Despite its unpredictable and diverse history, which affects people differently, this illness is initially marked by relapses (periods of neurological problems that are reversible) [1, 2, 11].

Doctors have a time-consuming and difficult task when diagnosing MS using MRI neuroimaging techniques. As a result, scientists are suggesting innovative approaches for precisely detecting these features. Deep Learning algorithms are one of the most recent disciplines of AI that have acquired huge traction for diagnosing a range of disorders using medical records. The capacity of Deep Learning networks in deducing and generating intrinsic and imperceptible portrayals of features from MRI records is one of its most notable capabilities. Another benefit of Deep Learning is its auto-feature detection stage, allowing for the incorporation of feature detection and classifying stages in Computer-Aided Diagnosis Systems utilizing Deep Learning. Since 2016, there has been a study on the topic of MS detection utilizing Deep Learning and MRI scans. The use of Deep learning for segmentation and classification applications was the focus of this field's research [2, 3].

2 Algorithm

2.1 Computer-Aided Diagnosis System

CAD systems are algorithms that are either fully automated or semi-automated and are designed to aid clinicians in the interpretation of medical pictures. The dataset, preprocessing, feature extraction, feature selection, classification, and model evaluation stages are all included [2].

Computer-aided diagnosis system (CADS) is becoming a next-generation tool for the diagnosis of human disease. Recently, Deep Learning (DL) techniques have been introduced to the medical image analysis domain with promising results on various applications, like the computerized prognosis for Alzheimer's disease, mild cognitive impairment, organ segmentations and detection, etc. In the context of CADS, most works focused on the problem of abnormality detection which is perfect for detecting Multiple Sclerosis lesions in the patients' MRI brain scans [2, 3].

Many applications of CADS in the diagnosis of MS using MRI scans with conventional ML and DL techniques exist currently. Deep learning may extract features directly from training data, reducing the need for explicit elaboration on feature extraction. The features may be able to compensate for, and potentially outperform, the discriminative capability of traditional feature extraction approaches. Deep learning techniques can shift the design paradigm of the CADS framework, resulting in a number of advantages over traditional frameworks. The two steps—feature extraction and selection—are the main difference between the two techniques. A downfall of CADS with ML in these two steps is that it uses trial and error methods and needs expertise in image processing and other AI techniques. Whereas, in CADS with DL the two steps are executed automatically in the neural network layers. The execution of CADS is not reduced by the augmentation of the input data with DL techniques. The three phases of feature extraction, selection, and supervised classification can all be accomplished within the same deep architecture's optimization. The performance may be tweaked more readily and in a methodical manner with such a design. Therefore, using CADS with DL techniques yields better results in the detection of MS using MRI [2].

The block diagram as shown in Fig. 1 represents how our CADS system looks overall. In the first module, Data Input, the MRI scans of patients with and without Multiple Sclerosis Lesions are fed into the system. To clean this raw data, the second module includes various preprocessing techniques applied to make the model more accurate and efficient. This preprocessed data is now divided into two datasets, Training Set and Testing Set, being our third module, to validate the model and predict its unidentified outcomes. Then we multiply the training data, in the Augmented Training Set module, for normalizing the data into a more machine-readable form. Going into our next module, this data is fed into a Convolution Neural Network model consisting of convolution, pooling, and fully connected layers along with more specific layers depending on the input data. Finally, we change the hyperparameters of the model based on the accuracy achieved in the Output and Performance Computation modules [2].

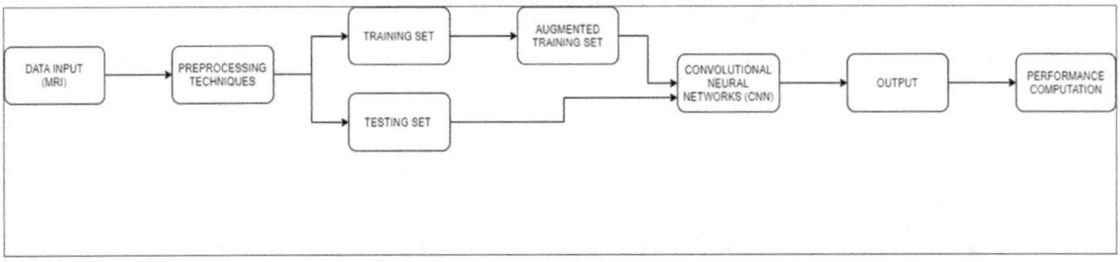

Fig. 1 CADS block diagram

2.2 Convolutional Neural Network

Convolution layers, pooling layers, and fully connected layers are the three principal layers in Convolutional Neural Networks [2, 3, 5]. Figure 2 depicts a model that includes the following elements: two convolutional layers, two pooling layers, and two fully linked layers. Owing to the fact that CNN has auto-feature selection, its execution is considerably superior to traditional AI methods. Higher-level characteristics are recovered when the models go deeper into the network's layers. Even though Convolutional Neural Networks have demonstrated adequate execution, this improvement has come at the cost of greater computational complexities, necessitating the use of more substantial and heavy processing hardware like Graphical Processing Unit (GPU) and Tensor Processing Unit (TPU) [2]. Convolutional layer, pooling layer, batch normalization, and fully linked layers are the primary components of a CNN architecture. One can study the temporal and spatial relations of an object by looking at the specific layers on CNN [3, 5].

Convolutional layers. Convolutional layers are the layers in a CNN where filters are applied to the input images or the resulting feature maps. In the convolutional layer, the essential parameters are the number of kernels and their sizes. Filters are used to extract features and the activation function is applied to each value in the feature map [3, 5, 8, 9]. The convolution layer feature extraction is displayed in Fig. 3 below.

Pooling Layer. Pooling layers, like convolutional layers, are employed in the lowering of the network's dimensionality [2, 3, 5, 8]. As shown in Fig. 4, this layer performs a certain method. The two most common methods are (i) max pooling, where in the filter zone only the maximum unit is taken and (ii) average pooling, where in the filter zone only the average unit is taken. Stride and window size are the parameters in this case. The pooling layer is mostly used for dimension reduction, but it also aids in achieving translation invariance [5]. Average pooling (AP) and maximum pooling (MP) are two popular pooling algorithms (MP).

AP—to determine the elements' average value in each pooling zone.

MP—to choose the pooling region's maximum value.

Fully Connected Layers. In a CNN model, fully connected layers, as seen in Fig. 5, are put prior to the output from classification. Also, they are utilized to

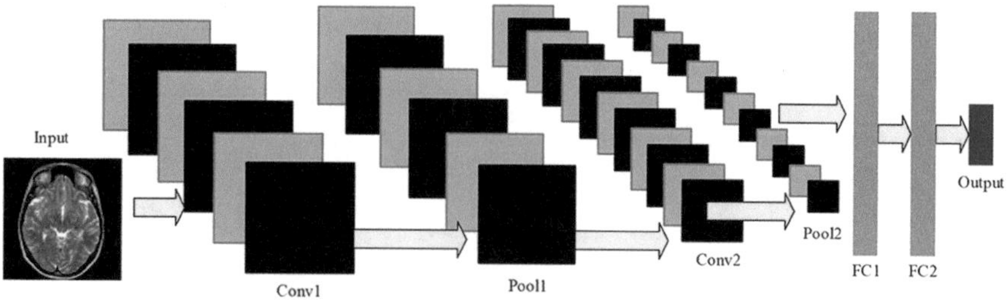

Fig. 2 CNN diagram

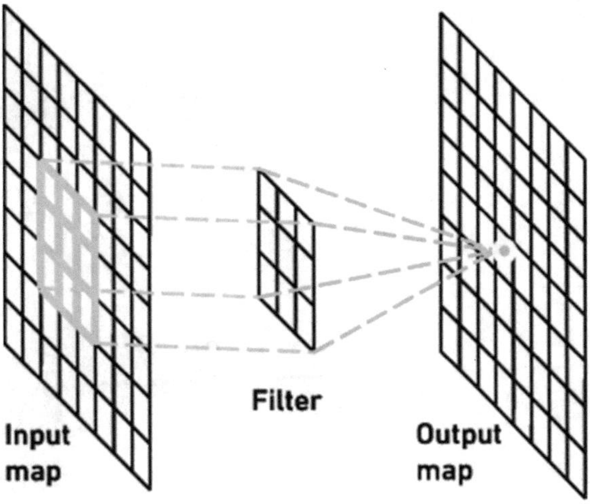

Fig. 3 Convolution layer diagram

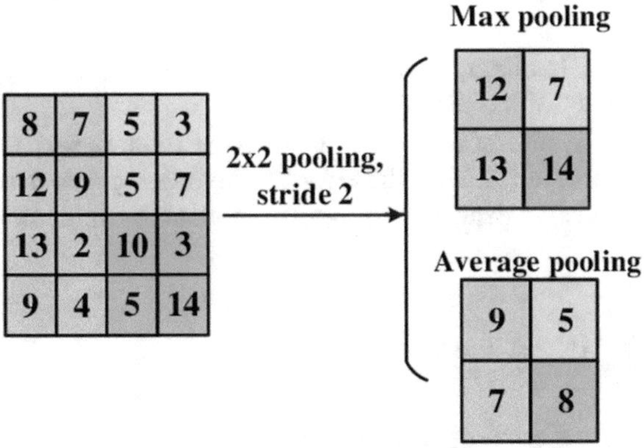

Fig. 4 Pooling method

combine information from the final feature maps. They flatten the results before classification. The parameters are the number of nodes and activation function [2, 8, 10].

Batch Normalization. By adding extra layers in a deep NN, batch normalization makes the network quicker and more stable [4, 8, 10]. The layers will execute standardization and normalization operations on the input from the previous layer. The parameter being updated in the previous layer causes the change in each layer's input distribution and leads to the model being trained slowly. This is called internal covariate shift. We solve this by using batch normalization. The input layer will now have a uniform distribution that was generated by normalizing the layers' input over a mini batch [8].

Dropout. Dropout is a method in which neurons are deleted at random. They "disappear" at random. This has the effect of making the layer look like and be treated like a layer with a different number of nodes and connectivity to the prior

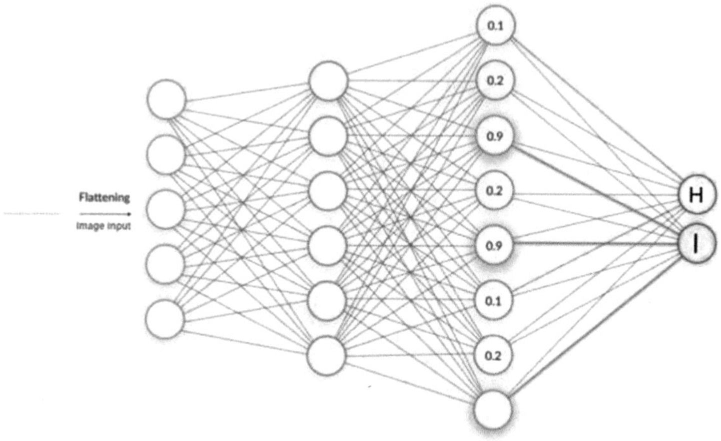

Fig. 5 Fully connected layers

layer. In effect, each update to a layer during training is performed with a different "view" of the configured layer [4, 8].

Overfitting is an issue that occurs often during neural network model training. Overfitting is when the error for the training data is minimal; however, the error on the testing data is large. We use dropout to overcome this problem [8, 10].

3 Preprocessing

3.1 Data augmentation

The practice of synthesizing new data from existing data is known as data augmentation. This could be used with any type of data, including numbers and images. Random (but realistic) transformations, such as picture rotation, scaling, and noise injection, are used to increase the diversity of the training set as shown in Fig. 6(1, 2, 3), respectively. Collecting biological data for producing additional data from insufficient data is a well-known difficulty. The reduction of overfitting and improvement of classification task accuracy have been proved to be achieved by data augmentation [6, 8, 9].

To each image, these preprocessing steps are being applied:

1. The brain, the most important image segment, is cropped.
2. As dataset images are usually of different dimensions and sizes, to make them uniform is essential to feed them into the neural network. The images are resized in the form of (image_width, image_height, and number of channels).
3. Normalization is applied for sizing pixel units in a 0–1 range.

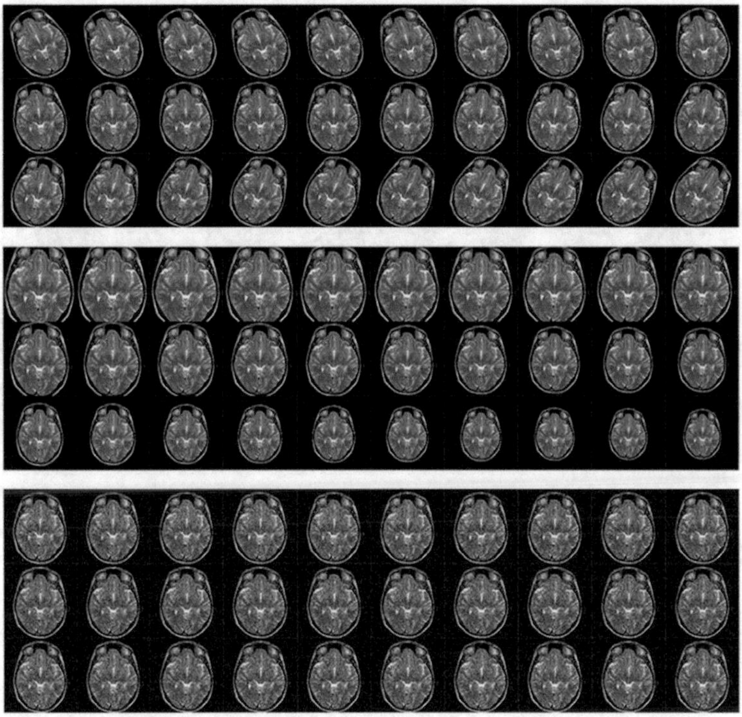

Fig. 6 1. Rotation. 2. Scaling. 3. Noise Injection

4 Dataset

A training set is used to create a model from a dataset, whereas testing (or validation) sets are utilized for model testing. Here, the testing data is employed for validating the model, while the training data is utilized to fit it. The test set is for predicting the unidentified outcomes using the models created [9]. We divide datasets into two sets, (i) training set and (ii) testing sets, to examine accuracies and precisions. The proportion to be shared isn't fixed and varies depending on the project. The 70/30 rule where the former is for the training set and the latter for the testing set is not mandatory. This is fully determined by the dataset and the types of projects. In our project, we are following the 20/80 rule—20% testing and 80% training of the dataset. The dataset as seen in Fig. 7 used by us is from an MRI Lesion Segmentation in Multiple Sclerosis Database from eHealthLab. There are 38 patients. For each patient, there are two folders—MRI scans from the 0th month and scans after 6–12 months. This leads to an average of 2000 MRI scans taken for the dataset. The age group varies from 15 to 51 [7].

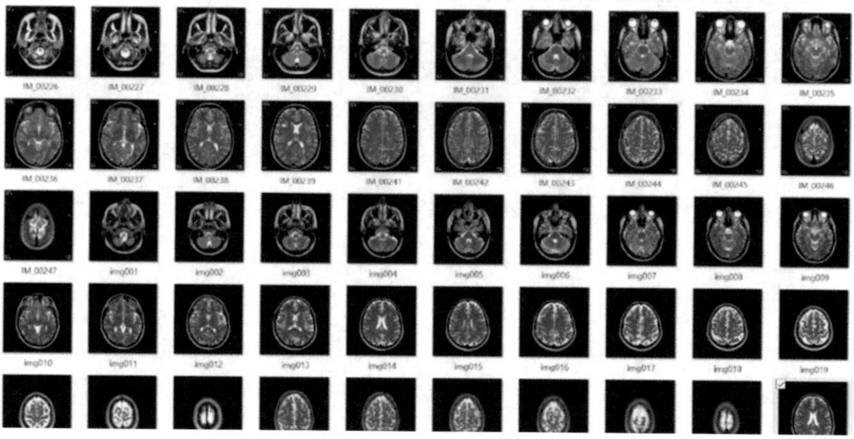

Fig. 7 Dataset

5 CNN Model

It's a seven-layer Convolutional Neural Network Model with Batch Normalization and Dropout for improved precision. The network's final layers are divided into two groups: Multiple Sclerosis diagnosed patients and non-diagnosed patients. The CNN model, as shown in Fig. 8, includes seven levels, which has five convolution layers, an input layer, batch normalized layers, and dense layers. A softmax activation function was utilized as the classification layer in the last layer to classify the images. MS patients and non-MS patients are separated by the features retrieved from the CNN model. Images have been utilized to detect and categorize Multiple Sclerosis diagnosed patients and non-MS patients to the CNN model after they have been preprocessed. The neural network was able to accurately recognize and categorize the photos.

5.1 Architecture Explanation

Every input MRI scan image of shape of (840,840, 3). These are input into the CNN model which then passes via the layers below.

1. Zero Padding layer—(2, 2) pool size.
2. Convolutional layers +32 layers, (7, 7) filter size +1—stride size.
3. Batch Normalization layers for normalization of pixel values to overcome the internal covariate shift problem and accelerate the processing.
4. ReLU activation layers.
5. Max Pooling layers, setting filter size as 4 and stride as 4.
6. A Flatten layer to flatten it to a 1D vector from the 3D matrix.

Fig. 8 CNN layers model

7. Dropout layer to eliminate the overfitting issue.
8. A fully connected Dense layer having 1 neuron and using Softmax activation function.

6 Results and Discussion

There are two types of AI methods for Multiple Sclerosis identification: traditional machine learning and deep learning. DL algorithms have taken a leading role in MS diagnosis for the past few years. DL techniques, unlike traditional machine learning algorithms, are extremely effective in diagnosing MS. Deep neural network layers have been used for feature detection in CADS based on DL techniques. This improved the accuracy of the Computer-Aided Diagnosis System in the diagnosing of multiple sclerosis. DL methods being used have raised hopes for reliable MS diagnosis utilizing MRI modalities [2].

Convolutional neural networks are the most well-known of the numerous deep learning models (CNN). CNN models have the benefit of auto-feature extraction. Higher-level characteristics are recovered as the network goes deeper in these models [2]. Solely, convolutional neural network methods are used for many sorts of classification and segmentation approaches using DL methodologies [2, 5].

The unavailability of existing MRI databases pertaining to additional participants and other methods, as well as datasets unavailability with functional neuroimaging methods that are used in the research of detection of Multiple Sclerosis, are two important obstacles to automated MS diagnosis [6]. The preprocessing techniques and data augmentation were used to overcome these challenges [6, 8, 9]. A method for increasing the variety of the training set is by using randomized changes such as picture rotation, scaling, and noise injection. Collecting biological data like MRI in order to create more data from small datasets is a well-known difficulty [6].

We solve internal covariate shifts by using batch normalization. The input layer will now have a uniform distribution that was generated by normalizing the layers' input over a mini batch [8]. When the error of only the training data is modest, but the error on the testing data turns out to be big, it is known as overfitting. Hence, we use dropout methods to try and overcome the trouble of overfitting [8, 10].

After compiling the model and training it with 24 epochs, on the 23rd epoch, we achieved the model with the best validation accuracy as shown in Fig. 9(1, 2). On this model with the best validation accuracy, the test set had an 89% accuracy and an 0.88 f1 score as shown in Fig. 9(3).

7 Conclusion

Multiple Sclerosis is a central nervous system chronic condition wherein your own immune system attacks the myelin protecting the nerve fibers. Early MS identification is critical because it can stop the illness from worsening and save lives. The CNN-based MRI detection technique proposed in this study adds to a quicker and more efficient MS identification model. Deep Learning methods have taken a prominent role in MS diagnosis in recent years.

Deep Learning methods, unlike traditional machine learning algorithms, are extremely successful in diagnosing MS. Deep neural network layers have been used for feature detection in CADS with DL techniques. This improved the accuracy of the Computer-Aided Diagnosis System in the diagnosing of multiple sclerosis. Deep Learning methods being used have given rise to the confidence for reliable MS diagnosis utilizing MRI modalities.

The capability of CNN has piqued the curiosity of radiology experts, and various studies in topics like image reconstruction, natural language processing, segmentation, classification, and lesion identification have already been published. CNN models have the benefit of auto-feature extraction. The unavailability of existing MRI databases pertaining to additional participants and other methods, as well as datasets unavailability with functional neuroimaging methods that are used in the research of detection of Multiple Sclerosis, are two important obstacles to automated MS diagnosis. To get the best classification results in automated MS diagnosis, large datasets are essential. As the datasets include a limited number of subjects, advanced Deep Learning models cannot be used to examine them.

This paper comprises of Computer-Aided Diagnosis System (CADS) that includes the techniques used to process data and the methods in the CNN layers like Dropout, Batch Normalization, and the like that make the neural network quicker and more stable, which also overcame overfitting of the dataset. This led to the implementation of our project with an accuracy of 89%.

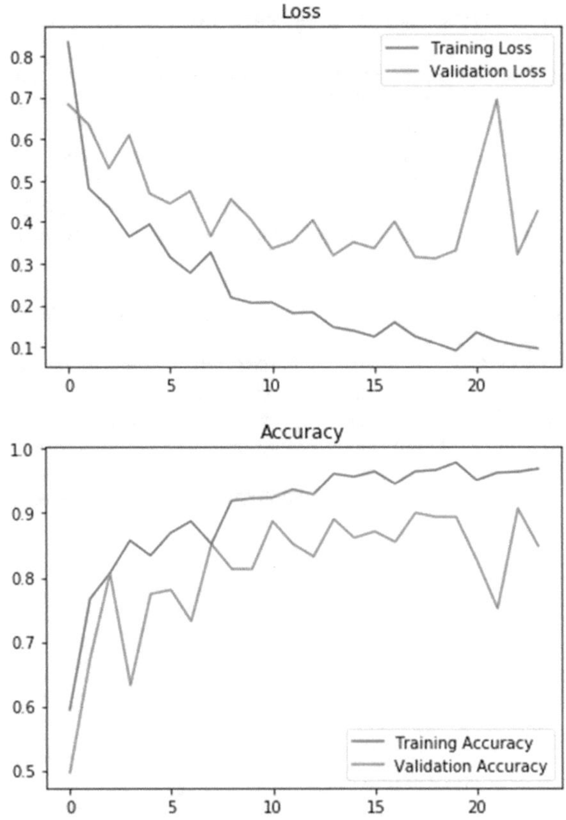

	Validation set	Test set
Accuracy	91%	89%
F1 score	0.91	0.88

Fig. 9. 1. Loss Graph. 2. Accuracy Graph. 3. Accuracy and f1 score table

References

1. Goldenberg MM (2012) Multiple sclerosis review. P & T: a peer-reviewed journal for formulary management 37(3):175–184
2. Shoeibi A, Khodatars M, Jafari M, Moridian P, Rezaei M, Alizadehsani R, Khozeimeh F, Gorriz JM, Heras J, Panahiazar M, Nahavandi S, Acharya UR (2021Sep) Applications of deep learning techniques for automated multiple sclerosis detection using magnetic resonance imaging: a review. Comput Biol Med 136:104697. https://doi.org/10.1016/j.compbiomed.2021.104697. Epub 2021 Jul 31 PMID: 34358994
3. Siar M, Teshnehlab M (2019) Diagnosing and classification tumors and MS simultaneous of magnetic resonance images using convolution neural network *. 1–4. https://doi.org/10.1109/CFIS.2019.8692148

4. Dai Y, Zhuang P (2019) Compressed sensing MRI via a multi-scale dilated residual convolution network
5. Maleki M, Teshnehlab PM, Nabavi M (2013) Diagnosis of multiple sclerosis (MS) using convolutional neural network (CNN) from MRIs
6. Loizou CP, Murray V, Pattichis MS, Seimenis I, Pantziaris M, Pattichis CS (2011) Multi-scale amplitude modulation-frequency modulation (AM-FM) texture analysis of multiple sclerosis in brain MRI images. IEEE Trans Inform Tech Biomed 15(1):119–129
7. MRI Lesion Segmentation in Multiple Sclerosis Database, in eHealth laboratory (2018) University of Cyprus. Available online at: http://www.medinfo.cs.ucy.ac.cy/index.php/facilities/32-software/218-datasets
8. Baldeon Calisto MG, Lai-Yuen SK (2020) Self-adaptive 2D–3D ensemble of fully convolutional networks for medical image segmentation. In: Proc. SPIE 11313, Medical Imaging 2020: Image Processing, 113131W
9. Pruenza C, Solano M, Díaz J, Arroyo R, Izquierdo G (2019) Model for prediction of progression in multiple sclerosis. Int J Interact Multimed Artif Intell
10. Valverde S, Salem M, Cabezas M, Pareto D, Vilanova JC, Ramió-Torrentà L, Rovira A, Salvi J, Oliver A, Llado X (2018) One-shot domain adaptation in multiple sclerosis lesion segmentation using convolutional neural networks
11. The Multiple Sclerosis International Federation, Atlas of MS, 3rd Edition (September 2020)

Rice Leaf Disease Identification Using Transfer Learning

Prince Rajak, Yogesh Rathore, Rekh Ram Janghel, and Saroj Kumar Pandey

1 Introduction

Agriculture plays an essential part in the Indian economy. It contributes to the second largest share of rice output [1]. Rice is harvested in nearly all of India's states, including West Bengal, Uttar Pradesh, Chhattisgarh, Punjab, Odisha, Bihar, Andhra Pradesh, Tamil Nadu, Assam and Karnataka, with West Bengal ranking first with total rice production of 146.05 lakh tons [2]. The agricultural industry contributes roughly 20.1% of the overall gross domestic output [3]. In India, rice is one of the most widely consumed grains. Diseases affect the development and quality of rice plants, lowering the profitability of the cultivation. Different illnesses can affect specific rice crops, making it difficult for farmers to identify them due to their limited expertise gathered through experience. An autonomous data processing specialised system is required for this accuracy and early detection of plant diseases detection. As a result, it is possible to cultivate a healthy and prosperous crop.

Deep learning (DL) is a powerful method that has been applied to agriculture to solve a variety of problems, including weed and seed detection, plant disease identification, plant recognition, fruit counting, root segmentation and so on. DL is a development of machine learning (ML) techniques that effectively trains a huge

P. Rajak (✉) · Y. Rathore · R. R. Janghel · S. K. Pandey
Department of Information Technology, National Institute of Technology Raipur, Raipur 492010, India
e-mail: rajakprince123@gmail.com

Y. Rathore
e-mail: yogeshrathore23@gmail.com

R. R. Janghel
e-mail: rrjanghel.it@nitrr.ac.in

S. K. Pandey
e-mail: saroj.pandey@gla.ac.in

P. Singh et al. (eds.), *Machine Learning and Computational Intelligence Techniques for Data Engineering*, Lecture Notes in Electrical Engineering 998,
https://doi.org/10.1007/978-981-99-0047-3_39

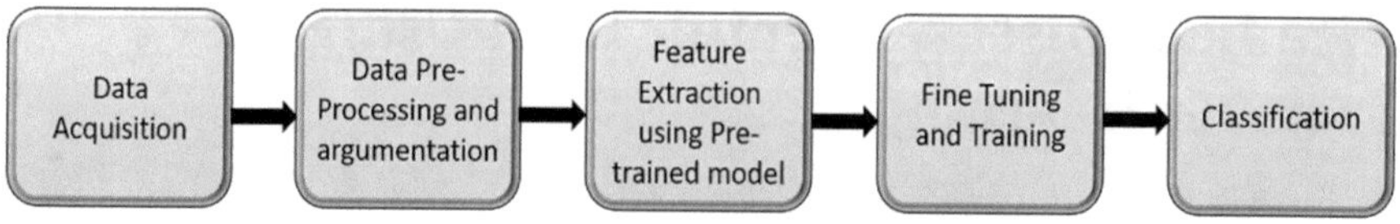

Fig. 1 Identification of disease using the transfer learning technique

quantity of data and automatically understands the characteristics of the input before producing an output based on decision regulation. The Convolution Neural Network (CNN) is a powerful tool for analysing visual data. It's a three-layer feed-forward ANN (Artificial Neural Network) with input layer, dense layer and output layer. The dense layer is made up of a convolutional layer, a pooling layer, a normalisation layer and a fully connected layer. It also has a set of automatic weights that it uses to learn the spatial connection of the input data and complete a classification job (Fig. 1).

Transfer learning (TF) [4]is a technique for repurposing a pre-trained CNN to solve a new issue. As a result, the training duration of the model may be lowered when compared to a model created from scratch, and the suggested model's performance can be improved, decreasing the computational power. By removing the last fully connected layers (bottleneck layer) [5] or fine-tuning the last few levels that will operate more specifically for the concerned dataset, TF may be used to produce a model that can be utilised as a fixed feature extractor. The various pre-trained models in our study integrate knowledge in the form of weights, which are transferred to our experiment for the feature extraction process utilising the TL technique. The DL model has been fine-tuned to increase the accuracy of identifying different types of rice leaf diseases.

The rest of this paper is organised as follows: Sect. 2 describes the literature review of related work; Sect. 3 describes the dataset used and DL method and TL applied to it after processing the data; Sect. 4 describes the brief discussion of the experimental results followed by concluding the paper in Sect. 5.

2 Literature Review

In this paper, authors [6] proposed and used the pre-trained InceptionResNetV2 model with TL for the identification of rice disease and have gained an accuracy of 95.67%. Authors in [7] used pre-trained Alexnet and GoogleNet for classifying test images from the datasets in which they included convolution, pooling, fully connected layers and have gained an accuracy of 98.67% for Alexnet and 96.25% for GoogleNet. Authors in [8] used two-stage simple CNN that has been constructed by VGG16. In the first stage, the entire dataset of 9 classes is divided into 17 classes. After they trained those datasets for better accuracy and after performing, they gained an accuracy of about 93%. Authors [9] have applied the pre-trained VGG16 model and do fine-tunned the model and gained the accuracy of about 60%.

Authors in [7] used TL to design their DL model because they developed their own dataset, which is minimal in size. The suggested CNN architecture is based on VGG-16, and it has been trained and tested using data from rice fields and the internet and gained the accuracy of 92.46%. Authors [10] concentrated on the diseases of the rice plant and have gathered 619 damaged rice plant images from the field, which were divided into four categories: sheath blight, bacterial leaf blight, rice blast and healthy leaves. They also used a support vector machine, a classifier and a deep CNN pre-trained as a characteristics extractor and achieve an accuracy of 95%. In [11], images of disease leaves are used to discover diseases in rice using a CNN written in the R programming language. BS (Brown Spot), BLB (Bacterial Leaf Blight) and LS (Leaf Smut) are the three diseases represented in the disease photos collected from the UCI ML Repository and train the model, achieving the accuracy of 86.66%. In [12] used images of disease leaves, CNN and the R programming language were utilised to discover diseases in rice. BS, LS and BLB are among the illnesses represented in the disease photos collected from the UCI ML Repository, performed training and achieved the accuracy of 92.46%.

Authors [13] proposed a colour feature extraction method for extracting the features and an SVM classifier is used for the classification of four different types of leaf diseases and achieve an accuracy of 94.65%. Authors [14] proposed a method that uses deep CNN and SVM classifiers in order to classify nine different rice leaf disease types. They used TL to improve the performance and gained the accuracy of 97.5%. Authors [15] described how they classified photos of damaged and healthy rice leaves using transfer learning. MobileNet-v2 is best suited for mobile applications with memory and computational constraints and gained an accuracy of 62.5%.

3 Proposed Methodology

In this work, we utilised the rice leaf disease (RLD) dataset and several TL models from TensorFlow Hub and the library Keras application. We imported these models and froze the final layers, then fine-tuned the model to train it and obtain the results shown below. The architecture of this work is displayed here (Fig. 2).

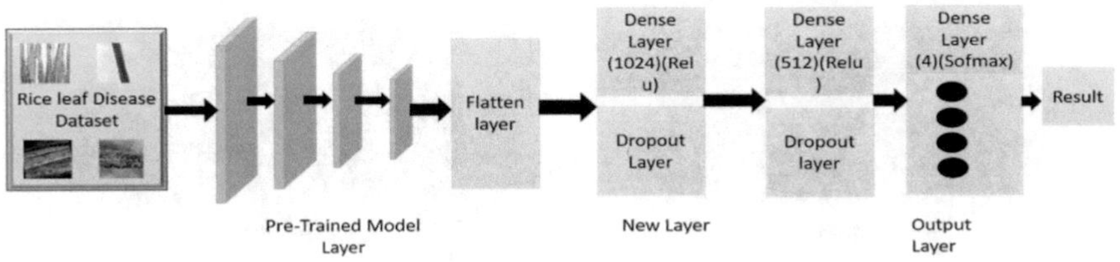

Fig. 2 Proposed work

3.1 Dataset Description

The primary obstacle to the identification of rice leaf diseases is the absence of appropriate datasets in the agricultural area. Increased access to rice leaf datasets might open up new avenues for study into rice leaf disease detection. Image datasets are required for ML/DL algorithms that need the identification of various rice leaf diseases. We present a quick overview of the rice leaf dataset in this part. The leaf dataset in this study includes four types of sick rice leaf images: brown spot, bacterial leaf blight, leaf-smut and healthy. The RLD Dataset, prepared by Caesar Arssetya [16] and uploaded to the personal drive, was utilised for the proposed task. This dataset comprises 16,092 images of separated rice leaves divided into four categories: healthy (4050), bacterial leaf blight (4050), brown spot (4050) and leaf smut (3942). From 200 by 300 pixels to 1080 by 1080 pixels, the images are available in a variety of sizes. Figure 3 shows sample photos from the dataset for each class (Table 1 and Fig. 3).

Bacterial leaf blight (BLB): It's a bacterial infection caused by the bacterium Xanthomonas oryzae. The diseased leaves turn a greyish-green tint, roll up and yellow before turning straw coloured and dying after withering. Wavy margins characterise the lesions, which advance towards the base. Bacterial slime-like morning dew drop might be seen on early lesions.

Table 1 Describe the RLD dataset used

Class	Training	Validation
Bacterial leaf blight	3510	540
Brown spot	3510	540
Healthy	3510	540
Leaf Smut	3510	432

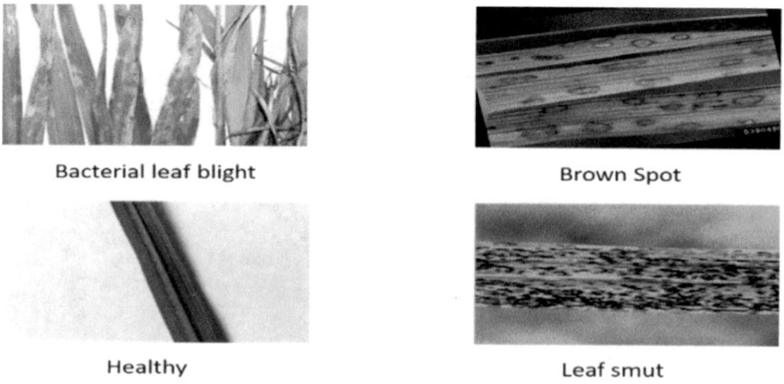

Fig. 3 Types of Rice Disease

Table 2 Some publicly available datasets of rice disease

Source	Total images	Classes	Link
Caesar Arssetya	16,092	4	[16]
Kaggle	120	3	[16]
Mendeley data	5932	4	[17]
Kaggle	5447	4	[18]
Kaggle	850	3	[19]

Brown spot (BS): It's a fungal infection. Infected leaves have a number of large spots on them that can harm the entire leaf. Small, round and dark brown to purple-brown lesions can be seen in the leaves in the early stages. The fully grown lesions are round to oval in shape, with a light brown to grey core and a reddish-brown edge generated by the fungi's toxin.

Leaf Smut (LS): LS, produced by the fungus Entyloma oryzae (EO), is a widespread but mild rice disease caused by the fungus EO. On both sides of the leaves, the fungus creates slightly elevated, angular black dots (sori). The fungus is dispersed by spores in the air and survives the winter on sick leaf litter in the soil. Leaf smut is a late season ailment that does minimal damage. The disease thrives in environments with high nitrogen levels (Table 2).

3.2 *Deep Learning Technique*

ML and DL are extremely powerful technologies as the field has evolved due to their ability to handle large amounts of data. Hidden layers have been found to help DL recognise patterns. A couple of the most often utilised deep learning techniques are listed below and describe some of the pre-trained models.

Convolutional Neural Network:

CNNs are made up of many layers of artificial neurons. Artificial neurons are mathematical functions that, like their biological counterparts, assess the weighted sum of numerous inputs and output an activation value. When you input a picture into a ConvNet, each layer develops several activation functions that are passed on to the next layer. The first layer usually extracts basic properties such as horizontal or diagonal edges. This data is sent on to the next layer, which is in charge of recognising more complex characteristics such as corners and combinational edges. As we get further into the network, it becomes capable of recognising increasingly complex items such as objects, faces and so on. The classification layer generates a series of confidence ratings based on the activation map of the final convolution layer, which indicates how likely the picture is to belong to which class (Fig. 4).

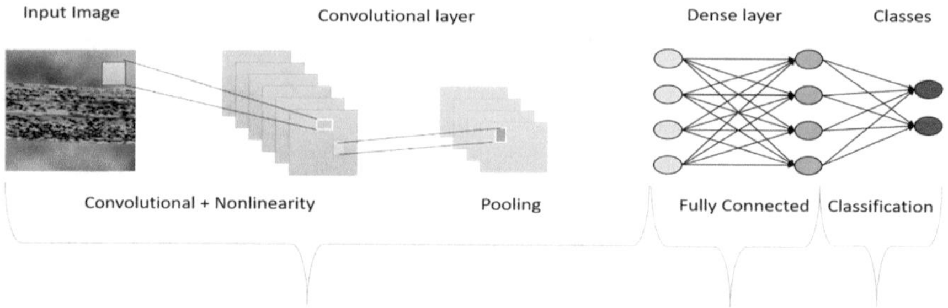

Fig. 4 Architecture of CNN

Transfer learning and Fine Tuning:

TL is a branch of ML that focuses on storing and transferring knowledge learnt while solving one issue to a similar but distinct problem. The theory behind TL for image classification is that if a model is trained on a large and general enough dataset, it may be used as a generic model of the visual world. The learned feature maps may then be used to train a complex model on a large dataset without having to start from scratch (Fig. 5).

Unfreeze some of the top layers of a frozen model base and train both the new classifier layers and the base model's final layers at the same time. This allows us to 'fine-tune' the underlying model's higher order feature representations to make them more relevant to the job at hand. Fine-tuning is the term for this procedure. Various pre-trained TL models are available, some of which are described here (Table 3).

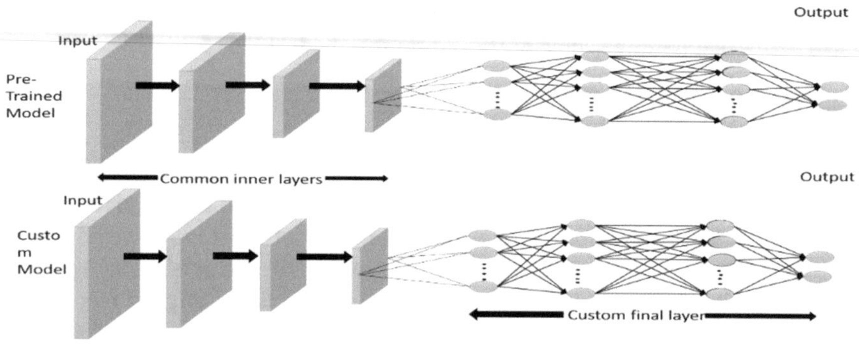

Fig. 5 Workflow of transfer learning

Table 3 shows different pre-trained model description

Model	Layers	Training dataset	Classes	Image size
Iception_V3	48	ImageNet	1000	299 by 299
MobileNet_V2	53	ImageNet	1000	224 by 224
EfficientNetB3	387	ImageNet	1000	300 by 300
DenseNet169	23	ImageNet	1000	244 by 244
ResNet50	50	ImageNet	1000	224 by 224
VGG-16	16	ImageNet	1000	244 by 244

4 Experimental Results

We present results for all situations in our models in this section. The model's performance with the RLD dataset is shown here, along with the model's name, epoch, loss, validation, training accuracy and validation, accuracy. There are eight models in total in the table, and their visualisation is detailed below with plots of training accuracy and validation accuracy, as well as plots of training loss and validation loss (Tables 4 and 5).

Visualisation: Plots shows the training and validation accuracy and loss of different model we have worked with (Figs. 6, 7, 8, 9, 10, 11, 12 and 13).

5 Conclusion

The automatic rice leaf disease identification helps farmers and industry to grow their profit. This work can be used for different disease identification like tomato, apple, etc. In this, we used the different pre-trained models using TL and frozen the final layer and performed fine-tuning. We trained different pre-trained models with our dataset and got validation accuracy with InceptionResnetV2 about 91%, Movilenet_v2 about 90%, InceptionV3 about 90%, Xception about 91% and, with DenseNet169, we achieved an accuracy of 94%. So, DenseNet169 performs much better with the proposed work, and the training accuracy of 98 and validation accuracy of 94 are obtained using DenseNet169. In future work, we are going to increase the dataset and classes to build a better model.

Table 4 Analysis of different pre-trained models implemented on the RLD dataset

Model	Epoch	Loss	Validation loss	Training accuracy	Validation accuracy
Inception_resnet_v2	10	0.3142	0.5629	0.9137	0.8945
	20	0.2206	0.4095	0.9315	0.8828
	30	0.1698	0.4114	0.9399	0.9141
Mobilenet_v2	10	0.1974	0.4276	0.9399	0.9043
	20	0.1557	0.5062	0.9518	0.9043
	30	0.1364	0.5211	0.9670	0.9082
DenseNet169	10	0.1024	0.2944	0.9679	0.9492
	20	**0.0526**	**0.3642**	**0.9805**	**0.9473**
	30	0.0825	0.4057	0.9772	0.9414
VGG16	10	0.3361	0.3692	0.8731	0.8711
	20	0.2387	0.4067	0.9095	0.8828
	30	0.2124	0.3998	0.9255	0.8750
Inception_v3	10	0.3171	0.5258	0.9010	0.8965
	20	0.2268	0.4327	0.9281	0.9043
	30	0.1942	0.5024	0.9349	0.8828
VGG19	10	0.4691	0.7155	0.8223	0.7500
	20	0.3488	0.4904	0.8672	0.8184
	30	0.3484	0.4516	0.8706	0.8633
Nasnet	10	0.2449	0.4133	0.9213	0.9023
	20	0.1384	0.3755	0.9560	0.8945
	30	0.1711	0.4310	0.9442	0.8867
Xception	10	0.2794	0.5886	0.9357	0.8965
	20	0.2132	0.6270	0.9560	0.9141
	30	0.1732	0.7296	0.9602	0.9102

Table 5 Comparison of previous work and proposed work

Author	Method	Dataset	Accuracy
[7]	TL+VGG16	Self	92.46
[12]	CNN+R language	UCI ML repository	92.46
[13]	SVM	Self	94.63
Proposed work	TL+DenseNet169	Hybrid	94.73 (20 epoch)

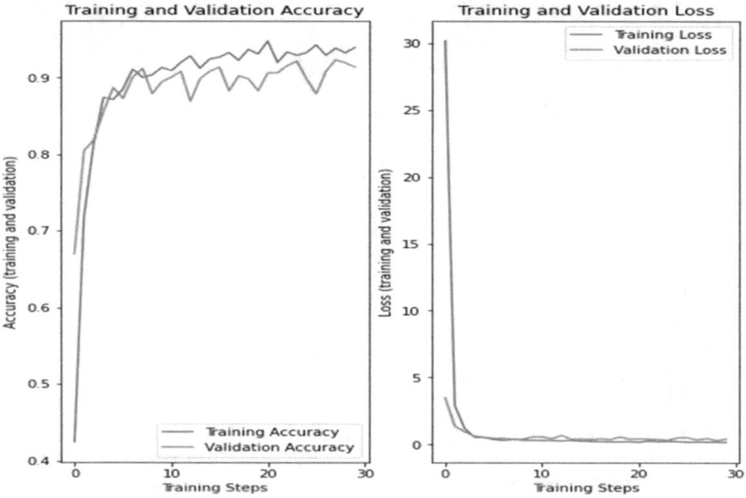

Fig. 6 Inception_Resnet_V2

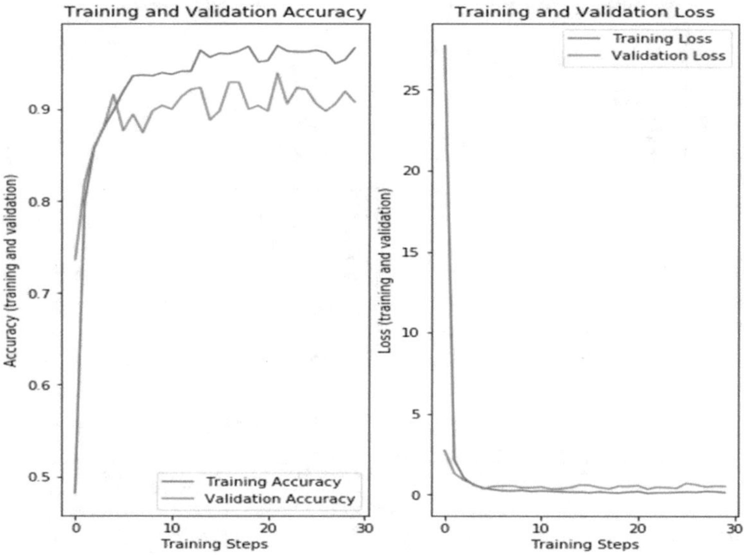

Fig. 7 MobileNet_v2

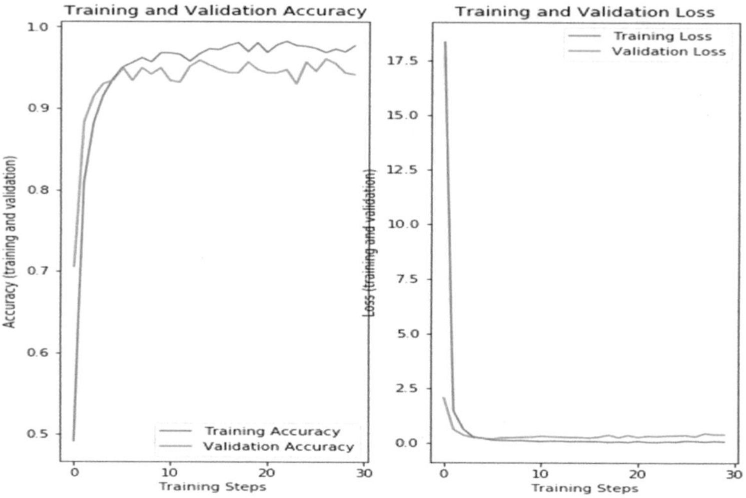

Fig. 8 DenseNet169This is the best result which we have gained from our work with densenet169

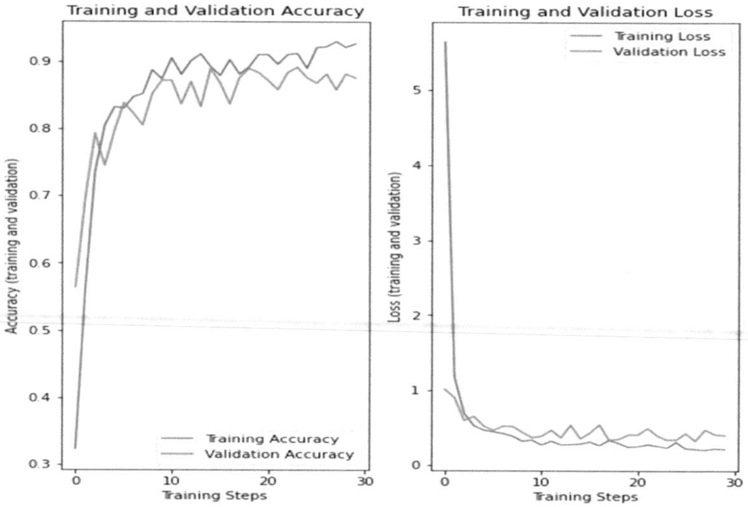

Fig. 9 VGG16

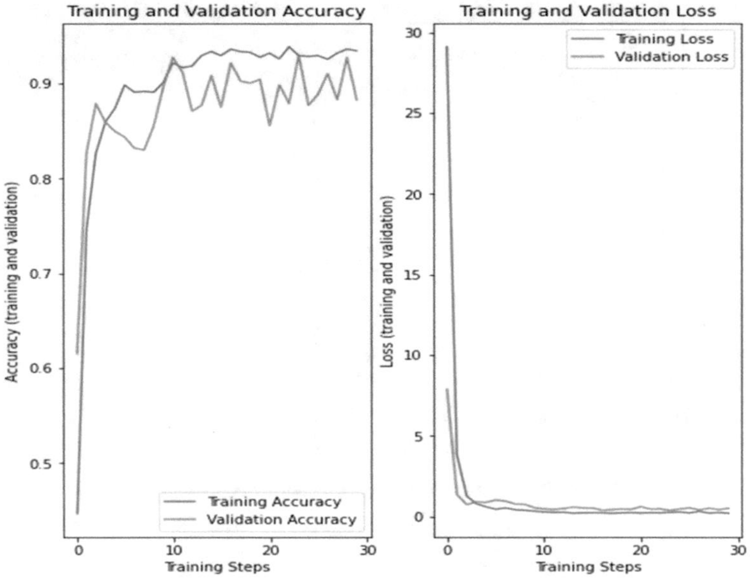

Fig. 10 Inception_V3

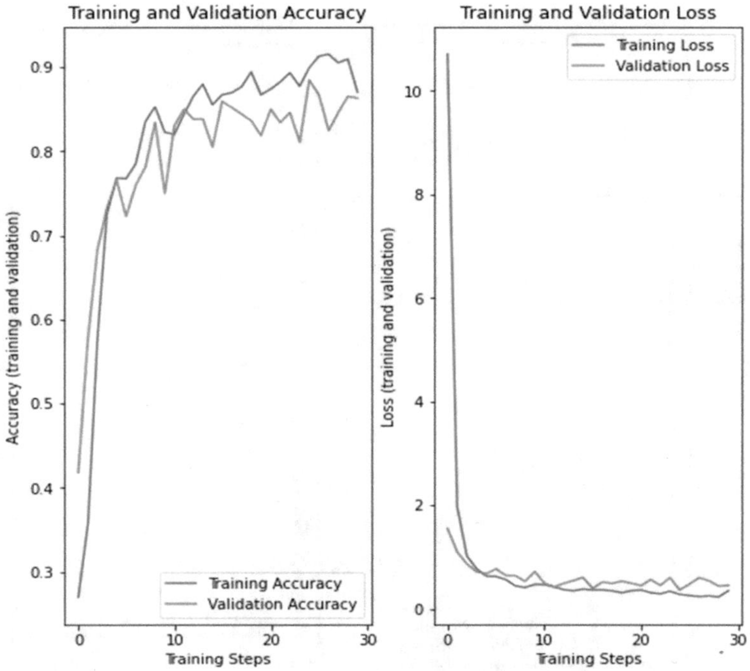

Fig. 11 VGG19

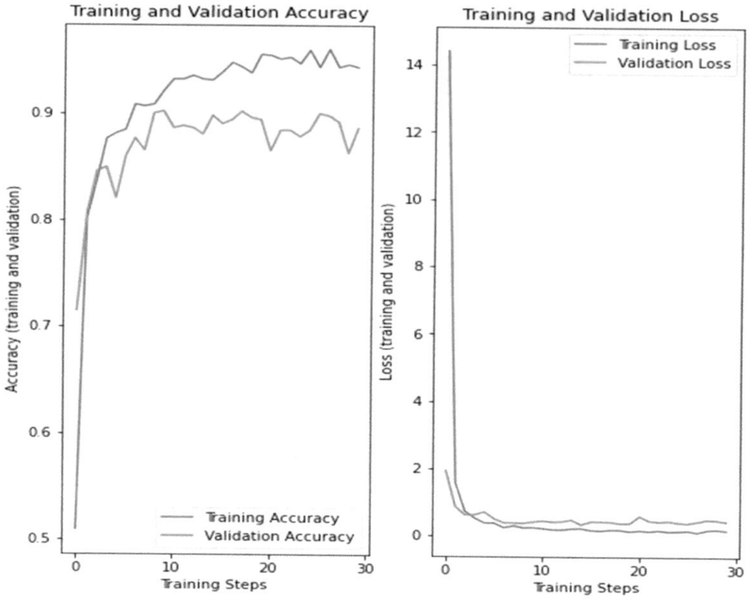

Fig. 12 Nasnet

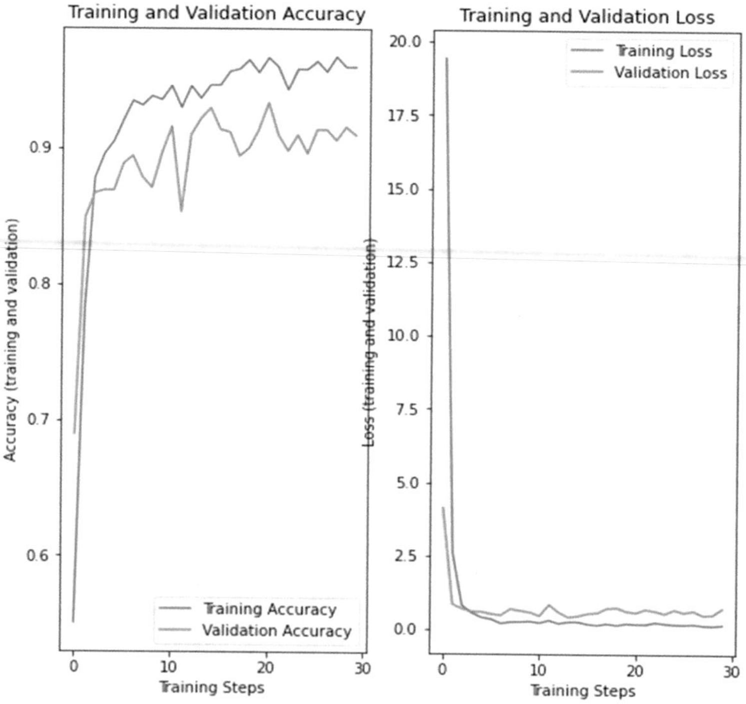

Fig. 13 Xception

References

1. "Top countries based on production of milled rice | Statista." https://www.statista.com/statistics/255945/top-countries-of-destination-for-us-rice-exports-2011/. Accessed 10 Jan 2022
2. "10 Largest Rice Producing States In India." https://www.fazlani.com/component/content/article/12-food-grains/36-10-largest-rice-producing-states-in-india. Accessed 10 Jan 2022
3. "Agriculture in India: industry overview, market size, role in development...| IBEF." https://www.ibef.org/industry/agriculture-india.aspx. Accessed 10 Jan 2022
4. Weiss K, Khoshgoftaar TM, Wang D (2016) A survey of transfer learning. J Big Data 3(1):1–40
5. "Transfer learning and fine-tuning | TensorFlow Core." https://www.tensorflow.org/tutorials/images/transfer_learning. Accessed 10 Jan 2022
6. Krishnamoorthy N, Prasad LN, Kumar CP, Subedi B, Abraha HB, Sathishkumar VE (2021) Rice leaf diseases prediction using deep neural networks with transfer learning. Environ Res 198:111275
7. Jadhav SB, Udupi VR, Patil SB (2020) Identification of plant diseases using convolutional neural networks. Int J Inf Technol 1–10
8. Rahman CR, Arko PS, Ali ME, Khan MAI, Apon SH, Nowrin F, Wasif A (2020) Identification and recognition of rice diseases and pests using convolutional neural networks. Biosys Eng 194:112–120
9. Andrianto H, Faizal A, Armandika F (2020) Smartphone application for deep learning-based rice plant disease detection. In: 2020 international conference on information technology systems and innovation (ICITSI), pp 387–392. IEEE.
10. Lu Y, Yi S, Zeng N, Liu Y, Zhang Y (2017) Identification of rice diseases using deep convolutional neural networks. Neurocomputing 267:378–384
11. Sony A (2019) Prediction of rice diseases using convolutional neural network (in Rstudio). Int J Innov Sci Res Technol 4(12):595–602
12. Ghosal S, Sarkar K (2020) Rice leaf diseases classification using CNN with transfer learning. In: 2020 IEEE Calcutta conference (CALCON), pp 230–236. IEEE
13. Shrivastava VK, Pradhan MK (2021) Rice plant disease classification using color features: a machine learning paradigm. J Plant Pathol 103(1):17–26
14. Hasan MJ, Mahbub S, Alom MS, Nasim MA (2019) Rice disease identification and classification by integrating support vector machine with deep convolutional neural network. In: 2019 1st international conference on advances in science, engineering and robotics technology (ICASERT), pp 1–6. IEEE
15. Shrivastava VK, Pradhan MK, Minz S, Thakur MP (2019) Rice plant disease classification using transfer learning of deep convolution neural network. In: International archives of the photogrammetry, remote sensing & spatial information sciences
16. "Rice Leaf Diseases Dataset | Kaggle.". https://www.kaggle.com/vbookshelf/rice-leaf-diseases. Accessed 10 Jan 2022
17. "Rice Leaf Disease Image Samples - Mendeley Data." https://data.mendeley.com/datasets/fwcj7stb8r/1. Accessed 10 Jan 2022
18. "Rice Diseases Image Dataset | Kaggle.". https://www.kaggle.com/minhhuy2810/rice-diseases-image-dataset . Accessed 10 Jan 2022
19. "Rice Disease Dataset | Kaggle.". https://www.kaggle.com/nischallal/rice-disease-dataset. Accessed 10 Jan 2022

Surface Electromyographic Hand Gesture Signal Classification Using a Set of Time-Domain Features

S. Krishnapriya, Jaya Prakash Sahoo, and Samit Ari

1 Introduction

Surface electromyography (sEMG) is the study of muscle activity based on the analysis of the electrical signals generated from the human skin surface using electrodes. A collection of signals are generated by all the muscle fibers of a single motor unit which is termed as motor unit action potential. The electromyographic signal is the aggregation of motor unit action potentials which are picked up by the sensor electrodes. Therefore, sEMG signals are the collection of information about the human hand gestures, movement of limb, and human intension [1]. EMG signals have a wide variety of applications [2–4] such as musculo-skeletal system, hand gesture recognition, interpretation of sign languages, prosthetics, biometric systems, human–machine interactions, etc.

Hand gesture recognition (HGR) is an important way to convey the information between deaf and dump people. It is also used as a human–machine interface, robot control and in many more applications due to the advantage of high flexibility and user-friendliness [5]. But the performance of a HGR system depends on the sensor used for data acquisition, the feature extraction technique and the classifier. Several sensors are used by the researchers to acquire the raw input data. These sensors are data glove, vision-based sensor, sEMG sensor, etc. The data glove is more accurate and robust but the user feels uncomfortable on wearing the glove. The vision-based sensors are very popular sensors due to its comfortability as there is no requirement to wear any device on the users hand. However, it's performance is affected by the complex environment and skin color noise [6]. In the past few years, the HGR using sEMG signals have gained popularity in the research society as they are physiological

S. Krishnapriya (✉) · J. P. Sahoo · S. Ari
Department of Electronics and Communication Engineering, National Institute of Technology, Rourkela, India
e-mail: krishnapriyaskumar@gmail.com

P. Singh et al. (eds.), *Machine Learning and Computational Intelligence Techniques for Data Engineering*, Lecture Notes in Electrical Engineering 998,
https://doi.org/10.1007/978-981-99-0047-3_40

signals closely related to human motion. The advantages of the sEMG systems are its low cost and portability.

In recent years, many researchers have proposed novel feature extraction techniques and have developed several classifiers to recognize the gesture class using the sEMG signals. Some researchers have proposed several time-domain features [7, 8] such as mean absolute value, waveform length, standard deviation, variance, etc., and some frequency-domain features [9] such as discrete wavelet transform, short-time Fourier transform, etc., are also proposed in the literature. From the several literature survey, we concluded that the performance of the recognition system mainly depends on the classification accuracy and the distinguishable features between the gesture classes in a dataset. Therefore in this work, a set of time-domain features (SoTF) is proposed for the recognition of gesture classes using the sEMG signal.

The contributions in this work are as follows:

- A set of time-domain features such as average, standard deviation, and waveform length (denoted as SoTF) are proposed to recognize the hand gesture using sEMG signals.
- Implementation of three classifiers such as kNN, SVM, and RF for the recognition process using the proposed SoTF feature.
- The performance of the recognition system is evaluated on publicly available 52 gesture classes of NinaPro DB1 dataset. In this study, exercise-wise the recognition performance of NinaPro DB1 dataset is also analyzed.

The rest of the paper is organized as follows. The recent works on sEMG-based hand gesture are described in the Sect. 2. The methodology of the proposed work is discussed in Sect. 3. The Ninapro DB1 dataset, validation techniques, and detailed experimental results and discussions are presented in Sect. 4. Section 5 concludes the paper and provides future scope of the work.

2 Related Works

In this section, a literature survey on recent existing techniques for the recognition of hand gestures using sEMG signals along with their limitations is described below. Several features such as mean absolute value (MAV), marginal discrete wavelet transform (mDWT), histogram (HIST), waveform length (WL), cepstral coefficients (CC), short-time fourier transform (STFT), and variance (VAR) were proposed by Atzori et al. [7] for the recognition of sEMG-based hand gesture signal. The combination of features were classified using four different classifiers, namely support vector machine (SVM), multi-layer perceptron (MLP), k-nearest neighbors (kNN), and linear discriminant analysis (LDA). Several of the feature-classifier combinations achieved a similar accuracy of around 76% for the NinaPro DB1 database. The authors found that advanced features like mDWT did not have any advantage over simpler features like MAV or WL. A feature combination of root mean square (RMS), HIST, mDWT, and time-domain statistics (TD) were put forward by Atzori et al. [9]. The features were analyzed individually as well as in combinations using four classifiers: kNN, SVM, LDA, and random forest (RF). The highest classification accuracy of 75.32% was obtained using all the feature combination and RF classifier on NinaPro DB1 dataset. Pizzolato et al. [10] applied the same set of features on two classifiers SVM and RF to compare six acquisition setups. The DB1 dataset in both the classifiers performed the best when trained with the combination of all the four features. The random forest classifier performed better than SVM giving an accuracy of 64.45%. Modified versions of the classifiers that are based on extreme learning machines (ELM) were introduced by Cene et al. [8]. A feature combination of RMS, Variance (VAR), MAV, and Standard Deviation (SD) were used. The reliable version of the standard ELM and regularized ELM, which are S-ELM and R-RELM produced accuracies of 73.13% and 75.03%, respectively, on NinaPro DB1 dataset. A model combining long short-term memory (LSTM) with multi-layer perceptron (MLP) to incorporate temporal dependencies along with the static characteristics of the sEMG signal was designed by He et al. [11]. The model on being evaluated on the NinaPro DB1 database produced an accuracy of 75.45%. Subsets of the NinaPro database consisting of 12 finger gestures and 8 isometric and isotonic hand gestures have been evaluated separately by Du et al. [12] and Saeed et al. [13]. Du [12] achieved an accuracy of 75% for the 12 finger gestures and 76% for the 8 hand gestures using random forest, while [13] achieved an accuracy of 85.41% for the 12 finger gestures and 76% for the 8 hand gestures with a feature combination of MAV, ZC, SSC, and WL using random forest classifier. Similar methods have also been evaluated on other sub-databases of NinaPro. Li et al. [14] has used a combination of MAV, RMS, and difference absolute standard deviation value (DASDV) giving an accuracy of around 68% for SVM and kNN classifiers on DB5 dataset. Several combinations of RMS, MAV, WL, slope sign change (SSC), integral absolute value (IAV), zero crossing (ZC), mean value of square root (MSR), maximum amplitude (MAX), and absolute value of the summation of square root (ASS) have been experimented by Zhou et al. [15] producing an average accuracy of 84% using random forest on DB4 dataset. In this work, SoTF are proposed to recognized the hand gesture signals.

3 Methodology

The major steps to be followed for the recognition of gesture classes using sEMG signal is shown in Fig. 1. These steps are data acquisition, feature extraction, and classification. In data acquisition, the ten channel sEMG signal is acquired using the MyoBock sensor. Then the time-domain features are extracted from each channel and finally, they are concatenated to represent a gesture class which is denoted as set of time-domain features (SoTF). The SoTF is classified using three different classifiers to find the best classification accuracy.

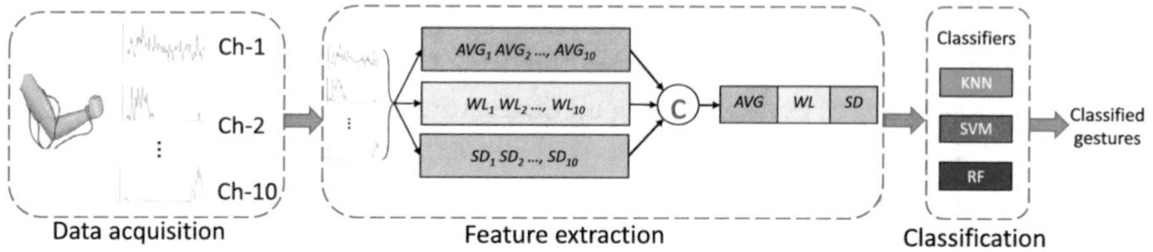

Fig. 1 Framework of the proposed sEMG-based hand gesture classification system

3.1 Data Acquisition and Pre-Processing

The sEMG signals have been taken from the NinaPro Database [7]. The gesture classes in the database are discussed in Sect. 4.2. Ten active double differential OttoBock MyoBock 13E200 sEMG electrodes are used to acquire sEMG signals. The electrode output is an amplified, bandpass-filtered and root mean square rectified version of the raw sEMG signal. The amplification factor is set to 14000 and the two filter cut off frequencies are 90–450 Hz. The electrodes acquire data at a frequency 100 Hz [7].

3.2 Proposed SoTF

Features are extracted from each of ten channels of the sEMG signal corresponding to a gesture. Time-domain features are used here since they are quick and easy to implement [13, 16]. The three time-domain features extracted from the sEMG signals are average (AVG_{1-10}), waveform length (WL_{1-10}), and standard deviation (SD_{1-10}).

Average It is the sum of all the signal amplitudes a_i in an interval with N points.

$$AVG = \frac{\sum_{i=1}^{N} a_i}{N} \tag{1}$$

Waveform Length It is the sum of the absolute differences between two adjacent samples in an interval with N points.

$$WL = \sum_{i=1}^{N-1} |a_{i+1} - a_i| \tag{2}$$

Standard Deviation It is the measure of variation of the signal values from the mean value $\overline{a}$ in an interval with N points.

$$SD = \sqrt{\frac{\sum_{i=1}^{N-1} |a_{i+1} - \overline{a}|}{N}} \tag{3}$$

The three features are concatenated to form a SoTF.

3.3 Classification

Three different classification models are used in this work. These are kNN, SVM, and RF.

K-Nearest Neighbor (kNN) kNN is a supervised machine learning algorithm that identifies the new data point as belonging to the majority class among its K-nearest neighbors [17, 18]. In this work, we have used the Euclidean distance to compute the nearest neighbors.

Support Vector Machine (SVM) SVM is a supervised machine learning algorithm that finds an optimal hyperplane that can separate the classes efficiently. Various kernels can be chosen based on the type of data to be classified. Since rbf kernel is suitable for classifying a nonlinear dataset in high-dimensional space, it is chosen as the kernel [19]. An rbf kernel-based SVM is used when the similarity between points in the transformed domain is gaussian and is given by the equation [4].

$$rbf(a, b) = exp\left(-\frac{(a-b)^2}{\gamma^2}\right) \tag{4}$$

where γ determines the training time and is related to the number of support vectors. a and b are two vectors in the input space. The regularization parameter C decides the extent of misclassification that is allowed. The more the value of C, the lesser the allowed misclassification.

Random Forest (RF) RF is a supervised machine learning algorithm that classifies the data based on the majority votes from N uncorrelated decision trees that together make up the larger random forest [18, 20]. It does not require the data to be normalized and is suitable for handling large data. It can perform well even with missing data and does not face over-fitting issues as it cancels out the bias on averaging all the predictions.

4 Experimental Evaluation and Results

4.1 Experimental Setup

The experiments were performed with an Intel(R) Core(TM) i5-9300H CPU @ 2.40 GHz on a 64-bit operating system with 8GB RAM. The programming was carried out in spyder (Python 3.8) integrated development environment (IDE).

4.2 NinaPro DB1 Dataset

The NinaPro DB1 dataset is a collection of 52 hand gestures which are repeated 10 times by 27 subjects. Five seconds are spent performing each repetition, followed by 3 seconds of rest. The gestures are classified into three exercises. The first exercise consists of 12 basic finger movements and 8 isometric and isotonic hand configurations. The second exercise comprises of 9 basic wrist movements, while the third exercise includes 23 functional and grasping movements as shown in Fig. 2 [7].

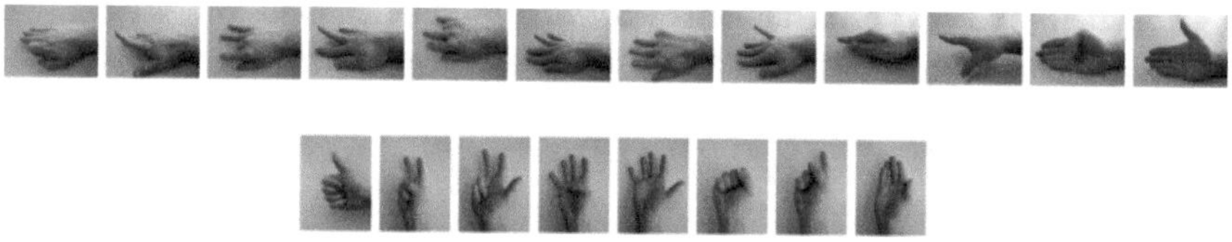

(a)

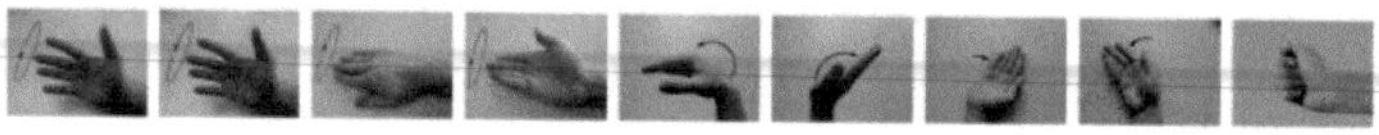

(b)

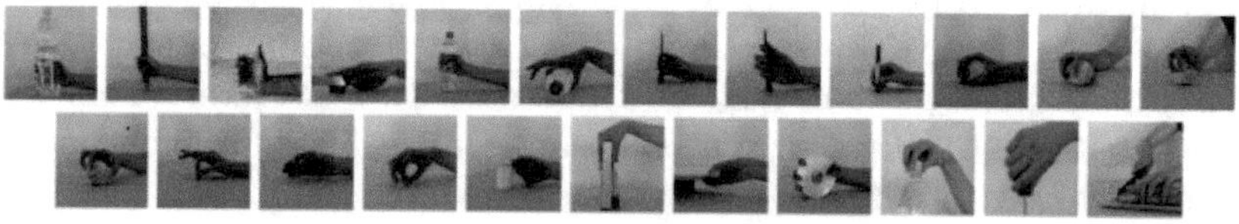

(c)

Fig. 2 52 gesture classes of the NinaPro DB1 dataset [7]. **a** 12 basic finger movements and 8 isometric and isotonic hand configurations; **b** 9 basic wrist movements; **c** 23 functional and grasping movements

4.3 Experimental Method

As an initial step, the data needs to be properly represented for easy access. Each subject's sEMG signals are available as separate folders in the database. Each exercise file is stored in a separate mat file within each subject folder. For simplicity, the entire data is represented gesture-wise in a single file. The data within the single file is arranged in such a way that the signals of all the repetitions of all subjects for the first gesture are followed by those for the second gesture. Now, each of the repetitions is filtered out for feature extraction. Figure 3 shows the signal extracted for the first gesture. The individual features extracted from the sEMG channels and SoTF are given to three different classifiers to compare the performances. The obtained features are of different scales which affects the modeling process and hence need to be standardized. On standardizing, the mean and the variance are converted to 0

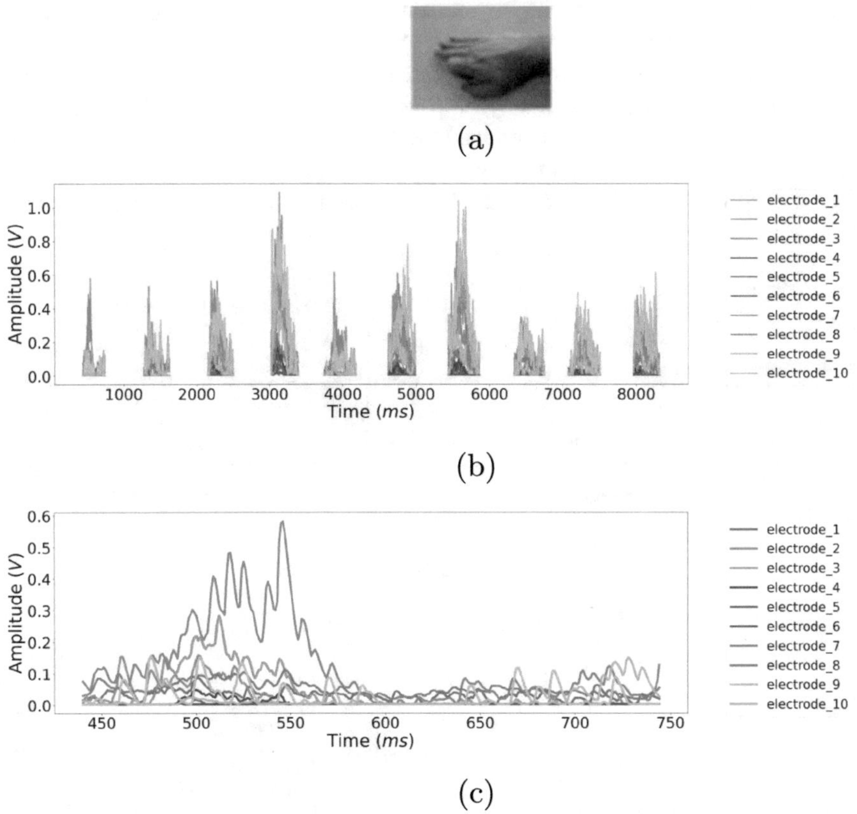

Fig. 3 **a** Gesture pose of the dataset "Gesture 1" [7], **b** Extracted signal for all the 10 repetitions of Subject 1, **c** Extracted signal for first repetition of Subject 1

and 1, respectively. Eighty percent of this data is randomly chosen for training the classifiers, while the rest 20% is used for testing [21].

4.4 Results and Discussions

In this work, a set of time domain features (SoTF) have been proposed that can categorize different hand gestures based on the sEMG signals from the forearm. To choose the best classifier for the SoFT, the classification accuracies are compared using kNN, rbf SVM, and random forest classifiers. Figure 4a offers the accuracies obtained for kNN classifier with a varying number of neighbors (K). On classifying the sEMG signals based on waveform length, standard deviation, and average features individually, they reach a maximum accuracy of 64%, 59%, and 69%, respectively, at $K = 3$. The SoTF gives a maximum accuracy of 82% at $K = 3$. The accuracies obtained for SVM classifier with a varying regularization parameter (C) are shown in Fig. 4b. Classification based on individual features, waveform length, standard deviation, and average features achieve maximum accuracies of 69%, 64%, and 77%, respectively, at $C = 100$. A maximum accuracy of 70% is achieved for the SoFT at $C = 10$. Figure 4c shows the accuracies obtained for random forest classifier with a varying number of trees (N). Maximum accuracy of 69.08%, 67.3%, 79.05%, and 86.07% are obtained on classifying the sEMG signals based on waveform length, standard

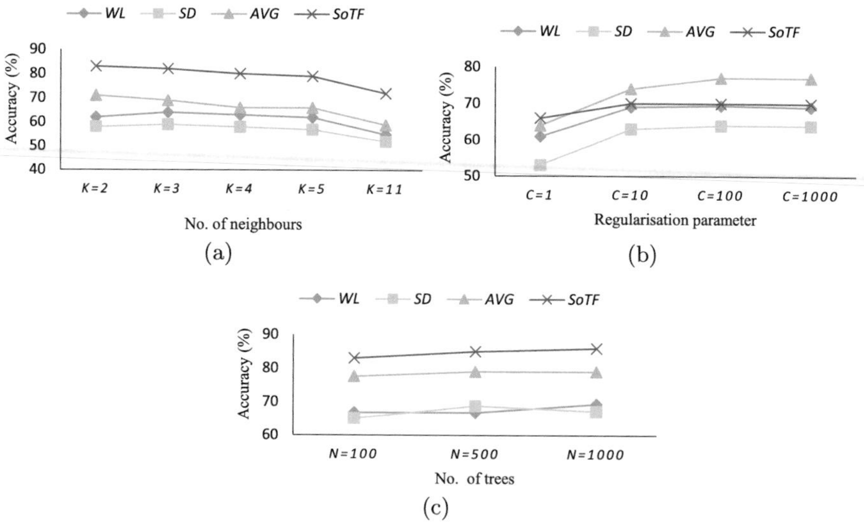

Fig. 4 Parameter study of different classifiers. **a** Classification accuracy obtained on varying the number of neighbors K in kNN classifier, **b** Classification accuracy obtained on varying regularization parameter C in SVM classifier, **c** Classification accuracy obtained on varying the number of trees N in RF classifier

Table 1 Best accuracies (in %) obtained for the SoTF using three classifiers for each exercise

Exercise	Classifiers		
	kNN ($K = 3$)	SVM ($C = 1000$)	RF ($N = 1000$)
1 (20 gestures)	86	73	88.24
2 (9 gestures)	91	74	90.53
3 (23 gestures)	81	69	84.45
All (52 gestures)	82	70	86

deviation, average features, and SoTF, respectively, at $N = 1000$. The best accuracies obtained with the SoTF are presented in Table 1 along with the exercise-wise accuracies obtained for the three classifiers at their best parameter values. The random forest achieves an accuracy of 86% which is the best among the three classifiers.

Figure 5 shows the confusion matrix obtained for the SoTF for each exercise. We can see that there are several gestures that show misclassification. Gestures in exercise 3 have more number of misclassifications than those in 1 and 2. Gesture 4 (middle finger extension) from exercise 1 is misclassified as gesture 6 (ring finger extension). Nine of the test samples of gesture 23 (Wrist supination with rotation axis through little finger) are classified as gesture 21 (Wrist supination with rotation axis through middle finger). Ball grasping gestures 40, 41, and 42 belonging to exercise 3 which are three finger, precision sphere, and tripod, respectively, show the highest misclassification in exercise 3 due to their high similarity. Also, gesture 32 (large diameter) gets misclassified as gesture 30 (small diameter) and 31 (fixed hook). The misclassification in the gestures is due to the activation of the same muscles due to similarity in gestures. Table 2 compares the classification accuracy of the proposed method with existing methods in literature. Furthermore, Table 3 compares the classification accuracy of 12 finger gestures and 8 isometric and isotonic hand gestures separately.

5 Conclusions

This paper has introduced a set of time domain features (SoTF) to classify 52 hand gesture classes of sEMG signals. The proposed SoTF is able to generate a distinguishable feature for each gesture class. Three different classifiers, namely kNN, SVM, and RF are implemented using the SoTF feature. Various parameter studies is presented to develop the best classifier using the proposed feature. The experimental results show that a recognition accuracy of 86% is achieved using the SoTF and RF classifier on 52 gesture classes of the NinaPro DB1 dataset which is superior to the earlier reported techniques. The exercise-wise recognition performance of the benchmarked dataset is also analyzed using the proposed feature. The confusion

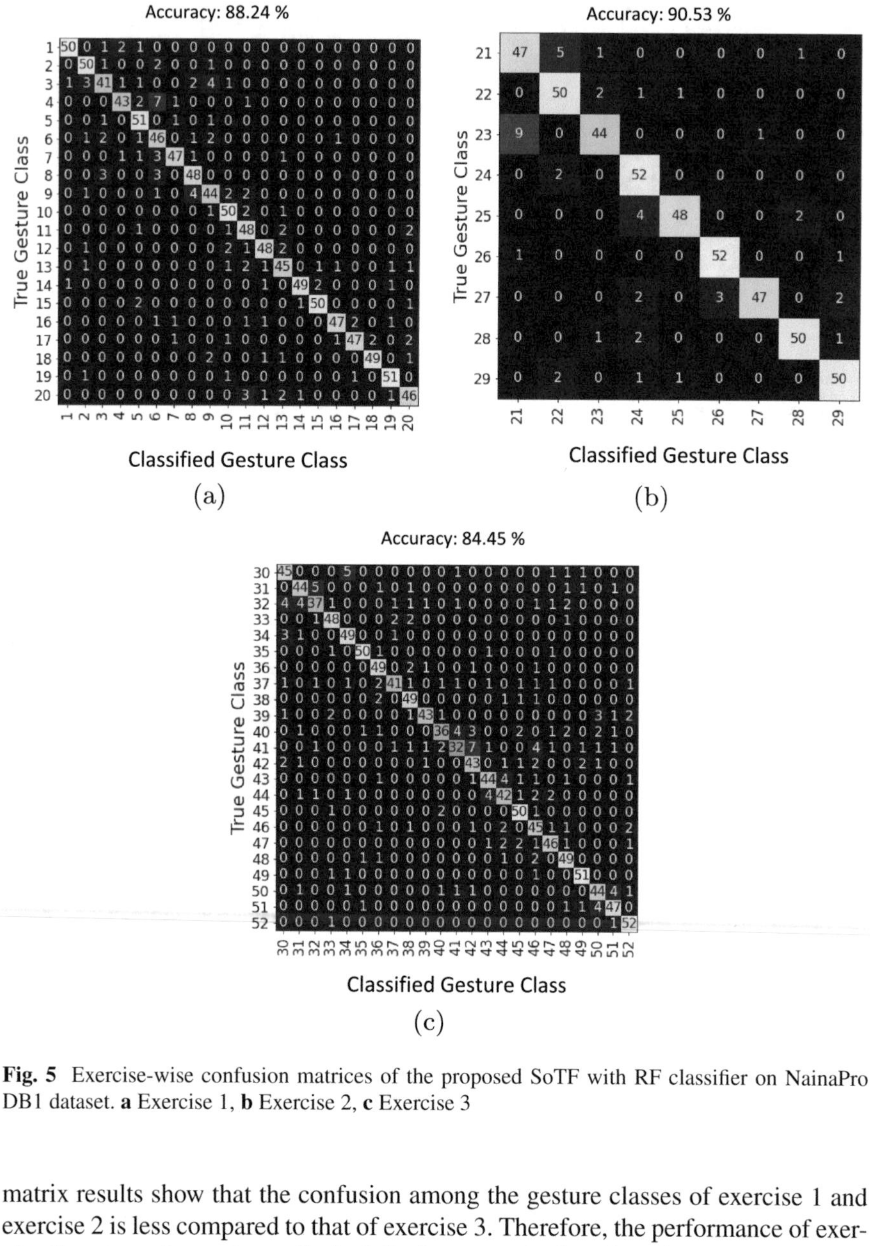

Fig. 5 Exercise-wise confusion matrices of the proposed SoTF with RF classifier on NainaPro DB1 dataset. **a** Exercise 1, **b** Exercise 2, **c** Exercise 3

matrix results show that the confusion among the gesture classes of exercise 1 and exercise 2 is less compared to that of exercise 3. Therefore, the performance of exercise 3 gesture classes is are limited. In future work, deep learning techniques with the current system may be introduced to overcome the above limitation.

Table 2 Comparison of classification accuracies obtained for the proposed SoTF with existing methods on 52 gestures of the NinaPro DB1 database

Author	Feature	Classifier	Accuracy (%)
Atzori et al. [7]	*WL*	kNN	73
Atzori et al. [7]	*WL*	SVM	75
Atzori et al. [9]	*RMS, mDWT*	kNN	65
Atzori et al. [9]	*MS + TD + HIST + mDWT*	RF	75
Pizzolato et al. [10]	*RMS + TD + HIST + mDWT*	SVM	60
Pizzolato et al. [10]	*RMS + TD + HIST + mDWT*	RF	65
Cene et al. [8]	*RMS + VAR + MAV + SD*	R-RELM	75.03
He et al. [11]	–	LSTM + MLP	75.45
Proposed work	*SoTF*	kNN	82
Proposed work	*SoTF*	SVM	70
Proposed work	*SoTF*	RF	86

Table 3 Comparison of classification accuracies obtained for the proposed SoTF with existing methods on a subset of NinaPro DB1 Database

Author	No. of gestures	Feature	Classifier	Accuracy (%)
Du et al. [12]	12	–	RF	75
Saeed et al. [13]	12	*MAV + ZC +SSC + WL*	LDA	85.41
Proposed work	12	*SoTF*	RF	87.65
Du et al. [12]	8	–	RF	76
Proposed work	8	*SoTF*	RF	88.42

References

1. Rodríguez-Tapia B, Soto I, Martínez DM, Arballo NC (2020) Myoelectric interfaces and related applications: current state of EMG signal processing-a systematic review. IEEE Access 8:7792–7805
2. Luo R, Sun S, Zhang X, Tang Z, Wang W (2019) A low-cost end-to-end sEMG-based gait sub-phase recognition system. IEEE Trans Neural Syst Rehabil Eng 28(1):267–276
3. Pancholi S, Joshi AM (2018) Portable EMG data acquisition module for upper limb prosthesis application. IEEE Sens J 18(8):3436–3443
4. Raurale SA, McAllister J, Del Rincón JM (2021) Emg biometric systems based on different wrist-hand movements. IEEE Access 9:12256–12266
5. Guo L, Lu Z, Yao L (2021) Human-machine interaction sensing technology based on hand gesture recognition: a review. IEEE Trans Hum-Mach Syst
6. Sahoo JP, Ari S, Ghosh DK (2018) Hand gesture recognition using DWT and F-ratio based feature descriptor. IET Image Process 12(10):1780–1787

7. Atzori M, Gijsberts A, Kuzborskij I, Elsig S, Hager AGM, Deriaz O, Castellini C, Müller H, Caputo B (2014) Characterization of a benchmark database for myoelectric movement classification. IEEE Trans Neural Syst Rehabil Eng 23(1):73–83
8. Cene VH, Tosin M, Machado J, Balbinot A (2019) Open database for accurate upper-limb intent detection using electromyography and reliable extreme learning machines. Sensors 19(8):1864
9. Atzori M, Gijsberts A, Castellini C, Caputo B, Hager AGM, Elsig S, Giatsidis G, Bassetto F, Müller H (2014) Electromyography data for non-invasive naturally-controlled robotic hand prostheses. Sci Data 1(1):1–13
10. Pizzolato S, Tagliapietra L, Cognolato M, Reggiani M, Müller H, Atzori M (2017) Comparison of six electromyography acquisition setups on hand movement classification tasks. PloS one 12(10), e0186,132 (2017)
11. He Y, Fukuda O, Bu N, Okumura H, Yamaguchi N (2018) Surface EMG pattern recognition using long short-term memory combined with multilayer perceptron. In: 2018 40th annual international conference of the IEEE engineering in medicine and biology society (EMBC), pp 5636–5639. IEEE (2018)
12. Du Y, Jin W, Wei W, Hu Y, Geng W (2017) Surface EMG-based inter-session gesture recognition enhanced by deep domain adaptation. Sensors 17(3):458
13. Saeed B, Gilani SO, ur Rehman Z, Jamil M, Waris A, Khan MN (2019) Comparative analysis of classifiers for EMG signals. In: 2019 IEEE Canadian conference of electrical and computer engineering (CCECE), pp 1–5. IEEE (2019)
14. Li Y, Zhang W, Zhang Q, Zheng N (2021) Transfer learning-based muscle activity decoding scheme by low-frequency sEMG for wearable low-cost application. IEEE Access 9:22804–22815
15. Zhou T, Omisore OM, Du W, Wang L, Zhang Y (2019) Adapting random forest classifier based on single and multiple features for surface electromyography signal recognition. In: 2019 12th international congress on image and signal processing, biomedical engineering and informatics (CISP-BMEI), pp 1–6. IEEE (2019)
16. Paul Y, Goyal V, Jaswal RA (2017) Comparative analysis between SVM & KNN classifier for EMG signal classification on elementary time domain features. In: 2017 4th international conference on signal processing, computing and control (ISPCC), pp 169–175. IEEE (2017)
17. Xing W, Bei Y (2020) Medical health big data classification based on KNN classification algorithm. IEEE Access 8:28808–28819. https://doi.org/10.1109/ACCESS.2019.2955754
18. Chethana C (2021) Prediction of heart disease using different KNN classifier. In: 2021 5th international conference on intelligent computing and control systems (ICICCS), pp 1186–1194 (2021). 10.1109/ICICCS51141.2021.9432178
19. Apostolidis-Afentoulis V, Lioufi KI (2015) SVM classification with linear and RBF kernels. July): 0-7. Classification with Linear and RBF kernels [21] (2015). http://www.academia.edu/13811676/SVM
20. Javeed A, Zhou S, Yongjian L, Qasim I, Noor A, Nour R (2019) An intelligent learning system based on random search algorithm and optimized random forest model for improved heart disease detection. IEEE Access 7:180,235–180,243. 10.1109/ACCESS.2019.2952107
21. Padhy S (2020) A tensor-based approach using multilinear SVD for hand gesture recognition from sEMG signals. IEEE Sens J 21(5):6634–6642

Supervision Meets Self-supervision: A Deep Multitask Network for Colorectal Cancer Histopathological Analysis

Aritra Marik, Soumitri Chattopadhyay, and Pawan Kumar Singh

1 Introduction

The uncontrolled cell division in certain epithelial tissues of the colon and rectum in the large intestine, due to mutation in certain genes, results in colorectal carcinoma or colorectal cancer (CRC) [27]. As of 2020, CRC is one of the leading causes of cancer-related deaths in the world, accounting for around 11% of cancer patients worldwide. The early detection of CRC can increase the survival rates by around 90% [28] which explains the need for its early detection.

Traditionally, the standard diagnosis procedures for CRC, which are carried out by pathologists, include faecal occult blood test (FOBT) and faecal immunochemical test (FIT) for detection of haemoglobin in the blood, followed by colonoscopy for studying the cause behind it. However, manual observation and diagnosis is susceptible to observer-based variations. The necessity of higher efficiency in colorectal cancer histopathological (CRCH) analysis from the tissue images along with the extraction of underlying features from those images has paved the way for deep learning-based methods in the literature.

Medical imaging-based histopathological analysis has been approached classically using traditional machine learning and handcrafted feature extraction [30, 32], which, however, fail to capture complex underlying patterns within the image data. As such, deep learning methods, particularly CNNs [13], have gained popularity due to their capability in recognizing salient and translationally invariant image features for robust classification and investigation. CNNs have been successfully applied to several facets of medical imaging including histopathology [31], chest X-rays [5] and CT scans [18].

A. Marik · S. Chattopadhyay · P. K. Singh (✉)
Department of Information Technology, Jadavpur University, Jadavpur University Second Campus, Plot No. 8, Salt Lake Bypass, LB Block, Sector III, Salt Lake City, Kolkata 700106, West Bengal, India
e-mail: pawansingh.ju@gmail.com

P. Singh et al. (eds.), *Machine Learning and Computational Intelligence Techniques for Data Engineering*, Lecture Notes in Electrical Engineering 998,
https://doi.org/10.1007/978-981-99-0047-3_41

In this paper, we propose a novel deep learning framework combining supervised and self-supervised techniques for feature learning for the purpose of CRCH analysis. The supervised learning framework is guided by deep metric learning using triplet loss for learning a discriminative embedding space along with self-supervised image reconstruction for learning pixel-level tissue image features. To the best of our knowledge, such a multitask pipeline has not been approached yet in regard to histopathological analysis. The downstream classification is done by feeding the extracted features into an SVM classifier [8].

The main contributions of this work may be summarized as follows:

1. A novel multitask training pipeline is proposed for learning robust representations of colorectal histopathological images.
2. Deep metric learning is used to learn a discriminative embedding space, augmented by a self-supervised image reconstruction module that enforces learning of pixel-level information for enhanced histopathological analysis.
3. Once trained, the encoder is used off-the-shelf to extract features which are used to train an SVM classifier [8] for the downstream classification. Upon comparison, the proposed framework outperforms several state-of-the-art works in literature on a publicly available CRCH dataset [16].

The rest of the paper is organized as follows. Section 2 discusses some recent works that are relevant to our proposed method. Section 3 describes the proposed multitask learning pipeline in detail. Section 4 outlines the experimental evaluation of the method on a public dataset. Finally, Sect. 5 concludes and also discusses future extensions of the present research.

2 Related Works

2.1 Colorectal Cancer Histopathology

The work in tissue-image classification in the past few years has primarily been under two categories, texture feature-based methods and deep learning methods. The first study on multi-texture-feature analysis for colorectal tissue classification was presented by Kather et al. [16], which obtained the highest accuracy of 87.4% in the multi-class classification task. The study also presented a new dataset of 5000 histological images of human CRC including 8 tissue classes (described in Table 1), the dataset we have used in our work. Among deep learning-based approaches, a bilinear CNN model was proposed by [31] which extracted and fused features from stain-decomposed histological images, achieving an accuracy of 92.6%. The CNN architecture proposed by [7] highlighted the importance of stain normalization in the literature, although achieving an accuracy of only 79.66%. Sabol et al. [25] proposed a method in which the classifier performance was enhanced by a fine-tuned CNN model to an accuracy of 92.74% on the multi-class classification task. Ohata et al. [21] first introduced the concept of fine-tuning a deep learning model which was pre-trained on ImageNet with respect to the current dataset while coping with the limited data available with respect to biomedical imaging. They used 108 different combinations of feature extractors and classifiers, out of which the pre-trained DenseNet-169 model and the SVM classifier obtained an accuracy of 92.08%. It is to be noted that in all of the aforementioned works, the CRCH dataset by [16] has been used. To this end, we bring to table a novel and efficient multitask network for robust CRCH classification.

2.2 Deep Metric Learning

Metric learning [17] is based on the principle of similarity between data samples. Specifically, deep metric learning utilizes deep architectures by learning embedded feature similarity through training on raw data. Metric learning has been extensively applied in three-dimensional modelling [9], medical imaging [2], facial recognition [15, 19, 26] and signature verification [4]. The supervised module of deep metric learning of the proposed architecture has been inspired by the triplet network proposed by [14] for learning distance-based metric embeddings of a multi-class dataset.

2.3 Self-supervised Learning

Self-supervised learning [22] has gained tremendous popularity in recent years due to its capability of learning very good quality representations without the need for explicit supervision. Such techniques include contrastive learning [6, 12], generation/reconstruction-based [3, 10], clustering-based [1] and so on. We take inspiration from self-supervised literature to augment our metric learning model with a reconstruction network to learn minute pixel-level visual information.

3 Methodology

3.1 Overview

We intend to combine supervised learning with self-supervision as a pre-training paradigm for robust CRCH image analysis. Specifically, we design a metric learning framework that learns to maximize intra-class similarity and simultaneously distinguish samples from a different class. Further, to enhance learning of image-level texture features we introduce an image reconstruction decoder that takes in the embedding of a corrupted version of the histopathological image and tries to output

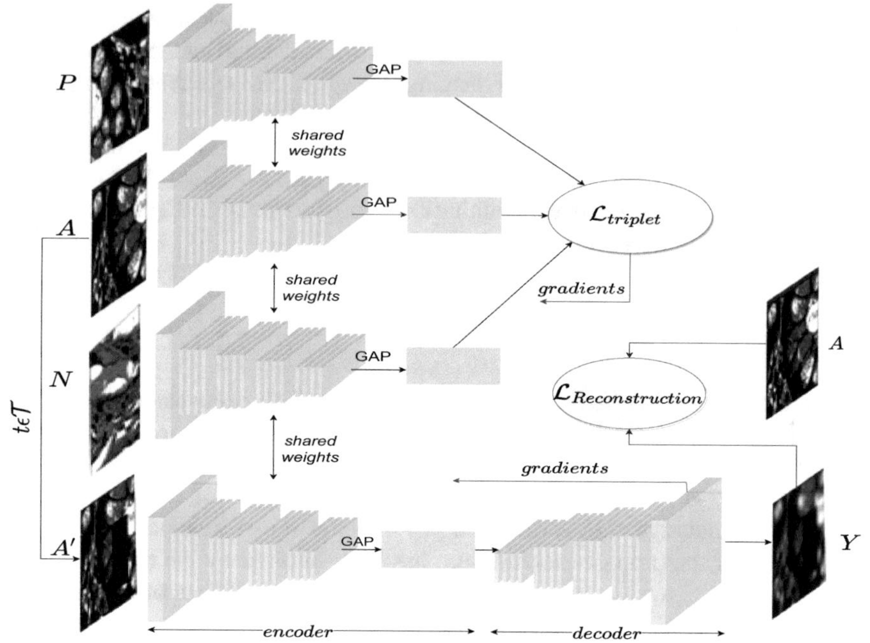

Fig. 1 Schematic representation of architecture proposed in this study

its original version. The corrupted version is produced by random spatial transformations such as blurring or dropping randomly distributed pixels. A reconstruction loss is used to train this branch. Note that the encoder weights are shared across the two modules and gradients flow throughout the architecture. Once the network is trained, the frozen encoder is used to extract image features, which are used to train an SVM classifier for the final classification. Figure 1 shows the overall architecture of the proposed pipeline.

3.2 Deep Metric Learning

The main purpose of metric learning is to learn a discriminative embedding space wherein samples belonging to the same class are close together, while those from different classes are farther apart. Typically, metric learning losses aim at minimizing the distance metric between intra-class or 'positive' pairs while maximizing the same between inter-class or 'negative' pairs. In this work, we have used the state-of-the-art Siamese convolutional networks [14] that comprise three CNN branches with shared weights, each corresponding to the anchor, positive and negative images comprising the triplet. The network is trained using a simple triplet loss objective. To keep our framework simple and lightweight, we have used an ImageNet pre-trained ResNet-18 [13] network as the encoder backbone, with the output embedding dimension being set to 512.

$$\mathcal{L}_{triplet} = \sum_{i=1}^{K} max(0, \|\mathcal{F}(A^{(i)}) - \mathcal{F}(P^{(i)})\|^2 - \|\mathcal{F}(A^{(i)}) - \mathcal{F}(N^{(i)})\|^2 + \mu) \quad (1)$$

where, (A, P, N) constitute a triplet, $\mathcal{F}(\cdot)$ denotes the shared encoder and μ is the permissible margin value of the triplet loss, set to 0.2 experimentally.

3.3 Image Reconstruction Network

Image reconstruction aims at learning pixel-level information so as to restore its original version from a corrupted one. We achieve this by introducing a decoder module that starts from an embedding vector into a series of upsampling layers to finally yield the output image dimensions. We have used the state-of-the-art U-Net decoder [24] network, *excluding* the skip connections from the encoder. A corrupted view of the original 'anchor' image is formed using random image transformations such as affine transformations, blurring or dropping pixels, and is passed through a shared encoder network, the output of which is then passed through the decoder. A reconstruction loss is employed between the output image (Y) and the original image (A) before corruption. In our work, we have used the simple mean squared error (MSE) as the reconstruction loss to train our network.

$$\mathcal{L}_{reconstruction} = \frac{1}{M} \sum_{i=1}^{K} \sum_{j=1}^{M} \|A_j^{(i)} - Y_j^{(i)}\|_2^2 \quad (2)$$

Combining Eqs. 1 and 2, the overall training loss objective can be written as follows:

$$\mathcal{L}_{train} = \lambda \cdot \mathcal{L}_{triplet} + \mathcal{L}_{reconstruction} \quad (3)$$

Here, λ is a hyperparameter that is used for the relative weighting of the losses. Experimentally, it has been set to 10.

3.4 Final Classification

Once the model is fully trained, we use the encoder while keeping its weights fixed and extract features from the images which are then fed into an SVM classifier [8] so as to train it for the final classification step. The SVM classifier is a supervised algorithm that aims at finding the optimal hyperplane(s) that separate the respective classes by mapping the samples onto a space such that the distance between class boundaries is maximized. The unseen sample features are likewise mapped on the sample space and thus, classes are predicted based on where they fall.

4 Results and Discussion

4.1 Dataset Description

We train and evaluate our proposed framework on a publicly available dataset by Kather et al. [16] comprising 5000 colorectal histopathological images uniformly distributed across 8 classes. Each image is of dimensions 150×150 px and belongs to exactly one category among those in Table 1. For our purpose, we split the dataset as 0.75/0.25 for train/test respectively.

4.2 Implementation Details

Our model has been implemented in PyTorch [23] on a 12GB K80 Nvidia GPU. The encoder–decoder framework was trained for 200 epochs using the Stochastic Gradient Descent (SGD) optimizer [29] with a learning rate of 0.01. To keep the pipeline fairly straightforward, we chose the backbone encoder as the ImageNet pre-trained ResNet-18 [13] that outputs an embedding of dimension 512. The decoder used is the state-of-the-art U-Net decoder [24] network *without* the skip connections from the encoder. Each image was resized to 256×256 *px* using bilinear interpolation before being passed through the encoder, the batch size being set to 32.

Table 1 Image categories in the CRCH dataset [16] used in this research. Each class contains 625 images, which are split as 0.75/0.25 for train/test, respectively

Label	Category
0	Tumour Epithelium
1	Simple Stroma
2	Complex Stroma
3	Immune Cells
4	Debris
5	Normal Mucosa
6	Adipose Tissue
7	Background

4.3 Evaluation Metrics

The four metrics used for evaluating our proposed method on the CRCH dataset [16] are, namely, *Accuracy*, *Precision*, *Recall* and *F1-Score*. The formulas for the metrics, derived from a confusion matrix, C, are given in the following Eqs. 4, 5, 6, 7.

$$Accuracy = \frac{\sum_{i=1}^{N} C_{ii}}{\sum_{i=1}^{N}\sum_{j=1}^{N} C_{ij}} \tag{4}$$

$$Precision_i = \frac{C_{ii}}{\sum_{j=1}^{N} C_{ji}} \tag{5}$$

$$Recall_i = \frac{C_tii}{\sum_{j=1}^{N} C_{ij}} \tag{6}$$

$$F1 - Score_i = \frac{2}{\frac{1}{Precision_i} + \frac{1}{Recall_i}} \tag{7}$$

Here, N signifies the number of classes in the respective dataset.

4.4 Qualitative Analysis

We analyse the discriminative embedding space learned by the joint encoder–decoder model by visualizing the embedding space of the extracted features in the two-dimensional plane using t-distributed stochastic neighbourhood embedding (t-SNE) [20] at intervals during the training process. The t-SNE plots have been put in Fig. 2. It is evident from the plots that over the training epochs, the embedding space gets more and more discriminative and the classes get fairly well separated, thus qualitatively suggesting that a robust representation has been learned by the pre-training paradigm.

4.5 Comparison with State of the Art

Performance comparison between our proposed method and other state-of-the-art methods for the CRCH dataset [16] has been provided in Fig. 2. It can be examined that our proposed framework outperforms all other existing state-of-the-art methods by a significant margin. It may be taken into consideration that some of the previously existing works in the literature reported accuracy as the only evaluation metric. It is insufficient and does not provide enough insights into false positives and true negatives. Since CRCH analysis is a multi-class classification task, the absence of

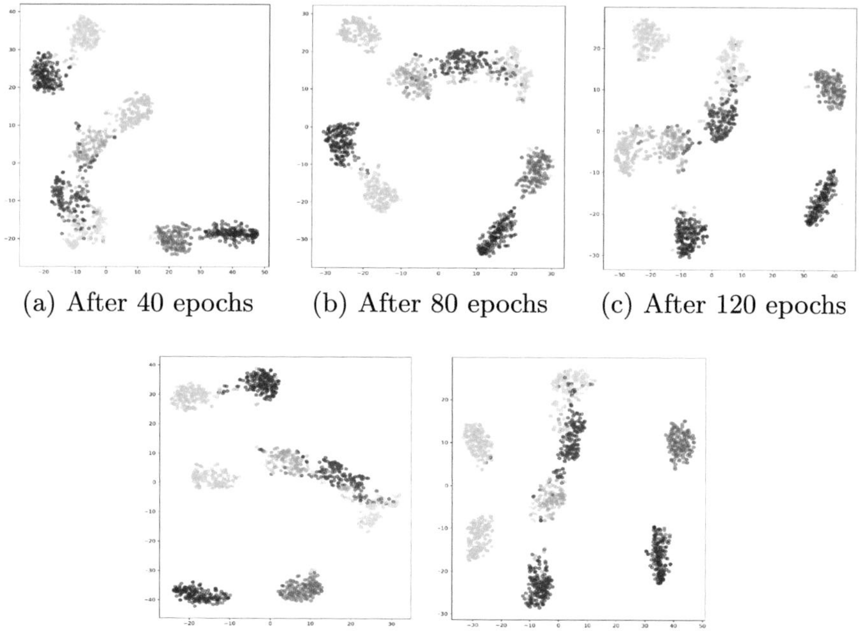

(a) After 40 epochs (b) After 80 epochs (c) After 120 epochs

(d) After 160 epochs (e) After 200 epochs

Fig. 2 t-SNE plots obtained by the encoder at different stages of the training process

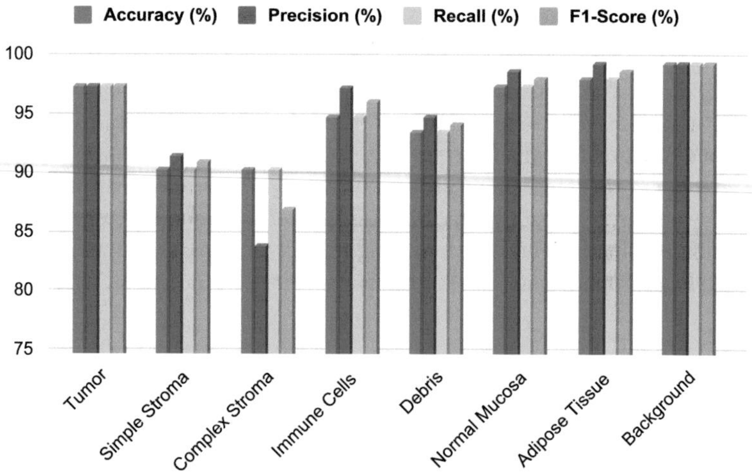

Fig. 3 Class-wise evaluation metrics obtained by the proposed method on the CRCH dataset

Table 2 Comparison of the proposed framework with state-of-the-art methods on publicly available CRCH dataset

Method	Accuracy (%)	Precision (%)	Recall (%)	F1-Score (%)
Ciompi et al. [7]	79.66	–	–	–
Kather et al. [16]	87.40	–	–	–
Ohata et al. [21]	92.08	–	–	92.12
Wang et al. [31]	92.60	–	92.80	–
Sabol et al. [25]	92.74	92.50	92.76	92.64
Ghosh et al. [11]	92.83	92.83	93.11	92.97
Proposed method	95.22	95.34	95.22	95.26

sufficient evaluation metrics makes the works unreliable. On the other hand, our proposed pipeline achieves commendable performance on all of the mentioned metrics considered.

4.6 Ablation Study

Since our proposed framework comprises two components, i.e. a supervised metric learning model and a self-supervised image reconstruction model, we quantitatively determine the contribution of each of them by performing an ablation study. We define the baselines for the same as follows:

- **Triplet**: This denotes the metric learning model alone excluding the decoder part, such that the encoder is trained using triplet loss only. All other experimental parameters are kept identical. Please refer to Sect. 3.2 for full details.
- **Reconstruction**: This denotes the encoder–decoder model which takes in a randomly transformed histopathological image and tries to reconstruct the original image from it, trained using reconstruction loss. All other experimental parameters are kept identical. Please refer to Sect. 3.3 for full details.

The overall performance results are shown in Table 3. From Table 3, it can be observed that the triplet model shows a better performance than the reconstruction baseline, thereby highlighting the importance of optimizing the embedding space for better discrimination. Furthermore, it can also be noticed that combining the two components together, i.e. our proposed framework improves the classification performance of the individual stand-alone baselines, affirming the contributions of the respective modules.

Table 3 Ablation study on the proposed pipeline

Method	Accuracy (%)	Precision (%)	Recall (%)	F1-Score (%)
Triplet	94.58	94.65	94.58	94.61
Reconstruction	93.39	93.50	93.39	93.41
Triplet + Reconstruction (Proposed)	95.22	95.34	95.22	95.26

5 Conclusion and Future Work

In this work, we present a novel training strategy that leverages supervised metric learning and self-supervised image reconstruction for robust representation learning of histopathological images. While metric learning optimizes the embedding space, the reconstruction module enforces pixel-level information learning, which aids the overall training pipeline, improving the downstream evaluation. We have analysed our framework qualitatively as well as compared with several existing state-of-the-art works in literature, along with a suitable ablation study to investigate the contributions of the respective modules. The results highlight the prowess of the proposed method for image classification on the CRCH dataset.

However, our work does have certain limitations, such as the relatively poor performance for the class 'Complex Stroma' as shown in Fig. 3. We intend to investigate this in future. Possible extensions of our work may be on the lines of improving the respective modules using alternate metric learning losses, or using VAE/GAN-based models for reconstruction and so on. Further, this work provides a foundation for CRCH representation learning and paves the way towards contrastive learning-based self-supervised approaches [6, 12] as large-scale supervised learning gradually becomes infeasible. We intend to explore these lines as well in our future works.

References

1. Alwassel H, Mahajan D, Korbar B, Torresani L, Ghanem B, Tran D (2019) Self-supervised learning by cross-modal audio-video clustering. arXiv:1911.12667
2. Annarumma M, Montana G (2018) Deep metric learning for multi-labelled radiographs. In: Proceedings of the 33rd annual ACM symposium on applied computing
3. Atito S, Awais M, Kittler J (2021) Sit: self-supervised vision transformer. arXiv:2104.03602
4. Bromley J, Bentz JW, Bottou L, Guyon I, LeCun Y, Moore C, Säckinger E, Shah R (1993) Signature verification using a "siamese" time delay neural network. World Scientific, IJPRAI
5. Chattopadhyay S, Kundu R, Singh PK, Mirjalili S, Sarkar R (2021) Pneumonia detection from lung x-ray images using local search aided sine cosine algorithm based deep feature selection method. Int J Intell Syst
6. Chen T, Kornblith S, Norouzi M, Hinton G (2020) A simple framework for contrastive learning of visual representations. In: ICML
7. Ciompi F, Geessink O, Bejnordi BE, de Souza GS (2017) The importance of stain normalization in colorectal tissue classification with convolutional networks. In: IEEE ISBI

8. Cortes C, Vapnik V: Support-vector networks. Mach Learn (1995)
9. Dai G, Xie J, Zhu F, Fang Y (2017) Deep correlated metric learning for sketch-based 3d shape retrieval. In: AAAI
10. Deepak P, Philipp K, Jeff D, Trevor D, Efros AA (2016) Context encoders: feature learning by inpainting. In: CVPR
11. Ghosh S, Bandyopadhyay A, Sahay S, Ghosh R, Kundu I, Santosh K (2021) Colorectal histology tumor detection using ensemble deep neural network. Elsevier, EAAI
12. He K, Fan H, Wu Y, Xie S, Girshick R (2020) Momentum contrast for unsupervised visual representation learning. In: CVPR
13. He K, Zhang X, Ren S, Sun J (2016) Deep residual learning for image recognition. In: IEEE CVPR
14. Hoffer E, Ailon N (2015) Deep metric learning using triplet network. In: International workshop on similarity-based pattern recognition
15. Hu J, Lu J, Tan YP (2014) Discriminative deep metric learning for face verification in the wild. In: IEEE CVPR
16. Kather JN, Weis CA, Bianconi F, et al (2016) Multi-class texture analysis in colorectal cancer histology. In: Scientific reports, nature
17. Kaya M, Bilge, HŞ (2019) Deep metric learning: a survey. Symmetry
18. Kundu R, Basak H, Singh PK, Ahmadian A, Ferrara M, Sarkar R (2021) Fuzzy rank-based fusion of CNN models using Gompertz function for screening COVID-19 CT-scans. In: Scientific reports. Nature
19. Liu J, Deng Y, Bai T, Wei Z, Huang C (2015) Targeting ultimate accuracy: face recognition via deep embedding. arXiv:1506.07310
20. Van der Maaten L, Hinton G (2008) Visualizing data using t-SNE. J Mach Learn Res
21. Ohata EF, Chagas JVSd, Bezerra GM, Hassan MM, de Albuquerque VHC, Filho PPR (2021) A novel transfer learning approach for the classification of histological images of colorectal cancer. J Supercomput Springer
22. Ohri K, Kumar M (2021) Review on self-supervised image recognition using deep neural networks. Knowl-Based Syst
23. Paszke A, Gross S, Massa F, Lerer A, Bradbury J, Chanan G, Killeen T, Lin Z, Gimelshein N, Antiga L et al (2019) Pytorch: an imperative style, high-performance deep learning library. In: NeurIPS
24. Ronneberger O, Fischer P, Brox T (2015) U-net: convolutional networks for biomedical image segmentation. In: MICCAI
25. Sabol P, Sinčák P, Hartono P, Kočan P et al (2020) Explainable classifier for improving the accountability in decision-making for colorectal cancer diagnosis from histopathological images. Elsevier, JBI
26. Schroff F, Kalenichenko D, Philbin J (2015) Facenet: a unified embedding for face recognition and clustering. In: CVPR
27. Society AC (2020) What is colorectal cancer? American Cancer Society. www.cancer.org/cancer/colon-rectal-cancer/about/what-is-colorectal-cancer.html
28. Society AC (2021) Survival rates for colorectal cancer. American Cancer Society. www.cancer.org/cancer/colon-rectal-cancer/detection-diagnosis-staging/survival-rates.html
29. Sutskever I, Martens J, Dahl G, Hinton G (2013) On the importance of initialization and momentum in deep learning. In: ICML
30. Takamatsu M, Yamamoto N, Kawachi H, Chino A, Saito S, Ueno M, Ishikawa Y, Takazawa Y, Takeuchi K (2019) Prediction of early colorectal cancer metastasis by machine learning using digital slide images. In: CMPB
31. Wang C, Shi J, Zhang Q, Ying S (2017) Histopathological image classification with bilinear convolutional neural networks. In: IEEE EMBC
32. Xu Y, Ju L, Tong J, Zhou CM, Yang JJ (2020) Machine learning algorithms for predicting the recurrence of stage IV colorectal cancer after tumor resection. In: Scientific reports, nature

Study of Language Models for Fine-Grained Socio-Political Event Classification

Kartick Gupta and Anupam Jamatia

1 Introduction

Text classification is one of the fundamental problems in Natural Language Processing (NLP), in which the objective is to label the textual data like phrases, sentences, and paragraphs. It is used in a wide range of applications such as sentiment analysis [7], spam detection [8], summarization [9], and even news classification. In this paper, we deal with the classification of socio-political events which directly impact the world on a day-to-day basis. However, in the field of text classification, their results are often limited by the quality of feature extraction. This phenomenon is particularly prominent in short text classification tasks such as news classification since the short text does not contain enough contextual information in contrast to paragraphs and documents. Natural languages refer not only to entities but essentially also to situations. Therefore, various aspects of situations are worth analyzing in modeling linguistic meaning. The event-based dimension of information is prominent for reasoning about why and how the world around us is evolving. The world is dynamic, and events are important aspects of everything that happens in this world.

In the modern world, an enormous amount of data is produced and stored every day thus challenging us to innovate more effective and efficient methodologies to store, process, and extract information. The task of text classification has been substantially benefited by the revival of the deep neural network due to their noteworthy achievement, with less essential requirement of feature engineering techniques. For such reasons, deep learning strategies have more preferred over machine learning approaches due to their robustness and ability to deal with large data. Therefore,

K. Gupta (✉)
Indian Institute of Information Technology Agartala, Tripura, India
e-mail: guptakartik99@gmail.com

A. Jamatia
National Institute of Technology Agartala, Tripura, India
e-mail: anupamjamatia@nita.ac.in

P. Singh et al. (eds.), *Machine Learning and Computational Intelligence Techniques for Data Engineering*, Lecture Notes in Electrical Engineering 998,
https://doi.org/10.1007/978-981-99-0047-3_42

deep learning models have minimized the needs of users and rapidly improve their performance in various NLP tasks.

The paper is organized as follows. Section 2 reviews the background work done. Section 3 describes the process of data collection and corpus creation: training and validation datasets. Next, the experimentation performed and models developed are described in Sect. 4. Subsequently, Sect. 5 presents the results obtained by these models, whereas Sect. 6 discusses the error analysis. We present the conclusions in Sect. 7.

2 Background Related Works

Various approaches and methodologies have been proven effective for the task of event detection and classification from the text corpus. [18] developed one of the earliest embedding models: Latent Semantic Analysis (LSA). It was a linear model which was trained on 200 K words, comprising less than 1M parameters. Martina Naughton has developed a methodology for event detection at the sentence level using a Support Vector Machine (SVM) classifier and a Language Modeling (LM) approach has been used to check the performance of the system [20]. Timothy Nugent compared various supervised classification methods for detecting a variety of different events and achieved good results with Support Vector Machines (SVM) and Convolutional Neural Networks (CNN) [21]. The key component of these approaches consists of a machine-learned embedding model that maps text into a low-dimensional continuous feature vector, thus hand-crafted features were not required.

A paradigm shift started when much larger embedding models were developed using gigantic amounts of training data. In 2013, Google developed a series of Word2Vec embeddings [19] that were trained on 6 billion words and straightaway became popular. A new type of deep contextualized word representation: ELMo (Embeddings from Language Models) [4] was jointly developed by the teams from AI2 and the University of Washington comprising of contextual embedding model based on a three-layer bidirectional LSTM [13] with 93 M parameters trained on 1B words, which performed much better than Word2Vec because they captured the contextual information. A benchmark was achieved with the development of Transformer [14] models like BERT [1] based on the bidirectional transformer which consists of 340 M parameters and trained on 3.3 billion words. Further development of models like XLNet [6] and A Robustly Optimized BERT Pre-training Approach (RoBERTa) [5] even surpassed BERT in performance and are the current state-of-the-art embedding models. Previous work on ACLED data done by Samantha Kent [17] and Benjamin Radford [11] gives great results but is limited only to `BERT` and `RoBERTa`, but does not explore different architectures like `XLNet` and `ELMo`, thus, do not explain the difference in the results due to different architectures, which could lead to better understanding and hence further enhancement in the field of NLP.

3 Corpus Acquisition and Annotations

To commence the project, the data required was collected from the Armed Conflict Location and Event Data Project (ACLED) database [3] which consists of news snippets of 25 event types pertaining to various political violence, demonstrations, and nonviolent and politically important events. Data was acquired from the ACLED website.[1]

To create the corpus, we first downloaded the data from all the regions available beginning from 1 January 2018 to 12 July 2021. The next step was to re-sample the data to counterbalance the corpus which contains 25 distinct event classes. Initially, the data was highly imbalanced.

The following step was to re-sample the data in order to balance the corpus based on the 25 fine-grained event classifications. The data was initially rather unbalanced. The data contained `1,034,527` samples, in which the largest class in the corpus `PEACE_PROT` contributed of `31.84%` of the data. On the contrary, the smallest class `CHEM_WEAPON` contributed as low as `0.00096%` to the corpus.

We used the following procedure for re-sampling our data: (1) All the classes containing more than 20 K unique samples were under-sampled (capped) to only 20 K samples per class. (2) All the classes that contained between 20 K and 5 K samples were retained as they are i.e. were neither under-sampled nor over-sampled. (3) All the classes that consisted of the number of samples between 5K and 1K were over-sampled to twice the number (100% oversampling) of samples. (4) Finally, all the classes that contained the number of samples below 1K were over-sampled by 500%. After re-sampling, the balanced data was stratified and split into train and validation data randomly. The train set contains (n = 257,967) samples and the validation set consists of (n = 28,664) samples. It was a 90%–10% split of the re-sampled corpus in which both training and test data contained same class distribution. Table 1 depicts the original data and resampled data.

Following this step, we created two different versions of the original corpus. The first version referred to as `CLEAN`, contains the cleaned text which is free of all dates, months, locations, punctuations, and symbols from the prepared corpus. The second version, referred to as `LEMM` contains the lemmatized text of the `CLEAN` set. The inspiration of the techniques for text normalization and cleaning of ACLED data and some baseline classification models trained using ACLED data are described in [15].

4 Experiments

This task includes a variety of experiments that were different from each other, either different model architecture was used or the way the input data used was processed before the model training. The standard method for all tests was the same, for each language model we prepared 2 models: one model with `CLEAN` corpus and the other

[1] www.acleddata.com/data-export-tool.

Table 1 Class distribution statistics of the corpus

Classes	Original data	Resampled data
PEACE_PROT	329,399 (31.84%)	20,000 (6.97%)
ARM_CLASH	239,828 (23.18%)	20,000 (6.97%)
ATTACK	132,202 (12.77%)	20,000 (6.97%)
ART_MISS_ATTACK	64,743 (6.25%)	20,000 (6.97%)
MOB_VIOLENCE	42,079 (4.06%)	20,000 (6.97%)
VIOLENT_DEMON	41,003 (3.96%)	20,000 (6.97%)
AIR_STR	34,020 (3.28%)	20,000 (6.97%)
PROT_WITH_INTER	31,784 (3.07%)	20,000 (6.97%)
REMOTE_EXPLOS	30,620 (2.95%)	20,000 (6.97%)
PROPY_DISTRUCT	24,315 (2.35%)	20,000 (6.97%)
ABDUC_FORCED_DISSAP	14,624 (1.41%)	14,624 (5.10%)
ARREST	8,668 (0.83%)	8,668 (3.02%)
CHANGE_TO_GROUP_ACT	7,925 (0.76%)	7,925 (2.76%)
DISR_WEAPON	7,499 (0.72%)	7,499 (2.61%)
GOVT_REG_TER	6,148 (0.59%)	6,148 (2.14%)
OTHERS	5,353 (0.51%)	5,353 (1.86%)
FORCE_AGAINST_PROT	3,098 (0.29%)	6,196 (2.16%)
GRENADE	2,948 (0.28%)	5896 (2.05%)
NON_STATE_ACTOR_OVERTAKE_TER	2,670 (0.25%)	5,340 (1.86%)
SEXUAL_VIOL	2,196 (0.21%)	4,392 (1.53%)
AGRMNT	1,460 (0.14%)	2,920 (1.01%)
NON_VIOL_TERR_TRANS	989 (0.09%)	5934 (2.07%)
SUICIDE_BOMB	635 (0.06%)	3,810 (1.32%)
HQ_EST	311 (0.03%)	1,866 (0.65%)
CHEM_WEAPON	10 (0.0009%)	60 (0.02%)

with LEMM corpus as input. To conduct the experiments, the system configuration used Interactive Development Sessions in Python 3.7 on Google Colab and Kaggle. We used TensorFlow-2 and TensorFlow-1.3 as our deep learning frameworks. The GPU used was a Nvidia Tesla P100-PCIE with a RAM capacity of 25.46 GB. The pretrained RoBERTa and XLNet models were downloaded from HuggingFace,[2] whereas BERT and ELMo embeddings were downloaded from TensorFlow-Hub.[3]

[2] www.huggingface.co/transformers.

[3] www.tfhub.dev/text.

4.1 BERT

For Model-1 $\texttt{BERT}_{\text{LEMM}}$ and Model-2 $\texttt{BERT}_{\text{CLEAN}}$, both of which used the fine-tuned the BERT base model [1] as the base model architecture, models were trained on the training corpus using a batch size of (64), and a learning rate of (5e-5). During our model training analysis, $\texttt{BERT}_{\text{LEMM}}$ was trained for (3) epochs, but we observed that the model tends to overfit if we continue to train the model for more than (2) epochs. So, Model-2 $\texttt{BERT}_{\text{CLEAN}}$ was trained only for (2) epochs using the same hyperparameters. class_weights were allotted to all event classes for both $\texttt{BERT}_{\text{LEMM}}$ and $\texttt{BERT}_{\text{CLEAN}}$ to counterbalance any residual discrepancies in class sizes due to an imbalance in the number of samples per class so that the models train evenly on all the classes. The contradistinction from $\texttt{BERT}_{\text{LEMM}}$ is, that the text corpus that was used during the training of $\texttt{BERT}_{\text{CLEAN}}$ model, which was CLEAN version of the corpus i.e. which was not lemmatized and free from pre-processing. This indicates that the text snippets right from the original corpus were fed into the system after they were cleaned (removal of symbols) and removal of locations, dates, and months. For both models $\texttt{BERT}_{\text{LEMM}}$ and $\texttt{BERT}_{\text{CLEAN}}$, we have used Adam optimizer [16] to reduce the categorical cross-entropy loss function [2]. We considered $\texttt{BERT}_{\text{LEMM}}$ as our baseline model.

4.2 ELMo

For Model-3 $\texttt{ELMo}_{\text{LEMM}}$ and Model-4 $\texttt{ELMo}_{\text{CLEAN}}$, we used fine-tuned the pre-trained ELMo embeddings [4], followed by a fully connected dense layer and finally, a dense layer with a softmax activation function. $\texttt{ELMo}_{\text{LEMM}}$ was trained using a batch size of (64) to feed the training data, a learning rate of (5e-5) was used. During our analysis, we found that we trained the model for (5) epochs; however, we noticed that $\texttt{ELMo}_{\text{LEMM}}$ overfits if we continue training the model for more than (3) epochs. Using these observations, Model-4 $\texttt{ELMo}_{\text{CLEAN}}$ was trained on the training corpus only for (3) using the same hyperparameters. $\texttt{ELMo}_{\text{LEMM}}$ used the LEMM version of the corpus as text input for training, whereas $\texttt{ELMo}_{CLEAN}$ used the CLEAN version. For both $\texttt{ELMo}_{\text{LEMM}}$ and $\texttt{ELMo}_{\text{CLEAN}}$, we used Adam optimizer [16] in order to reduce the categorical cross-entropy loss [2].

4.3 RoBERTa

In Model-5 $\texttt{RoBERTa}_{\text{LEMM}}$ and Model-6 $\texttt{RoBERTa}_{\text{CLEAN}}$, we used fine-tuned the pre-trained RoBERTa base model [5] as the base architecture. The last hidden states were passed through a fully connected dense layer followed by dense layer with softmax activation function. For $\texttt{RoBERTa}_{\text{LEMM}}$, we trained the model on the training corpus using a batch size of (64), and a learning rate of (5e-5). During analysis, we found that we trained the model for (3) epochs; however, it was noticed that $\texttt{RoBERTa}_{\text{LEMM}}$ overfits if we continue to train it for more than (2) epochs. On finding this we trained $\texttt{RoBERTa}_{\text{CLEAN}}$ only for (2) epochs. We used the LEMM version of the corpus as text input which was also used for $\texttt{RoBERTa}_{\text{LEMM}}$ on the contrary $\texttt{RoBERTa}_{\text{CLEAN}}$ used CLEAN version of the training corpus. For both $\texttt{RoBERTa}_{\text{LEMM}}$ and $\texttt{RoBERTa}_{\text{CLEAN}}$, we used Adam optimizer.

4.4 XLNet

For Model-7 $\texttt{XLNet}_{\text{LEMM}}$ and Model-8 $\texttt{XLNet}_{\text{CLEAN}}$, we used the pre-trained XLNet-base-cased model [6]. After collecting the last hidden states, they were passed through a fully connected dense layer followed by dense layer with softmax activation function. $\texttt{XLNet}_{\text{LEMM}}$ was trained with a learning rate of (5e-5), a batch size of (64) for feeding the training corpus and trained the model for (3) epochs. For $\texttt{XLNet}_{\text{LEMM}}$, we used the LEMM corpus. We observed that the model overfits after (2) epochs. So, $\texttt{XLNet}_{\text{CLEAN}}$ was trained only for (2) epochs. We used Adam optimizer [16] to reduce the categorical cross-entropy loss [2].

5 Results Analysis

For evaluating the performance of the models for fine-grained event classification, we used *micro, macro* and *weighted* F1 score(s) as the performance metrics. The average is calculated using the *weighted* F1 method, which takes into account the contributing portion of each class in the sample. The *macro* version is also similar however, the performance of the model is calculated separately for each class, and then the mean is calculated. This guideline was inspired by [10].

Table 2 depicts the results of our eight experiments. Models were tested on an unseen test set provided by ACL, which consisted of 829 samples for the Subtasks of Task 2, *Fine-Grained Classification of Socio-Political Events*, first presented at **ACL-IJCNLP 2021**'s workshop: *Challenges and Applications of Automated Extraction of Socio-political Events from Text (CASE)* [10]. Test data is also accessible from the CASE-2021 site.[4]

We observed that Model-6 $\texttt{RoBERTa}_{\text{CLEAN}}$, which is using the RoBERTa-base model [5] architecture and CLEAN corpus for model training, performs the best which achieves weighted F1-score of 0.81, macro-F1-score of 0.78, and micro-F1-score of 0.80. Moreover, the results we achieved are very similar to the top-performing models of the CASE @ ACL-IJCNLP 2021 [12, 17].

[4] http://piskorski.waw.pl/resources/case2021/data.zip.

Table 2 Evaluated results

Model	Evaluation metric	Precision	Recall	F1-score
$\texttt{BERT}_{\text{LEMM}}$	Micro Avg.	0.75	0.75	0.75
	Macro Avg,	0.74	0.74	0.72
	Weighted Avg.	0.79	0.75	0.75
$\texttt{BERT}_{\text{CLEAN}}$	Micro Avg.	0.77	0.77	0.77
	Macro Avg,	0.77	0.77	0.74
	Weighted Avg.	0.82	0.77	0.76
$\texttt{ELMo}_{\text{LEMM}}$	Micro Avg.	0.53	0.53	0.53
	Macro Avg,	0.59	0.53	0.51
	Weighted Avg.	0.64	0.53	0.54
$\texttt{ELMo}_{\text{CLEAN}}$	Micro Avg.	0.59	0.59	0.59
	Macro Avg,	0.58	0.59	0.56
	Weighted Avg.	0.63	0.59	0.58
$\texttt{RoBERTa}_{\text{LEMM}}$	Micro Avg.	0.74	0.74	0.74
	Macro Avg,	0.73	0.73	0.71
	Weighted Avg.	0.78	0.74	0.74
$\texttt{RoBERTa}_{\text{CLEAN}}$	Micro Avg.	0.80	0.80	**0.80**
	Macro Avg,	0.78	0.80	**0.78**
	Weighted Avg.	0.83	0.80	**0.81**
$\texttt{XLNet}_{\text{LEMM}}$	Micro Avg.	0.76	0.76	0.76
	Macro Avg,	0.75	0.73	0.73
	Weighted Avg.	0.79	0.76	0.76
$\texttt{XLNet}_{\text{CLEAN}}$	Micro Avg.	0.79	0.79	0.79
	Macro Avg,	0.78	0.78	0.76
	Weighted Avg.	0.82	0.79	0.79

Followed by Model-8 $\texttt{XLNet}_{\text{CLEAN}}$, which uses the XLNet base model [6] architecture which also uses CLEAN corpus as input data for model training, achieves a weighted F1-score of 0.79, micro-F1-score of 0.79, and macro-F1-score of 0.76 as mentioned in Table 2. From Table 2, we can observe that all the models which have used CLEAN as input data for training have better performance in contrast to all the models using the same architecture and similar model configurations but using LEMM corpus as the input data. Another inference that we can draw is that Transformer [14] and self-attention models tend to perform much better on raw or natural data in comparison to data which is pre-processed, lemmatized text with the removal of stop words.

6 Error Analysis

To get a better understanding of the results, this section gives a short qualitative analysis of the misclassifications and hypotheses for the potential reasons for them. To study the misclassification, we analyzed it in two verticals: analysis with individual classes and for the architecture of the language models.

6.1 Error(s) Due to Redundancy in Corpus

As shown in Table 3, classes like NON_STATE_ACTOR_OVERTAKE_TER, OTHERS, and CHEM_WEAPON have consistently underperformed of all the classes from the corpus on the unseen test data. A few misclassified examples are mentioned below:

Text: *On August 11, ISIS militants captured a port in the town of Mocimboa da Praia.*
ISIS forces took control of an airfield located near the port
Correct: NON_STATE_ACTOR_OVERTAKE_TER
Prediction: GOVT_REG_TER

Text: *Residents return to Syria's Dabiq after Turkish-backed rebel fighters seized the town from the Islamic State group*
Correct: OTHERS
Prediction: NON_VIOL_TERR_TRANS

Text: *A rocket struck a hospital after dozens of people were killed and scores more were injured in a suspected chemical attack in Jinji. The suspected chemical attack in the rebel-held capital killed 28 people on Sunday including 7 children, opposition activists said.*
Correct: CHEM_WEAPON
Prediction: AIR_STR

The redundancy in the corpus also contributes significantly to the misclassification produced by the models which could be explained be due to the following reasons:

- Minority classes in an imbalanced dataset lack in terms of samples thus, it may lead to poor performance of the class since the model(s) are not much trained on minority classes. Therefore, very few unique features make it difficult to generalize.
- Classes could be very closely related to each other thus, having many common features making it complex for the model to accurately classify the correct label. As shown, classes NON_STATE_ACTOR_OVERTAKES_TER, GOV_REG_TERRIT, and HQ_EST are so closely related to each other that samples of GOV_REG_TERRIT are classified as NON_STATE_ACTORS_OVERTAKES_TER as seen in

Table 3 Worst-performing event types

Model	Class name	f1-score
$BERT_{LEMM}$	NON_STATE_ACTOR_OVERTAKE_TER	0.16
	OTHERS	0.16
	CHEM_WEAPON	0.36
$BERT_{CLEAN}$	CHEM_WEAPON	0.28
	OTHERS	0.40
$ELMo_{LEMM}$	OTHERS	0.13
	CHEM_WEAPON	0.20
	NON_STATE_ACTOR_OVERTAKE_TER	0.25
$ELMo_{CLEAN}$	CHEM_WEAPON	0.10
	GOVT_REG_TER	0.17
	OTHERS	0.26
$RoBERTa_{LEMM}$	OTHERS	0.29
	NON_STATE_ACTOR_OVERTAKE_TER	0.36
$RoBERTa_{CLEAN}$	OTHERS	0.41
	NON_STATE_ACTOR_OVERTAKE_TER	0.51
$XLNet_{LEMM}$	OTHERS	0.29
	NON_STATE_ACTOR_OVERTAKE_TER	0.30
$XLNet_{CLEAN}$	PROP_DISTRUCT	0.45
	OTHERS	0.50

Confusion matrix in Fig. 1vis-a-vis samples of NON_STATE_ACTORS_OVERTAKES_TER are predicted as HQ_EST class as shown in Confusion Matrix in Fig. 2a.

- The sample might contain text pertaining to multiple classes thus making it difficult for model to classify the correct labels. As shown above, the text sample-3 contains news related to both CHEM_WEAPON and AIR_STRIKE thus, the model failed to predict label as expected. Error trends due to this can be easily observed in Confusion Matrices for RoBERTa and XLNet in Fig. 1d, and Fig. 2d respectively.
- Human error(s) also contribute to errors produced by the language models.

6.2 *Error(s) Due to Model Architecture*

To understand the misclassification, Figs. 1b and 2b represents, the Confusion Matrices of $ELMo_{LEMM}$ and $BERT_{LEMM}$ which depict that these two models have consistently misclassified on the test data. According to the results of model predictions as found in Table 3, the overall descending order of model-wise performance follows RoBERTa, XLNet, BERT, and ELMo. The misclassification and the performance of the models could be justified by understanding their architecture and associated limitations. The

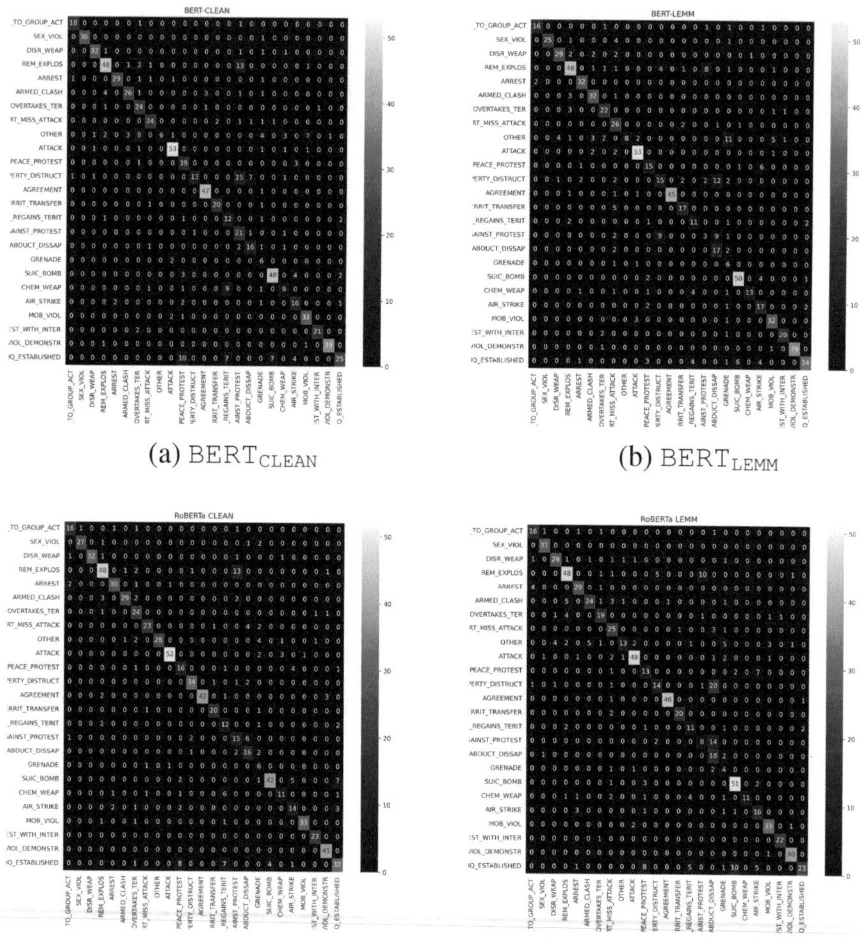

(a) $\texttt{BERT}_{\texttt{CLEAN}}$ (b) $\texttt{BERT}_{\texttt{LEMM}}$

(c) $\texttt{RoBERTa}_{\texttt{CLEAN}}$ (d) $\texttt{RoBERTa}_{\texttt{LEMM}}$

Fig. 1 Confusion matrices for the best results on different corpus used in the tenfold experiment

poor performance by the ELMo embeddings [4] is because only uses bidirectional LSTM [13], simply concatenated left-to-right and right-to-left, thus, it could not take advantage of both the contexts simultaneously. Therefore, ELMo embeddings are not truly bidirectional, lacking ability to capture the full context and all the associated features of the input text in order to give comparable performance to Transformer [14] based language models: RoBERTa, XLNet and BERT. ELMo models have produced the highest number of FPs and FNs and lowest number of TPs and TNs which can be seen in Confusion Matrix in Fig. 2a and b.

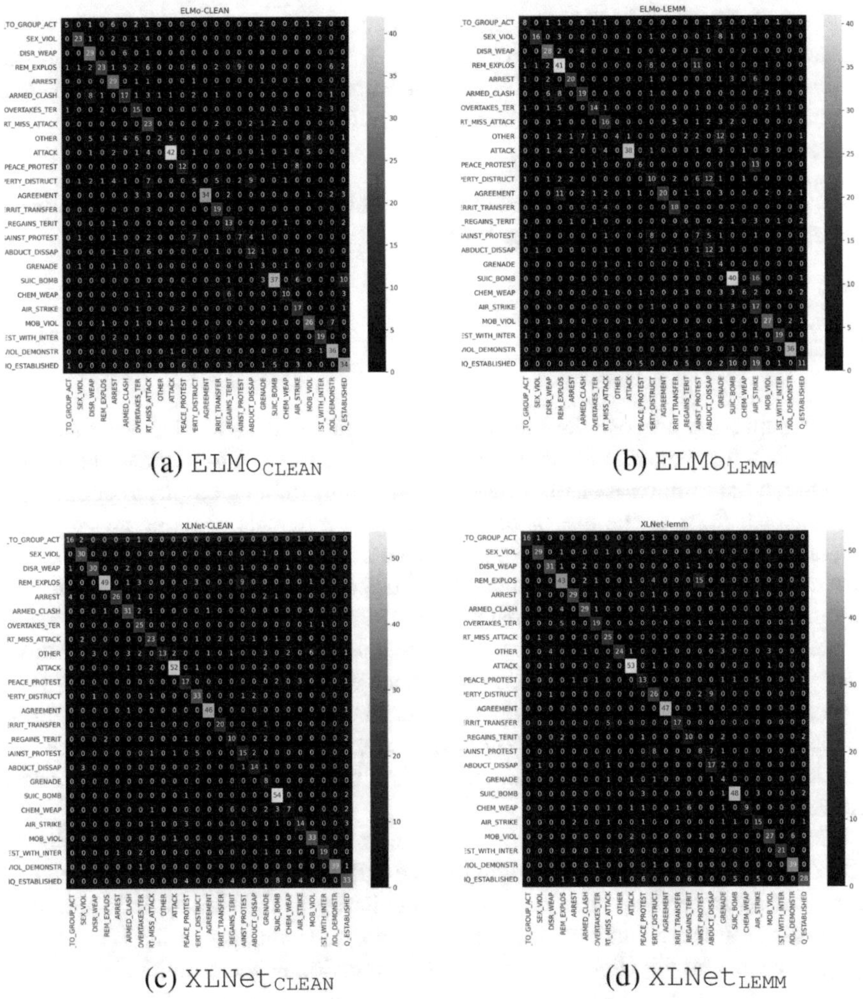

(a) $\texttt{ELMo}_{\texttt{CLEAN}}$ (b) $\texttt{ELMo}_{\texttt{LEMM}}$

(c) $\texttt{XLNet}_{\texttt{CLEAN}}$ (d) $\texttt{XLNet}_{\texttt{LEMM}}$

Fig. 2 Confusion matrices for the best results on different corpus used in the tenfold experiment

Our baseline model architecture BERT base model [1] gets outperformed by other Transformer-based models XLNet [6] and RoBERTa [5] due to the following reasons:

- BERT base is highly under-trained in comparison to RoBERTa and XLNet which is also argued in [5, 6] which also mention that BERT failed to achieve it's potential due. Results can be compared in Fig. 1a, c and Fig. 2c.

- BERT architecture ignores the interdependence of the masked positions, which results in a pretrain-finetune disparity, whereas XLNet architecture eliminates this independent assumption. Thus, XLNet has a higher number of TPs and less number of FPs and FNs which can be seen in Fig. 2d. Thus, having better performance.

We propose the aforementioned reasons could be possible explanations for misclassification by our models on the test data.

7 Conclusion

In this paper, we propose the use of various language models using deep learning approaches for the task of fine-grained classification of socio-political events. We resampled the original ACLED data for creating our training corpus to counter the imbalanced classes. The use of different language models gives us an empirical understanding of the architecture of language models and their impact on performance for a particular task. In comparison to the baseline figures provided by the organizers [10] and the top performers of the task, we achieved nearly similar results while only using a small fraction of the original data. This signifies that our methodology of re-sampling the training corpus to handle imbalanced class distribution had a positive effect and boosted the model performance. We also observe that attention models tend to perform better with unprocessed or raw data in comparison to corpus applied with preprocessing like lemmatization, stopword removal, and lower case conversion.

Consistent with previous work, we find that it is difficult to classify text snippets that contain event instances belonging to more than one class. We also found that a similar issue persists when due to high similarity in the classes and sharing of linguistic features. Situations like these give rise to challenges to the task of fine-grained event classification. Fine-grained classification of socio-political events has proven to be a more layered issue than was initially anticipated, but with radical developments in research, the task seems surmountable. Future work must focus on building upon previous endeavors, to find methodologies to accurately classify text containing events related to multiple classes and closely related classes. CASE @ ACL-IJCNLP 2021 Shared Task is a prominent step forward in achieving the goal of fine-grained classification of socio-political events and we look forward to seeing how future research will be affected by the work that has been done here.

References

1. Devlin J, Chang M, Lee K, Toutanova K (2018) BERT: pre-training of deep bidirectional transformers for language understanding. CoRR, arXiv:abs/1810.04805
2. Zhang Z, Sabuncu MR (2018) Generalized cross entropy loss for training deep neural networks with noisy labels. In: Bengio S, Wallach HM, Larochelle H, Grauman K, CesaBianchi N, Garnett R (eds) Advances in neural information processing systems 31: annual conference on neural information processing systems 2018, NeurIPS2018, December 3–8, 2018, Montréal Canada, pp 8792–8802
3. Raleigh C, Linke A, Hegre H, Karlsen J (2010) Introducing ACLED: an armed conflict location and event dataset: Special data feature. J Peace Res 47(5):651–660
4. Peters ME, Neumann M, Iyyer M, Gardner M, Clark C, Lee K, Zettlemoyer L (2018) Deep contextualized word rep-resentations
5. Liu Y, Ott M, Goyal N, Du J, Joshi M, Chen D, Levy O, Lewis M, Zettlemoyer L, Stoyanov V (2019) Roberta: a robustly optimized BERT pretraining approach. CoRR, arXiv:abs/1907.11692
6. Yang Z, Dai Z, Yang Y, Carbonell J, Salakhutdinov RR, Le QV (2019) Xlnet: generalized autoregressive pretraining for language understanding. In: Wallach H, Larochelle H, Beygelzimer A, d'Alché-Buc F, Fox E, Garnett R (eds) Advances in neural information processing, vol 32, pp 5754–5764. Curran Associates Inc.
7. Medhat W, Hassan A, Korashy H (2014) Sentiment analysis algorithms and applications: a survey. Ain Shams Eng J 5(4):1093–1113
8. Jindal N, Liu B (2007) Analyzing and detecting review spam. In: Seventh IEEE international conference on data mining (ICDM 2007), pp 547–552
9. Lloret E, Palomar M (2012) Text summarisation in progress: a literature review. Artif Intell Rev 37(1):1–41
10. Haneczok J, Jacquet G, Piskorski J, Stefanovitch N (2021) Fine-grained event classification in news-like text snippets - shared task 2, CASE 2021. In: Proceedings of the 4th workshop on challenges and applications of automated extraction of socio-political events from text (CASE 2021), pp 179–192. Association for Computational Linguistics
11. Radford BJ (2021) Case 2021 task 2: zero-shot classification of fine-grained sociopolitical events with transformer models. In: Proceedings of the 4th workshop on challenges and applications of automated extraction of socio-political events from text (CASE 2021), pp 203–207
12. Hürriyetoğlu A (ed) (2021) Proceedings of the 4th workshop on challenges and applications of automated extraction of socio-political events from text (CASE 2021). Association for Computational Linguistics
13. Schuster M, Paliwal K (1997) Bidirectional recurrent neural networks. IEEE Trans Signal Process 45:2673–2681
14. Vaswani A, Shazeer N, Parmar N, Uszkoreit J, Jones L, Gomez AN, Kaiser L, Polosukhin I (2017) Attention is all you need. In: Guyon I, von Luxburg U, Bengio S, Wallach HM, Fergus R, Vishwanathan SVN, Garnett R (eds) Advances in neural information processing systems 30: annual conference on neural information processing systems 2017, December 4–9, 2017, Long Beach, CA, USA, pp 5998–6008
15. Piskorski J, Haneczok J, Jacquet G (2020) New benchmark corpus and models for fine-grained event classification: to BERT or not to BERT? In: Proceedings of the 28th international conference on computational linguistics, pp 6663–6678, Barcelona, Spain (Online). International Committee on Computational Linguistics
16. Kingma DP, Ba J (2015) Adam: a method for stochastic optimization. In: Bengio Y, LeCun Y (eds) 3rd international conference on learning representations, ICLR 2015, San Diego, CA, USA, May 7–9, 2015, conference track proceedings
17. Kent S, Krumbiegel T (2021) CASE 2021 task 2 socio-political fine-grained event classification using fine-tuned RoBERT a document embeddings. In: Proceedings of the 4th workshop on challenges and applications of automated extraction of socio-political events from text (CASE 2021), pp 208–217. Online. Association for Computational Linguistics

18. Deerwester SC, Dumais ST, Landauer TK, Furnas GW, Harshman RA (1990) J Am Soc Inf Sci 41(6):391–407
19. Mikolov T, Sutskever I, Chen K, Corrado G, Dean J (2013) Distributed representations of words and phrases and their compositionality. In: Neural and information processing system (NIPS)
20. Naughton M, Stokes N, Carthy J (2008) Investigating statistical techniques for sentence-level event classification. In: 22nd international conference on computational linguistics, proceedings of the conference,18–22 August 2008, Manchester, UK, pp 617–624
21. Nugent T, Petroni F, Raman N, Carstens L, Leidner JL (2017) A comparison of classification models for natural disaster and critical event detection from news. In: 2017 IEEE international conference on big data, big data 2017,Boston, MA, USA, December 11–14, 2017, pp 3750–3759. IEEE Computer Society

Fruit Recognition and Freshness Detection Using Convolutional Neural Networks

R. Helen, T. Thenmozhi, R. Nithya Kalyani, and T. Shanmuga Priya

1 Introduction

In our daily routine, fruit sorting is one of the most tedious processes. It requires a lot of effort and manpower and consumes a lot of time as well. In modern times, industries are approving automation and smart machines to make their work easier and more efficient in the field of fruit sorting. So, a smart fruit sorting system has been introduced to identify the type of fruit and then tag the name of a specific fruit. In the supermarket, the fruits are collected from storage and sorted to identify the fresh fruits such as banana, apple, and orange as per the choice of the user. The improper handling and storage of fruits might cause food poisoning. To prevent fruit decay and to consume fresh fruits, many methods have been analyzed by sorting fruits automatically. Sorting fruit one by one using hands is one of the most difficult jobs. Ultimately, it consumes a lot of effort, manpower, and time as well. The color and shape of fruit are identified through Machine Learning algorithms [1–3]. The quality of fruits and vegetables is classified through computer vision applications [4]. The fruit freshness is identified through machine learning methods [13]. The quality of fruits and vegetables has been classified through deep neural networks [11, 12]. The

R. Helen (✉) · T. Thenmozhi · R. Nithya Kalyani · T. Shanmuga Priya
Department of Electrical and Electronics Engineering, Thiagarajar College of Engineering, Madurai, Tamil Nadu, India
e-mail: rheee@tce.edu

T. Thenmozhi
e-mail: thenmozhithangaraj22@gmail.com

R. Nithya Kalyani
e-mail: nithyameena75@gmail.com

T. Shanmuga Priya
e-mail: shanmugapriya.pct@gmail.com

P. Singh et al. (eds.), *Machine Learning and Computational Intelligence Techniques for Data Engineering*, Lecture Notes in Electrical Engineering 998,
https://doi.org/10.1007/978-981-99-0047-3_43

fruit type and freshness have been recognized through deep learning techniques [14–16]. The quality control and type of fruit are detected through Convolutional Neural Networks [17]. The relative accuracy improvement upto 38% is obtained for small datasets by novel assisted Artificial Neural Networks [18]. Artificial Neural Networks are used for solving small-scale problems by spatial transformer Networks as input [19]. Features of fruits and vegetables were preprocessed, segmented, and extracted and quality was classified based on color, texture, size, and shape through computer vision [20]. In this paper, a fruit sorting machine is introduced to differentiate between various fruits. Using TensorFlow and OpenCV, an apple is detected in front of the Raspberry pi camera and diagnosed the freshness of the fruit within a limited time.

2 Materials and Methods

The system consists of mainly five phases as follows:

i. Image Acquisition
ii. Image pre-processing
iii. Image Segmentation
iv. Feature extraction
v. Classification

The block diagram of the Digital Image processing system is shown in Fig. 1.

2.1 *Image Acquisition*

The image is captured with the help of a pi camera which is connected to the raspberry pi 3 shown in Fig. 2a. The blocks of Raspberry Pi Model B are represented in Fig. 2b. The fruit must be placed in front of the camera with a plain white background for easy capturing of the image without any distortion. The fruit may be placed at any degree of angle or elevation. The pi camera can capture directly any object which supports

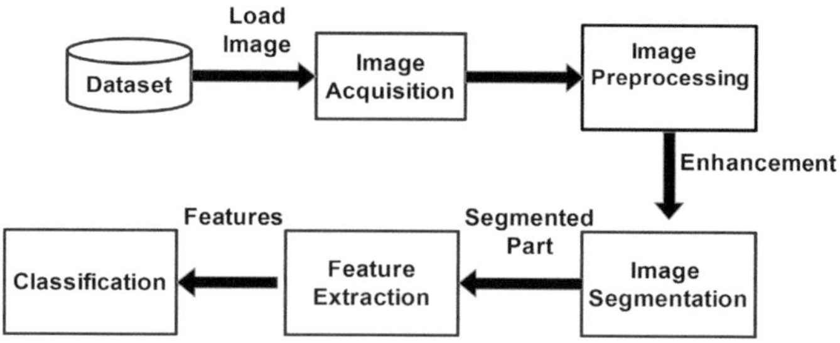

Fig. 1 Block diagram of digital image processing system

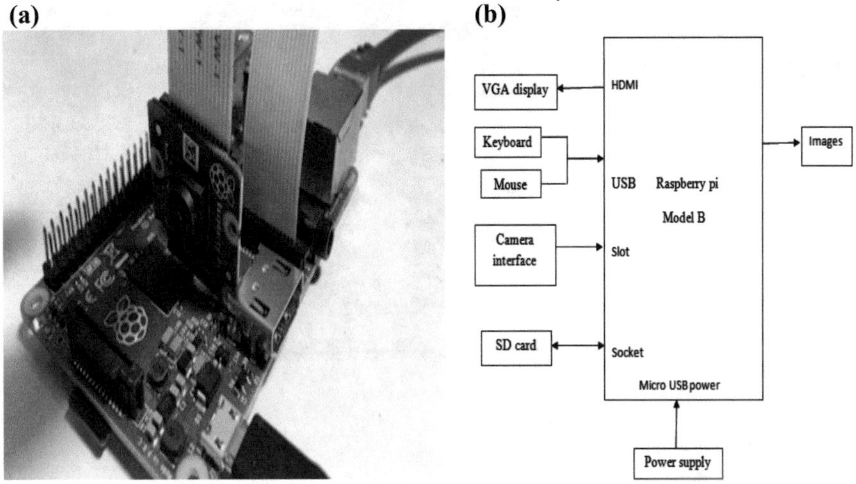

Fig. 2 **a** Pi camera setup with Raspberry Pi 3 and **b** Block Representation of Raspberry Pi

Python's buffer protocol. The protocol helps to pass the object to the destination and the image data will be written directly to the object. The fruit and vegetable grading system has been done through computer intelligence using Raspberry Pi [6, 7].

2.2 *Image Pre-Processing*

OpenCV is a library of 2500 programming algorithms mainly designed for real-time computer vision applications [5]. It uses NumPy arrays as images and colors as default in planar BGR. The image is then converted from BGR format to 8*8 image format. The image is preprocessed to remove unwanted disturbances and provide a better resolution of an image.

2.3 *Image Segmentation*

The digital image is subdivided into various segments which helps to reduce the complexity of the image to make the processing easier. This process is known as Image segmentation. Although different types of Image segmentation are available, Threshold-based image segmentation is preferred in this paper [9, 10]. It helps to create a binary image based on the threshold value of the pixel intensity of an image. It is one of the best and most efficient methods for fruit segmentation.

Fig. 3 Pictorial representation of fresh and rotten fruits

2.4 *Feature Extraction*

2.4.1 Training Data

The Input data is obtained from Kaggle competitions which have three types of fruits—apples, bananas, and oranges with two classes (fresh and rotten) as targets for each fruit. Each fruit is taken from Pi's camera at different angles. Images of fresh bananas, apples, and oranges taken in all kinds of angles are stored in three folders. Similarly, Images of rotten bananas, apples, and oranges taken in all kinds of angles are stored in another three folders. So, from six folders, a total of 5989 images are considered as input data. Out of which, 3596 images are utilized for training, 596 images for validation, and 1797 images for testing the data. The fruit samples per class are demonstrated in Fig. 3. The fruit images are the input data, extracted as features that feed into deep learning algorithms for classification. The fruit names are categorized as a Target. The Training data is utilized for training the model. This trained data is then fed into machine learning and deep learning models so that model can provide better results.

2.4.2 Testing Data

After training the model, testing data helps to evaluate the performance of the model. For testing the model, the new images are captured using the raspberry pi camera module. This camera module produces a 5MP clear resolution image, or 1080p HD video at a recording speed of 30fps. At first, the predicted data is received from the trained model without giving the labels, then the true labels are compared with the predicted data and the performance of the model is obtained. While training the model, Validation data will be used to check the performance during training and test data will be used after training the model. The input images need to be converted into array form for the training and testing process because Deep Learning models access numerical data only. After converting the images into array form, the features/pixels range from 0 to 255. Further, the Feature scaling process helps to change the range of features/pixels from 0–255 to 0–1 range. This scaling reduces the training time. The validation data is used to evaluate the performance of the model during training time.

2.5 Classification

In the Classification phase, the type of fruit and freshness (i.e., rotten or good) of fruit are classified by deep learning algorithms [8]. The Pictorial representation of fresh and rotten fruits is displayed in Fig. 3.

3 Proposed Methodology

This work is implemented in Python V3.7 using TensorFlow V2.4.0. The image is obtained from the pi camera, extracted, resized, and scaled using OpenCV. The images are converted into array variables for the training process. Figure 4 demonstrates the block diagram of the proposed model.

Base Sequential model is used for training and further tuning of parameters. Conv2D is a Two-Dimensional convolutional layer (where filters are applied to the

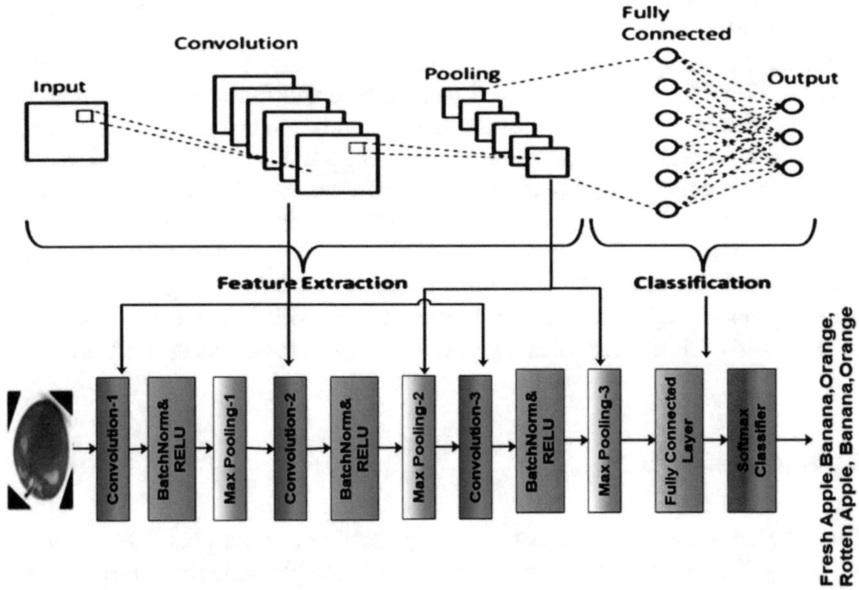

Fig. 4 Block diagram of the proposed model

original image with specific features map to reduce the number of features), creating the convolution kernel with a fixed size of boxes applied on the image which is shown in Fig. 4. Sixteen filters are preferred which help to produce a tensor of outputs. The input of the image is given as a size of 100 width and 100 height and 3 channels for RGB. An activation function is a node placed at the end or in between the layers of the neural network model. Activation function helps to pass the right neuron or fire the neuron. So, the activation function of the node defines the output of the specific node by giving a set of inputs. Max pooling is a pooling operation that calculates the maximum value in each batch of each feature map shown in Fig. 5b. It takes the value from input vectors and prepares the vector for the next layers. The dropout layer drops irrelevant neurons from previous layers shown in Fig. 5a. This helps to avoid overfitting problems. In overfitting, the model gives better accuracy during training time than testing time. Flatten layer converts the 2D array into a 1D array of all features which is shown in Fig. 5c. The dense layer reduces the count of neurons obtained from the flattened layer. The dense layer uses all the inputs of previous layer neurons and performs calculations and sends outputs. Adam Optimizer is a function that helps to reduce losses. Here, Categorical Loss entropy is calculated to find losses. So, to reduce the losses, the learning rate of the neural network is defined by the optimizer. Now reduceLROnplateau is used to reduce the learning rate when there is no improvement in the accuracy. The Number of CNN layers and their parameters (neurons) is shown in Table 1. Here, the performance is evaluated using accuracy by calculating the percentage of correct predictions and overall predictions on the validation set.

This method makes the model less dependent on pre-processing of an image which decreases the necessity of human effort. Efficient filters are used to exploit spatial locality between neurons.

In this paper, three layered sequential two-dimensional Convolutional Neural Network (2D-CNN) is proposed to detect the type and freshness of the fruit. The parameter designed in terms of the number of neurons is mentioned in Table 1. Three dropout layers are also included to avoid overfitting problems. The proposed methodology helps to reduce the computational time and minimizes the losses occurred while training the model. This method helps to share the parameters and reduce the dimensions of the image which makes the model easy to deploy and higher classification accuracy is obtained. In this paper, Input dataset of 6000 fruit images was captured in all degrees of angle. This method successfully classified the images without any external paid OCR or NLP software's help that makes the model cost-effective.

4 Hardware Setup

Here, the hardware setup consists of the pi camera and Raspberry Pi Model B. Raspberry Pi cameras are capable of high-resolution photographs, along with full HD 1080p video which can be fully controlled programmatically. The Raspberry Pi Camera module cable is inserted into the Raspberry Pi camera port. The cable slots

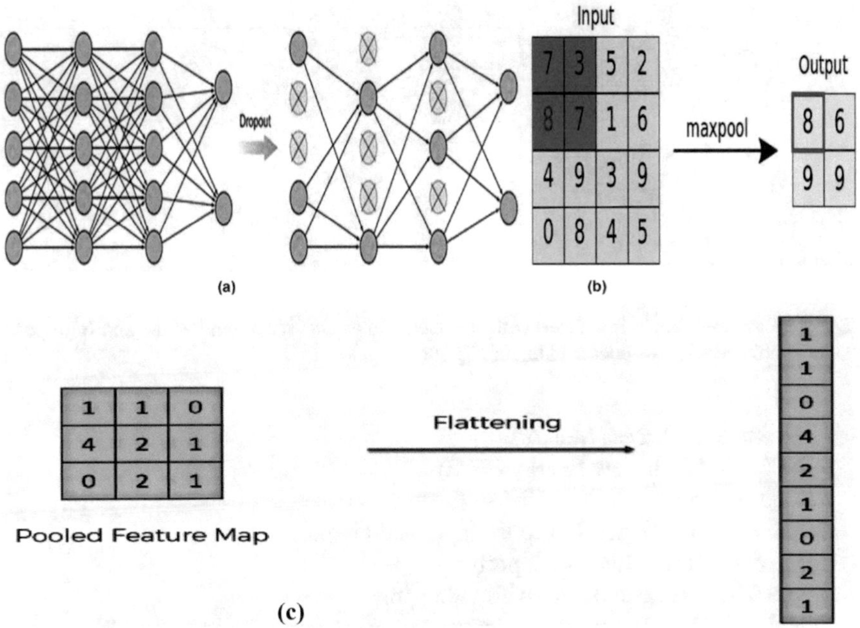

Fig. 5 **a** Dropout, **b** Pooling, and **c** Flattening process of CNN

Table 1 Convolutional neural network layers and their parameters

Layers	Parameters
CNN-1(CON2D)	(32,3,3)
Dropout_1	0.3
CNN-2(CON2D)	(64,3,3)
Dropout_2	0.4
CNN-3(CON2D)	(128,3,3)
Dropout_3	0.5

into the connector are situated between the USB and micro-HDMI ports, with the silver connectors facing the micro-HDMI ports shown in Fig. 6a and b. Command-line application is used to allow the camera module to capture the picture of the fruit. The image to be captured is to be placed in front of the camera. This captured image is fed into the proposed model and the type and condition of the fruit are displayed as an output on the screen.

4.1 Hardware Specifications

- ARM processor with 64bit Quad-Core at 1.2 GHz

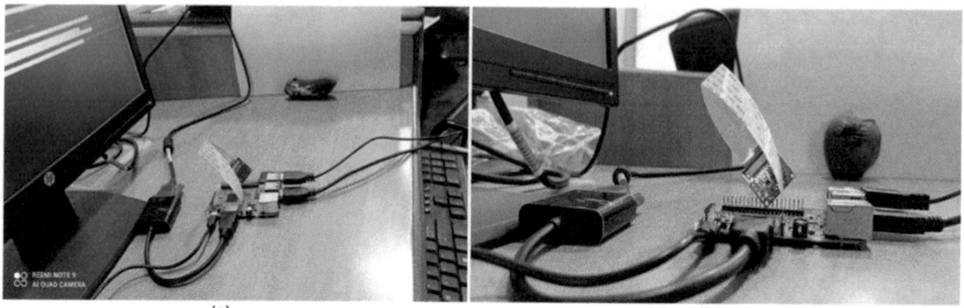

Fig. 6 **a** Proposed model interfaced with Raspberry Pi to classify rotten banana and **b** proposed model interfaced with Raspberry Pi to classify rotten apple

- 1 GB Random Access Memory
- In-built Bluetooth Low Energy (BLE)
- 100 Base Ethernet
- 40-pin extended General Purpose Input and Output
- 4 Universal Serial Bus with 2 ports
- 4 Pole 3.3 mm stereo output with video port
- Full-size HDMI Camera Serial Interface camera port for connecting Pi camera
- Display Serial Interface display port for connecting touchscreen display
- Micro SD port for loading customized OS and storing data

5 Results and Discussion

In this paper, Performance metrics are determined through accuracy. Accuracy is the number of correctly recognized images from all the input images. This proposed model achieves 94% accuracy during the fruit sorting process. Similarly, the model achieves 96% accuracy during the identification of the freshness of the fruit. Training Loss and validation losses for fruit sorting are shown in Fig. 7a. The Training and validation losses for fruit freshness classification are shown in Fig. 7b.

The Accuracy varies with the number of epochs which is shown in Fig. 8. The testing loss and accuracy are explained in Table 2.

The results of carrying out these experiments demonstrate that the proposed approach is capable to recognize the fruit name automatically with a high degree of accuracy of 96%. This real-time project consumes an overall 10 s for the whole process as it has less training time with minimum testing losses. This proposed technique is outperformed when compared with other machine learning algorithms in sorting and grading the good quality of a fruit and it was shown in Table 3. Hence overall machine learning techniques gave an accuracy of 83%.

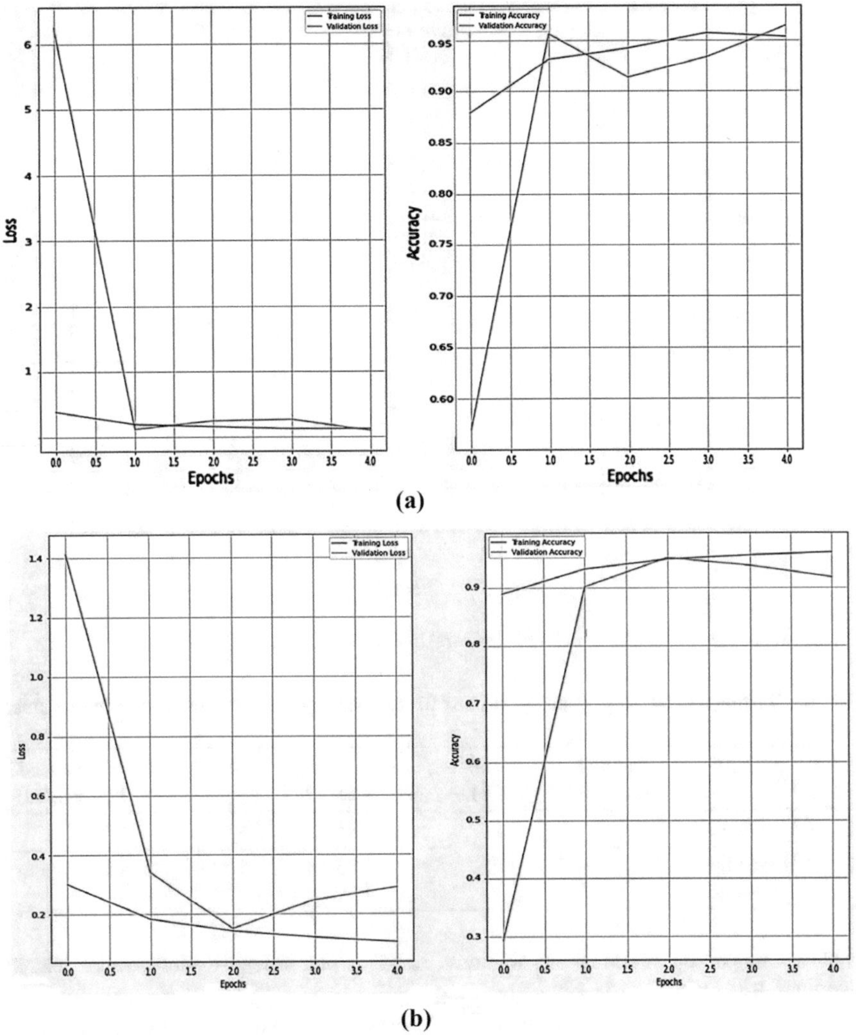

Fig. 7 **a** Training Accuracy and Loss for Fruit sorting and **b** Training loss and Accuracy for fruit freshness classification

6 Conclusion

In this project, Images of three types of fruit such as apple, oranges, and bananas are taken from Pi's camera. These fruits are correctly classified and sorted. The freshness condition of these fruits (rotten or fresh) is also determined by the Sequential convolutional Neural Networks model. This method gives an accuracy of 94% for sorting and 96% for freshness detection. This real-time project consumes less training time with minimum testing losses.

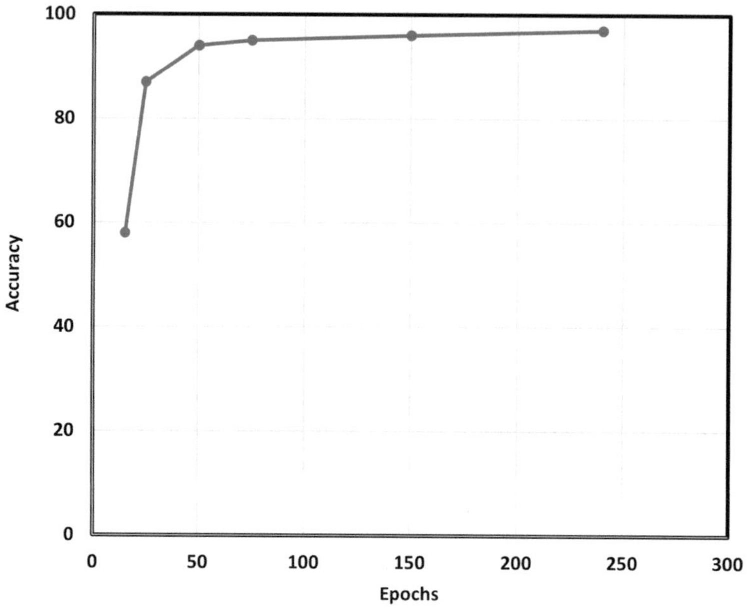

Fig. 8 Accuracy varies with epochs for fruit freshness classification

Table 2 Testing and training accuracy and loss for fruit sorting and freshness condition classification

S.no	Activity	Training accuracy (%)	Training loss (%)	Testing accuracy	Testing loss	Consumed time (Min)
1	Sorting	89	11	94	6	5
2	Condition (Freshness)	93	7	96	4	3

Table 3 Comparison of sorting and freshness detection with state-of-the-art machine learning algorithms with the proposed CNN

Accuracy	Artificial neural network	K-nearest neighbor	Support vector machine	Logistic regression	Convolutional neural networks
Sorting accuracy (%)	90	80	83	84	**94**
Freshness condition accuracy	92	85	89	86	**96**

References

1. Rege S, Memane R, Phatak M, Agarwal P (2013) 2D geometric shape and color recognition using digital image processing. Int J Adv Res Electr, Electron Instrum Eng 2(6):2479–2487
2. Zawbaa HM, Abbass M, Hazman M, Hassenian AE (2014) Automatic fruit image recognition system based on shape and color features. In: International Conference on Advanced Machine Learning Technologies and Applications). Springer, Cham, pp. 278–290
3. Rocha A, Hauagge DC, Wainer J, Goldenstein S (2008) Automatic produce classification from images using color, texture and appearance cues. In: 2008 XXI Brazilian Symposium on Computer Graphics and Image Processing. IEEE, pp. 3–10
4. Bhargava A, Bansal A (2021) Fruits and vegetables quality evaluation using computer vision: a review. J King Saud Univ-Comput Inf Sci 33(3):243–257
5. Patil MSV, Jadhav MVM, Dalvi MKK, Kulkarni MB (2014) Fruit quality detection using opencv/python. system 1722:1730
6. Pandey R, Naik S, Marfatia R (2013) Image processing and machine learning for automated fruit grading system: a technical review. Int J Comput Appl 81(16):29–39
7. Mhaski RR, Chopade PB, Dale MP (2015) Determination of ripeness and grading of tomato using image analysis on Raspberry Pi. In: 2015 Communication, Control and Intelligent Systems (CCIS). IEEE, pp. 214–220
8. Ertam F, Aydın G (2017) Data classification with deep learning using Tensorflow. In: 2017 international conference on computer science and engineering (UBMK). IEEE, pp. 755–758
9. Nandhini P, Jaya J (2014) Image segmentation for food quality evaluation using computer vision system. Int. J. Eng. Res. Appl. 4(2), 01–03
10. Girshick R, Donahue J, Darrell T, Malik J (2014) Rich feature hierarchies for accurate object detection and semantic segmentation. In: Proceedings of the IEEE conference on computer vision and pattern recognition. pp. 580–587
11. Zeng G (2017) Fruit and vegetables classification system using image saliency and convolutional neural network. In: 2017 IEEE 3rd Information Technology and Mechatronics Engineering Conference (ITOEC). IEEE, pp. 613–617
12. Toshev A, Szegedy C (2014) Deeppose: Human pose estimation via deep neural networks. In: Proceedings of the IEEE conference on computer vision and pattern recognition. pp. 1653–1660
13. Mudaliar G, Priyadarshini RK (2021) A machine learning approach for predicting fruit freshness classification. In: International research journal of engineering and technology (IRJET) vol 08, Issue: 05. e-ISSN: 2395-0056
14. Valentino F, Cenggoro TW, Pardamean B (2021) A design of deep learning experimentation for fruit freshness detection. In: IOP Conference Series: Earth and Environmental Science, vol. 794, No. 1. IOP Publishing, p. 012110
15. Chung DTP, Van Tai D (2019) A fruits recognition system based on a modern deep learning technique. In: Journal of physics: conference series, vol 1327, No. 1. IOP Publishing, p. 012050
16. Fu Y (2020) Fruit freshness grading using deep learning. Doctoral dissertation, Auckland University of Technology
17. Naranjo-Torres J, Mora M, Hernández-García R, Barrientos RJ, Fredes C, Valenzuela A (2020) A review of convolutional neural network applied to fruit image processing. Appl Sci 10(10), 3443
18. Hijazi A, Al-Dahidi S, Altarazi S (2020) A novel assisted artificial neural network modeling approach for improved accuracy using small datasets: application in residual strength evaluation of panels with multiple site damage cracks. Appl Sci 10(22):8255
19. Sharma S, Shivhare SN, Singh N, Kumar K (2019). Computationally efficient ann model for small-scale problems. In: Machine intelligence and signal analysis. Springer, Singapore, pp. 423–435
20. Bhargava A, Bansal A (2018) Fruits and vegetables quality evaluation using computer vision: a review. J King Saud Univ Comput Inf Sci

Modernizing Patch Antenna Wearables for 5G Applications

T. N. Suresh Babu and D. Sivakumar

1 Introduction

The intensive research into fifth-generation (5G) technology is a strong insignia of the methodological revolution mandatory to report the emergent demand and requirements for wireless broadband and Internet of Things (IoT)-based applications. The researchers can refine their study objectives and contribute to progress thanks to the timely improvement in 5G technology. Not just smartphones, but also numerous IoT devices will leverage 5G technologies to enable faster transmission speeds, lower latency, and therefore increased remote execution capacity [1].

Wearables will be able to be deployed in large numbers per macrocell or picocell thanks to 5G's increased capacity and throughput, reduced latency, and network densification. New higher-frequency spectrums, such as millimetre wave (mmW) bands, are projected to be supported by future wearables. As a result, a separate model for mmW propagation, communication topologies, and component design paradigms is required. Materials, fabrication methods, and assessment procedures for wearable antennas should all be considered during the design and evaluation process. The state-of-the-art review on wearable antennas operating across various frequencies demonstrated a considerable movement away from basic topologies, materials, and manufacturing methods in the mmW frequencies, in preparation for 5G-centric applications. The fabrication process has improved as the antenna has progressed from simple microstrip or planar monopole-based topologies to more advanced arrays

T. N. Suresh Babu (✉)
Department of Electronics and Communication Engineering, Adhiparasakthi Engineering College, Melmaruvathur, Tamilnadu 603 319, India
e-mail: sureshbabu@apec.edu.in

D. Sivakumar
Department of Computer Science and Engineering, AMET University, Kanathur, Tamilnadu 603 112, India

P. Singh et al. (eds.), *Machine Learning and Computational Intelligence Techniques for Data Engineering*, Lecture Notes in Electrical Engineering 998,
https://doi.org/10.1007/978-981-99-0047-3_44

and metamaterial-based topologies. Laser cutting, embroidery, metallized eyelets, inkjet printing, and other techniques are projected to enable the accurate production of increasingly complex structures with smaller dimensional tolerances for higher-frequency operation. In addition to the standard assessments, wearable antenna evaluations include SAR analysis, deformation studies on detailed human phantoms, reliability studies, functionality tests, and operation under a variety of environmental circumstances [2–4].

Wearable technology is growing more desirable for a variety of applications, ranging from the military to health care to consumer electronics. They'll also play a crucial role in the next 5G networks, which are expected to provide higher bit rates, fewer outages, and a broader coverage area than previous technologies. Because of the limited accessible spectrum and the need for high-speed data transfer, 5G is expected to utilize higher-frequency spectra in the mmW bands. This necessitates a new mmW propagation model, as well as new communication architectures and component design paradigms, including for wearable devices.

For 5G Wearable Applications, the reported multiple-input multiple-output (MIMO) antenna [5] covers sub-6 GHz frequencies. A new tri-band compact coplanar waveguide (CPW)-fed liquid crystal polymer (LCP)-based antenna [6] for WLAN, Worldwide Interoperability for Microwave Access (WiMAX), and 5G systems is described. For sub-6 GHz 5G wireless applications, a high-performance bus-shaped flexible tri-band antenna has been proposed [7]. An inductive ground plane and a slots-loaded patch make up a novel seven-band miniaturized antenna for wearable applications [8]. For WLAN applications at 2.45 GHz (2.34–2.93 GHz) or 3.1 GHz (2.81–3.77 GHz), a small 5G monopole reconfigurable antenna is proposed [9]. An ultra-wideband (UWB) MIMO antenna is proposed for 4.5G/5G wearable devices that cover the 3.3–10.9 GHz band with a reflection coefficient of less than −10 dB [10]. An UWB antenna with an artificial magnetic conductor (AMC) as a meta surface [11] is exhibited for IoT applications operating at 5G (10.125 to 10.225 GHz) operating frequencies to demonstrate bandwidth improvement. Based on the defected ground surface (DGS) concept, a T-shaped patch antenna and right-angled-shaped slots assembly [12] are proposed to generate numerous resonances in the Ka-band (26–40 GHz). A unique 3D printed patch antenna for 5G wearable applications has been created, which operates in the 28 GHz frequency band recommended for future communications systems [13]. For millimetre wave applications, a wearable microstrip jeans 5G antenna [14] is proposed, which spans the frequency range of 21.18 GHz to 36.59 GHz. For wearable applications of 5G and beyond 5G networks, a flexible microstrip patch antenna array [15] constructed on a liquid crystal polymer (LCP) substrate is proposed. For high-speed 5G applications, an improved fabric wearable antenna [16] has been reported. The article [17] presents wearable circular polarized antennas for health care, 5G, energy harvesting, and IoT systems. However, not much information is available on meta surface-enabled multiband planar antenna for sub-6 GHz and beyond-6 GHz 5G applications in a wearable scenario.

In this study, a compact tetra-band wearable textile (jeans) patch antenna is developed for sub-5G and beyond 5G applications, consisting of an inductive Meta surface ground plane and a slots filled patch. The antenna design, parameter analysis, and simulation results, including S_{11}, radiation patterns, and peak gain, are described and shown. In addition, the suggested antenna is examined under various bending conditions. Furthermore, the antenna performance is evaluated at various body locations, and the specific absorption ratio (SAR) performance is modeled using CST Microwave StudioTM software on a human body model.

2 Antenna Design

The envisioned tetra-band antenna's configuration is shown in Fig. 1. The antenna is developed on a 1.8 mm thick jeans substrate with a permittivity of 1.6. The overall dimensions are 30 mm × 42 mm. An inductive Meta surface ground is made of a free 5 × 4 array of periodic square copper patches, each measuring 5 × 5 mm on the substrate.

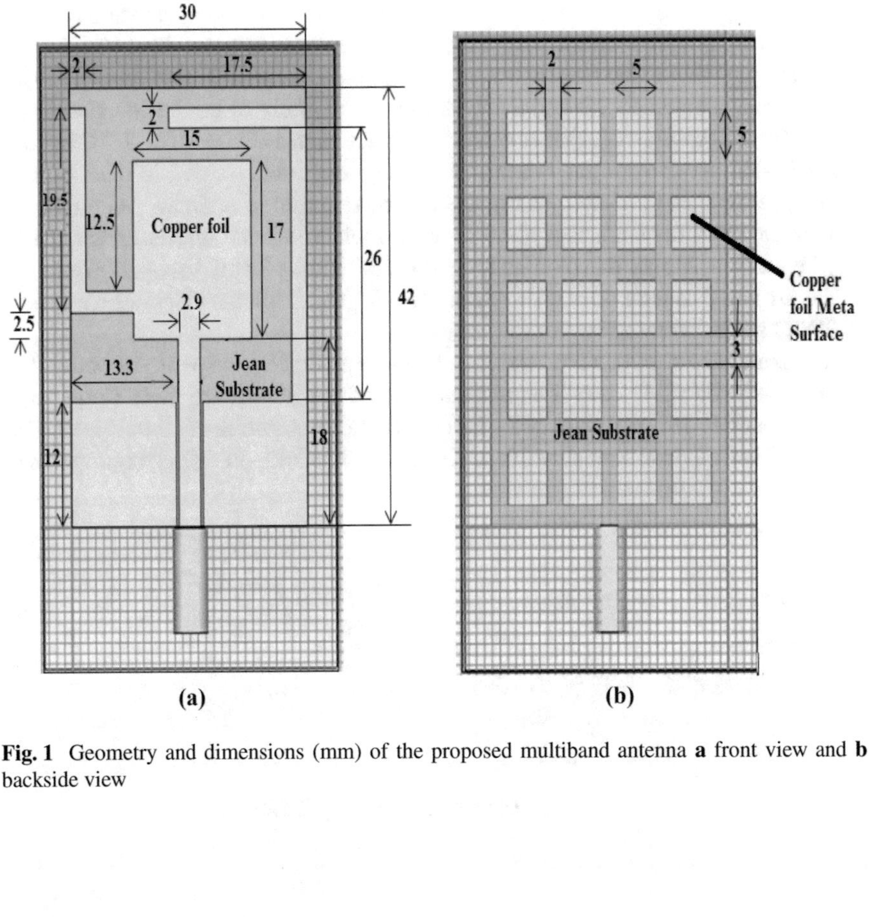

Fig. 1 Geometry and dimensions (mm) of the proposed multiband antenna **a** front view and **b** backside view

3 Antenna Performance Analysis

The proposed antenna's performance is assessed using free space return loss (S_{11}), gain radiation pattern, directivity radiation pattern, and surface current distribution characteristics. The proposed antenna is designed and simulated using Computer Simulation Technology (CST). The antenna design starts with a simple rectangular patch antenna with a coaxial-fed. Then, by attaching L-shaped strips with patches and also modifying the copper ground to 5×4 array of periodic square copper patches, each of size 5×5 mm, the meta surface ground is created.

3.1 *Simulation Results*

5G will require spectrum in three essential frequency ranges to provide comprehensive coverage and serve all use cases in the future. Sub-1 GHz, 1–6 GHz, and over 6 GHz are the three ranges. Broad coverage in urban, suburban, and rural regions, as well as support for Internet of Things (IoT) services, is possible at sub-1 GHz. 1–6 GHz provides a decent balance of coverage and capacity. This covers a band of frequencies between 3.3 and 3.8 GHz, which is likely to be the foundation for many early 5G services. It also covers frequencies such as 1800 MHz, 2.3 GHz, and 2.6 GHz that may be assigned to or reformed by operators for 5G. In the long run, more spectrums in the 3–24 GHz range will be required to maintain 5G quality of service and boost demand. To achieve the ultra-high broadband speeds envisioned for 5G, frequencies higher than 6 GHz are required. Figures 2 and 3 show the response of the proposed antenna design's return loss (S_{11}) and voltage standing wave ratio (VSWR) against frequency.

The simulated results were carried out and show that S_{11} response of better than -10 dB is achievable over four resonance bands: 1.909–2.426 GHz (impedance bandwidth of 24.7%), 3.498–4.211 GHz (impedance bandwidth of 18.03%), 19.838–21.196 GHz (impedance bandwidth of 6.62%), and 24.343–30.687 GHz (impedance

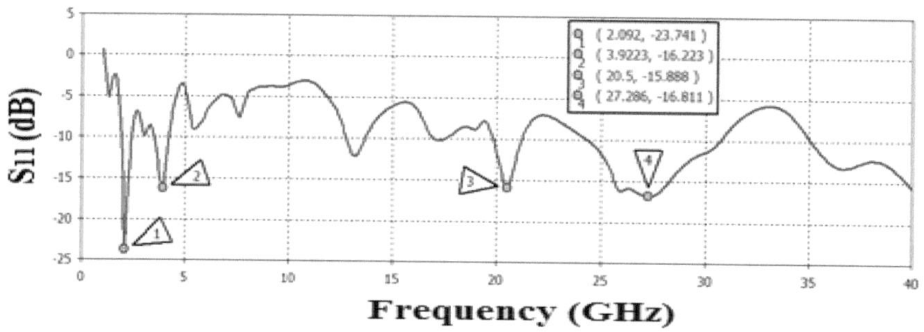

Fig. 2 Return loss (S_{11}) versus frequency

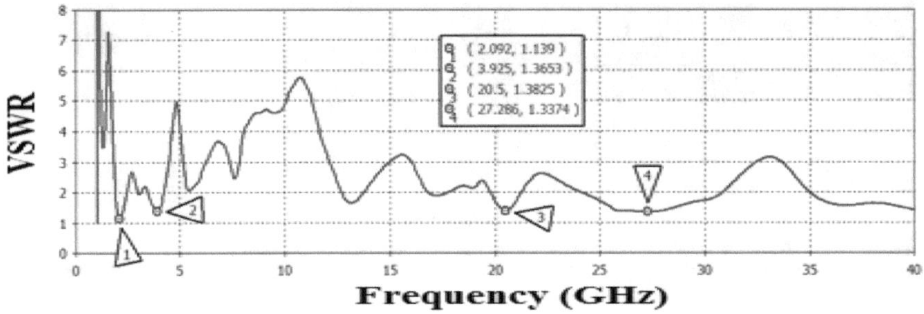

Fig. 3 VSWR versus frequency

bandwidth of 23.25%) with VSWR less than 2. Although more return loss is desirable in this case, there is no gain over 10 dB return loss because the antenna already receives more than 90% of the available power. When the VSWR is less than 2, the antenna is properly matched to the transmission line, and it receives more power. Hence, the antenna sufficiently supports the communication operation in the sub-(1–6) GHz band and also beyond the 6 GHz millimetre wave frequency range.

The simulated far-field radiation patterns at frequencies of 2.092, 3.925, 20.5, and 27.286 GHz are depicted in Fig. 4. It can be concluded that the H-plane (xy-plane) of the antenna has an almost omnidirectional radiation pattern and the obtained average gains are approximately 2.42 dBi, 4.353 dBi, 8.05 dBi, and 9.157 dBi for 2.092, 3.925, 20.5, and 27.286 GHz, respectively.

Figure 5 shows the simulated directivity for frequencies of 2.092, 3.925, 20.5, and 27.286 GHz. The surface current distribution of the antenna can be used to analyze the antenna's resonant properties. At frequencies of 2.092, 3.925, 20.5, and 27.286 GHz, Fig. 6 depicts the antenna's simulated current distribution. The parameter simulation study, shown in Fig. 6, is carried out on the length of both strips and the length of the rectangular patch based on the present distribution of the aforesaid four frequencies.

3.2 Bending Performance

Bending tests are essential because the bent nature of the jeans dielectric substrate allows the antenna to be integrated into various electrical devices in a curved shape.

Three bending models with radii of $R = 8$ mm, $R = 10$ mm, and $R = 50$ mm were simulated to test the antenna's bending performance. Figure 7 depicts the proposed antenna bending model with S_{11} versus frequency. Although the return loss under limit bending varies substantially in the S_{11} test, it is still less than -10 dB in the operational frequency band, which matches the actual engineering requirements. In summary, the antenna's desired bandwidth is preserved in both circumstances, and the antenna's performance remains stable under bending conditions. The working

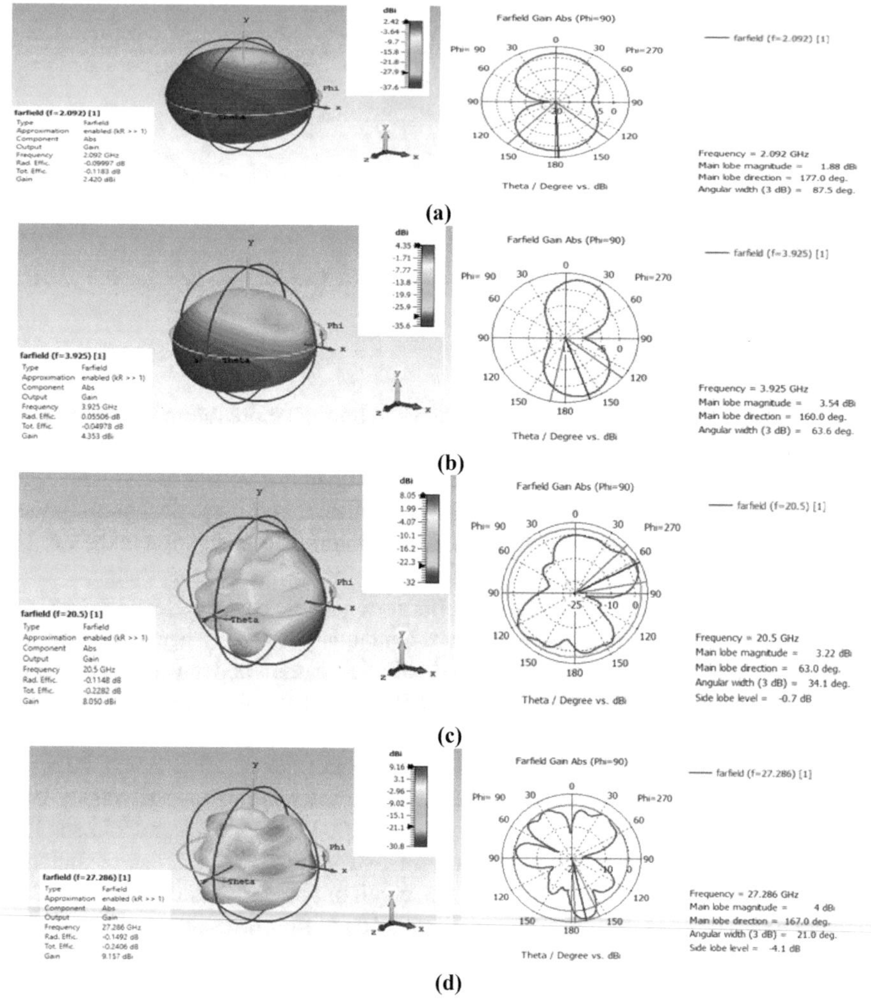

Fig. 4 3D and 2D far-field gain radiation pattern for **a** 2.092 GHz, **b** 3.925 GHz, **c** 20.5 GHz, and **d** 27.286 GHz

frequency band for various bending models with radii of $R = 8$ mm, $R = 10$ mm, and $R = 50$ mm is shown in Table 1.

3.3 On-Body Performance

SAR is one of the most essential metrics for assessing antenna security. A precise international standard for comparative absorptivity has been established to ensure

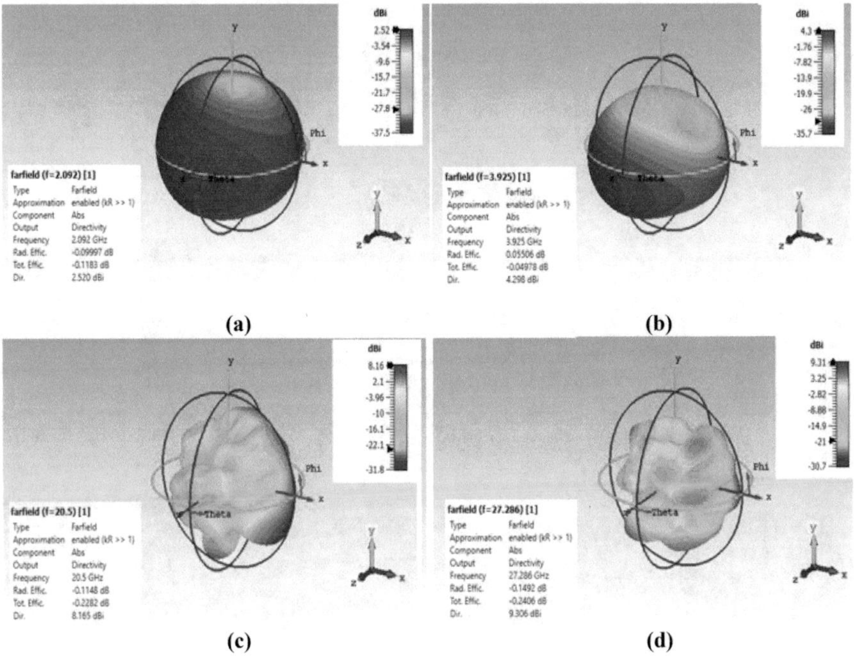

Fig. 5 Directivity pattern for **a** 2.092 GHz, **b** 3.925 GHz, **c** 20.5 GHz, and **d** 27.286 GHz

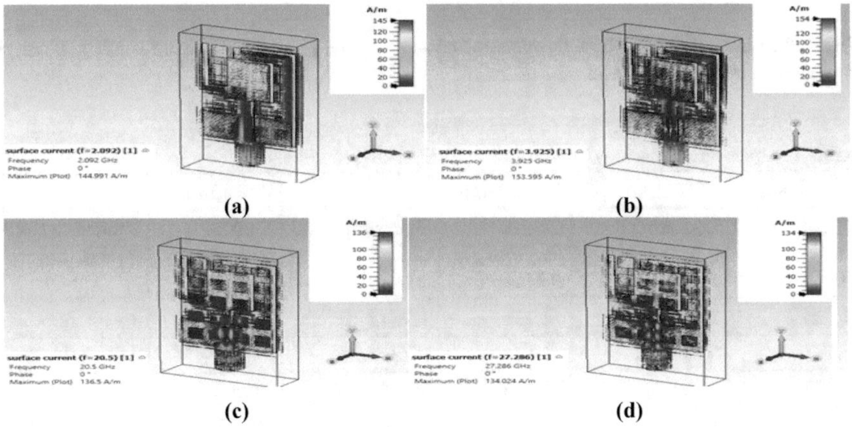

Fig. 6 Surface current distribution for **a** 2.092 GHz, **b** 3.925 GHz, **c** 20.5 GHz, and **d** 27.286 GHz

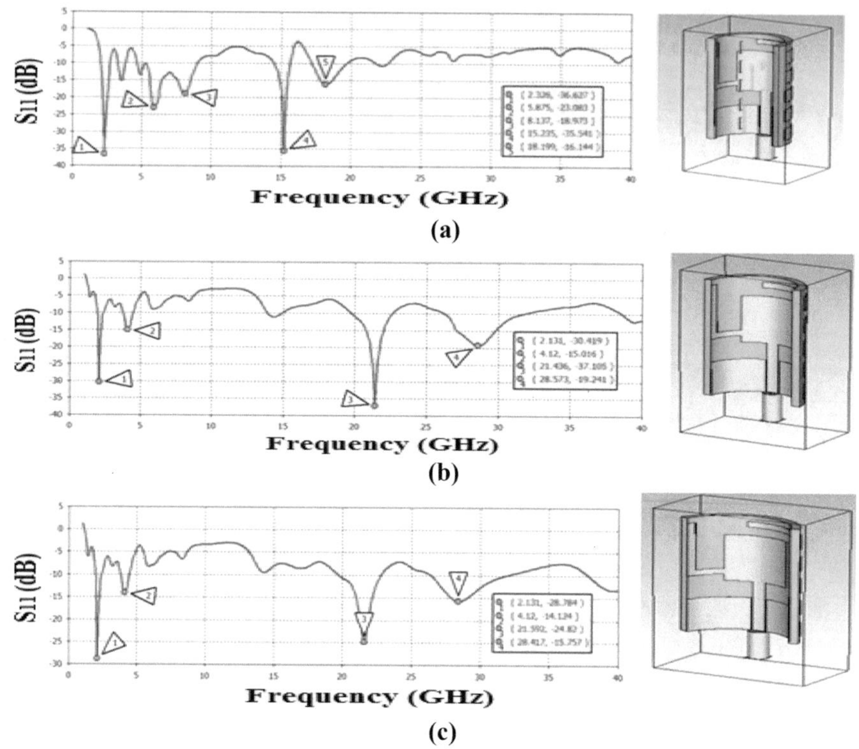

Fig. 7 The bending model with S_{11} versus frequency of the proposed antenna for **a** $R = 8$ mm, **b** $R = 10$ mm, and **c** $R = 12$ mm

Table 1 Operating frequency band for different bending models

$R = 8$ mm		$R = 10$ mm		$R = 12$ mm	
Center frequency (GHz)	Operating band (GHz)	Center frequency (GHz)	Operating band (GHz)	Center frequency (GHz)	Operating band (GHz)
2.326	2.04–2.73	2.131	1.96–2.42	2.131	1.97–2.45
5.875	5.27–9.01	4.12	3.79–4.49	4.12	3.81–4.46
15.235	14.51–15.63	21.436	19.78–22.60	21.592	19.72–22.78
18.199	17.22–19.69	28.573	26.26–31.78	28.417	26.41–31.99

the safety of antenna radiation to the human body. Wearable antennas are designed to be worn close to the physical body, and the SAR is utilized to address the risks posed to the human body by wearable specialized electronics. SAR is a critical criterion for determining the amount of electromagnetic field consumed by human tissues. Figure 8 simulates the basic model of human tissues with their electrical properties such as relative permittivity (ε_r), conductivity (σ), mass density (ρ), and thickness

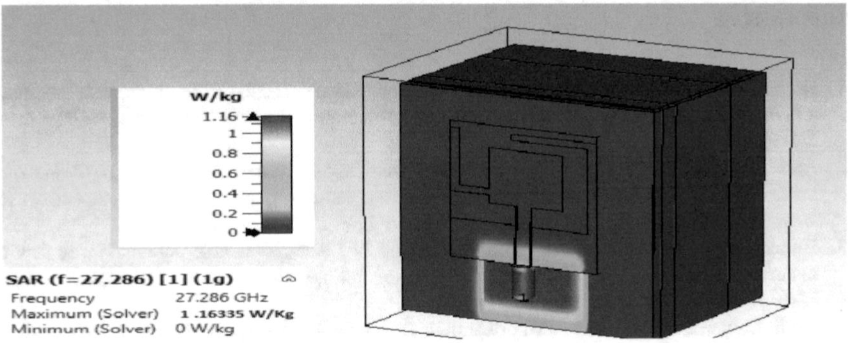

Fig. 8 Simulated SAR distribution in 1 g tissues on the human tissue model

(*d*), consisting of skin ($\varepsilon_r = 38.0067$, $\sigma = 1.184$, $\rho = 1001$ kg/m^3, $d = 1$ mm), fat ($\varepsilon_r = 10.8205$, $\sigma = 0.58521$, $\rho = 900$ kg/m^3, $d = 2$ mm), and muscle ($\varepsilon_r = 55$, $\sigma = 1.437$, $\rho = 1006$ kg/m^3, $d = 10$ mm) with SAR distributions at 27.28 GHz with an input power of 0.1 W. At a height of 2 mm from the tissue surface, the antenna is positioned. The peak 1 g SAR value for the patch alone is 1.16 W/kg, according to the simulated SAR distributions. As a result, the SAR restriction is satisfied with a value that is substantially below the permitted limit of 1.6 W/kg.

4 Conclusion

A wearable textile Meta surface ground-based patch antenna for millimetre wave applications is presented, explored, evaluated, and simulated in this work. The topic of metric optimization for various antenna settings has been thoroughly studied. With VSWR less than 2, S_{11} response of better than -10 dB can be achieved over four resonance bands: 1.909–2.426 GHz (impedance bandwidth 24.7%), 3.498–4.211 GHz (impedance bandwidth 18.03%), 19.838–21.196 GHz (impedance bandwidth 6.62%), and 24.343–30.687 GHz (impedance bandwidth 23.25%). The antenna design shown in this article has the potential to be very useful in 5G technology. The antenna has a fair gain and good radiation characteristics, according to the simulation findings. In addition, the suggested antenna is put to the test under various bending conditions. The measured results reveal that, in addition to meeting SAR criteria, the antenna still has good matching and radiation patterns at the desired operating frequencies.

References

1. Kumar S, Dixit AS, Malekar RR, Raut HD, Shevada LK (2020) Fifth generation antennas: a comprehensive review of design and performance enhancement techniques. IEEE Access 8:163568–163593
2. Aun NFM, Soh PJ, Al-Hadi AA, Jamlos MF, Vandenbosch GAE, Schreurs D (2017) Revolutionizing wearables for 5G: 5G technologies: recent developments and future perspectives for wearable devices and antennas. IEEE Microwave Mag 18(3):108–124
3. Paracha KN, Abdul Rahim SK, Soh PJ, Khalily M (2019) Wearable antennas: a review of materials, structures, and innovative features for autonomous communication and sensing. IEEE Access 7:56694–56712
4. Boric-Lubecke O, Lubecke VM, Jokanovic B, Singh A, Shahhaidar E, Padasdao B (2015) Microwave and wearable technologies for 5G. In: 12th International Conference on Telecommunication in Modern Satellite, Cable and Broadcasting Services (TELSIKS). IEEE, Nis, Serbia, pp. 183–188
5. Zou H, Li Y, Wang M, Peng M, Yang G (2018) Compact frame-integrated MIMO antenna in the LTE Band 42/4.9-GHz Band/5.2-GHz WLAN for 5G wearable applications. In: 2018 international conference on microwave and millimeter wave technology (ICMMT). IEEE, Chengdu, China, pp. 1–3
6. Du C, Li X, Zhong S (2019) Compact liquid crystal polymer based tri-band flexible antenna for WLAN/WiMAX/5G applications. IEEE Access 1–1 (2019)
7. Kamran A, Tariq F, Amjad Q, Karim R, Hassan A (2020) A bus shaped tri-band antenna for Sub-6 GHz 5G wireless communication on flexible PET substrate. In: 22nd International Multitopic Conference (INMIC). IEEE, Islamabad, Pakistan, pp. 1–6
8. Alam M, Siddique M, Kanaujia BK, Beg MT, Rambabu SKK (2018) Meta-surface enabled hepta-band compact antenna for wearable applications. IET Microwaves Antennas Propag 13(13):2372–2379
9. Ullah S, Ahmad I, Raheem Y, Ullah S, Ahmad T, Habib U (2020) Hexagonal shaped CPW Feed Based Frequency Reconfigurable Antenna for WLAN and Sub-6 GHz 5G applications. In: 2020 International Conference on Emerging Trends in Smart Technologies (ICETST), IEEE, Karachi, Pakistan, pp. 1–4
10. Li Y, Shen H, Zou H, Wang H, Yang G (2017) A compact UWB MIMO antenna for 4.5G/5G wearable device applications. In: 2017 Sixth Asia-Pacific Conference on Antennas and Propagation (APCAP). IEEE, Xi'an, China, pp. 1–3
11. Kassim S, Rahim HA, Malek F, Sabli NS, Salleh MEM (2019) UWB nanocellulose coconut coir fibre inspired antenna for 5G applications. In: 2019 International Conference on Communications, Signal Processing, and their Applications (ICCSPA). IEEE, Sharjah, United Arab Emirates, pp. 1–5
12. Jilani SF, Abbasi QH, Alomainy A (2018) Inkjet-printed millimetre-wave pet-based flexible antenna for 5G wireless applications. In: 2018 IEEE MTT-S International Microwave Workshop Series on 5G Hardware and System Technologies (IMWS-5G). IEEE, Dublin, Ireland, pp. 1–3
13. Fawaz M, Jun S, Oakey WB, Mao C, Elibiary A, Sanz-Izquierdo B, Bird D, McClelland A (2018) 3D printed patch Antenna for millimeter wave 5G wearable applications. In: 12th European Conference on Antennas and Propagation (EuCAP 2018). IEEE, London, UK, pp. 1–5
14. Sharma D, Dubey SK, Ojha VN (2018) Wearable antenna for millimeter wave 5G communications. In: 2018 IEEE Indian Conference on Antennas and Propagation (InCAP). IEEE, Hyderabad, India, pp. 1–4
15. Saeed MA, Ur-Rehman M (2019) Design of an LCP-based antenna array for 5G/B5G wearable applications. In: 2019 UK/ China Emerging Technologies (UCET). IEEE, Glasgow, UK, pp. 1–5

16. Rubesh Kumar T, Madhavan M (2021) Hybrid fabric wearable antenna design and evaluation for high speed 5G applications. Wireless Personal Commun
17. Sabban A (2022) Wearable circular polarized antennas for health care, 5G, energy harvesting, and IoT systems. Electronics 11(427):1–27

Veridical Discrimination of Expurgated Hyperspectral Image Utilizing Multi-verse Optimization

Divya Mohan, S. Veni, and J. Aravinth

1 Introduction

Hyperspectral images [1] can capture more accurate spectral responses when they have hundreds of bands, making them best suited to tracking subtle variations in ground covers and their changes over time. Due to insufficiency in the number of reference samples in comparison to the number of input target features, the Hughes phenomenon occurs [2]. With an increased number of spectral channels in digital images, there happens a decrease in the precision of data and as a result information dimensionality increases. In addition, the smile effect can directly affect Hyperion images in a variety of ways around the signal spectrum, and it can differ from scene to scene. Since a spectral smile is not apparent to the naked eye, researchers must use a predictor to make it recognizable. Thus, this condition affects the effective and optimized discrimination of the hyperspectral image. A further difficult problem with hyperspectral images appears to be endmember extraction, which is complicated by the poor spatial resolution [3] and can result in mixed pixels. The inclusion of multiple mixed pixels in hyperspectral images makes endmember extraction difficult. Even if the image has been perfectly pre-processed and a highly precise function is extracted,

D. Mohan (✉)
Research Scholar, Department of Electronics and Communication Engineering,
Amrita School of Engineering, Amrita Vishwa Vidyapeetham, Coimbatore 641112, India
e-mail: divyamohan19@gmail.com

S. Veni
Department of Electronics and Communication Engineering, Amrita School of Engineering,
Amrita Vishwa Vidyapeetham, Coimbatore 641112, India
e-mail: s_veni@cb.amrita.edu

J. Aravinth
Department of Electronics and Communication Engineering, Amrita School of Engineering,
Amrita Vishwa Vidyapeetham, Coimbatore 641112, India
e-mail: j_aravinth@cb.amrita.edu

P. Singh et al. (eds.), *Machine Learning and Computational Intelligence Techniques for Data Engineering*, Lecture Notes in Electrical Engineering 998,
https://doi.org/10.1007/978-981-99-0047-3_45

a single small pixel can contain high-level detail in the hyperspectral image. This clearly shows the need to come up with an optimized algorithm which can withstand the complexities of hyperspectral images [4]. So, this makes to deliberate about developing a novel solution to easily process the hyperspectral image without losing its versatile resources.

With this motivation, in this work, a method is attempted to eliminate the Hughes phenomenon issue using a filter wrapper semi-supervised band selection technique where only the appropriate bands are specifically selected and also the difficulties created by a lack of labeled dataset or an overfitness attribute are removed. Secondly, the smile effect in the hyperspectral image is tried to minimize with the help of stabilized smile frown technique by means of spectral sharpening. Thirdly, optimized resource utilization is carried out since it is a novel method to prevent excess spectral data removal. The proposed system is expected to give better computation time and accuracy compared to many other existing methods.

The organization of the proposed work is as follows. Sect. 2 describes the related work. Sect. 3 covers the methodology and workflow in detail followed by the results and discussions in Sect. 4. Sect. 5 concludes the presented work.

2 Related Work

Hyperspectral imaging [1] has recently been praised as a cutting-edge and exciting analytical tool for use in research, monitoring, and business. It is a method for creating a spatial chart of spectral variation that can be used for a number of purposes and is a relatively new and promising area of study for both automatic target analysis [3] and assessing its analytical composition.

Cao et al. [2] suggested two independent semi-supervised band selection approaches so that the spectral and spatial information can be fused together to give better classification accuracy for hyperspectral images. This fusion can also be effectively utilized in improving the quality of band selection. Though they offered better performance, huge computation time remains one of the drawbacks. Hong et al. [12] found that it may be difficult to distinguish the same products in spatially distinct scenes or locations. They suggested a solution to this problem by using a technique called invariant attribute profiles to derive invariant features from hyperspectral imagery (HSI) in both the spatial and frequency domains (IAPs). However, a singularity in most statistical processing methods is hindering process efficiency. Das et al. [9]. The research performed multi-way sparsity estimation in hyperspectral image processing since hyperspectral images have an underlying multi-way structure. The research provides an effective criterion for estimating multi-way sparsity, which will make signal inversion and noise reduction activities easier. Even though sometimes, hyperspectral images have a low spatial resolution [4], which makes their interpretation difficult.

Although the existing research works have performed well in implementing the techniques, it was observed that there is a high demand to correct the inaccuracy produced due to the complex nature of the hyperspectral images. A novel methodology using semi-supervised filter wrapper band selection and stabilized smile frown technique is developed to overcome the complexities which are briefed in the following section.

3 Proposed Methodology

The steps of the proposed methodology are shown in Fig. 1 and are elaborated as follows: (i) The first step is to reduce the Hughes phenomenon using a filter wrapper semi-supervised band selection technique that selects the appropriate band using an independent filter based on Bhattacharya distance, Kullback–Leibler divergence, Jeffries–Matusita distance, etc. [6]. The wrapper method is then executed using the logistic regression variable. Labeled propagation and projection matrix [4] are used to perform a semi-supervised band selection, removing the difficulties created by a lack of labeled dataset or an overfitness attribute. (ii) The second step is, by using stabilized smile frown technique, the smile effect in the hyperspectral image is minimized. This technique uses the spectrum's reflectance readings to detect the smile-affected signal's max and min noise fraction [8] values to prevent excessive spectrum data removal. (iii)Thirdly, the end member extraction [10] is carried out by Volume Shrunk Pure Pixel Actualize Method. This divides the input along the specimen and finds the pixel purity index, after which the dimension is reduced by orthogonal subspace projection and max noise fraction transformation [5]. Then a volume matrix determi-

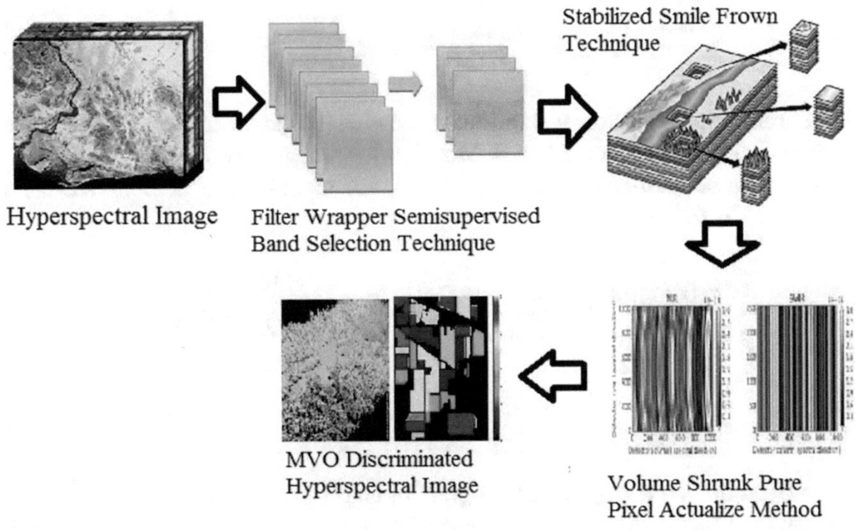

Fig. 1 Flow diagram of the proposed methodology

nation is also used to remove the mix pixel spectrum. (iv) Finally, the most important variables are chosen for segmentation [6, 7]. Then optimizing the use of such minute data is effectively done by incorporating a multi-verse optimization algorithm [10]. This employs a nature-inspired concept to work with extremely structured data and make the most use of everything that's viable.

3.1 Filter Wrapper Semi-Supervised Band Selection Technique

Hyperspectral imaging is a method for creating a spatial map of spectral variance that can be utilized for a variety of purposes. The Hughes phenomenon occurs when the number of training sites or reference samples is insufficient in relation to the number of input target features. As the number of spectral channels in digital images grows, the precision of data suffers as a result of the increased information dimensionality [11]. Hence, there is a high need to reduce the bands and select the appropriate bands needed for the segmentation. The first method is to reduce the Hughes phenomenon by utilizing a filter wrapper semi-supervised band selection technique, which uses an independent filter to pick the suitable band based on Bhattacharya distance, Kullback–Leibler divergence, Jeffries–Matusita distance, and other factors. The wrapper method is then implemented with the logistic regression variable as a parameter. A semi-supervised band selection is performed using labeled propagation and a projection matrix, removing the challenges caused by a lack of labeled dataset or an overfitness attribute.

Let the hyperspectral data cube be written as

$$I \in R^{(H \times W \times N)} \tag{1}$$

where H and W are the hyperspectral data cube's height and width, respectively, and N is the total number of spectral bands. Mutual information between two bands, Bi and Bj, with joint probability distribution P(bi, bj) and marginal probability distribution M(bi) and M(bj), is written as

$$I\left(B_i, B_j\right) \sum_{b_i \epsilon B_i, b_j \epsilon B_j} P\left(b_i, b_j\right) \mathbf{log} \frac{P\left(b_i, b_j\right)}{M\left(b_i\right), M\left(b_j\right)} \tag{2}$$

$$P\left(b_i, b_j\right) = \frac{H\left(b_i, b_j\right)}{i \times j} \tag{3}$$

where H(bi, bj) is the gray-level histogram of bands Bi, Bj. The grayscale image is taken and the Bhattacharya distance of the image between bands Bi and Bj is found using

$$B_{i,j} = \frac{1}{B}\left(\alpha_i, \alpha_j\right)^T \left(\frac{\beta_i + \beta_j}{2}\right)^{-1} \left(\alpha_i - \alpha_j\right) + \frac{1}{2}\ln\left[\frac{\left|\beta_i + \frac{\beta_j}{2}\right|}{|\beta_i|^{\frac{1}{2}}\,\left|\beta_j\right|^{\frac{1}{2}}}\right] \tag{4}$$

Here, α_i and α_j are band means, while β_i and β_j are band covariance matrices.Then the Bhattacharya distance between bands Bi and Bj is determined and the Jeffries–Matusita distance is evaluated between bands Bi and Bj using

$$JM = \sqrt{\mathbf{2}\left(\mathbf{1} - e^{B_{i,j}}\right)} \tag{5}$$

where $B_{i,j}$ is the Bhattacharya distance obtained from the equation (4). After evaluating the band distances, a projection matrix P is to be created which can be used to classify X using Y = XP, which can be acquired through semi-supervised approaches. Based on the distance the band is chosen on the selected spectrum and the propagation and projection matrix is found out using

$$I_x{}^2 + I_y{}^2 \tag{6}$$

Then various sub-bands for image Im are selected with wavelet transform, histogram distribution and image $\mathbf{I_m}'$ is found. The process is repeated, filtered, and reconstructed image [11] is obtained. The reconstructed image with a necessary number of bands is then processed for smile reduction in the next technique.

3.2 *Stabilized Smile Frown Technique*

Low-frequency events that change spectro-image over tracks are referred to as spectral smile. Spectral smile [13] refers to low-frequency phenomena that alter spectro-images across tracks. The sensor response is affected in two ways by a spectral grin. First, the spectral PSF's central wavelength varies with column number inside a specific band, i.e., central wavelength increases as it gets closer to the image's edges. Second, when moved out from the core columns, the PSF width grows, implying that spectral resolution decreases. The smile correction step seeks to partially compensate for a band's non-homogeneous spectral resolution. This spectral heterogeneity leads to a given sharp spectral feature to become gradually over-smoothed when it is convolved by increasingly larger PSFs, contributing to the smile brightness gradient. This stabilized smile frown technique looks at a spectral sharpening [14] influenced by spatial sharpening techniques used in picture processing. By boosting the local contrast of the "smiled" spectra, this approach seeks to replicate an improvement in spectral resolution up to the reference. The following equation can be used to sharpen:

$$L_{\mathbf{sharp}} = \frac{L(\theta, \lambda) - \frac{\omega_{\theta,\lambda}}{\mathbf{2}}\left(L(\theta, \lambda - V)\right) + \left(L(\theta, \lambda + V)\right)}{\mathbf{1} - \omega_{\theta,\lambda}} \tag{7}$$

For any line number, $L\ (\theta, \lambda)$ is the reflectance value for column and band. V specifies the neighborhood, and $\omega_{\theta,\lambda}$, the level of sharpness within the range [0...1], $\omega_{\theta,\lambda} = 1$ indicating infinite sharpening. Both the local spectral structure and the instrument characteristics are taken into consideration in this formula. The local scale is defined by parameter V, which is set to unity to correspond to thin absorption characteristics represented by three neighboring specters [15]. The parameter refers to the ratio of the current point spread function width. The relationship equation could be written as follows:

$$\omega_{\theta,\lambda} = \omega_{max} \frac{W\ (\theta, \lambda) - \mathbf{min} W\ (\lambda)}{\mathbf{max} W\ (\lambda) - \mathbf{min} W\ (\lambda)} \tag{8}$$

where $\omega_{max}$is the most sharpening, and min and max(W(θ, λ) are the minimum and maximum of the full width at half maximum (W) of all detection elements within band, respectively. As a result, for the smile, is near to zero, and for the image edges, is close to maximum.

The maximum noise fraction (MNF) transformation can be used to determine the amount of energy in a grin. After two cascaded principal component transformations and noise whitening, an MNF rotation on hyperspectral data provides new components sorted by the signal-to-noise ratio.

3.3 Volume Shrunk Pure Pixel Actualize Method

Endmember extraction [15] looks to be a difficult task with hyperspectral pictures, which is aggravated by the low spatial resolution, which results in mixed pixels. End member extraction is an important process in feature selection that is carried out by volume shrunk pure pixel actualize method, which divides the input along the specimen and finds the pixel purity index, after which the dimension is reduced by orthogonal subspace projection and max noise fraction transformation, and finally, a volume matrix determination is used to remove the mix pixel spectrum. As a result, the most important variables [13] are chosen for segmentation. Even if the image has been perfectly preprocessed and a highly precise function is extracted from the hyperspectral image, a single small pixel contains high-level detail. The smile-reduced image is converted to HSV image in which H refers to hue, S refers to saturation, and V refers to value counts. The end members are extracted based on HSV properties and unwanted pixels are removed. When all of the pixels have been examined and no replacement is required, the process is terminated. Thus, the endmembers required for segmentation are carried out efficiently and the extracted features are segmented using the next process.

3.4 Multi-verse Optimization Algorithms

It is critical to develop an optimal algorithm that can handle the complexities of hyperspectral images. As a result, it's time to think about coming up with a new way to interpret hyperspectral images quickly without compromising their diversity. Optimizing the use of such minute data is imperative. As a result, this research incorporated a multi-verse optimization algorithm, which employs a nature-inspired concept to work with extremely structured data and make the most use of everything that's viable. The MVO [14] algorithm was inspired by three primary notions from multiverse theory: white holes, black holes, and wormholes. The MVO algorithm's detailed steps are listed below:

(i) Take M * N input image that needs to be split.
(ii) Determine the image's histogram.
(iii) Configure MVO's control parameters, such as the number of universes, the number of iterations (Max), the cost function (between-class variance), and the threshold levels.
(iv) The best thresholds can be found by maximizing the cost function with the MVO pseudo-code.
(v) The best universe is a collection of ideal threshold values that have the best maximum cost function value.
(vi) Using the corresponding threshold values, the input image is segmented. his technique uses the advantages of the MVO algorithm in the segmentation of hyperspectral image and a very optimized discrimination is carried with high accuracy.

This methodology postulates that universes with high inflation rates are extremely likely to have white holes in order to improve the overall inflation rate of the universes. On the other hand, universes with low inflation rates are more likely to have black holes. As a result, moving objects from a universe with a high inflation rate to a universe with a low inflation rate is always a possibility. This can ensure that the average inflation rates of all universes improve over the iterations. Also, this algorithm uses a roulette wheel method to mathematically describe the white/black hole tunnels and interchange the items of universes. The universe is arranged based on their inflation rates in each iteration and chooses one of them to have a white hole using the roulette wheel. The performance of this methodology is discussed in detail in the upcoming section.

4 Results and Discussions

This segment provides a detailed description of the implementation results as well as the performance of the proposed system. This work has been implemented in the working platform of MATLAB with the following system specification and the results are discussed below, Platform: MATLAB, OS: Windows 7, Processor: 64-bit Intel processor, RAM: 8GB RAM. This research utilizes a hyperspectral image taken over Pavia University with the ROSIS (Reflective Optics System Imaging Spectrometer). The implementation also covers Salinas and Indian pine datasets and the results of each dataset are compared and discussed in the following sections.

4.1 Experimental Results and Analysis

The hyperspectral image from the dataset is given as input to the proposed methodology and a discriminated image is obtained as an output from the novel approach with increased accuracy and low computation time. The following images depict the results obtained at various steps of the proposed framework. Initially, in the filter wrapper semi-supervised band selection technique, the input image is converted into a grayscale image for further processing. This grayscale image is then processed using the novel technique filter wrapper semisupervised band selection technique for band selection process which selects the needed bands and eliminates the unwanted bands

After the band selection process, the smile effect in the image due to hyperspectral sensors must be eliminated. This is carried out by stabilized smile frown technique which sharpens the image between minimum and maximum noise values. This output image contains highly needed information which must be extracted for the segmentation process.

From the smile-reduced image the end members needed for segmenting the image accurately must be extracted. Volume shrunk pure pixel actualize method extracts all the end members based on pixel purity index and volume transformation from the hyperspectral image in which each pixel has useful information. Finally, the multiverse optimization algorithm optimizes the end member selected by the previous process based on black holes, white holes, and worm holes and efficiently segments the needed information from the image and produces a discriminated output as in the following Fig. 2.

4.2 Performance Metrics of Proposed Method

The performance metrics of the proposed method based on various datasets is depicted and explained above. The accuracy of the proposed framework based on the computation time on three different datasets is evaluated in Table 1.

The proposed framework has a percentage accuracy of 96 with a very low computational time of 0.1 s for the Pavia dataset as plotted in Fig. 3. Similarly, it shows an accuracy of 96.2% with a very low computational time of 0.1 s for the Indian Pines dataset as shown in Fig. 4. The proposed framework's accuracy is assessed using the

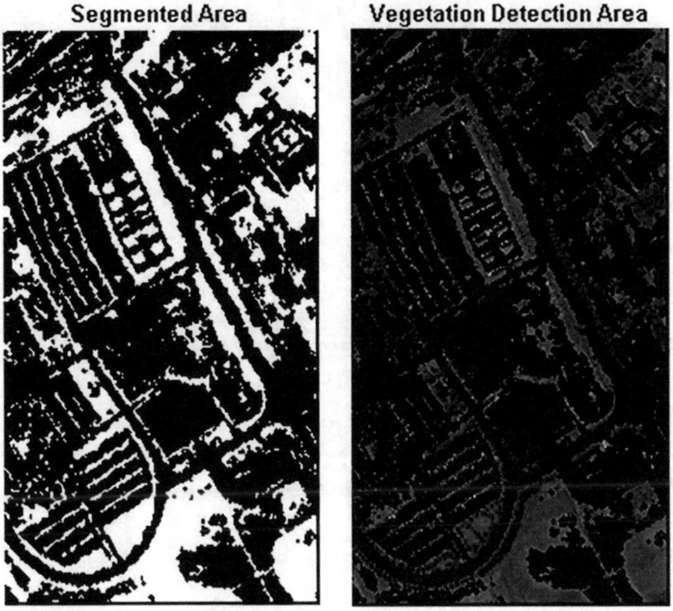

Fig. 2 Discriminated image using MVO

Table 1 Performance evaluation based on computation time and accuracy with 3 datasets

University of Pavia dataset		Indian Pines dataset		Salinas dataset	
Computation time (in s)	Accuracy (%)	Computation time (in s)	Accuracy (%)	Computation time (in s)	Accuracy (%)
0.1	96	0.1	96.2	0.1	96.4
0.2	96.2	0.2	96.4	0.2	96.8
0.3	97	0.3	97.2	0.3	97
0.4	97.2	0.4	97.4	0.4	97.5
0.5	97.4	0.5	97.5	0.5	97.6
0.6	97.8	0.6	97.6	0.6	97.8
0.7	98	0.7	98	0.7	98
0.8	98.1	0.8	98.4	0.8	98.5
0.9	98.3	0.9	98.5	0.9	98.7
1.0	98.5	1.0	98.6	1.0	98.8

Salinas dataset, and it has an accuracy of 96.4% with a very low computing time of 0.1 s, and as the computational time grows, the accuracy improves as well, reaching 98.8% accuracy with 1 s computational time as plotted in Fig. 5.

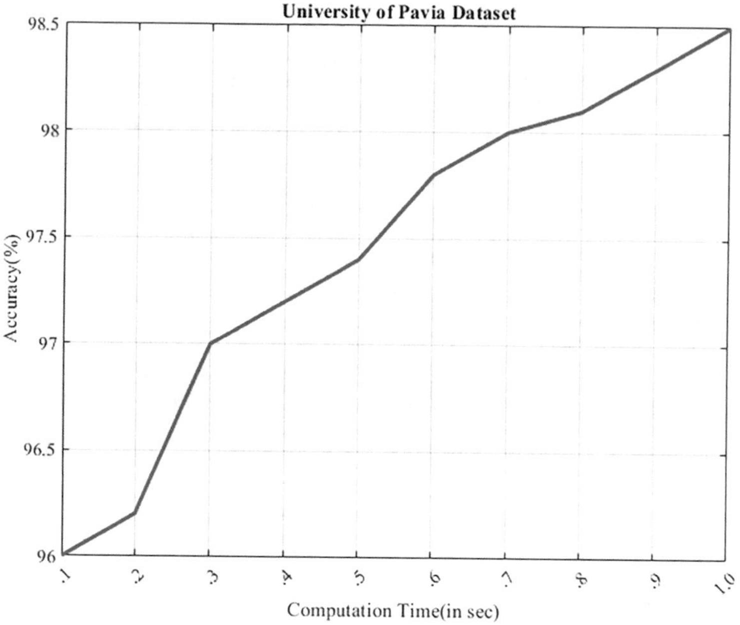

Fig. 3 Performance metrics based on computation time and accuracy with Pavia dataset

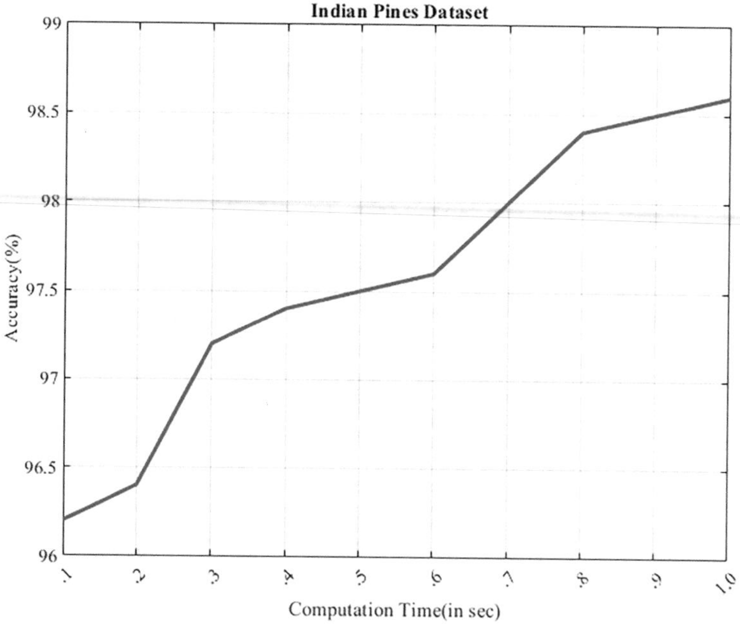

Fig. 4 Performance metrics based on computation time and accuracy with Indian pines dataset

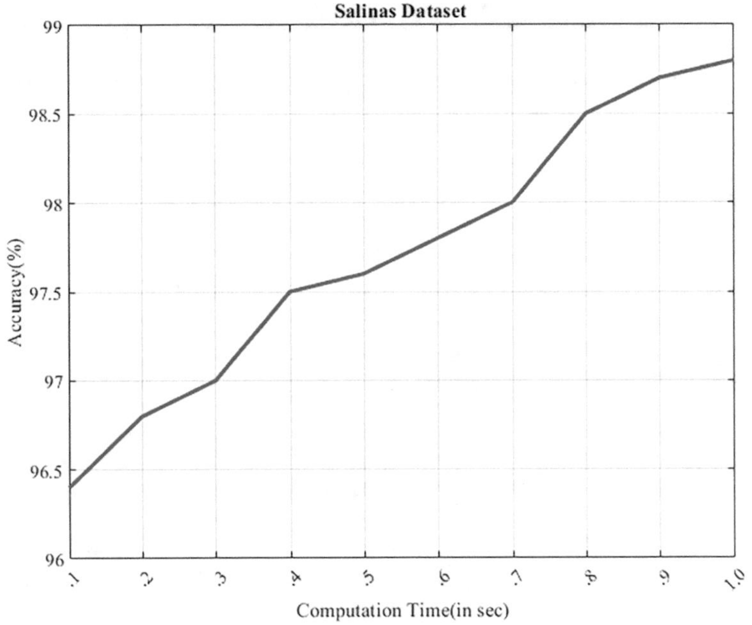

Fig. 5 Performance metrics based on computation time and accuracy with Salinas dataset

4.3 Comparison Results of the Proposed Method

The proposed method is compared with existing techniques [16] based on its various performances and evaluated. The results of the comparative study of the novel approach to segment the hyperspectral image are depicted and explained below using Table 2.

Table 2 Comparison metrics based on computational time with different datasets

Methods	Computation time (in s)		
	Pavia dataset	Salinas dataset	Indian Pines dataset
SVM	7.24	6.20	8.68
JSRC	66.09	145.67	23.03
EPF	8.45	7.68	8.88
LBP	737.51	301.68	69.33
EMAP	216.11	50.17	19.83
IFRF	8.75	4.95	2.83
IID	1608.51	139.67	155.45
SPCA-GP	5.20	3.88	1.28
Proposed method	1.08	0.76	0.60

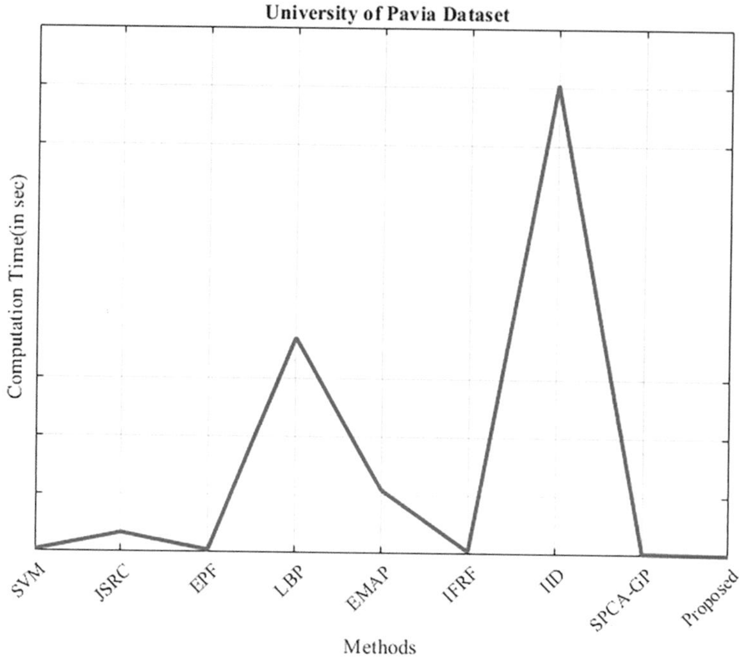

Fig. 6 Comparison metrics based on computational time with Pavia dataset

The computation time of the proposed method is calculated. All the three datasets are used for comparison and the computational time is evaluated (Figs. 6 and 7).

The proposed method performs in 1.08 s which is a much reduced time while comparing with other techniques the timing shows to be the most perfect. The results show that the proposed method has the lowest computational time of 0.76 s of all the other compared methodologies. Finally, the proposed method's computation time is once again compared with 8 other different existing methods with the Indian pines dataset as plotted in Fig. 8. In this dataset also, the proposed method computes in a very low time than all eight other techniques [16]. Thus, this illustrates that the computation time of the proposed method is 0.60 s which is comparatively lower than other datasets.

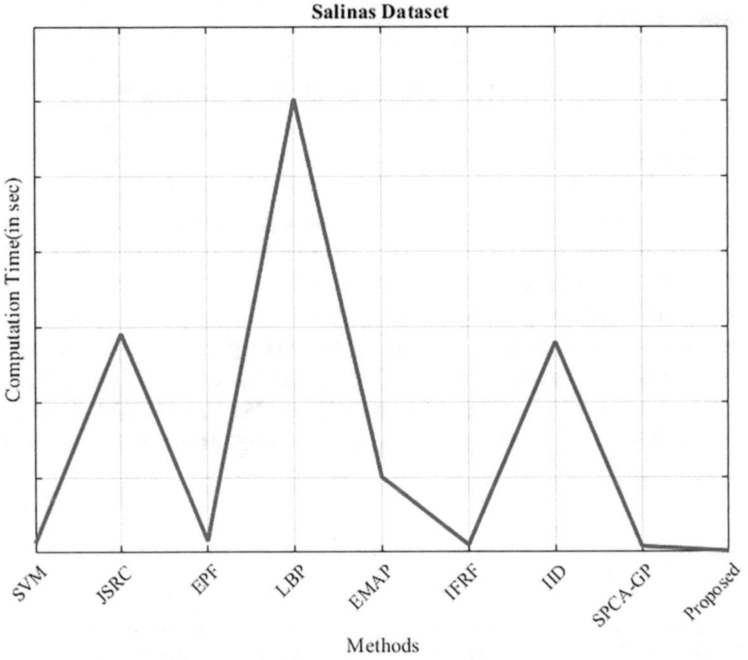

Fig. 7 Comparison metrics based on computational time with Salinas dataset

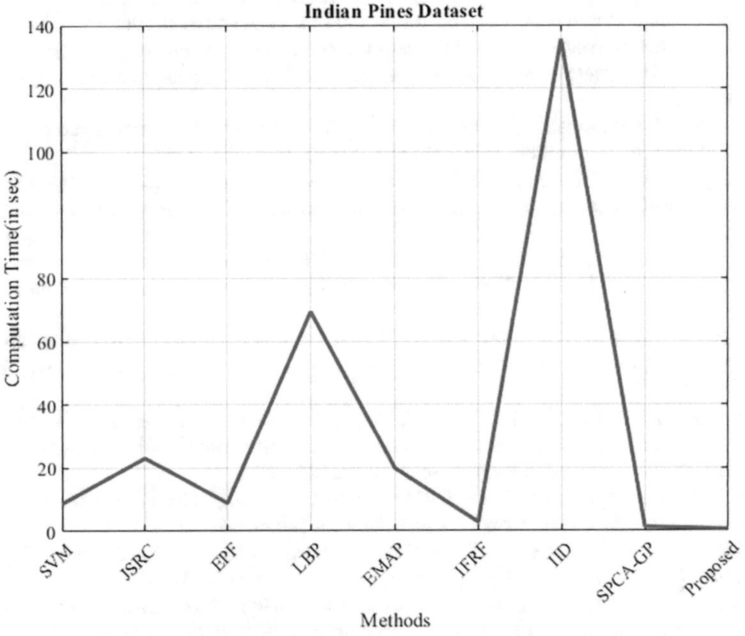

Fig. 8 Comparison metrics based on computational time with Indian Pines dataset

5 Conclusions

The novel proposed methodology eliminates all kinds of complexities in a hyperspectral image and provides a well-processed image by which the discrimination is carried out accurately. The bands in the hyperspectral image are reduced initially using a filter wrapper semi-supervised band selection technique and then the smile effect produced due to hyperspectral imaging sensors is eliminated by stabilized smile frown technique. Then the end members of the hyperspectral image are selected using volume shrunk pure pixel actualize method. The expurgated image produced from the above process with accurately necessitating features and useful pixels is segmented using the MVO algorithm. This algorithm helps to segment the image in a much-optimized manner with high accuracy and low computational time. Then the proposed method is compared with SVM, JSRC, EPF, LBP, EMAP, IFRF, IID, and SPCA-GP for three various datasets and computational time is noted.

References

1. Anand A, Pandey PC, Petropoulos GP, Palvides A, Srivastava PK, Sharma JK, Malhi RKM et al (2020) Use of hyperion for mangrove forest carbon stock assessment in bhitarkanika forest reserve: a contribution towards blue carbon initiative. Remote Sens 12:597. https://doi.org/10.3390/rs12040597
2. Cao X, Ji Y, Liang T, Li Z, Li X, Han J, Jiao L (2018) A semi-supervised spatially aware wrapper method for hyperspectral band selection. Int J Remote Sens 39(12):4020–4039
3. Anand R, Veni S, Aravinth J (2021) Robust classification technique for hyperspectral images based on 3D-discrete wavelet transform. Remote Sens 13:1255. https://doi.org/10.3390/rs13071255
4. Aravinth J, Sb Valarmathy (2015) A natural optimization algorithm to fuse scores for multimodal biometric recognition. Int J Appl Eng Res 10(8):21341–21354
5. Sowmya V, Renu RK, Soman KP (2015) Spatio-spectral compression and analysis of hyperspectral images using tensor decomposition. In: 24th national conference on communications, NCC
6. Sowmya V, Soman KP, Kumar S, Shajeesh KU (2012) Computational thinking with spreadsheet: convolution, high-precision computing and filtering of signals and images. Int J Comput Appl 60(19)
7. Cui B, Ma X, Xie X, Ren G, Ma Y (2017) Classification of visible and infrared hyperspectral images based on image segmentation and edge-preserving filtering. Infrared Phys Technol 81:79–88
8. Dadon A, Ben-Dor E, Karnieli A (2010) Use of derivative calculations and minimum noise fraction transform for detecting and correcting the spectral curvature effect (smile) in Hyperion images. IEEE Trans Geosci Remote Sens 48(6):2603–2612
9. Das S, Bhattacharya S, Pratiher S (2021) Efficient multi-way sparsity estimation for hyperspectral image processing. J Appl Remote Sens 15(2):026504
10. Geng X, Sun K, Ji L, Zhao Y, Tang H (2015) Optimizing the endmembers using volume invariant constrained model. IEEE Trans Image Process 24(11):3441–3449
11. Głomb P, Romaszewski M, (2020) Anomaly detection in hyperspectral remote sensing images. In: Pandey PC, Srivastava PK, Balzter H, Bhattacharye B, Petropoulos G (eds) Hyperspectral remote sensing: theory and applications. Elsevier

12. Hong D, Wu X, Ghamisi P, Chanussot J, Yokoya N, Zhu XX (2020) Invariant attribute profiles: a spatial-frequency joint feature extractor for hyperspectral image classification. IEEE Trans Geosci Remote Sens 58(6):3791–3808
13. Khan U, Sidike P, Elkin C, Devabhaktuni V (2020) Trends in deep learning for medical hyperspectral image analysis. arXiv:2011.13974
14. Tien-Heng H, Jean-Fu K (2020) Comparison of CNN algorithms on hyperspectral image classification in agricultural lands. Sensors 20(6):1734
15. Riese FM, Keller S (2020) Supervised, semi-supervised, and unsupervised learning for hyperspectral regression. In: Hyperspectral image analysis. Springer, Cham, pp 187–232
16. Zhu X, Li G (2019) Rapid detection and visualization of slight bruise on apples using hyperspectral imaging. Int J Food Prop 22(1):1709–1719

Self-supervised Learning for Medical Image Restoration: Investigation and Finding

Jay D. Thakkar, Jignesh S. Bhatt, and Sarat Kumar Patra

1 Introduction

Medical imaging is the process of acquiring images in form of photos or video frames to study and diagnose various diseases. The development in this field is expeditious while leading to many computer vision-based challenges including in detection/monitoring of a disease [8, 13, 19], classification [3, 16], segmentation [20, 21], registration [5], denoising [14, 18], and restoration of medical imagery. Image restoration is the process of considering a degraded image and estimating a clean, non-degraded version of it. Applications of restoration can be seen in many different areas specifically in medical imaging, remote sensing, and forensic science. Degradations in images are due to many reasons including noise, blur, and haze. In medical imaging, the main types of degradation are due to sensor noise, camera misfocus, variations in the ambient temperature of the scanning room, and the inevitable movements of patients while imaging [14]. Hence, it is required to estimate a clean (undegraded) image, i.e., restore the image by inverting the unknown degradations and decreasing noise from the image.

After the rise of convolution neural networks (CNNs), computer vision systems have made drastic progress [15]. The credit for the success of CNN highly goes to supervised learning methods. However, such methods rely upon a large amount of ground truth information that is needed for supervised training. The lack of such vital information in fields like medical imaging is a challenge because of the lack

J. D. Thakkar · J. S. Bhatt (✉) · S. K. Patra
Indian Institute of Information Technology Vadodara, Vadodara, India
e-mail: jignesh.bhatt@iiitvadodara.ac.in

J. D. Thakkar
e-mail: 201961006@iiitvadodara.ac.in

S. K. Patra
e-mail: skpatra@iiitvadodara.ac.in

P. Singh et al. (eds.), *Machine Learning and Computational Intelligence Techniques for Data Engineering*, Lecture Notes in Electrical Engineering 998,
https://doi.org/10.1007/978-981-99-0047-3_46

of domain knowledge in human annotators including, the scarcity of training data itself. On the other hand, the use of unsupervised learning is many times avoided in the apprehension of undetected features for final decision and diagnosis. As a result, the use of alternate learning methods, say self-supervised learning (SSL), has emerged [9]. To some extent, the SSL is an effective combination of supervised and unsupervised learning paradigms. There are two main parts in SSL, i.e., the first part is called the "pretext task", while the second is the "downstream task" used for fine-tuning the network on a specific small labeled dataset. The broad philosophy of the SSL is to learn a pretext task on huge, unlabeled data followed by fine-tuning this partially trained network on small labeled data in a downstream task. One major benefit of SSL is that one can use the same pretext task on different downstream tasks like classification, Detection, and segmentation. Interestingly, the idea of self-supervision is widely adapted in natural language processing problems [7] but found very limited usage in the field of computer vision problems. In the medical image restoration case, we investigate the SSL as the downstream task to undo the degradation and finally restore the image. With this, one could exploit the unlabeled medical image data to train/improve the pretext neural model.

In this work, a generalized method with an option to incorporate multiple pretext tasks is suggested. Note that the choice of pretext task significantly impacts the results of the self-supervised learning model. Investigation and finding with detailed comparative experiments on MRI and CT datasets followed by ablation analysis using different combinations of loss functions are reported. Further, such investigation would also help to understand better interpretability of neural models [10, 11].

2 Methodology

In this section, a general methodology (Fig. 1) is proposed for medical image restoration using self-supervised learning. We first discuss the overall strategy and then discuss how self-supervision works with different pretext tasks for MRI/CT restoration using different combinations of loss functions.

As shown in Fig. 1, medical image Iref is degraded with Gaussian blur of different variances. It is then corrupted by Rician noise for MRI and Poisson noise for CT to simulate a given MRI/CT image as available in practice. This is then given to a network that is built like an autoencoder-type architecture for SSL (Fig. 1). The encoder network works as a pretext task. The goal here is to learn features only using the given data. Generally, pretext learning is done separately on large unlabeled datasets. It is noticeable that the pretext learning part plays a key role in medical image restoration. Therefore, the experiments are performed on different pretext learning tasks to get better insights into the restoration process. Two of the major pretext learning tasks are rotation learning (rotnet) [6] and models genesis [24]. In our work, the unlabeled ImageNet dataset for rotation learning pretext task and unlabeled medical datasets for models genesis pretext task for the training are used. Thus, this part generates pseudo-mappings (pretext task weights) on the unlabeled data. The

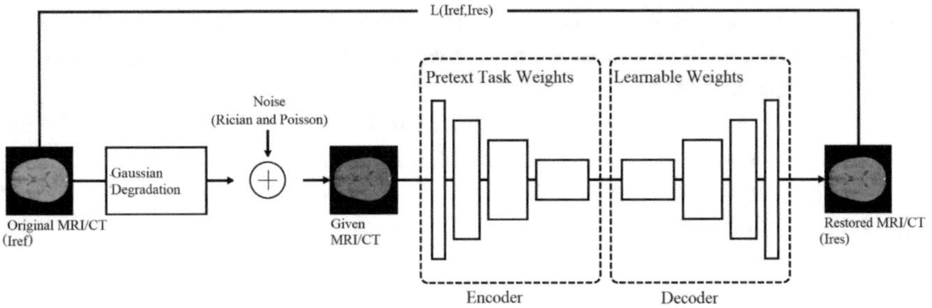

Fig. 1 Method for investigating restoration of medical images using self-supervised learning. Here, we employ two different pretext tasks, i.e., rotation learning and models genesis, against a downstream task, i.e., restoration for the learnable weights. Combinations of ℓ_1, RMSE, and SSIM are constructed as L(Iref, Ires). Gaussian blur with different variances is introduced as the degradation while, Rician and Poisson noises are considered to represent different statistical properties of MRI and CT datasets, respectively. We use AlexNet for rotation pretext, while we use modified AlexNet for downstream restoration, and we employ 3D U-Net for model genesis encoder–decoder

second part is the decoder part, where the task is to estimate a clean (restored) version of the given image. It is the downstream restoration task that performs fine-tuning with a small labeled dataset. Note that the pseudo-mappings learned within the pretext encoder network are now fine-tuned at the decoder network using this relatively small ground truth. This way, we effectively alleviate the practical issues of large training data and scarcity of ground truth in medical imaging.

Rotation learning uses geometric transformations to alter the images and use those transformed images to learn features in the encoder network (Fig. 1). In this work, the pretext/rotation learning part is trained using AlexNet architecture on ImageNet dataset [4]. Here, we train the encoder network using a set of training images to predict the rotation angles of the input images. These are randomly rotated with 0, 90, 180, and 270 degrees of rotation as geometric transformations, i.e., input images are randomly rotated versions of the input image. The aim is to train the pretext encoder network to predict the degrees with which the reference images are rotated. Therefore, the pretext is trained on the four-way classification task to predict one of the four rotations. The knowledge gained thus is later used for medical image restoration downstream task in the decoder network. The decoder is trained using modified AlexNet architecture. To achieve the restoration, we add seven layers of transpose convolution with ReLU after five layers of AlexNet.

Models genesis is a pretext task that learns the underlying features of medical images by applying known degradations on patches of an image and then forcing the pretext encoder network to restore those patches within the image. In this work, we apply different known degradations. An image is cropped in different arbitrary-sized patches, and different degradation/transformations are applied to the patches. We consider local shuffling, non-linear distortion, inpainting, and outpainting, as well as different combinations of these methods. These degraded patches are given to the pretext encoder network as inputs, and the network aims to learn visual representation by restoring the patches within an image. In our work, both the pretext/models genesis and the downstream/restoration learning are trained on 3D U-net architecture with the skip connections [15].

In this work, the downstream task is the restoration of medical images. The network takes the knowledge of the already learned encoder networks (pretext tasks) and uses them to further train the downstream restoration in the decoder part with small labeled MRI and CT datasets.

2.1 Combinations of Different Loss Functions

Different loss functions yield different learning machines [23]. In this work, we experiment with different combinations of loss functions to better appreciate the learning curve of self-supervised learning for medical image restoration. It includes differentiable loss functions, i.e., root-mean-squared error (RMSE), structural similarity index metric (SSIM), and ℓ_1 loss function separately as well as in combinations i.e., RMSE+SSIM, SSIM+ℓ_1, and ℓ_1+RMSE. We now define them.

$$\mathrm{RMSE}(\mathbf{I_{ref(i)}}, \mathbf{I_{res(i)}}) = \sqrt{\frac{1}{n}\sum_{i=1}^{n}[\mathbf{I_{ref(i)}} - \mathbf{I_{res(i)}}]^2}, \tag{1}$$

where $\mathbf{I_{ref(i)}}$ is reference and $\mathbf{I_{res(i)}}$ is restored images, respectively, and n is total number of pixels in an image.

$$\mathrm{SSIM}(\mathbf{I_{ref(i)}}, \mathbf{I_{res(i)}}) = \frac{(2\mu_{I_{ref(i)}}\mu_{I_{res(i)}} + C_1)(2\sigma_{I_{ref(i)}I_{res(i)}} + C_2)}{(\mu^2_{I_{ref(i)}} + \mu^2_{I_{res(i)}} + C_1)(\sigma^2_{I_{ref(i)}} + \sigma^2_{I_{res(i)}} + C_2)}, \tag{2}$$

where $\mu_{I_{ref(i)}}$ and $\mu_{I_{res(i)}}$ are means; and $\sigma^2_{I_{ref(i)}}$ and $\sigma^2_{I_{res(i)}}$ are variances of reference and restored image, respectively. Here, C_1 and C_2 are constants to ensure stability when the denominator becomes zero.

$$\ell_1(\mathbf{I_{ref(i)}}, \mathbf{I_{res(i)}}) = \sum_{i=1}^{n}|\mathbf{I_{ref(i)}} - \mathbf{I_{res(i)}}|. \tag{3}$$

For testing purposes, we rely on the well-known peak signal-to-noise ratio (PSNR). It is the ratio between the maximum possible power of an image and the power of corrupting noise that affects the quality of its representation. In this work, it is defined as

$$\mathrm{PSNR}(\mathbf{I_{ref(i)}}, \mathbf{I_{res(i)}}) = 10\log_{10}\frac{[\max(\max(\mathbf{I_{ref(i)}}), \max(\mathbf{I_{res(i)}}))]^2}{\mathrm{MSE}(\mathbf{I_{ref(i)}}, \mathbf{I_{res(i)}})}, \tag{4}$$

where max(.) indicates maximum value of the argument and RMSE(.) is as defined in Eq. (1).

3 Experiments and Result Analysis

In this section, we present results obtained by employing the proposed methodology as shown in Fig. 1 on medical datasets of different modalities. First, qualitative (visual) comparisons are done by incorporating different combinations of loss functions and then we conduct ablation experiments along with quantitative analysis.

3.1 Datasets, Specifications, and Parameter Settings

Brain MRI medical images for the supervised and rotation learning (rotnet) pretext are used in the experiments. Publicly available "Brain Tumor Segmentation Challenge (Brats - 2020)" dataset [1, 2, 12] is considered for the experiments; and publicly available "Lung Nodule Analysis (LUNA - 2016)" [17] dataset is used for models genesis as the pretext task. All the algorithms are implemented and trained on a Supercomputer with Intel(R) Xeon(R) Gold 6139, Nvidia Quadro GP100 16GB accelerator card, and 96 GB RAM. We have further used Intel(R) Core(TM) i7-7700HQ CPU with 16 GB RAM and Nvidia GeForce GTX 1050 Ti Graphics card for testing. The programming is done in python and PyTorch. The loss functions are optimized using the Adam optimizer. The networks are trained with 200 epochs with the early stopping of 25 epochs.

3.2 Restoration of Brain MRI Dataset

Here, we show the results of the rotation learning approach on the brain MRI dataset and compare it with the supervised ImageNet approach for the restoration of MRI images. The dataset consists of multimodal scans of 369 patients of size $512 \times 512 \times 155$ voxels [1, 2, 12]. Each image cube has 25 slices, which gives a total of 9225 images and we resize each of them to 224×224 pixels. For validation purpose, we have the total data of 125 patients, which makes a total of 3125 images. A learning rate of 0.002 with a weight decay of 1e−5 is employed, and a batch size of 64 is kept throughout the downstream task training process. The training is done by gradient descent algorithm and then verified against validation data. With this setup, the downstream task training took approximately 100 minutes per epoch.

Figure 2 shows the visual results of a challenging case of supervised pretext and rotation learning pretext, where the images are degraded by applying Rician noise of $\mu = 0, \sigma = 1$, and Gaussian blur of $\mu = 0, \sigma = 0.01$. The experiments are done using

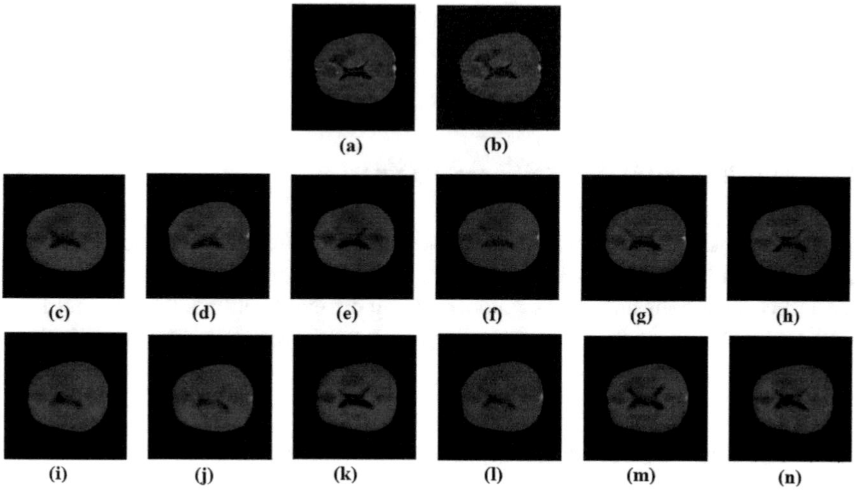

Fig. 2 Visual result of brain MRI restoration obtained by supervised pretraining ((**c**) to (**h**)) and corresponding rotnet pretraining ((**i**) to (**n**)): **a** reference image; **b** degraded image by Rician noise of $\mu = 0$, $\sigma = 1$ and Gaussian blur of $\mu = 0$, $\sigma = 0.01$; restoration **c** and **i** by ℓ_1; **d** and **j** by RMSE; **e** and **k** by SSIM; **f** and **l** by ℓ_1+RMSE; **g** and **m** by SSIM+RMSE; and **h** and **n** by ℓ_1+SSIM

different combinations of loss functions as shown in Fig. 2. From the figure, one can see that when trained on normal autoencoder-type architecture, both the rotnet and supervised pretraining fail to restore the details of the medical images. However, it works better at restoring the edges of the medical images. Figure 2h, n show that the combination SSIM+ℓ_1 gives better results when compared to the other combinations of loss functions. It is also evident that the combinations of loss functions give better results compared to training on individual loss functions.

3.3 *Restoration of Lung CT Dataset*

The lung LUNA16 CT dataset consists of 1,186 lung nodules annotated in 888 CT scans of different patients [17]. The dataset has a total of 28279 images and for our experiment, we divide them into three different parts. The training subset consists of 14159 images, the validation has 5640 images, and we test with 8480 images. The input image cubes are of size $64 \times 64 \times 32$ voxels. A learning rate of 0.01 and batch size of 16 throughout our downstream task training process is employed. The training is done by gradient descent algorithm and then tested against a testing subset of the LUNA16 dataset. With this setup, the downstream restoration task training took approximately 150 minutes per epoch.

Figure 3 shows visual results of a few challenging models genesis pretext task examples. The images in the figure are degraded by Poisson noise of $\lambda = 5$ and

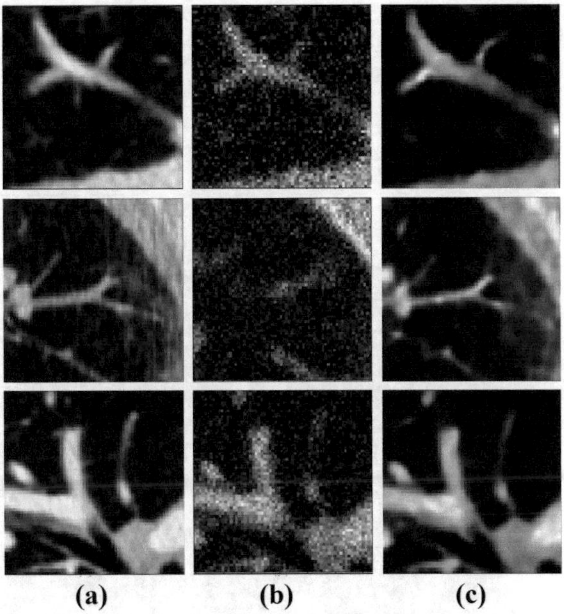

Fig. 3 Restoration of lung CT using models genesis as pretext task: **a** given images, **b** degraded versions with $\lambda = 5$ of Poisson noise and $\mu = 0$, $\sigma = 0.01$ of Gaussian blur in (**a**), and **c** restored versions of (**b**) with fine-tuning of images by SSIM+ℓ_1

Gaussian blur of $\mu = 0$, $\sigma = 0.01$. As in the case of MRI brain (Fig. 2), the lung CT restoration (Fig. 3) also gives better results when trained on SSIM+ℓ_1 compared to the individual training on different loss functions. One can also note that self-supervised models genesis, which is trained on medical data, gives better results compared to ImageNet-trained rotnet. Thus, we note that a pretext task trained on medical image data gives better results compared to a supervised or rotation-learning task trained on the ImageNet data.

3.4 Ablation Experiments and Quantitative Analysis

An ablation study refers to removing or interchanging parts of a neural model to analyze how it affects a specific task. In this subsection, the ablation experiments are carried out in order to get better insight into the pretext and downstream parts of the network for medical image restoration. Here, the performance of the models with different pretext tasks is analyzed, and we also analyze by freezing different layers in the downstream task. Table 1 shows the comparison of our experiments with a state-of-the-art restoration method [22].

Pretext tasks versus combination of loss functions Here, it is described how different pretext tasks affect the medical image restoration task. Figure 4 show the comparison of experimenting on the test dataset (BRATS20) with 3125 images using the supervised approach and rotation learning approach. The images are degraded by

Table 1 A quantitative result on the restoration of medical images using different restoration techniques degraded by Gaussian noise of $\mu = 0$, $\sigma = 0.01$ and Gaussian blur of $\mu = 0$, $\sigma = 0.01$

Restoration technique	PSNR (dB)
Rotnet learning	18.33
Models genesis	32.34
Deep image prior [22]	38.22

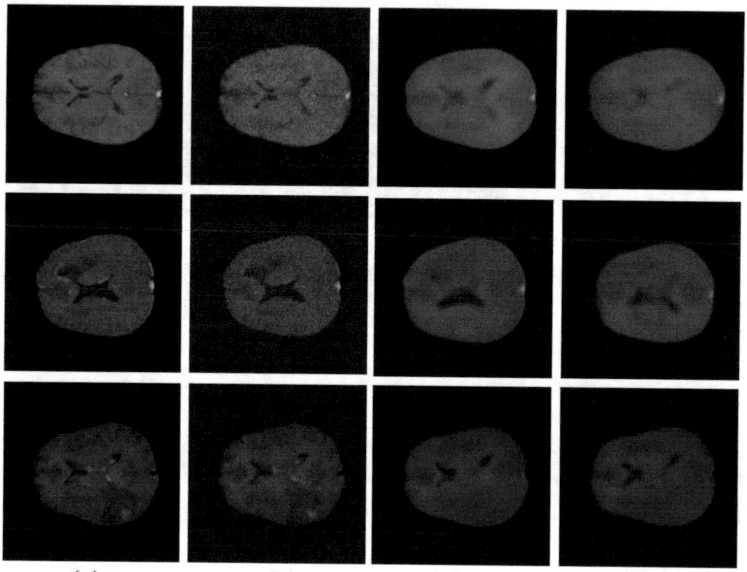
(a) (b) (c) (d)

Fig. 4 Ablation experiment-1 of different pretext tasks for restoration of BRATS20 MRI dataset with 3125 MRI images: **a** given three representative images, **b** degraded versions of images in (**a**) by Rician noise of $\mu = 0$, $\sigma = 1$ and Gaussian blur of $\mu = 0$, $\sigma = 0.01$, **c** restored versions of images by supervised pretext, and **d** restored versions of images by rotnet pretext

Rician noise of $\mu = 0$, $\sigma = 1$ and Gaussian blur of $\mu = 0$, $\sigma = 0.01$. Table 2 lists the PSNR values of degrading the testing images. One can observe from Table 2 that for supervised learning, training on RMSE loss gives slightly better results compared to the rotation learning approach.

Downstream tasks versus fine-tuning Here, we show how fine-tuning the downstream task affects medical image restoration and also highlight that the use of medical data is better than using ImageNet or any other generalized data for medical image restoration. In two different ablation experiments, Figs. 5 and 6 show the comparison of how fine-tuning the downstream task work when compared to using the pretext task without any fine-tuning. As shown in Fig. 5, the images are degraded by Rician noise of $\mu = 0$, $\sigma = 1$ and Gaussian blur of $\mu = 0$, $\sigma = 0.01$. Table 3 denotes

Table 2 A quantitative result on restoration of MRI dataset (BRATS20) degraded by Rician noise of $\mu = 0, \sigma = 1$ and Gaussian blur of $\mu = 0, \sigma = 0.01$ using different combinations of loss functions

Pretext tasks	Performances of different learning machines					
	RMSE	SSIM	ℓ_1	RMSE+SSIM	RMSE+ℓ_1	SSIM+ℓ_1
Supervised learning	18.98	19.12	18.22	19.03	19.09	18.44
Rotnet learning	18.29	18.30	17.92	18.33	18.25	17.98

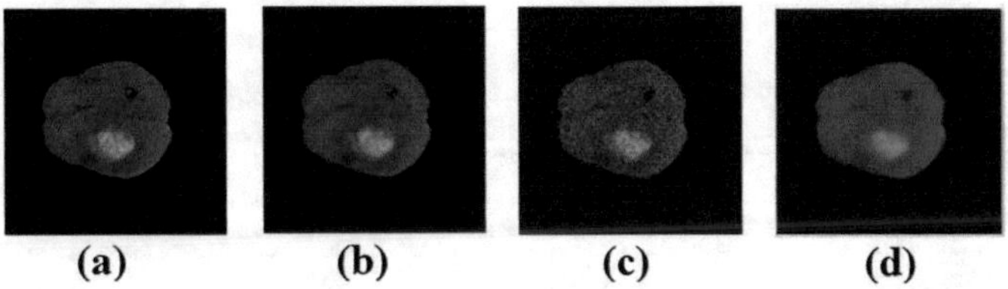

Fig. 5 Ablation experiment-2 of fine-tuning vs. without fine-tuning for restoration of MRI dataset (BRATS20): **a** given image, **b** degraded version by Rician noise of $\mu = 0, \sigma = 0.01$ and Gaussian blur of $\mu = 0, \sigma = 0.01$ in (**a**), **c** restored version using fine-tuning in (**a**), and **d** restored version without fine-tuning in (**a**)

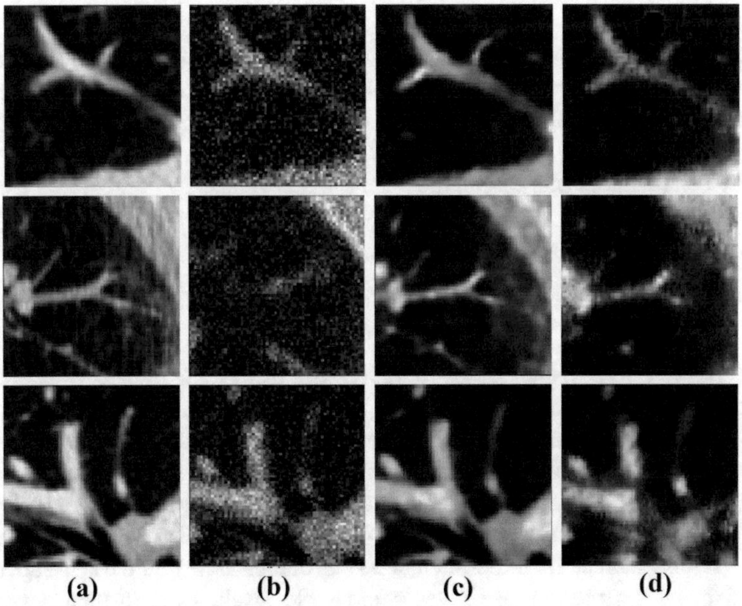

Fig. 6 Ablation experiment-3 of using models genesis as pretext task learning for restoration of lung CT (LUNA16 dataset) image: **a** given representative images, **b** degraded versions of images using Poisson noise $\lambda = 5$ and Gaussian blur $\mu = 0, \sigma = 0.01$ in (**a**), and **c** restored versions with fine-tuning of images in (**b**), **d** restored versions without fine-tuning of images in (**b**)

Table 3 MRI restoration (Ablation Experiment-2, Fig. 5): Fine-tuning versus without fine-tuning on MRI dataset (BRATS20) with 3125 images using supervised and rotnet approaches

MRI Restoration				PSNR (dB)
Pretext	Gaussian blur (%)	Rician noise (%)	Finetuning	
Supervised learning	5	5	No	12.11
Supervised learning	5	5	Yes	19.03
Rotnet learning	5	5	No	12.01
Rotnet learning	5	5	Yes	18.33

Table 4 CT restoration (Ablation experiment-3, Fig. 6): Fine-tuning versus without fine-tuning on CT dataset (LUNA16) with 8480 images for models genesis approach

CT restoration				PSNR (dB)
Pretext	Gaussian blur (%)	Noise (%)	Finetuning	
Models genesis	5	Gaussian—5	No	20.53
Models genesis	5	Gaussian—5	Yes	25.11
Models genesis	5	Poisson—5	No	27.88
Models genesis	5	Poisson—5	Yes	32.34

the PSNR values for the rotnet and supervised approach. Figure 6 shows the experiments performed on the LUNA16 CT test dataset of 8480 images. The images are degraded by Poisson noise $\lambda = 5$ and Gaussian blur $\mu = 0$, $\sigma = 0.01$. It is noticeable from Fig. 6, and Table 4 that when trained on medical images, the restoration works decently even without fine-tuning the data.

4 Conclusion and Discussion

This work presents a methodology equipped with different self-supervised learning strategies for medical image restoration. We discuss our investigation and finding on different modalities of medical image restoration. Unlike existing deep learning approaches, self-supervision helps neural networks to learn valuable semantics with a fairly less amount of explicitly labeled data. Hence, the approach is convenient for practical applications in medical imagery, with the scarcity of ground truth as well as limited training data pairs. Experimental results demonstrate that the self-supervised medical image restoration task gives comparable results even to their supervised counterparts. The ablation experiments have given better insights into better appreciating different counterparts of self-supervised learning including, pretext tasks, downstream tasks, different combinations of loss functions, and with/without fine-tuning. It would certainly help build more interpretable neural models based on SSL for medical image restoration. In the future, one can conduct experiments to solve different downstream tasks like super-resolution and segmentation. Similar investigations could also be conducted in order to get better interpretability of such a practically useful neural model for other vision-based tasks.

Acknowledgements Authors are thankful to Gujarat Council on Science and Technology (GUJ-COST) for providing PARAM Shavak GPU-based supercomputer to IIIT Vadodara for conducting the exhaustive experiments in this research work.

References

1. Bakas S et al (2018) Identifying the best machine learning algorithms for brain tumor segmentation, progression assessment, and overall survival prediction in the brats challenge. ArXiv:abs/1811.02629
2. Bakas S et al (2017) Advancing the cancer genome atlas glioma mri collections with expert segmentation labels and radiomic features. Sci Data 4
3. Challa U, Yellamraju P, Bhatt J (2019) A multi-class deep all-cnn for detection of diabetic retinopathy using retinal fundus images. In: Deka B, Maji P, Mitra S, Bhattacharyya DK, Bora PK, Pal SK (eds) Pattern recognition and machine intelligence. Springer International Publishing, Cham, pp 191–199
4. Deng J et al (2009) Imagenet: a large-scale hierarchical image database. In: 2009 IEEE conference on computer vision and pattern recognition, pp 248–255 (2009)
5. Deshpande VS, Bhatt JS (2019) Bayesian deep learning for deformable medical image registration. In: Deka B, Maji P, Mitra S, Bhattacharyya DK, Bora PK, Pal SK (eds) Pattern recognition and machine intelligence. Springer International Publishing, Cham, pp 41–49
6. Gidaris S, Singh P, Komodakis N (2018) Unsupervised representation learning by predicting image rotations. ArXiv:abs/1803.07728
7. Howard J, Ruder S (2018) Fine-tuned language models for text classification. ArXiv:abs/1801.06146
8. Jeberson K, Kumar M, Jeyakumar L, Yadav R (2020) A combined machine-learning approach for accurate screening and early detection of chronic kidney disease. In: Agarwal S, Verma S, Agrawal DP (eds) Machine intelligence and signal processing. Springer Singapore, Singapore, pp 271–283
9. Kolesnikov A, Zhai X, Beyer L (2019) Revisiting self-supervised visual representation learning. In: Proceedings of the IEEE/CVF conference on computer vision and pattern recognition, pp 1920–1929
10. Mahapatra D, Poellinger A, Shao L, Reyes M (2021) Interpretability-driven sample selection using self supervised learning for disease classification and segmentation. IEEE Trans Med Imaging
11. Manaswini P, Bhatt JS (2021) Towards glass-box cnns. arXiv:2101.10443
12. Menze B et al (2014) The multimodal brain tumor image segmentation benchmark (brats). IEEE Trans Med Imaging 99
13. Rai S, Bhatt JS, Patra SK (2022) Accessible, affordable, and low-risk lungs health monitoring in covid-19: deep reconstruction from degraded lr-uldct. Accepted, IEEE International Symposium on Biomedical Imaging (ISBI)
14. Rai S, Bhatt JS, Patra SK (2021) Augmented noise learning framework for enhancing medical image denoising. IEEE Access 9:117153–117168
15. Ronneberger O, Fischer P, Brox T (2015) U-net: Convolutional networks for biomedical image segmentation. ArXiv:abs/1505.04597

16. Savitha G, Jidesh P (2019) Lung nodule identification and classification from distorted ct images for diagnosis and detection of lung cancer. In: Tanveer M, Pachori RB (eds) Machine intelligence and signal analysis. Springer, Singapore, pp 11–23
17. Setio A et al (2017) Validation, comparison, and combination of algorithms for automatic detection of pulmonary nodules in computed tomography images: The luna16 challenge. Med Image Anal 42:1–13
18. Sharma A, Chaurasia V (2019) A review on magnetic resonance images denoising techniques. In: Tanveer M, Pachori RB (eds) Machine intelligence and signal analysis. Springer, Singapore, pp 707–715
19. Sharma M, Bhatt JS, Joshi MV (2017) Early detection of lung cancer from CT images: nodule segmentation and classification using deep learning. In: Verikas A, Radeva P, Nikolaev D, Zhou J (eds) Tenth international conference on machine vision (ICMV 2017), vol 10696, pp 226–233. International Society for Optics and Photonics, SPIE. https://doi.org/10.1117/12.2309530
20. Shivhare SN, Sharma S, Singh N (2019) An efficient brain tumor detection and segmentation in mri using parameter-free clustering. In: Tanveer M, Pachori RB (eds) Machine intelligence and signal analysis. Springer, Singapore, pp 485–495
21. Soni P, Chaurasia V (2019) Mri segmentation for computer-aided diagnosis of brain tumor: a review. In: Tanveer M, Pachori RB (eds) Machine intelligence and signal analysis. Springer, Singapore, pp 375–385
22. Ulyanov D, Vedaldi A, Lempitsky V (2020) Deep image prior. Int J Comput Vis 128:1867–1888
23. Zhao H, Gallo O, Frosio I, Kautz J (2017) Loss functions for image restoration with neural networks. IEEE Trans Comput Imaging 3(1):47–57
24. Zhou Z, Sodha V, Pang J, Gotway M, Liang J (2021) Models genesis. Med Image Anal 67:101840

An Analogy of CNN and LSTM Model for Depression Detection with Multiple Epoch

Nandani Sharma and Sandeep Chaurasia

1 Introduction

"Depression is a mood disturbance which typically includes feelings of apprehension, gloom, helplessness, and worthlessness" [2]. In the words of the World Health Organisation, across the globe 350 million people are affected by depression [3]. In a prediction by the World Health Organisation (WHO) 322 million people are estimated to be in hardship with the pain of depression by the year 2030 [4]. The mild effects of depression on humans are Loneliness, sadness,emptiness, feelings of self guilt, and no enthusiasm for the future. At its worst end the depression may lead to the suicide. According to the WHO 7,00,000 people death cause is suicide every year. With these many mild and major health effects on life depression is the least talked about topic. Reasons being least talked about are : unavailability of the resources, scarcity of healthcare workers experts for depression, and most effect is social stigma attached with mental disorder.

Social media has become a non-detachable need of human life. Social media platforms are used mostly for self-disclosure of feelings such as closeness of friendships, patch up, the feeling of sadness, breakup, heartbreak, breakdown, burnout. People are more interested in self-disclosure of feelings on social media as compared to the sharing to the people in actuality. Morgan and Cotten [1]; research suggests that the online communication platform is associated with reduction of depression symp-

N. Sharma · S. Chaurasia (✉)
Manipal University Jaipur, Jaipur, India
e-mail: sandeep.chaurasia@jaipur.manipal.edu

N. Sharma
e-mail: nandani.219351015@muj.manipal.edu; nandani@poornima.org

N. Sharma
Poornima Institute of Engineering and Technology, Jaipur, India

P. Singh et al. (eds.), *Machine Learning and Computational Intelligence Techniques for Data Engineering*, Lecture Notes in Electrical Engineering 998,
https://doi.org/10.1007/978-981-99-0047-3_47

Fig. 1 Word cloud of the tweet data on depression

toms. This research supports depression symptoms that can be detected by social media text.

Twitter is a microblogging site which is mostly used for the purpose of self-disclosure of emotion and feeling. The microtext in twitter is named as tweets. Twitter has 221million, Monetizable Daily Active Users till Q3 2021 [6]; this number is approximately equal to the population of some country of the world. Twitter with this much acknowledged platform of social media is a good source for natural language processing classification. Boyed et al. state that twitter is fundamentally the means of general communication of individual, group and complete world [7] (Fig. 1).

With given world health organisation medical statistics for depression there is no laboratory test to inspect the depression disorder [4]. The universal approach to detection of depression in person are, questionnaires, interviews [3, 14, 15]. Questionnaires are most of the time subjective, Twitter is a descriptive platform to express emotions. Another problem with this universal approach is that people may be unwilling to talk about depression [5]; Whereas on twitter people are expressing their emotion and feeling with their own enthuthiasm.

The process of the Sentimental analysis is classification of the text into the categories to the polarity positive, negative and neutral.The depression detection is a more fine grained classification under the umbrella of sentimental analysis classification. Depression detection is fine grained sentiment analysis because the indications of the depression in the tweet text are often subtle [5]; and thus not immediately obvious to the human reader so the machine learning approach is one of the possible solutions.

2 Related Work

Kim et al. [8]; classify the fine grained shade of mental health such as depression, anxiety, bipolar, BPD, Schizophrenia and autism using the autonomous binary models for each mental health issue. Data is collected from Reddit for CNN with max pooling and XGBoost model classification.

Ahmed Husseini Orabi et al. [9] Implemented a different deep learning model as CNN, with max, multi channelCNN, Multichannelpooling CNN, BiLSTM for depression text classification on twitter data. Once the optimization of word embedding technique is performed on the random trainable, skip-gram, and CBOW.

Abdulqader Almars [3] worked on depression detection for tweets in the Arabic language. Author combines the attention technique along with Bidirectional Long Short term memory model for detection of tweet dataset in Arabic language. Accuracy achieved 0.83% for Arabic language tweets.

Wen [16]; worked on previously scraped tweets data for depression detection with a model created with embedding bags and linear layer using word look up table.With this model accuracy achieved is 99.04 and F1 score 85%.

Diveesh Sings and Aileen Wang [5] in their research work first scrap the twitter for depression the instead of labelling tweets manually label them with data with polarity score. Then training following deep learning models: RNN, CNN and GRU. with variation of character base and word base in parallel the variation of learned and pretrained embedding. They have observer the word GRU achieve accuracy of 98% amog other variations.

Harleen Kaur et al. [17] work on the Covid-19 tweet data using some deep learning model Recurrent neural network, Heterogeneous Euclidean overlap metric, Hybrid heterogeneous support vector machine.

Amit Kumar Sharma et al. [20] classifies the IMDB dataset for sentiment analysis of the viewer with the word2vec with the CNN model of deep learning. The CNN model achieved 99% accuracy for training data and 82.19% of testing data.

Convolutional Neural Network is a deep learning approach. A Large extent of research is performed with convolutional neural networks in image processing [12]. In the state of art the convolutional neural network is used for text classification and showed successful results. In analogy to other deep learning models, convolutional neural networks can be trained with few parameters as well as few connections [11]. The training of the Convolutional Neural Network is not that much convolution [11]. In text processing the Convolutional Neural Network is used to reduce the dimension of the text while preserving the feature of the text and then classify. The reduction is done in such a way that it preserves the feature of the text, for the purpose of the classification. The input matrix is convolute with the filter size and stride for output. The convolutional neural network is divided in two steps: feature learning and classification.

In conclusion, it has been observed that least work is done on the comparison of the deep learning model Convolutional neural network, long short term memory and combination of convolutional neural network and long short term memory with multiple epochs analysis.

3 Experimental Work

This research is using different flavours of Convolutional neural network, long short term memory and combination of convolutional neural network and long short term memory. The research work is comparing the five different combination of Convolutional neural network with max pooling, Convolutional neural network with average pooling, Convolutional neural network with multiple layer, Long short term memory and Convolutional neural network combined with Long short term memory. Measure the performance of each model at 5, 10 and 15 epoch.

3.1 Dataset

There is a lack of public dataset availability for the depression tweet. The reason being lack of availability is due to the ethics of treatment and patient privacy policy bind with the health related data. Keeping this in mind, a random dataset of twitter data available publicly is taken as a label of the non depression [18]. For the depression label of the dataset the publicly available twitter data with keyword depression is scraped for a clock of 24 h [19]. The non depression tweets are 12000 in count and for depression tweets 2345 are used. Total tweet data is 14,345. Attributes of the tweeter to be considered for the classification are Ids, tweet text and labels (Table 1).

3.2 Preprocessing

It has been observed by Ahmed Husseini Orabi et al. [9]; that language of the tweet data is unstructured in nature tweets are filled with misspelt, #tags, url and character limitation words. As the language of tweet text mainly does not follow natural language grammar rules as the. With this observation some cleaning and preprocessing is required for the data before sending it to train the models.English language words

Table 1 Depression dataset summary

Sentiment	Tweet count
Non depression	12000
Depression	2345

are a contraction as an example "I will" is contracted with "I'll". It may affect the model so the contraction of English words will expand.

In the cleaning process of tweet data firstly all the hashtags are removed. Then mention, associated with the tweets are cleaned. Emoji of different types are brought out. The image URL text associated with the link to next information that does not add the sentiment of the tweet text so are also withdrawn. In addition the punctuation is also removed. Then the word of each tweet is stemmed using porter stemmer. Once the sanitization of the tweet data is completed it's time to convert the text into numbers and create a vocabulary index based on frequency. For generalising the length of each tweet is 140 post padding performed for the short tweets and prune the long tweets. Now the text is converted to the number it's time to split the data in 60% training, 20% validation and 20% for testing.

3.3 Experiment

The numerical value converted from text in data preprocessing, if provided directly to any model for training, will not be helpful as the semantic of the word is not saved during the conversion process. For maintaining the semantic information each word of the complet vocabulary is represented in the vector space. In word embedding words are represented in some N specific dimensional space. The representation is designed for the purpose of maintaining the semantics of the words. In this research, word2vec pretrained model is used with the input dimension as 20000, output dimension as 300 and input length as 140.

Activation function when the data is non linearly separable then there is requirement of activation function for classification. Sigmoid function squeezes the value in 0 and 1 For performing the Sigmoid function formula is:

$$F(z) = \frac{1}{1 + e^{-z}} \tag{1}$$

ReLU is another activation function with the full form Rectified linear unit. Relu is mostly used for the hidden layer of the deep learning models. The ReLU formula:

$$F(z) = \max(0, z) \tag{2}$$

For the purpose of curtailment of the dimensionality and deformation compensation of convolutional layers the pooling operation is performed [13]. For this purpose the global max pooling and average max polling is used. The global max pooling for summarising the feature maximum value is selected. The Global average pooling the average complete feature matrix is performed.

LSTM is a deep learning model mostly used for sequential data classification. Long short term memory is not only to maintain the previous state (Sn-1) learning, but also to maintain all the previous state learning (L0-n-1) for the purpose of the current state (State). LSTM works on two main principles: Forget Principle: according to this principle insignificant information is forgotten. This principle ensures that unnecessary information is not remembered for long run. Saving principle : Save all the important information for later use. The use of this principle is in maintaining long term memory. LSTM is a combination of 3 gates: input gate, output gate and forget gate. The equation of these gate are as following:

Input Gate Equation:

$$i_t = \sigma(w_i[h_{t-1}] + b_i) \tag{3}$$

Output Gate Equation:

$$f_t = \sigma(w_i[h_{t-1}] + b_o) \tag{4}$$

Output Gate Equation:

$$f_t = \sigma(w_i[h_{t-1}] + b_f) \tag{5}$$

The components explained above can be used to create different models of combination.

To train the model, the first model created is CNN with Global MaxPooling with multiple layers. In the Sequential model the first model is pre trained word2vec embedding.In next step convolution 1d layer is added with filter size 5 and activation function ReLU. Then the global max layer is used. Further dense layer is added to reduce the overfitting with the activation function ReLU, followed by the dropout laye with rate 0.2. At last we output is given by the dense layer with activation layer sigmoid function.

To train the model, the second model created is CNN with Global Average Pooling with multiple layers. In the Sequential model the first model is pre trained word2vec embedding. Second layer in the model is dropout with the rate 0.2. In the next step convolution 1d layer is added with filter size 3 and activation function ReLU. Then the global average layer is used. Further dense layer is added to reduce the overfitting with the activation function ReLU, followed by the dropout laye with rate 0.2. At last we output is given by the dense layer with activation layer sigmoid function.

To train the model, the third model created is a Multi layer CNN with multiple layers. In the Sequential model the first model is pre trained word2vec embedding. Second layer in the model is dropout with the rate 0.2. In the next step we add three convolution 1d layers one after the other with filter size 3, 4 and 5 respectively and activation function ReLU. Then the global average layer is used. Further dense layer is added to reduce the overfitting with the activation function ReLU, followed by the dropout laye with rate 0.2. At last we output is given by the dense layer with activation layer sigmoid function.

Now it is time to train the model number forth using a long term short memory. In the Sequential model the first model is pre trained word2vec embedding. Second layer in the model is Spatial dropout one dimensional with the rate 0.25. In the next step we add Long short term memory layers with dropout 0.5. Then a dropout layer with rate 0.2 is added. Further dense layer is added to reduce the overfitting with the activation function sigmoid which gives output.

Finally we are at last model name as convolution neural network with long term short term with following layers. In the Sequential model the first model is pre trained word2vec embedding. Second layer is Long short term memory layers. In the next step we add Then a dropout layer with rate 0.2 is added. Then the model is combined with a convolution layer with filter size 3 and activation function ReLU.Then the global average layer is used and a dropout layer with rate 0.2. Further dense layer is added to reduce the overfitting with the activation function sigmoid which gives output.

4 Result

Five models created above are now evaluated with the help of prominent metrics such as precision, recall and F1-score. Evaluation formula can be given by:

$$\text{Accuracy} = \frac{TP + TN}{TP + FP + TN + FN} \tag{6}$$

$$\text{Precision} = \frac{TP}{TP + FP} \tag{7}$$

$$\text{Recall} = \frac{TP}{TP + FN} \tag{8}$$

$$\text{F1 score} = 2 * \frac{(\text{Precision} * \text{Recall})}{(\text{Precision} + \text{Recall})} \tag{9}$$

In the above formula TP is true positive, FP is false negative, TN is true negative and finally FN is false negative which represents the state of confusion matrix.

Accuracy of CNN with global max Pooling can be observed in Fig. 2 and Table 3 with 5 epoch, 10 epoch and finally 15 epoch. In Fig. 2. the growth of the accuracy during the training of CNN with global max pooling at 5, 10, 15 epoch is visualised. In the training phase the following 0.9984, 0.9984 and 0.9982 accuracy is achieved by the end of 5, 10 and 15 epochs. The accuracy of 97.86, 97.86, and 97.54% is achieved by the model tested for 5, 10 and 15 epochs Table 3. It has been observed that with epoch 5 and 10 the accuracy is the same and higher 97.86%. Precision for non depressive disorder with 5 epoch and 10 epoch is 0.98 is highest among all the three epochs. Precision for depressive disorder with 15 epoch is 0.99 is highest among all the three epochs.Recall is higher with epoch 10 and 15 with value 1.00 for non depressive class and for depressive disorder with epoch 5 and 10 with value 0.98. F1 score 0.99 for all the epochs for non depressive whereas for depressive with epoch 5 and 10 with value 0.93.

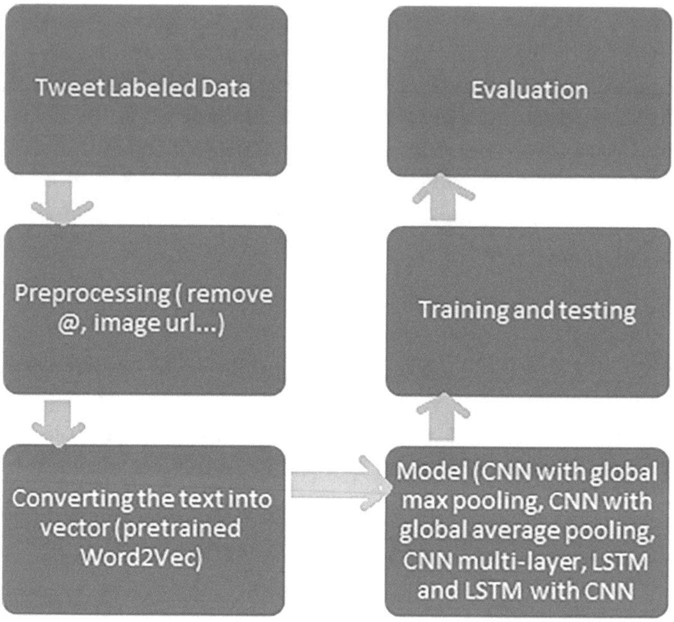

Fig. 2 Flow diagram of the model

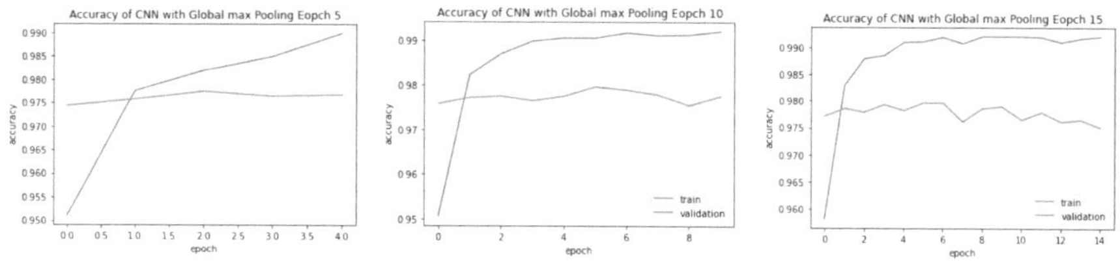

Fig. 3 Accuracy of CNN with global max pooling with epoch 5,10 and 15

Accuracy of CNN with global Average Pooling can be observed in Fig. 3 with 5 epoch, 10 epoch and finally 15 epoch. In Fig. 3. the growth of the accuracy during the training of CNN with global Average pooling at 5, 10, 15 epoch is visualised. In the training phase the following 0.9843, 0.9883 and 0.9905 accuracy is achieved by the end of 5, 10 and 15 epochs. The accuracy of 95.99, 96.59 and 97.08% is achieved by the model tested for 5, 10 and 15 epochs Table 3. It has been observed that this model achieves best accuracy with 97.08% (Table 2).

In Fig. 4. the growth of the accuracy during the training of CNN with multi-layer at 5, 10, 15 epoch is visualised. In the training phase the following 0.9926, 0.9967 and 0.9982 accuracy is achieved by the end of 5, 10 and 15 epochs. The accuracy of 97.89, 97.86 and 97.75% is achieved by the model tested for 5, 10 and 15 epochs Table 3. Accuracy of CNN multi-layer can be observed in Fig 4 with 5 epoch, 10

Table 2 Precision, Recall, F1-Score of five different applied models

Model	Label 5	Precision 10	Recall	F1-Score
Epoch 5				
CNN with Global MaxPooling	DN	0.98	0.99	0.99
	N	0.97	0.89	0.93
CNN with Global Average Pooling	DN	0.97	0.98	0.98
	N	0.90	0.85	0.87
Multi layer CNN	DN	0.98	0.99	0.99
	N	0.97	0.90	0.93
LSTM	DN	0.99	1.00	1.00
	N	1.00	0.95	0.97
LSTM with CNN	DN	0.99	1.00	0.99
	D	0.98	0.96	0.97
Epoch 10				
CNN with Global MaxPooling	DN	0.98	1.00	0.99
	N	0.97	0.89	0.93
CNN with Global Average Pooling	DN	0.97	0.99	0.98
	N	0.94	0.84	0.89
Multi layer CNN	ND	0.98	0.99	0.99
	D	0.96	0.90	0.93
LSTM	ND	0.99	1.00	0.99
	D	0.99	0.95	0.97
LSTM with CNN	DN	0.99	0.99	0.99
	N	0.94	0.97	0.96
Epoch 15				
CNN with Global MaxPooling	ND	0.97	1.00	0.99
	D	0.99	0.86	0.92
CNN with Global Average Pooling	ND	0.97	0.99	0.98
	D	0.95	0.87	0.91
Multi layer CNN	ND	0.98	0.99	0.99
	D	0.96	0.89	0.93
LSTM	ND	0.99	1.00	0.99
	D	0.99	0.96	0.97
LSTM with CNN	DN	0.99	1.00	0.99
	N	0.99	0.95	0.97

epoch and finally 15 epoch. It has been observed that model with epoch 5 gave best accuracy of three that is 97.89%

Accuracy of LSTM can be observed in Fig. 5 and Table 3 5 with 5 epoch, 10 epoch and finally 15 epoch. In Fig. 5. the growth of the accuracy during the training

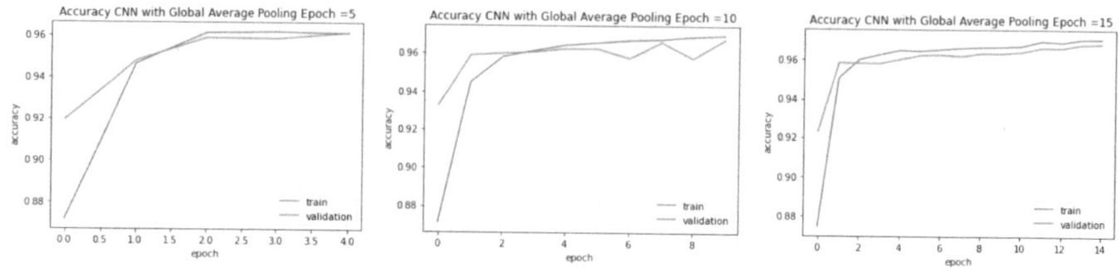

Fig. 4 Accuracy of CNN with global average pooling with epoch 5,10 and 15

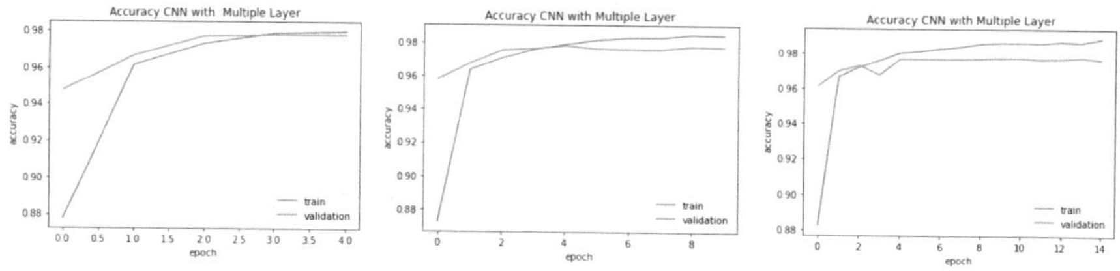

Fig. 5 Accuracy of CNN with Multi Layer with epoch 5,10 and 15

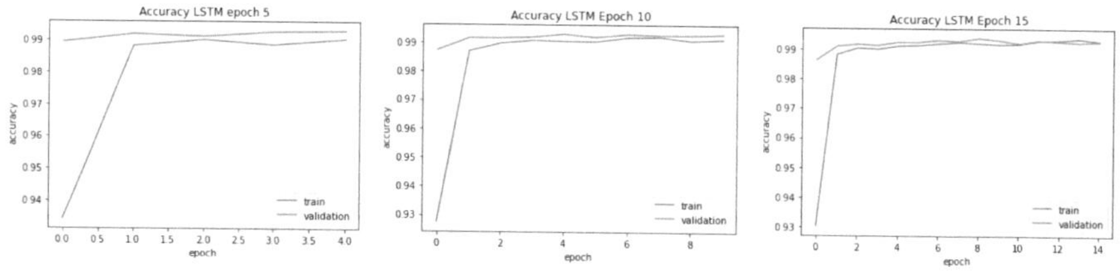

Fig. 6 Accuracy of LSTM with epoch 5,10 and 15

of LSTM at 5, 10, 15 epoch is visualised. In the training phase the following 0.9943, 0.9979 and 0.9982 accuracy is achieved by the end of 5, 10 and 15 epochs. The accuracy of 99.19 ,99.12 and 99.12% is achieved by the model tested for 5, 10 and 15 epochs Table 3. It has been observed that models with epoch 5 gave best accuracy of three that is 99.19%

Accuracy of LSTM with CNN can be observed in Fig. 6 and Table 3 with 5 epoch, 10 epoch and finally 15 epoch. In Fig. 6. the growth of the accuracy during the training of LSTM followed by CNN with global max pooling at 5, 10, 15 epoch is visualised. In the training phase accuracy progress is shown in Fig. 6 by the end of 5, 10 and 15 epochs. The accuracy of 98.98, 98.56 and 99.16% is achieved by the model tested for 5, 10 and 15 epochs Table 3. It has been observed that models with epoch 15 gave the best accuracy of three that is 99.16% (Fig. 7).

One of the research objectives is to compare the results of the five models named CNN with global max pooling, CNN with global average pooling, CNN with multiple Layer, LSTM and CNN with LTSM. As shown in Table 3, it has been observed that the LSTM Model with the epoc 5 performs best as compared with the other models

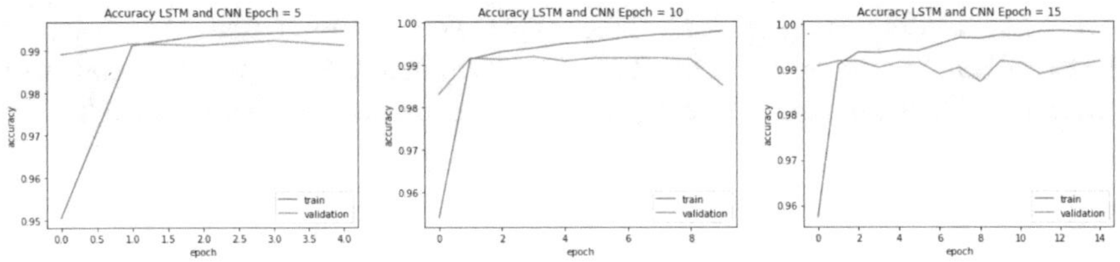

Fig. 7 Accuracy of LSTM with CNN with epoch 5, 10 and 15

with accuracy 99.19%. Whereas the CNN with global average pooling with epoch 5 performs lowest with the accuracy 95.99%.

From Table 2 the precession, recall and F1 of the model LSTM is best among the other models trained in the research with 5, 10, 15 epoch. LSTM model achieved the F Score of 1.0 for non depression class. Performance of CNN with Global Average Pooling is lowest among all other models.

5 Conclusion and Future Work

From Table 3 the LSTM model achieves the accuracy of 99.19% followed by LSTM with CNN with 99.16% accuracy.

From the Table 2 on if the performance of all the models is evaluated on the basis of precession the performance is of LSTM for non depressive class score 1 with epoch 5 and also for the depressive class also the LSTM performance is appreciable with the score 0.99. Then the result is followed by the LSTM with CNN non depressive class score 0.99. And the CNN with Global Average Pooling model performance is with the depressive class with 5 epoch with score 0.9.

With accuracy of upto 95.99–99.19% the model can be implemented in real life depression detection. The models can be used for the purpose of the early detection of depression in clinical depression detection and it is a good tool for treatment of depression.

Table 3 Accuracy of 5 model

Model	Epoch 5 (%)	Epoch 10 (%)	Epoch 15 (%)
CNN with global max pooling	97.86	97.86	97.54
CNN with global average pooling	95.99	96.59	97.08
Multi layer CNN	97.89	97.86	97.75
LSTM	99.19	99.12	99.12
LSTM with CNN	98.98	98.56	99.16

In present work demographic attributes of the tweet text is not used as the feature of the model training, according to the WHO report women are more prone to depression as compared to the men.In future depression text detection can be applied to the real time twitter text.

References

1. Morgan C, Cotten SR (2003) The relationship between Internet activities and depressive symptoms in a sample of college freshmen. Cyber Psychol Behav 6:133–142
2. Battle J (1978) Relationship between self-esteem and depression. Psychol Rep 42(3):745–746
3. Almars AM (2022) Attention-based Bi-LSTM model for arabic depression classification. CMC-Comput Mater Contin 71(2):3091–3106
4. Mustafa RU, Ashraf N, Ahmed FS, Ferzund J, Shahzad B, Gelbukh A (2020) A multiclass depression detection in social media based on sentiment analysis. In: Proceedings of the 17th IEEE international conference on information technology-new generations. Springer, Berlin, pp 659–662
5. Sings D, Wang A, Depression Detection Through Tweets in Stanford University, Stanford CA 94304 [online]. Available https://web.stanford.edu/class/archive/cs/cs224n/cs224n.1184/reports/6879557.pdf
6. https://www.statista.com/statistics/970920/monetizable-daily-active-twitter-users-worldwide/
7. Boyd D, Golder S, Lotan G (2010) Tweet, tweet, retweet: conversational aspects of retweeting on twitter. IEEE Computer Society, Los Alamitos, CA, USA, pp 1–10.
8. Kim J, Lee J, Park E, Han J (2020) A deep learning model for detecting mental illness from user content on social media. Sci Rep 10(1):1–6
9. Ahmed Husseini Orabi AH, Buddhitha P, Orabi MH, Inkpen D (2018) Deep learning for depression detection of twitter users. In: Proceedings of the fifth workshop on computational linguistics and clinical psychology: from keyboard to clinic, pp 88–97
10. Reece AG, Reagan AJ, Lix KL, Dodds PS, Danforth CM, Langer EJ (2017) Forecasting the onset and course of mental illness with Twitter data. Sci Rep 7(1):1–11
11. Kim H, Jeong YS (2019) Sentiment classification using convolutional neural networks. Appl Sci 9(11):2347
12. Krizhevsky A, Sutskever I, Hinton G E (2012) Imagenet classification with deep convolutional neural networks. Neural Inf Process Syst 1097–1105
13. Otsuzuki T, Hayashi H, Zheng Y, Uchida S (2020) Regularized pooling. In: International conference on artificial neural networks. Springer, Cham, pp 241–254
14. Whooley O (2014) Diagnostic and statistical manual ofmental disorders (dsm). Wiley Blackwell Encyclo-Pedia Ofhealth, Illn, Behav, Soc 5:381–384
15. Park M, Cha C, Cha M (2012) Depressive moods of users portrayed in twitter. In: Proceedings of the ACM SIGKDD workshop on healthcare informatics (HI-KDD). Beijing, China, pp 1–8
16. Wen S, Detecting depression from tweets with neural language processing. J Phys: Conf Ser 1792: 012058
17. Kaur H, Ahsaan SU, Alankar B, Chang V (2021) A proposed sentiment analysis deep learning algorithm for analyzing COVID-19 tweets. Inf Syst Frontiers 1–3
18. https://www.kaggle.com/ywang311/twitter-sentiment/data
19. https://github.com/eddieir/Depression_detection_using_Twitter_post
20. Sharma AK, haurasia S, Srivastava DK (2020) Sentimental short sentences classification by using CNN deep learning model with fine tuned Word2Vec. Procedia Comput Sci 167: 1139–1147

Delaunay Tetrahedron-Based Connectivity Approach for 3D Wireless Sensor Networks

Ramesh Kumar and Tarachand Amgoth

1 Introduction

In Wireless Sensor Networks (WSN), sensor nodes gather data from their surroundings and transmit it to a base station. To perform this operation, network connectivity is essential. Each connected nodes have their own area of interest where they gather data. Their position is such that the coverage area has the minimum amount of overlapping while maintaining connectivity. Failure of nodes and sometimes environmental conditions abruptly cut off the coverage and connectivity. This affects the overall network performance. Sometimes network partitions are also created. In a 3D environment where the surface is not plane, the radio signal of nodes is not regularly available. Mountaineer areas, densely populated buildings, etc. are examples of scenarios where radio signal strength is diminished due to obstacles. For this, different deployment techniques are explored. We have shown a functioning network in a 3D scenario in Fig. 1. Here, nodes are installed on hills using the tower. The presence of large and dense trees also affects radio strength and connectivity.

In real-life problems, finding the optimal number and position of sensors for maximum coverage and connectivity is very challenging. Proper deployment of nodes assures good coverage and long-lasting connectivity. There are mainly two types of deployment techniques: random and deterministic. For 3D terrain, random deployment is not suitable, as it creates an unnecessary loop in the data forwarding path. So we focus on a deterministic approach where nodes' positions are calculated, and then devices are installed there. There are multiple factors that are affected by the deployment pattern, like energy consumption, network lifetime, network traffic, data

R. Kumar (✉) · T. Amgoth
Department of Computer Science & Engineering, Indian Institute of Technology (Indian School of Mines), Dhanbad, Jharkhand, India
e-mail: ramesh.iitism@gmail.com

T. Amgoth
e-mail: tarachand@iitism.ac.in

P. Singh et al. (eds.), *Machine Learning and Computational Intelligence Techniques for Data Engineering*, Lecture Notes in Electrical Engineering 998,
https://doi.org/10.1007/978-981-99-0047-3_48

reliability, deployment cost, etc. In 2D space, deployment needs a position of location with two coordinates only. Assuming the direction of radio signal traversal in a spherical manner, it cannot reach beyond the particular range. However, in 3D, a third component, height, is also considered. A random deployment has certain limitations like redundancy, network congestion, delay, energy wastage, etc. To overcome these, a deterministic approach is adopted. The improper deployment also creates a coverage hole in the region. It optimizes the number of sensors required, coverage area, and network lifetime.

For the proper functioning of the network, connectivity is checked, and the nodes are added if required. For connection establishment among network partitions, relay nodes are more suitable. Relays are a type of node that has a larger transmission range and more energy. Compared to dynamic deployment, static deployment is less complex and more cost-efficient. For connectivity using a static relay is presented in CRP [1] algorithm. In dynamic deployment, mobile nodes are used, which change their position based on the network requirement. But automatic movement requires nodes with high computational power and energy. The use of mobile nodes for collecting data is also considered for connectivity in the presence of obstacles [2]. But, in high-hill areas, mobile nodes may get stuck and run out of battery. There is a high probability of physical damage to mobile devices. In 3D terrain, the energy depletion rate is also very high. So the deployment of nodes is preferable to using mobile devices for data collection. Since the terrain surface is not planar, we cannot deploy nodes in a straight line.

Most of the deployment methods in the 2D plane consider the ideal plane surface, which is not the same as in the 3D plane. Compared to a 2D surface, a 3D surface coverage area is more complex in the real-world scenario. Here, objects and terrain obstruct the radio signal. The optimal sensor deployment problem in the 3D plane is NP-complete [3]. A random deployment is not optimal due to the presence of 3D terrain. Using a mobile device is also not suitable for 3D deployment. In this paper, we propose a Delaunay tetrahedron-based deployment in a 3D environment. Sensor nodes are deployed at predetermined positions to ensure connectivity. 3D space is sectorized into a tetrahedron shape into which nodes can be placed. We compare this scheme with random 3D-grid-based deployment, demonstrating that the proposed technique is more efficient.

2 Related Work

There are many deployment techniques that exist in the literature using the deterministic approach. Some of them are based on computational geometry to determine the position of sensors. A Delaunay triangle-based deployment scheme for the 2D network is discussed in [4]. Others use different optimization techniques to minimize the number of nodes required. Evolutionary algorithms like PSO and GA are also tested and reported efficiently. But, the deployment strategy adapted for the 2D environment does not fit well for the 3D environment in each of the scenarios. We present a brief discussion about those methods.

Here, we specifically discuss techniques for 3D networks. In paper [5], the sensor deployment approach employs a DT score, which comprises two phases: contour-based deployment and Delaunay triangulation-based deployment. The probabilistic sensor deployment model scores each potential position, and the site with the highest score is chosen for deployment. This method is restricted to 2D deployment only. For 2D deployment, many studies are available. But in recent days, 3D deployment techniques are being explored. We discuss a few of them below. Paper [6] discusses a 3D deployment approach based on the distributed particle swarm optimization (DPSO) algorithm and a suggested 3D virtual force (VF) methodology. It overcame the problems associated with PSO, like local optima. In paper [7], authors discussed a deployment strategy (WTDS) for 3D terrains that is based on guided wavelet transform (WT). Here the sensor moves within the mutation phase of the genetic algorithms (GAs) with the primary focus of maximizing the quality of coverage (QoC). The number of sensors on a 3D surface is optimized using a probabilistic sensing model and Bresenham's line of sight (LOS) algorithm. 3D-UWSN-Deploy [8] applies a "divide and conquer" strategy where the whole space is partitioned into a collection of sub-cubes with the objective of maximizing sub-cube size. Here, a tree is formed where sensors are placed at the node using breadth-first search. Vertex Coloring-based Sensor Deployment for 3D terrain [VC-SD (3D)] applies the concept of vertex coloring to the graph. It determines the sensor requirements and its positions [9]. The authors in [10] presented a Bresenham line-of-sight-based realistic coverage model for 3D environments, which they used to re-formalize the 3D WSN deployment problem and incorporate a realistic spatial model of the environment. This work presents a multi-objective genetic algorithm that incorporates novel adaptive and guided genetic operators. Additionally, it used two optimization techniques to boost performance: search space reduction as well as sampling-based evaluation. In the paper, [11], the WVSN deployment problem for 3D indoor monitoring is attempted for coverage, connectivity, and obstacle awareness. A continuous 3D space is discretized using 3D grids such that a more precise version can be achieved by adjusting the grid granularity. The approach discussed in [12] used movable robotic sensors mounted on the vertices of a truncated octahedral grid to provide coverage in three dimensions. There are different types of deployment methods like prism sensor node deployment strategy, Cube Node Deployment, Hexagonal Prism Node Deployment, and Pyramid Node Deployment. Boundary detection for 3D environments is discussed in [13, 14] which can be used to place the sensor nodes in the target area. These algorithms provide the uncovered area and also determine the probable positions of sensor nodes. A local search-based method is discussed in [15].

3 Proposed Method

Delaunay triangulation is a very efficient technique for creating triangles of high quality. From a collection of points on the plane, it generates a weave of continuous, non-overlapping triangles. For a 3D surface, a tetrahedron is formed by four triangles, six straight lines, and four vertices. It is the most basic structure of all regular convex polyhedra. Compared with hexahedra and pentahedra structures, the tetrahedron has a lot of flexibility. In the 3D domain, the Delaunay tetrahedron is more effective for quality structure. Watson proposed an algorithm for Delaunay triangulation known as Watson's algorithm [16]. But, for a 3D environment, it has certain limitations.

Suppose we have N partitions, P_i for $i = 1, 2, ..., N$. Each partition is represented as a node in a polyhedron. Among these $P - i$, we select any four nodes to form a tetrahedron. The selection of nodes is such that no other node, out of the four selected, lies within the tetrahedron. For each node, there is a tetrahedron that consists of closer nodes only compared with the remaining nodes. It is written mathematically as in Eq. 1 where T_i is the set of any four nodes, $P_{i+k}|k = 0, 1, 2, 3$ and r is the any point inside tetrahedron. A conversion from polyhedron to tetrahedron is shown in Fig. 2. To define the quality of a tetrahedron, a value of $\frac{3r}{R}$ is calculated, where r is the radius of the tetrahedron's inscribed sphere and R is of the circumscribed spheres, respectively [17]. This value is 1 for the equilateral tetrahedron and 0 for the flat tetrahedron. In other words, we can elaborate the idea of Delaunay tetrahedron as follows: Four non-coplanar points s_i, s_j, s_k, s_l form a Delaunay tetrahedron if there lies a point s_x which is more closer to s_i, s_j, s_k, s_l than any other points existing. Here, s_x is center of the sphere passing through the points s_i, s_j, s_k, s_l. Points s_i, s_j, s_k, s_l form a Delaunay tetrahedron (T_D) and its passing sphere is the circumsphere of T_D, i.e. $T_D \Rightarrow \exists s_x$ such that $\| s_x - s_i \| = \| s_x - s_j \| = \| s_x - s_k \| = \| s_x - s_l \|$.

$$T_i = \{r \in R^n | dist(r, P_k) \leq dist(r, P_j); \ j = 1, 2...N, j \neq k\} \tag{1}$$

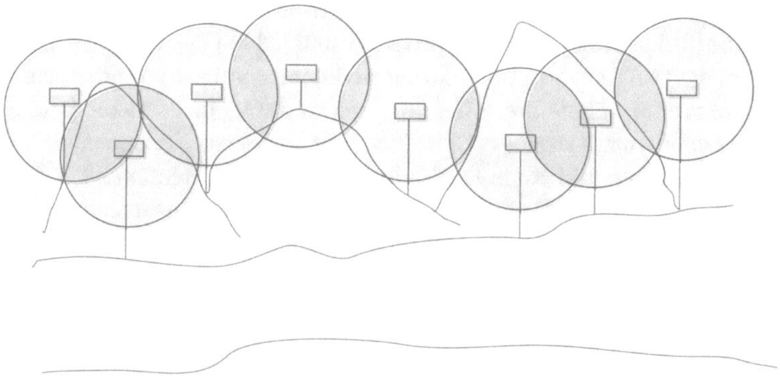

Fig. 1 WSN in 3D terrain

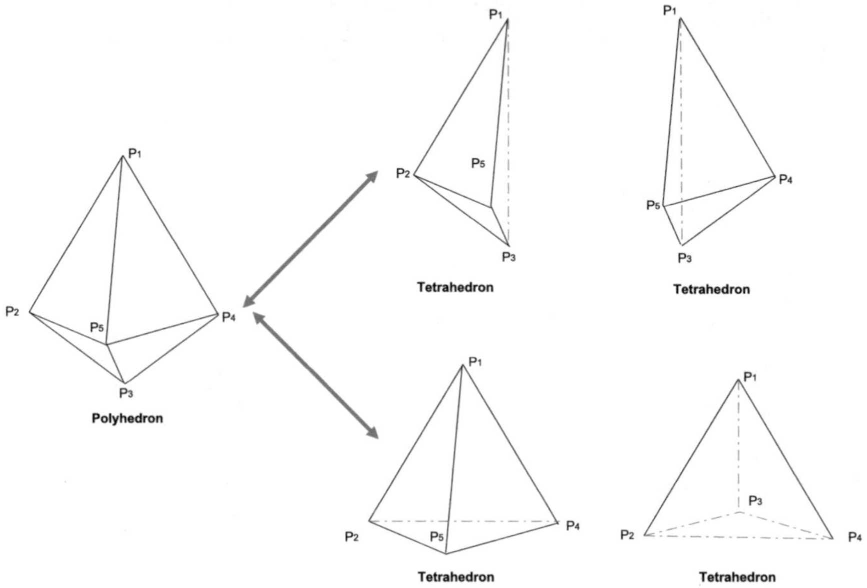

Fig. 2 Polyhedron to tetrahedron

Recovering link process is as below:

- Triangulate the area considering each partition as vertices.
- Intermediate points are inserted on the edge and surface of each triangle.
- Intermediate points are moved inward for optimal tetrahedron finding
- Based on the conditions, intermediate points are selected or rejected for the Delaunay tetrahedron.
- Length of a triangle is such that it fits inside a sphere of radius R_s where R_s is the range of sensor nodes.
- Place relay nodes at each circumcenter of newly found triangles.

For a set of planar polygons P, 3D constrained Delaunay tetrahedron is the set of vertices V, and any polygon is obtained as the union of faces of T_V where T_V is the triangulation of vertices [18]. To triangulate the uncovered area, we consider the intermediate points. These are the points on the edge and surface of the polyhedron. The intermediate points are selected such that they optimally triangulate the region without affecting the geometry of the polyhedron. Their positions can be changed to optimize the number of tetrahedrons. It helps in creating the missing links among nodes, shown in Fig. 3. Adding intermediate points is also needed in the case of the Schönhardt polyhedron because such polyhedrons cannot be triangulated without adding intermediate points. To find the need for the addition of intermediate points is an NP-complete problem[14].

Algorithm 1 3D network connectivity

INPUT: P
OUTPUT: T

Triangulate P
for $i, j \leftarrow 1$ to $| P |$ **do**
  Find intermediate points P_k on line $\overline{P_i P_j}$
  if $| \overline{P_k P_{k+1}} | < \frac{2}{3} \star \pi \star R$ **then**
    append P_k in set P_{int}
    Triangulate $P \cup P_{int}$
  end if
end for

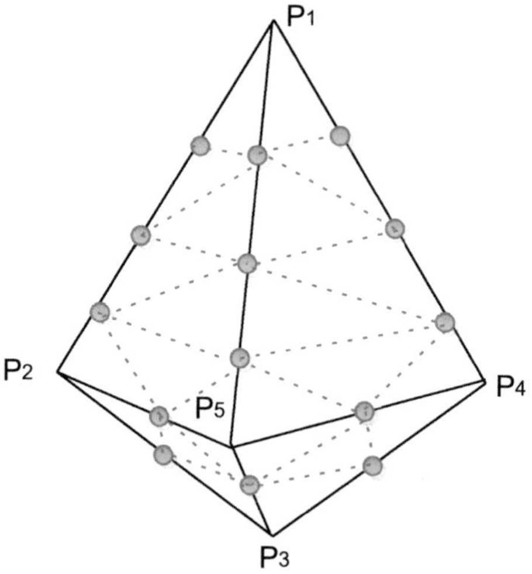

Fig. 3 Triangulation inside tetrahedron

$$\sum_{i=1}^{3} s_i < 2 \times \pi \times R \tag{2}$$

For the possible intermediate points, we use the circle theorem, which relates the side of an equilateral triangle to its circumradius. The side of equilateral triangle d_{int} is equal to $\sqrt{3} \times R$. So, we select the intermediate points at a distance of $\sqrt{3} \times R$ only. Thus, we get multiple intermediate points to form an internal tetrahedron. To optimize the number of tetrahedrons, these intermediate points are slightly shifted with unit length. After the final optimization of intermediate points, the distance between two consecutive intermediate points must be less than one-third of the circumcircle perimeter, i.e. $| \overline{P_k P_{k+1}} |^3 < 8\sqrt{2} \times \pi \times R^3$. This is based on geometric condition: Volume of tetrahedron < Volume of Circum-sphere.

The circumcenter of each tetrahedron is a possible location for the relay node. The x and y coordinates are possible geographic locations, while the z coordinate is the height where the node should be kept. This can be achieved using a tower, pole, or flying balloon tied with rope.

Algorithm 2 Tetrahedron selection

```
INPUT: P, T
OUTPUT: C_s
SP = shortest_path(P)
for i ← 1 to | T | do
    if T_i intersect SP.edges() then
        Find C_i = T_i.Circumcenter()
        append C_i in set C_s
        place relay at C_s
    end if
end for
```

4 Performance Evaluation

Connectivity is the prime requirement for data delivery accurately. Base station/Sink ensures all sensor node connectivity in the network. We compare our method with 3D- grid-based approaches. We calculate the number of relay nodes required for connectivity in the network, such that data reaches the base station/sink. The locations for relay deployment have been calculated using the methods discussed above. To compare the results, we have considered two sizes of the target field. One field is of size 200 m × 200 m, and the other field size is 400 m× 400 m. The reason behind considering different field sizes is that we want to show the effect on the performance of the algorithm when the field is smaller as well as larger. The transmission range of the relay node is 50 m. The objective of the proposed method is to optimize the number of relays required. In Fig. 4, we have shown the comparison of the proposed

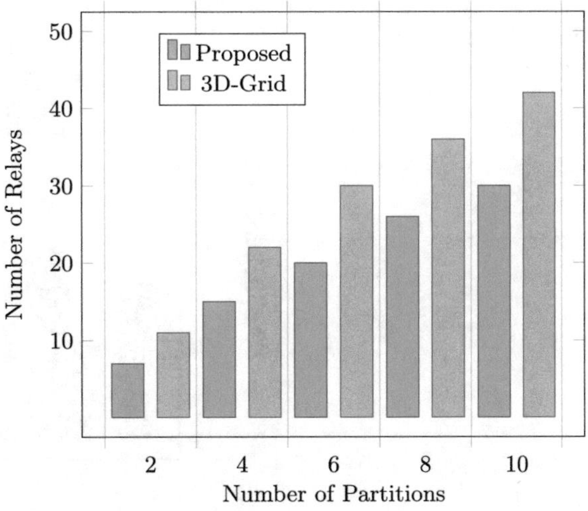

Fig. 4 Partitions versus relays for field size of 200 m × 200 m

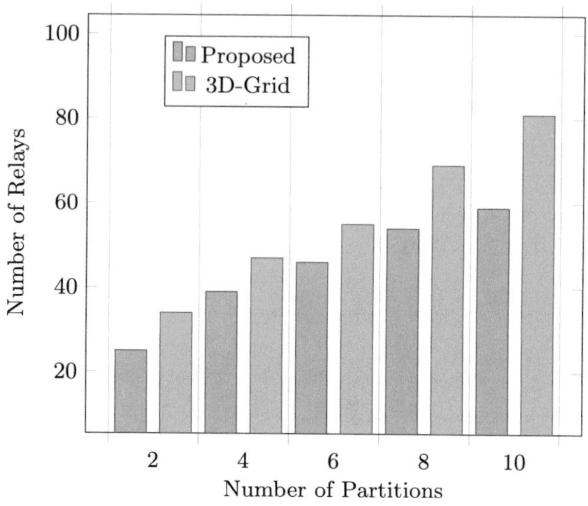

Fig. 5 Partitions versus relays for field size of 400 m × 400 m

method and the 3D-Grid-based approach. We can notice that, as the number of partitions increases, there are more relay nodes required. This is because when the number of partitions increases, each partition needs a set of relays to connect with the other partition. But this requirement is comparatively lower in our proposed method. Similarly, in Fig. 5, we depicted the correlation between the number of partitions and the required number of relays needed for the field of size 400 × 400. Similar to Fig. 4, here too, the number of relays needed increases for more partitions. But, when there is an increase in field size, the number of required relays also increases. In both situations, there is a variation in the height of the relays deployed.

In Fig. 6, we have shown the performance over the network lifetime and compared it with that of 3D-grid-based deployment. When an optimal number of nodes are placed, it reduces the intermediate hop counts. Since at each hop, data is received,

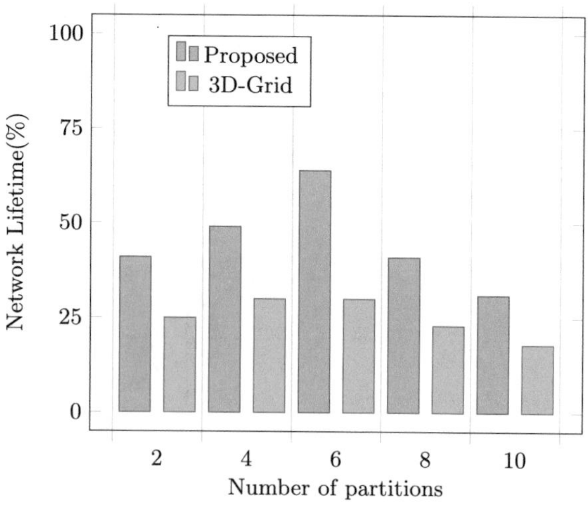

Fig. 6 Partitions versus Network lifetime

processed, and forwarded. It consumes energy at each step. Decreasing the hop count reduces the intermediate node energy consumption. When nodes save energy, the overall network lifetime is increased. In Fig. 6, we find that the network lifespan of the proposed method is greater when compared with the other one.

5 Conclusion

A 3D-WSNs is an emerging sub-field of WSN. In this work, we propose a 3D node placement for the connectivity of WSNs. We focus on the networks that have been partitioned into multiple sub-networks due to several reasons. The 3D space between partitions is triangulated using the Delaunay tetrahedron. The circumcenter of each tetrahedron is the probable location for node placement. We simulated the proposed approach and compared it with a 3D-grid-based deployment strategy. We observe that our method outperforms the other one. We may further develop this work by taking into account nodes that have multichannel interfaces.

References

1. Kumar R, Amgoth T (2020) Adaptive cluster-based relay-node placement for disjoint wireless sensor networks. Wirel Netw 26(1):651–666
2. Kumar R, Amgoth T, Das D (2020) Obstacle-aware connectivity establishment in wireless sensor networks. IEEE Sens J 21(4):5543–5552
3. Kong L, Zhao M, Liu X-Y, Jialiang L, Liu Y, Min-You Wu, Shu Wei (2014) Surface coverage in sensor networks. IEEE Trans Parallel Distrib Syst 25(1):234–243
4. Kumar R, Amgoth T, Sah DK (2021) Deployment of sensor nodes for connectivity restoration and coverage maximization in wsns. In: 2021 sixth international conference on wireless communications, signal processing and networking (WiSPNET). IEEE, pp 209–213
5. Chun-Hsien W, Lee K-C, Chung Y-C (2007) A delaunay triangulation based method for wireless sensor network deployment. Comput Commun 30(14–15):2744–2752
6. Yanzhi D (2020) Method for the optimal sensor deployment of wsns in 3d terrain based on the dpsovf algorithm. IEEE Access 8:140806–140821
7. Unaldi N, Temel S, Asari VK (2012) Method for optimal sensor deployment on 3d terrains utilizing a steady state genetic algorithm with a guided walk mutation operator based on the wavelet transform. Sensors 12(4):5116–5133
8. Khalfallah Z, Fajjari I, Aitsaadi N, Rubin P, Pujolle G (2016) A novel 3d underwater wsn deployment strategy for full-coverage and connectivity in rivers. In: 2016 IEEE international conference on communications (ICC), pp 1–7
9. Arivudainambi D, Pavithra R (2020) Coverage and connectivity-based 3d wireless sensor deployment optimization. Wirel Pers Commun 1–20
10. Saad A, Senouci MR, Benyattou O (2020) Toward a realistic approach for the deployment of 3d wireless sensor networks. IEEE Trans Mobile Comput 1
11. Brown T, Wang Z, Shan T, Wang F, Xue J (2017) Obstacle-aware wireless video sensor network deployment for 3d indoor monitoring. In: GLOBECOM 2017—2017 IEEE global communications conference, pp 1–6

12. Nazarzehi V, Savkin AV (2018) Distributed self-deployment of mobile wireless 3d robotic sensor networks for complete sensing coverage and forming specific shapes. Robotica 36(1):1–18
13. Qiang D, Wang D (2004) Boundary recovery for three dimensional conforming delaunay triangulation. Comput Methods Appl Mech Eng 193(23–26):2547–2563
14. Chen J, Zhao D, Huang Z, Zheng Y, Gao S (2011) Three-dimensional constrained boundary recovery with an enhanced steiner point suppression procedure. Comput Struct 89(5–6):455–466
15. Tam NT, Thanh Binh HT, Dat VT, Lan PN et al (2020) Towards optimal wireless sensor network lifetime in three dimensional terrains using relay placement metaheuristics. Knowl-Based Syst 206:106407
16. Watson DF (1981) Computing the n-dimensional delaunay tessellation with application to voronoi polytopes. The Comput J 24(2):167–172
17. Golias NA, Dutton RW (1997) Delaunay triangulation and 3d adaptive mesh generation. Finite Elements Anal Design 25(3–4):331–341
18. Cavalcanti PR, Mello UT (1999) Three-dimensional constrained delaunay triangulation: a minimalist approach. In: IMR, pp 119–129

CNN Based Apple Leaf Disease Detection Using Pre-trained GoogleNet Model

Sabiya Fatima, Ranjeet Kaur, Amit Doegar, and K. G. Srinivasa

1 Introduction

In Computer-Vision (CV) one of the current challenges is in the identification and efficient recognition of leaf diseases of fruit because these are a very important application for Agronomy [1]. Apple is a very important fruit plant and it is very famous worldwide due to its nutrient importance. Kashmir is the largest apple producer in India, so far. Because Kashmir produces 77.7% of apples, India becomes the world's sixth-biggest apple producer. Then Kashmir becomes the 11th largest apple grower, after Russia and beyond Brazil [2].

Apple is the main fruit of the world's cooler climates zones in India. In 1930 apple scab disease was detected in Kashmir valley, then in 1973, the disease of Apple was introduced as an epidemic and it infected 70,000 acres area so, there was a loss of 5,400,000 Rupees only in a season, so it became one of the main problems for the nation declared by the Indian Government. In 1977, it is found in Himachal Pradesh, and in 1983, there was a loss of Rs. 15,000,000 then this amount was compensated to the farmers by the HP-state-Government. So the total loss was 50,000,000 which is equal to 10.0% of the total income of India. So according to [3], because of disease and pests on crops, India losses Rs. 50 thousand crores annually.

S. Fatima (✉) · R. Kaur · A. Doegar · K. G. Srinivasa
Department of CSE, National Institute of Technical Teachers Training and Research (NITTTR), Chandigarh, India
e-mail: sabiya1990fatima@gmail.com

R. Kaur
e-mail: ranjeet.cse@nitttrchd.ac.in

A. Doegar
e-mail: amit@nitttrchd.ac.in

K. G. Srinivasa
e-mail: kgsrinivasa@nitttrchd.ac.in

P. Singh et al. (eds.), *Machine Learning and Computational Intelligence Techniques for Data Engineering*, Lecture Notes in Electrical Engineering 998,
https://doi.org/10.1007/978-981-99-0047-3_49

Plant diseases have historically been recognized by expert eye inspection. However, there is a possibility of inaccuracy due to subjective perception [4]. In this regard, a variety of spectroscopic and imaging techniques for diagnosing plant diseases have been investigated [5, 6], but they are very expensive and less efficient. Autonomous plant disease diagnostics using machine learning has become a realistic option in recent years. K-means clustering and support vector machine (SVM) are traditional approaches [7–11] that require complex preprocessing and extracting features from processed images, which reduce the efficiency of disease diagnostics. Recently, convolutional networks (CNNs) have been a research area in computer vision because they have achieved significant success due to their automatic feature extraction from images directly without any preprocessing. The CNN model, on the other hand, maybe utilized as an automated approach for identifying plant diseases, allowing agricultural specialists and technicians to establish a robust apple leaf disease classification [12–16]. Thus we have focused on the detection of Apple leaf disease using deep CNN which is described further in the given sections.

The contribution of this work is to utilize the GoogleNet [17] model for automatic feature extraction and detection of whether the Apple leaf image is diseased or healthy. This work also explains the working of this CNN and its application to the dataset containing diseased and healthy Apple leaves. The rest of the paper is structured as follows: Sect. 2 presents the related existing approaches in this domain using machine learning and deep learning. Section 3 details the proposed approach by explaining each step including Dataset acquisition and preprocessing, GoogleNet architecture and its working, and measures of performance evaluation are used in the next section. Section 4 explains the experimental results of the proposed approach and its performance comparison with other existing works and finally, Sect. 5 presents the conclusion and future scope.

2 Related Work

Many works on the identification of plant leaf disease using leaf images have been widely discussed and presented by researchers worldwide. Some research studies for various plant leaf diseases are addressed using machine learning or deep learning in the following.

Rumpf et al. [5] proposed an early detection method for the disease in sugar beet plants using SVM. However, it may not achieve good accuracy. Al Hiary et al. [18] proposed a classification technique for plant leaf diseases using K-mean clustering along with the ANN (artificial neural network) technique. First, identify the green color. Second, Otsu's technique has been used to cover the image, and the last two steps have been used to completely remove the infected boundaries in the leaf. It performed well on five different leaves named early scorch, cotton mold, ashen mold, late scorch, and tiny whiteness, but had poor detection accuracy with noisy data. Kulkarni and Patil [19] utilized ANN with a Gabor filter for extracting the features of the system; it provides improved outcomes. However, the performance can be improved by using color and shape-based features of input images. Akhtar et al. [20] applied Otsu Thresholding and morphology algorithms for segmentation of disease spots, then statistical Haralick texture features (GLCM), DCT, DWT classified using KNN, SVM with linear kernel, Decision Tree, Nave Bayesian, and RNN on a total of 40 images of rose leaves and Found SVM is better with DCT and DWT features. However, it is based on a small dataset.

In [21], the authors first converted alfalfa RGB images to HSV and L*a*b color spaces. The K-means and Fuzzy C-mean clustering algorithms are used for the segmentation of lesion spots. Color, shape, and texture features are obtained from the lesion spots. Qin et al. concluded the K-median, ReliefF, and SVM found the best accuracy of 80% on average. However, there is a need for better techniques. Vijayalakshmi [22] utilized K-means segments optimized through genetic algorithms and color-texture features calculated by the CCM method. These features are well classified by SVM for only the dataset of 106 images, which is small in size. Wang et al. [13] introduced a system for estimating the disease severity by using Deep Learning on images of only the black rot disease of apples from the plantVillage [23] dataset. CNN with VGG16, VGG19, Inception v3, ResNet50 transfer learning was implemented and concluded that VGG16 has resulted in better accuracy, 90.4% among them. However, it shows lower performance in the single apple disease category. Khan et al. [12] contributed to the segmentation of the ailment/lesion spot and selection of the most important features using VGG16, Café alexnet model max-pooling, and GA optimized those features, then classified them into SVM and KNN with different kernels. Researchers have concluded that M-SVM produced 98.6% accuracy but [12] utilized only two diseases, i.e., Rot and Scab of Apple. Agrawal et al. [24] utilized the grape disease images from the [23] dataset. Convert the RGB to LAB and HIS color spaces, then apply K-means and extract color features. LAB and HIS then SVM with different kernels like linear, RBF, quadratic, and polynomial. They found polynomial SVM as the best with 90% with LAB and HIS features, but this was recorded on a small dataset. Ghazi et al. [25] Made use of K-mean clustering. Color and texture features are classified by SVM on different plant leaves and found to be 85.65%, which means they need to improve.

Kumar et al. [26] have proposed a novel optimization algorithm for the selection of features in plant disease named ESMO (Exponential Spider Monkey) with SVM and reported 92.12% accuracy. It used 1000 different plant leaf images from the [23] dataset. Joshi and Shah [27] proposed a hybrid fuzzy C-mean technique along with a neural network approach with 80% accuracy only. Khan et al. [28] mainly contributed to enhancement, disease segments, and the selection of important features. Color features by the color histogram, texture features by LBP were extracted and optimized by GA. In [29], Chang et al. first converted the RGB image to luminance and HIS space, and then color texture features were estimated by the CCM matrix. I implemented ANN, SVM, and KNN with different kernels, and concluded that ANN is the best with an accuracy of 93.81% for small cropped leaf images. Febrinanto and Dewi [30] have mainly focused on citrus diseases. The image is transformed into L*a*b* color-space then Resizing, rescaling, and K-means has been done based on a*, b* color variables by ignoring L (i.e., luminosity/brightness) as classified by KNN. A total of 120 images of diseases were used and the recorded accuracy was 90.83%, but healthy leaves were not accurately identified.

Mohanapriya and Balasubramani [31] mainly focused on PCA for the diseased part, extracted texture features, and then classified them by Nave Bayes with an average accuracy of 90%. Yadav et al. [32] for feature extraction, the author used Alexnet, and for feature subset selection, he employed the particle swarm optimization approach (PSO) and classified by SVM with an accuracy of 97.3%. Overfitting arises due to the minimal number of classes. By showing the advantages of fuzzy logic decision-making rules, Sibiya and Sumbwanyambe [33] suggested the "Leaf Doctor" application, which can be used to assess the severity of the plant leaf disease. Kaur et al. [34] extracted texture features by the GLCM method from the grayscale image, which was extracted by K-means segmentation, and these features were classified by the KNN classifier and recorded accuracy of 92.25% and FDR of 0.94 by using a small dataset of only 25 images for training purposes.

The literature shows that in terms of high classification rate and low error probability, there is still scope for improvement.

3 Proposed Approach

This section describes how CNN is used to classify apple plant leaves disease in this work. The four essential elements of the proposed methodology are represented in Fig. 1 that are data acquisition and preprocessing, retraining GoogleNet CNN, disease classification, then evaluation of performance.

3.1 Data Acquisition and Preprocessing

There are 54,323 images of 14 crops and 38 plant diseases in the PlantVillage database [23]. Only Apple leaf images are gathered from this dataset. An overview of our

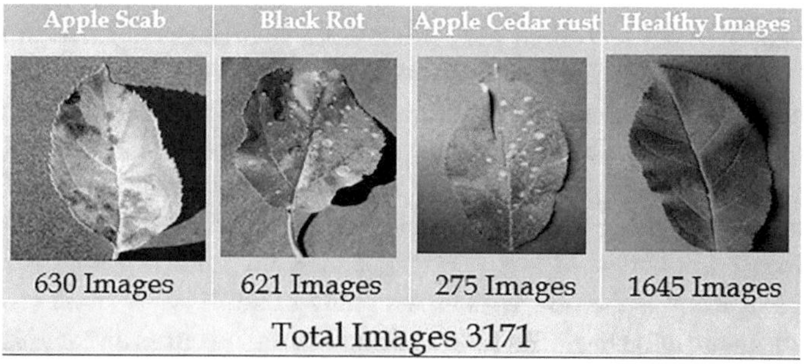

Fig. 1 Apple leaf samples from PlantVillage dataset

dataset as well as a representative example of each class is shown in Fig. 1. In our dataset, there are 3171 images of three diseases and a healthy category. We used a 70:15:15 ratio of random images for training, validation, and testing in our case. As a result, there are 2221 training images, 475 validation images, and 475 testing images. To be used as input for the GoogleNet approach, all of the images were resized to 224 × 224 × 3 pixels while the original size was 256 × 256 into 3 color channels; an input image is 224 × 224 pixels in size and is supplied to the first convolution layer.

3.2 *Retrain GoogleNet CNN*

Szegedy et al. [17] presented GoogleNet in their study, and it was the winner of the ILSVRC in 2014. GoogleNet features nine inception modules for its major auxiliary classifiers, four convolutional layers, five fully-connected layers, three average-pooling, four max-pooling, three softmax layers and seven million parameters [25]. Additionally, the fully-connected layer employs dropout regularization, and all convolutional layers use ReLU activation. On the other hand, this network is substantially deeper, with a total of 22 layers.

DeepCNN image classification techniques are divided into two stages: feature extraction and classification module. In an end-to-end learning architecture, the feature is extracted by providing training images, and then the training images are sent into the softmax layer. In contrast to handcrafted features, deep-learning algorithms rapidly learn low-level, mid-level, and abstract features from images. To extract the deep features, a training set of images is fed into the pre-trained GoogleNet method. We used a total of 144 layers, including convolutional and fully connected layers. We can relocate features of the GoogleNet approach by utilizing our [23] dataset for both training and validation. The objective of the softmax function is to do re-learning using the dataset's four classes. GoogleNet was previously trained for 1000 categories using features of huge datasets. Using stochastic gradient descent, the effective learning rate is 0.0003 to start the training. Transfer learning is a deep network approach that enables us to retrain a network by fine-tuning parameters to new levels. The flowchart of proposed work is shown in Fig. 2 and the proposed model's training and validation results are shown in Fig. 3.

We modified a fully connected and the output classification layers in the pre-trained GoogleNet. The final fully connected layer is the same in size as the number of classes in our dataset, which is four. The fully connected layer's learning rate parameters are increased to enhance GoogleNet's learning process. The learning technique is stochastic gradient descent, with a starting learning rate of 0.0003 and a maximum of 6 epochs. After optimizing the model, it is trained on Apple image classification.

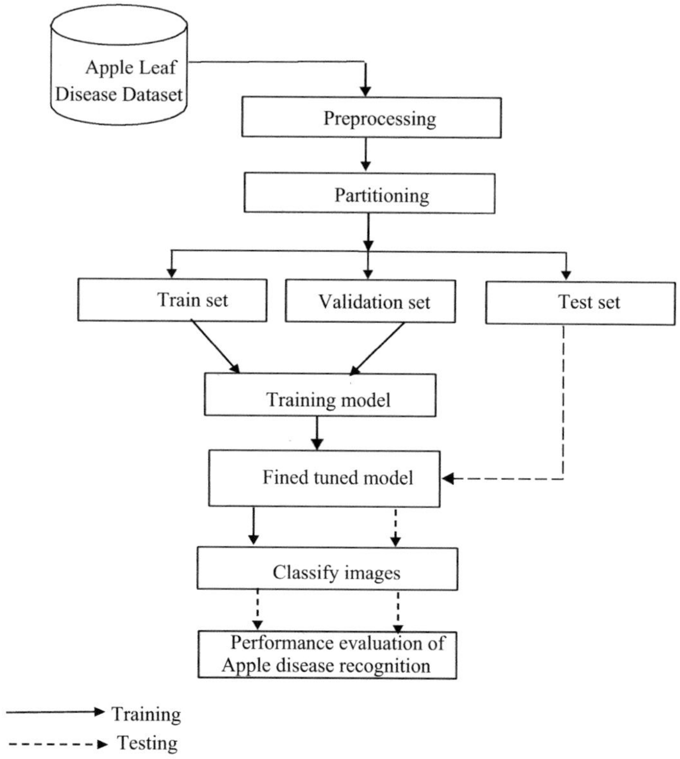

Fig. 2 Flow chart of proposed work

3.3 *Disease Detection and Classification Process*

The number of output classification layers is the same in size as the number of classes in our dataset, which is four.

Due to the potential for these models to learn features automatically during the training stage, each output has a different probability for the input image; the model then picks the output with the highest probability as its class prediction. Finally, this step uses the pre-trained set to identify the disease present on the leaf.

4 Experimental Results and Discussion

This section details an experiment that was carried out to test the proposed GoogleNet CNN classification on the Apple Leaf image. The proposed model is implemented in the MATLAB2020b environment using a laptop with Windows 10 OS, 8 GB of

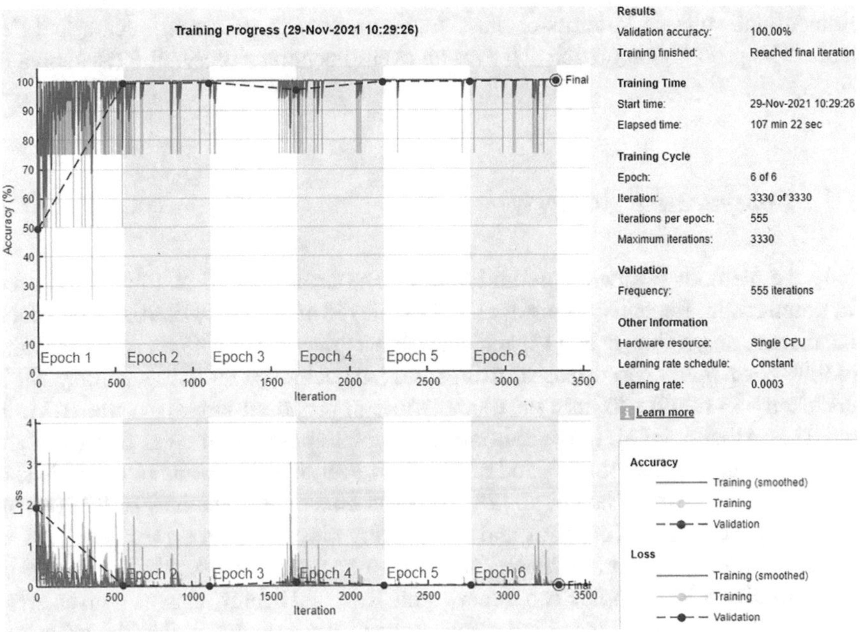

Fig. 3 Results of retraining GoogleNet over PlantVillage Apple-dataset

RAM, and a 2.90 GHz Intel Core i7 CPU. The validation of the proposed approach is completed in 107 min and 22 sec.

The proposed deep CNN for detecting apple leaf diseases was developed and trained using a popular and widely used deep learning technique. In this context, image preprocessing was performed first, then fine-tuning to improve classification accuracy while reducing the training time. Because we are employing the CPU-based framework as indicated in Fig. 3, the training time in our work is a little longer. We used six epochs, with the network performing 555 iterations for each epoch. So, the training process completes in 3330 iterations. The validation process demonstrates that our proposed method is 100% accurate.

In transfer learning, the parameters are used to avoid network overfitting in the training phase, as over-fitting reduces the classification accuracy of the deep networks [23]. A deep network can be retrained using techniques like rotation and rescaling. We just resized images in our case. Table 1 compares the proposed method to existing methods in terms of accuracy, precision, recall, and F1-score using the PlantVillage Data set [23] of Apple Leaf images; illustrates the results of class-wise classification accuracy obtained using a 70:15:15 training-validation-testing ratio and shows the class wise classification accuracy; we achieved 100% accuracy for Black Rot, Apple Scab, and Apple Rust, indicating that DCNN properly divided the images into their respective classes, and 99.6% for Healthy leaves. With other methods, the proposed method achieves maximum accuracy. The proposed GoogLeNet deep CNN is more efficient and reliable in terms of classification accuracy, precision, recall, and F1-score. The proposed approach achieves an overall accuracy of 99.79% (as shown in Fig. 4).

4.1 Comparative Analysis

Only the methods evaluated on the Plant Village dataset listed in Table 1 are used for comparison. Recently, Wang et al. [13] achieved an accuracy of 90.4% on Black Rot disease only. Khan et al. [12] achieved an accuracy of 98.10% on Apple Rot and 96.90% for Apple Scab diseases, Atila et al. [35] achieved 99.91% accuracy but it has taken 643 min 3 s to train the model while in proposed work it's only 107 min and 22 s, Albattah et al. [14] used only 1645 total apple leaf images of Rot, Scab, Rust and healthy and reported good results. However, it needs to add more apple leaf images in their work. Khan et al. [28] achieved an average accuracy of 97.20% on three (Rot, Scab, Rust) diseases and one healthy class with an error rate of 2.8%. While the proposed work achieved an average accuracy of 99.79% (as shown in Fig. 4) on Black Rot, Apple Scab, cedar Apple Rust and healthy classes with an error rate of 0.2% only. From the comparison analysis, it is concluded that the proposed method outperforms the existing methods on the same Apple leaf dataset.

5 Conclusion

GoogleNet is used in this work to classify the apple leaf images. Because it is a powerful pre-trained network to extract features and obtain a higher rate of classification. Apple Plant Village dataset having cedar_apple_rust, Black_rot, Apple_Scab, and healthy leaves are used to train the pre-trained GoogLeNet. For this purpose, it is only the last feature learner layer and the classification layer modified and retrained the modified model to classify the Apple leaf images into four classes more accurately than the other existing models. The GoogleNet model with 144 layers is trained on 70% of Apple leaf images, achieved a validation accuracy of 100%, while 99.79% is in the testing process. So, it is concluded that the proposed method outperforms the existing methods on the same Apple leaf dataset. In the future, our aim is to investigate huge datasets with various types of Apple Diseases for detection and classification method.

Table 1 Analysis of the proposed work against existing works

Method	Dataset	Disease type	Accuracy (%)	Overall accuracy (%)	Precision (%)	Recall (%)	F1 score (%)
Wang et al. [13]	PlantVillage dataset	Black rot Healthy	–	90.4	–	–	–
Khan et al. [12]	PlantVillage dataset	Apple rot Apple scab	98.10 96.90	98.6	98.5	98.5	–
Geetharamani and Pandian [16]	PlantVillage dataset	Black rot Apple scab Apple rust Healthy	98 96 98 96	96.46	96.47	99.89	98.15
Atila et al. [35]	PlantVillage dataset	Black rot Apple scab Apple rust Healthy	–	99.91	98.42	98.31	–
Albattah et al. [14]	PlantVillage dataset	Black rot Apple scab Apple rust Healthy	–	–	99.4	98.8	99.15
Khan et al. [28]	PlantVillage dataset	Black rot Apple scab Apple rust Healthy	98.10 97.30 94.62 98.0	97.20	97.0	97.15	–
Proposed	PlantVillage dataset	Black rot Apple scab Apple rust Healthy	100 100 100 99.6	99.79	99.79	99.9	99.82

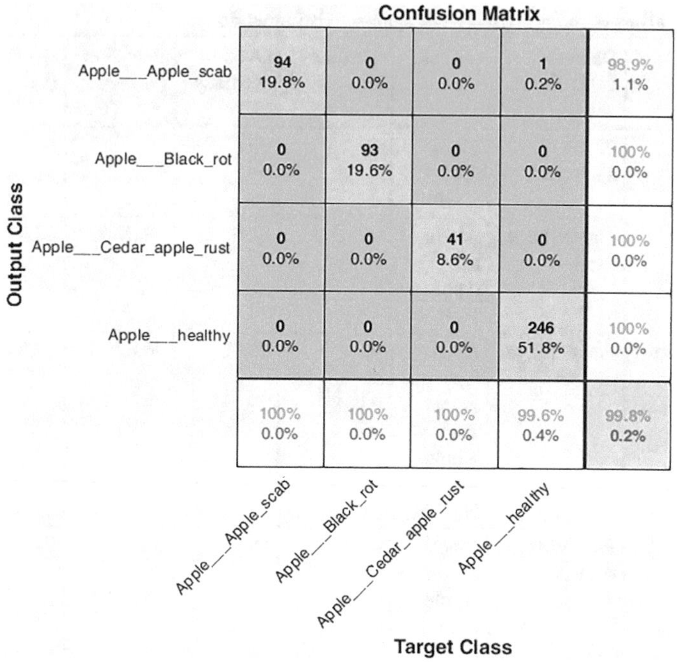

Fig. 4 Confusion-matrix for proposed work

References

1. Savary S, Ficke A, Aubertot JN, Hollier C (2012) Crop losses due to diseases and their implications for global food production losses and food security. Food Secur 4(4):519–537
2. Why India should start taking apples seriously. The Economic Times. https://economictimes.indiatimes.com/news/economy/agriculture/why-india-should-start-taking-apples-seriously/articleshow/71767025.cms?from=mdr. Accessed 18 June 2020
3. Crops worth Rs 50,000 crore are lost a year due to pest, disease: study. The Economic Times. https://economictimes.indiatimes.com/news/economy/agriculture/crops-worth-rs-50000-crore-are-lost-a-year-due-to-pest-disease-study/articleshow/30345409.cms. Accessed 18 June 2020
4. Dutot M, Nelson LM, Tyson RC (2013) Predicting the spread of postharvest disease in stored fruit, with application to apples. Postharvest Biol Technol 85:45–56
5. Rumpf T, Mahlein AK, Steiner U, Oerke EC, Dehne HW, Plümer L (2010) Early detection and classification of plant diseases with Support Vector Machines based on hyperspectral reflectance. Comput. Electron. Agric. 74(1):91–99
6. Yuan L, Huang Y, Loraamm RW, Nie C, Wang J, Zhang J (2014) Spectral analysis of winter wheat leaves for detection and differentiation of diseases and insects. Field Crop Res 156:199–207
7. Dubey SR, Jalal AS (2016) Apple disease classification using color, texture and shape features from images. Signal, Image Video Process 10(5):819–826
8. Omrani E, Khoshnevisan B, Shamshirband S, Saboohi H, Anuar NB (2014) Potential of radial basis function-based support vector regression for apple disease detection. Measurement 55:512–519

9. Shuaibu M, Lee WS, Hong YK, Kim S (2017) Detection of apple Marssonina blotch disease using particle swarm optimization. Trans ASABE 60(2):303–312
10. Singh V, Misra AK (2017) Detection of plant leaf diseases using image segmentation and soft computing techniques. Inf Process Agric 4(1):41–49
11. Ali H, Lali MI, Nawaz MZ, Sharif M, Saleem BA (2017) Symptom based automated detection of citrus diseases using color histogram and textural descriptors. Comput Electron Agric 138:92–104
12. Khan MA et al (2018) CCDF: automatic system for segmentation and recognition of fruit crops diseases based on correlation coefficient and deep CNN features. Comput Electron Agric 155:220–236
13. Wang G, Sun Y, Wang J (2017) Automatic image-based plant disease severity estimation using deep learning. Comput Intell Neurosci 2017:8
14. Albattah W, Nawaz M, Javed A, Masood M, Albahli S (2021) A novel deep learning method for detection and classification of plant diseases. Complex Intell Syst 1–18
15. Dumais S, Cutrell E, Cadiz J, Jancke G, Sarin R, Robbins DC (2003) Stuff I've seen: a system for personal information retrieval and re-use. In: Proceedings of the 26th annual international ACM SIGIR conference on research and development in information retrieval, pp 72–79
16. Geetharamani G, Pandian A (2019) Identification of plant leaf diseases using a nine-layer deep convolutional neural network. Comput Electr Eng 76:323–338
17. Szegedy C et al (2015) Going deeper with convolutions. Proc IEEE Conf Comput Vis Pattern Recognit 7:1–9
18. Al Hiary H, Ahmad S, Reyalat M, Braik M, ALRahamneh Z (2011) Fast and accurate detection and classification of plant diseases. Int J Comput Appl 17(1):31–38
19. Kulkarni AH, Patil A (2012) Applying image processing technique to detect plant diseases. Int J Mod Eng Res 2(5):3661–3664
20. Akhtar A, Khanum A, Khan SA, Shaukat A (2013) Automated plant disease analysis (APDA): performance comparison of machine learning techniques. In: 11th international conference on frontiers of information technology. IEEE, pp 60–65
21. Qin F, Liu D, Sun B, Ruan L, Ma Z, Wang H (2016) Identification of Alfalfa leaf diseases using image recognition technology. PLoS ONE 11(12):e0168274
22. Vijayalakshmi S (2019) Schematic intelligent image processing techniques for leaf disease detection. Manonmaniam Sundaranar University
23. Mohanty SP, Hughes DP, Salathé M (2016) Using deep learning for image-based plant disease detection. Front Plant Sci 7:1419
24. Agrawal N, Singhai J, Agarwal DK (2018) Grape leaf disease detection and classification using multi-class support vector machine. In: International conference on recent innovations in signal processing and embedded systems (RISE). IEEE, pp 238–244
25. Ghazi MM, Yanikoglu B, Aptoula E (2017) Plant identification using deep neural networks via optimization of transfer learning parameters. Neurocomputing 235:228–235
26. Kumar S, Sharma B, Sharma VK, Sharma H, Bansal JC (2020) Plant leaf disease identification using exponential spider monkey optimization. Sustainable Computing: Informatics and systems 28:100283
27. Joshi M, Shah D (2018) Hybrid of the fuzzy c means and the thresholding method to segment the image in identification of cotton bug. Int J Appl Eng Res 13(10):7466–7471
28. Khan MA et al (2019) An optimized method for segmentation and classification of apple diseases based on strong correlation and genetic algorithm based feature selection. IEEE Access 7:46261–46277
29. Chang YK, Mahmud MS, Shin J, Price GW, Prithiviraj B (2019) Comparison of image texture based supervised learning classifiers for strawberry powdery mildew detection. AgriEngineering 1(3):434–452
30. Febrinanto FG, Dewi C (2019) The implementation of k-means algorithm as image segmenting method in identifying the citrus. IOP Conf Ser: Earth Environ Sci 243(1):012024
31. Mohanapriya K, Balasubramani M (2019) Recognition of unhealthy plant leaves using Naive Bayes classifier. IOP Conf Ser: Mater Sci Eng 561(1):012094

32. Yadav R, Rana Y, Nagpal S (2019) Plant leaf disease detection and classification using particle swarm optimization. In: International conference on machine learning for networking. Springer, pp 294–306
33. Sibiya M, Sumbwanyambe M (2019) An algorithm for severity estimation of plant leaf diseases by the use of colour threshold image segmentation and fuzzy logic inference: a proposed algorithm to update a 'Leaf Doctor' application. AgriEngineering 1(2):205–219
34. Kaur S, Pandey S, Goel S (2018) Semi-automatic leaf disease detection and classification system for soybean culture. IET Image Proc 12(6):1038–1048
35. Atila Ü, Uçar M, Akyol K, Uçar E (2021) Plant leaf disease classification using EfficientNet deep learning model. Eco Inform 61:101182

Adaptive Total Variation Based Image Regularization Using Structure Tensor for Rician Noise Removal in Brain Magnetic Resonance Images

V. Kamalaveni, S. Veni, and K. A. Narayanankuttty

1 Introduction and Related Work

In image processing, total variation filtering and total variation denoising both are known as total variation regularization. This method is used for noise removal in images. This method is based on the fact that images with maximum and possibly erroneous details have maximum total variation. Total variation is the sum of gradients computed at every pixel position across the entire image that is the integral of the absolute image gradient across the image domain [1–3]. By reducing the total variation of the image subject to the condition that the image must be a close match to the original image it will remove unwanted noise present in the image whilst preserving important details such as edges. This concept was pioneered by Rudin-Osher-Fatemi in 1992. Total variation denoising compared to median filtering or linear smoothing filters enable edge preserving denoising. Linear filtering reduces noise but at the same time smooths away edges to a greater extent [4]. Followed by the ROF model different researchers proposed the adaptive total variation models [3, 5–9]. Adaptive TV models enable efficient edge preservation. Next we present popular the existing adaptive total variation models here.

Chambolle and Lions [10] proposed the Adaptive TV model which combines the features of the two convex functionals $|\nabla u|$ and $|\nabla u|^2$ described by the following functional given in Eq. 1.

V. Kamalaveni (✉) · S. Veni
Department of Electronics and Communication Engineering, Amrita School of Engineering, Amrita Vishwa Vidyapeetham, Coimbatore, India
e-mail: vvkamalaveni@gmail.com

S. Veni
e-mail: s_veni@cb.amrita.edu

K. A. Narayanankuttty
Amrita School of Engineering, Amrita Vishwa Vidyapeetham, Coimbatore, India

P. Singh et al. (eds.), *Machine Learning and Computational Intelligence Techniques for Data Engineering*, Lecture Notes in Electrical Engineering 998,
https://doi.org/10.1007/978-981-99-0047-3_50

$$F(u) = \frac{1}{2\varepsilon} \int\limits_{|\nabla u| \langle \varepsilon} |\nabla u|^2 + \int\limits_{|\nabla u| \geq \varepsilon} \left(|\nabla u| - \frac{\varepsilon}{2}\right) + \int\limits_{\Omega} |u - u_0|^2 \quad (1)$$

where $|\nabla u|$ is the gradient of an image, u_0 is the noisy initial image, u is the observed image, ε is a parameter that must be chosen a prior. In Chambolle and Lions model all those image gradients which are greater than or equal to ε are considered as edges and those image gradients which are less than ε are considered to be belonging to flat inner noisy region. The model behaves like ROF model at the edges so the edges are preserved well. The model behaves like the Tikhonov model inside the flat noisy inner region.

Bollt et al. [11] proposed a graduated adaptive image denoising which uses a local compromise between total variation and isotropic diffusion which is described by Eq. 2.

$$\min_u J(u) = \frac{1}{2\varepsilon} \int\limits_{\Omega} |\nabla u|^{P(|\nabla u|)} + +\frac{\lambda}{2} \int\limits_{\Omega} |u - u_0|^q \quad q = 1 \text{ or } q = 2 \quad (2)$$

$$P(|\nabla u|) = 2 - 10\frac{|\nabla u|^3}{M^3} + 15\frac{|\nabla u|^4}{M^4} - 6\frac{|\nabla u|^5}{M^5} \quad \text{if } |\nabla u| \leq M \quad (3)$$

$$P(|\nabla u|) = 0 \quad \text{if } x > M \quad (4)$$

where $|\nabla u|$ is the gradient of an image, u_0 is the noisy image, u is the observed image, M decides which are the image gradients to be considered as edge.

The variable exponent adaptive model proposed by Chen et al. [12] has the following form shown in Eq. 5.

$$\min_u E(u) = \min_u \int\limits_{\Omega} \left[|\nabla u|^{\alpha(|\nabla u|)} + \frac{\lambda}{2}(u - u_0)^2 \right] dx\, dy \quad (5)$$

Here

$$\alpha(|\nabla u|) = 1 + \frac{1}{1 + \left(\frac{|\nabla u|}{k}\right)^2} \quad (6)$$

In the Chen model the regularization term uses the variable exponent. The variable exponent value depends on the value of the edge function. At image edges, the edge function value is 0, so the value of the variable exponent becomes one. Therefore the model behaves like the ROF model. This leads to strong edge preservation. In flat noisy inner region the value of the edge function is 1, so the variable exponent becomes 2. Therefore the model behaves like a Tikhonov model. This provides strong smoothing [8, 13, 14].

Chen et al. [15] proposed an adaptive TV algorithm making use of new edge stopping function based on difference of curvature. In this algorithm the regularization term as well as the fidelity term change adaptively depending on whether the pixel belongs to an edge or flat region. This model is described by Eq. 7.

$$\min_u E(u) = \min_u \iint_\Omega \left[|\nabla u|^{P(D)} + \frac{1}{2}\lambda(D)(u - u_0)^2 \right] dxdy \tag{7}$$

Here the functions $P(D)$ and $\lambda(D)$ depend on difference of curvature which is defined as follows.

$$D = \|u_{nn}\| - \|u_{ee}\| \tag{8}$$

Here u_{un} is the directional second derivative of image u along gradient direction and u_{ee} is the directional second derivative of image u along the edge direction. Now analysing the behaviour of this adaptive TV model the following facts are found. At the object boundaries the regularization term adapts to TV norm and the weight of the fidelity term assumes a large value. So the model behaves like an ROF model which enables efficient edge preservation. In the constant smooth region the regularization term adapts to L2 norm and the weight of the fidelity term is small. So the model behaves like a Tikhonov model which enables strong smoothing.

The observation from the literature survey is as follows. Though the adaptive total variation method enables edge preserving denoising may change important information contained in the images such as fine details, contrast and texture. The reason for all these problems is the error in distinguishing between the edges and the flat noisy inner region. In most of the adaptive total variation models, gradient based edge stopping function is used as a tool to determine the edges. It is proved in our previous work that in locating edges, an edge stopping function based on the trace of the structure tensor matrix is more accurate than an edge stopping function based on the gradient of an image [16]. This fact gives new direction in the research work that is to develop an adaptive total variation model using the trace of the structure tensor matrix based edge stopping function which more accurately locates edges and ramp. This new adaptive total variation model will totally stop diffusion across edges and permit strong diffusion in the flat inner noisy region. The proposed new adaptive total variation based regularization algorithm removes the Rician noise present in the brain MR images very effectively in addition to eliminating the above mentioned drawbacks associated with the conventional adaptive total variation regularization.

This paper is organized as follows. Section 3 explains in detail the structure tensor matrix and its applications. In addition the section explains in detail the new proposed adaptive variable exponent based model using a trace of the structure tensor matrix in edge function. Section 4 compares the performance of the proposed model with existing adaptive total variation models. Section 5 concludes the paper.

2 Materials and Methods

2.1 Structure Tensor Matrix

The structure tensor matrix is which is computed at every pixel position in an image as per Eq. 9. The eigenvalue decomposition can be applied to structure the tensor matrix to find out eigenvalues (l_1, l_2) and eigenvectors (e_1, e_2). Here the eigenvector e_1 is a unit vector normal to the gradient edge while the eigenvector e_2 is the tangent.

$$T = \begin{bmatrix} u_x^2 & u_x u_y \\ u_x u_y & u_y^2 \end{bmatrix} = \nabla u . \nabla \mathrm{u}^{\mathrm{T}} \tag{9}$$

$$T = \begin{bmatrix} G_\rho * u_x^2 & G_\rho * u_x u_y \\ G_\rho * u_x u_y & G_\rho * u_y^2 \end{bmatrix} = \begin{bmatrix} a & b \\ c & d \end{bmatrix} = \nabla u . \nabla \mathrm{u}^{\mathrm{T}} \tag{10}$$

$$l_{1,2} = \frac{1}{2}\left(a + d \pm \sqrt{4b^2 + (a-d)^2}\right) \tag{11}$$

$$e_1 = \begin{pmatrix} e_{1x} \\ e_{1y} \end{pmatrix} = \begin{bmatrix} \dfrac{2b}{\sqrt{\left(d-a+\sqrt{(a-d)^2+4b^2}\right)^2+4b^2}} \\ \dfrac{d-a+\sqrt{(a-d)^2+4b^2}}{\sqrt{\left(d-a+\sqrt{(a-d)^2+4b^2}\right)^2+4b^2}} \end{bmatrix} \tag{12}$$

$$e_2 = \begin{pmatrix} e_{2x} \\ e_{2y} \end{pmatrix} = \begin{bmatrix} -e_{1y} \\ e_{1x} \end{bmatrix} \tag{13}$$

In Eq. 9 ∇u is the gradient of an image, u_x is the derivative of image u along x direction. u_y is the derivative of image u along y direction. The component wise smoothing of tensor matrix with Gaussian filter having standard deviation ρ is to be done as given in Eq. 10. The merits of the structure tensor matrix are mentioned here. (i) The error in locating edges is minimized due to element by element Gaussian smoothing and also the cancellation effect of opposite gradients is avoided. (ii) By comparing the eigen values it is possible to determine whether the pixel is part of an edge or corner or constant area. If $l_1 = l_2 \approx 0$ then the pixel is a part of the flat region. If $l_1 >> l_2 \approx 0$ then the pixel is part of an edge. If $l_1 \geq l_2 >> 0$ then the pixel is part of a corner [17–19].

2.2 Proposed Adaptive Total Variation Based Image Regularization Using Structure Tensor

The authors proposed a new adaptive local feature driven total variation based regularization model using the structure tensor matrix which is described by the Eq. 14.

$$\min_u\{E(u)\} = \min_u \iint_{(x,y)\in\Omega} \left[|\nabla u|^{\alpha(h)} + \frac{1}{2}\beta(h)(u-u_o)^2\right]dxdy \tag{14}$$

In the above mentioned model the variable exponent $\alpha(h)$ and regularization parameter $\beta(h)$ uses an edge stopping function based on the trace of structure tensor matrix which is described by Eq. 15. In Eq. 14 u_0 is the initial noisy image and u is the observed image.

$$c(h) = \exp\left(-\left(\frac{h}{k}\right)^2\right) \tag{15}$$

$$\alpha(h) = 1 + c(h) \tag{16}$$

$$\beta(h) = 0.2(1 - c(h)) \tag{17}$$

Here parameter h is the trace of the tensor matrix, the sum of eigen values that is $h = l_1 + l_2$, k is the gradient threshold parameter. Weight of fidelity term is also known as the regularization parameter. In the noisy inner region $l_1 = l_2 \approx 0$, the function $c(h) = 1$ therefore the variable exponent $\alpha(h)$ becomes two and the model becomes the Tikhonov model. This results in efficient noise removal in the inner region. When h becomes larger the value of the function $c(h)$ becomes equal to 0. At the image edges $l_1 >> l_2 \approx 0$, the edge function $c(h) = 0$ therefore the variable exponent $\alpha(h)$ becomes one and the model becomes an ROF model, so the edges are preserved well. At image edges $\beta(h)$ becomes 0.2 since $c(h)$ is zero, this value of the regularization parameter enables strong edge preservation which is proved in our previous work [20]. In the inner region $c(h)$ is either exactly one or nearly one example 0.999. This results in a small value of the regularization parameter either 0 or 0.0002 and this value enables strong smoothing [21–24]. The basic steps in the proposed adaptive total variation regularization algorithm using structure tensor is described by the flowchart shown in Fig. 1.

The proposed model described by Eq. 14 is solved using the steepest descent method [15, 25] and the following partial differential Eq. 18 is obtained as a solution of the proposed model.

$$\frac{\partial u}{\partial t} = \nabla(g(|\nabla u|)\nabla u) + \beta(h)(u_0 - u) \tag{18}$$

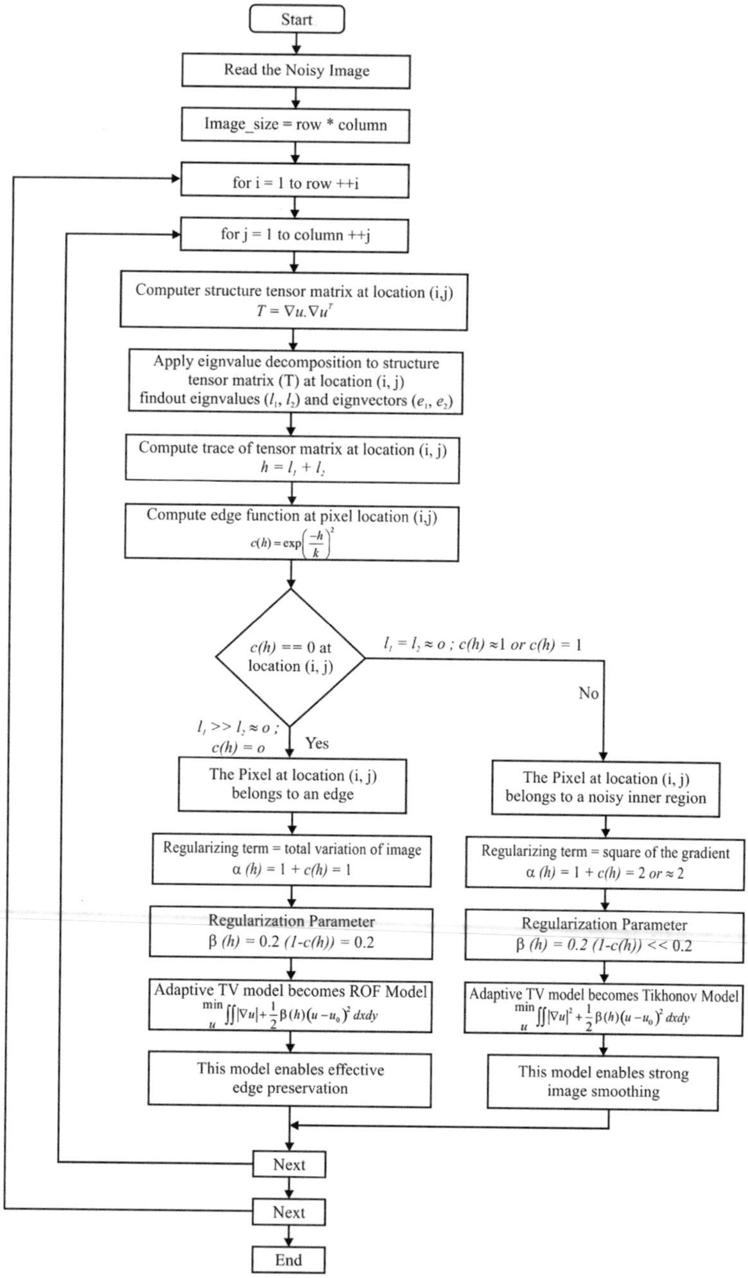

Fig. 1 Flowchart showing steps of adaptive total variation regularization algorithm using structure tensor matrix

Here $u = u_0$ the original noisy image at time $t = t_0$ and the diffusivity function $g(|\nabla u|)$ is given below.

$$g(|\nabla u|) = \phi'(|\nabla u|)/|\nabla u| \tag{19}$$

$$\phi(|\nabla u|) = |\nabla u|^{\alpha(h)} \tag{20}$$

The orthogonal decomposition of Eq. 18 is given in Eq. 21.

$$\frac{\partial u}{\partial t} = \phi''(|\nabla u|)u_{nn} + g(|\nabla u|)u_{ee} + \beta(h)(u_0 - u) \tag{21}$$

In Eq. 21

$$\phi''(|\nabla u|) = \alpha(h)(\alpha(h) - 1)|\nabla u|^{\alpha(h)-2} \tag{22}$$

$$g(|\nabla u|) = \alpha(h)|\nabla u|^{\alpha(h)-2} \tag{23}$$

$$u_{vv} = e_2^{\mathrm{T}}\, H\, e_2 = \begin{pmatrix} e_{2x} & e_{2y} \end{pmatrix} \begin{pmatrix} u_{xx} & u_{xy} \\ u_{xy} & u_{yy} \end{pmatrix} \begin{pmatrix} e_{2x} \\ e_{2y} \end{pmatrix} \tag{24}$$

$$u_{ww} = e_1^{\mathrm{T}}\, H\, e_1 = \begin{pmatrix} e_{1x} & e_{1y} \end{pmatrix} \begin{pmatrix} u_{xx} & u_{xy} \\ u_{xy} & u_{yy} \end{pmatrix} \begin{pmatrix} e_{1x} \\ e_{1y} \end{pmatrix} \tag{25}$$

$$u_{nn} = \frac{1}{(u_x^2 + u_y^2)}\left(u_{yy}u_y^2 + u_{xx}u_x^2 + 2u_x u_y u_{xy}\right) \tag{26}$$

$$u_{ee} = \frac{1}{(u_x^2 + u_y^2)}\left(u_{xx}u_y^2 + u_{yy}u_x^2 - 2u_x u_y u_{xy}\right) \tag{27}$$

The proposed model described by Eq. 21 can be implemented by means of a numerical iterative algorithm given in Eq. 28. In Eq. 28 n denotes the number of iterations. In Eq. 21 replace u_{nn} by u_{ww} and replace u_{ee} by u_{vv}.

$$\begin{aligned} u_{i,j}^{n+1} &= u_{i,j}^n + \Delta t\phi''(|\nabla u|)_{i,j}^n (u_{ww})_{i,j}^n + \Delta t g(|\nabla u|)_{i,j}^n (u_{vv})_{i,j}^n \\ &+ \Delta t\beta(h)_{i,j}^n \left(u_{i,j}^n - (u_o)_{i,j}^n\right) \end{aligned} \tag{28}$$

Table 1 Comparison of quality metric SSIM for denoised images generated by Chen model, Erik model, Chambolle, TV-difference of curvature model and the proposed model using structure tensor

Modality	Slice/noise level	Chen model	Erik model	Chambolle model	TV-diffcur	Proposed model—tensor
T1-weighted	Transverse-55 (10%)	0.9650	0.9161	0.4740	0.8240	0.9897
	Transverse-55 (15%)	0.9612	0.9414	0.5275	0.8559	0.9892
T1-weighted	Transverse-69 (10%)	0.9710	0.9132	0.4739	0.7598	0.9879
	Transverse-69 (15%)	0.9612	0.9409	0.4254	0.8448	0.9863
T2-weighted	Coronal-67 (10%)	0.9735	0.9170	0.511	0.7714	0.9899
	Coronal-67 (15%)	0.9631	0.9353	0.4600	0.8323	0.9897
T2-weighted	Sagittal-84 (10%)	0.9644	0.9233	0.5444	0.8812	0.9902
	Sagittal-84 (15%)	0.9738	0.9410	0.5010	0.8457	0.9902
PD-weighted	Transverse-64 (10%)	0.9680	0.9125	0.5165	0.8018	0.9881
	Transverse-64 (15%)	0.9597	0.9358	0.4363	0.8144	0.9894
PD-weighted	Sagittal-64 (10%)	0.9650	0.9215	0.4892	0.8405	0.9880
	Sagittal-64 (15%)	0.9625	0.9357	0.4244	0.8433	0.9892
Mean	SSIM	0.9657	0.9278	0.4819	0.8262	0.98898

3 Experimental Results and Discussion

The performance of the proposed adaptive TV model using a trace of the tensor matrix in the edge function is compared with the existing adaptive variable exponent based total variation models such as Chen model, Erik model, Chambolle model and TV model using difference curvatures. In the performance comparison the quality metrics SSIM, FSIM, PSNR, MAE are computed and the results are presented in the Tables 1, 2, 3, and 4. The visual quality of denoised images produced by the above mentioned algorithms is compared in Figs. 2, 3, 4, and 5. Rician noise corrupted MRI slices are denoised and the average value of the quality metrics computed are presented [26–30]. It is found that the proposed algorithm performs extremely best compared to all other algorithms from the value of the quality metrics experimentally computed. The SSIM, PSNR, FSIM and MAE values of the denoised image are generated by different algorithms such as Chen Model, Erik Model, Chambolle Model, TV Model

Table 2 Comparison of quality metric PSNR for denoised images generated by Chen model, Erik model, Chambolle, TV-difference of curvature model and proposed model using structure tensor

Modality	Slice/noise level	Chen model	Erik model	Chambolle model	TV-diffcur	Proposed model—tensor
T1-weighted	Transverse-55 (10%)	39.7373	37.2551	22.8851	19.4111	44.7362
	Transverse-55 (15%)	37.3672	36.5702	26.4126	23.8043	42.8219
T1-weighted	Transverse-69 (10%)	40.2485	37.4829	26.3495	17.7472	44.3789
	Transverse-69 (15%)	37.6674	36.6993	22.9390	23.6176	42.8227
T2-weighted	Coronal-67 (10%)	39.4025	37.1587	25.7693	17.3413	44.9378
	Coronal-67 (15%)	37.6350	36.5866	22.8934	21.4896	43.7684
T2-weighted	Sagittal-84 (10%)	37.6217	37.3078	25.8276	21.5006	44.7783
	Sagittal-84 (15%)	39.0915	36.7418	23.9579	22.3598	43.6229
PD-weighted	Transverse-64 (10%)	38.9516	36.8058	23.6685	18.7264	43.9873
	Transverse-64 (15%)	37.0377	36.2772	19.2914	20.9401	42.9877
PD-weighted	Sagittal-64 (10%)	38.9610	36.8549	22.4483	20.069	43.9203
	Sagittal-64 (15%)	34.3960	36.3628	19.2061	22.6528	42.4932
Mean	PSNR	38.176	36.8419	23.4707	20.8049	43.7713

using difference curvature and the proposed adaptive TV model using the trace of the structure tensor matrix based edge function are compared and the results are shown in Figs. 6, 7, 8, and 9. The bar chart proves that the proposed algorithm using structure tensor has larger magnitudes of PSNR, FSIM, MAE and SSIM compared to all other adaptive variable exponent based total variation algorithms.

Table 1 shows that the proposed algorithm generates higher SSIM values in comparison to other models, which means that the denoised image is highly similar to the original image. Table 2 shows the value of PSNR quality metric for denoised images generated by different models. The proposed algorithm performs best since it achieves maximum PSNR values compared to other adaptive models. This finding agrees with the visual quality of images generated by the proposed algorithm. Table 3 proves that the proposed algorithm can generate a visually enhanced image compared to other models which is proved from the higher values of FSIM quality metric. An observation from Table 4 is that the mean absolute error is very minimum for the

Table 3 Comparison of quality metric FSIM for denoised images generated by Chen model, Erik model, Chambolle model, TV-difference of curvature model and proposed model using structure tensor

Modality	Slice/noise level	Chen model	Erik model	Chambolle model	TV-diffcur	Proposed model—tensor
T1-weighted	Transverse-55 (10%)	0.9829	0.9728	0.8062	0.8828	0.9960
	Transverse-55 (15%)	0.9852	0.9868	0.8786	0.9230	0.9966
T1-weighted	Transverse-69 (10%)	0.9864	0.9653	0.8338	0.8350	0.9941
	Transverse-69 (15%)	0.9874	0.9830	0.7548	0.9160	0.9947
T2-weighted	Coronal-67 (10%)	0.9890	0.9729	0.8771	0.8324	0.9963
	Coronal-67 (15%)	0.9891	0.9880	0.8125	0.8979	0.9973
T2-weighted	Sagittal-84 (10%)	0.9902	0.9780	0.8993	0.9068	0.9969
	Sagittal-84 (15%)	0.9898	0.9881	0.8434	0.9076	0.9975
PD-weighted	Transverse-64 (10%)	0.9830	0.9735	0.8350	0.8678	0.9940
	Transverse-64 (15%)	0.9829	0.9487	0.7570	0.8813	0.9951
PD-weighted	Sagittal-64 (10%)	0.9829	0.9567	0.8352	0.8993	0.9943
	Sagittal-64 (15%)	0.9833	0.9691	0.7472	0.9190	0.9945
Mean	FSIM	0.9860	0.9735	0.8233	0.8890	0.9956

images generated by the proposed algorithm compared to other models. This means that the proposed model is very efficient in edge preserving denoising than other models.

The visual quality of denoised images produced by various adaptive total variation models is compared in Figs. 2, 3, 4, and 5. The value of the gradient threshold parameter (k) used in the proposed new algorithm is 12. In the experimental analysis different gradient threshold parameters are tried starting from $k = 1$ up to 15 and $k = 12$ gives proper balancing between edge preservation and noise removal. The number of iterations used in the proposed algorithm is 10. The proposed model performs best in producing sharper edges and preserving weak edges and fine details than those of other methods which could be noticed in the denoised images shown. The amount of noise present in the denoised image of the proposed model could be removed further by increasing the gradient threshold parameter k. The proposed model is able

Table 4 Comparison of quality metric MAE for denoised images generated by Chen model, Erik model, Chambolle, TV-difference of curvature model and proposed model using structure tensor

Modality	Slice/noise level	Chen model	Erik model	Chambolle model	TV-diffcur	Proposed model—tensor
T1-weighted	Transverse-55 (10%)	1.6395	1.8568	15.5004	7.0806	0.5637
	Transverse-55 (15%)	1.9270	1.4742	16.3329	5.8291	0.5334
T1-weighted	Transverse-69 (10%)	1.3841	1.8923	10.45481	7.7202	0.5897
	Transverse-69 (15%)	2.0927	1.5602	11.0514	6.0966	0.6702
T2-weighted	Coronal-67 (10%)	1.4398	1.8840	11.0903	8.3962	0.5390
	Coronal-67 (15%)	1.8097	1.6177	15.4983	6.8273	0.5270
T2-weighted	Sagittal-84 (10%)	1.7802	1.7572	10.9683	5.6478	0.5197
	Sagittal-84 (15%)	1.4055	1.5327	15.2187	6.3314	0.5069
PD-weighted	Transverse-64 (10%)	1.7201	1.6931	14.4087	7.1438	0.7531
	Transverse-64 (15%)	2.0751	2.1893	23.5607	7.4384	0.6207
PD-weighted	Sagittal-64 (10%)	1.6395	2.0952	16.5524	6.0546	0.5965
	Sagittal-64 (15%)	3.5621	1.7165	23.9566	8.3579	0.8484
Mean	MAE	1.8729	1.7724	15.3827	6.9103	0.6057

to bring an appropriate balance between noise removal and preservation of weak edges as well as fine details whereas other adaptive total variation algorithms could not.

4 Conclusion

To eliminate Rician noise present in the magnetic resonance images effectively a new adaptive total variation based image regularization algorithm using structure tensor matrix is proposed which uses a variable exponent in the regularization term and an adaptive fidelity term that is the weight of the fidelity term that adaptively changes depending on whether a pixel is part of an edge or a constant region. In the proposed algorithm the edge function value is computed at every pixel position using trace of

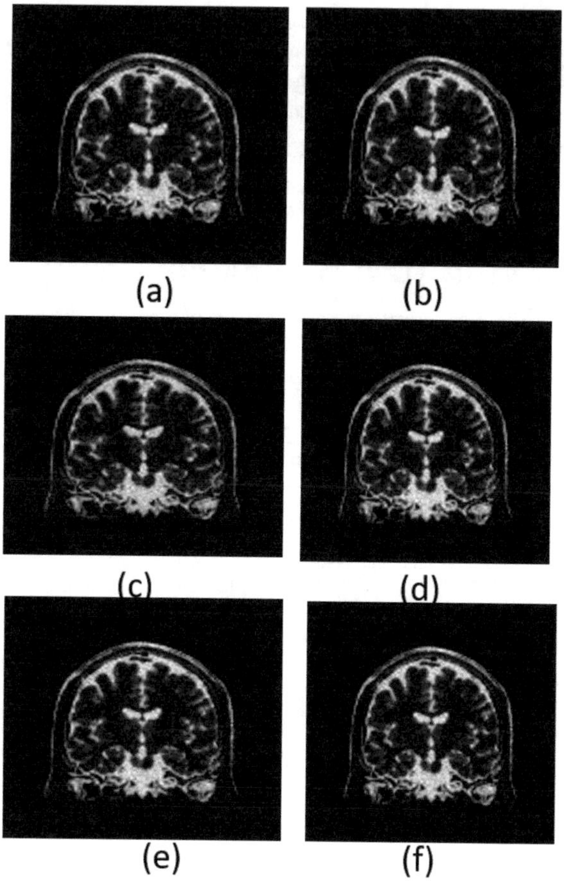

Fig. 2 Comparison of denoised images generated by different models. **a** T2-weighted coronal MRI slice-67 with 15% noise level. **b** Denoised image by Chen model. **c** Denoised image by Chambolle model. **d** Denoised image by Erik model. **e** Denoised image by TV-difference of curvature. **f** Denoised image by proposed algorithm using structure tensor

the structure tensor matrix. The performance of the proposed model using an edge function based on structure tensor is analysed in detail. The experimental results show that the proposed algorithm using structure tensor based edge function performs best compared to the existing adaptive variable exponent based total variation models such as Chen model, Erik model, Chambolle model and TV model using difference curvature both qualitatively as well as quantitatively. The proposed algorithm is tested on brain MR image data set for different noise levels and the performance is evaluated in terms of MAE, PSNR, FSIM and SSIM for Rician noise.

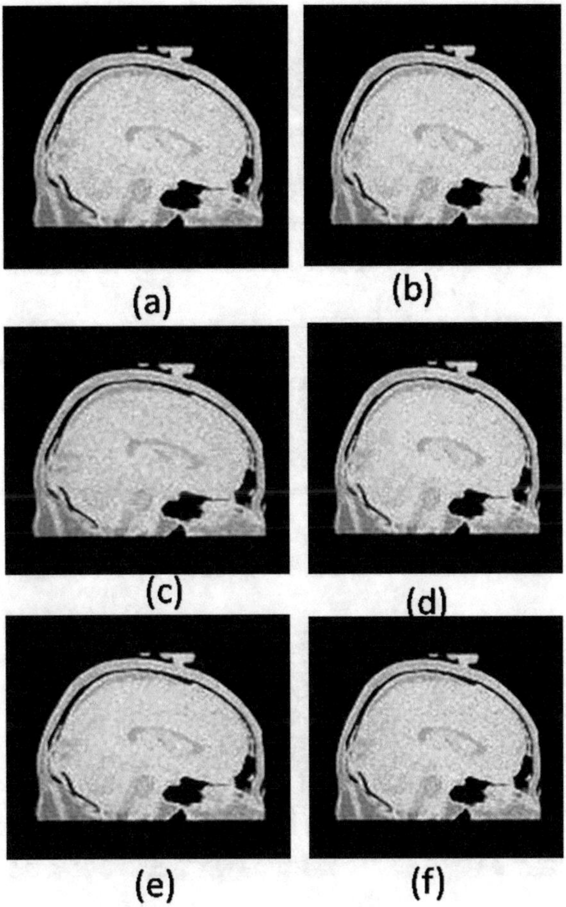

Fig. 3 Comparison of denoised images generated by different models. **a** PD-weighted sagittal MRI slice-64 with 10% noise level. **b** Denoised image by Chen model. **c** Denoised image by Chambolle model. **d** Denoised image by Erik model. **e** Denoised image by TV-difference of curvature. **f** Denoised image by proposed algorithm using structure tensor

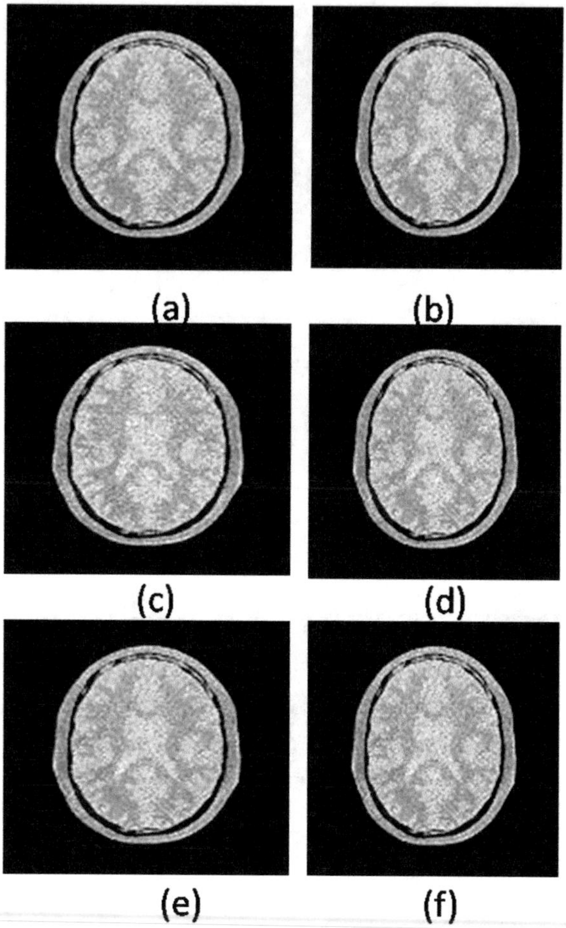

Fig. 4 Comparison of denoised images generated by different models. **a** PD-weighted sagittal MRI slice-64 with 15% noise level. **b** Denoised image by Chen model. **c** Denoised image by Chambolle model. **d** Denoised image by Erik model. **e** Denoised image by TV-difference of curvature. **f** Denoised image by proposed algorithm using structure tensor

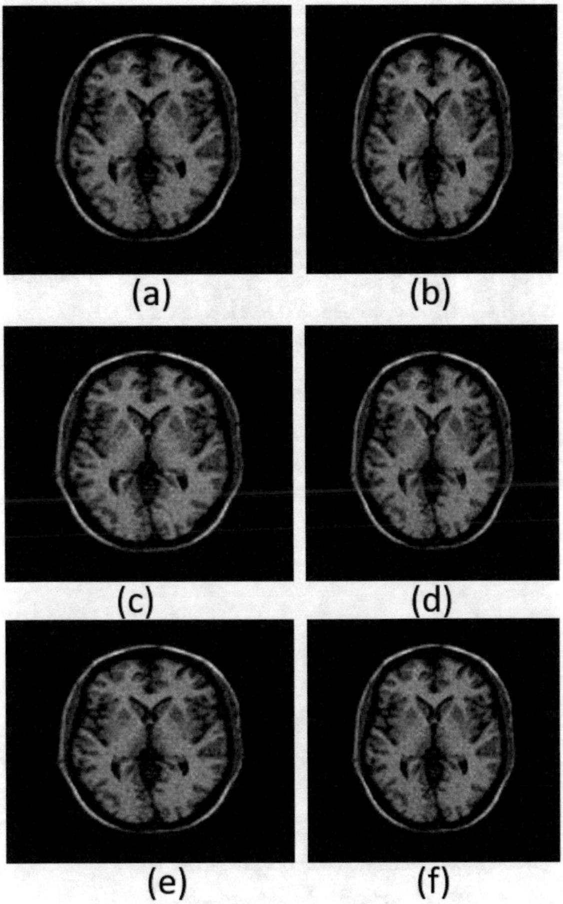

Fig. 5 Comparison of denoised images generated by different models. **a** T1-weighted transverse MRI slice-55 with 10% noise level. **b** Denoised image by Chen model. **c** Denoised image by Chambolle model. **d** Denoised image by Erik model. **e** Denoised image by TV-difference of curvature. **f** Denoised image by proposed algorithm using structure tensor

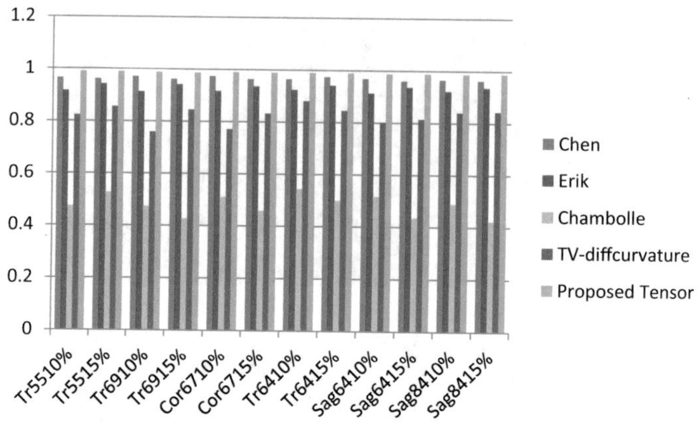

Fig. 6 Comparison of quality metric SSIM for denoised images generated by **a** Chen model. **b** Erik model. **c** Chambolle model. **d** TV-difference of curvature. **e** Proposed algorithm using structure tensor

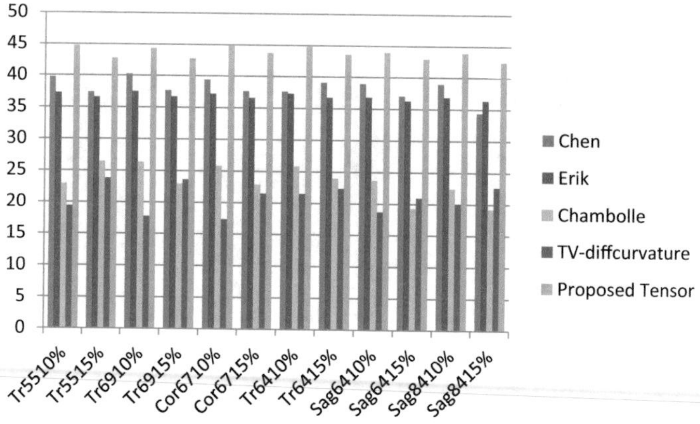

Fig. 7 Comparison of quality metric PSNR for denoised images generated by **a** Chen model. **b** Erik model. **c** Chambolle model. **d** TV-difference of curvature. **e** Proposed algorithm using structure tensor

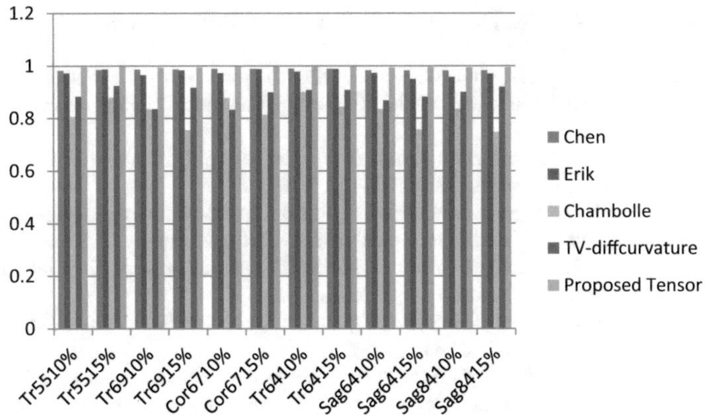

Fig. 8 Comparison of quality metric FSIM for denoised images generated by **a** Chen model. **b** Erik model. **c** Chambolle model. **d** TV-difference of curvature. **e** Proposed algorithm using structure tensor

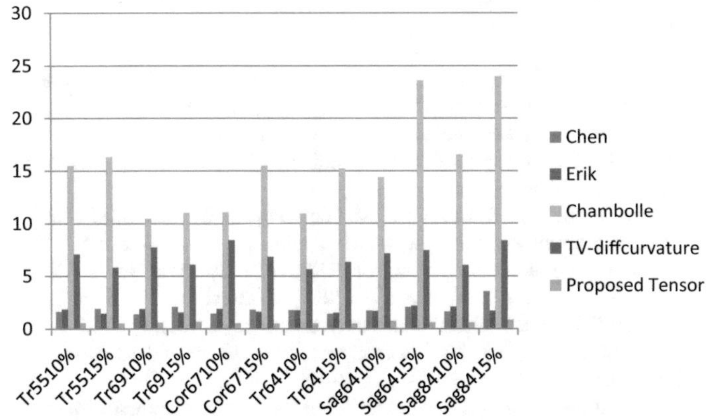

Fig. 9 Comparison of quality metric MAE for denoised images generated by **a** Chen model. **b** Erik model. **c** Chambolle model. **d** TV-difference of curvature. **e** Proposed algorithm using structure tensor

References

1. Vogel CR, Oman ME (1998) Fast robust total variation based reconstruction of noisy, blurred images. IEEE Trans Image Process 7
2. Karthik S, Hemanth VK, Soman KP, Balaji V, Sachin Kumar S, Sabarimalai Manikandan M (2012) Directional total variation filtering based image denoising method. Int J Comput Sci Issues 9(2, No 1). ISSN 1694-0814
3. Beck A, Teboulle M (2009) Fast gradient-based algorithms for constrained total variation image denoising and deblurring problems. IEEE Trans Image Process 18(11):2419–2434
4. Fan L, Zhang F, Fan H, Zhang C (2019) Brief review of image denoising techniques. Vis Comput Ind, Biomed Art 2:7. https://doi.org/10.1186/s42492-019-0016-7

5. Chambolle (2004) An algorithm for total variation minimization and applications. J Math Imaging Vis 20:89–97
6. Zhao Y, Liu JG, Zhang B, Hongn W, Wu Y-R (2015) Adaptive total variation regularization based SAR image depeckling and despleckling evaluation index. IEEE Trans Geosci Remote Sens 53(5):2765–2774
7. Liu et al (2016) Hybrid regularizers-based adaptive anisotropic diffusion for image denoising. SpringerPlus. https://doi.org/10.1186/s40064-016-1999-6
8. Zheng H, Hellwich O Adaptive data-driven regularization for variational image restoration in the BV space. Computer Vision & Remote Sensing, Berlin University of Technology, Berlin
9. Bresson X, Laurent T, Uminsky D, von Brecht JH (2013) An adaptive total variation algorithm for computing the balanced cut of a graph. Math Commons
10. Chambolle A, Lions P-L (1997) Image recovery via total variation minimization and related problems. Numer Math 76:167–188
11. Bollt EM, Chartrand R, Esedoglu S, Schultz P, Vixie KR (2009) Graduated adaptive image denoising: local compromise between total variation and isotropic diffusion. J Adv Comput Math 31(1–3):61–85
12. Chen, Levine, Rao (2004) Variable exponent, linear growth functionals in image processing. SIAM J Appl Math 66:1383–1406
13. Sachinkumar S, Mohan N, Prabaharan P, Soman KP (2016) Total variation denoising based approach for R-peak detection in ECG signals. Procedia Comput Sci 93:697–705
14. Soman KP, Poornachandran P, Athira S, Harikumar K (2015) Recursive variational mode decomposition algorithm for real time power signal decomposition. Procedia Technol 21:540–546
15. Chen Q, Montesinos P, Sun QS, Heng PA, Xia DS (2010) Adaptive total variation denoising based on difference curvature. Image Vis Comput 28(3):298–306
16. Kamalaveni V, Veni S, Narayanankutty KA (2017) Improved self-snake based anisotropic diffusion model for edge preserving image denoising using structure tensor. Multimed Tools Appl 76(18):18815–18846
17. Brox T, Van den Boomgaard R, Lauze FB, Van de Weijer J, Weickert J, Mrazek P, Kornprobst P (2006) Adaptive structure tensors and their applications. In: Weickert J, Hagen H (eds) Visualization and image processing of tensor fields, vol 1. Springer, pp 17–47
18. Wang H, Chen Y, Fang T, Tyan J, Ahuja N (2006) Gradient adaptive image restoration and enhancement. In: IEEE international conference on image processing (ICIP), Atlanta, GA
19. Wu J, Feng Z, Ren Z (2014) Improved structure-adaptive anisotropic filter based on a nonlinear structure tensor. Cybern Inf Technol 14(1):112–127
20. Kamalaveni V, Narayanankutty KA, Veni S (2016) Performance comparison of total variation based image regularization algorithms. Int J Adv Sci Eng Inf Techol 6(4). Scopus and Elsevier indexed
21. Prasath S, Thanh DNH (2019) Structure tensor adaptive total variation for image restoration. Turk J Electr Eng Comput Sci 27(2):1147–1156. https://doi.org/10.3906/elk-1802-76
22. Li C, Liu C, Wang Y (2013) Texture image denoising algorithm based on structure tensor and total variation. In: 5th international conference on intelligent networking and collaborative systems (INCoS)
23. Faouzi B, Amiri H (2011) Image denoising using non linear diffusion tensors. In: 2011 8th international multi-conference on systems, signals & devices
24. Wu X, Xie M, Wu W, Zhou J (2013) Nonlocal mean image denoising using anisotropic structure tensor. Adv Opt Technol 3013:794728. https://doi.org/10.1155/2013/794728
25. You YL, Kavesh M (2000) Fourth-order partial differential equations for noise removal. IEEE Trans Image Process 9(10):1723–1730
26. AksamIftikhar M et al (2014) Robust brain MRI denoising and segmentation using enhanced non-local means algorithm. Int J Imaging Syst Technol 24:54–66
27. Getreuer P, Tong M, Vese LA (2011) A variational model for the restoration of MR images corrupted by blur and Rician noise. In: Bebis G et al (eds) Advances in visual computing. ISVC 2011. Lecture notes in computer science, vol 6938. Springer, Berlin, Heidelberg

28. Yang J et al (2015) Brain MR image denoising for Rician noise using pre-smooth non local mean filter. BioMed Eng OnLine
29. Mohan J, Krishnaveni V, Guo Y (2013) A new neutrosophic approach of Wiener filtering for MRI denoising. Meas Sci Rev 13(4)
30. Singh, Ghanapriya, and Rawat T (2013) Color image enhancement by linear transformations solving out of gamut problem. Int J Comput Appl 67(14):28–32

Survey on 6G Communications

Rishav Dubey, Sudhakar Pandey, and Nilesh Das

1 Introduction

The year 2019 ushers in a new age of 5G wireless connectivity. As we write this survey, a number of countries have begun to make 5G services available to the masses, and more countries are preparing to join this in near future. 5G will pervade all sectors of life as a breakthrough technology, resulting in enormous economic and societal benefits. However, from the standpoint of technical development, in order to meet the growing demand for communications and networking in the next decade, we have to explore what will be the future beyond 5G communication networks. Now it's time to think about what new development will bring about evolution in the 5G network which would be implemented in the 6G communication network. In this survey, we suggest 6G visions to pave the way for the development of 6G and beyond. We begin by describing the current level of 5G technology, as well as the importance of ongoing 6G development. Based on current and forthcoming wireless communications development, we expect 6G will include three key features: mobile ultra-broadband, super Internet-of-Things (IoT), and artificial intelligence (AI) [1].

The 5G system is an innovative wireless communication architecture that uses a service-based design instead of a communication-oriented design to accomplish "connected things" [2]. Unlike earlier generations, 6G will change the progression of wireless communication from "connected objects" to "connected intelligence". Sixth-generation technology will transform three areas: new media, new services,

R. Dubey · S. Pandey · N. Das (✉)
National Institute of Technology, Raipur, Raipur, India
e-mail: nilesh101das@gmail.com

R. Dubey
e-mail: er.rishavdubey@yahoo.com

S. Pandey
e-mail: spandey.it@nitrr.ac.in

P. Singh et al. (eds.), *Machine Learning and Computational Intelligence Techniques for Data Engineering*, Lecture Notes in Electrical Engineering 998,
https://doi.org/10.1007/978-981-99-0047-3_51

and new infrastructures. The 6G system is intended to use powerful artificial intelligence (AI) technology in various domains, collecting, transmitting, and learning data quickly and effectively anytime, anywhere to develop a huge variety of novel applications and intelligent services. Distributed and decentralized AI, in particular, will support the upcoming 6G, which will be highly flexible and secure, and applying to network systems which will be more human-centric. So 6G communications will have more robust security and protection of personal information. Traditional Machine Learning enabled networks relying on a central server, on the other hand, have serious privacy and security issues, such as single points of failure, and so are unable to provide ubiquitous and safe AI for 6G [3]. Furthermore, standard centralized ML approaches may not be ideal for ubiquitous ML because of the high overhead incurred by centralized data collecting and processing. As a result, decentralized machine learning solutions are becoming increasingly important for 6G, where all private data is stored locally on training devices. Federated Learning (FL) has recently received a lot of interest from academia and business as an emergent decentralized ML solution [4]. In FL, devices collaborate to train a shared model using their local data, and so only send model changes to centralized parameter servers rather than raw data.

2 Use Case Scenario for 6G Communication

Sensors and actuators make up the physical devices [5]. Sensors and actuators are used to sense a variety of environmental characteristics, including temperature, wetness, and movement. They take measurements, store data, process it, and send it to other nodes and cloud computers. These physical devices are where IoT big data is collected. It is critical to comprehend the collected data in order to make IoT devices intelligent and make context aware processing. We must use some smart processes to understand the data using a light weight algorithm because the power of these gadgets is limited by the battery inside them. The fact that data is growing in size as a result of these devices implies that data growth is directly proportionate to the number of IoT devices. The accuracy and efficiency with which data is processed can have a significant impact on IoT security decisions. In comparison to previous generations, the 5G service model has been transformed into a service-based architecture [6]. The new use cases which will be available with 6G are depicted in Table 1.

2.1 New Media

In a decade, with the high-paced development of wireless communication technologies, the methods of information communication will increasingly shift from augmented/virtual reality (AR/VR) to high-fidelity extended reality (XR) interaction and even wireless holographic communication will be realized [8, 9]. Users can

Table 1 Network use cases of 6G communication network [7]

Use case	Typical applications	Key requirements
Extended reality (ER)	Immersive gaming, remote surgery, remote industrial control	High data rate (>1.6 Gbps/device), low latency, high reliability
Holographic telepresence	Online education, collaborative working, deep-immersive gaming	Ultra-high data rate (terabits per second)
Multi-sense experience	Remote surgery, tactile Internet, remote controlling and repairing	Stringently low latency
Tactile Internet for Industry 4.0	Industrial automation, smart energy consumption	<1 ms E2E latency
Intelligence transport and logistics	Automated road speed enforcement, real-time parking management	Stringently high reliability and low latency
Ubiquitous global roaming	World-wide roaming services for UE, portable devices, industrial apps	Low-cost fully global coverage
Pervasive intelligence	Computer vision, SLAM, speech recognition, NLP, motion control	High decision accuracy and transparency, complex data privacy

make use of novel holographic communication and display services such as virtual travel, virtual gaming, virtual teaching and other completely realistic holographic interactions at any time from any location.

2.2 New Services

6G communication will give customers a wide range of new services which will be beyond current teleport technology [10]. The traditional service model has been advanced by visible light communication, holographic teleport, quantum communication and other communication technologies. So, the newer service model will provide better services to the users by enabling sectors like remote surgery, cloud PLC, and intelligent transportation systems [11]. These new technologies are intended to give better high-precision and trustworthy services.

2.3 New Infrastructure

The 6G communication system will produce various rising infrastructures such as deep learning networks, integrated terrestrial and space, decentralized and trustable infrastructure [12]. In particular, the deep learning network has brought many

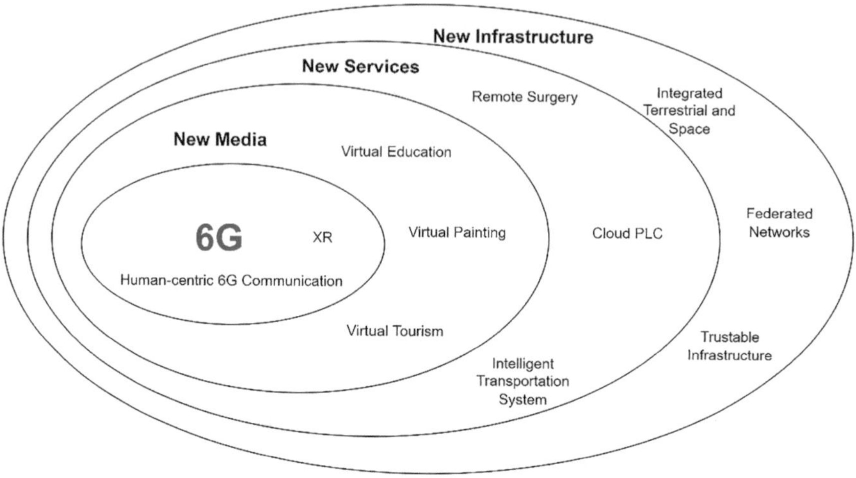

Fig. 1 Applications for 6G communication network

emerging intelligent applications to industry and people because it benefits from the low latency and high bandwidth of the 6G network (Fig. 1).

3 Requirements and Infrastructure for 6G Communication

6G combines important features that allow for ubiquitous, high-performance and automated communication networks while maintaining safety and confidentiality. Below, we'll go through the 6G communication performance standards in further depth.

3.1 High Performance Networking

6G is often regarded as a complicated networking system involving numerous diverse space, air, ground, and undersea communication networks. Through networks and systems, such as undersea-land communications and satellite communication networks, three-dimensional super-connection networks provide various sorts of network services and dense coverage by providing global connectivity and integrated networking [13]. 6G communications would achieve about few terabytes per second, data throughput per user, with the use of large-scale distributed networks, giving benefits such as increased end-to-end reliability, ultra-low latency and energy performance networking. In comparison to 5G communications, 6G communications enable efficient and low-overhead networking and connectivity not just in urban locations but also in less dense areas, such as the undersea environment. 6G communications make use of innovative communication networks to enable very diversified data such as augmented reality and virtual reality (AR/VR), resulting in a novel communication experience that includes virtualization.

3.2 Higher Energy Efficiency

Wireless devices with limitations in battery life have higher energy efficiency requirements in the 6G future. As a result, two significant 6G research topics increased battery life and reduced energy usage. According to research, energy conservation technology, green communication and wireless power transmission are suggested ways to improve the working hours and energy efficiency of wireless devices [14]. Using various energy conservation methods, wireless devices may collect energy from wind, solar and thermal energy, and ambient radio-frequency, extending the battery life. Wireless devices with rechargeable hardware can also employ wireless energy transfer technologies to replenish their energy supply from charging stations or dense networks. Symbiotic radio is a new technology that combines passive backscatter devices with a primary transmission system, and was recently launched to help wireless devices save energy [15]. Ambient backscatter communication is a popular example of SR, which lets network devices communicate via ambient RF waves rather than active RF transmission, which eliminates the need for batteries [16]. AI based solutions are important for green communication plans because of energy efficiency and implementation. Machine learning or deep learning algorithms may be utilized to optimize a wireless device's compute job offloading option as well as the optimal working and the best operating and resting time scheduling solution, increasing efficiency and decreasing consumption of energy.

3.3 High Security and Privacy

The majority of current research in 4G and 5G communications focuses on network throughput, reliability, and latency. Privacy and security issues in wireless communication, on the other hand, have been largely ignored in recent decades. Since data privacy is closely related to users' activity, maintaining data security and privacy has become a key component of intelligent 6G communications. Internet service providers are legally able to collect a large quantity of personal information from their customers, resulting in frequent privacy data leaks, so ML approaches would be used to provide security-optimized deep learning to overcome this issue in 6G networks.

3.4 High Intelligence

Users will benefit from 6G's high intelligence by receiving high-quality, tailored, and intelligent services. High-intelligent 6G includes operational intelligence, application intelligence, and service intelligence.

Operational Intelligence. Performance optimization and resource allocation issues are common in traditional network operations. Optimization approaches based on game theory, and other theories are frequently utilized to reach a sufficient level of network functioning. However, in large-scale and time-varying settings, these optimization theories may fail to deliver the best solution. These complex problems can be solved using advanced machine learning approaches.

Application Intelligence. At the moment, 5G network applications are getting increasingly smarter. Intelligent apps are one of the application criteria for 6G networks. Wireless communication technologies provided by federated learning are used to link devices to 6G networks, allowing them to execute a variety of sophisticated applications [17]. Users may, for example, require sophisticated voice assistants in the future to complete their daily schedules. Users will benefit from highly intelligent applications thanks to the 6G network's pervasive AI.

Service Intelligence. The 6G network being a human-centric network will provide intelligent services that are both satisfying and individualized. Federated Learning, for example, uses distributed learning to give users with individualized healthcare, personalized advice, and personalized intelligent voice services [18]. In the future, 6G networks will be strongly integrated with intelligent services.

3.5 Increased Device Density

As compared to 5G, 6G has substantially higher device density and AI integration, as well as significantly higher bandwidth speeds and lower latency. Network capacity issues are becoming increasingly significant as device density rises and data traffic grows at a rapid pace. One alternative is to deploy a growing number of tiny radio cells capable of transmitting data fast and efficiently. Through high-performance power grids, these cells must be seamlessly connected to fiber-optic infrastructures. Through flexible and extensive wireless networks, fiber optics networks may provide very large bandwidth capacity and reliability for big devices with low latency [12].

3.6 Green Communication

Green communication necessitates making informed decisions about resource usage and communication efficiency. In 6G communication, green communication optimization and offloading decisions are some complex optimization problems which

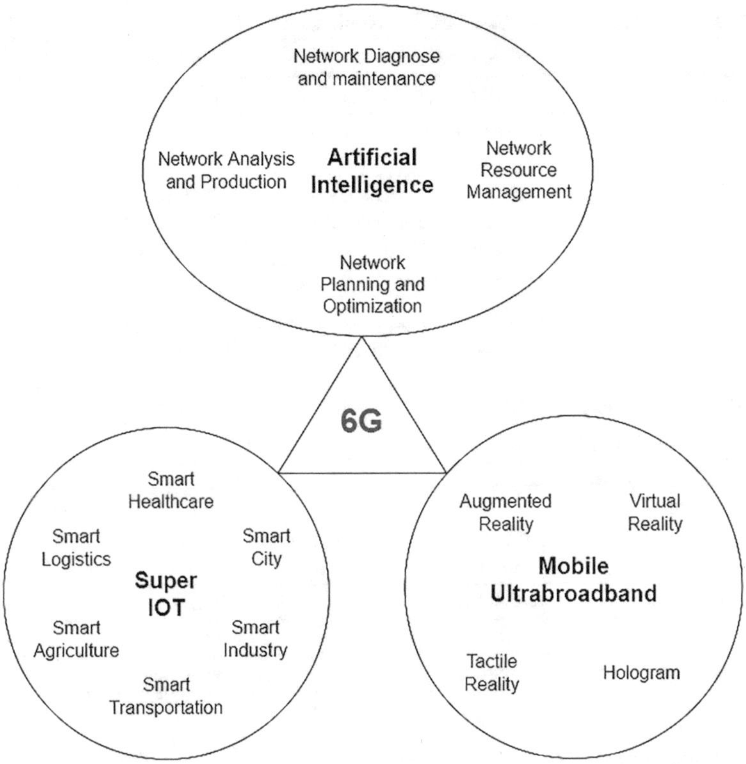

Fig. 2 Important aspects of 6G communication network

cannot be solved using traditional programming and mathematical models. This is due to high network traffic, changing network conditions and a huge number of network devices (Fig. 2).

4 5G to 6G Comparison

Higher frequencies in the wireless spectrum are used by both 5G and 6G to deliver more data at a quicker rate. 5G, on the other hand, runs at frequencies of less than 6 gigahertz (GHz) and greater than 25 GHz, referred to as low band and high band, respectively. 6G will have a frequency range of 1–3 terahertz (THz). 6G will give 1000 times faster speeds than 5G at certain frequencies. The comparison between 5G and 6G based on KPI requirements is summarized in Table 2.

Table 2 5G and 6G comparison based on KPI requirements [7]

KPI	5G requirement	6G requirement
Peak data rate	20/10 Gbps (DL/UL)	1 Tbps
User-experienced data rate	100/50 Mbps (DL/UL, dense urban)	>1 Gbps
Latency	UP: 4/1 ms (eMBB/URLLC) CP: 10 ms (eMBB/URLLC) E2E: not defined	UP: 10 μs to 100 CP: remarkably improved E2E: considered
Mobility	Up to 500 km/h (high-speed trains)	Up to 1000 km/h (airlines)
Connection density	10^6 per km^2 (with relaxed QoS)	10^7 per km^2
Network energy efficiency	Not defined	10–100 times better than that of 5G
Peak spectral efficiency	30/15 bps/Hz (DL/UL)	90/45 bps/Hz (DL/UL)
Area traffic capacity	10 $Mbps/m^2$	1 $Gbps/m^2$ (e.g. indoor hot spots)
Reliability	>99.999% (URLLC: 32 bytes within 1 ms, urban macro)	>99.99999%
Signal bandwidth	>100 MHz	>1 GHz
Positioning accuracy	<10 m	cm level
Timeliness	Undefined	Considered

5 Challenges for 6G Communication

There are numerous challenges obstructing the 6G communication network. We will brief about three main challenges i.e., THz communication, path loss and channel capacity in this section.

5.1 THz Sources

The THz range is too low for photonic devices that produce optical signals and too high for electronic devices that produce microwave signals, since it is placed between the optical and microwave frequency bands [19]. There are two forms of THz signals: continuous signals and pulsed signals. Since it can be made with a transmitter or antenna of reasonable complexity, the pulsed signal is heavily researched in the existing studies on THz communications [20].

5.2 *Path Loss*

THz signals have a significant free-space path loss, which is caused by both molecule absorption and dispersion losses. The potential energy of THz signals is translated into the internal kinetic energy of the molecule in the air, and the proportion of molecules of water vapor determines the molecular absorption loss. The dispersion loss is caused by the extension of the electromagnetic wave in space.

5.3 *Channel Capacity*

The newest developments in graphene-based electronics have paved the way for terahertz-band electromagnetic communication between nanodevices. Depending on the path length and chemical characteristics of the channel, this frequency spectrum has the potential to produce very high bandwidths, ranging from the full band to several gigahertz-wide channels. The molecular absorption noise is governed by the types and amounts of molecules that the THz wave beam encounters, since different molecules may produce different amounts of absorption noise.

6 Conclusion

We presented 6G concepts in this article, and we discussed the use case scenarios and covered the areas of new media, services, and infrastructure. We also discussed the requirements and infrastructure needed for 6G communication, and talked about performance networking, energy efficiency, security and privacy, intelligence, device density and green communication. We also represented 5G to 6G comparison on the basis of key performance indicators which gave us the key differences between them. We also discussed the challenges faced by the 6G Communication system covering areas such as THz sources, path loss and channel capacity. We look at important technologies to realize each component. THz communications, in particular, is a good option for supporting mobile ultra-broadband, while artificial intelligence can be utilized to accomplish intelligent IoT.

References

1. Zhang L, Liang YC, Niyato D (2019) 6G visions: mobile ultra-broadband, super internet-of-things, and artificial intelligence. China Commun 16(8):1–14
2. Han B, Jiang W, Habibi MA, Schotten HD (2021) An abstracted survey on 6G: drivers, requirements, efforts, and enablers. arXiv preprint arXiv:2101.01062

3. Kaur J, Khan MA, Iftikhar M, Imran M, Haq QEU (2021) Machine learning techniques for 5G and beyond. IEEE Access 9:23472–23488
4. Liu Y, Yuan X, Xiong Z, Kang J, Wang X, Niyato D (2020) Federated learning for 6G communications: challenges, methods, and future directions. China Commun 17(9):105–118
5. De Silva CW (2007) Sensors and actuators: control system instrumentation. CRC Press
6. Ji H et al (2017) Introduction to ultra reliable and low latency communications in 5G. arXiv preprint arXiv:1704.05565
7. Han B et al (2021) An abstracted survey on 6G: drivers, requirements, efforts, and enablers. arXiv preprint arXiv:2101.01062, Tables II and III
8. David K, Berndt H (2018) 6G vision and requirements: is there any need for beyond 5G? IEEE Veh Technol Mag 13(3):72–80
9. Dang S, Amin O, Shihada B, Alouini M-S (2020) What should 6G be? Nat Electron 3(1):20–29
10. Zhang Z, Xiao Y, Ma Z, Xiao M, Ding Z, Lei X, Karagiannidis GK, Fan P (2019) 6G wireless networks: vision, requirements, architecture, and key technologies. IEEE Veh Technol Mag 14(3):28–41
11. Liu Y, Yu JJQ, Kang J, Niyato D, Zhang S (2020) Privacy-preserving traffic flow prediction: a federated learning approach. IEEE Internet Things J 1
12. Giordani M, Polese M, Mezzavilla M, Rangan S, Zorzi M (2020) Toward 6G networks: use cases and technologies. IEEE Commun Mag 58(3):55–61
13. Xiao Y, Shi G, Krunz M (2020) Towards ubiquitous AI in 6G with federated learning. arXiv preprint arXiv:2004.13563
14. Huang T, Yang W, Wu J, Ma J, Zhang X, Zhang D (2019) A survey on green 6G network: architecture and technologies. IEEE Access 7:175758–175768
15. Long R, Guo H, Zhang L, Liang Y-C (2019) Full-duplex backscatter communications in symbiotic radio systems. IEEE Access 7:21597–21608
16. Yang G, Zhang Q, Liang Y-C (2018) Cooperative ambient backscatter communications for green internet-of-things. IEEE Internet Things J 5(2):1116–1130
17. Li T, Sahu AK, Talwalkar A, Smith V (2020) Federated learning: challenges, methods, and future directions. IEEE Signal Process Mag 37(3):50–60
18. Kang J, Xiong Z, Niyato D, Zou Y, Zhang Y, Guizani M (2020) Reliable federated learning for mobile networks. IEEE Wirel Commun 27(2):72–80
19. Akyildiz I, Jornet J, Han C (2014) Teranets: ultra-broadband communication networks in the terahertz band. IEEE Wirel Commun 21(4):130–135
20. Josep MJ, Akyildiz IF (2014) Femtosecond-long pulse-based modulation for terahertz band communication in nanonetworks. IEEE Trans Commun 62(5):1742–1754

Human Cognition Based Models for Natural and Remote Sensing Image Analysis

Naveen Chandra and Himadri Vaidya

1 Introduction

Cognitive science is a multidisciplinary study of cognition and intelligence. It concentrates on characterizing the nature of human knowledge i.e. how that knowledge is represented, processed, and transformed within human nervous systems and intelligent systems/machines. Cognitive science explains human cognitive skills such as visual perception, attention, cognition, language, memory, thinking, decision making, reasoning, and cognitive development. Cognitive science implies that the process of thinking consists of representational structures in the human mind and some computational methods/procedures that operate on those structures.

According to William James (a well-known philosopher), the study of attention is divided into two categories i.e., selective attention and divided attention [1]. Memory is defined as the mental storage of data, information, and the processes involved in acquiring, storing, and retrieving that information. Memory serves as the center stage around which both elementary (categorization, encoding) and advanced (reasoning, problem-solving) form of cognition revolves. Memory can be of two types firstly; long-term memory in which information can be stored for days, weeks, and years secondly, short-term memory in which information is stored for either seconds or minutes. Memory can also be categorized as declarative and procedural forms. Declarative/explicit memory refers to our memory for facts, specific knowledge, meanings, and experiences. Procedural/implicit memory allows us to remember actions and motor sequences [1].

N. Chandra (✉)
Wadia Institute of Himalayan Geology, Dehradun, India
e-mail: naveenchandra@wihg.res.in

H. Vaidya
Graphic Era Hill University, Dehradun, India

P. Singh et al. (eds.), *Machine Learning and Computational Intelligence Techniques for Data Engineering*, Lecture Notes in Electrical Engineering 998,
https://doi.org/10.1007/978-981-99-0047-3_52

Perception is the complex sequence of processes by which we can organize and interpret information about the world that has been collected by our sensory receptors which in turn allows us to see and hear the world around us. Perception serves to interpret sensation in all of its various forms which include vision (eyes), audition (ears), olfaction (nose), tactile senses (skin), and gustation (tongue). Vision and hearing are two major senses that allow us to perceive the environment. Some well-known facts about perception are that perception is limited, it is selective, it refers to the distal stimulus and not the proximal stimulus, it requires time, it is not entirely veridical, it requires memory, it, requires internal representations and it is not influenced by context [1].

The reasoning is defined as a mental process by which any given information can be transformed into new or more useful forms. It is an important cognitive capacity of human life. Reasoning can be of two type's namely inductive reasoning and deductive reasoning. A computational approach towards cognition has developed the concept of reasoning [1].

Cognition is defined as a psychological, mental action or process of acquiring knowledge and understanding through thought, experience, and the senses which result in perception, sensation, or intuition. In simple words, cognition involves how we acquire, store, retrieve and use knowledge. Whenever cognition is used, some information is acquired, placed in storage, and uses that information for further decision-making processes. Cognition is a base for the processing of information, applying knowledge, and changing preferences. Cognition or cognitive processes are interrelated with one another, and they don't exist in isolation. There are complex interactions and coordinations among various components of cognition. According to the behaviorist approach, humans were viewed as passive organisms therefore cognitive processes are passive, rather than active. Cognitive processes are generally not directly observable and are efficient and accurate. They can be natural or artificial, conscious or unconscious [2].

2 Cognition Based Model for Natural Images

2.1 Attention-Based Model

Visual attention is an important cognitive parameter that is commonly used by human beings to interpret and understand the surroundings by concentrating on some focused objects. In this method objects with the maximum attention, and values are popped out iteratively from an image [3]. When a person visualizes the world infinite visual information is sensed by him but only particular objects from that environment are used for further analysis and the rest of the objects are ignored from that scene [4] and this process of selection is termed as visual attention. This process involves various complex processes in the retina and cortex and uses human cognitive parameters such as attention, visual perception, and memory.

Attention is one of the major cognitive skills in human beings and it is an important parameter for image interpretation. This method is carried out in two steps. Initially, the original image is segmented into different regions. To reduce the computational time the input image is downsized.

In the first step, remarkable regions which might be the real objects in the human mind are extracted. By using the adjacent matrices the relationship between the regions is defined [3]. Secondly, the program reverts to the input/original image to extract features of an object/target derived in the first step. In image segmentation, similar pixels are grouped into specific regions. It is defined as a pre-clustering process [5] in which the meaning is not assigned to the segmented region. Further JSEG [5], considers texture and color information for the segmentation of an image. Adjacent matrices for the segmented region are defined which includes boundary and feature matrix to define the relationships amongst various segments. Suppose a color image C is segmented into M regions, i.e., $C = \{r_i\}, i = 1 \ldots M$ where r_i is the ith region. Then boundary matrix B is given by (1).

$$B = \left\{1_{i,j}\right\}, i, j = 1 \ldots M \quad (1)$$

The feature matrix is given by (2)

$$A^k = \{a^k_{i,j}\}, k = 1 \ldots K \quad (2)$$

where, A^k is the kth feature matrix, and the difference in terms of the kth feature between region i and j is denoted by $a^k_{i,j}$. However, three feature matrices (one for texture and the other two for texture) measures the difference between a region and its respective surroundings.

Suppose S is the set of total combinations of R regions, then the size of S is given as 2^R. For instance, S contains eight subsets for $R = 3$, which are defined as $\{\}$, $\{1\}$, $\{1, 2\}$, $\{1, 2, 3\}$, $\{1, 3\}$, $\{2\}$, $\{2, 3\}$, $\{3\}$. Assuming that $M \in S$ is one merged (valid) region satisfying the following constraints:

(1) $M \neq \varnothing$
(2) $M \neq \{1 \ldots S\}$
(3) M is a combination of neighboring regions. An object is attractive if it is different from its surroundings [4] and the attention value of the merged region is given by (3).

$$F(M) = D\left(M, M'\right) - h\lceil D(M) - D(M')\rceil \quad (3)$$

where $h\lceil D(M) - D(M')\rceil$ is the mean function of $D(M)$ and $D(M')$. The popping out process is implemented by finding the valid combination of (M^t) whose attention value $F(M)$ achieves to be maximum, i.e., $M^t = \operatorname{argmax}\lceil F(M)\rceil$.

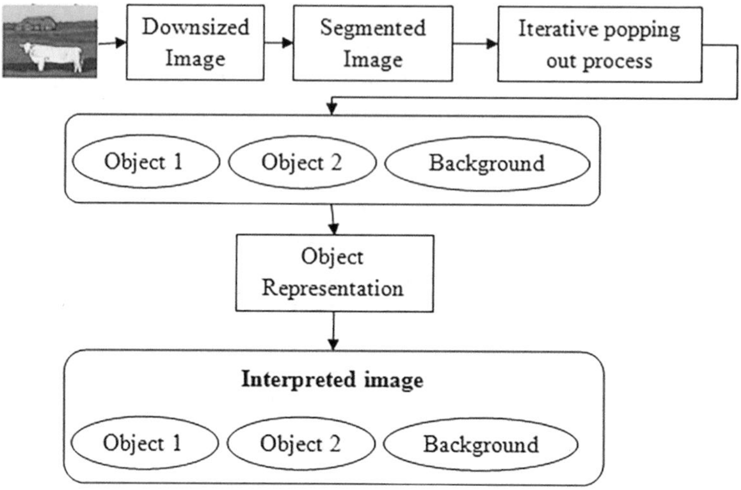

Fig. 1 Process of attention based interpretation [3]

The attention-based image interpretation method showcases better results when compared to the global-feature-based approach because the various sources of information are well-defined based on human attention. Here the process of image interpretation is executed using attentive objects and background. The flow process of the model is shown in Fig. 1.

2.2 *Understanding Based Model*

It is a model which analyses and interprets the data which is in the form of images. It is used to interpret and understand the complex visual pattern of medical images. It is a cognitive model because it operates by using reasoning and thought processes that take place in the human mind [6]. A new class of cognitive categorization is described using interpretation, analysis, and reasoning process which can perform deep analyses of data being interpreted. This model uses human cognitive parameters such as reasoning and thought processes.

In this model interpretation and analysis are carried out using cognitive resonance. It is the process of distinguishing similarities and differences between analyzed data and a set of data represented by a knowledge base. During image interpretation main stress is on cognitive resonance which leads to an interpretation of semantic information and analysis of data [7]. During analysis, the content, form, shape, and meaning of data are analyzed which is used further for extracting significant features from the image. Simultaneously, knowledge collected by the system is used to generate some expectations. Later, these expectations are compared with the features of the analyzed data. Then cognitive resonance identifies the similarity and differences

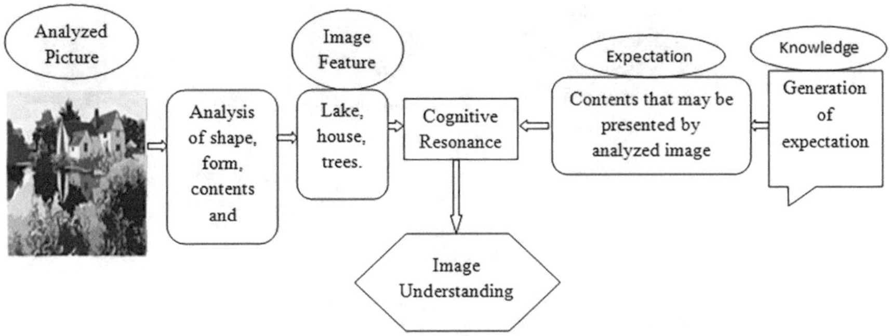

Fig. 2 Process of cognitive interpretation of image data [6]

between analyzed data and generated expectation. This model uses cognitive techniques for the interpretation of medical images. The main focus of the model is on cognitive reasoning which is used to understand the input image. UBIAS is capable of processing deep semantic analysis of the input image. Semantic information in the input image is composed of different data units (ontologies) which are easy to interpret and provide a meaning which is used for the further decision-making process. The interesting thing about this model is that it combines the study of psychological and philosophical subjects. The flow process of the model is shown in Fig. 2.

2.3 Koch and Ullman Model

One of the well-known models for visual interpretation based on attention skills was introduced by Koch and Ullman. This model provided the platform for many attention-based models. Many features in the image are collected in saliency maps. Salient regions in the image are identified using the Winner-take-all (WTA) algorithm [8]. This model is based on feature integration theory by Treisman and Gelade. This model uses human cognitive parameters such as attention, reasoning, recognition, and perception.

This model uses the WTA approach in which a neural network determines the most salient objects in the image and its detailed description for further interpretation. WTA network describes how the maximum salient objects are interpreted and implemented by the neural network. This is a biological method that is processed in the human brain [9]. WTA is still an overhead because there are many other methods to determine saliency maps.

After determining the saliency map the object is routed into a central representation that contains the properties of a single location in the image therefore this approach is also referred to as the selective routing model. This model doesn't describe further processing of the information. This model suggests a method for processing the selected regions called inhibition of return (IOR).

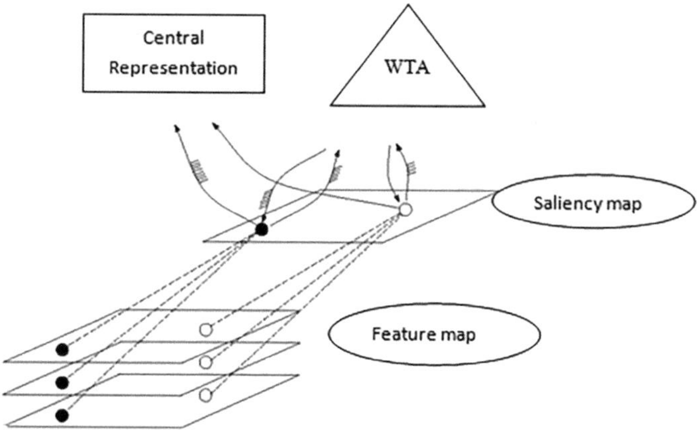

Fig. 3 The architecture of Koch–Ullman model [8]

The concept of central representation is a biological point of view that involves the simple and complex processing of information in the human mind. This is one of the important cognitive models for interpretation as it uses many cognitive parameters such as attention for object recognition and memory for storing and processing information. Different features in the image are processed in parallel and their saliency maps are represented in different feature maps. It is observed that this method uses the bottom approach for image interpretation and analysis. The flow process of the model is shown in Fig. 3.

2.4 Bayesian Model

The Bayesian model is an attention-based method for searching and interpreting objects in an image. A Top-down guided search algorithm has been used for searching houses in an image. An experiment was conducted using forty satellite images. It is difficult to recognize objects when an interpreter has to consider all possible views of an object [10]. Visual search of an object becomes faster when an interpreter knows the exact object to be searched. Recognition and interpretation behavior in human beings can be understood by the integration of bottom-up information which is perceived by our senses from the surroundings and top-down information from the descriptive knowledge of the objects/targets in the surroundings. In the past, some visual recognition methods have been proposed namely, feature integration which processes the low-level visual features in separate feature maps. Later these feature maps are combined to form a master map which is used for guided attention. The guided search method uses top-down approach knowledge for creating an activation map. Then by using the bottom-up approach feature maps are created which contribute to weighing features such as location (e.g., the bottom-right corner) and

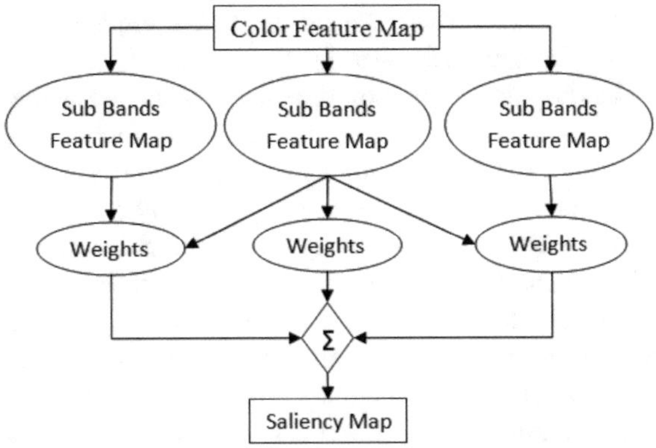

Fig. 4 The architecture of feature integration theory [10]

(e.g., a green color) which characterize the objects in an image. By using the combination of these weighted maps attention can be shifted between the different locations in an image. This model uses human cognitive parameters such as attention, visual perception, and recognition. This model used the theories of the Bayesian framework for searching and interpreting in a probabilistic way. It is a combination of attention and recognition and was developed using top-down biasing towards desired features and a representational framework for desired targets [10].

Firstly, variations in the features of different objects are identified by which the target in the given image can be searched secondly in a visual interpretation task, the probability of different targets is computed in different feature maps lastly, the objects with the highest probability will be interpreted first and the location of those objects are also identified.

The main advantage of this approach is in its simplicity and less computational time for recognition algorithms [10]. This model was based on two parameters i.e., description and the location of target/object [10]. The Bayesian model was evaluated through images containing different objects and diverse scenes. The use of the probability distribution function amended the accuracy of the model. The flow process of the model is shown in Fig. 4.

3 Cognition Based Model for Satellite Images

3.1 Damage Assessment from High-Resolution Satellite Image

This model uses QuickBird and IKONOS images and proposes a cognitive method for detecting the damage caused by an earthquake in the town of Xiang's in DuJiangyan [11]. High-resolution satellite images were being used for interpreting the damaged buildings [12]. There are many semi-automated algorithms for damage detection caused by natural hazards [13].

In this model, an experiment is conducted using morphological anomalies for detecting damages through convex hulls of a building and their original shape after the hazard [14]. This model uses human cognitive parameters such as thinking, reasoning, and perception. Many researchers are working to improve automated object extraction by focusing on human perception [15]. Humans can identify and classify images based on contextual information or by using object recognition algorithms. Some applications introduced cognitive along with semantic knowledge for understanding/interpretation of remote sensing images [4].

There are several methods/processes/techniques for object detection, and image understanding, based on fuzzy and cognitive studies, similarly this model also attempts to simulate the way human beings interpret satellite imagery. This model uses fuzzy theory to perform the fusion of low-level features and semantic features.

In the first step, low-level features like texture, shape color, and tone are extracted by using object recognition/detection and image processing-based procedures. There are two methods of getting low-level features first, the original image is segmented which consists of object attributes and features second, extraction of features from image pixels through filtering [11]. In the second step, semantic features are obtained using the top-down approach, and thirdly, integration of these features is performed using fuzzy logic approach via membership function for image understanding and object recognition in a way more like humans.

Fuzzy logic is a form of mathematical logic in which truth can assume a range of values between 0 and 1. In general, it quantifies the ambiguous statements, where strictly logical statements, either no or yes replaced by a range of [0…1]. Therefore, the fuzzy logic approach is capable of emulating the human mind and considers all linguistic rules. The fuzzy classification method is widely used for information extraction from images. Here, every object of each class is given a membership degree which is the result of fuzzy-based classification [11].

It is a hybrid model which combines the top-down and bottom-up approaches. This approach involves semantic and low-level features resulting in improved accuracy for various image understanding approaches. As this is a cognitive model therefore rules and knowledge can be reused. The flow process of the model is shown in Fig. 5.

3.2 *Polsar Image Interpretation Model*

A hierarchical cognitive system was developed for the interpretation of high-resolution-PolSAR images. In past, cognitive methods have also been proposed for detecting objects like buildings, roads from satellite images [16–22]. Here, interpretation of the PolSAR image is carried out using three different layers [23]. Firstly, the human's visual cognition system is further divided into two parts i.e., low-level and senior-level visual cognition. The low level is used for the image segmentation using different scales and the senior level is used for feature binding and feature

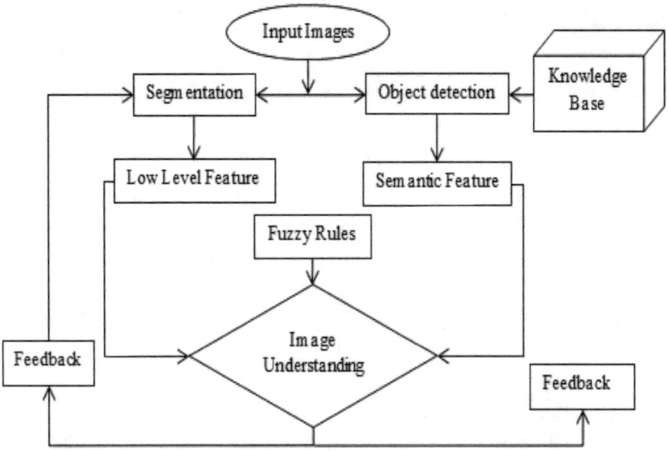

Fig. 5 Model for damage assessment [11]

extraction. During visual cognition, the visual features from the image are extracted and combined using the prior knowledge to obtain the preliminary results. Secondly, a human's logical cognition system during this different classifier such as neural network and fuzzy logic is used to detect the target from an image. Fuzzy C-means clustering (FCM) algorithm and pulse coupled neural network (PCNN) were used for determining the target pixel of an image [23]. Thirdly, during the human psychology cognition system, the unknown targets are identified based on the context features.

Psychological cognition removes the uncertain parameters and increases the overall accuracy of object/target detection based on the background. The quantitative evaluation of the method was done using detected probability, undetected probability, and false alarm [23]. The framework of the model is shown in Fig. 6.

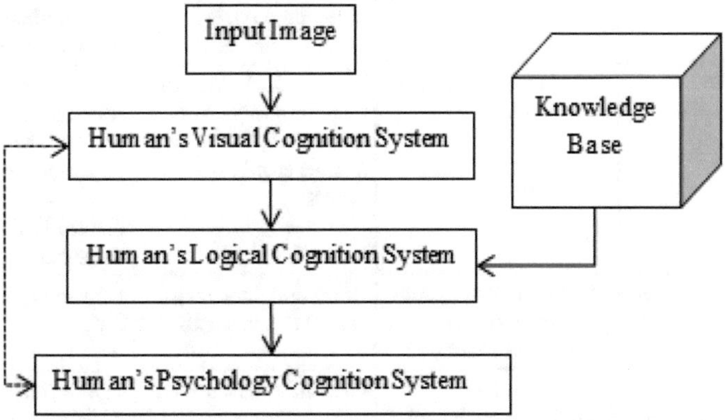

Fig. 6 The overall framework of the system [16]

4 Comparison of Models

Cognitive models discussed above have some capabilities and limitations. In the attention-based model objects with lower attention values are difficult to interpret therefore a method can be proposed so that the objects which are neglected by the human sensors can also be interpreted easily. UBIAS system is dedicated to medical imaging but it needs to be tested for other images. The performance of the Bayesian model needs to be improved for illumination conditions. The routing mechanism of information to central representation is still an overhead in the Koch and Ullman model. A comparative summary of the models has been given in Table 1.

Table 1 Summary of models

Model	Interpretation mechanism	Capabilities	Limitations
Attention-based model	Interpretation is carried out using a popping out algorithm	This model interprets the objects in the image along with the background. It uses the matrix representation to reduce the complexity	Objects with lower attention values are difficult to interpret and it is computationally very costly
UBIAS	Interpretation is performed using cognitive resonance	It uses advanced graph techniques for the diagnosis of medical images. By using cognitive resonance semantic information from the image is easily interpreted	It is only dedicated to semantic analysis of selected medical visualization. Without using the reasoning process the entire interpretation process is difficult to conduct
Koch and Ullman	Interpretation is carried out using WTA	This model served as a base for many visual attention models implemented later	The process of routing salient region to central representation needs to be explained
Bayesian model	Based on the Bayesian framework	This model is used for dual purposes i.e., for identifying the location and description of an object	The model needs to be tested using natural scenes. The performance of the model was not well under illumination color circumstances
Cognitive model for damage assessment	Extraction of semantic features from the recognized classes	Rules and knowledge can be reused easily	High computational time
Polsar image interpretation based on human cognition mechanism	Using a human cognitive mechanism	Fast and robust system	The overall accuracy of the system is needed to be improved

5 Conclusion

Image interpretation models have been using three methodological approaches: Decision theory-based interpretation, Case-based interpretation, and Appearance-based interpretation. One of the main challenges of cognitive image interpretation is to develop a flexible system that is capable of performing complex image analysis tasks and interpreting information from various scenes and images.

An image interpretation model inherits the advantage from human attention systems and applies to every image and scene, e.g., natural images, outdoor images, office environment, or medical images. A training image can give good results. There are certain limitations such as salient objects are easier to interpret rather than inconspicuous ones. There is a complexity in the mechanism of the human brain.

Cognitive image interpretation is a highly interdisciplinary field and the disciplines investigate the area from different perspectives. Psychologists can investigate human behavior on image interpretation strategies to understand the internal processes in the brain which can result in psychophysical theories or models. Advancement in the field of cognitive image interpretation field could greatly help in solving the challenges in vision and image interpretation problems. However, several factors are needed to be explored. These factors can bridge the gap between human interpreters and computational models. There are many probabilistic models which are used for image interpretation mechanisms and by modeling the cognitive processes in the human brain these models can be used for cognitive image interpretation.

References

1. Nadel L (2003) Encyclopedia of cognitive science, vols 1–4. Nature Publishing Group, New York
2. Matlin M (1983) Cognition. CBS College Publishing, Japan
3. Fu H, Chi Z, Feng D (2006) Attention-driven image interpretation with application to image retrieval. Pattern Recogn 39(9):1604–1621
4. Theeuwes J (1993) Visual selective attention: a theoretical analysis. Acta Psychol 93–154
5. Deng Y, Manjunath B (2001) Unsupervised segmentation of color-texture regions in images and video. IEEE Trans Pattern Anal Mach Intell 23(8):800–810
6. Ogiela L (2009) UBIAS systems for cognitive interpretation and analysis of medical images. Opto-Electron Rev 17(2):166–179
7. Ogiela MR, Ogiela L, Tadeusiewicz R (2010) Cognitive reasoning UBIAS & E-UBIAS systems in medical informatics. In: 2010 6th international conference on networked computing (INC). IEEE, pp 1–5
8. Frintrop S, Rome E, Christensen HI (2010) Computational visual attention systems and their cognitive foundations: a survey. ACM Trans Appl Percept (TAP) 7(1)
9. Frintrop S (2006) VOCUS: a visual attention system for object detection and goal-directed search, vol 3899. Springer
10. Elazary L, Itti L (2010) A Bayesian model for efficient visual search and recognition. Vision Res 50(14):1338–1352
11. Zhang C, Wang T, Liu X, Zhang S (2010) Cognitive model based method for earthquake damage assessment from high-resolution satellite images: a study following the WenChuan

earthquake. In: 2010 sixth international conference on natural computation (ICNC), vol 4. IEEE, pp 2079–2083
12. Fumio Yamazaki YY, Matsuoka M (2005) Visual damage interpretation of buildings in Bam City using QuickBird images following the 2003 Bam, Iran, Earthquake. Earthq Spectra 21:S329–S336
13. Yamazaki F, Matsuoka M (2007) Remote sensing technologies in post-disaster damage assessment. J Earthq Tsunami 1:193–210
14. Andre G, Chiroiu L, Mering C, Chopin F (2003) Building destruction and damage assessment after earthquake using high resolution optical sensors. The case of the Gujarat earthquake of January 26, 2001. In: Proceedings of the IEEE international geoscience and remote sensing symposium, vols I–VII, pp 2398–2400
15. Mallinis DKG, Karteris M, Gitas I (2007) An object-based approach for the implementation of forest legislation in Greece using very high resolution satellite data. In: Lang S, Blaschke T, Hay GJ (eds) Object-based image analysis—spatial concepts for knowledge-driven remote sensing applications. Springer, Berlin
16. Chandra N, Vaidya H, Ghosh JK (2020) Human cognition based framework for detecting roads from remote sensing images. Geocarto Int 1–20
17. Chandra N, Ghosh JK (2018) A cognitive viewpoint on building detection from remotely sensed multispectral images. IETE J Res 64(2):165–175
18. Chandra N, Ghosh JK (2017) A cognitive method for building detection from high-resolution satellite images. Curr Sci 1038–1044
19. Chandra N, Ghosh JK, Sharma A (2019) A cognitive framework for road detection from high-resolution satellite images. Geocarto Int 34(8):909–924
20. Chandra N, Ghosh JK (2016) A cognitive perspective on road network extraction from high resolution satellite images. In: 2016 2nd international conference on next generation computing technologies (NGCT), October. IEEE, pp 772–776
21. Chandra N, Ghosh JK, Sharma A (2016) A cognitive based approach for building detection from high resolution satellite images. In: 2016 international conference on advances in computing, communication, & automation (ICACCA) (Spring), April. IEEE, pp 1–5
22. Chandra N, Sharma A, Ghosh JK (2016) A cognitive method for object detection from aerial image. In: 2016 international conference on computing, communication and automation (ICCCA), April. IEEE, pp 327–330
23. Kang L, Zhang Y, Zou B, Wang C (2015) High-resolution PolSAR image interpretation based on human images cognition mechanism. In: 2015 IEEE international conference on geoscience and remote sensing symposium (IGARSS), July. IEEE, pp 1849–1852

Comparison of Attention Mechanisms in Machine Learning Models for Vehicle Routing Problems

V. S. Krishna Munjuluri Vamsi, Yashwanth Reddy Telukuntla, Parimi Sanath Kumar, and Georg Gutjahr

1 Introduction

The Vehicle Routing Problem (VRP) is a classic NP-Hard problem and one of the most studied combinatorial optimization problems. In a VRP, the objective is to select a set of minimum-cost vehicle routes through customer locations so that each route starts and ends at a common location (depot). There are different variants for VRP; in this manuscript, we consider the Capacitated Vehicle Routing Problem (CVRP), in which each of the vehicles is restricted to a certain load capacity.

VRPs are very general in the way they are defined and can aptly represent a wide variety of combinatorial optimization problems. The VRP is a very abstract problem and many real-world problems can be modeled using the idea of tour planning. It may include vaccination drives (as in [8]), catastrophe relief distribution (as in [9]), and farming and cultivation community commutes (as in [2]). Multiple methods have been proposed that take various approaches in solving VRPs [3, 14, 15, 18].

To solve VRPs, a large number of exact and heuristic algorithms have been proposed in the literature. Some of the classic exact algorithms are Dijkstra's method, Branch-and-Bound, and Dynamic Programming. Latest advancements in the indus-

V. S. Krishna Munjuluri Vamsi (✉) · P. S. Kumar
Department of Computer Science and Engineering, Amritapuri, India
e-mail: munjulurivkrishna@am.students.amrita.edu

P. S. Kumar
e-mail: psanathkumar@am.students.amrita.edu

Y. R. Telukuntla
Department of Electronics and Communication, Amritapuri, India
e-mail: yashwanthreddy@am.students.amrita.edu

G. Gutjahr
Center for Research in Analytics & Technologies for Education, Amrita Vishwa Vidyapeetham, Amritapuri, India
e-mail: georgcg@am.amrita.edu

P. Singh et al. (eds.), *Machine Learning and Computational Intelligence Techniques for Data Engineering*, Lecture Notes in Electrical Engineering 998,
https://doi.org/10.1007/978-981-99-0047-3_53

try saw big data-aided heuristics being used to solve the VRPs [23] and a few have also proposed a better constraint relaxation strategy to improve the robustness of heuristics and meta-heuristics to solve the CVRP [12], and we guide the readers to Sect. 2 for the traditional CVRP constraints. Although the exact algorithms provide a guarantee that the optimal solution is obtained, such algorithms are computationally very demanding and will usually not scale to problem instances of realistic size. Therefore, heuristic algorithms are often the only available option to achieve acceptable running times in practical applications.

With the recent improvements in deep learning methods such as end-to-end neural networks, it has become possible to use machine learning techniques to solve the VRP problem. Such algorithms find a solution as a permutation of nodes at any instance by making a sequence of decisions. In this way, solving the VRP can be represented by a Markov Decision Process (MDP). MDPs, in general, require To solve such an MDP, we need a framework where the natural idea is to make one decision after the other. Reinforcement Learning (RL) [16] is one such framework where an "agent" is constantly interacting with some "environment" and gains "reward" based on its "actions". In our context, the agent will be the actual vehicle that goes and serves the customers, the environment can be the customers and their demands, and we define our reward will be the inverse of the distance traveled by taking that action. And by actions, we mean the steps taken by the vehicle to serve its customers.

Now, for the RL agent to perform better, it needs to keep track of its current environment and make informed decisions about its customers. This is where the Attention Mechanisms come in. The vehicle attends to different customer nodes differently based on how the attention layers modify the node embeddings. After a route is complete, Attention Mechanism re-computes the node embeddings because the graph structure changes considerably. So, for choosing customers to be served next, the vehicle has the best possible representation of the current distribution of customer nodes. Each attention layer has multiple heads and we can have multiple such layers (discussed in detail in Sect. 4) which naturally brings the need for optimization of these hyperparameters of the model.

The remaining of the paper is organized as follows. Section 2 gives a problem definition of the VRP. Section 3 gives some background on VRPs and how a VRP can be solved by methods from machine learning. Section 4 describes AMs in more detail and the need for comparison. In the Results section, Sect. 5, we talk about the comparison that has been performed and our findings from it. In conclusion, we summarized our findings.

2 Problem Definition

VRPs can be modeled in different ways. One of the common models is the vehicle flow formulation where an integer value is used to count the number of times a particular edge is traversed by the vehicle. Two indices are used to represent the edges and therefore the formulation is known as a two-index formulation. Additionally, three-index formulations also exist where a third index is used to identify which vehicle is used by a particular edge. VRPs can also be modeled as set partitioning problems.

A two-index formulation extended from the Dantzig et al. [7] is given. Let V be the vertex set or the set of customer locations. Let the cost of going from node i to node j be represented by c_{ij}. Here, x_{ij} is a binary variable that has value 1 if the arc going from i to j is considered as part of the solution and 0 otherwise. Let the number of available vehicles be denoted by K and $r(S)$ corresponds to the minimum number of vehicles needed to serve a set $S \subseteq V$ of customers. Also assume that $0 \in V$ is the depot node. VRP can be formulated as

$$\text{Minimize} \sum_{i \in V} \sum_{j \in V} c_{ij} x_{ij} \tag{1}$$

subject to

$$\sum_{i \in V} x_{ij} = 1 \quad \forall j \in V \setminus \{0\} \tag{2}$$

$$\sum_{j \in V} x_{ij} = 1 \quad \forall i \in V \setminus \{0\} \tag{3}$$

$$\sum_{i \in V} x_{i0} = K \tag{4}$$

$$\sum_{j \in V} x_{0j} = K \tag{5}$$

$$\sum_{i \notin S} \sum_{j \in S} x_{ij} \geq r(S), \quad \forall S \subseteq V \setminus \{0\}, S \neq \emptyset \tag{6}$$

Constraints (2) and (3) ensure that exactly one arc enters and exactly one leaves each vertex associated with a customer. Constraints (4) and (5) ensure that the number of vehicles leaving the depot is equal to the number entering. Constraint (6) ensures that the routes should be connected and that the demand on each route should not exceed the capacity of the vehicle. The capacity constraint (6) can be converted into a subtour elimination constraint (7) to get an alternative formulation. This new constraint ensures that at least $r(s)$ edges leave each customer set S.

$$\sum_{i \in S} \sum_{j \in S} x_{ij} \leq |S| - r(s) \tag{7}$$

As discussed, VRPs can be solved by many sophisticated methods (e.g., solving VRPs by using the learning mechanism provided by neural networks), but solving VRPs was all started with a classic algorithm [11]. The algorithm proposed by them iteratively matches vertices to form a set of vehicle routes, where each iteration is called a "stage of aggregation" (This algorithm was only able to solve small problem instances). Over the years many heuristics have been proposed to solve large problem instances (for example, the edge exchanges method by Kindervater and Savelsbergh [11]). Meta-heuristics are also methods for solving VRPs, which eventually overtook problem-specific solution methods.

Another approach that will be considered in this paper is to represent the VRP as a sequence-to-sequence problem in a pointer network, where the input is a sequence of nodes and the output is a permutation of this sequence. Methods from machine learning are used to learn how to permute sequences in an optimal way. This approach was originally developed by Google for use in machine translation and improved by Google in 2020, which is very much used in Google's recent chatbot Meena [1]. We will now discuss this approach in more detail.

3 Sequence-to-Sequence Model for Solving VRPs

Pointer networks [21] free us with the limitation of output length generally depending on the input given. The labels assigned by the approximate solver to the sequence-to-sequence models [19] are trained in a supervised manner. But the performance of the pointer network is only as good as the quality of the training dataset, which in turn might give us expensive solutions. To train a policy modeled by PN without supervised signals, we can use reinforcement learning (RL) algorithms. Element-wise projections will be replacing Long short-term memory (LSTM) [13] of the pointer network, which are invariant to the input order and will not introduce redundant sequential information.

Recurrent Neural Networks (RNN) is a type of Neural Network, where the output of the previous layer is taken as input to the current step/layer. In an encoder-decoder architecture, both encoder and decoder use multiple RNN layers [22] each. At first, Encoder takes the entire input sequence and churns out a fixed-length internal representation/vector; this helps the model to understand more about the context and internal dependencies associated with it. This fixed-length internal representation/vector is then passed to the decoder's first layer. The decoder layer's role is to convert the fixed-length internal representation/vector to an output sequence. The Encoder-Decoder architecture (where we encode the problem to a fixed dimension vector in some way and then decode the vector while solving the problem) generally fails to remember (short-term memory) if the given input sequence is long enough. This implies that carrying information from the previous step to the next step is challenging, and during back-propagation, RNN suffers from the vanishing gradient problem (where Gradients are used to update a neural network's weights in the learning process). The vanishing gradient problem occurs when the gradient shrinks/becomes extremely small (during back-propagation); it doesn't contribute to efficient learning. So because of this, the RNN's layers stop learning for longer sequences, thus short-term memory.

To overcome the above limitation, we can try implementing an LSTM network, where every LSTM node has a cell state/vector, which gets passed down like a chain. The state of the cell can be modified (both add and remove). From both RNNs Sequence-to-Sequence model and Pointer Networks and LSTM's (encoder-decoder/vanilla architecture), we expect them to have the capability to save/access the information seen thus far, which in turn means that the final encoder should be having the whole input data/sequence and which might cause data loss because of having the complete input sequence as a compressed vector. Passing that single compressed vector to a decoder for deciphering is in itself a highly complex task and which again is a bottleneck. This is where Attention comes in. Every hidden state from each encoder node in attention [16] at every time step makes predictions after deciding which one is more informative and passes it on to the decoder.

4 Attention Mechanisms

As discussed in Sect. 1, we need a more robust way of selecting the next customers. The attention mechanism can be thought of as follows. The vehicle, after servicing a customer, needs to decide where to go next for the overall optimality of the entire fleet. At the serviced customer node, the vehicle creates a query asking its neighbors. The neighbors reply to the vehicle by passing their key and value vectors (The computation of these vectors is explained in the later paragraphs). Now, the vehicle compares the product of the query and key to the value vector, the idea of query, key and values is derived from [20]. Based on the resulting distribution, the vehicle decides which customer to serve next. This is essentially the "neural message passing mechanism [10]" which is a good way of thinking about the attention mechanisms.

In the architecture proposed by [17], the attention mechanism is included in the ENCODE (refer Fig. 1) step of the encoder-decoder architecture and it helps in computing the node embeddings effectively (The Encoder has been explained at a

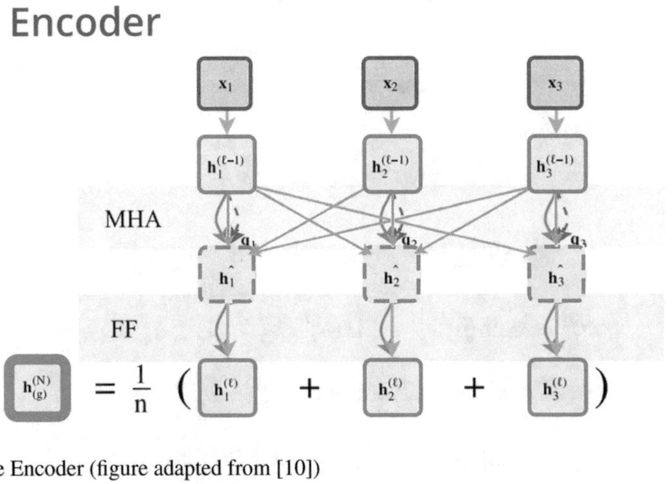

Fig. 1 The Encoder (figure adapted from [10])

greater depth in Sect. 3). These final node embeddings are used, later in the DECODE stage, to compute the context vector. Now, the keys are generated according to the embeddings (just as the case with the encoding), but the queries are based on the compatibilities (computed from the context vector). In this work, we restrict ourselves to discussing the specifics of the Attention Mechanism in the ENCODE step.

The basic building block of the Multi-Head Attention (MHA) is the scaled dot-product attention. Each head of the MHA learns a different type of information from the graph embeddings and the node embeddings (updated by the previous layer of MHA). The scaled-dot-product compares how similar the query (Q) is relative to the key (K) of the neighboring node, and the output vector is compared against the values (V) vector of the same neighbor. And to achieve this, we generate the Q, K, V for the current node and, from that, calculate the compatibility vector. This compatibility vector, aptly named, tells us how compatible the neighbor is with respect to the query that is generated. From the compatibility vector, we take a softmax of the compatibility vector (scaling) and multiply it with V (dot-product). Essentially, this is the scaled-dot-product attention. Now, for MHA, we do this as many times as the number of heads in the model and concatenate all the outputs to finally project it back onto the embedding space.

Each attention layer consists of an MHA and a fully connected Feed Forward (FF) layer. And we can have many layers of attention but it can quickly get really expensive in terms of computing power. Notionally, the more the number of heads and the number of layers, the better the model should perform because it has a better chance to take into account the surrounding nodes and design routes based on REINFORCE policy (a reference to RL algorithm).

In the equations that follow, we just give the formulation for the ideas that are described in the above paragraphs.

$$q_{im}^{(l)} = W_m^Q h_i^{(l-1)}, k_{im}^{(l)} = W_m^K h_i^{(l-1)}, v_{im}^{(l)} = W_m^V h_i^{(l-1)}, \tag{8}$$

$$u_{ijm}^{(l)} = (q_{im}^{(l)})^T k_{jm}^l, \tag{9}$$

$$a_{ijm}^{(l)} = \frac{e^{u_{ijm}^{(l)}}}{\Sigma_{j'=0}^h e^{u_{ij'm}^{(l)}}}, \tag{10}$$

$$h_{im}^{'(l)} = \sum_{j=0}^{n} a_{ijm}^{(l)} v_{jm}^{(l)}, \tag{11}$$

$$\text{MHA}_i^{(l)}(h_0^{(l-1)}, \ldots, h_n^{(l-1)}) = \sum_{m=1}^{M} W_m^O h_{im}^{'(l)}, \tag{12}$$

$$\hat{h}_i^{(l)} = \tanh(h_i^{(l-1)} + \text{MHA}_i^{(l)}(h_0^{(l-1)}, \ldots, h_n^{(l-1)})), \tag{13}$$

$$FF(\hat{h}_i^{(l)}) = W_1^F \text{ReLu}(W_0^F \hat{h}_i^{(l)} + b_0^F) + b_1^F, \tag{14}$$

$$\hat{h}_i^{(l)} = \tanh(h_i^{(l-1)} + FF(\hat{h}_i^{(l)})). \tag{15}$$

In Eq. (8), we compute query, keys, and values at node i based on the node embeddings provided by the previous layer $(l-1)$. Here, W^Q, W^K, and W^V are the parameters that learn how to compute queries, keys, and values, respectively. And, these parameters stay independent (different for different layers of attention).

In Eq. (9), $u_{ijm}^{(l)}$ is the compatibility vector that is obtained by looking at how similar the query generated by node i at the mth head in the lth layer (which is $q_{im}^{(l)}$) is to the values with respect to the key of the neighboring node j at the mth head in the lth layer (which is k_{jm}^{l}). This computes compatibility to all the surrounding nodes j in some neighborhood of i and can hence be expensive.

In Eq. (10), the attentions, for the nodes j in some neighborhood of node i at the mth head in lth layer, $a_{ijm}^{(l)}$ are obtained by taking the softmax of the compatibilities.

In Eq. (11), $h_{im}^{'(l)}$ is obtained by taking the product of the attentions with the values vector $v_{jm}^{(l)}$ of the node j at the mth head in the lth layer.

In Eq. (12), the final MHA output is obtained by taking the product with the weights matrix W_m^O with the output of the previous equation. This is the step where all the learnings from the heads are merged and projected back onto the embedding space that the customer nodes are in.

As we have seen in Eqs. (8) through (12), we can have as many layers and as many heads as we like. This is the Multi-Head Attention sublayer. In MHA, one way of thinking about the heads is that each of the heads gets better at identifying a certain aspect of the neighborhood. And we concatenate all of this and project the vector back onto the embedding space. And each of the layers refines the learnings from the previous layer because the next layer takes its output and computes all the compatibilities, attentions, and products using a different set of weights for each layer. From this, we can already see that with these two variables m and l, there are already a lot of possible architectures for our Attention Mechanism. And apparently, the more their number is, the better the Attention Mechanism is at attending to its neighboring nodes.

However, when applying this approach with real data, the number of heads and layers is limited. This is because the number of parameters in the layers increases exponentially, and after a certain point, training the network becomes infeasible. Also, the more parameters we have, the less time can be spend on finding optimal values for them. So, there is a need to compare and look for optimal values of m and l.

While we mainly worked on MHA, the Attention Mechanism is incomplete without the FF sublayer which is described in Eqs. (13) through (15). This constitutes the ENCODE part which tries to get the best node embeddings by stacking multiple attention layers.

As it was already established, there is a trade-off when selecting the number of layers and heads: the more layers and heads we use, the more complex our model becomes and the harder it gets to train the model. This is also related to the so-called bias-variance trade-off in machine learning, where increasing the flexibility of a model will decrease the bias but will increase the variance. And since the performance of the model can usually be represented as a combination of bias and variance, there exists an optimal level of complexity for the model where neither the bias nor the variance is too large.

In other words, when designing attention mechanisms for the VRP, multiple topologies are possible and it is not a prior obvious which of them is best. The different topologies can be described in terms of the numbers m and l and we would like to explore which values for m and l lead to good performance.

5 Simulation Results

Since we are trying to explore the performance of the models if we about the same time to train them, we fix the training epochs to be a constant for each of the models. For this comparison, the number of epochs is fixed to be 400 in batches of 128 for all of the models irrespective of the number of customer nodes. We trained the models to solve CVRPs of small (approximately 20 and 30 customer nodes) and medium (50 customer nodes) sizes. For each problem size, we trained the models with 1 head, 4 heads, and 8 heads. The testing data is the benchmark datasets from Christofides [5, 6] and Augerat [4]. Even though the models can handle dynamic inputs (i.e., with changing demands and the customer locations), the datasets are static and don't change with time. In other words, all the customer locations and their demands are known beforehand.

All the simulations were done on Google's Colab Notebooks with one Tesla K80 GPU. The implementation was done in Python. Tensorflow and Keras were the primarily used frameworks.

The results are shown in Table 1. The comparison is done with constant epochs for models with 1 head, 4 heads, and 8 heads. The cost of the solution generated by the models is recorded in the table and the optimal cost is shown next to it (the optimal cost is known for the selected benchmark datasets [4–6]).

6 Discussion and Conclusion

The OR community is showing much interest in solving VRPs. This is done to utilize the workforce maximally while reducing the costs of travel or transportation in a variety of applications. Usually, it is not feasible to apply exact methods that take too much time which should be started from scratch even if the problem instance changes slightly. Machine learning methods have much potential to approximate the costs of the routes efficiently and to get good generalizations if the distribution of the customers at testing is similar to that of the customers at training. In particular, research has shown that RNNs and LSTMs on their own are too naive for the job and that introducing RL can lead to better solutions.

To improve those RL models, advanced attention mechanisms have been proposed. We have motivated how there are different topologies possible for the AMs

Table 1 Comparison results

DataSet	1 Head	4 Heads	8 Heads	Optimal cost	Nodes
Augerat Set P (P-n21-k8)	369	400	267	211	20
Chirstofides Set E (E-n22-k4)	722	589	583	375	21
Augerat Set P (P-n22-k2)	359	397	270	216	21
Augerat Set P (P-n22-k8)	862	747	666	603	21
Augerat Set B (B-n31-k5)	751	971	814	672	30
Augerat Set A (A-n33-k6)	1217	1157	1079	742	32
Chirstofides Set E (E-n33-k4)	1237	1201	1285	835	32
Augerat Set P (P-n51-k10)	1009	1220	1068	741	50
Christofides Set E (E-n51-k5)	771	863	813	521	50
Augerat Set B (B-n51-k7)	1466	1242	1294	1032	50
Chirstofides et al. (CMT 1)	771	863	813	524	50
Augerat Set B (B-n52-k7)	1419	1620	1917	747	51
Augerat Set P (P-n55-k7)	887	1037	826	568	54

and why there is a need for a thorough comparison. In our work, we have tested the model by varying different hyperparameters like the number of heads and the number of attention layers that make up the entire AM. We observe that small CVRP instances are already benefiting from more heads. The larger instances do not show a clear trend, possibly due to the limited training we could provide. The smaller CVRP instances section is a clear demonstration of the potential benefits the Attention Mechanisms can add to the Deep Reinforcement Learning-based solutions to VRPs, in general.

In the future, we would like to perform a more exhaustive testing that takes into account a lot more values for the number of heads and for the number of layers. Depending upon the availability, we would love to test the model for CVRP instances with 100 customer nodes and above, which is where the more advanced architectures is expected to significantly outperform more simple models.

References

1. Adiwardana D, Luong MT, So DR, Hall J, Fiedel N, Thoppilan R, Yang Z, Kulshreshtha A, Nemade G, Lu Y et al (2020) Towards a human-like open-domain chatbot. arXiv:2001.09977
2. Ajayan S, Dileep A, Mohan A, Gutjahr G, Sreeni K, Nedungadi P (2019) Vehicle routing and facility-location for sustainable lemongrass cultivation. In: 2019 9th international symposium on embedded computing and system design (ISED). IEEE, pp 1–6
3. Anbuudayasankar S, Ganesh K, Lee TR (2011) Meta-heuristic approach to solve mixed vehicle routing problem with backhauls in enterprise information system of service industry. In: Enterprise information systems: concepts, methodologies, tools and applications. IGI Global, pp 1537–1552
4. Augerat P (1995) Approche polyèdrale du problème de tournées de véhicules. PhD thesis, Institut National Polytechnique de Grenoble-INPG
5. Christofides N, Eilon S (1969) An algorithm for the vehicle-dispatching problem. J Oper Res Soc 20(3):309–318
6. Christofides N, Mingozzi A, Toth P (1981) Exact algorithms for the vehicle routing problem, based on spanning tree and shortest path relaxations. Math Program 20(1):255–282
7. Dantzig G, Fulkerson R, Johnson S (1954) Solution of a large-scale traveling-salesman problem. J Oper Res Soc Am 2(4):393–410
8. Gutjahr G, Krishna LC, Nedungadi P (2018) Optimal tour planning for measles and rubella vaccination in Kochi, South India. In: 2018 international conference on advances in computing, communications and informatics (ICACCI). IEEE, pp 1366–1370
9. Gutjahr G, Viswanath H (2020) A genetic algorithm for post-flood relief distribution in Kerala, South India. In: ICT systems and sustainability. Springer, Berlin, pp 125–132
10. Kool W, Van Hoof H, Welling M (2018) Attention, learn to solve routing problems! arXiv:1803.08475
11. Laporte G (2009) Fifty years of vehicle routing. Transp Sci 43(4):408–416
12. Letchford AN, Salazar-González JJ (2019) The capacitated vehicle routing problem: stronger bounds in pseudo-polynomial time. Eur J Oper Res 272(1):24–31
13. Libovický J, Helcl J (2017) Attention strategies for multi-source sequence-to-sequence learning. arXiv:1704.06567
14. Malairajan R, Ganesh K, Punnniyamoorthy M, Anbuudayasankar S (2013) Decision support system for real time vehicle routing in Indian dairy industry: a case study. Int J Inf Syst Supply Chain Manag (IJISSCM) 6(4):77–101
15. Mohan A, Dileep A, Ajayan S, Gutjahr G, Nedungadi P (2019) Comparison of metaheuristics for a vehicle routing problem in a farming community. In: Symposium on machine learning and metaheuristics algorithms, and applications, Springer, Berlin, pp 49–63
16. Nazari M, Oroojlooy A, Snyder LV, Takáč M (2018) Reinforcement learning for solving the vehicle routing problem. arXiv:1802.04240
17. Peng B, Wang J, Zhang Z (2019) A deep reinforcement learning algorithm using dynamic attention model for vehicle routing problems. In: International symposium on intelligence computation and applications. Springer, Berlin, pp 636–650
18. Rao TS (2019) An ant colony and simulated annealing algorithm with excess load VRP in a FMCG company. In: IOP conference series: materials science and engineering, vol 577. IOP Publishing, p 012191
19. Sutskever I, Vinyals O, Le QV (2014) Sequence to sequence learning with neural networks. In: Advances in neural information processing systems, pp 3104–3112
20. Vaswani A, Shazeer N, Parmar N, Uszkoreit J, Jones L, Gomez AN, Kaiser Ł, Polosukhin I (2017) Attention is all you need. In: Advances in neural information processing systems, pp 5998–6008
21. Vinyals O, Fortunato M, Jaitly N (2015) Pointer networks. arXiv:1506.03134
22. Yu Y, Si X, Hu C, Zhang J (2019) A review of recurrent neural networks: Lstm cells and network architectures. Neural Comput 31(7):1235–1270
23. Zheng S (2019) Solving vehicle routing problem: A big data analytic approach. IEEE Access 7:169565–169570

Performance Analysis of ResNet in Facial Emotion Recognition

Swastik Kumar Sahu and Ram Narayan Yadav

1 Introduction

Facial expression recognition is one of the most important tasks in image processing. It can be used in public places for business development and for security purposes. Malls and restaurants can use it to get customer feedback on their emotion. It can also be used for security purposes in some special locations. Various models have already been implemented for the emotion recognition. But every model have their own pros and cons. Convolutional neural networks (CNN) perform quite well in facial emotion classification [1]. But while training the CNN using the backpropagation algorithm, the CNN model suffers from the vanishing gradient problem [2–5]. As the gradient becomes vanishingly small so the weights don't get updated anymore and the model stops further training. To avoid this problem ResNet was proposed where an additional skip connection layer is used, which solves the vanishing gradient problem [6, 7]. As there is no vanishing gradient problem in ResNet so a deep neural network can be constructed using ResNet which can classify the facial emotion with very high accuracy. Convolutional neural networks have some major problems such as long training time and low recognition rates in the complex background [2, 3, 8, 9]. Along with these problems the vanishing gradient problem also prevents the CNN to create a deep network structure, which decreases the accuracy of the model [6, 7]. On the other hand ResNet can be used to create a deep network with high accuracy [8]. This ResNet model detected facial emotions with an accuracy of 85.96% in 50 epochs. ResNet can be used to create a deep neural network with can further increase the accuracy without facing the vanishing gradient problem.

S. K. Sahu (✉) · R. N. Yadav
Maulana Azad National Institute of Technology, Bhopal, India
e-mail: swastikism@gmail.com

P. Singh et al. (eds.), *Machine Learning and Computational Intelligence Techniques for Data Engineering*, Lecture Notes in Electrical Engineering 998,
https://doi.org/10.1007/978-981-99-0047-3_54

2 Related Works

Generally the basic CNN model has an input layer, many hidden layers, and an output layer [1]. The hidden layer can be made up of one or many layers [10]. Mainly the CNN has a convolutional layer, Relu, pulling, and fully connected layers. In the convolutional layer a convolution operation is applied to the input image for feature extraction [1, 9]. After that pooling operation is performed by which the size of the image decreases which again decreases the number of calculations so that the speed of the model can be increased. The Relu is used before the fully connected layer. As the CNN works basically by extracting the features of the input image so there is no need for manual feature extraction. The CNN model learns the emotion recognition skill for a set of input images called the training images set and after that the performance of the model is tested using the test images set. But while training the model using gradient-based methods and backpropagation algorithm the CNN suffers from the vanishing gradient problem [6, 7]. To solve this problem ResNet was proposed. In this paper a ResNet model has been implemented using tensorflow and python for face emotion detection and performance analysis of the model.

3 Methodology

In this paper a ResNet model has been implemented for the facial emotion recognition and the performance of the model has been analyzed by plotting graphs and confusion matrix. The block diagram of the proposed model is given in Fig. 1 which basically consists of two residual blocks (RES blocks), where each RES block contains one convolution block and two identity blocks.

In the proposed ResNet model first zero padding is applied to the input data and then convolution operation is applied. The convolution layer is mainly used for facial feature extraction. After that batch normalization and ReLU operations are performed. Then max pooling is used to decrease the size of the image to reduce the number of calculations.

After the max pooling layer two RES blocks are there, where each RES block contains one convolution block and two identity blocks. In the convolution block the skip connection layer contains convolution, max pooling and batch normalization operation, where as in the identity block the skip connection is just an identity connection without any convolution or pooling layer.

After the RES blocks average pooling is used. Then the data is transferred to a flattened layer. After that 3 dense layers are used and finally the emotion class is detected using the Adam optimizer.

The data set is taken from kaggle [11] and it contains 5 emotion classes, those are anger, disgust, sad, happiness, and surprise. Each emotion class is assigned an integer from 0 to 4. The data set has been divided to make separate training and testing set, so that it can be confirmed that the model has learned and not just memorized.

Fig. 1 The proposed model

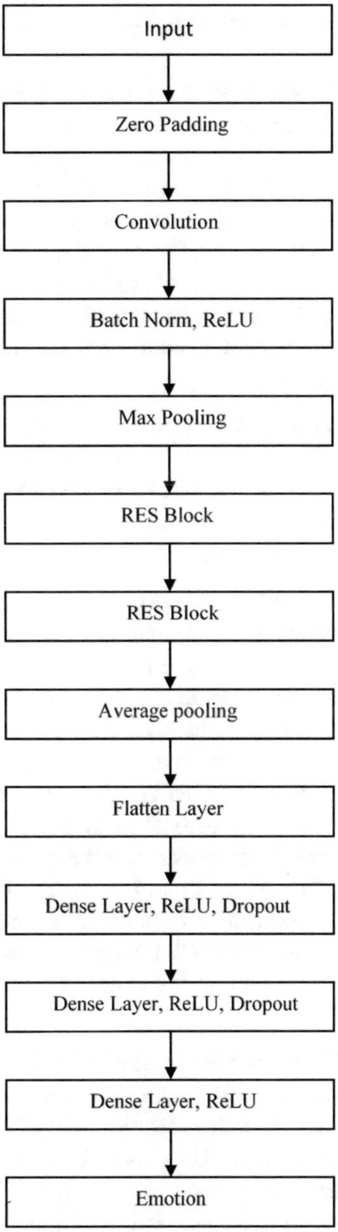

4 Experiment

In this paper the proposed ResNet model has been implemented using python and tensorflow. The data set has been taken from kaggle [11] where the facial emotions belong to five different emotion classes such as anger (0), disgust (1), sad (2), happiness (3), and surprise (4). The sample images of the dataset used in this experiment are 48 × 48 pixels in size. Each pixel has a value from zero to 255. The dataset contains a total of 24,568 sample images which are divided into training, testing, and validation sets where the training set contains 22,111 images, the testing set contains 1229 images, and the validation set contains 1228 images. After training the model for 50 epochs with batch size 128 the performance of the proposed model has been analyzed.

Zero padding. In this experiment each input image has 48 × 48 pixels where each pixel has a value from 0 to 255. Zero padding is used to make the size of the input sequence equal to a power of two [12]. In zero padding, zeros are added to the end of the input sequence so that the total number of samples becomes equal to the next higher power of two.

Convolution layer. The convolutional layer applies a filter to the input image to create a feature map that summarizes the presence of detected features in the input [1]. This layer performs the 2D convolution operation [1, 6, 7]. The input images are 48 × 48 pixels where each pixel has a value from 0 to 255. The convolution is a linear operation that involves the multiplication of a set of weights with the input. As this method is designed for two-dimensional input, so the multiplication is performed between an array of input data which is basically the input image and a two-dimensional array of weights, which is called a filter or a kernel. The filter should be smaller than the input data. A dot product is applied between a filter-sized patch of the input image and the kernel [1]. The dot product is the element-wise multiplication between the filter-sized patch of the input image and the kernel, which is then summed. Basically convolutional layer identifies the different features of an image so that it can detect the face in an image irrespective of the position of the face.

Batch Normalization. Normalization is the data pre-processing method used to get the numerical data to a common scale without distorting its shape. When data is input to a deep learning algorithm. It is desired to change the values to a balanced scale [1, 3]. The reason for using normalization is partly to ensure that the model can generalize appropriately. Batch normalization is used to make the neural networks faster and more stable through normalization of the layers' inputs by re-centering and re-scaling.

ReLU. ReLU stands for a rectified linear unit which is a piecewise linear function that will output the input directly if it is positive, otherwise, it will output zero [1, 3, 6, 8]. It is one of the most popular activation functions because it is easier to train. The ReLu function is defined as,

$$f(x) = \max(0, x) \tag{1}$$

Pooling. Pooling is used to reduce the size of the image which reduces the number of calculations. Max Pooling calculates the maximum value for patches of a feature map, and average pooling calculates the average value for patches of a feature map [1, 6]. Pooling is used after the convolution layer.

RES Block. In this proposed model 2 RES blocks have been used. The block diagram of the RES block is given in Fig. 2. Each RES lock consists of one convolution block and two identity blocks. In the identity block after performing the convolution and batch norm three times in the main path; the input is directly added using an identity connection in the short path, and after that ReLU function is applied [6]. But in the convolution block the input is not directly added in the short path, as the short path contains the convolution, max pooling, and the batch normalization operation, because to add two images the size of the image should be the same, and after max pooling the size of the image has been decreased. So the input can't be just added to the end of the main path. The block diagram of the identity block and convolution block are given in Figs. 3 and 4 respectively.

Softmax classifier. Softmax is a nonlinear activation function which also acts as a squashing function. The role of the squashing function is to convert the output to a range of 0 to 1. Therefore it can give the probability of a data point belonging to a particular class [5, 6, 12]. The softmax classifier gives a single output which basically represents a particular emotion, as every emotion has been assigned to a different number. The softmax classifier is defined as

$$softmax(z) = \frac{e^{z_i}}{\sum_{j=1}^{K} e^{z_j}} \tag{2}$$

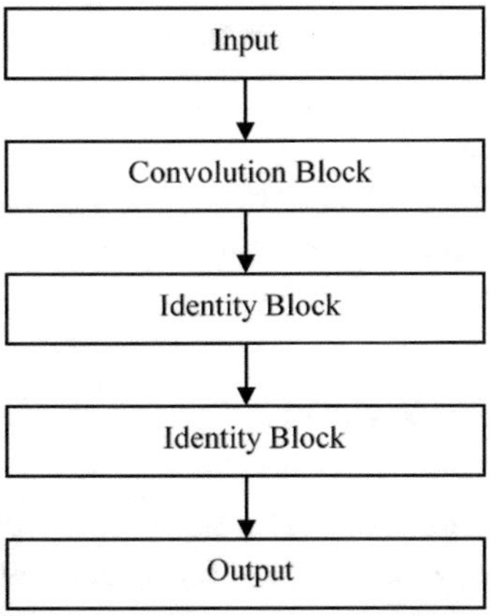

Fig. 2 The RES Block

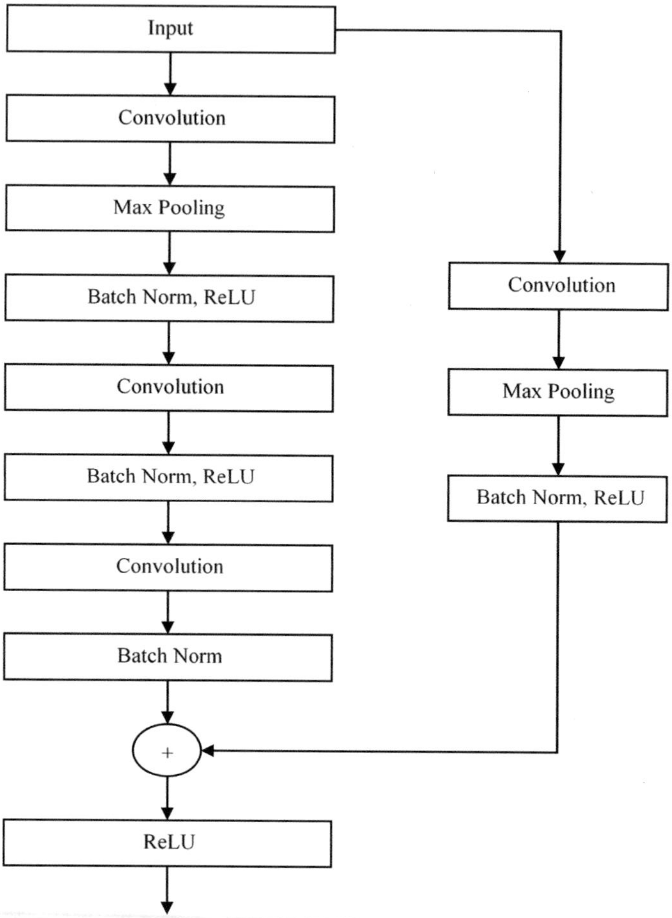

Fig. 3 The convolution Block

where z_i values are the elements of the input vector to the softmax function and K is the number of classes in the multi-class classifier. In this equation the denominator represents the sum across every value passed to a node in the layer.

5 Results

A graph has been plotted to compare the training and validation performance which is given in Fig. 5a, b for accuracy and loss respectively. It is observed that the validation graph follows the training graph. So it is clear that the proposed model has learned to predict the emotions and not just memorized the images. A confusion matrix has

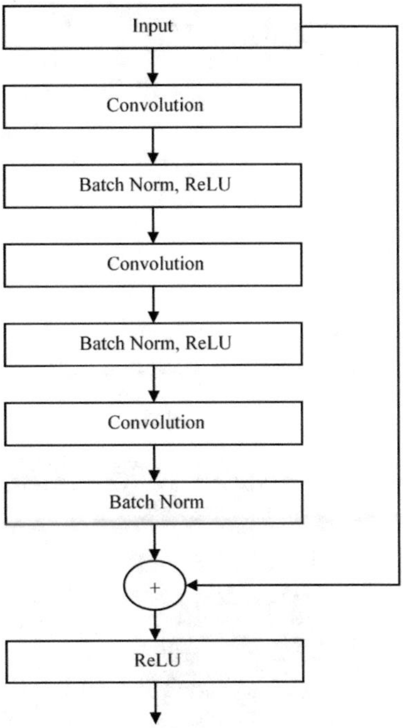

Fig. 4 Block diagram of the identity block

also been plotted in Fig. 5c, which represents the relation between predicted emotion and actual emotion.

A few images were plotted with their actual emotion value and predicted emotion values in Fig. 7. From the confusion matrix it can be seen that the detection of the emotion class 'disgust' was low as compared to other classes because that emotion has very less sample images in the data set which can be seen from the bar plot in Fig. 5, which represents the number of images per each emotion class. After training this model for 50 epochs its accuracy was found to be 85.96% (Fig. 6).

The accuracy of the ResNet model has been compared with 2 other methods (Table 1), and it was found that ResNet performed better than the other 2 methods.

The ResNet overcomes the vanishing gradient problem faced by the CNN with the help of RES blocks, where the RES blocks contain an additional skip connection layer. Therefore this ResNet model can be used to construct a deep neural network model with better performance. In this experiment pooling layer decreased the number of calculations by decreasing the size of the image and ReLU introduced nonlinearity. The convolution layer was used for feature extraction. And finally softmax classifier gave the probability of an image belonging to a certain category. Image augmentation techniques can be used for under-represented data. The performance of the ResNet model can also be increased by adding more RES blocks. ResNet can be used to construct a deep neural network for accurate facial emotion recognition.

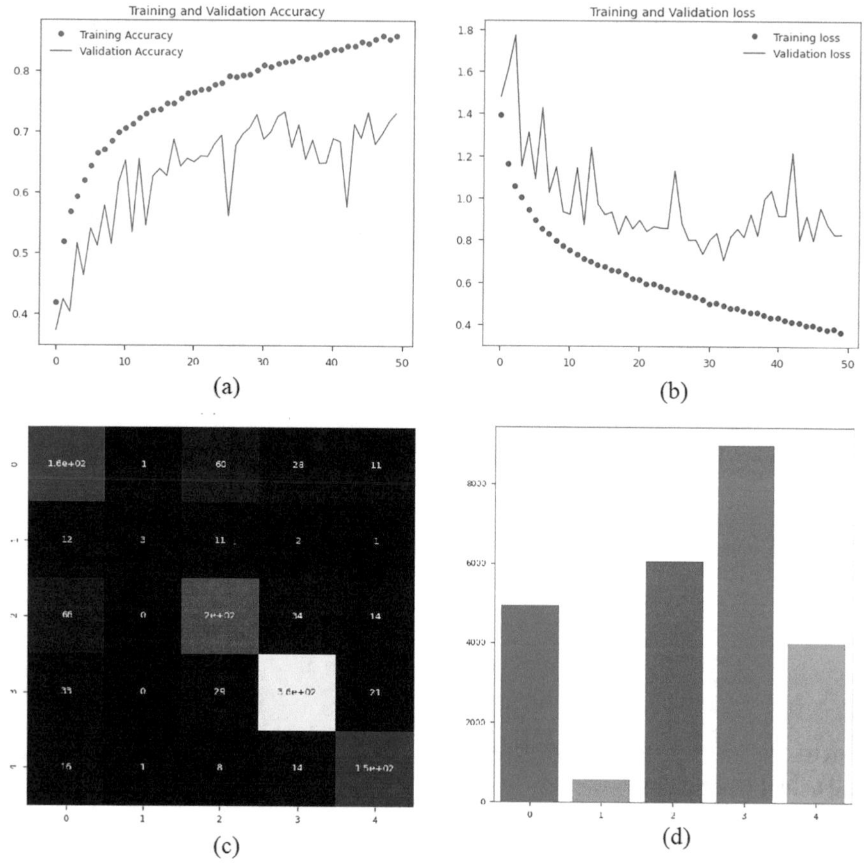

Fig. 5 **a** Graph of training and validation accuracy. **b** Graph of training and validation loss. **c** The confusion matrix. **d** Bar graph showing number of sample images per each emotion class

Fig. 6 Accuracy of the model after 50 epochs

Fig. 7 Comparison of the predicted and actual emotions class of a few sample face images after running the proposed ResNet model for 50 epochs conclusion

Table 1 Performance comparison of three algorisms

Method	Accuracy
The proposed ResNet model	85.96
R-CNN [1]	79.34
FRR-CNN [1]	70.63

References

1. Zhang H, Jolfaei A, Alazab M (2019) A face emotion recognition method using convolutional neural network and image edge computing. IEEE Access 7:159081–159089
2. Al-Abri S, Lin TX, Tao M, Zhang F (2020) A derivative-free optimization method with application to functions with exploding and vanishing gradients. IEEE Control Syst Lett 5(2):587–592
3. Zhu N, Yu Z, Kou C (2020) A new deep neural architecture search pipeline for face recognition. IEEE Access 8:91303–91310
4. Zhang Y, Wang Q, Xiao L, Cui Z (2019) An improved two-step face recognition algorithm based on sparse representation. IEEE Access 7:131830–131838
5. Liu C, Hirota K, Ma J, Jia Z, Dai Y (2021) Facial expression recognition using hybrid features of pixel and geometry. IEEE Access 9:18876–18889
6. Ou X, Yan P, Zhang Y, Tu B, Zhang G, Wu J, Li W (2019) Moving object detection method via ResNet-18 with encoder–decoder structure in complex scenes. IEEE Access 7:108152–108160
7. Wang L, Xiang W (2019) Residual neural network and wing loss for face alignment network. In: 2019 IEEE 14th International conference on intelligent systems and knowledge engineering (ISKE) (pp. 1093–1097). IEEE
8. Yang H, Han X (2020) Face recognition attendance system based on real-time video processing. IEEE Access 8:159143–159150
9. Rathgeb C, Dantcheva A, Busch C (2019) Impact and detection of facial beautification in face recognition: an overview. IEEE Access 7:152667–152678
10. Arya R, Kumar A, Bhushan M (2021) Affect recognition using brain signals: a survey. In: Singh V, Asari VK, Kumar S, Patel RB (eds) Computational methods and data engineering. Advances in intelligent systems and computing, vol 1257. Springer, Singapore. https://doi.org/10.1007/978-981-15-7907-3_40
11. www.kaggle.com/c/challenges-in-representation-learning-facial-expression-recognition-challenge/data
12. Kiliç Ş, Askerzade İ, Kaya Y (2020) Using ResNet transfer deep learning methods in person identification according to physical actions. IEEE Access 8:220364–220373

Combined Heat and Power Dispatch by a Boost Particle Swarm Optimization

Raghav Prasad Parouha

1 Introduction

Nowadays, optimization algorithms inspired by nature (called NI algorithms, i.e. Nature Inspired algorithms) are created for solving optimization problems with extreme nonlinearity and complexity [1]. They are famous due to their avoidance of local optima, capability of global search, derivation-free mechanisms and flexibility. Such methods use the ideas from insects and animals behaviors living in this nature entitled as swarm intelligence (SI) algorithms such as particle swarm optimization (PSO) [2], firefly algorithm (FA) [3], artificial bee colony (ABC) [4], cuckoo search (CS) [5], salp swarm optimizer (SSO) [6], etc. Despite the fact that a large number of SI algorithms have been introduced in the literature, they have not been able to solve a wide range of problems [7]. Particularly, for some problems an algorithm can produce satisfactory outcomes but not for others. As a result, for solving a variety of optimization problems there is a necessity to develop some efficient algorithms.

Recently, PSO and its variants have been effectively used to solve difficult optimization problems among several SI algorithms. It also resolves several practical problems of different areas [8–14], due to its positive points like implementation is easy, parameters to be controlled are few, and it has fast convergence. However, the PSO algorithm suffers with premature/slow convergence and easily stuck in local optima problems. To deal with these problems, so many PSO alternatives were introduced in literature [15].

Shi and Eberhart [16] improved the performance of traditional PSO through linearly time-varying inertia weight which may be efficient in solving many optimization problems. One recently made strategy for updating the particle's velocity

R. P. Parouha (✉)
Indira Gandhi National Tribal University, Amarkantak, M.P, India
e-mail: rparouha@igntu.ac.in

P. Singh et al. (eds.), *Machine Learning and Computational Intelligence Techniques for Data Engineering*, Lecture Notes in Electrical Engineering 998,
https://doi.org/10.1007/978-981-99-0047-3_55

of traditional PSO suggested in [17], which enhances diversity and premature convergence of swarm. Zhan et al. [18] offered new position/velocity updates (orthogonal culture strategy based) of PSO for well diversity maintenance. A novel velocity updating system based on perturbed global best was introduced by Xinchao [19] to avoid diversity loss. To avoid premature convergence, Wang et al. [20] proposed a changed velocity process for traditional PSO. Tsoulos [21], provided an update pattern (included local search, similarity check and stopping rule) of velocity.

Ratnaweera et al. [22] suggested a fresh PSO (HPSO-TVAC) which is generated by time-varying control factors to the velocity. A dynamically changed inertia weight for velocity update formula advised by Yang et al. [23]. For guiding globally direction to particle, Mendes et al. [24] recommended a novel PSO using different parameters. Different neighborhood topological planning is applied in PSO by Kennedy and Mendes [25]. To maintain diversity, Liang and Suganthan [26] used a multi-swarm approach. Beheshti et al. [27] developed FGLT-PSO based on global-local-neighborhood topology. Parouha [28] designed MTVPSO for solving nonconvex economic load dispatch problems. Parouha and Verma [29] developed an advanced hybrid algorithm (*h*aDEPSO) with the combination of suggested advanced DE (aDE) and PSO (aPSO). These three algorithms were successfully applied to different complex unconstrained optimization functions. Further, Verma and Parouha [30] successfully applied *h*aDEPSO, aDE and aPSO to solve constrained functions with engineering optimization problems. Also, Parouha and Verma [31] presented a systematic overview of developments in DE and PSO with their advanced suggestion for optimization problems.

Noticeably, there are few parameters in PSO and parameter tuning in it is quite simple, boosting the quality of the algorithm and therefore it is now a broadly used method. In PSO inertia weight is a significant parameter to enhance PSO performance. When inertia weight is large and small then it makes particles tend to encouraging global exploration and local exploitation correspondingly. Likewise, the cognitive and social parameters balance the global and local search for PSO effectively. The PSO is largely dependent on its parameters (which direct particles to the optimum) and position update (to balance diversity). As a result, numerous investigators have attempted to improve the accuracy and speed of PSO by modifying its control parameters and position update equation.

Motivated by the above facts of PSO (like parameter influences, advantages and disadvantages) and collected works, an improved PSO algorithm (IPSO) suggested in this paper. It contains gradually varying acceleration coefficients and inertia weight. The proposed factor makes IPSO more random nature which may increase local and global search balance and ability to provide a quality optimal solution.

The remaining parts of the paper are structured as follows: The second section presents a brief summary of the classical PSO, the third section discusses the introduced technology, the fourth section shows the outcomes and discussion, and at last the fifth section discusses the learnings of this paper and some research ideas which can be used in future.

2 Classical PSO

Eberhart and Kennedy [2] introduced PSO in 1995. There are two main processes of PSO, firstly updating velocity and secondly position update. In *j*-dimensional search space, it starts with random swarms (population). Particle '*i*' at iteration '*t*' has a velocity $v_i^t = (v_{i,1}^t, v_{i,2}^t, \ldots, v_{i,j}^t)$ and position $x_i^t = (x_{i,1}^t, x_{i,2}^t, \ldots, x_{i,j}^t)$ (which is a vector). Until the current iteration (t), the best solution achieved by i^{th} particle is defined as $p_{best_i}^t = (p_{best_{i,1}}^t, p_{best_{i,2}}^t, \ldots, p_{best_{i,j}}^t)$ and g_{best}^t (that is, global best) amongst best. Updated velocity and position obtained as follows.

$$v_{i,j}^{t+1} = wv_{i,j}^t + c_1r_1\left(p_{best_{i,j}}^t - x_{i,j}^t\right) + c_2r_2\left(g_{best_j}^t - x_{i,j}^t\right) \tag{1}$$

$$x_{i,j}^{t+1} = x_{i,j}^t + v_{i,j}^{t+1} \tag{2}$$

where $x_{i,j}^t$: *i*th particle present position, $v_{i,j}^t$: *t*th existing velocity; t and t_{max} ($1 \le t \le t_{max}$): present and maximum iteration of the algorithm, $i\,1 \le i \le np$(swarmsize): *i*th particle of the swarm, c_1 and c_2: cognitive and social acceleration coefficients, r_1 & r_2: : random uniformly distributed in (0, 1), w: inertia weight. Mostly, $c_1$and$c_2 \in [0, 2]$ and $w = 1$ [32].

3 Proposed Methodology

Because PSO depends on the population, so it is said as a population-based optimization algorithm. It has some advantages (such as easy implementation, simplicity and robustness) and disadvantages (such as premature convergence) due to the role of control parameters (w—effectively controlled previous velocity and c_1 and c_2—affected more either globally or locally).

Motivated by above remarks, improved PSO (IPSO) is introduced. It is established on new gradually varying (increasing and/or decreasing) c_1, c_2 & w. Mathematically, these are presented below

$$w(\text{inertia weight}) = \begin{cases} w_{max} \times \left(\dfrac{\left(\frac{w_{min}+w_{max}}{2}\right)}{w_{max}}\right)^{\left(\frac{t}{t\max/2}\right)^2} & ; t \le \frac{t_{max}}{2} \\ \left(\frac{w_{min}+w_{max}}{2}\right) \times \left(\dfrac{w_{min}}{\left(\frac{w_{min}+w_{max}}{2}\right)}\right)^{\left(\frac{t-t\max/2}{t\max/2}\right)^2} & ; t > \frac{t_{max}}{2} \end{cases}$$

$$c_1(\text{cognitive acceleration coefficients}) = c_{1min} \times \left(\frac{c_{1max}}{c_{1min}}\right)^{\left(\frac{t}{t_{max}}\right)^2}$$

$$c_2(\text{social acceleration coefficients}) = c_{2min} \times \left(\frac{c_{2max}}{c_{2min}}\right)^{\left(\frac{t}{t_{max}}\right)}$$

where max & min represented maximum and minimum values respectively.

In IPSO, for the ith particle velocity and position updated by (3) and (4) respectively.

$$v_{i,j}^{t+1} = \left\{ \begin{array}{ll} w_{max} \times \left(\frac{\left(\frac{w_{min}+w_{max}}{2}\right)}{w_{max}}\right)^{\left(\frac{t}{t_{max}/2}\right)^2} & ; t \le \frac{t_{max}}{2} \\ \left(\frac{w_{min}+w_{max}}{2}\right) \times \left(\frac{w_{min}}{\left(\frac{w_{min}+w_{max}}{2}\right)}\right)^{\left(\frac{t-t_{max}/2}{t_{max}/2}\right)^2} & ; t > \frac{t_{max}}{2} \end{array} \right\} v_{i,j}^{t} + c_{1min} \times \left(\frac{c_{1max}}{c_{1min}}\right)^{\left(\frac{t}{t_{max}}\right)^2} r_1\left(p_{best_{i,j}}^{k} - x_{ij}^{k}\right) + c_{2min} \times \left(\frac{c_{2max}}{c_{2min}}\right)^{\left(\frac{t}{t_{max}}\right)} r_2\left(g_{best_j}^{t} - x_{i,j}^{t}\right) \quad (3)$$

$$x_{i,j}^{t+1} = x_{i,j}^{t} + v_{i,j}^{t+1} \quad (4)$$

In the search process of the introduced IPSO: (i) the proposed w decreased gradually maintaining global exploration and local exploitation abilities because when it increases exploration ability is strong and exploitation is strong when it decreases (ii) c_1 and c_2 decreased and increased gradually respectively to improvise the PSO's global (comprehensive) search competency in the initial stage because when c_1 increases and c_2 decreases it pushes elements to passage in the whole solution whereas the value of c_1 decreases and c_2 increases (as iteration increases) then it twitches the elements to the inclusive solution. Then, by a vast analysis of unconstrained benchmark function and also on engineering design optimization by the proposed IPSO algorithm the control parameter, $c_{1min} = c_{2max} = 2.5$; $c_{1max} = c_{2min} = 0.5$, $w_{max} = 0.9$ and $w_{min} = 0.4$ are recommended for optimization problems. During the search process for IPSO influences and/or behavior of proposed w, c_1 and c_2 are illustrated in Fig. 1a, b.

4 Simulation Results and Analysis

Before solving the CHPED problem [33] (consists of 4 power, 1 heat and 2 cogenerations only unit), suggested IPSO validated on following two complex functions.

$$\text{Rosenbrock function } f_1(x) = \sum_{i=1}^{n-1}\left(100\left(x_{i+1} - x_i^2\right) + (x_i - 1)^2\right)$$

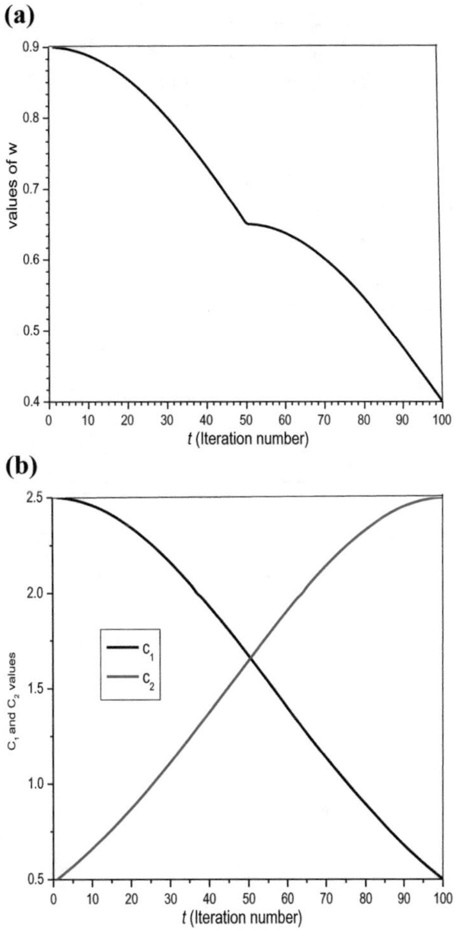

Fig. 1 **a** Distinction of w throughout iterations. **b** Distinction of c_1 and c_2 throughout iterations

$$\text{Rastrigin function } f_2(x) = 10n + \sum_{i=1}^{n}\left(x_i^2 - 10\cos(2\pi x_i)\right)$$

The program is accompanied by C language on PC-i7 Intel®, 8 GB RAM, 2.20 GHz. Parameter: Population sizes (np) = 40, dimension = 10, independent runs (stopping criteria) = 20 for benchmark function, which is absolutely identical to the PSODE [28], remain same as mentioned above section. The produced results considered benchmark function compared with traditional DE [34] and PSODE [35]. Among all the algorithms, the best results are indicated by italic bold standards in each table. The explanation of results and discussion of several investigations are as below.

The numerical results (mean and standard deviation) of the finest objective function value listed in Table 1 over 20 trail runs and graphical results in terms of convergence plot are plotted in Fig. 2a, and b. From these results it can be said that IPSO best performer in comparison to others and should be further recommended to apply real-life problems.

Further, the proposed system is implemented in one of the systems of CHPED (consists of 4 power, 1 heat and 2 cogenerations-only units). During transmission the power units are considered along with its power loss. The demanded power is 600 MW, and heat units are 150MWth. The feasible sets for power and heat-only units are given below.

Table 1 Numerical results on benchmark function

Function	Standards	IPSO	DE [34]	PSO [35]	PSODE [35]
Rosenbrock	Mean	3.417E-256	2.8662E+000	1.6916E+000	2.2864E-051
	Standard deviation	5.251E-268	1.6892E+000	1.7038E+003	2.3016E-101
Rastrigrin	Mean	2.072E-019	2.0894E+000	3.3311E+000	3.1011E-002
	Standard deviation	1.801E-021	1.4189E+000	2.9464E+000	2.8418E-004

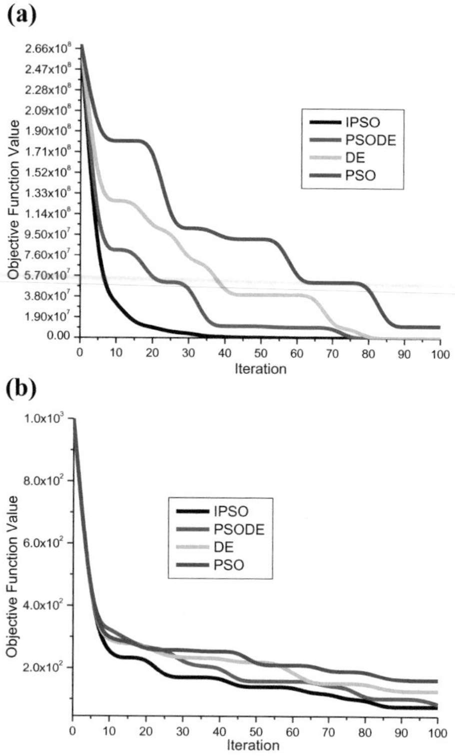

Fig. 2 **a** Convergence plot on rosenbrock function. **b** Convergence plot on rastrigrin function

- Four pure power units

$$F_{P_1}(P_{P_1}) = 25 + 2P_{P_1} + 0.008P_{P_1}^2 + \left|100 * sin\left(0.042 * \left(P_{P_{1,min}} - P_{P_1}\right)\right)\right| : 10 \le P_{P_1} \le 75$$

$$F_{P_2}(P_{P_2}) = 60 + 18P_{P_2} + 0.003P_{P_2}^2 + \left|140 * sin\left(0.04 * \left(P_{P_{2,min}} - P_{P_2}\right)\right)\right| : 20 \le P_{P_2} \le 125$$

$$F_{P_3}(P_{P_3}) = 100 + 2.1P_{P_3} + 0.0012P_{P_3}^2 + \left|160 * sin\left(0.038 * \left(P_{P_{3,min}} - P_{P_3}\right)\right)\right| : 30 \le P_{P_3} \le 175$$

$$F_{P_4}(P_{P_4}) = 120 + 2P_{P_4} + 0.001P_{P_4}^2 + \left|180 * sin\left(0.037 * \left(P_{P_{4,min}} - P_{P_4}\right)\right)\right| : 40 \le P_{P_4} \le 250$$

- Cogeneration Units

$$F_{C_1}(P_{C_1}.H_{C_1}) = 2650 + 14.5P_{C_1} + 0.034P_{C_1}^2 + 4.2H_{C_1} + 0.03H_{C_1}^2 + 0.031P_{C_1}.H_{C_1}(\$/h)$$

- Heat-only unit

$$F_{K_1}(H_{K_1}) = 950 + 2.0109H_{K_1} + 0.038H_{K_1}^2(\$/h) 0 \le H_{K_1} \le 2695.2$$

The power loss coefficient table is also given as follows.

$$PC = 10^{-6} * \begin{vmatrix} 39 & 10 & 12 & 15 & 15 & 16 \\ 10 & 40 & 14 & 11 & 15 & 20 \\ 12 & 34 & 35 & 17 & 20 & 18 \\ 15 & 11 & 17 & 39 & 25 & 19 \\ 15 & 15 & 20 & 25 & 49 & 14 \\ 16 & 20 & 18 & 19 & 14 & 15 \end{vmatrix}$$

The feasible region for cogeneration units is depicted in Fig. 3a and combined heat and power units are given in Fig. 3b.

Parameters for CHPED problem: $np = 50$, $t_{max} = 500$ and compared with 2 existing algorithms BCO [36] and NSGA-II [37] also comparative results are numerically listed in Table 2 and graphically presented in Fig. 4a, b.

From Table 2, it can be decided that the proposed IPSO is efficient and comparatively better than other algorithms. From Fig. 3a, it can be perceived that the proposed IPSO reduce the loss by 9% when compared with BCO and 100% reduction when compared with NSGA-II. From Fig. 3b, the comparison of the total cost is done for BCO, NSGA-II and this proposed IPSO algorithm. On comparing these obtained results, the IPSO algorithm reduces the cost by 27$ when compared to BCO and 422$ on comparing NSGA-II.

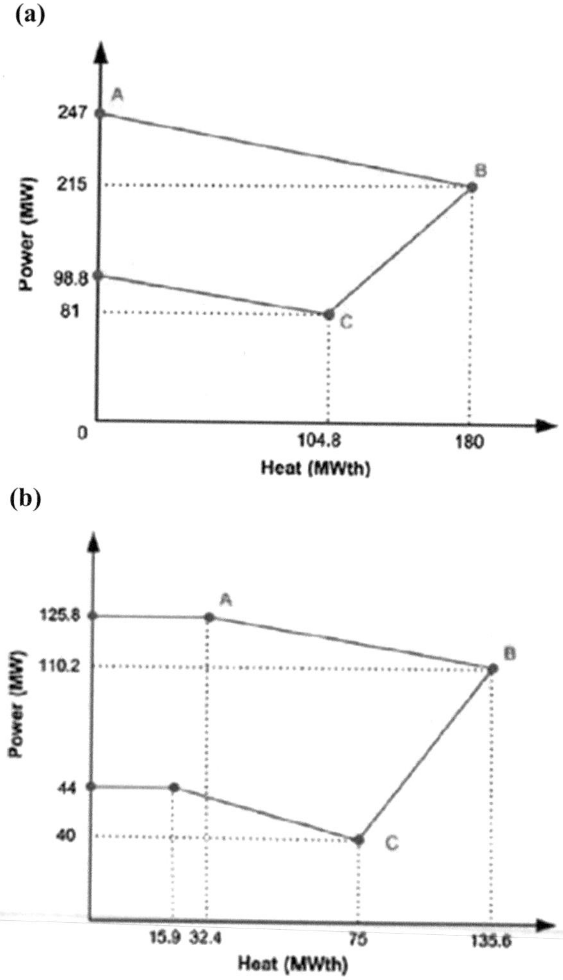

Fig. 3 **a** Heat power reasonable region of cogeneration unit 1. **b** Heat power possible region of cogeneration unit 2

5 Conclusion and Prospect Advice

Since classical PSO suffers from a few major shortcomings, so to remove these drawbacks and improve the PSO algorithm, an improved PSO (viz. IPSO) described in the presented paper, on basis of increasing and/or decreasing parameters. The proposed w decreased gradually which maintains global exploration and local exploitation abilities because when it increases exploration ability is strong and exploitation is strong when it decreases and the suggested c_1 and c_2 decreased and increased gradually respectively which improves global search competency in the initial stage because as c_1 increases and c_2 decreases it pushes elements to passage in the whole solution

Table 2 Numerical results on the considered system of the CHPED problem

Units/criteria	Algorithm		
	BCO [36]	NSGA-II [37]	IPSO
P1	43.9457	74.5357	40.6311
P2	98.5888	99.3518	100.828
P3	112.932	174.719	114.782
P4	209.771	211.017	210.049
PC1	98.8000	100.936	57.5209
PC2	44.0000	44.1036	17.9663
HC1	12.0974	24.3678	98.5196
HC2	78.0236	72.5270	74.5127
H1	59.8790	53.1052	42.5569
Total power	608.038	704.664	**607.368**
Total heat	150	150	150
Total cost	10,317	10,712.86	**10,290.82**
Total loss	8.0384	104.664	**7.3681**

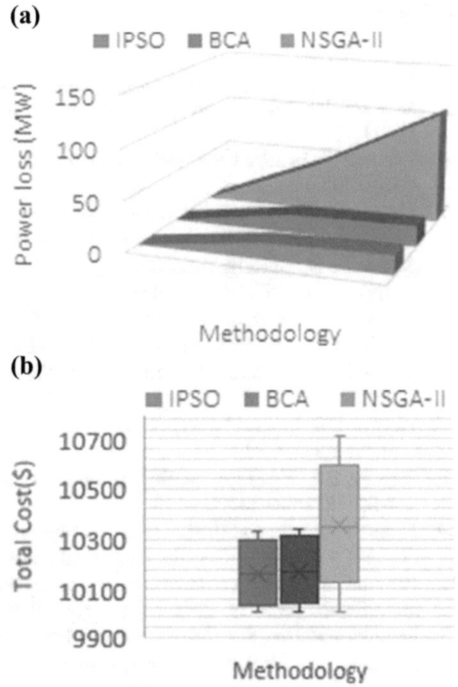

Fig. 4 **a** Comparison of power loss (MW). **b** Comparison of total cost ($)

whereas the value of c_1 decreases and c_2 increases (as iteration increases) then it twitched the elements for the inclusive solution.

For measuring the effectiveness of the proposed IPSO, two multifaceted functions are simulated and compared with traditional PSO, classical DE, and the hybrid variant of PSO and DE (named PSODE). After this, IPSO was implemented in one of the systems of the CHPED problem (consists of 4 power, 1 heat and 2 cogenerations-only units) and two existing methods namely BCO and NSGA-II.

On the basis of graphical and numerical results analysis, it can be observed that the proposed IPSO is much reliable, accurate, efficient, effective and capability of avoiding fall into local optimal solution and much better than other algorithms.

As a future scope, to deal with multi-objective real-world optimization problems IPSO may be applied.

Acknowledgement This research supported by DST-SERB Govt. of India; grant number CRG/2020/000817.

References

1. Verma K, Bhardwaj S, Arya R, Islam MS Ul, Bhushan M, Kumar A, Samant P (2019) Latest tools for data mining and machine learning. Int J Innov Technol Explor Eng (IJITEE) 8(9S), 18–23
2. Kennedy J, Eberhart R (1995) Particle swarm optimization. In: Proceedings of ICNN'95—International conference on neural networks, vol 4, pp 1942–1948
3. Yang X-S (2009) Firefly algorithms for multimodal optimization. In: International symposium on stochastic algorithms. Springer, pp 169–78
4. Karaboga D, Basturk B (2007) A powerful and efficient algorithm for numerical function optimization: artificial bee colony (ABC) algorithm. J Glob Opt 39, 459–471
5. Gandomi AH, Yang X-S, AH Alavi (2013) Cuckoo search algorithm: a metaheuristic approach to solve structural optimization problems. Eng Comput 29:17–35
6. Mirjalili S, Gandomi AH, Mirjalili SZ, Saremi S, Faris H (2017) Salp swarm algorithm: a bioinspired optimizer for engineering design problems. Adv Eng Softw 114:163–191
7. Wolpert DH, Macready WG (1997) No free lunch theorems for optimization. IEEE Trans Evol Comput 1(1):67–82
8. Parouha RP (2018) Economic load dispatch using memory based differential evolution. Int J Bio-Insp Comput 11:159–170
9. Parouha RP, Das KN (2016) DPD: an intelligent parallel hybrid algorithm for economic load dispatch problems with various practical constraints. Expert Syst Appl 63:295–309
10. Parouha RP, Das KN (2016) A robust memory based hybrid differential evolution for continuous optimization problem. Knowl-Based Syst 103:118–131
11. Parouha RP, Das KN (2016) A novel hybrid optimizer for solving economic load dispatch problem. Int J Electr Power Energy Syst 78:108–126
12. Parouha RP, Das KN (2015) An efficient hybrid technique for numerical optimization and applications. Comput Ind Eng 83:193–216
13. Das KN (2015) An ideal tri-population approach for unconstrained optimization and applications. Appl Math Comput 256:666–701
14. Parouha RP (2014) An effective hybrid DE with PSO for constrained engineering design problem. J Int Acad Phys Sci 18:01–15

15. Suri RS, Dubey V, Kapoor NR, Kumar A, Bhushan M (2021) Optimizing the compressive strength of concrete with altered compositions using Hybrid PSO-ANN. In: 4th International conference on information systems and management science (ISMS 2021), Springer, December 14–15, 2021, The faculty of ICT, University of Malta, Msida, Malta
16. Shi Y, Eberhart R (1998) A modified particle swarm optimizer. In: The 1998 IEEE International conference on evolutionary computation proceedings, Anchorage, AK, pp 69–73
17. Liang JJ, Qin AK, Suganthan PN (2006) Comprehensive learning particles warm optimizer for global optimization of multimodal functions. IEEE Trans Evolut Comput 10:281–295
18. Zhan Z-H, Zhang J, Li Y (2011) Orthogonal learning particle swarm opti-mization. IEEE Trans Evol Comput 15:832–846
19. Xinchao Z (2010) A perturbed particle swarm algorithm for numerical optimization. Appl Soft Comput 10, 119–124
20. Wang Y, Li B, Weise T, Wang J, Yuan B (2011) Self-adaptive learning basedparticle swarm optimization. Inf Sci 181:4515–4538
21. Tsoulos IG (2010) Enhancing PSO methods for global optimization. Appl Math Comput 216:2988–3001
22. Ratnaweera A, Halgamuge SK (2004) Self-organizing hierarchical particle swarm optimizer with time-varying acceleration coefficients. IEEE Trans Evol Comput 8:240–255
23. Yang X, Yuan J, Yuan J (2007) A modified particle swarm optimizer with dynamic adaptation. Appl Math Comput 189:1205–1213
24. Mendes R, Kennedy J (2004) The fully informed particle swarm: simpler, may be better. IEEE Trans Evol Comput 8:204–210
25. Kennedy J, Mendes R (2002) Population structure and particle swarm performance. In: IEEE Congress on evolution of computers, Honolulu, pp 1671–1676
26. Liang JJ, Suganthan PN (2005) Dynamic multi-swarm particle swarm optimizer. In: Swarm intelligence symposium, California, pp 124–129
27. Beheshti Z, Shamsuddin SM, Sulaiman S (2014) Fusion global-local-topology particle swarm optimization for global optimization problems. Math Probl Eng
28. Parouha RP (2019) Non-convex/non-smooth economic load dispatch using modified time varying particle swarm optimization. Computational Intelligence Wiley. https://doi.org/10.1111/coin.12210
29. Parouha RP (2021) Design and applications of an advanced hybrid meta-heuristic algorithm for optimization problems. Artif Intell Rev 54(8):5931–6010
30. Verma P (2021) An advanced hybrid algorithm for constrained function optimization with engineering applications. J Amb Intell Human Comput. https://doi.org/10.1007/s12652-021-03588-w
31. Parouha RP (2022) A systematic overview of developments in differential evolution and particle swarm optimization with their advanced suggestion. Appl Intell. https://doi.org/10.1007/s10489-021-02803-7
32. Huynh DC (2010) Parameter estimation of an induction machine using advanced particle swarm optimization algorithms. IET J Electr Power App 4:748–760
33. Neyestania M, Hatami M (2019) Combined heat and power economic dispatch problem using advanced modified particle swarm optimization. J Renew Sustain Energy 11:015302. https://doi.org/10.1063/1.5048833
34. Niu B, Li L (2008) A novel PSO-DE-based hybrid algorithm for global optimization. In: Huang D-S, Wunsch II DC, Levine DS, Jo K-H (eds) ICIC 2008. LNCS (LNAI), vol 5227, pp 156–163
35. Storn R (1997) Differential evolution—a simple and efficient heuristic for global optimization over continuous spaces. J Glob Optim 11:341–359
36. Basu M (2011) Bee colony optimization for combined heat and power economic dispatch. Expert Syst Appl 38(11):13527–13531
37. Basu M (2013) Combined heat and power economic emission dispatch using non dominated sorting genetic algorithm-II. Int J Electr Power Energy Syst 53:135–141

A QoE Framework for Video Services in 5G Networks with Supervised Machine Learning Approach

K. B. Ajeyprasaath and P. Vetrivelan

1 Introduction

In today's world, MPEG Dash plays a significant role in network traffic since it gives the end user an easily accessible video streaming. According to a recent study by the cisco Index, Video traffic contributes to 70% of entire Internet traffic. In 2022, it is predicted that video traffic will contribute to 82% of total Internet traffic [1]. Hence the Immense growth has become a vital issue for both video content providers and network operators to give the best quality video to end users by managing the network traffic. Content providers or network operators focus on giving better proficiency to end users known as QoE. End users might encounter various experiences based on available bandwidth, their necessity, and network condition. All the above-mentioned factor has an active or passive impact on QoE. To examine the QoE for any particular multimedia service, active users take part in subjective feedback to value the offered service [2]. Alternatively, Objective methods applied to define the objective QoE from a set of standards represent the subjective assessment [3]. In recent days, both QoE and QoS are taken into account to examine if there is some default in the QoE prediction process [4]. By employing the supervised machine learning classification model, user-level QoE can be forecasted depending on the QoS [5, 6]. To gauge the quality, the content providers have complete access to the video quality but the network operators only have admission to network traffic. The network traffic gathered and studied from the user devices or network elements or both sides are used to evaluate the network quality. The entire data collected is used to know the user expectations and helps to enhance the QoE in the near future. The network

K. B. Ajeyprasaath · P. Vetrivelan (✉)
School of Electronics Engineering, Vellore Institute of Technology, Chennai, India
e-mail: vetrivelan.p@vit.ac.in

K. B. Ajeyprasaath
e-mail: ajeyprasaath.kb2019@vitstudent.ac.in

P. Singh et al. (eds.), *Machine Learning and Computational Intelligence Techniques for Data Engineering*, Lecture Notes in Electrical Engineering 998,
https://doi.org/10.1007/978-981-99-0047-3_56

operators have no admission to encrypted or the QoE at the user end. So, the current user-level QoE monitoring or the indirect network monitoring by DPI cannot be used by network operators. Alternatively, 5G is expected to support high bandwidth and low delay contributing to enhanced QoE expectations [7]. One of the key challenges in 5G is adapting to the increased growth of video streaming traffic and maintaining smooth QoE [8]. The objective of this work is to forecast the QoE based on the mean opinion score of the end user observing the indirect traffic conditions at the edge facilities adjacent to the user. By using a predictive mechanism, the QoS metrices are gathered and analyzed. After that a real-time mechanism is used to map the QoS to QoE correlation i.e., a supervised machine learning classification algorithm is used to forecast QoE (MOS) for every single video segment in contradiction to QoS metrices. The supervised machine learning classification algorithms are examined and a suitable model is picked in the ML part. These QoE forecasts help network operators relate network traffic with video quality to overpower QoE degradation.

2 Background Work

In delivering better quality service to the end user the most important point to be considered is to analyze their satisfaction level from the network operator view. The most popular method to determine user satisfaction depends on QoS metrics gathered from the network. QoS is described in parameters such as jitter, bandwidth, throughput, packet loss, and delay therefore in QoS, the user's opinion and view are not considered since it only concentrates on the systems technical performance. So, to take the user's opinion into account, the QoE is developed based on User's perception and views which may Vary corresponding to User's needs. The QoE chain is driven by many components that actively or passively impact the user insight for any video services. The components which influence the QoE chain are related to one another which is represented in Fig. 1. In many scenarios objective QoE named stall, representation stream, bitrate, and visual quality have a very robust effect on user QoE. The QoS metrics are the most essential parameters considered to measure user satisfaction since they adversely impact the QoE metrices consequently an important point to be considered is that measuring QoS parameters is easy in comparison with monitoring the QoE metrices.

Nowadays, QoE forecasts have moved to a data-driven approach and subsequently they depend on machine learning (ML) models [9]. Thus, this method gives objective assessment techniques more adeptness to provide more precise values. The main objective of ML techniques is to construct supervised learning models and to forecast based on the given input data. In the first step, QoS parameters and user feedbacks gathered using a system known as a probe from the client and server-side which is used to train and validate the model subsequently, in the second step, real-time QoS metrices gathered using a probe is given in the trained model to forecast the user experience without changing the absolute network traffic. Dynamic adaptive streaming over HTTP(DASH) is governing network traffic in the recent era. DASH supplies

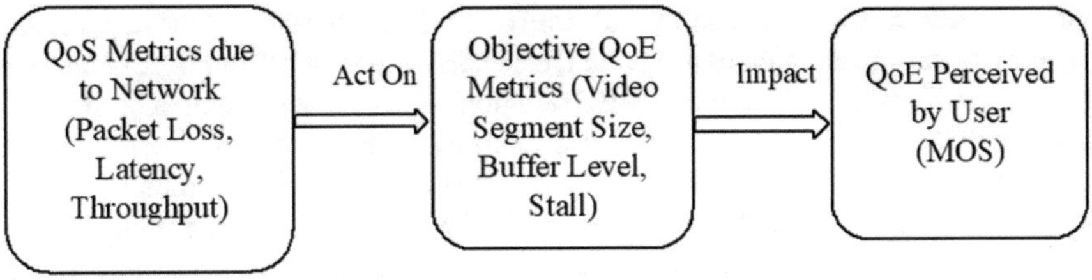

Fig. 1 QoS and QoE correlation

layouts to deliver the best quality of video streaming service in the internet. Dash adheres to ABS standards [10]. The adaptive bit rate streaming (ABS) algorithm aims to detect a favorable quality of video streaming. ABS algorithms decide on the quality of the segments to be downloaded based on the network's available resources. The choice of the ABS algorithm plays a significant role in end-user satisfaction (Fig. 2). Dynamic Adaptive Streaming Over HTTP denotes DASH it works on the principle of dividing a larger video file into a number of very tiny segments of equal intervals. Every segment is encrypted in varying bitrates and resolutions the illustration of the individual segment is defined in the media presentation description (MPD) file. Based on the current network scenario user determines the video segments which can be played the user chooses the segment with the most efficient bitrate feasible which can be downloaded in that duration without causing any interruption.

DASH users are permitted to swap between a variety of video qualities to enhance the viewing experience based on the available network. Thus, it is necessary to know about network conditions for enhanced video distribution. To make a very efficient video distribution and to satisfy the client demand a new architecture named MEC

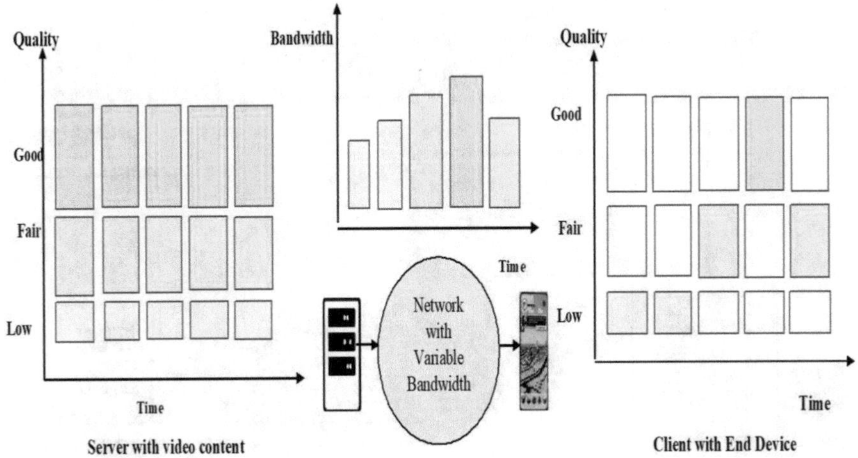

Fig. 2 Dynamic adaptive streaming over HTTP

with 5G network emerges. It permits to store of the data at the edge of the network this paper aims at network functioning and QoE enhancement MEC with a 5G network uses a scenario where a systematic MEC function will analyze the network traffic data and give existing data on network traffic necessities of the front or backhaul network. Later on, this data will go to the cloud where the network traffic can be remodeled and stabilized to enhance the client QoE to achieve this we need to employ a probe mechanism at the client side which can analyze the network traffic this mechanism helps us forecast the QoE at the client side.

3 Design and Analysis

This work suggests the technique to forecast user QoE using a probe mechanism at edge facilities with 5G data's Fig. 3. Architecture for QoE Framework describes the entire use for the QoE forecast mechanism with the client, Access point (AP), network, switch, server. The AP will also act as an edge computer node that can be employed to monitor the network, our analysis is set up on an NS-3 network simulator, packet capturing, DASH player, and ML-supervised model. In this, packet capturing is used to capture the QoS metrics at the AP indirectly by using the DASH player, we can get the QoE parameter log file. ML-supervised classification model helps us to forecast QoE based on QoS parameters. NS-3 network simulator-based simulation to mimic real-time systems to analyze how network work.

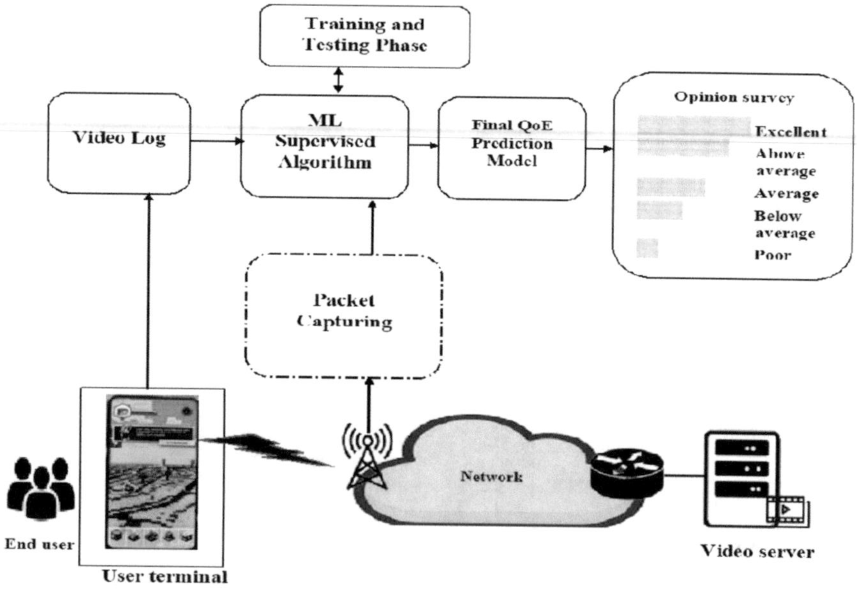

Fig. 3 Architecture for QoE framework

The geometrical arrangement of the system is described in Fig. 3. Architecture for QoE Framework using NS-3 network simulator combined with a variety of network conditions for the whole video streaming the layout comprises of video server, switch, network, Access point, and video client. On the server side video is divided into small chunks of 2-s duration with varying bitrates and resolution connected to the switch the link between the switch and AP performs as a bottle neck varying network conditions are mimicked between AP and switch connection through a Linux traffic controller employing downlink throughput parameters from gathered 3G, 4G, 5G network in per second duration for every distinct network the user employs a DASH player to play the similar video of a same duration from the video server. The log file created while playing the video content gets saved with ample chunk QoE parameters. Based on the resolution, codec, frame rate, bit rate, duration of frame, MOS can be estimated by using ITU-T P.1204 in a similar way the network traffic can be obtained and saved at an AP to get network level QoS parameters. To construct a supervised ML classification model that can forecast the QoE of the client more exactly about the video performance, the created data needs more processing to make it suitable to a supervised ML classification algorithm. Video log data and packet capturing data are preprocessed to make it apt for giving in the supervised ML classification model. In the network traffic we analyze that paradigm of each and every segment is consistent the downloading of each segment takes place by instigating an HTTP get request subsequently, the download takes place till the next segment request emerges based on this finding per segment-related QoS parameter can be pre-required. The dataset employed to create a supervised ML classification model comprises per segment QoS parameters as an aspect and QoE components as a base. The video log file gives the necessary data of each segment, which comprises of the stall, bitrate, resolution, and respective segment's MOS values. In this paper MOS value of each distinct segment is considered to construct a supervised ML classification model. Figure 4 illustrates the Imbalance Distribution of Mean Opinion Score.

To construct a supervised ML classification model following phases are considered removing the unnecessary data, evaluating whether a column needs to be dropped or not, fill out the missing values or reject rows with missing values, and filtering unnecessary outliers. Figure 5 describes the Distribution of the Mean Opinion Score after cleaning the dataset. It is divided into training and testing in the ratio of 70% and 30%.

The analysis of the supervised ML classification algorithm is made and five of the algorithms are selected and their parameters are tuned which is represented in Table 1.

The training and testing of the dataset are carried out with all five algorithms and from that, algorithm with the best accuracy is finally selected to forecast the values for the supervised ML classification model from the analysis which is represented in Fig. 6.

It is noted that the random forest classifier (RFC) is identified as the best model with the highest accuracy of 76% so, the RFC is opted. The trained model does not undergo underfitting or overfitting from the analysis.

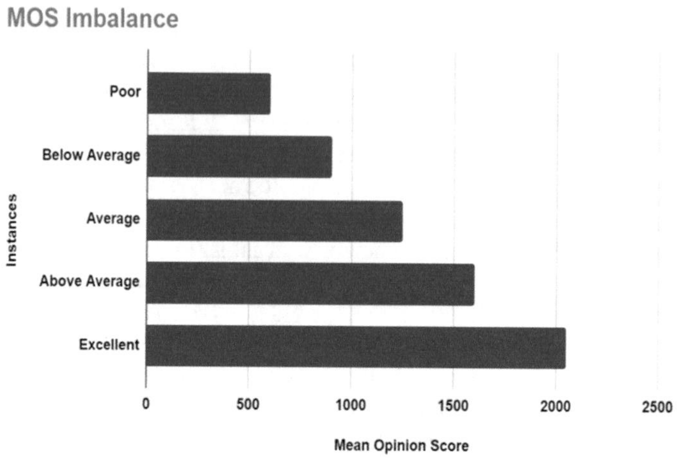

Fig. 4 Imbalance distribution of mean opinion score

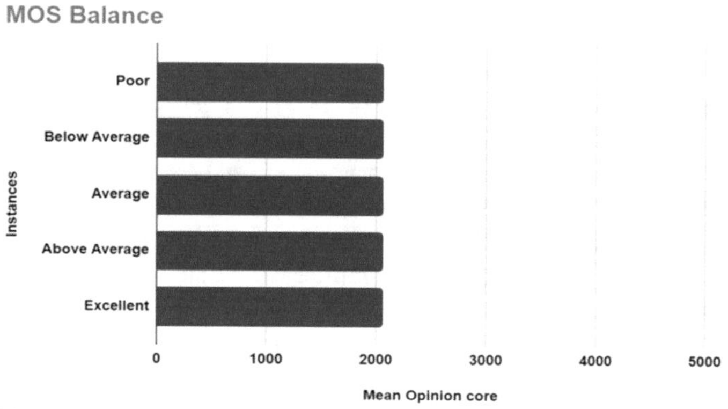

Fig. 5 Balance distribution of mean opinion score

Table 1 Supervised ML classification parameters

Machine learning supervised classification models	Parameters
Logistic regression (LR)	Penalty = 12
Support vector classifier (SVC)	Degree = 3
K-nearest neighbors (KNN)	Neighbors = 3
Decision tree classifier (DTC)	Criterion = Entropy
Random forest classifier (RFC)	Criterion = Gini

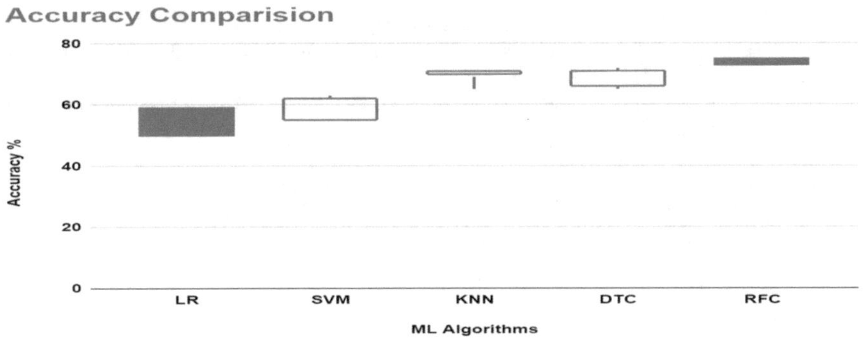

Fig. 6 Result of analyzed ML algorithms

4 Conclusion

An outline of 5G video services framework is specified in this work. This framework can be employed in various arenas of wireless networks. The representation of DASH video streaming in static and mobility scenarios for QoS-QoE correlation and prediction method based on supervised ML classification is done. In this work different algorithms are analyzed among that RFC is opted. Since it has the highest accuracy 76% in comparison with other models. Hence, in real-time scenario RFC can be suggested to predict QoE which helps content providers and network operators to satisfy the end user needs.

References

1. Barakabitze AA et al (2020) QoE management of multimedia streaming services in future networks: a tutorial and survey IEEE Commun Surv Tutor **22**(1), 526–565. https://doi.org/10.1109/COMST.2019.2958784
2. Zhao T, Liu Q, Chen CW (2017) QoE in video transmission: a user experience-driven strategy. In: IEEE Commun Surv Tutor **19**(1), 285–302. https://doi.org/10.1109/COMST.2016.2619982
3. Barman N, Zadtootaghaj S, Schmidt S, Martini MG, Möller S (2018) An objective and subjective quality assessment study of passive gaming video streaming. Int J Netw Manage 2054
4. Huang R, Wei X, Zhou L, Lv C, Meng H, Jin J (2018) A survey of data-driven approachon multimedia qoe evaluation. Front Comp Sci 12:08
5. Petrangeli S, Wu T, Wauters T, Huysegems R, Bostoen T, Turck FD (2017) A machine learning-based framework for preventing video freezes in HTTP adaptive streaming. J Netw Comput Appl 94:78–92
6. Orsolic I, Pevec D, Suznjevic M, Skorin-Kapov L (2017) A machine learning approach to classifying YouTube QoE based on encrypted network traffic. Multimed Tools Appl 76:22267–22301
7. Swetha S, Raj D (2017) Optimized video content delivery over 5G networks. In: 2017 2nd International conference on communication and electronics systems (ICCES). IEEE, pp 1000–1002. https://doi.org/10.1109/CESYS.2017.8321232

8. Wang R, Yang Y, Wu D (2017) A QoE-aware video quality guarantee mechanism in 5g network. In: International conference on image and graphics. Springer, Cham, pp 336–352. https://doi.org/10.1007/978-3-319-71589-6_30
9. Ul Mustafa R, Ferlin S, Rothenberg CE, Raca D, Quinlan JJ (2020) A supervised machine learning approach for dash video qoe prediction in 5g networks. In: Proceedings of the 16th ACM symposium on QoS and security for wireless and mobile networks, July 2017, Washington, DC, USA. https://doi.org/10.1145/3416013.3426458
10. Tan X, Xu L, Zheng Q, Li S, Liu B (2021) QoE-driven DASH multicast scheme for 5G mobile edge network. J Commun Inf Netw 6(2), 153–165. https://doi.org/10.23919/JCIN.2021.9475125

A Survey of Green Communication and Resource Allocation in 5G Ultra Dense Networks

Dhanashree Shukla and Sudhir D. Sawarkar

1 Introduction

The 5G network allows reliable communication with increased network capability and uniform quality of service to more number of users. The 5G NR (New Radio) technology was introduced, which aims at fast, scalable, and responsive mobile broadband communication. Because of the reduced physical range due to higher frequencies, 5G divides the geographical areas into small cells. There is always a trade-off between the speed and the distance to which service can be provided. The lowest 5G band ranges from 600–850 MHz, the medium 5G band uses 2.5–3.7 GHz, and the higher frequency band of 5G uses 25–39 GHz. The higher band has the capacity to provide speeds of Gigabits per second but suffers from non-transmission through walls. 5G wireless devices use low power and local antenna array for their transmission and reception. The channels of 5G are generally reused from cell to cell. The cells in the 5G are typically connected through an optical fiber high bandwidth network and a wireless backhaul connection can be used in-between the cells. For mmWave communication, a direct transmitter must be present near the user without any wall. There are 3 main applications for 5G networks namely—enhanced Mobile Broadband (eMBB), massive Machine Type Communication (mMTC), and Ultra Reliable Low Latency Communication (URLLC). Key use cases for each of the main applications are shown in Fig. 1.

The performance of the 5G band is defined on four main parameters namely—Speed, Latency, Error Rate, and Range. It is expected that 5G will provide 1 Gbps speed. Typical latency in 5G equipment should be in the range of 10–20 ms. Adoptive signal coding is used to reduce the error rate in 5G but at the cost of bandwidth. It is proven that 5G NR software deployment on 4G hardware is at least 15% better in terms of performance. For the 5G networks, tall towers are not required to mount the antennas whereas a small region-based distribution is possible with the help of microcells, femtocells, and picocells. The high-level components in a base station and the power consumption for each of these components are shown in Fig. 2c. As

D. Shukla (✉) · S. D. Sawarkar
Datta Meghe College of Engineering, Navi Mumbai 400703, India
e-mail: dhanashree.shukla@dmce.ac.in

P. Singh et al. (eds.), *Machine Learning and Computational Intelligence Techniques for Data Engineering*, Lecture Notes in Electrical Engineering 998,
https://doi.org/10.1007/978-981-99-0047-3_57

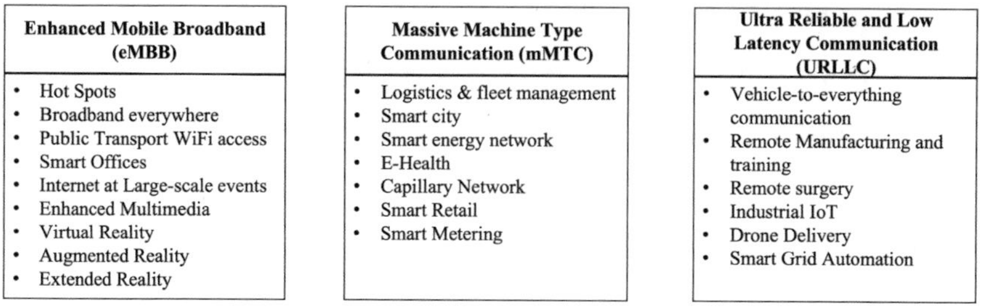
Enhanced Mobile Broadband (eMBB)
Hot Spots
Broadband everywhere
Public Transport WiFi access
Smart Offices
Internet at Large-scale events
Enhanced Multimedia
Virtual Reality
Augmented Reality
Extended Reality
Massive Machine Type Communication (mMTC)
Logistics & fleet management
Smart city
Smart energy network
E-Health
Capillary Network
Smart Retail
Smart Metering
Ultra Reliable and Low Latency Communication (URLLC)
Vehicle-to-everything communication
Remote Manufacturing and training
Remote surgery
Industrial IoT
Drone Delivery
Smart Grid Automation

Fig. 1 Key pillars use cases

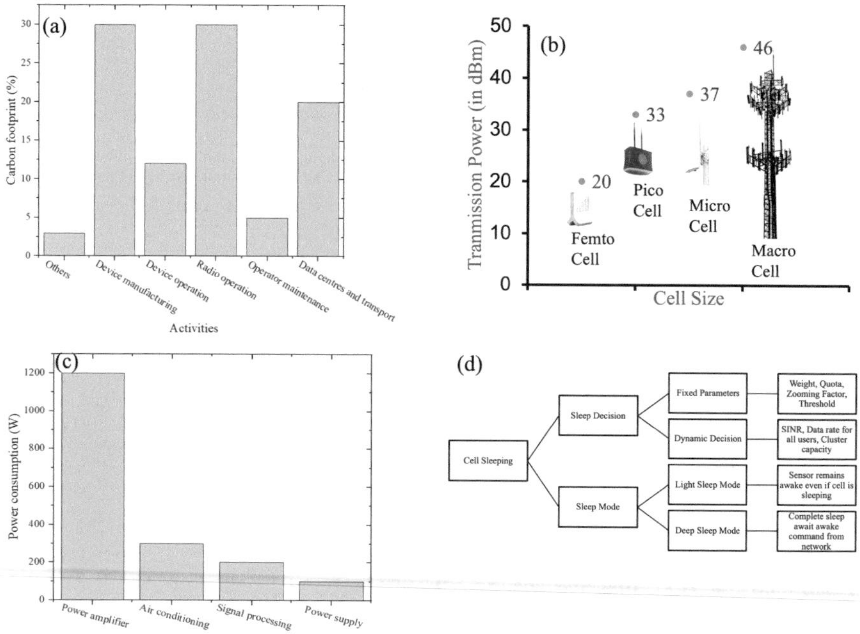

Fig. 2 **a** Carbon Footprint in Mobile Network. **b** Cell size versus transmission power. **c** BS Components versus Power consumption. **d** Overview of cell sleeping

the cell size increases, the transmission power also increases as shown in Fig. 2b. A brief comparison of Microcell, Picocell, and femtocell on various parameters is mentioned in Table 1.

According to the Ericsson Mobility Report 2021, Mobile Data traffic is expected to increase from 65 EB/ month to 288 EB/ month by 2027. 5G subscribers are expected to increase from 1.13 Bn to 5.75 Bn by 2027. This drastic increase in connected devices and corresponding data traffic is expected to result in an alarming increase in carbon footprint. Radio Operations contribute to 30% of overall communication field carbon footprint as shown in Fig. 2. To overcome this challenge, green com-

Table 1 Small cell types comparison

Types of small cell various aspects	Microcell	Picocell	Femtocell	Relay nodes	Remote radio head (RRH)
Area covered by a cell	250 m-1 km	<100 m–300 m	< 10 m–50 m	300 m	Few km
Implementation spot	Outdoor/Exterior	Outdoor/indoor	Indoor	Outdoor/Indoor	Outdoor
Implementation layout	Planned	Planned	Unplanned	Planned	Planned
Access mode	Open access	Open access	Open/closed/hybrid access	Open access	Open access
Frequency parameters	Centrally planned	Centrally planned	Locally determined	Centrally planned	Centrally planned
Backhaul connection	Fiber	X2 interface	Internet protocol	Wireless	Fiber
Transmit power	30–43 dBm	23–30 dBm	< 23 dBm	30 dBm	46 dBm

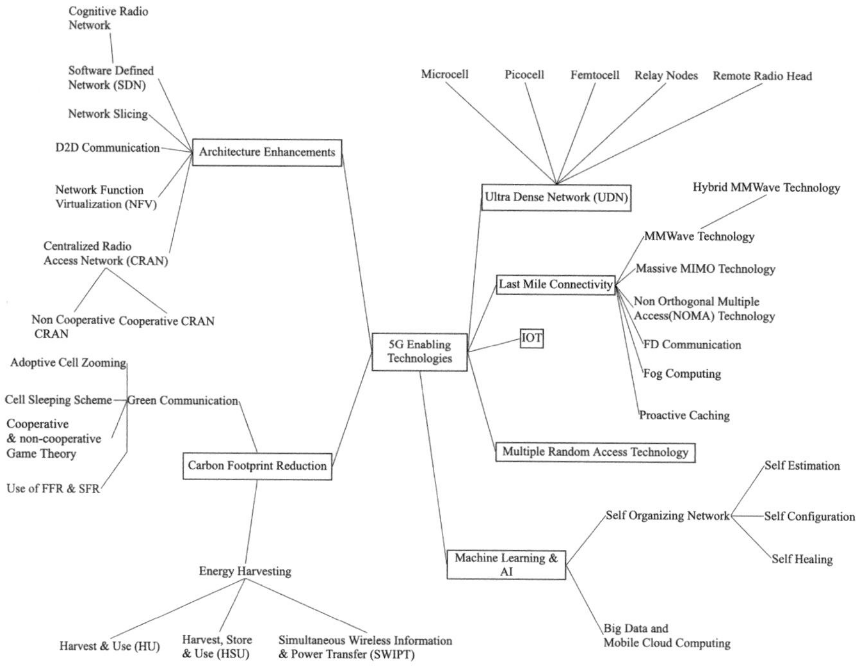

Fig. 3 5G Key research areas

munication and resource allocation have come into focus. As shown in Fig. 3, 5G research can be divided into many areas. This paper discusses the current research for Green Communication in Ultra Dense Networks, with an emphasis on resource allocation using machine learning.

2 Review of Recent Literature

2.1 *Green Communication: The Advancement*

Green communication means decreasing the power consumption in wireless networks and eventually decreasing the carbon footprint. It also covers the strategy for planning the network in an energy-efficient manner which includes minimization of the density of base stations for certain coverage area. If we compare features of 4G and 5G as given in Table 2, we can observe carbon footprint going up in 5G versus 4G.

Abrol et al. [1] proposed a strategy called green communication. In it, a small cell becomes the transmission station reducing the power required by a traditional base station covering hundreds of miles. A Simultaneous Wireless Information and

Table 2 Green Communication perspective: 4G versus 5G

Feature	4G	5G
Average power utilization per site	1.3 kW	1.1 kW
Base station density	8–10 BS/km^2	40–50 BS/km^2
Use of green technology	Phantom cells, soft cells, liquid cells, green base station	D2D, massive MIMO, spectrum sharing
Carbon impression per mobile subscriber	23 kg	31 kg
CO^2 discharge	170 Mto	235 Mto
Number of base stations	7.6 millions	11.2millions
Radio access network electricity consumption	77 TWh	86 TWh
Femtocell power utilization	6 W	5 W(expected)

Power Transfer (SWIPT) is also explained which saves up to 30% of energy. They mentioned three major small cell access types: closed, open, and hybrid access. A mobile relay station can be used in 5G, it serves as a base station on demand. It increases the coverage efficiency resulting in energy saving. Instead of a single antenna if four antennas are used in the beam-steering mode it saves up to 55% energy. A small cell implementation can effectively save 11.1% energy. A massive Multiple-Input Multiple-Output (MIMO) deployment with zero forced processing can achieve a transfer rate of 30.7 Mb/Joule of energy. A relay deployment compared to Long-Term Evolution (LTE) deployment can save up to 9.7 dB of power. Typically there are three techniques deployed for energy-efficient green communication namely Energy-efficient architecture, Energy-efficient resource management, and Energy-efficient radio technologies.

Arevalo et al. [3] designed an algorithm, Optical Topology Search (OTS), which conducts a techno-economic analysis. The variables in their algorithms were cell size and bit rate per user. They found that femtocells are the best solution for delivering ultra-high bit rates. Tsai et al. [27] proposed metaheuristic deployment of 5G network with a brief review on the hyper-dense deployment problem. A study conducted by Gandotra et al. [10] shows that carbon footprint is directly proportional to the number of cellular towers installed. According to their study, the total CO emissions around the end of 2020 were 345 million tons. The telecommunication sector contributed around 2% of the total carbon emissions. They proposed that, to decrease power consumption focus must be given to resource allocation, optimal network planning, and renewable energy. Liu et al. [16] investigated the impact of dynamically switching on/off base stations on problems caused by the random and dense deployment of small cells. Their simulation shows that a combination of Small Cell Base Stations(SBSs) sleep and optimum spectrum allocation coefficient can optimize network. Bouras et al. [5] proposed a better dense deployment strategy for femtocells which can result in better energy efficiency and interference management. They proposed two sleep modes as shown in Fig 2d. **Light sleep mode** (discontinuous transmission mode): In this mode, sensors of the cell remain awake measuring an increase in received power for uplink. This sleep mode can give power saving up to 40%. **Deep sleep mode**: In this mode, the entire cell is switched off and core network tracks users through Mobility Management Entity (MME) resulting in 70% energy saving. A congestion control scheme incorporating SINR and cell zooming is proposed by Mukherjee et al. [18] They have achieved 20% power saving with the proposed scheme and an SINR improvement of 18%. There are two types of interference, inter-tier and intra-tier interference, as we have two different types of cells, Macro cell and small cell in the network. There will be offloading between these two types of cells. This offloading can cause load imbalance. Yang et al. [30] have proposed two schemes, dedicated channel deployment and partially shared channel deployment using cooperative game theory. They have tried to use various concepts like cell range expansion, cognitive radio, and self-organizing networks for reducing the power consumption. Ghosh et al. [11] have designed small cell zooming using a weighted 5G mobile network for maximum energy. They found that if a higher number of femtocells are chosen, then 35% of the power is saved, SINR is increased by 30%, and spectral efficiency by 60%. The weight is defined as the number of devices connected to the base station, and quota is the maximum number of devices that can be connected to the base station as shown in Fig. 2d. They proposed the two players of the game are adjacent femtocells. The higher weighted femtocell wins the game and zooms its coverage area. Xu et al. [28] have proposed a method of adaptive cell zooming. They proposed that cells with lower traffic loads can switch off and save power. Their proposed methodology was dependent on sleep threshold and dependent on traffic requirements. They have also used Cell Zooming Factor (CZF) which expresses cell coverage as shown in Fig. 2. When the number of users connected to a cell is less than a threshold, the cell will try to offload users to Macro Cell and go to sleep. The threshold is derived from maximizing small cell sleeping probability and optimizing macro cell service capacity. Ganame et al. [9] has developed an algorithm that allows optimization of deployment in 5G base stations. They have used a Monte-Carlo simulation. They could achieve an outage rate of around 11.63% and maximum coverage of 98%. They could achieve roughly the removal of 5% base station with the proposed algorithm. Zhou et al. [35] have proposed a green cell planning and deployment strategy for small cell networks. They modeled different traffic patterns using stochastic geometry and made an energy-efficient scheme based on traffic patterns received. A summary of improvements achieved by various scholars is given in Table 3.

3 Resource Allocation

As more number of users connect to the 5G network, efficient allocation of resources like time, frequency, and space becomes an important factor for delivering satisfactory QoS. In D2D communication in a multicell context, Jiang et al. [14] achieved a Maximum SINR of 32 dB & throughput up to 55 megabits per second by using

Table 3 Comparison of Green Communication Methods

Method	Refs.	Throughput	Power consumption	Improvement in SINR	Spectral efficiency
SFR/FFR	[26]	~38 Mbps	~	~	~88%
Zoom Game	[11]	~	−35%	~30%	~60%
Swarm Intelligence	[9]	~	~	~	~
Heuristic resource Pairing	[14]	~55 Mbps	~	~22%	~
Small Base Station sleep	[16]	~	−40%	~	~
Adoptive Cell Zooming	[28]	~	−41%	~	~
Hybrid Strategy for sleep mode	[5]	~	−25%	~	~
Congestion control scheme for cell zooming	[18]	~	−20%	18%	~

various solutions. They have used Farthest Distance-Based Resource Allocation (FDRA), Heuristic Optimization Resource Allocation Algorithm (HORA), Dynamic Fractional Power Control Scheme (DFPC), Full Compensation Power Control, and Half Compensation Power Control. Bouaziz et al. [4] have proposed two-stage algorithm: the first stage involves femtocell selection, and the second stage involves resource allocation using a QoS-RAS(Resource Allocation Scheme) and a QoS-aware resource allocation. They have considered parameters such as Resource Utilization Ratio, Request Dropped Probability, Fairness Index, and Total Average Throughput. Saddoud et al. [24] proposed two algorithms that make use of linear programming and Lagrangian duality along with scheduling. Khan et al. [15] proposed a Maximum SINR association strategy which connects devices with the BS that provides the best downlink SINR. They also proposed a maximum received power approach, where devices are linked to the BS that has the highest received power. Saddoud et al. [23] proposed a Dynamic Borrowing Scheduler (DBS) for IoT communications. DBS used roughly 70% of the bandwidth for M-M flows and 30% for H-H flows. Thus, in comparison to the 'Without Borrowing' scheduler, the bandwidth is well utilized. Yoshino et al. [31] proposed a new algorithm in modem management. Cell quality of connection-destination candidate is extracted from communication logs based on UE area ID. These measurements are directly used to make a connection decision. In terms of throughput, the proposed method outperforms the conventional method by 3.2 times for a UE traveling at 50 km/hr. Hasabelnaby et al. [13] proposed an approach based on two links, FSO and mmWave, with an optimal resource allocation scheme. Because FSO is impacted in foggy weather, whereas mmWave is impacted in rainy weather. Each RRH is directly linked to the BBU via one FSO link and one mmWave link. Omran et al. [20] attempted to maximize the number of MC (Macro Cells) offloaded users served using SCs. The authors investigated the situation in which the Macro Cell (MC) is congested and the Small Cells (SCs) are not fully loaded. The MC attempts to offload some of its users who are already located outside the range of SCs to these SCs. The authors believe that idle users within the range of each SC are willing to assist other users and act as relays. The authors have used User Relay Link Matrix, Utility Matrix, and Optimal assignment algorithm. Saddoud et al. [22] proposed a radio resource management (RRM) approach based on the UL IoT flows' Quality of Service (QoS) requirements. The authors used the Radio Resource Scheme in conjunction with traffic classifiers (M2M and H2H). Zhou et al. [36] created an intelligent traffic control policy by utilizing a deep learning algorithm known as long short-term memory (LSTM). LSTM considers not only current but also historical data. It uses past and current datasets to make localized predictions about future traffic characteristics. They were able to achieve 9.5 Gbps throughput with an average packet loss rate of 11%. Elkourdi et al. [7] proposed a cell selection/user association algorithm based on a Bayesian game. The algorithm takes into account the capabilities of the access nodes as well as the traffic type of the user equipment (UE). They were able to achieve a latency of up to 0.02 ms compared to 7 ms when using the maximum SINR approach. Al-Dulaimi et al. [2] proposed two spectrum coexistence frameworks for small cells: time filling and space filling (non-overlapping or overlapping). The highest throughput of 78% is achieved by non-overlapped filling with 76 small cells. Similarly, in non-overlapped filing, the highest number of transmitted packets of 85 % is obtained with 76 small cells. The smallest End-to-End delay of 0.56 msec is achieved with the 76 number of small cells in the non-overlapped filling. Guanding et al. [32] investigated the use of FD communications in UDN. The authors propose a novel ICIC (Inter-Cell Interference Coordination) scheme with four steps: base station association, user pairing and mode selection, power control, and resource block allocation. In-band full-duplex (FD) communication, which allows a device to transmit and receive on the same frequency spectrum at the same time, has recently been proposed to improve spectrum utilization. The simulation achieved a 25% throughput gain with 100 users in their study. Farooq et al. [8] have proposed a novel proactive load balancing scheme based on a semi-Markov process called OPERA. It takes advantage of the fact that daily human movement is highly predictable. They have been able to reduce the percentage of unsatisfied users from 19% in Real Deployment Settings to 0.35% (Users who cannot get the desired bit rate). Dai et al. [6] have proposed a two-tier approach for better network planning. Step 1: Training a model for predicting RSS at the end user location. Step 2: Use Genetic algorithm and greedy algorithm to optimize the BS deployment and its parameters. They have been able to achieve a 7.68% gain by the greedy algorithm and 18.5% gain by using Genetic Algorithm. Nguven et al. [19] have used Apollonian circle and straight line to analyze handover performance. They have proposed optimum handover settings to minimize both handover failure and ping pong effect. Masood et al. [17] have proposed Deep Neural Networks-based model to augment ray-tracing tools which are used for modeling various obstacles when designing networks. Authors have achieved 25% increase in prediction accuracy as compared to state-of-the-art empirical models and a 12x decrease in prediction time as compared to ray tracing. Zhang et al. [34] has shown Adaptive Interference Aware (AIA) Virtual Network Function (VNF) placement for 5G network. They have considered two scenarios for 5G communication, i.e. autonomous driving vehicles and 4K HD video transmission. Their AIA approach achieved an improvement of 24%. Xue et al. [29] has modeled a cell capacity of a 5G cellular network with interbeam interference. Their simulation predicts the number of beams required to maximize the cell capacity for a given network. Sarma et al. [25] have given Game Theory Models, Machine Learning Models, and other miscellaneous models to alleviate the interference. Zambianco et al. [33] designed a binary quadratic non-convex optimization that minimizes interslice interference which occurs due to multiple base stations and is generated due to the multiplexing of the spectrum. Soultan et al. [26] proposed a combination of Fast Frequency Reuse (FFR) and Soft Frequency Reuse (SFR) for both regular and irregular cellular networks. Gonzalez et al. [12] have proposed a method to combine the base station location at optimal places with optimal power allocation. They proposed instead of using a small number of high-power macro cells one can use a large number of microcells. They also took into consideration non-uniform distribution across the coverage area. They could reach a quality of service (QoS) of 0.15 bps per Hz with a power of 13.3 dB. Summary of all the Resource Management research covered above is given in Table 4.

Future Scope

Our study indicates a few challenges and limitations with the current research in the area of green communication and resource allocation in 5G. The below area can serve as a direction for future research. The models and algorithms are run on a small number of cells or users. In reality, many wireless networks are very large, with significant variations in parameters such as the number of users, obstacles, movement, and so on. Some studies take into account a fixed combination of cells (like Microcell—Picocell, Femtocell—Macro cell). To be useful in the real world, the methods/scheme must work across all possible combinations, as conditions can change due to new cell deployment. One of the most important aspects of 5G rollout will be non-standalone rollout, which will take advantage of existing 4G networks to quickly roll out 5G. There is scope for research on resource allocation and green communication in non-standalone 5G. Many of the research papers assume that the parameters are fixed in advance; however, there is room for improvement in the models/schemes proposed, where the parameters can be changed dynamically based on changes in external or internal conditions. A key area of focus for future research could be the computational effort required for the models and schemes proposed. As many wireless networks are already carrying a large amount of traffic which is expected to increase multi-fold with 5G, putting additional computational requirements on them might impact QoS.

Table 4 Resource Management research summary

Refs. No.	Analytical tool	Scheme/model	Work done	Used for
[4]	MATLAB	QoS -Resource allocation scheme algorithm	Improved resource utilization ratio, dropped request probability, throughput	Femtocell selection and resource allocation
[24]	MATLAB	Schedulers:static, dynamic, Borrowing	Improvement in rejection ratio, accepted flows	Resource allocation in H-H, M-M,IoT
[15]	–	Cell association strategy	SINR and Throughput improvement	Cell Association and selection
[23]	MATLAB	Dynamic borrowing scheduler and traffic classifier	Improvement in rejection ratio, accepted flows, Bandwidth utilization	Resource allocation in M-M flows, IoT Flows, H-H Flows
[31]	–	New modem manager model	Improvement in throughput	Cell selection method
[13]	–	Lexicographic Optimization	Avg BER, Avg Transmitted power	Resource allocation for optical & mmwave link
[20]	–	Optimal assignment algorithm	Total number of served user versus offloaded users, offloading efficiency versus offloaded users	Cell selection and Offloading
[22]	MATLAB	Radio resource management Scheme	Improvement in accepted flows and rejected flows	Resource Allocation in H-H, M-M, IoT
[36]	–	Long short-term memory algorithm	Throughput, Packet Loss rate	Radio Resource Assignment
[7]	–	Bayesian cell selection algorithm	Latency reduction	Cell selection
[2]	–	Non-overlapped & Overlap space filling scheme	Improvement in Throughput	Planning of small cell deployment
[32]	–	Inter-cell interference coordination scheme	Base station association, user pairing and mode selection, power control, and resource block allocation	interference reduction
[8]	MATLAB	Proactive load balancing framework 'OPERA'	Reduction in unsatisfied Users percentage	Load imbalance in HetNets
[6]	–	Greedy algorithm	RSS improvement	Base station deployments
[19]	NS3	Apolinion circle and straight line method	Handover failure and ping pong effect reduction	Handover
[17]	–	Deep neural network algorithm	RSS Prediction	Network design and performance optimization
[21]	–	Long short-term memory	Handover prediction	Radio resource wastage and overhead reduction

References

1. Abrol A, Jha RK (2016) Power optimization in 5g networks: a step towards green communication. IEEE Access 4:1355–1374
2. Al-Dulaimi A, Al-Rubaye S, Cosmas J, Anpalagan A (2017) Planning of ultra-dense wireless networks. IEEE Netw 31(2):90–96
3. Arévalo GV, Gaudino R (2019) Optimal dimensioning of the 5g optical fronthaulings for providing ultra-high bit rates in small-cell, micro-cell and femto-cell deployments. In: 2019 21st international conference on transparent optical networks (ICTON), pp 1–4. IEEE
4. Bouaziz A, Saddoud A, Chaouchi H et al (2021) QoS-aware resource allocation and femtocell selection for 5g heterogeneous networks
5. Bouras C, Diles G (2017) Energy efficiency in sleep mode for 5g femtocells. In: 2017 wireless days, pp 143–145. IEEE
6. Dai L, Zhang H (2020) Propagation-model-free base station deployment for mobile networks: Integrating machine learning and heuristic methods. IEEE Access 8:83375–83386
7. Elkourdi M, Mazin A, Gitlin RD (2018) Towards low latency in 5g hetnets: a Bayesian cell selection/user association approach. In: 2018 IEEE 5G world forum (5GWF), pp 268–272. IEEE
8. Farooq H, Asghar A, Imran A (2020) Mobility prediction based proactive dynamic network orchestration for load balancing with QoS constraint (opera). IEEE Trans Veh Technol 69(3):3370–3383
9. Ganame H, Yingzhuang L, Ghazzai H, Kamissoko D (2019) 5g base station deployment perspectives in millimeter wave frequencies using meta-heuristic algorithms. Electronics 8(11):1318
10. Gandotra P, Jha RK, Jain S (2017) Green communication in next generation cellular networks: a survey. IEEE Access 5:11727–11758
11. Ghosh S, De D, Deb P, Mukherjee A (2020) 5g-zoom-game: small cell zooming using weighted majority cooperative game for energy efficient 5g mobile network. Wirel Netw 26(1):349–372
12. González-Brevis P, Gondzio J, Fan Y, Poor HV, Thompson J, Krikidis I, Chung PJ (2011) Base station location optimization for minimal energy consumption in wireless networks. In: 2011 IEEE 73rd vehicular technology conference (VTC Spring), pp 1–5. IEEE
13. Hasabelnaby MA, Selmy HA, Dessouky MI (2019) Optimal resource allocation for cooperative hybrid FSO/mmW 5g fronthaul networks. In: 2019 IEEE photonics conference (IPC), pp 1–2. IEEE
14. Jiang F, Wang Bc, Sun Cy, Liu Y, Wang X (2018) Resource allocation and dynamic power control for D2D communication underlaying uplink multi-cell networks. Wirel Netw 24(2):549–563
15. Khan MF (2020) An approach for optimal base station selection in 5g hetnets for smart factories. In: 2020 IEEE 21st international symposium on a world of wireless, mobile and multimedia networks(WoWMoM), pp 64–65. IEEE
16. Liu Q, Shi J (2018) Base station sleep and spectrum allocation in heterogeneous ultra-dense networks. Wirel Pers Commun 98(4):3611–3627
17. Masood U, Farooq H, Imran A (2019) A machine learning based 3D propagation model for intelligent future cellular networks. In: 2019 IEEE global communications conference (GLOBECOM), pp 1–6. IEEE
18. Mukherjee A, Deb P, De D (2017) Small cell zooming based green congestion control in mobile network. CSI Trans ICT 5(1):35–43
19. Nguyen MT, Kwon S (2020) Geometry-based analysis of optimal handover parameters for self-organizing networks. IEEE Trans Wirel Commun 19(4):2670–2683
20. Omran A, Sboui L, Rong B, Rutagemwa H, Kadoch M (2019) Joint relay selection and load balancing using d2d communications for 5g hetnet mec. In: 2019 IEEE international conference on communications workshops (ICC Workshops), pp 1–5. IEEE (2019)

21. Ozturk M, Gogate M, Onireti O, Adeel A, Hussain A, Imran MA (2019) A novel deep learning driven, low-cost mobility prediction approach for 5g cellular networks: the case of the control/data separation architecture (CDSA). Neurocomputing 358:479–489
22. Saddoud A, Doghri W, Charfi E, Fourati LC (2018) 5g radio resource management approach for internet of things communications. In: International conference on Ad-Hoc networks and wireless, pp 77–89. Springer (2018)
23. Saddoud A, Doghri W, Charfi E, Fourati LC (2019) 5g dynamic borrowing scheduler for IoT communications. In: 2019 15th international wireless communications & mobile computing conference (IWCMC), pp 1630–1635. IEEE
24. Saddoud A, Doghri W, Charfi E, Fourati LC (2020) 5g radio resource management approach for multi-traffic IoT communications. Comput Netw 166:106,936
25. Sarma SS, Hazra R (2020) Interference mitigation methods for d2d communication in 5g network. In: Cognitive informatics and soft computing, pp 521–530. Springer
26. Soultan EM, Nafea HB, Zaki FW (2021) Interference management for different 5g cellular network constructions. Wirel Pers Commun 116(3):2465–2484
27. Tsai CW, Cho HH, Shih TK, Pan JS, Rodrigues JJPC (2015) Metaheuristics for the deployment of 5g. IEEE Wirel Commun 22(6):40–46
28. Xu X, Yuan C, Chen W, Tao X, Sun Y (2017) Adaptive cell zooming and sleeping for green heterogeneous ultradense networks. IEEE Trans Veh Technol 67(2):1612–1621
29. Xue Q, Li B, Zuo X, Yan Z, Yang M (2016) Cell capacity for 5g cellular network with inter-beam interference. In: 2016 IEEE international conference on signal processing, communications and computing (ICSPCC), pp 1–5. IEEE
30. Yang C, Li J, Guizani M (2016) Cooperation for spectral and energy efficiency in ultra-dense small cell networks. IEEE Wirel Commun 23(1):64–71
31. Yoshino M, Shingu H, Asano H, Morihiro Y, Okumura Y (2019) Optimal cell selection method for 5g heterogeneous network. In: 2019 IEEE 89th vehicular technology conference (VTC2019-Spring), pp 1–5. IEEE
32. Yu G, Zhang Z, Qu F, Li GY (2017) Ultra-dense heterogeneous networks with full-duplex small cell base stations. IEEE Netw 31(6):108–114
33. Zambianco M, Verticale G (2020) Interference minimization in 5g physical-layer network slicing. IEEE Trans Commun 68(7):4554–4564
34. Zhang Q, Liu F, Zeng C (2019) Adaptive interference aware VNF placement for service customized 5g network slices. In: IEEE INFOCOM 2019-IEEE conference on computer communications, pp 2449–2457. IEEE (2019)
35. Zhou L, Sheng Z, Wei L, Hu X, Zhao H, Wei J, Leung VC (2016) Green cell planning and deployment for small cell networks in smart cities. Ad Hoc Netw 43:30–42
36. Zhou Y, Fadlullah ZM, Mao B, Kato N (2018) A deep-learning-based radio resource assignment technique for 5g ultra dense networks. IEEE Netw 32(6):28–34

A Survey on Attention-Based Image Captioning: Taxonomy, Challenges, and Future Perspectives

Himanshu Sharma, Devanand Padha, and Arvind Selwal

1 Introduction

Computer Vision (CV) is an interdisciplinary branch of Artificial Intelligence (AI) that delivers approaches to devise effective digital systems that can perceive, comprehend, and interpret visual data in the same way humans do. Digital images are one of the most common visual resources generated over the Internet. The exponential growth of image-structured data over the internet has created a slew of insurmountable problems, one of which is how to search images over the web. Earlier, text-based metadata conveyed the visual contents of images. However, with massive amounts of images being generated every day, it becomes impossible to precisely describe all of them. As a result, automated strategies for implicitly describing image contents are required. When digital systems generate image descriptions, they become clever enough to narrate images and perform a multi-modal search. Therefore, there is a growing demand for systems that can describe images with promising results.

Image captioning (IC), a new sub-field of CV, is dedicated to automated image caption production. The generated description is extensive enough to describe all of the image' key features. A generic architecture of the IC model consists of a serial pipeline with two components [1]: a visual component and a description generation component. The visual component extracts and encodes a collection of semantic (objects, attributes, scenes) and spatial information from the query image [2]. The

H. Sharma (✉) · D. Padha · A. Selwal
Department of Computer Science and Information Technology, Central University of Jammu, Samba 181143, India
e-mail: himanshusharma.csit@gmail.com

D. Padha
e-mail: devanand.csit@cujammu.ac.in

A. Selwal
e-mail: arvind.csit@cujammu.ac.in

P. Singh et al. (eds.), *Machine Learning and Computational Intelligence Techniques for Data Engineering*, Lecture Notes in Electrical Engineering 998,
https://doi.org/10.1007/978-981-99-0047-3_58

description generation component receives the encoded details and expresses them in textual descriptions. Previously, visual components were made of object detectors only [1]. Object detectors are later substituted with more reliable global and local region-based image features like GIST, SIFT, HOG, and others [3]. The development of convenient and accurate object detection models like Convolutional Neural Networks (CNN) and more complex natural language decoders such as Recurrent Neural Networks (RNN) and Long Short-Term Memories (LSTM) further influenced the advent of IC. The classical IC architecture is thus transformed into a novel encoder-decoder form [4], wherein CNN encodes visual inputs and RNN decodes this visual context to produce captions.

Template-based IC, retrieval-based IC, and end-to-end IC are the three general types of IC approaches. Template-based IC methods are also known as generative methods. These techniques integrate the detected semantic content into predefined templates using some grammatical rules. The retrieval-based IC models, also known as ranking approaches, combine both image and text modalities into a multi-modal space in which semantically similar items are clustered together [1, 3]. The retrieval-based IC techniques are useful for both multi-modal search as well as for novel description generation. Succeeding a breakthrough in the accuracy of machine translation systems using an end-to-end architecture, researchers propose to build a similar end-to-end framework for IC by using a combination of CNN and RNN.

With recent developments in deep learning, the efficiency of IC models has improved dramatically. Despite these advancements, IC approaches cannot focus only on the visual subject of our interest. The introduction of the attention mechanism in IC addresses this limitation. Even though several studies in the literature have examined AIC approaches [5, 6], no empirical analysis has been conducted. This inspires us to carry out an analytical AIC survey. In this work, we comprehensively review AIC techniques (Table 1), taxonomy, IC dataset, challenges and future horizons. The following are the significant contributions of our research:

i. An in-depth investigation of AIC approaches.
ii. A contrast of the benchmark datasets used for training and evaluation of AIC models.
iii. Most recent developments and future directions in AIC.

The components are listed below in the order in which they appear. Section 2 covers AIC and its various types. Section 3 is a comprehensive review of the AIC literature. Section 4 categorizes and highlights the AIC dataset. Sections 5 and 6 explore open research issues and conclusions, respectively.

Table 1 A comparison of attention-based image captioning models

Year	Author(s)	Attention type	Image encoder	Language decoder	Optimization	Datasets	Evaluation metrics	Performance (BLEU-4)%
2015	Xu et al. [4]	Region	VGGNet	LSTM	RMSProp, Adam	Flickr 30K, MS COCO	BLEU, METEOR	25.0
2015	Jin et al. [9]	Semantic	VGGNet	LSTM	Adam	Flickr 30K, MS COCO	BLEU, METEOR, CIDEr	28.2
2015	Yang et al. [15]	Spatial	RNN, VGG Net	LSTM	AdaGrad	MS COCO	BLEU, METEOR, CIDEr	29.0
2017	Lu et al. [11]	Semantic, Spatial	ResNet	LSTM	Adam	Flickr30K, MS COCO	BLEU, METEOR, CIDEr	33.2
2017	Chen et al. [29]	Spatial, Channel-wise	VGG Net, ResNet	LSTM	Adadelta	Flickr 8K, Flickr 30K, MS COCO	BLEU, METEOR	30.4
2017	Gan et al. [10]	Semantic	ResNet	LSTM	Adam	Flickr 30K, MS COCO	BLEU, METEOR, CIDEr	34.1
2017	Pedersoli et al. [7]	Region	VGG Net	RNN	Adam	MS COCO	BLEU, METEOR, CIDEr	28.8
2017	Tavakoli et al. [31]	Semantic, Spatial	VGG Net	LSTM	–	MS COCO	BLEU, METEOR, ROUGE, CIDEr	28.7
2017	Liu et al. [8]	Region	VGG Net	LSTM	Adam	Flickr 30K, MS COCO	BLEU, METEOR	27.6
2017	Wu et al. [16]	Spatial	VGG Net	LSTM	–	Flickr 30K, MS COCO	BLEU, METEOR, ROUGE, CIDEr	40.0
2018	Yao et al. [32]	Semantic, Spatial	R-CNN	LSTM	Adam	MS COCO	BLEU, METEOR, ROUGE, CIDEr, SPICE	37.1
2018	Ye et al. [19]	Hybrid	ResNet	LSTM	Adam	MS COCO, Flickr 30K	BLEU, METEOR, ROUGE, CIDEr, SPICE	35.5
2018	Anderson et al. [33]	Variational	R-CNN	LSTM	–	MS COCO	BLEU, METEOR, ROUGE, CIDEr, SPICE	36.2
2018	Lu et al. [34]	Semantic	R-CNN	LSTM	Adam	MS COCO, Flickr 30K	BLEU, METEOR, ROUGE, CIDEr, SPICE	34.9

(continued)

Table 1 (continued)

Year	Author(s)	Attention type	Image encoder	Language decoder	Optimization	Datasets	Evaluation metrics	Performance(BLEU-4)%
2019	Ke et al. [12]	Semantic	R-CNN	LSTM	–	MS COCO	BLEU, METEOR, ROUGE, CIDEr, SPICE	36.8
2019	Qin et al. [20]	Hybrid	R-CNN	LSTM	Adam	MS COCO	BLEU, METEOR, ROUGE, CIDEr, SPICE	37.4
2019	Huang et al. [13]	Semantic	R-CNN	LSTM	Adam	MS COCO	BLEU, METEOR, ROUGE, CIDEr, SPICE	38.1
2019	Gao et al. [35]	Semantic, Spatial	ResNet	LSTM	Adam	MS COCO, Flickr 30K	BLEU, METEOR, ROUGE, CIDEr, SPICE	37.5
2019	Chen and Zhao[36]	Variational	ResNet	LSTM	Adam	MS COCO, Flickr 30K	BLEU, METEOR, ROUGE, CIDEr	35.4
2019	Wang et al. [37]	Semantic, Spatial	R-CNN, ResNet	LSTM	Adam	MS COCO	BLEU, METEOR, ROUGE, CIDEr, SPICE	37.6
2019	Yao et al. [38]	Semantic, Spatial	R-CNN	LSTM	Adam	MS COCO	BLEU, METEOR, ROUGE, CIDEr, SPICE	38.0
2019	Yang et al. [21]	Hybrid	ResNet	LSTM	Adam	MS COCO	BLEU, METEOR, ROUGE, CIDEr, SPICE	38.9
2019	Wang et al. [39]	Semantic, Spatial	R-CNN	LSTM	Adam	MS COCO	BLEU, METEOR, ROUGE, CIDEr, SPICE	32.5
2019	Yang et al. [40]	Semantic, Spatial	R-CNN	LSTM	Adam	MS COCO	BLEU, METEOR, ROUGE, CIDEr, SPICE	36.9
2019	Li et al. [22]	Adaptive	R-CNN	LSTM	Adam	MS COCO	BLEU, METEOR, ROUGE, CIDEr, SPICE	39.5
2019	Herdade et al. [17]	Spatial	R-CNN	Transformer	Adam	MS COCO	BLEU, METEOR, ROUGE, CIDEr, SPICE	35.5
2019	Li et al. [41]	Hybrid	VGG Net	Transformer	Adam	MS COCO	BLEU, METEOR, ROUGE, CIDEr, SPICE	37.8
2020	Pan et al. [42]	Spatial, Channel-wise	R-CNN, ResNet	LSTM	Adam	MS COCO	BLEU, METEOR, ROUGE, CIDEr, SPICE	40.3

(continued)

Table 1 (continued)

Year	Author(s)	Attention type	Image encoder	Language decoder	Optimization	Datasets	Evaluation metrics	Performance(BLEU-4)%
2020	Huang et al. [43]	Adaptive	R-CNN	LSTM	Adam	MS COCO	BLEU, METEOR, ROUGE, CIDEr, SPICE	37.0
2020	Liu et al. [44]	Semantic, Spatial	R-CNN	LSTM	Adam	MS COCO	BLEU, METEOR, ROUGE, CIDEr	37.9
2020	Wang et al. [23]	Hybrid	R-CNN	Bi-LSTM	Adam	MS COCO, Visual Genome	BLEU, METEOR, ROUGE, CIDEr, SPICE	36.6
2020	Sammani et al. [24]	Hybrid	R-CNN	LSTM	Adam	MS COCO	BLEU, METEOR, ROUGE, CIDEr, SPICE	38.0
2020	Zhou et al. [14]	Semantic	R-CNN	LSTM	Adam	MS COCO, Flickr 30K	BLEU, METEOR, ROUGE, CIDEr, SPICE	38.0
2020	Yu et al. [25]	Hybrid	R-CNN	Multimodal Trans-former	Adam	MS COCO	BLEU, METEOR, ROUGE, CIDEr	40.7
2020	Cornia et al. [26]	Hybrid	R-CNN, ResNet	Transformer	Adam	MS COCO	BLEU, METEOR, ROUGE, CIDEr	39.1
2020	Guo et al. [27]	Hybrid	R-CNN	Transformer	Adam	MS COCO	BLEU, METEOR, ROUGE, CIDEr, SPICE	39.9
2021	Wang et al. [28]	Hybrid	R-CNN	LSTM	–	MS COCO	BLEU, METEOR, ROUGE, CIDEr, SPICE	–
2021	Hossian et al. [45]	Semantic	R-CNN	Bi-LSTM	Adam	MS COCO	BLEU, METEOR, ROUGE, CIDEr	33.6

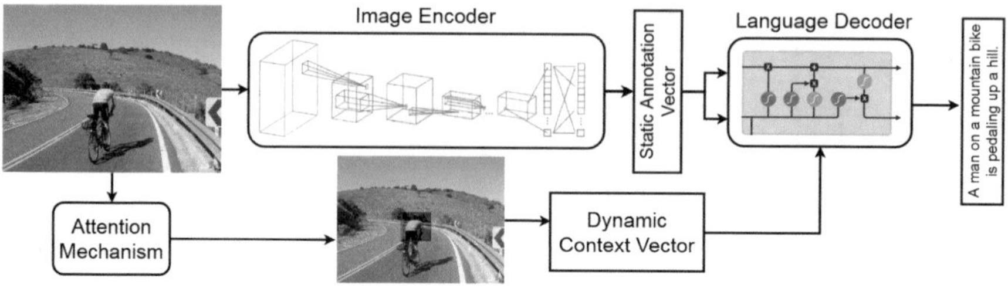

Fig. 1 The architecture of an attention-based image captioning model [4]

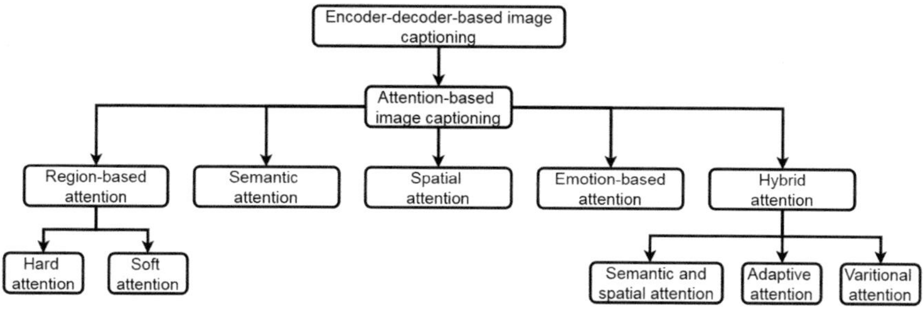

Fig. 2 A proposed taxonomy of attention-based image captioning systems

2 Attention-Based Image Captioning

A CNN-based image encoder and an LSTM-based language decoder form an AIC model's general architecture [4]. The CNN analyzes the query image and abstracts the required visual information to create static annotation vectors. Using this vector, the decoder guesses the words that will be appended to the caption. As shown in Figure 1, the model accomplishes the attention mechanism by altering the static annotation vector collected by the CNN. The attention mechanism analyzes the entire image for the previously created words and determines which portions of the image deserve attention in the present time frame. Succeeding a selection, the attention model collects visual inputs from this region and combines them with the static annotation vector to generate a dynamic context vector. The dynamic context is used by the language decoder to anticipate the next word close to the visual region currently selected. The types of attention used by the model to produce captions might take many different forms. Figure 2 depicts a taxonomy of different types of attention in IC.

2.1 Region-Based Attention

One of the early attempts in AIC is region-based attention. In region-based attention, an input image is split into a series of equal areas [4, 7, 8]. These regions direct the decoder component to generate captions at each time frame. The region-based attention is further classified into two kinds depending on the number of regions that receive attention at any given time frame.

i. ***Hard attention***: In hard attention, the decoder only pays attention to one of the "n" visual areas at any given moment. Using the LSTM hidden states, the model determines which image region requires attention at the "t+ 1" time frame. The specified attention region is used to create a new dynamic context vector. This context vector is used by the LSTM to construct the caption's next word.
ii. ***Soft attention***: Soft attention takes into account both previously attended image areas $(r^1, r^1, ...r^t)$ as well as the newly picked region (r^{t+1}). At the "t + 1" time frame, all of these areas are combined to create a dynamic context vector. Based on this dynamic context, the language decoder creates the next word of the caption.

2.2 Semantic Attention

Semantic attention is a more sophisticated version of region-based attention in which a collection of non-regular semantic regions of the query image are discovered using a mix of global and local features [9–14]. The decoder visits a single or a group of such semantic regions at a specified moment to generate captions.

2.3 Spatial Attention

An image has both semantic and spatial details that must be appropriately described. The region-based attention and semantic attention regardless focus only on the semantic contents. However, the global linkages must also be accurately conveyed. Spatial attention is concerned with implementing these global relations in the generated caption [15–17]. The spatial attention extracts the global context of the semantic entities and assists the decoder in expressing the interconnection between them.

2.4 Emotion-Based Attention

Individual emotions can be attended to while forming captions [18]. Emotion-based AIC systems strive to detect the mood and state of individuals present in the query image. Emotional attention is one of the most important attributes that the human eye pays attention to. Emotion-based attention strives to decipher the emotions of the current living entities. Emotional attention is the focus of the AIC subsection known as Stylized Captioning (SC).

2.5 Hybrid Attention

Implementing only one type of attention might cause the system to become biased and ineffective. As a result, different attention approaches are frequently combined to build a hybrid attention model [19–28]. Some of the most typical kinds of hybrid attention are outlined here.

i. ***Hybrid semantic-spatial attention***: This hybrid attention model combines the abilities of both semantic and spatial attention. The model attends to both the global as well as local attributes of the image.
ii. ***Adaptive attention***: The adaptive attention model itself decides the type of attention needed by the decoder at each time frame.
iii. ***Variational attention***: There are certain cases when one visual entity needs more attention than the other. The variational attention mechanism is suited for such situations as it can vary the level of attention between objects.

3 Literature Survey

Xu et al. [4] established one of the earliest attention-enabled image captioning techniques. As illustrated in Figure 1, the model extracts several high-level feature vectors known as annotation vectors from the query image. The context vector is dynamically constructed using the annotation vector at each time frame. To calculate attention using annotation vectors, the model employs a multilayered neural network directed by the values of prior hidden states. This model computes attention through the use of both hard and soft attention processes. Jin et al. [9] advocated that the scene-specific attention context be captured. A huge number of non-regular visual areas are used to represent the query image. A CNN encoder is used to extract a feature vector for each visual area. The global context vector is extracted in addition to the feature vector. To create captions, the feature vectors and context vectors are supplied in the LSTM decoder. Yang et al. [15] proposed an attention method based on the reviewer module. The model employs a review module to perform several review steps on the context vector and generate a thought vector that contains information about the regions requiring attention. Based on the thought vector and previous hidden states, the decoder constructs the next word of the caption. Sugano and Bulling [2] enhanced the gaze-assistance-based AIC even further by adding gaze information into the model's LSTM design. They developed a new semantic attention-based algorithm that combines the advantages of top-down and bottom-up techniques. The model learns to selectively pay attention to semantic information and fuse them into LSTM hidden states.

The image captioning model, according to Lu et al. [11], does not need attention to anticipate all of the words in the caption. Conjunctions, prepositions, and other non-semantic words can be predicted using previously created words. As a result of this research, a model based on selective attention is offered. The model implicitly determines whether or not to pay attention to the word currently being generated. Chen et al. [29] drew a similar conclusion and devised a hybrid attention method that incorporates both spatial and channel-wise attention. Gan et al. [10] developed a semantic attention-based compositional network in which the LSTM parameters are constructed using the attended semantic notions. Gan et al. [10] extended their previous work in Gan et al. [18] by presenting a novel IC framework for creating stylized captions. The model utilizes a new LSTM component that filters the stylized words in the lexicon for each visual input. Pedersoli et al. [7] devised an attention-based paradigm that establishes a direct connection between an image and its semantic phrases. The first IC model for personalized attention is developed by Park et al. [30]. The model generates hashtags and posts phrases for the Instagram dataset. Tavakoli et al. [31] studied human scene description abilities and created a salience-based semantic attention model. Liu et al. [8] produced a quantitative evaluation score for the agreement between the generated attention map and human attention by matching image regions to words. A framework for visual question answering and image captioning is created by Wu et al. [16], which directly learns high-level semantic structures.

Yao et al. [32] developed an attention model based on Graph Convolutional Networks (GCNs) that incorporates semantic and spatial objects. The model creates a graph of the observed visual content and then uses GCN to generate the context of the LSTM decoder. Learning a high-dimensional transformation matrix of the query image is proposed by Ye et al. [19]. The model implements a variety of attention, such as spatial, channel-wise, and so on, using the transformation matrix. Pan et al. [42] use a unified attention block that makes use of both spatial and channel-wise attention. Instead of concentrating on the visual background, Ke et al. [12] recommended paying attention to the words to improve captioning. Qin et al. [20] introduced the Look Back (LB) approach, which incorporates prior attention input into the current time frame. Huang et al. [13] introduced a module that extends the traditional attention technique to identify the relevance between attention outcomes and queries. Huang et al. [43] extend this work by constructing an adaptive AIC model in which attention may be varied depending on the requirements. To improve captioning reliability, Gao et al. [35] presented a two-pass AIC model that leverages an intentional residual attention network. Liu et al. [44] suggested a global and local information exploration and distilling strategy for word selection that abstracts scenes, spatial data, and attribute level data. Wang et al. [23] employed a memory mechanism-based attention method to simulate human visual interpretation and captioning skills. Sammani et al. [24] proposed a unique IC strategy that uses the current training captions instead of creating novel captions from scratch. Zhou et al. [14] provide improved multi-modeling using parts of speech (POS).

Anderson et al. [33] deploy attention at several levels by combining top-down and bottom-up techniques. The model decodes all scene-specific characteristics by attending to the semantic content at both the global and local levels. Chen and Zhao [36] also implemented a similar level-specific AIC model. Wang et al. [37] and Yao et al. [38] built a hierarchical attention-based IC model that computes attention at multiple levels. Wang et al. [28] went on to design a dynamic AIC model that generates captions without ignoring function words. Lu et al. [34] and Yang et al.

[21] constructed a traditional slot filling-based attention technique, where semantic contents filled template slots for generating captions.

Wang et al. [39] implemented a scene-graph-based image captioning model that incorporates all the visual information into the language decoder using a graph. Yang et al. [40] also implemented a similar auto-encoder model that incorporates the language inductive bias into the encoder-decoder framework.

Li et al. [22] proposed an augmented transformer model to achieve both vision-guided attention and concept-guided attention in a single framework. Herdade et al. [17] developed an object-relational transformer that directly contains information about the spatial relations between items identified through attention. Yu et al. [25] developed a multi-modal transformer for image captioning that captures both intra-modal and inter-modal interactions in a single attention block. Li et al. [41] used an entangled attention technique to allow the transformer to acquire both semantic and visual information at the same time. Cornia et al. [26] created a meshed memory transformer-based model for attention that employs both low- and high-level visual information.

Guo et al. [27] synthesize captions using a self-attention extension to overcome the limits of the transformer-based IC model. Hossain et al. [45] employed Generative Adversarial Networks (GANs) to develop a unique image and text description synthesis technique. To label images, the model leverages an attention mechanism that has been trained on both real and synthetic data.

4 Benchmark Datasets

One of the dominant studies in CV is AIC, which uses a variety of deep learning models like CNN, RNN, and LSTMs to create a mapping between images and captions. Deep learning-based models require large amounts of training data to achieve a reliable level of model training. A wide range of benchmark AIC datasets is available for training and assessment. Table 2 contains a list of benchmark datasets for AIC training and validation. Table 2 details the AIC dataset, including the number of images, captions per image, scene types, and image sources.

5 Open Research Challenges

Based on the comprehensive literature survey of AIC, we conclude the following areas that need focus in the future.

i. **Limited accuracy and efficiency in real-time situations**: Since AIC is still in its early stages, there is a lot of room for development in terms of quality and reliability [28, 45]. The testing efficiency of the AIC technique is now sufficient on images in the training dataset. However, when it comes to real-time scenarios, the efficiency plummets substantially. This is because AIC models are only trained on a restricted set of visual content. One of the key research

Table 2 A comparative summary of benchmark attention-based IC datasets

Year	Author(s)	Dataset	Images	Captions	Scenes	Image source
2006	Grubinger et al. [46]	IAPR TC-12	20000	1–5	Mixed	Primary collected
2010	Rashtchian et al. [47]	PASCAL 1K	1000	5	Mixed	PASCAL VOC
2011	Ordonez et al. [1]	SBU dataset	1000000	1	Mixed	Flickr.com
2013	Hodosh et al. [3]	Flickr 8K	8092	1–5	People and animals	Flickr.com
2014	Gong et al. [48]	Flickr 30K	31783	5	People and animals	Flickr.com
2014	Lin et al. [49]	MS COCO	328000	5	Mixed	Web
2015	Lin et al. [50]	NYU-v2	1449	5	Indoor scenes	NYU-RGBD

problems in AIC is adding more and more visual infusions for good real-time performance.

ii. **Scarcity of domain-specific datasets**: Considering that AIC is a developing discipline, there is a great need for unique and efficient domain-specific datasets. Only a few of these datasets are available for now [18, 30]. The dynamic and demanding quality of IC, on the other hand, demands huge training data with relevant descriptions. Consequently, the research community must concentrate on the development of these domain-specific datasets.

iii. **Inefficiency when generating extended sentences**: RNNs suffer from the vanishing gradient problem when generating longer phrases. The problem is solved by storing the temporal context in the hidden states of the LSTM model. A similar issue arises in attention-based systems when the AIC model loses its attention context while producing lengthier sentences. Future studies will focus on maintaining attention context over extended periods utilizing hybrid attention strategies.

iv. **Attention misinterpretation**: Human attention is a very intelligent mechanism that can discriminate between background and foreground information with pinpoint precision. However, AIC systems have not yet reached this level. The attention mechanism used in IC systems sometimes considers the background, resulting in descriptions that include non-salient features as well [19–22, 27, 28, 45]. Consequently, when reviewed by a human expert, the generated captions are inadequate. Thus, the degree of attention that should be paid to semantic content is an open research topic for AIC.

v. **Limited use of modern deep learning techniques**: Advanced deep learning-based models like Transformers and Graph Neural Networks (GNN) still find a limited application in AIC. [17, 25–27, 41] have begun using them in IC with encouraging results. As a result, adopting more of these sophisticated models into AIC remains an ongoing research problem.

6 Conclusions

In this study, we examined various cutting-edge AIC techniques. According to the findings, hybrid AIC models outperform other strategies, with hybrid AIC accounting for 60 % of top-performing techniques. In addition, R-CNN and LSTM are used as image encoder and language decoder in 80 % and 70 % of the top approaches, respectively, indicating that encoder-decoder architecture predominates other AIC strategies. Yu et al. [25] have achieved the highest BLUE-4 score of 40.7 % among all AIC techniques using a hybrid attention mechanism. The transformer-based language model, even though a novel introduction in AIC, yields a promising BLUE-4 of 39.9 % indicating that they may produce better results in the future. In addition to these findings, we observe that attention-based methods do not perform fairly in real-time situations. Moreover, these approaches suffer from several other problems such as limited vocabulary, high complexity, and misinterpretation of attention. The limited availability of personalized datasets is another important issue that needs to be addressed. Recently, the field of IC is reporting more success with the application of data augmentation techniques. The futuristic approaches can take advantage of the robustness of hybrid approaches, where the advantages of both traditional and recent models are combined.

References

1. Vicente O, Girish K, Tamara B (2011) Im2text: Describing images using 1 million captioned photographs. Adv Neural Inf Process Syst 24:1143–1151
2. Sugano Y, Bulling A (2016). Seeing with humans: gaze-assisted neural image captioning. arXiv: 1608.05203 [cs]
3. Micah H, Peter Y, Julia H (2013) Framing image description as a ranking task: data, models and evaluation metrics. J Artif Intell Res 47:853–899
4. Xu K, Ba J, Kiros R, Cho K, Courville A, Salakhudinov R, Zemel R, Bengio Y Show, attend and tell: neural image caption generation with visual attention, p 10
5. Hossain MZ, Sohel F, Shiratuddin MF, Laga H (2019) A comprehensive survey of deep learning for image captioning. ACM Comput Surv (CsUR) 51(6):1–36
6. Zohourianshahzadi Z, Kalita JK (2021) Neural attention for image captioning: review of outstanding methods. Artif Intell Rev 1–30
7. Pedersoli M, Lucas T, Schmid C, Verbeek J (2017) Areas of attention for image captioning. arXiv:1612.01033 [cs]
8. Liu C, Mao J, Sha F, Yuille A Attention correctness in neural image captioning, p 7
9. Jin J, Fu K, Cui R, Sha F, Zhang C (2015) Aligning where to see and what to tell: image caption with region-based attention and scene factorization. arXiv:1506.06272 [cs, stat]
10. Gan Z, Gan C, He X, Pu Y, Tran K, Gao J, Carin L, Deng L. Semantic compositional networks for visual captioning. arXiv:1611.08002 [cs]
11. Lu J, Xiong C, Parikh D, Socher R (2017) Knowing when to look: adaptive attention via a visual sentinel for image captioning. In: Proceedings of the IEEE conference on computer vision and pattern recognition, pp 375–383

12. Ke L, Pei W, Li R, Shen X, Tai YW Reflective decoding network for image captioning. arXiv:1908.11824 [cs]
13. Huang L, Wang W, Chen J, Wei XY (2019) Attention on attention for image captioning. arXiv:1908.06954 [cs]
14. Zhou Y, Wang M, Liu D, Hu Z, Zhang H (2020) More grounded image captioning by distilling image-text matching model. In: 2020 IEEE/CVF conference on computer vision and pattern recognition (CVPR), pp 4776–4785, Seattle, WA, USA, June 2020. IEEE
15. Yang Z, Yuan Y, Wu Y, Cohen WW, Salakhutdinov RR Review networks for caption generation, p 9
16. Wu Q, Shen C, van den Hengel A, Wang P, Dick A (2016) Image captioning and visual question answering based on attributes and external knowledge. arXiv:1603.02814 [cs]
17. Herdade S, Kappeler A, Boakye K, Soares J (2020) Image captioning: transforming objects into words. arXiv:1906.05963 [cs]
18. Gan C, Gan Z, He X, Gao J, Deng L (2017) StyleNet: generating attractive visual captions with styles. In: 2017 IEEE conference on computer vision and pattern recognition (CVPR), pp 955–964, Honolulu, HI, July 2017. IEEE
19. Senmao Y, Junwei H, Nian L (2018) Attentive linear transformation for image captioning. IEEE Trans Image Process 27(11):5514–5524
20. Qin Y, Du J, Zhang Y, Lu H Look back and predict forward in image captioning, p 9
21. Yang X, Zhang H, Cai J (2019) Learning to collocate neural modules for image captioning. In: 2019 IEEE/CVF international conference on computer vision (ICCV), pp 4249–4259, Seoul, Korea (South), October 2019. IEEE
22. Jiangyun L, Peng Y, Longteng G, Weicun Z (2019) Boosted transformer for image captioning. Appl Sci 9(16):3260
23. Li W, Zechen B, Yonghua Z, Hongtao L (2020) Show, recall, and tell: image captioning with recall mechanism. Proc AAAI Conf Artif Intell 34(07):12176–12183
24. Sammani F, Melas-Kyriazi L (2020) Show, edit and tell: a framework for editing image captions. In: 2020 IEEE/CVF conference on computer vision and pattern recognition (CVPR), pp 4807–4815, Seattle, WA, USA, June 2020. IEEE
25. Jun Yu, Jing L, Zhou Yu, Qingming H (2020) Multimodal transformer with multi-view visual representation for image captioning. IEEE Trans Circuits Syst Video Technol 30(12):4467–4480
26. Cornia M, Stefanini M, Baraldi L, Cucchiara R (2020) Meshed-memory transformer for image captioning. arXiv:1912.08226 [cs]
27. Guo L, Liu J, Zhu X, Yao P, Lu S, Lu H (2020) Normalized and geometry-aware self-attention network for image captioning. In: 2020 IEEE/CVF Conference on Computer Vision and Pattern Recognition (CVPR), pp 10324–10333, Seattle, WA, USA, June 2020. IEEE
28. Wang C, Gu X (2021) An image captioning approach using dynamical attention. In: 2021 international joint conference on neural networks (IJCNN), pp 1–8
29. Chen L, Zhang H, Xiao J, Nie L, Shao J, Liu W, Chua T-S (2017) SCA-CNN: spatial and channel-wise attention in convolutional networks for image captioning. arXiv:1611.05594 [cs]
30. Chunseong Park C, Kim B, Kim G (2017) Attend to you: personalized image captioning with context sequence memory networks. arXiv:1704.06485 [cs]
31. Tavakoli HR, Shetty R, Borji A, Laaksonen J (2017) Paying attention to descriptions generated by image captioning models. arXiv:1704.07434 [cs]
32. Yao T, Pan Y, Li Y, Mei T (2018) Exploring visual relationship for image captioning. arXiv:1809.07041 [cs]
33. Anderson P, He X, Buehler C, Teney D, Johnson M, Gould S, Zhang L (2018) Bottom-up and top-down attention for image captioning and visual question answering. arXiv:1707.07998 [cs]
34. Lu J, Yang J, Batra D, Parikh D (2018) Neural baby talk. arXiv:1803.09845 [cs]
35. Gao L, Fan K, Song J, Liu X, Xu X, Shen HT Deliberate attention networks for image captioning, p 8
36. Chen S, Zhao Q (2019) Boosted attention: leveraging human attention for image captioning. arXiv:1904.00767 [cs]

37. Weixuan W, Zhihong C, Haifeng H (2019) Hierarchical attention network for image captioning. Proc AAAI Conf Artif Intell 33:8957–8964
38. Yao T, Pan Y, Li Y, Mei T (2019) Hierarchy parsing for image captioning. arXiv:1909.03918 [cs]
39. Wang D, Beck D, Cohn T (2019) On the role of scene graphs in image captioning. In: Proceedings of the beyond vision and language: in tegrating real-world knowledge (LANTERN), pp 29–34, Hong Kong, China, 2019. Association for Computational Linguistics
40. Yang X, Tang K, Zhang H, Cai J (2019) Auto-encoding scene graphs for image captioning. In: 2019 IEEE/CVF conference on computer vision and pattern recognition (CVPR), Long Beach, CA, USA, June 2019. IEEE, pp 10677–10686
41. Li G, Zhu L, Liu P, Yang Y Entangled transformer for image captioning, p 10
42. Pan Y, Yao T, Li Y, Mei T X-linear attention networks for image captioning, p 10
43. Huang L, Wang W, Xia Y, Chen J (2020) Adaptively aligned image captioning via adaptive attention time. arXiv:1909.09060 [cs]
44. Liu F, Ren X, Liu Y, Lei K, Sun X (2020) Exploring and distilling cross-modal information for image captioning. arXiv:2002.12585 [cs]
45. Hossain MZ, Sohel F, Shiratuddin MF, Laga H, Bennamoun M (2021) Text to image synthesis for improved image captioning. IEEE Access 9:64918–64928
46. Grubinger M, Clough P, Müller H, Deselaers T (2006) The iapr tc-12 benchmark: a new evaluation resource for visual information systems. In: International workshop onto image, vol 2
47. Rashtchian C, Young P, Hodosh M, Hockenmaier J (2010) Collecting image annotations using amazon's mechanical turk. In: Proceedings of the NAACL HLT 2010 workshop on creating speech and language data with amazon's mechanical Turk, pp 139–147
48. Gong Y, Wang L, Hodosh M, Hockenmaier J, Lazebnik S (2014) Improving image-sentence embeddings using large weakly annotated photo collections. In: European conference on computer vision, pp 529–545. Springer
49. Lin T-Y, Maire M, Belongie S, Hays J, Perona P, Ramanan D, Dollár P, Zitnick CL (2014) Microsoft coco: common objects in context. In: European conference on computer vision, pp 740–755. Springer
50. Lin D, Kong C, Fidler S, Urtasun R (2015) Generating multi-sentence lingual descriptions of indoor scenes. arXiv:1503.00064

QKPICA: A Socio-Inspired Algorithm for Solution of Large-Scale Quadratic Knapsack Problems

Laxmikant, C. Vasantha Lakshmi, and C. Patvardhan

1 Introduction

Evolutionary computation has gone through various inspirational ideas in the past few decades. Inspiration from natural selection, behaviours of animals, birds, and fish, physical and chemical reactions have been successfully proposed in the literature. Recently, algorithms inspired by human interaction and knowledge transfer have been proposed. The applicability of these meta-heuristics has been well-established on a wide range of problems concerning planning, design, simulation, identification, control, and classification and is gradually increasing with their utilization in solving untried optimization problems [2]. These meta-heuristics do not guarantee to find the optimal global solution. They only provide a satisficing solution or a "near-optimal" solution within "reasonable" computation time for otherwise intractable problems for finding optimal solutions.

In these meta-heuristic algorithms, exploration (diversification) and exploitation (intensification) are the fundamental operations to search for the optimum solution in the solution space. Exploration is used to explore the feasible solution space to find better solutions. Exploitation is used to explore the solution space around the solution in hand, to get more promising solutions nearby. A proper balance in the computational effort devoted to these two operations is critical to the success of any such meta-heuristic [2].

Laxmikant (✉) · C. Vasantha Lakshmi · C. Patvardhan
Dayalbagh Educational Institute, Dayalbagh, Agra 282005, India
e-mail: lkrdkrishnan@gmail.com

C. Vasantha Lakshmi
e-mail: vasanthalakshmi@dei.ac.in

C. Patvardhan
e-mail: cpatvardhan@dei.ac.in

P. Singh et al. (eds.), *Machine Learning and Computational Intelligence Techniques for Data Engineering*, Lecture Notes in Electrical Engineering 998,
https://doi.org/10.1007/978-981-99-0047-3_59

Genetic Algorithms (GA) [15], Evolution Strategies (ES) [30, 32], and Differential Evolution (DE) [33] are some widely studied evolutionary computation meta-heuristics inspired by natural selection theory [8]. Particle Swarm Optimization (PSO), Ant Colony Optimization (ACO), Artificial Bee Colony (ABC), Cuckoo Search (CS), and Firefly Algorithm (FA) are inspired by the social and individual behaviours of insects, birds, and animals [5]. Simulated Annealing (SA) [9, 20], Gravitational Search Algorithm (GSA), and Chemical-Reaction Optimization Algorithm (CROA) are some physics- and chemistry-inspired evolutionary computation meta-heuristics [34].

It is well-established that the actual real-life evolution of individuals is not biological only; it has social evolution, to a more considerable extent. This social evolution is much faster than the biological evolution [31]. Therefore, algorithms inspired by the interaction of individuals and societies are expected to perform better than algorithms inspired by biological evolution. Numerous such algorithms mimicking human behaviour and knowledge sharing have been proposed in the literature. Society and Civilisation Algorithm [29], Parliamentary Optimization Algorithm [4], and Imperialist Competitive Algorithm [1] are some prominent algorithms belonging to this class. These algorithms follow grouping mechanisms to divide the entire solution space into different discrete or overlapping solution groups. These groups provide two-layered influence or inspiration—individuals in a group mimic the behaviour of their local best while local bests in each group follow the global best.

Imperialist Competitive Algorithm (ICA) has been applied to many engineering search and optimization problems [18]. It is inspired by the socio-political process of imperialism and imperialistic competition. The operators in ICA are designed to solve continuous optimization problems. These operators cannot be directly applied to those problems with binary nature. Towards this end, a few discrete versions of ICA have also been developed and reported in the literature viz., Discrete Binary ICA (DB-ICA) [24], Discrete ICA (DICA) [12], Modified ICA (MICA) [23], and Binary ICA (BICA) [22] to name a few. BICA uses nine transfer functions for binary assimilation and outperforms DB-ICA. This paper presents an enhanced ICA dubbed as QKPICA to solve large-scale Quadratic Knapsack Problems (QKPs). The computational performance of this new algorithm is compared with the existing binary version of ICA—Binary ICA (BICA). QKPICA easily outperforms BICA on the large-scale Quadratic Knapsack benchmark problems tested.

The remainder of the paper is organized as follows. The Quadratic Knapsack problem description is given in Sect. 2. ICA and BICA are described in Sect. 3. The proposed QKPICA is explained in Sect. 4. The benchmark dataset and parameters used in the computational experiments are given in Sect. 5. The performance of BICA and QKPICA on large-scale QKP benchmark instances are reported and discussed in Sect. 6. Section 7 concludes the paper.

2 Quadratic Knapsack Problems (QKPs)

Quadratic Knapsack Problem is a quadratic counterpart of the binary or 0-1 knapsack problem. In a QKP, each item pair—i and j—has an associated positive integer value p_{ij} that gets added to the knapsack profit whenever items i and j are selected together in a solution along with their individual profits. These pairs of values form a matrix $Q_{n\times n} = (p_{ij})$ of positive integers, where $i, j \in \{1, 2, \ldots, n\}$. The diagonal values p_{ii} in this matrix represent the selection values of ith item, while other values p_{ij} represent bonus profit when both items i and j are selected. The objective of QKP is to search for a subset of items that maximizes the overall sum of values Eq. (1), and at the same time, the sum of their weights do not exceed the knapsack capacity C, Eq. (2).

$$\text{Maximize profit: } \sum_{i=1}^{n}\sum_{j=1}^{n} p_{ij}x_i x_j \tag{1}$$

$$\text{Subject to: } \sum_{n=1}^{n} w_i x_i \le C \tag{2}$$

$$\text{where } x_i, x_j \in \{0, 1\},\ i, j = 1, 2, \ldots, n$$

QKP has been applied in the fields of finance [21], cluster analysis [28], clique problem [11, 25], compiler design [16], and VLSI design [13] to name a few. QKP has been reported as one of the hardest combinatorial optimization problems in NP-hard class [6, 27]. The hardness of the QKP stems from the fact that even for 1000 items, the input or problem matrix size is 1000 × 1000, i.e., 1M and for 2000 items it becomes 4M.

Gallo et al. [14] introduced QKP and derived upper bounds using upper planes. Various exact algorithms like Lagrangian relaxation [6], Lagrangian decomposition [3], semi-definite programming [17], and others are also been given for QKP. Julstrom [19] provided a greedy Genetic Algorithm (GGA) with the use of three heuristics for QKP—Absolute Value Density (AVD), Relative Value Density (RVD), and Dual Heuristic, and reported their performance on QKP instances of 100 and 200 variable objects. Patvardhan et al. [26] presented a novel Quantum-inspired Evolutionary Algorithms that showed better performance than GGA over a wide range of benchmark instances in terms of consistency in finding the optimal solution. This work presents an enhanced socio-inspired meta-heuristic QKPICA for solving large-scale QKPs of sizes 1000 and 2000. Some other salient literature references on this problem include [7, 35].

3 ICA and BICA

Imperialist Competitive Algorithm (ICA) is a widely studied and applied socio-inspired evolutionary algorithm. It simulates the imperialistic competition among countries to improve themselves by taking control of other weaker ones.

The initial population generated in ICA is called a population of countries; each country corresponds to an individual solution. The cost of a country is the counterpart of fitness in canonical EAs. When evaluated, these countries are divided into two groups according to their cost—imperialists and colonies. Each imperialist country forms an empire by possessing some colonies in the proportion of its cost, i.e., an imperialist with a higher cost acquires a higher number of colonies. The total power of an empire is calculated using its imperialist's and colonies' power.

After building empires, ICA starts moving colonies in an empire towards its imperialist. This movement of colonies is done by two variation operators: Assimilation and Revolution. In Assimilation, colonies move towards their relevant best imperialist. However, in Revolution, a random deviation is added to the direction of this movement. Whenever a colony exhibits better fitness than its imperialist, ICA exchanges its role with the imperialist. As the algorithm progresses, empires start an imperialistic competition by improving their power by taking over the colonies of other weaker empires. In this competition, a more powerful empire has more chances of possessing the weakest colony. Hence, powerful empires become more powerful as the weaker ones collapse into them. Eventually, there comes a state when all empires collapse into one. This empire contains all the countries as its colonies. Thus, the imperialistic competition ends, and the ICA terminates. Pseudocode of ICA is provided in Algorithm 1.

Algorithm 1 Pseudo-code of Imperialist Competitive Algorithm

```
Generate random countries
Form empires
repeat
    Assimilate: move the colonies in an empire toward their imperialist
    Revolution: move the colonies with some deviation toward their imperialist
    If a colony outperforms its imperialist, then exchange their roles
    Imperialistic competition: better empires fight for colonies from weaker empires
    Collapse empires having no colonies to the stronger empires
until only one empire remains
```

Some combinatorial optimization problems are binary, i.e., solutions to these problems have a canonical binary representation. Examples are the set covering problem, maximum-independent-set problem, and knapsack problem. Binary strings can represent solutions to these problems by including (excluding) a set, vertex, or item in (from) a candidate solution according to the corresponding entry in the binary string being 1 (0).

Table 1 Transfer functions [22]

Number	Name	Transfer function
1	TF1	$F(d) = \frac{1}{1+e^{-2d}}$
2	TF2	$F(d) = \frac{1}{1+e^{-d}}$
3	TF3	$F(d) = \frac{1}{1+e^{-\frac{d}{2}}}$
4	TF4	$F(d) = \frac{1}{1+e^{-\frac{d}{3}}}$
5	TF5	$F(d) = \left\lvert erf\left(\frac{\sqrt{\pi}}{2}d\right)\right\rvert = \left\lvert \frac{2}{\sqrt{\pi}} \int_0^{\frac{\sqrt{\pi}}{2}d} e^{-t^2} dt \right\rvert$
6	TF6	$F(d) = \tanh(d)$
7	TF7	$F(d) = \left\lvert \frac{d}{\sqrt{1+d^2}} \right\rvert$
8	TF8	$F(d) = \left\lvert \frac{2}{\pi} \arctan(\frac{\pi}{2}d) \right\rvert$
9	TF9	$F(d) = 2 \times \left\lvert \frac{1}{1+e^{-d}} - 0.5 \right\rvert$

BICA [22] provides nine different binary versions of ICA. Each time the assimilation process is designed using a different transfer function. BICA converts the distance between a colony and its imperialist, say d, into a probability value using a transfer function $F(d)$ (see Table 1). The colonies in BICA move towards their imperialist by changing 0s and 1s according to this probability value. To update the position of colonies, Eq. 3 is utilized for transfer functions TF1 through TF4, and Eq. 4 is utilized for TF5 through TF9.

$$x_i(t+1) = \begin{cases} 0 & \text{if } rand < F(d_i(t+1)) \\ 1 & \text{if } rand \geq F(d_i(t+1)) \end{cases} \tag{3}$$

$$x_i(t+1) = \begin{cases} Not(x_i(t)) & \text{if } rand < F(d_i(t+1)) \\ x_i(t) & \text{if } rand \geq F(d_i(t+1)) \end{cases} \tag{4}$$

Here, $x_i(t)$ represents ith bit of a colony at time t.

4 The Proposed QKPICA

Heuristics from [19] are utilized to seed the initial population consisting of N countries in QKPICA. Each country C in this population is represented as a binary vector in n-dimensional search space as Eq. 5.

$$C = (c_1, c_2, \ldots, c_n), \text{ where } c_i \in \{0, 1\} \tag{5}$$

Here, each binary value c_i in a country C represents inclusion (1) or exclusion (0) of ith item in the knapsack.

The best M countries in this population are designated imperialists. This selection is based on their cost $f(C_i)$ as given in Eq. 6.

$$f(C) = \sum_{i=1}^{n} \sum_{j=1}^{n} p_{ij} c_i c_j, \ \forall c_i, c_j \in \{C\} \tag{6}$$

These M imperialists form M empires by acquiring remaining $(N - M)$ countries in the proportion of their cost. Hence, a better imperialist acquires more colonies and creates a bigger empire. The assignment of colonies to these empires is done using the roulette-wheel selection process [15]. That is, the initial number of colonies $C.E_i$ in an empire E_i is calculated in proportion to the empires' relative fitness, as in Eq. 7.

$$C.E_i = round\left\{\left|\frac{f(M_i)}{\sum_{j=1}^{M} f(M_j)}\right| \cdot (N - M)\right\} \tag{7}$$

Colonies in empires try to improve with assimilation and revolution processes. Assimilation is implemented by moving the colonies towards their corresponding imperialist. This movement of colonies is done based upon the Hamming distance between the imperialist and the corresponding colonies. Equation 8 is used to find the moving bits (m) for a colony with the Hamming distance D; U represents the integer uniform distribution function. Out of the D different bits, m bits are chosen randomly, and such selected bits are copied from imperialist to the corresponding bits in the colony. This assimilation process is listed in pseudocode given in Algorithm 2.

$$m \approx U(0, D) \tag{8}$$

Algorithm 2 Pseudo-code of proposed binary assimilation process

```
Calculate the distance D between colony and its imperialist (M_i)
Calculate the number of bits m to move toward imperialist using Eq. 8
Randomly select colony's m bits out of D bits and change according to M_i
if generated new_colony outperforms the colony then
    Keep the new_colony, discard the colony
else
    Discard the new_colony
end if
```

A random change in the colony is made in the revolution process. If this change improves the colony, then it is kept; otherwise, it is discarded. For revolution in QKPICA, a simple mutation operator [15] is used.

After assimilation and revolution, competition within the empire starts, making the best country the imperialist of the empire. This process updates the best local

Table 2 Parameter values used in BICA and QKPICA

Parameters	BICA	Proposed QKPICA
β	1.5	–
γ	0.8	0.8
ξ	0.77	0.77
N	100	100
M	10	10

solution. The newly updated imperialist is the next moving position in the subsequent assimilation process.

Empire's hunger for power starts a competition between them called imperialistic competition. In this competition, empires try to improve their power by acquiring colonies from the weaker empires. The power of an empire $(P.E_i)$ is calculated as the sum of the imperialist's cost and a fraction (ξ) of its colonies' mean cost, Eq. 9.

$$P.E_i = f(M_i) + \xi.\text{mean}\{f(\text{colonies of empire } E_i)\} \tag{9}$$

In this imperialistic competitive process, weaker empires become weaker and collapse into stronger ones. Eventually, when only one empire remains, QKPICA terminates.

5 Computational Experiments

For this study, a benchmark dataset of 80 large-scale QKP instances of 1000 and 2000 objects is taken from [35]. Both BICA and QKPICA are implemented using the C++ programming language on a Windows 10 operating system. This machine was configured with Intel® Core™ i7-7700 CPU (3.60GHz) with 8M Cache and 8GB RAM. The best parameters for BICA as suggested in [22]—assimilation rate $(\beta) = 1.5$, revolution rate $(\gamma) = 0.8$, and value of $\xi = 0.77$—are used in these experiments. However, QKPICA omits the assimilation rate (β) parameter; it is a parametric improvement in QKPICA over BICA and other ICAs. The parameters used in these experiments are summarized in Table 2. The maximum number of iterations is taken as 100.

6 Results and Discussion

Results of best BICA and QKPICA on QKP instances of 1000 objects are given in Table 3. The solution quality of BICAs and QKPICA on a problem instance is compared in Figure 1. QKPICA provided the near-optimal solution with a maximum

Table 3 Results of best BICA and QKPICA on 40 QKP instances of size 1000

Instance	Best	Best BICA				QKPICA			
	[7, 35]	BS	MBS	WS	$\%_{gap}$	BS	MBS	WS	$\%_{gap}$
1000_25_1.dat	6172407	2134501	2042651.4	2006858	65.42	**6165895**	**6165895**	**6165895**	0.11
1000_25_2.dat	229941	24270	20208.4	18542	89.45	**229933**	**229933**	**229933**	0.00
1000_25_3.dat	172418	15484	12465.7	11392	91.02	**172362**	**172362**	**172362**	0.03
1000_25_4.dat	367426	49645	41256.1	39467	86.49	**367426**	**367426**	**367426**	0.00
1000_25_5.dat	4885611	2127097	2023034.8	1986929	56.46	**4884243**	**4884243**	**4884243**	0.03
1000_25_6.dat	15689	1467	1057.8	923	90.65	**15689**	**15689**	**15689**	0.00
1000_25_7.dat	4945810	2117375	2040555.4	2016643	57.19	**4943010**	**4943010**	**4943010**	0.06
1000_25_8.dat	1710198	583966	561269.6	547112	65.85	**1710073**	**1710073**	**1710073**	0.01
1000_25_9.dat	496315	70621	65712.9	62152	85.77	**496315**	**496315**	**496315**	0.00
1000_25_10.dat	1173792	315169	295934.2	287267	73.15	**1173607**	**1173607**	**1173607**	0.02
1000_50_1.dat	5663590	3149736	3036959.5	2988560	44.39	**5663547**	**5663547**	**5663547**	0.00
1000_50_2.dat	180831	16048	13929.4	12895	91.13	**180725**	**180725**	**180725**	0.06
1000_50_3.dat	11384283	4266789	4059769.5	3996590	62.52	**11384216**	**11384216**	**11384216**	0.00
1000_50_4.dat	322226	31593	26928.4	25415	90.20	**322226**	**322226**	**322226**	0.00
1000_50_5.dat	9984247	4337463	4082879.5	4008561	56.56	**9983765**	**9983765**	**9983765**	0.00
1000_50_6.dat	4106261	1640327	1586601.2	1569140	60.05	**4105798**	**4105798**	**4105798**	0.01
1000_50_7.dat	10498370	4203466	4070508.5	3989437	59.96	**10498286**	**10498286**	**10498286**	0.00
1000_50_8.dat	4981146	2611823	2527737	2495028	47.57	**4979004**	**4979004**	**4979004**	0.04
1000_50_9.dat	1727861	390882	363678.7	347176	77.38	**1727861**	**1727861**	**1727861**	0.00
1000_50_10.dat	2340724	607481	574499.5	563434	74.05	**2339279**	**2339279**	**2339279**	0.06
1000_75_1.dat	11570056	6322683	6101388.7	6041021	45.35	**11568349**	**11568349**	**11568349**	0.01
1000_75_2.dat	1901389	338122	303192.3	291996	82.22	**1900524**	**1900524**	**1900524**	0.05
1000_75_3.dat	2096485	390718	367111.5	355203	81.36	**2089683**	**2089683**	**2089683**	0.32
1000_75_4.dat	7305321	3765610	3623303.3	3570479	48.45	**7302458**	**7302458**	**7302458**	0.04
1000_75_5.dat	13970240	6423191	6116594.5	6003690	54.02	**13959904**	**13959904**	**13959904**	0.07
1000_75_6.dat	12288738	6649445	6143169.5	6010454	45.89	**12288040**	**12288040**	**12288040**	0.01

(continued)

Table 3 (continued)

Instance	Best	Best BICA				QKPICA			
	[7, 35]	BS	MBS	WS	$\%_{gap}$	BS	MBS	WS	$\%_{gap}$
1000_75_7.dat	1095837	124364	117150.1	111536	88.65	**1093132**	**1093132**	**1093132**	0.25
1000_75_8.dat	5575813	2167368	2086158.3	2059177	61.13	**5571537**	**5571537**	**5571537**	0.08
1000_75_9.dat	695774	77195	66879.3	64039	88.91	**695484**	**695484**	**695484**	0.04
1000_75_10.dat	2507677	548723	527185.7	511347	78.12	**2507677**	**2507677**	**2507677**	0.00
1000_100_1.dat	6243494	2190235	2128520.5	2084381	64.92	**6238137**	**6238137**	**6238137**	0.09
1000_100_2.dat	4854086	1326937	1258723.4	1229441	72.66	**4849993**	**4849993**	**4849993**	0.08
1000_100_3.dat	3172022	696853	654800.3	638013	78.03	**3167690**	**3167690**	**3167690**	0.14
1000_100_4.dat	754727	66308	59413.8	55688	91.21	**753941**	**753941**	**753941**	0.10
1000_100_5.dat	18646620	8611169	8105766.3	8009500	53.82	**18640211**	**18640211**	**18640211**	0.03
1000_100_6.dat	16020232	8543792	8128405.9	7982271	46.67	**16014039**	**16014039**	**16014039**	0.04
1000_100_7.dat	12936205	7761290	7520678.9	7408694	40.00	**12935577**	**12935577**	**12935577**	0.00
1000_100_8.dat	6927738	2579841	2465329.8	2405037	62.76	**6923970**	**6923970**	**6923970**	0.05
1000_100_9.dat	3874959	992735	887901	853804	74.38	**3874959**	**3874959**	**3874959**	0.00
1000_100_10.dat	1334494	155265	142425.4	137150	88.37	**1331725**	**1331725**	**1331725**	0.21

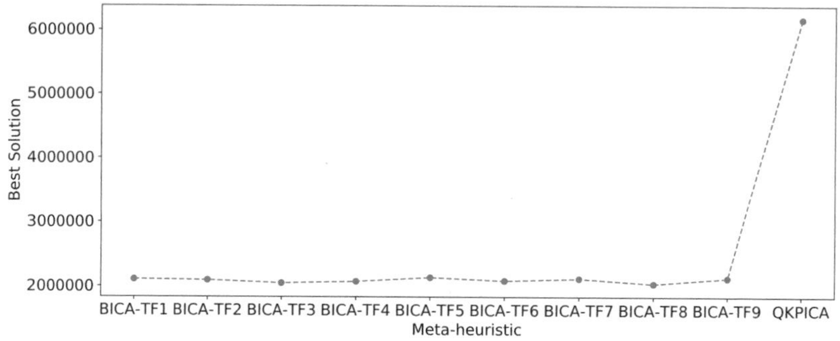

Fig. 1 Comparison of solution quality for a QKP instance of size 1000

0.32% gap and performed far better than BICA in all the instances. BS and WS are the best and worst solutions found in a run, and MBS is the average of best solutions in 10 runs.

Results of best BICA and QKPICA on QKP instances of 2000 objects are given in Table 4. QKPICA also performed better on these instances on all four parameters of solution quality viz., BS, MBS, WS and %gap from optimal.

The results shown in Tables 3 and 4 clearly show the superiority of QKPICA on the tested QKP instances. Further, a non-parametric statistical hypothesis analysis between BICA and QKPICA is conducted to compare the performance of both meta-heuristics. Such statistical results are necessary due to the stochastic nature of meta-heuristics [10]. In this paper, the Wilcoxon signed-rank test is utilized to compare both meta-heuristics pairwise. Further details about non-parametric statistical analysis for meta-heuristics can be found in [10].

The results of the Wilcoxon signed ranked test are provided in Table 5. It is evident that the QKPICA significantly improves over BICA, even with a significance level of 0.01.

7 Conclusion

QKP is one of the most challenging combinatorial problems in the NP-hard class. In this paper, an enhanced binary version of ICA dubbed as QKPICA is proposed to solve large-scale QKP instances of sizes 1000 and 2000. A new assimilation process is devised for this purpose. The results show that QKPICA outperforms each version of BICA by a large margin. A non-parametric statistical test, Wilcoxon signed-rank test, is also performed to support this fact, which shows the superiority of QKPICA even on a significance level of 1%. The structure of QKPICA is such that it can easily be adapted to other hard combinatorial optimization problems. Attempts are being made in this direction.

Table 4 Results of Best BICA and QKPICA on 40 QKP instances of size 2000

Instance	Best	Best BICA				QKPICA			
	[7, 35]	BS	MBS	WS	$\%_{gap}$	BS	MBS	WS	$\%_{gap}$
2000_25_1.dat	5268188	1389844	1348384.6	1326483	73.62	**5268172**	**5268172**	**5268172**	0.00
2000_25_2.dat	13294030	7332983	7251981.4	7215950	44.84	**13285189**	**13285189**	**13285189**	0.07
2000_25_3.dat	5500433	1602689	1541291.4	1510233	70.86	**5500169**	**5500169**	**5500169**	0.00
2000_25_4.dat	14625118	7810571	7587162.8	7537127	46.59	**14625118**	**14625118**	**14625118**	0.00
2000_25_5.dat	5975751	1817804	1786600.4	1775189	69.58	**5970585**	**5970585**	**5970585**	0.09
2000_25_6.dat	4491691	1063522	1013728.2	991802	76.32	**4491630**	**4491630**	**4491630**	0.00
2000_25_7.dat	6388756	1933111	1851172	1838766	69.74	**6388756**	**6388756**	**6388756**	0.00
2000_25_8.dat	11769873	6371427	6324303.6	6281623	45.87	**11765730**	**11765730**	**11765730**	0.04
2000_25_9.dat	10960328	5586158	5507282.4	5442300	49.03	**10960313**	**10960313**	**10960313**	0.00
2000_25_10.dat	139236	4438	3994.2	3704	96.81	**139236**	**139236**	**139236**	0.00
2000_50_1.dat	7070736	1535267	1453725.6	1428426	78.29	**7064595**	**7064595**	**7064595**	0.09
2000_50_2.dat	12587545	3980026	3895083.8	3840818	68.38	**12587419**	**12587419**	**12587419**	0.00
2000_50_3.dat	27268336	15272759	14910309.8	14801066	43.99	**27268336**	**27268336**	**27268336**	0.00
2000_50_4.dat	17754434	7403009	7251118.4	7177922	58.30	**17747606**	**17747606**	**17747606**	0.04
2000_50_5.dat	16805490	6747207	6612774.2	6522577	59.85	**16799771**	**16799771**	**16799771**	0.03
2000_50_6.dat	23076155	11983979	11827047.4	11725759	48.07	**23073040**	**23073040**	**23073040**	0.01
2000_50_7.dat	28759759	15365935	15102604.4	14980933	46.57	**28747831**	**28747831**	**28747831**	0.04
2000_50_8.dat	1580242	99820	95677	93414	93.68	**1580242**	**1580242**	**1580242**	0.00
2000_50_9.dat	26523791	14945207	14666392.6	14605587	43.65	**26521300**	**26521300**	**26521300**	0.01
2000_50_10.dat	24747047	13200048	13161610	13123172	46.66	**24747047**	**24747047**	**24747047**	0.00
2000_75_1.dat	25121998	10169048	10018515.2	9931889	59.52	**25118380**	**25118380**	**25118380**	0.01
2000_75_2.dat	12664670	2939354	2839667.2	2796624	76.79	**12661651**	**12661651**	**12661651**	0.02
2000_75_3.dat	43943994	23128639	22690198.8	22405280	47.37	**43941918**	**43941918**	**43941918**	0.00

(continued)

Table 4 (continued)

Instance	Best	Best BICA				QKPICA			
	[7, 35]	BS	MBS	WS	$\%_{gap}$	BS	MBS	WS	$\%_{gap}$
2000_75_4.dat	37496613	20585171	20173129	20013854	45.10	**37492420**	**37492420**	**37492420**	0.01
2000_75_5.dat	24834948	9668029	9390403.6	9269499	61.07	**24821916**	**24821916**	**24821916**	0.05
2000_75_6.dat	45137758	23202332	22794489.2	22582170	48.60	**45137758**	**45137758**	**45137758**	0.00
2000_75_7.dat	25502608	10331643	10190092.8	10051199	59.49	**25501501**	**25501501**	**25501501**	0.00
2000_75_8.dat	10067892	1890095	1826302.4	1792006	81.23	**10067600**	**10067600**	**10067600**	0.00
2000_75_9.dat	14171994	3599403	3482344.4	3447584	74.60	**14167730**	**14167730**	**14167730**	0.03
2000_75_10.dat	7815755	1315226	1286328.4	1262822	83.17	**7808983**	**7808983**	**7808983**	0.09
2000_100_1.dat	37929909	17773543	17339066.4	17137609	53.14	**37926035**	**37926035**	**37926035**	0.01
2000_100_2.dat	33648051	13939986	13812649.2	13709134	58.57	**33635306**	**33635306**	**33635306**	0.04
2000_100_3.dat	29952019	10862349	10756118.2	10656491	63.73	**29944350**	**29944350**	**29944350**	0.03
2000_100_4.dat	26949268	8955942	8743586.6	8646147	66.77	**26941888**	**26941888**	**26941888**	0.03
2000_100_5.dat	22041715	6586045	6487375.2	6444642	70.12	**22027758**	**22027758**	**22027758**	0.06
2000_100_6.dat	18868887	4849914	4704093.2	4672845	74.30	**18864880**	**18864880**	**18864880**	0.02
2000_100_7.dat	15850597	3503823	3410649	3355748	77.89	**15845991**	**15845991**	**15845991**	0.03
2000_100_8.dat	13628967	2427538	2354869	2298352	82.19	**13622153**	**13622153**	**13622153**	0.05
2000_100_9.dat	8394562	1163752	1103581.6	1073845	86.14	**8389997**	**8389997**	**8389997**	0.05
2000_100_10.dat	4923559	398067	387450	376832	91.92	**4919481**	**4919481**	**4919481**	0.08

Table 5 Non-parametric statistical analysis of BICA and QKPICA on QKP instances of sizes 1000 and 2000

Comparison	R^+	R^-	z-value	p-value
QKPICA vs Best BICA	80	0	−5.510932	3.5694E-8

References

1. Atashpaz-Gargari E, Lucas C (2007) Imperialist competitive algorithm: an algorithm for optimization inspired by imperialistic competition, pp 4661–4667. https://doi.org/10.1109/CEC.2007.4425083
2. Bäck T, Fogel DB, Michalewicz Z (1997) Handbook of evolutionary computation. CRC Press
3. Billionnet A, Faye A, Soutif É (1999) A new upper bound for the 0–1 quadratic knapsack problem. Eur J Oper Res 112(3):664–672
4. Borji A (2007) A new global optimization algorithm inspired by parliamentary political competitions. In: Gelbukh A, Kuri Morales ÁF (eds) MICAI 2007: advances in artificial intelligence, pp 61–71. Springer, Berlin
5. Brownlee J (2011) Clever algorithms: nature-inspired programming recipes, 1st edn. Lulu.com
6. Caprara A, Pisinger D, Toth P (1999) Exact solution of the quadratic knapsack problem. INFORMS J Comput 11(2):125–137. https://doi.org/10.1287/ijoc.11.2.125
7. Chen Y, Hao JK (2017) An iterated "hyperplane exploration" approach for the quadratic knapsack problem. Comput Oper Res 77:226–239
8. Darwin C (2009) The origin of species: by means of natural selection, or the preservation of favoured races in the struggle for life, 6 edn. Cambridge library collection - darwin, evolution and genetics. Cambridge University Press (2009). https://doi.org/10.1017/CBO9780511694295
9. Delahaye D, Chaimatanan S, Mongeau M (2019) Simulated annealing: from basics to applications. In: Handbook of metaheuristics, pp 1–35. Springer (2019)
10. Derrac J, García S, Molina D, Herrera F (2011) A practical tutorial on the use of nonparametric statistical tests as a methodology for comparing evolutionary and swarm intelligence algorithms. Swarm Evol Comput 1(1):3–18
11. Dijkhuizen G, Faigle U (1993) A cutting-plane approach to the edge-weighted maximal clique problem. Eur J Oper Res 69(1):121–130. https://doi.org/10.1016/0377-2217(93)90097-7
12. Emami H, Lotfi S (2013) Graph colouring problem based on discrete imperialist competitive algorithm. arXiv:1308.3784
13. Ferreira CE, Martin A, de Souza CC, Weismantel R, Wolsey LA (1996) Formulations and valid inequalities for the node capacitated graph partitioning problem. Math Program 74(3):247–266
14. Gallo G, Hammer PL, Simeone B (1980) Quadratic knapsack problems. In: Combinatorial optimization, pp 132–149. Springer (1980)
15. Goldberg DE (1989) genetic algorithms in search, optimization and machine learning, 1st edn. Addison-Wesley Longman Publishing Co., Inc, Boston, MA, USA
16. Helmberg C, Rendl F, Weismantel R (1996) Quadratic knapsack relaxations using cutting planes and semidefinite programming. In: Cunningham WH, McCormick ST, Queyranne M (eds) Integer programming and combinatorial optimization. Springer, Berlin, pp 175–189
17. Helmberg C, Rendl F, Weismantel R (2000) A semidefinite programming approach to the quadratic knapsack problem. J Comb Optim 4(2):197–215
18. Hosseini S, Al Khaled A (2014) A survey on the imperialist competitive algorithm metaheuristic: implementation in engineering domain and directions for future research. Appl Soft Comput 24:1078–1094

19. Julstrom BA (2005) Greedy, genetic, and greedy genetic algorithms for the quadratic knapsack problem. In: Proceedings of the 7th annual conference on Genetic and evolutionary computation, pp 607–614
20. Kirkpatrick S, Gelatt CD, Vecchi MP (1983) Optimization by simulated annealing. Science 220(4598):671–680
21. Laughhunn D (1970) Quadratic binary programming with application to capital-budgeting problems. Oper Res 18(3):454–461
22. Mirhosseini M, Nezamabadi-pour H (2018) BICA: a binary imperialist competitive algorithm and its application in CBIR systems. Int J Mach Learn Cybern 9(12):2043–2057
23. Mousavirad S, Ebrahimpour-Komleh H (2013) Feature selection using modified imperialist competitive algorithm. In: ICCKE 2013, pp 400–405. IEEE
24. Nozarian S, Soltanpoora H, Jahanb MV (2012) A binary model on the basis of imperialist competitive algorithm in order to solve the problem of knapsack 1-0. In: Proceedings of the international conference on system engineering and modeling, pp 130–135
25. Park K, Lee K, Park S (1996) An extended formulation approach to the edge-weighted maximal clique problem. Eur J Oper Res 95(3):671–682. https://doi.org/10.1016/0377-2217(95)00299-5
26. Patvardhan C, Prakash P, Srivastav A (2012) Novel quantum-inspired evolutionary algorithms for the quadratic knapsack problem. Int J Math Oper Res 4(2):114–127
27. Pisinger D (2007) The quadratic knapsack problem-a survey. Discret Appl Math 155(5):623–648
28. Rao M (1971) Cluster analysis and mathematical programming. J Am Stat Assoc 66(335):622–626
29. Ray T, Liew K (2003) Society and civilization: an optimization algorithm based on the simulation of social behavior. IEEE Trans Evol Comput 7:386–396
30. Rechenberg I (2018) Optimierung technischer systeme nach prinzipien der biologischen evolution
31. Reynolds RG (1994) An introduction to cultural algorithms. In: Proceedings of the third annual conference on evolutionary programming. World Scientific, pp 131–139
32. Schwefel HPP (1993) Evolution and optimum seeking: the sixth generation. Wiley, New York
33. Storn R, Price K (1997) Differential evolution - a simple and efficient heuristic for global optimization over continuous spaces. J Glob Optim 11(4):341–359. https://doi.org/10.1023/A:1008202821328
34. Xing B, Gao WJ (2014) Innovative computational intelligence: a rough guide to 134 clever algorithms (2014)
35. Yang Z, Wang G, Chu F (2013) An effective grasp and tabu search for the 0–1 quadratic knapsack problem. Comput Oper Res 40(5):1176–1185. https://doi.org/10.1016/j.cor.2012.11.023

Balanced Cluster-Based Spatio-Temporal Approach for Traffic Prediction

Gaganbir Kaur, Surender K. Grewal, and Aarti Jain

1 Introduction

An inefficient transportation system may tremendously impact the environment and economy of a country. According to the 2015 Urban Mobility Report, congestion has grown substantially over the past 20 years [1]. Urban commuters are facing a lot of stress due to traffic congestion [2]. Intelligent transportation systems (ITS), with better traffic prediction, are expected to have an efficient solution to this traffic congestion problem. Monitoring of traffic parameters like flow and speed helps reduce travel time and increases road safety. Wireless Sensor Networks (WSN) provide better technology for traffic monitoring and prediction using low-cost sensors. WSN is a self-organized monitoring network in which numerous sensors are deployed at roadways and intersections. This enhances the efficiency of traffic prediction by measuring traffic flow in real time [3]. ITS has an integrated system for traffic management and control, combining data communication, sensor technology, and electronic control technology to form a modern transport management system. Traffic Flow Prediction (TFP) uses historical traffic information to predict the nature of traffic, allowing commuters to avoid congestion and make better travel plans. Traffic sensors like cameras, radars, etc. deployed throughout the traffic network collect data and transmit it to a base station, where the data is aggregated to draw parameters for traffic analysis. Various centralized machine learning approaches like ARIMA

G. Kaur (✉) · S. K. Grewal
Department of Electronics and Communication, DCRUST, Haryana, India
e-mail: kgaganbir@gmail.com

S. K. Grewal
e-mail: skgrewal.ece@dcrustm.org

A. Jain
Department of Electronics and Communication, NSUT East Campus, Delhi, India
e-mail: aarti.jain@nsut.ac.in

P. Singh et al. (eds.), *Machine Learning and Computational Intelligence Techniques for Data Engineering*, Lecture Notes in Electrical Engineering 998,
https://doi.org/10.1007/978-981-99-0047-3_60

(AutoRegressive Integrated Moving Average), Neural networks, and Non-regression models have been used for traffic forecasting [5].

Traffic prediction is a complex time-variant process due to dependence on both space and time domains. Since the flow of traffic on connected roads is closely related, prediction is dependent on the spatial dimension. Traffic volume is largely affected by the topology of the road networks. In the temporal dimension, traffic changes dynamically with time. Recurrent neural networks (RNNs), GRUs, and Long Short-Term Memory (LSTM) have been extensively used for traffic prediction considering the temporal features only. Convolutional neural networks (CNNs) and graph convolutional network (GCN) have been used in literature for characterizing the spatial features of traffic. To accurately predict the traffic data, spatial and temporal characteristics of the traffic must be captured simultaneously. Zhao et al. [4] combine GRU and GCN to form Temporal-GCN. GCN captures the topology of the road networks, characterizing the spatial features and GRU captures the dynamic behavior of traffic, characterizing the temporal features.

Generally, in WSN the number of detectors or sensor nodes is large but resources are limited. Thus, it becomes crucial to reduce energy consumption for increasing the network lifetime. Centralized WSN approaches in which all the detectors transmit information to the Base Server(BS) increase the computation load on the BS. Deploying clustering techniques, in which a group of detectors send their collected data to a detector designated as Cluster Head (CH) and CH further aggregates and transmits information to the Base Server, which can effectively reduce the computation burden on the BS. Clustering also reduces the energy consumption of the network by decreasing the number of remote transmissions by a detector. Detectors in WSN are randomly deployed resulting in different node densities in different areas. Clustering must consider node density for optimal cluster formation [6]. The major contributions of this paper are as follows:

1. Fuzzy c means algorithm is used for designing a clustered WSN. Clusters thus formed are balanced by minimizing the variance within a cluster.
2. Traffic prediction algorithm using GCN and GRU with clustering is proposed which decreases the computational cost and up-link communication cost.
3. The proposed algorithm is assessed using real-world dataset to demonstrate the performance enhancement.

The rest of the paper is organized as follows: Sect. 2 outlines the associated work in this area, the problem definition is described in Sect. 3, and the methodology is explained in Sect. 4. Section 5 describes the dataset used for experiments and results based on evaluation metrics. Finally, Sect. 6 concludes the paper with a future direction for research.

2 Related Work

Recurrent Neural Networks (RNNs) have been extensively used for predicting the temporal characteristics of traffic data.RNN networks have a problem of vanishing gradient and gradient explosion with the increase in time lag and thus cannot be used for long-term traffic prediction. Variants of RNN like LSTM and GRU have been adopted to solve the above problems. Various models like bidirectional LSTM [8] and deep LSTM [9] have been designed for traffic prediction by combining single LSTM models. LSTM models take a longer time due to their complex structure. GRU models are simpler with only two gates and thus speed up training[10]. Fu et al. [10] first applied GRU for short-term traffic flow prediction and proposes a GRU NN model using LSTM and GRU neural networks. The model outperforms ARIMA, while the GRU model shows marginally better performance than the LSTM model. GRU NN has a lesser mean absolute error as compared to ARIMA.

Lv et al. [11] consider the nonlinear Spatio-temporal correlations using the stacked autoencoder (SAE) model for learning generic features of traffic. A greedy layerwise technique is used for training the model. This model shows better performance than Support Vector Machine (SVM) and backpropagation neural networks. Zheng et al. [12] combine two predictors, backpropagation and radial basis function, to form Bayesian Combined Neural Network (BCNN) model for prediction since a single prediction model shows better performance only for some fixed duration. The predictions from the two models are combined using the credit assignment algorithm. The credit values depend on the cumulative prediction performance for a given time period. The combined (hybrid) model outperforms the single predictors in 85 percent time intervals. The model can pick the best predictor models by tracking their performance online. Thus, the performance of the model is dependent on the predictors selected. If very high performing predictors are selected at a particular instant, the accuracy of the combined BCNN model will be high.

Traffic prediction becomes a challenging problem due to the complex spatial dependency of road networks and multiple temporal dependencies due to repeated time patterns. Many deep learning techniques like the combination of CNN and RNN have been widely used by researchers for traffic prediction. This combination cannot extract the connectivity and global nature of traffic networks. Thus, a Residual Recurrent Graph Neural Network (Res-RGNN)has been proposed in [13]. A novel hop scheme has also been integrated with Res-RGNN which overcomes the vanishing gradient problem in RNNs. Wu et al. propose a Deep Neural Network(DNN)-based TFP model (DNN-BTF), which uses the periodicity of traffic and Spatio-temporal characteristics of traffic data [14]. The spatial features are captured by the CNN and the temporal features by a gated RNN. They also introduced an attention model which determines the significance of traffic data in the past. The researchers also performed visualizations of the proposed model.

Given the complex spatio-temporal dependency of traffic, Bai et al. propose an attention temporal graph convolutional network (A3T-GCN) using GRU, GCN along with attention mechanism in [15]. The attention model learns the importance of traffic information at any particular time. The model outperforms the baseline models. Chen et al. use support vector regression (SVR) for predicting traffic parameters in the WSN and IoT networks [16]. Researchers use time-series data processed by logarithmic function to remove any fluctuations in traffic data. This processed data trains an SVR model which predicts future traffic. The method achieves significantly less mean square error in traffic prediction. Yi et al. in [17] used Tensorflow Deep Neural Network (DNN) architecture for predicting traffic conditions using real-time transportation data. The model differentiates congested and non-congested conditions using Traffic Performance Index (TPI) in a logistic regression analyser. Further, a hyper tangent activation function is applied at every layer of DNN. Traffic Performance Index is calculated as the ratio of the difference between maximum traffic speed and average speed at a particular time instant and maximum speed. TPI indicates the extent of congestion. 1 value means the highest congestion where vehicles cannot move and 0 value means congestion-free movement. 0.5 is taken as a critical value to differentiate between congested and non-congested conditions. However, only 1 percent of the daily data could be used due to memory limitations. This could be overcome with the use of multi GPUs.

3 Problem Definition

This paper focuses on predicting future traffic based on past traffic data considering the Spatio-temporal features of traffic information while decreasing the computation cost by clustering. Traffic information may be defined in terms of traffic speed, traffic flow, and traffic density. In this paper, traffic speed is used as a parameter for traffic information prediction.

Adjacency Matrix A^{NXN}: The road network is depicted as an unweighted graph G=(D, E). $D=\{d_1, d_2, d_3, ...d_N\}$ represents detectors or sensor nodes installed on roads of the network, and E is the edge connecting two detectors. The Adjacency Matrix A represents the connectedness of nodes in G. It is calculated from distance $(dist(d_i, d_j)$ between detectors d_i and d_j, $d_{i,j} = exp(-\frac{dist(d_i,d_j)^2}{\sigma^2})$, σ is the standard deviation. Adjacency matrix is then computed as $A_{i,j}=d_{i,j}$, if $d_{i,j} > \epsilon$ else 0. ϵ is the threshold and is set to 0.1.

Feature Matrix X^{NXn}: Features or attributes of the road network are represented as feature matrix X^N where N is the number of detectors or sensor nodes and n is the number of node attributes or the length of time series. Traffic speed is used as the attribute of nodes. Thus, the Spatio-temporal problem is to find a mapping function f() $[Y_{t+1}, ..., Y_{t+T}]= f(G : (Y_{t-n}, ...Y_t))$.

4 Methodology

4.1 Balanced Clustering-Based Traffic Prediction

To reduce the communication overhead, detectors on the road network are clustered and one cluster head (CH) is elected per cluster. CH collects information from all the detectors within the cluster, performs some form of aggregation on the data, and transmits it to the base server. This reduces the energy consumption and amount of data transmitted to the Base Server. Fuzzy C means clustering algorithm [18] is a popular soft clustering technique. The algorithm assigns a membership value (degree of belonging to the cluster) to each detector according to the distance between the detectors and the cluster center. The algorithm aims to minimize the Sum of the Squared distance between detectors and the cluster center:

$$J_m = \sum_{i=1}^{n} \sum_{j=1}^{N} u_{ij}^m \|y_i - c_j\|^2, 1 \leq m < \infty \quad (1)$$

N = Number of clusters, n = Number of data points, m=fuzziness parameter > 1

Membership function and cluster centers are updated according to Eqs. (2) and (3) [18]

$$u_{ij} = \frac{1}{\sum_{k=1}^{N} \frac{\|y_i - c_j\|}{\|y_i - c_j\|}^{\frac{2}{m-1}}} \quad (2)$$

$$C_j = \frac{\sum_{i=1} N u_{ij}^m x_i}{\sum_{i=1} N u_{ij}^m} \quad (3)$$

For each cluster, one node is elected as CH which aggregates collected information from all the member nodes and transmits it to the BS. The node nearest to the centroid of the cluster is elected as CH for that cluster.

Quality of clustering relies on the choice of the number of clusters and the process of formation of clusters. The performance of WSN is largely dependent on these parameters. Thus, the clusters formed must be balanced so that the variance within the clusters is minimized. To calculate the optimal number of clusters, total normalized variance is computed for different cluster numbers. Normalized variance is computed as the ratio of variance after clustering and variance without clustering as per equation (4) [19]. Thus, the number of clusters yielding minimum variance is evaluated.

$$NormalizedVariance = \frac{\sum_{i=1}^{C} N_i * var(C_i)}{N * var(C)} \quad (4)$$

C is the number of clusters and N is the number of nodes. N_i is the number of nodes in the ith cluster. The number of clusters is varied from 2 to 10 and normalized

Fig. 1 Normalized variance versus the number of clusters

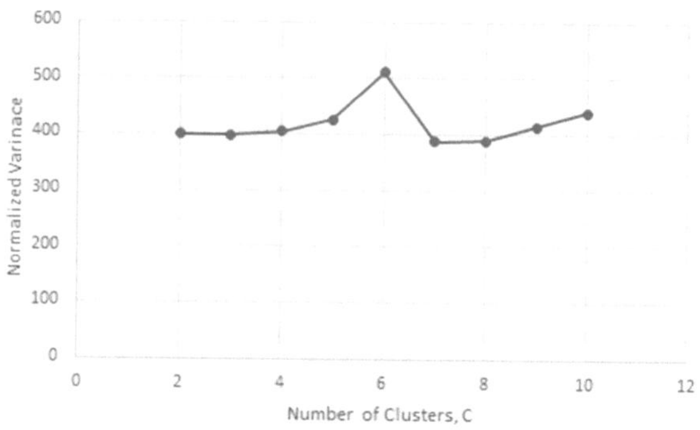

variance is calculated as per equation (4). Thus, the optimal number with minimum variance is selected. The plot of normalized variance and the number of clusters is shown in Fig. 1.

4.2 Spatio-Temporal Approach with GCN and GRU

Gated Recurrent Unit(GRU) is a form of standard RNN which uses two types of gates: reset gate and update gate. The reset gate holds the short-term memory of the network and the update gate holds the long-term memory. The reset gate equation is expressed as [4]

$$r_v^t = \sigma(W_r y_t, U_r h_{t-1}) \tag{5}$$

W_r and U_r are weight matrices for the reset gate, y_t is the input and h_{t-1} is the hidden state. Values of W_r and U_r are aggregated through a sigmoid function, thus, r_t gives a value ranging between 0 and 1. The update gate equation is expressed as

$$\text{Update gate,}u_v^t = \sigma(W_u x_t, U_u h_{t-1}) \tag{6}$$

GRU cell has a memory content that stores only the relevant past data. It generates a candidate hidden state as per the below equation:

$$C_v^t = tanh(W_c x_t + (r_v^t \circledast h_v^{t-1})U_c) \tag{7}$$

Update gate and reset gate control how much each hidden unit remembers or forgets the past information. When the reset gate is 0 the data from the earlier state is completely ignored. If the reset gate is set then the entire information from the previous hidden state is considered. The update gate decides how much data from the earlier hidden state will be transferred to the future state. Thus, the problem of the vanishing gradient is eliminated. The tanh activation function maps the data to (−1, 1), reducing the computations and thus eliminating the gradient explosion problem.

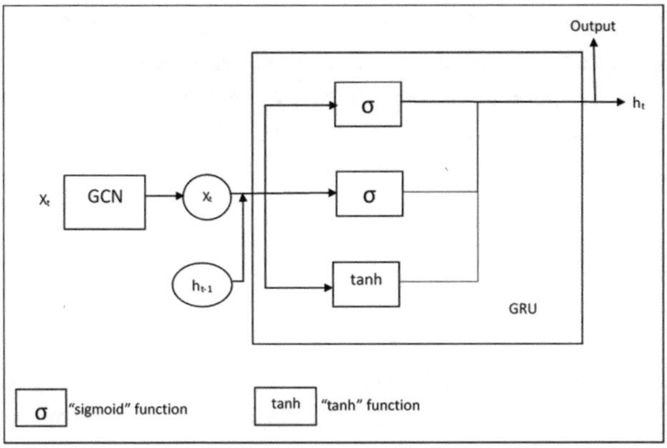

Fig. 2 GCN-GRU combined cell

Graph Convolution Networks (GCNs) are the neural networks that work on graphs and use convolutional aggregation. An undirected graph can be represented by $G = (D, A)$ where D represents a set of nodes $\{d_1, d_2, ...d_n\}$ and A is the adjacency matrix. $A_{i,j}$ is equal to the weight of the edge between the nodes d_i and d_j. If the edge is missing then $A_{i,j} = 0$. There is a degree matrix of A, defined as a diagonal matrix $M_{(i,i)} = \Sigma_{j=1}^{n} A_{i,j}$. Every node has a feature vector x_i. A feature matrix X_t is formed by stacking feature vectors $[x_1, x_2, ..., x_n]$ [20, 21].

$f(A, X_t)$ is graph convolution function. If X_t represents the feature matrix and W represents weights then GCN-GRU process is governed by the following equations:

$$\text{Update gate,} u_v^t = \sigma(W_u[f(A, X_t), h_v^{t-1}]) \tag{8}$$

$$r_v^t = \sigma(W_r[f(A, X_t), h) \tag{9}$$

$$C_v^t = tanh(W_c[f(A, X_t), r_v^t \circledast h_v^{t-1}]) \tag{10}$$

$$h_v^t = u_v^t \circledast h_v^{t-1} + (1 - u_v^t) \circledast c_v^t \tag{11}$$

GCN-GRU cell is shown in Fig. 2.

5 Experiments

5.1 Data Description

The performance of the clustered Spatio-temporal approach used in this paper is assessed using a real-world dataset from the Caltrans Performance Measurement System (PeMS) [23] for the state of California. Around 39,000 detectors installed on the roads collect traffic information in real time to form PeMS database. In this paper traffic speed data collected by 100 loop detectors for the first week of January 2021 is used. Traffic speed is aggregated every five minutes. The Adjacency matrix is computed from the distance between the detectors. Data is normalized to range (0,1). 80% data is used for training and 20% data for testing.

5.2 Evaluation Metrics

Root Mean Square Error (RMSE), Mean Absolute Error (MAE), and Accuracy are used as metrics for evaluating the difference between the actual traffic value, y_i, and predicted value, $\tilde{y_p}$, as follows:

$$RMSE = \sqrt{\frac{1}{n}\Sigma_{i=1}^{n}\left(y_i - \tilde{y_p}\right)^2} \tag{12}$$

$$MAE = \sum_{i=1}^{D} |y_i - \tilde{y_p}| \tag{13}$$

$$Accuracy = 1 - \frac{|y_i - \tilde{y_p}|}{|y_i|} \tag{14}$$

5.3 Results

For the experiments, the learning rate is set to 0.001 and 64 hidden units are used. The algorithm is run for 1000 epochs. The results of clustered approach are compared with the baseline Temporal-GCN (TGCN) model [4]. Table 1 clearly shows that the proposed clustered approach gives comparable results as TGCN with a considerable reduction in computation time. Figure 3 compares the RMSE values for training and testing, and the plot in Fig. 4 shows the MAE of prediction. Test RMSE equals training RMSE as the number of epochs increases. The accuracy of the two models is plotted in Figs. 5 and 6. Clustered approach predicts traffic information with almost the same accuracy as TGCN. The computation time is reduced by almost 50 times by using clustering.

Table 1 Traffic forecasting results of the proposed approach and TGCN model

Metrics	Clustered approach	TGCN
RMSE	3.268	3.05
MAE	2.395	2.19
Accuracy	0.953	0.9539
Computation Time	946.498s	51118.09s

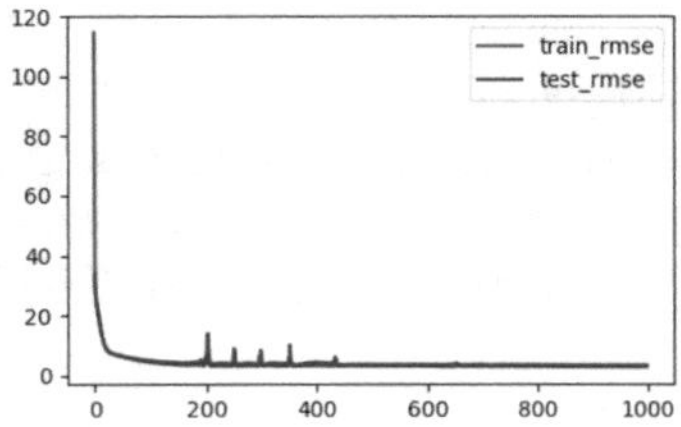

Fig. 3 RMSE plot for clustered approach

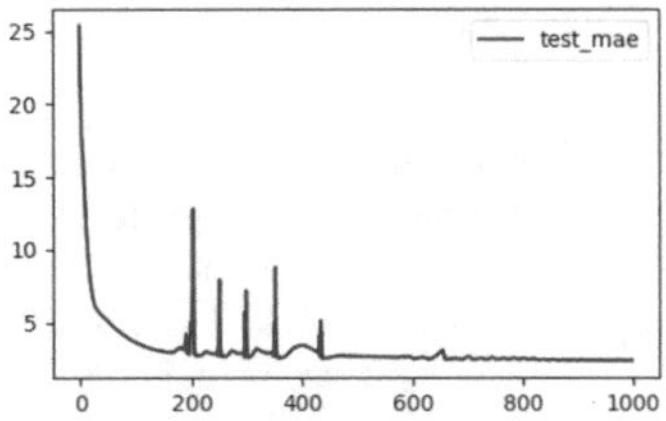

Fig. 4 MAE plot

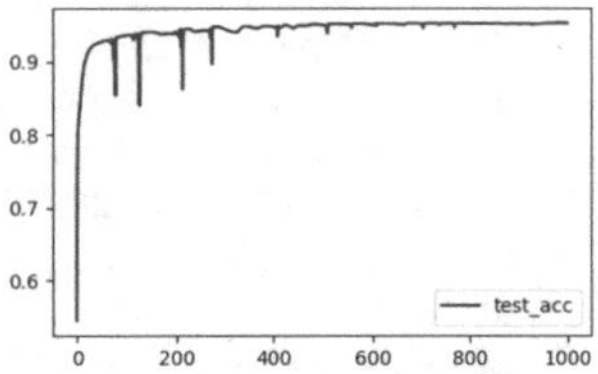

Fig. 5 Accuracy of clustered approach

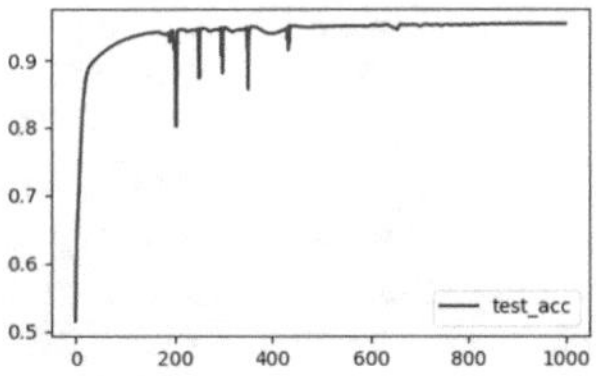

Fig. 6 Accuracy of TGCN model

6 Conclusion

In this paper, cluster-based traffic prediction approach is proposed which trains the model with both GCN and GRU. GCN captures the spatial features of traffic and GRU captures the temporal features. Clustering reduces the computational time and also decreases the total data transmitted to the base server by nodes, thus reducing the communication load. Clustering technique based on fuzzy c means is used and the optimal number of clusters is determined based on normalized variance of the clusters. The proposed model predicts traffic information with 95% accuracy. Computation time is reduced by almost 50 times as compared to the model without clustering. The work focuses on predicting future traffic information at the cost of the privacy of users' data. In future work, a prediction scheme considering the privacy of users could be developed thus eliminating possible security breaches.

References

1. Urban Congestion Trends Report (2020) https://ops.fhwa.dot.gov/congestionreport
2. IBM, Frustration rising: IBM 2011 Commuter Pain Survey (2011). https://newsroom.ibm.com/2011-09-08-IBM-Global-Commuter-Pain-Survey-Traffic-Congestion-Down-Pain-Way-Up, 1
3. Pascale A, Nicoli M, Deflorio F, Dalla Chiara B, Spagnolini U (2012) Wireless sensor networks for traffic management and road safety. IET Intell Transp Syst 6(1):67–77
4. Zhao L, Song Y, Zhang C, Liu Y, Wang P, Lin T, Deng M, Li H (2019) T-GCN: a temporal graph convolutional network for traffic prediction. IEEE Trans Intell Transp Syst 21(9):3848–3858
5. Xiao L, Peng X, Wang Z, Xu B, Hong P (2009) Research on traffic monitoring network and its traffic flow forecast and congestion control model based on wireless sensor networks. In: International conference on measuring technology and mechatronics automation, vol 1, pp 142–147. 11–12 April 2009
6. Su S, Zhao S (2018) An optimal clustering mechanism based on Fuzzy-C means for wireless sensor networks. Sustain Comput: Inform Syst 18:127–134
7. Zhu G, Liu D, Du Y, You C, Zhang J, Huang K (2020) Toward an intelligent edge: wireless communication meets machine learning. IEEE Commun Mag 58(1):19–25
8. Cui Z, Ke R, Pu Z, Wang Y (2020) Stacked bidirectional and unidirectional LSTM recurrent neural network for forecasting network-wide traffic state with missing values. Transp Res Part C: Emerg Technol 118:102674
9. Yu R, Li Y, Shahabi C, Demiryurek U, Liu Y (2017) Deep learning: a generic approach for extreme condition traffic forecasting. In: Proceedings of the SIAM international conference on data mining, Texas, USA, pp 777–785. Accessed 27–29 April 2017
10. Fu R, Zhang Z, Li L (2016) Using LSTM and GRU neural network methods for traffic flow prediction. In: 31st youth academic annual conference of Chinese association of automation (YAC), Wuhan, China, pp 324–328. Accessed 11–13 Nov 2016
11. Lv Y, Duan Y, Kang W, Li Z, Wang FY (2014) Traffic flow prediction with big data: a deep learning approach. IEEE Trans Intell Transp Syst 16(2):865–873
12. Zheng W, Lee DH, Shi Q (2006) Short-term freeway traffic flow prediction: Bayesian combined neural network approach. J Transp Eng 132(2):114–121
13. Chen C, Li K, Teo SG, Zou X, Wang K, Wang J, Zeng Z (2019) Gated residual recurrent graph neural networks for traffic prediction. In: Proceedings of the AAAI conference on artificial intelligence, vol 33(01), pp 485–492, July 2019

14. Wu Y, Tan H, Qin L, Ran B, Jiang Z (2018) A hybrid deep learning based traffic flow prediction method and its understanding. Transp Res Part C: Emerg Technol 90:166–180
15. Bai J, Zhu J, Song Y, Zhao L, Hou Z, Du R, Li H (2021) A3T-GCN: attention temporal graph convolutional network for traffic forecasting. ISPRS Int J Geo-Inf 10(7):485
16. Chen X, Liu Y, Zhang J (2021) Traffic prediction for internet of things through support vector regression model. Internet Technol Lett e336
17. Yi H, Jung H, Bae S (2017) Deep neural networks for traffic flow prediction. In: IEEE international conference on big data and smart computing (BigComp), pp 328–331. Accessed 13–16 Feb 2017
18. Bezdek JC, Ehrlich R, Full W (1984) FCM: the fuzzy c-means clustering algorithm. Comput Geosci 10(2–3):191–203
19. Saeedmanesh M, Geroliminis N (2017) Dynamic clustering and propagation of congestion in heterogeneously congested urban traffic networks. Transp Res Procedia 23:962–979
20. Zhang S, Tong H, Xu J, Maciejewski R (2019) Graph convolutional networks: a comprehensive review. Comput Soc Netw 6(1):1–23
21. Wu F, Souza A, Zhang T, Fifty C, Yu T, Weinberger K (2019) Simplifying graph convolutional networks. In: International conference on machine learning, pp 6861–6871
22. Jiang M, Chen W, Li X (2021) S-GCN-GRU-NN: a novel hybrid model by combining a spatiotemporal graph convolutional network and a gated recurrent units neural network for short-term traffic speed forecasting. J Data Inf Manag 3(1):1–20
23. Chen C (2002) Freeway performance measurement system (PeMS). University of California, Berkeley
24. Aggarwal C (2018) Neural networks and deep learning. Springer International Publishing AG, part of Springer Nature, 2018. https://doi.org/10.1007/978-3-319-94463-0
25. Fahmy HMC (2016) Wireless sensor networks. Springer Science, Business Media Singapore. https://doi.org/10.1007/978-981-10-0412-4

HDD Failure Detection using Machine Learning

I. Gokul Ganesh, A. Selva Sugan, S. Hariharan, M. P. Ramkumar, M. Mahalakshmi, and G. S. R. Emil Selvan

1 Introduction

Over the last few decades, many studies have focused on the task of predicting hard disc failures. A threshold-based method was employed in previous approaches. These, on the other hand, were only 3–10% accurate in forecasting drive failures. Failure prediction methods examine current and historical data describing system states, events, and processes. Machine learning, which enables the training of a prediction model from time-series data, evaluation of the model's performance, and deployment in a productive environment, is becoming a more widely utilized technology for failure prediction. In the dataset, when the records of one class outnumber those of the other, our classifier may become biased toward the prediction. To avoid this data imbalance, the ADASYN approach is performed. The primary idea behind ADASYN is to use a weighted distribution for different minority class examples

I. G. Ganesh · A. S. Sugan · S. Hariharan · M. P. Ramkumar (✉) · M. Mahalakshmi · G. S. R. E. Selvan
Department of Computer Science and Engineering, Thiagarajar College of Engineering, Madurai, Tamilnadu, India
e-mail: ramkumar@tce.edu

I. G. Ganesh
e-mail: gokulganesh@student.tce.edu

A. S. Sugan
e-mail: selvasugan@student.tce.edu

S. Hariharan
e-mail: shariharan@student.tce.edu

M. Mahalakshmi
e-mail: mahalakshmim@student.tce.edu

G. S. R. E. Selvan
e-mail: emil@tce.edu

P. Singh et al. (eds.), *Machine Learning and Computational Intelligence Techniques for Data Engineering*, Lecture Notes in Electrical Engineering 998,
https://doi.org/10.1007/978-981-99-0047-3_61

based on their learning difficulty, with more synthetic data created for more difficult minority class instances than for easier minority class examples. The selection of attributes is a basic and important thing in Machine Learning. Feature extraction is the process of selecting a collection of features from a variety of effective approaches to lower the dimension of feature space. PCA and LDA are feature extraction methods. PCA reduces noise by condensing a large number of characteristics into a few main components. LDA is a method for determining a linear combination of attributes that distinguishes or distinguishes two or more classes of objects or occurrences. Apache Spark is used to reduce the processing time. The goal of this research is to increase prediction accuracy by using machine learning techniques. This study provides hard disk failure prediction models based on Logistic Regression, Random Forest, and Decision Tree, all of which perform admirably in terms of performance, prediction, interpretability, and stability.

2 Literature Survey

The goal of this research was to forecast water line failure based on age, length, material, and month of failure. Both datasets were subjected to three different machine learning algorithms: Random Forest Classifier, Logistic Regression Classifier, and Decision Tree. A similar approach was applied at Waterloo to evaluate classifier performance, and the results are available. While the models' overall accuracy is rather good, their predictive ability has drastically diminished following cross-validation [1].

This thesis has studied in depth the fields of machine learning and, in particular, supervised learning and dimensional reduction. The study was undertaken with the aim of combining machine learning solutions for time series manipulation with predictive maintenance in the context of equipment condition assessment and maintenance requirements forecasts. The main challenge was integrating the LIME neighborhood generator into the iPCA method [2].

The study proposed a default diagnostic algorithm based on SVM and RF algorithms to classify several classes of normal and abnormal conditions and defect prognosis for the RUL classes in predictive maintenance of the EPFAN with sensor data. Each algorithm has good accuracy for the diagnostic and prognosis pattern. But the utmost accuracy, precision and recall have been achieved with random forest [3].

In this paper, a hybrid strategy for multiclass defect classification was created and assessed by integrating ICEEMDAN, PCA, and LSSVM enhanced by CSA-NMS. The suggested hybrid model of the ICEEMDAN PCA-LSSVM gives accumulator status findings utilizing all of these evaluation variables and a high geometric mean value of 0.9960, implying a virtually flawless categorization. The categorization of the accumulator state was previously identified as the most challenging of the four components evaluated [4].

The purpose of this study is to see whether utilizing the content of the tender documents as a data source, machine learning, and text extraction algorithms might detect signals of corruption during the public procurement process. The computation of numerous indices, as well as signs of data corruption, is extremely difficult to come by. All CPV groups have the same mean accuracy as the original model's results, but all other metrics and recalls have improved, resulting in better identification of positive discoveries in the dataset [5].

3 Proposed Methodology

The proposed method makes use of machine learning techniques to classify whether a hard disk is normal or contains some fault. The data is collected, processed, and relevant features are selected, and a suitable supervised machine learning algorithm is applied to build the classifier. The built classifier is then evaluated against the validation parameters. The proposed detection method is then executed with the Apache PySpark to reduce the time consumption (Fig. 1).

3.1 Dataset

All data is acquired via S.M.A.R.T readings of BackBlaze's storage hard drives to operate with a set of relevant data from a real-time environment. SMART stands for Self-Monitoring, Analysis, and Reporting Technology and is a monitoring system included in hard drives that reports on various attributes such as the model of the hard drive, capacity bytes of the drive, Reallocated sector count, reported uncorrectable errors, Command Timeout, Current Pending Sector Count, Offline Uncorrectable, etc. Backblaze counts a drive as failed when it is removed from a Storage Pod and replaced because it has totally stopped working, or because it has shown evidence of failing soon. To determine if a drive is going to fail soon, SMART statistics can be used as evidence to remove a drive before it fails catastrophically or impedes the operation of the Storage Pod volume [6] (Fig. 2.)

Data set:

- Total disks: 110,700
- Total Normal disks: 110,692
- Total failed disks: 8

3.2 Data Preprocessing

Data cleansing and processing is a very important process as the collected dataset cannot be assured of its suitability in building the classifier[7]. As a first step the missing values in each feature were detected and were replaced with the most

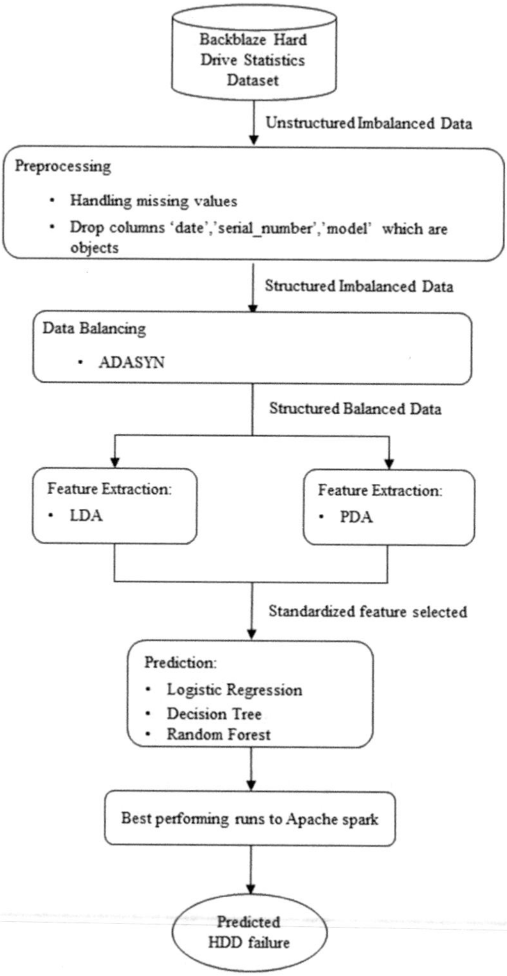

Fig. 1 Flow of execution

	date	serial_number	model	capacity_bytes	smart_1_normalized
0	7/5/2019	Z305B2QN	ST4000DM000	4.000790e+12	118.0
1	7/5/2019	ZJV0XJQ4	ST12000NM0007	1.200010e+13	77.0
2	7/5/2019	ZJV0XJQ3	ST12000NM0007	1.200010e+13	75.0
3	7/5/2019	ZJV0XJQ0	ST12000NM0007	1.200010e+13	75.0
4	7/5/2019	PL1331LAHG1S4H	HGST HMS5C4040ALE640	4.000790e+12	100.0

Fig. 2 Backblaze data set in a nutshell

commonly occurring value. And then some features which holds string values like the serial number and the model's name are not relevant in the task of prediction. So, these features were removed.

3.3 Data Balancing

The way the target variable is distributed affects how efficiently a model classifies a given input in the case of supervised learning. For instance, the dataset chosen here consists of samples representing normal drive after the application of preprocessing techniques 1,10,586 and the samples of the failed drive include 8. This proposed methodology uses ADASYN as the balancing algorithm. The model built using this data may achieve a higher accuracy but the limitation is that the built classifier always predicts the given drive as normal because the normal class is the majority class[8].

ADASYN:

ADASYN has been implemented in the imbalanced-learn module of sklearn. The technique depends on the Synthetic Minority Oversampling Technique (SMOTE), which is commonly used to solve Imbalanced Classification Problems in Machine Learning. By using ADASYN from imblearn to our sample data, we are able to create synthetic samples [9].

3.4 Feature Selection

PCA and LDA are two such techniques which have been tested in the proposed methodology to choose the best feature selection algorithm. PCA learns from the features supplied and classifies an unknown input [10]. Proper feature selection has to take place as there may be the presence of some irrelevant or correlated features which results in the poor quality of the model.

3.4.1 Principal Component Analysis (PCA)

PCA is an algorithm for lowering the dimensionality of huge datasets while minimizing information loss and boosting interpretability It accomplishes this by generating new statistically independent variables that optimize variance in a sequential manner. The purpose of PCA is to find patterns in a data set and then condense the variables down to their most significant characteristics [11].

3.4.2 Linear Discriminant Analysis (LDA)

LDA is an attempt to reduce the dimensionality of the data while maintaining the relationship between the features and the target variable. LDA uses the concept of feature projection to reduce the curse of dimensionality. The chosen dataset here has a large feature space and hence the data is projected in a low dimensional space [12].

3.5 Fault Detection Without Cloud Computing Resources

The data is divided 80:20 between training and testing for feature selection. Three algorithms have been explored on the chosen dataset to find the final best classifier. After that, the training data is used to fit the model, and the test data to validate it [13].

Logistic Regression:

The classification of a hard disk drive as normal or faulty belongs to the binary type of logistic regression. To deal with target variable imbalance, it is a viable mechanism to assign the class weights to the defining logistic regression algorithm.

Decision Tree:

The purpose of employing a Decision Tree is to develop a training model that can utilize basic decision rules inferred from training data to predict the class or value of the target variable.

Random Forest:

Random forest algorithm employs comprehensive learning, which is an approach for solving complicated problems that integrates a number of classifiers.

3.6 Fault Detection with Cloud Computing Resources

Traditional machine learning is not only complicated and difficult to set up, but it's also expensive. It needs pricey GPU cards to train and deploy massive machine learning models, such as deep learning. With the use of cloud computing, it reduces the managing cost of IT resources. Cloud provides automatic backup for the code. It greatly improves scalability.

Spark:

PySpark accelerates the processing of large data sets by dividing the work into chunks and allocating those chunks to different computational resources. It can manage thousands of real or virtual computers and handle up to petabytes of data. This helps in the execution of machine learning models with less time overhead.

4 Experimental Implementation and Evaluation

4.1 Experimental Setup

The project has been carried out using the google collab platform with python 3.0 as the programming language. Machine learning is supported by Python 3.0, which provides simple and legible code, as well as complicated algorithms and flexible processes. NumPy is a Python-based data analysis and high-performance scientific computing environment, Pandas for general-purpose data analysis, SciPy is a Python package for advanced computation, Seaborn for data visualization, and Scikit-learn for machine learning has been used. PySpark delivers Spark applications using Python APIs, and it also provides the PySpark shell for interactive data analysis in a distributed environment.

4.2 Evaluation Matrix

Evaluation metrics which are referred to by any Machine Learning system are based on the values of the confusion matrix.

Confusion Matrix:

A n x n square matrix which represents the number of instances correctly or incorrectly classified in n classes. Generally, more often it is a 2-dimensional matrix, indicating the information about binary class classification instances.

- True positives (TP): The number of failure class instances that are correctly classified as a failure by the classifier.
- False Negatives (FN): The number of failure class instances that are incorrectly classified as normal by the classifier.
- False Positives (TP): The number of failure class instances that are incorrectly classified as a normal class by the classifier.
- True Negatives (TN): The number of normal class instances that are correctly classified as a normal class by the classifier.

The diagonal elements of the confusion matrix indicate the number of instances that are correctly classified in n classes.

The non-diagonal elements of the confusion matrix indicate the number of instances that are incorrectly classified in n classes.

Accuracy: A metric to evaluate the overall performance of the model, evaluates the ratio of correctly classified instances to the overall total number of instances in the dataset as mentioned in the Eq. (1).

$$\text{Accuracy} = \frac{TP + TN}{TP + FP + TN + FN} \tag{1}$$

Precision: Precision evaluates the ratio of correctly classified instances of the failure class to overall predicted instances of the failure class as mentioned in Eq. (2).

$$\text{Precision} = \frac{TP}{TP + FP} \tag{2}$$

Recall: Recall evaluates the ratio of correctly classified instances of the failure class to overall instances of the failure class in the entire dataset under evaluation. It is also known as Sensitivity or True Positive Rate as mentioned in Eq. (3).

$$\text{Recall} = \frac{TP}{TP + FN} \tag{3}$$

F-measure: F-measure is a statistical technique which examines the accuracy of the AI model built by evaluating the weighted mean of precision and recall as mentioned in the Eq. (4).

$$\text{F} - \text{Measure} = 2 * \frac{Precision.Recall}{Precision + Recall} \tag{4}$$

4.3 Performance Evaluation on Data Balancing

This dataset used for the work is Predicting Hard Disk Drive Failure Dataset which consists of around 130 features and 110,700 samples. Among which around 110,586 Samples represent the failure class and the remaining 8 samples represent the normal class.

Figure 3 is more obvious that the application of the ADASYN algorithm has overcome the class imbalance issue by which the minority class instances have been raised from 8 to 110,586 samples. Thus, after the application of the algorithm, the number of samples representing the failure and normal class are distributed evenly. Then on the balanced dataset, we built logistic regression, decision tree, and random forest models and evaluated our model with the validation parameters. Table 1 shows the consolidated results of the performance of the models built using above mentioned three algorithms on the dataset before and after the application of the class distribution balancing algorithm.

Without any data balancing approach, it is more obvious that all three algorithms have achieved high accuracy with an average Precision, Recall, and F1-Score. This is mainly because an ml model will work more biased in an imbalanced dataset. This obviously leads to an increased accuracy. Thus, this clearly depicts accuracy as a wrong choice of metrics.

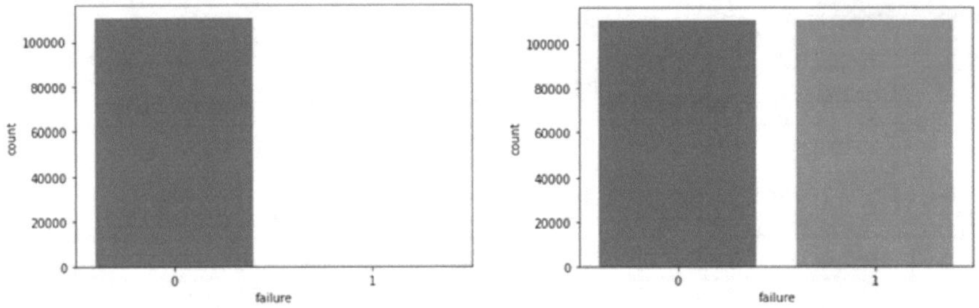

Fig. 3 Distribution of target variable analysis on data balancing

Table 1 Performance analysis on data balancing

S. no.	Evaluation metrics	Logistic regression		Decision tree		Random forest	
		Before ADASYN	After ADASYN	Before ADASYN	After ADASYN	Before ADASYN	After ADASYN
1	Accuracy (%)	96	61	99	78	100	81
2	Precision	0.00	0.63	0.00	0.98	0.00	0.97
3	Recall	0.00	0.56	0.00	0.60	0.00	0.62
4	F1-Score	0.00	0.59	0.00	0.74	0.00	0.76

Whereas the results observed while using ADASYN show a decreased accuracy but a well-balanced Precision-Recall ratio and with the increased accuracy. This proves that the model predicts both the classes at a balanced rate.

4.4 Performance Evaluation on Feature Selection

In the Dataset nearly 130 features have been represented but it does not mean that all features will help in building the best model. And some features will have correlation so those features will not help in the prediction. So, the right features have to be selected before building the model. Hence, the technique used for feature selection plays a vital role. In [10] the features have been selected randomly which devalues the quality of the model. This paper has explored the techniques of Principal Component Analysis (PCA) and Linear Discriminant Analysis (LDA) on the dataset in the feature selection process.

From Table 2, it is observed that, in this dataset, the performance of LDA is better.

PCA performs better in the case when the number of samples per class is less. Whereas LDA works better with large datasets having multiple classes and thus class separability is an important factor while reducing dimensionality. So, Linear Discriminant Analysis is chosen for the feature selection.

Table 2 Performance analysis on feature selection

S. no.	Evaluation metrics	Logistic regression		Decision tree		Random forest	
		Before LDA	LDA	Before LDA	LDA	Before LDA	LDA
1	Accuracy (%)	61	92	78	94	81	97
2	Precision	0.63	0.90	0.98	0.99	0.99	0.99
3	Recall	0.56	0.93	0.60	0.90	0.62	0.94
4	F1-Score	0.59	0.91	0.74	0.94	0.76	0.96

From Tables 1 and 2 it is inferred that a proper feature selection algorithm improves the quality of the built model. The use of LDA has improved the built model effectively in terms of evaluation parameters. On comparing the evaluation metrics like precision, recall, accuracy, and F1 score, the classifier built using the Random Forest algorithm performs well than the other chosen algorithms.

4.5 *Performance Evaluation Using Apache Spark*

Though validation parameters prove that the model built effectively predicts whether a hard disk drive has a chance of failure or not. But there occurs some limitation due to the time consumption taken by the algorithm during the process of fitting. So, to reduce the time factor, the proposed detection algorithm is executed in Apache PySpark which takes less processing time due to the feature of in-memory computation.

The Table 3 infers that the processing time taken has been drastically reduced from 5 min to 194 ms when executed in the Spark environment. This is because PySpark provides high processing speed as the data gets stored in cache thereby reducing the number of fetch operations from the disk. So, this reduction of time consumption will help in the effective prediction of failure.

Table 3 Performance measure of random forest in spark

S. no.		In spark	RF
1	Accuracy (%)	78	97
2	Recall	0.77	0.99
3	Precision	0.74	0.94
4	F1 score	0.75	0.96
5	Time taken	194 ms	5 min

5 Conclusion and Future Work

In this study, a machine learning model is proposed to evaluate Hard Disk Drive failures. The proposed approach has overcome the problem of imbalance distribution of the target variable, and a proper subset of features has been selected. The model presented here has shown good performance in terms of evaluation metrics. In future the performance of the model will be increased in the aspect of validation metrics and at the same time overhead will be reduced.

References

1. Amini M, Dziedzic R (2021) Comparison of machine learning classifiers for predicting water main failure (No 5579). EasyChair
2. Ramkumar MP, Narayanan B, Selvan GSR, Ragapriya M (2017) Single disk recovery and load balancing using parity de-clustering. J Comput Theoret Nanosci 14(1), 545–550. ISSN 15461955
3. Kusumaningrum D, Kurniati N, Santosa B (2021) Machine learning for predictive maintenance. In: Proceedings of the international conference on industrial engineering and operations management. IEOM Society International, pp 2348–2356
4. Buabeng A, Simons A, Frimpong NK, Ziggah YY (2021) Hybrid intelligent predictive maintenance model for multiclass fault classification
5. Ramkumar MP, Balaji N, Rajeswari G (2014) Recovery of disk failure in RAID-5 using disk replacement algorithm. Int J Innov Res Sci Eng Technol 3(3):2358–2362
6. Srijha V, Ramkumar M (2018) Access time optimization in data replication. In: 2018 2nd International conference on trends in electronics and informatics (ICOEI), pp 1161–1165. IEEE
7. Farooq B, Bao J, Li J, Liu T, Yin S (2020) Data-driven predictive maintenance approach for spinning cyber-physical production system. J Shanghai Jiaotong Univ (Science) 25(4):453–462
8. Giommi L, Bonacorsi D, Diotalevi T, Tisbeni SR, Rinaldi L, Morganti L, Martelli B (2019) Towards predictive maintenance with machine learning at the INFN-CNAF computing centre. In: Proceedings of international symposium on grids & clouds
9. Ramkumar MP, Balaji N Vinitha Sri S (2013) Stripe and disk level data redistribution during scaling in disk arrays using RAID 5. In: PSG—ACM Conference on intelligent computing, 26–27 April, vol 1, Article no 74, pp 469–474
10. Paraschos S (2021) Elevating interpretable predictive maintenance through a visualization tool (Doctoral dissertation, Master's thesis, Aristotle University of Thessaloniki, School of Informatics, MSc. Artif Intell 3:2021
11. Meena SM, Ramkumar MP, Asmitha RE, Emil Selvan GSR (2020) Text summarization using text frequency ranking sentence prediction. In: 2020 4th International conference on computer, communication and signal processing (ICCCSP), pp 1–5. IEEE
12. Andriani AZ, Kurniati N, Santosa B (2021) Enabling predictive maintenance using machine learning in industrial machines with sensor data. In: Proceedings of the international conference on industrial engineering and operations management
13. Tomer V, Sharma V, Gupta S, Singh DP (2021) Hard disk drive failure prediction using SMART attribute. Mater Today Proc

Energy Efficient Cluster Head Selection Mechanism Using Fuzzy Based System in WSN (ECHF)

Nripendra Kumar, Ditipriya Sinha, and Raj Vikram

1 Introduction

Wireless Sensor Network (WSN) consists of a large number of sensors, which are deployed in large areas to provide a variety of applications such as remote health monitoring, disaster management (forest fire), etc. In WSN, saving the energy of nodes is highly demanding due to the huge amount of energy drainage. To save energy from sensor nodes researchers have proposed many techniques. Cluster head selection is one of them. A hierarchical cluster-based approach is used to improve the network performance and save the battery power of sensor nodes. The nodes, which are sensed the event are called active and the rest of the nodes in the network are inactive. All active nodes constitute the cluster. Considering the demands of energy-efficient cluster formation and overhead reduction has motivated authors to design energy-aware cluster formation techniques. In this scheme a fuzzy-based cluster head selection scheme is designed. The scheme is designed on the basis of 4 parameters such that residual energy, node to base station distance node to node distance, and degree of nodes. This way energy consumption of sensor nodes is managed in an efficient way and the network lifetime is enhanced.

Our main contributions are listed as follows:

1. A fuzzy-based CH selection algorithm.
2. Enhanced the lifetime of the network.

N. Kumar (✉) · D. Sinha · R. Vikram
Department of Computer Science and Engineering, National Institute of Technology Patna, Patna, India
e-mail: nripendrak.pg20.cs@nitp.ac.in

D. Sinha
e-mail: ditipriya.cse@nitp.ac.in

R. Vikram
e-mail: raj.cs17@nitp.ac.in

P. Singh et al. (eds.), *Machine Learning and Computational Intelligence Techniques for Data Engineering*, Lecture Notes in Electrical Engineering 998,
https://doi.org/10.1007/978-981-99-0047-3_62

3. Compared the proposed approach with FBECS [1], BCSA [2] and LEACH [3] algorithm.

The rest of the paper is organized as follows: Sect. 2 describes the related work on the clustering problem, Sect. 3 identifies the scope of the work, and Sect. 4 gives the simulation results. At last we conclude the paper in Sect. 5.

2 Related Work

In [1], the author proposed a BCSA protocol. The network is partitioned into four equal sub-networks based on their distance from the base station in this technique. The sensor node closest to the base station has a greater associated probability value than the SN in the region farthest from the BS. In [4] the author proposed a probability-based cluster routing protocol for wireless sensor networks. Here, residual energy and initial energy are only considered for the cluster head selection [2]. The author proposed a fuzzy-based enhanced cluster head selection (FBECS) for WSN. The network is divided into subnetworks, and the probability is allocated to every node as per the separation distance. Here, three parameters: energy, distance to the base station, and density of sensors are considered for CH selection [5]. The author proposed an enhancement of network lifetime by fuzzy-based secure CH clustered routing protocol for mobile wireless sensor networks. In this paper, an energy resourceful clustering technique is established for mobile wireless sensor networks utilising Fuzzy-logic ideas. In [6], the author proposed a fuzzy-based novel clustering technique by exploiting spatial correlation in wireless sensor networks. The authors are trying to find a cluster head on the basis of chance and prediction of the required energy for transmission. They are not considering any distance parameters within the network. In [7], the author proposed a LEACH-C protocol, which enhances the network lifetime of the leach. [8] The author talks about a stochastic and egalitarian cluster-based energy-efficient approach addressing bi-level heterogeneity. The CH selection is based on dynamic likelihood calculated by ignoring uniform power use. For selecting the most relevant nodes as cluster head nodes from the nodes present in each cluster, a fuzzy logic-based deductive inference system was built and applied in [9].

3 Proposed Work

The sensor node (SN) collects data from the target region and relays it to the base station (BS). Every sensor node can function as both a sensing node and a CH. The CH combines the data gathered from its members and sends it to the BS. The focus is on energy optimization of the deployed SN because a large amount of energy is depleted during information transmission. This proposed work describes the selection of the finest eligible sensor node as a cluster head. The overall workflow of the proposed model depicts in Fig. 1.

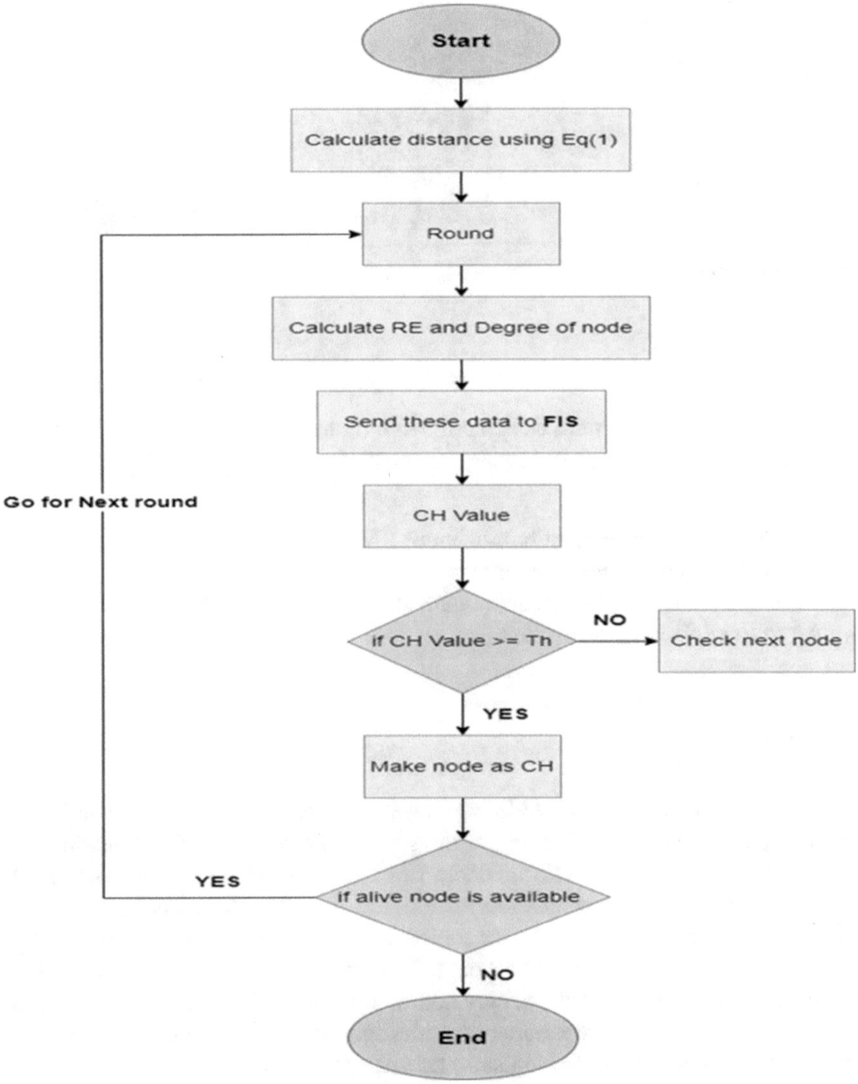

Fig. 1 Work flow diagram of proposed model

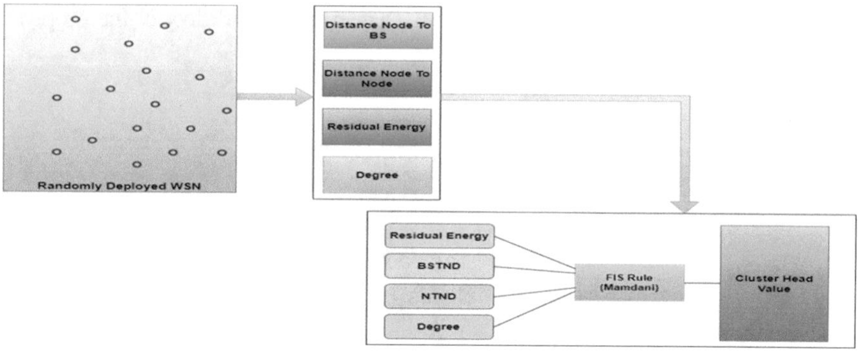

Fig. 2 Proposed model

3.1 Assumption

The suggested work's system design aims to monitor the environment by installing SN in the target region. The following are some of the assumptions that are made in this suggested work:

1. The network is made up of homogenous SN with the same energy level.
2. SN deployment is random in nature.
3. Once deployed, the BS and SN remain static, that is, they do not move.
4. All of the SN are GPS-enabled.

3.2 Proposed Model

The Fuzzy Logic based enhanced Energy-aware clustering scheme is depicted in this paper. Figure 2 shows the working of the proposed model. The CH selection uses fuzzy logic to choose the best candidate from the network's accessible nodes. As fuzzy input, four network parameters are employed in the model. The first consideration is the distance between the base station and the sensor nodes. Second, the SN's residual power level is taken into account. Third, the average distance between each sensor node and the rest of the sensor nodes. The fourth factor is the node degree, which is determined by the number of nodes in their sensing range. The proposed work is divided into the following sub-phases.

3.2.1 Deployment Phase

The SN is strewn around the target area in a haphazard manner. The suggested procedure kicks into effect after the SN is deployed in the target region. The establishment of clusters occurs after the start of each round. Before gathering information from the target location, BS broadcasts a packet (LOC BS) into the target region after the deployment. This packet provides critical information such as BS's position, as well as a time frame for SN to prevent a collision. To avoid collisions, the SN now broadcasts a message (hi message) throughout the network in line with the time window supplied by BS. The SN is estimated for all network parameters such as their distance from BS, residual energy, distance from neighboring nodes, and the degree of node based on their sensing range once all of the broadcasts have been finished. These are derived in the following way:

Distance: The distance between two points (x, y) and (x_1, y_1) is computed using the Euclidian distance equation (1).

$$Distance = \sqrt{(x - x_1)^2 + (y - y_1)^2} \tag{1}$$

Residual Energy: Energy is computed with the standard radio energy model [10]

Degree: degree is defined as the number of nodes present in their sensing region of a node.

3.2.2 Cluster Head Selection Phase

The FL technique is used in this proposed study to choose the best candidate for the post of CH. The phrase Fuzzy refers to anything a little hazy. When a scenario is unclear, the computer may be unable to generate a True or False response. The number 1 denotes True in Boolean logic, while 0 denotes False. A Fuzzy Logic algorithm, on the other hand, takes into account all of a problem's uncertainties, including the possibility of other answers besides True or False.

Fuzzification

For fuzzification, we utilized four variables as inputs mentioned in Table 1. Crisp input from the fuzzifiers in Fig. 3. The whole amount of energy left over is referred to as residual energy. The Euclidean distance between each SN and the BS is the Distance to BS. The average distance between a node and each other in the network is determined. The degree refers to the number of nodes in their sensing range. The FIS receives these crisp values (discrete values). Fuzzified membership values are provided in the rule base for IF-THEN situations. Because there are four input variables, each with three linguistic levels, the total number of potential fuzzy rules is $3^4 = 81$. Some if and then rules are given in pictorial format in Fig. 4 and matrix representation in Table 2. A value is derived by applying the fuzzy AND and OR operators to the inputs. All of the output is aggregated, and the maximum value is picked. Among the set of fuzziness we use fuzzy logic is used to calculate the CH value.

Defuzzification

Min of max Defuzzification also known as the maximum method is applied for this purpose. The output in this case is given by Eq. (2).

$$Z* = \frac{\sum_{i=1}^{n} x^{i}}{n} \tag{2}$$

The linguistic variable for output is mentioned in Table 3. Membership functions come in a variety of forms, including triangular, trapezoidal, Gaussian, and generalized bell. Because of its simplicity and speed of computation, we adopted the Triangular membership function (MF) in this suggested study. The triangular MF that is integrated into our FIS provided in Eq. (3).

$$\text{triangle}(x; a, b, c) = \max\left(\min\left(\frac{x-a}{b-a}, \frac{c-x}{c-b}\right), 0\right) \tag{3}$$

Three parameters, a, b, and c, define a triangular MF. We have an alternate formulation for the previous Eq. (2) by utilizing min and max. The x coordinates of the three corners of the underlying triangle MF are determined by the parameters a, b, and c (with a, b, and c). Once the cluster head value for each node has been calculated the defuzzifier turns the supplied input into a crisp set and calculates the cluster head value for each node. In Table 4 cluster head value of some sensors node with respect to rounds. After analyzing the cluster head value, we define a threshold value of 0.56

Table 1 Input function for the fuzzifier

Input	Linguistic variable
Residual energy	Low, medium, high
BS to node distance	Close, medium, far
Node to node distance	Close, medium, far
Degree of node	Spars, medium, dense

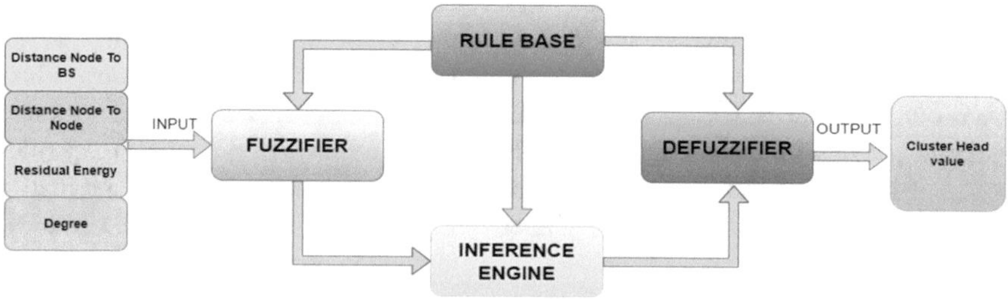

Fig. 3 The components of a fuzzy system

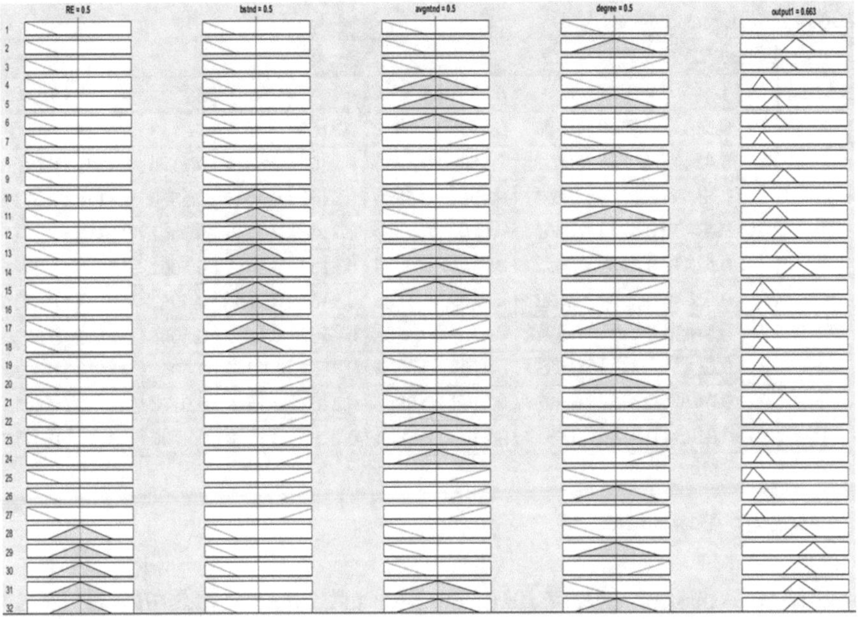

Fig. 4 Fuzzy rule set view

Table 2 If then rules

Rule no.	Residual energy	BS to node distance	Node to node distance	Degree of node	Cluster head value
1	Low	Close	Close	Sparse	Far better
2	Medium	Close	Close	Sparse	Better
3	Medium	Medium	Close	Medium	Far better
–	–	–	–	–	–
81	High	Close	Medium	Sparse	Far better

Table 3 Output function for defuzzifier

Output	Linguistic variables
Cluster head values	Worst, bad, fair, good, better, far better, best

for becoming a particular node's cluster head. If a node has a cluster head value greater than 0.56, that node will be chosen as the cluster head for the current round.

Table 4 Cluster head values

	Sensor ID										
Number of rounds		1	2	3	4	5	6	7	8	9	10
	1	0.6143	0.5329	0.6927	0.5958	0.5882	0.2162	0.5317	0.5513	0.6624	0.6901
	2	0.6143	0.6545	0.6927	0.6544	0.5885	0.2162	0.5317	0.6170	0.4858	0.6106
	3	0.6143	0.3664	0.6927	0.5067	0.5885	0.2162	0.5317	0.5550	0.3178	0.3715
	4	0.6143	0.3134	0.6927	0.5109	0.5885	0.2162	0.5317	0.5062	0.3178	0.3715
	5	0.6143	0.3134	0.6927	0.5109	0.5885	0.2162	0.5317	0.5062	0.3178	0.3715
	6	0.6143	0.3134	0.6927	0.5109	0.5885	0.2162	0.5806	0.5062	0.3178	0.3715
	7	0.6143	0.3134	0.6927	0.5109	0.5885	0.2162	0.6362	0.5062	0.3178	0.3715
	8	0.6143	0.3134	0.6927	0.5109	0.5885	0.2162	0.6952	0.5062	0.3178	0.3715
	9	0.6120	0.3134	0.6927	0.5109	0.5885	0.2162	0.7518	0.6170	0.3178	0.3715
	10	0.6143	0.3134	0.6927	0.5109	0.5885	0.2162	0.7518	0.5550	0.3178	0.3715

4 Result Analysis

To confirm its performance, ECHF is simulated and compared to FBECS [1], BCSA [2], and LEACH [3]. The simulation work is done in MATLAB since it supports all sorts of fuzzy MFs and is easy to build. The simulation parameter is given in Table 5.

It is depicted in Fig. 5 that when we increase the number of rounds in the network average residual energy of nodes decreases. The proposed ECHF gives better results when compared to FBECS [1], BCSA [2], and LEACH [3].

Figure 6 shows that when the number of rounds increases the number of dead nodes increases in the network. The proposed ECHF gives better results when compared to FBECS [1], BCSA [2], and LEACH [3].

Figure 7 shows that the proposed algorithm select more number of cluster head when we increase the number of the round. It gives better performance when compared to FBECS [1], BCSA [2], and LEACH [3].

Table 5 Simulation parameter

Parameters	Value
Number of nodes	100
Network Dimension	1000*1000
Initial energy of sensor node	200 J
Sensor node sensing range	250
Simulation Simulator	Matlab

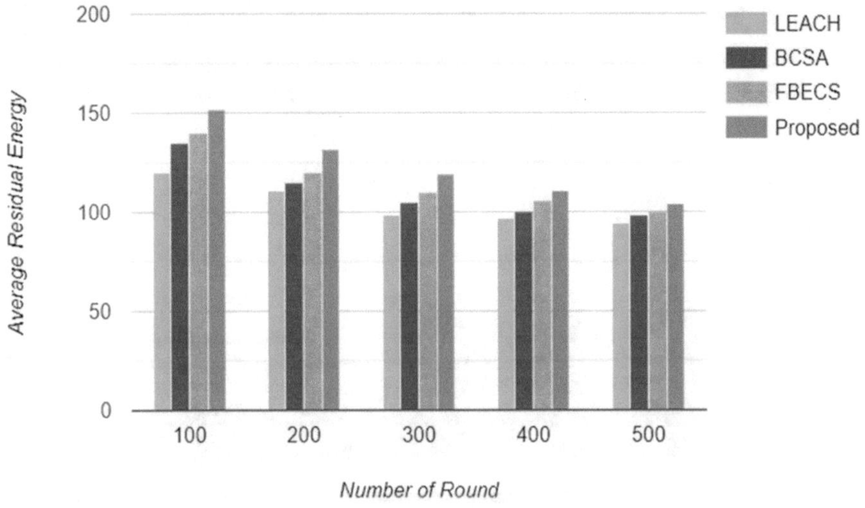

Fig. 5 Number of round versus average residual energy

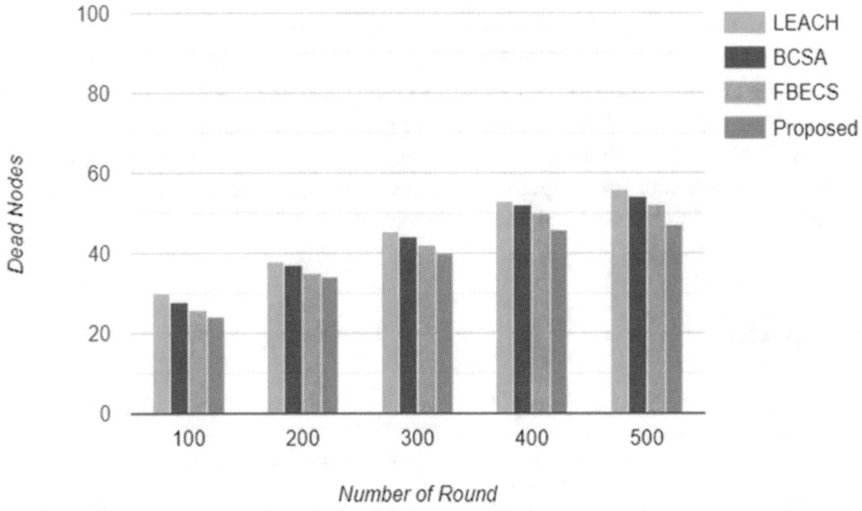

Fig. 6 Number of rounds versus dead nodes

5 Conclusions

Cluster head selection problems are an open challenge for researchers in WSN. In this paper, a fuzzy-based technique has been proposed for selecting the best cluster head in deployed sensors. For any sensor node, four parameters such as distance from the base station, residual energy, distance from neighboring nodes, and the degree of

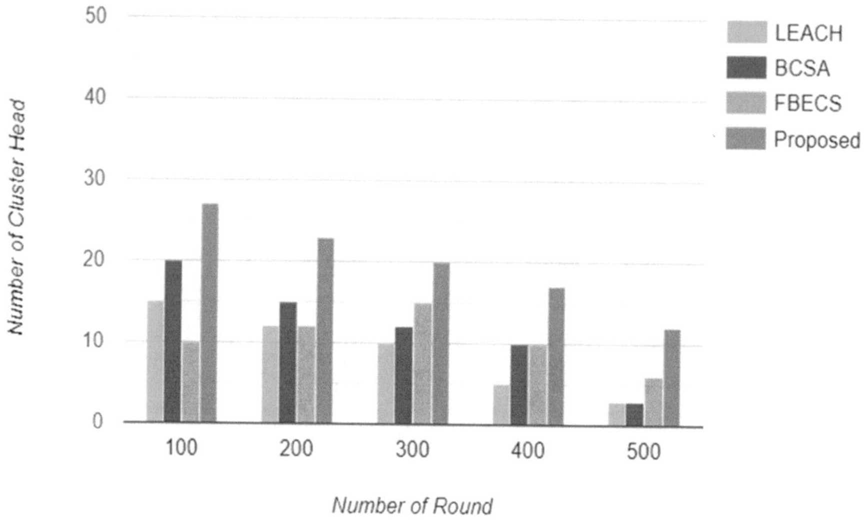

Fig. 7 Number of rounds versus number of cluster head

the node are computed and a fuzzy model is built with the Mamdani fuzzy rule model for the cluster head selection. The fuzzy interface system calculates the cluster head value using all four parameters and 81 rules. Every round, the CH value may change owing to changes in residual energy and node degree, since certain sensor nodes may become dead causing a change in the node degree. The simulation results show that the proposed protocol has outperformed the existing protocols, FBECS [1], BCSA [2], and LEACH [3] in terms of energy conservation and CH selection.

References

1. Zytoune O, Fakhri Y, Aboutajdine D (2009) A balanced cost cluster-heads selection algorithm for wireless sensor networks. Int J Comput Sci 4(1):21–24
2. Mehra PS, Doja MN, Alam B (2020) Fuzzy based enhanced cluster head selection (FBECS) for WSN. J King Saud Univ Sci 32(1):390–401
3. Singh SK, Kumar P, Singh JP (2017) A survey on successors of LEACH protocol. IEEE Access 5:4298–4328
4. Rawat P, Chauhan S (2021) Probability based cluster routing protocol for wireless sensor network. J Amb Intell Humaniz Comput 12(2):2065–2077
5. Rajesh D, Jaya T (2021) Enhancement of network lifetime by fuzzy based secure CH clustered routing protocol for mobile wireless sensor network. J Amb Intell Humaniz Comput 1–11
6. Singh M, Soni SK (2019) Fuzzy based novel clustering technique by exploiting spatial correlation in wireless sensor network. J Amb Intell Humaniz Comput 10(4):1361–1378
7. Heinzelman WB, Chandrakasan AP, Balakrishnan H (2002) An application-specific protocol architecture for wireless microsensor networks. IEEE Trans Wirel Commun 1(4):660–670
8. Elbhiri B, Saadane R, Aboutajdine D (2011) Stochastic and equitable distributed energy-efficient clustering (SEDEEC) for heterogeneous wireless sensor networks. Int J Ad Hoc Ubiquitous Comput 7(1):4–11

9. Balasubramaniyan R, Chandrasekaran M (2013) A new fuzzy based clustering algorithm for wireless mobile Ad-Hoc sensor networks. Int Conf Comput Commun Informat 2013:1–6. https://doi.org/10.1109/ICCCI.2013.6466313
10. Zytoune O, Fakhri Y, Aboutajdine D (2009) A balanced cost cluster-heads selection algorithm for wireless sensor networks. Int J Electron Commun Eng 3:761–764. http://scholar.waset.org/1307-6892/8896

Choosing Data Splitting Strategy for Evaluation of Latent Factor Models

Alexander Nechaev, Vasily Meltsov, and Dmitry Strabykin

1 Introduction

Recommender systems (RS) are software products that solve the task of users' interest prediction. The goal of these systems is to create a list of items with which a user is most likely to interact. In many domains, the only data available to make such a prediction are numerical ratings that have been either explicitly or implicitly given to existing items [1]. Consequently, the task of an RS is often stated as a task of rating prediction. In this case, matrix factorization algorithms show state-of-the-art performance in terms of prediction errors [2, 3].

As well as for most of the other machine learning methods, a dataset is split into the train, validation, and test subsets. It has been shown that the choice of a data splitting technique may largely impact the results of the model evaluation for complex deep learning methods [4]. This fact negatively affects one's ability to make an unbiased comparison of different models in both academic and business areas. For example, for research purposes, it may be hard to state which method performs better even if the same dataset and target metric are employed. One can either wilfully or unintentionally manipulate the results using different splits. At the same time, if one uses simple random splits in production, a model that actually performs better than previous ones may not be preferred over them due to biases in chosen validation set or even its size [5].

A. Nechaev (✉) · V. Meltsov · D. Strabykin
Vyatka State University, Kirov, Russia
e-mail: dapqa@yandex.ru

V. Meltsov
e-mail: meltsov@vyatsu.ru

D. Strabykin
e-mail: strabykin@vyatsu.ru

P. Singh et al. (eds.), *Machine Learning and Computational Intelligence Techniques for Data Engineering*, Lecture Notes in Electrical Engineering 998,
https://doi.org/10.1007/978-981-99-0047-3_63

However, the question of how much the choice of data splitting strategy affects the actual performance of a latent factor model remains unanswered. In production, the same model is often trained on new data, and there is a need to split data properly to make the most performant model. It is important that not only the accuracy should be taken into account, but also the amount of time needed to train a model. For example, the usage of cross-validation is usually considered as profitable in terms of the model's generalization performance, but it drastically increases evaluation time. The goal of our research is to experimentally estimate how much the choice of the data splitting strategy impacts the performance and evaluation times of latent factor models, and try to determine what strategy should be used in different cases.

2 Related Work

Recommender system datasets naturally contain biased data. Models trained on these data should take these biases into account to make more accurate predictions. However, in real-world pipelines, training is often performed not on a full dataset, but only on its part (to validate and compare different models, in particular—for hyper-parameter optimization). In many areas, simple random splits are used. This strategy is acceptable in some domains but may distort recommender systems data. If a model leverages existing biases, but used subsets do not preserve initial biases or even create new ones, given performance estimations will not be true. For machine learning models in common, there are two solutions to this problem.

The first solution is K-Fold cross-validation [6]. The usage of this technique allows smoothing the biases created by data splitting and making more reasonable evaluations but the overall computation time grows together with the number of folds. Such growth may be critical for complex models that require large amounts of computational resources for a single train-test run. Besides the presence of such a consequence, one should contemplate that there are cases when cross-validation may not improve evaluation accuracy. For example, Westerhuis et al. [7] showed how an actual performance might largely differ between three variants of cross-validation for their model. In our research, the focus is set on matrix factorization algorithms. Since some of them may heavily leverage existing biases [1, 8], there will be no difference in how many folds are in use if biases are distorted in every split. So, the need for cross-validation in our case is questionable.

The second common solution is to try to split a dataset in a such way that important characteristics of point distributions are preserved in all parts. Xu and Goodacre [9] conducted a comparative study that tried to help solve which splitting strategy was the best to estimate a model's generalization performance on synthetic datasets. According to the results, all the strategies were viable, but the choice could not be made apriori because observations were very data-dependent. Moreover, the problem was observed that the most representative points were sampled to validation sets and models were poorly fit. A possible solution, that also creates only a single split, may be found in the work by Joseph and Vakayil [10]. They present SPlit—a method of data splitting that relies on the most representative points of a continuous distribution. Experiments show that the usage of SPlit stably decreases the variance of errors shown by trained models in comparison to other existing algorithms.

However, a problem arises when it comes to matrix factorization datasets. SPlit, as well as other algorithms from its family, treats dataset rows as points of a multidimensional space. At the same time, recommender system datasets for which latent factor models are suitable usually contain only users' and items' identifiers, ratings, and timestamps. To apply such an algorithm to them, one needs to transform their rows to representative, comparable feature vectors. Unfortunately, the common solution to this problem is in the usage of feature vectors [11]. So, here is an impasse: to make feature vectors viable, data must be properly split. Moreover, even if there is a technique that can make sought vectors in a fast and unbiased manner, the length of latent factor vectors is usually much larger than the width of datasets for which methods like SPlit have been evaluated, so the splitting process may become unacceptably resource-consuming.

Consequently, both cross-validation methods and distribution-preserving splitting algorithms do not solve the problem stated at the start of the current section. At this point, the solution may lie in the area of special techniques for recommender system data. A comprehensive review of splitting methods used in recent works is presented by Meng et al. [4]. According to them, the following special splitting strategies may be used in our case (there are several others, but they require additional data):

- *Temporal Global*: all transactions are sorted by their timestamps, and a specified amount of the most recent ones is used for validation, while all others are used for training;
- *Temporal User*: the same as *Temporal Global*, but with stratification on users (i.e., given amounts of the latest transaction are calculated for each user individually).

It is important that the usage of temporal strategies solves the problem of 'transaction leakage', i.e., the usage of information 'from the future' in the model training process. According to the findings of Meng et al., the choice of data splitting strategy noticeably influences evaluation metric values and even may make the results of different investigations on the same dataset and metric incomparable. At the same time, there is still no known way to determine apriori which strategy would be the best for a given dataset; it conforms to the results from [9].

These findings help understand how the results may be distorted in research areas, but they do not answer the question about how the actual performance of models changes depending on which strategy is applied for training and hyperparameter optimization. Also, there is a need to evaluate splitting methods using matrix factorization models, on a larger variety of datasets, which have different characteristics and represent transactions from different domains. These points substantiate the goal of our study stated above.

3 Methodology

3.1 Datasets and Matrix Factorization Algorithms in Use

To perform the experiments, 15 datasets have been chosen from different domains. Movielens datasets [12] are de-facto standard ones for the evaluation of matrix factorization recommender models. They contain ratings that users have given to movies they watched. Amazon Reviews datasets [13] are other popular ones, containing ratings of products from Amazon online shops (divided by product categories). LibraryThing and Epinions datasets [14, 15] are initially used for sentiment analysis and user's cross-influence investigation, but they also contain ratings and may be used for training latent factor models. LibraryThing contains books' ratings, while Epinions contain ratings from general e-commerce service. Another book dataset is GoodRead Reviews [16], which is also primarily aimed at sentiment analysis but has much higher numbers of ratings per user and per item in comparison to LibraryThing.

In addition, to examine non-standard data for recommender systems, we use the Drug Recommendations dataset [17]. It contains ratings of drugs prescribed for several diseases. If one considers drugs as items and diseases as users, one can make an RS that predicts which drugs may be suitable for a given disease.

The summary description of used datasets is presented in Table 1.

There are a couple of matrix factorization algorithms that can be applied to build latent factor models of these data. To reach the goal of the current research, one needs to conduct a large number of experiments on different datasets. So, chosen models must be both simple enough to be trained in several minutes and generic enough to make conclusions about more complex ones. According to findings presented in [18], SVD and SVD++ algorithms [1] are suitable in this case, since many modern algorithms are built on their basis and have the same characteristics of their loss function's shape. Therefore, SVD and SVD++ algorithms are employed to perform the experiments.

3.2 Experiment Scheme

Experiments in our research are conducted as follows.

A dataset $\mathcal{K}$ is split into two parts, a test $\mathcal{K}_{test} \subset \mathcal{K}$ and non-test $\mathcal{K} \setminus \mathcal{K}_{test}$ ones, respectively. Next, the non-test part is split in some way, providing a pair of sets $\mathcal{K}_{train}$ and $\mathcal{K}_{valid}$. A model is trained on $\mathcal{K}_{train}$ and validated on $\mathcal{K}_{valid}$, thus giving a target metric value. Hyperparameters of a model are optimized using Bayesian Optimization to minimize (or maximize) the target metric value. If cross-validation is applied, there are multiple pairs of train and validation sets, and the target metric value is averaged across them. After hyperparameter optimization, a model with given hyperparameter values is trained on $\mathcal{K} \setminus \mathcal{K}_{test}$ and evaluated on $\mathcal{K}_{test}$ thus giving the final metric value that practically estimates the model performance. For splitting a

Table 1 Summary description of used datasets

Dataset	# of Ratings	# of Users	# of Items
Movielens 100 k	100000	943	1682
Movielens 1 M	1000209	6040	3706
Movielens 10 M	10000054	69878	10677
Epinions	188478	116260	41269
LibraryThing	1387125	70618	385251
GoodRead Reviews (w/ spoilers)	1330981	18868	25469
Drug Recommendations	53471	708	2635
Amazon Ratings (Software)	459436	21663	375147
Amazon Ratings (Amazon Fashion)	883636	186189	749233
Amazon Ratings (All Beauty)	371345	32586	324038
Amazon Ratings (Appliances)	602777	30252	515650
Amazon Ratings (Gift Cards)	147194	1548	128877
Amazon Ratings (Luxury Beauty)	574628	12120	416174
Amazon Ratings (Magazine Subscriptions)	89689	2428	72098
Amazon Ratings (Prime Pantry)	471614	10814	247659

dataset to test and non-test parts, the same strategy *must* be used in all cases, while for splitting the non-test part to train and validation ones different strategies are applied. This allows comparing the actual performance of models fit using different strategies.

The important attribute of a splitting strategy is how it behaves over time. It has been mentioned in [9] that the dataset size may be crucial for a proper choice. In a production environment, the amount of available data always grows, and if a chosen strategy performs well with some small count of rows, it does not mean that it will perform such well after a couple of months. To take this peculiarity into account, we suggest making 5 versions of each dataset $\mathcal{K}$ (noted as $\mathcal{K}^{(1)}, \ldots, \mathcal{K}^{(5)}$), corresponding to different states of the dataset in time ($\mathcal{K}^{(5)}$ is the smallest, while $\mathcal{K}^{(1)}$ is the largest and the latest one). For each $\mathcal{K}^{(5)} \ldots \mathcal{K}^{(2)}$, $|\mathcal{K}^{(i)}| = 0.9|\mathcal{K}^{(i-1)}|$. These versions are employed instead of using only full datasets.

To split a dataset to test and non-test parts, we always use the temporal global strategy (test size = 0.2) to avoid the 'transaction leakage' problem and make the fairest final estimations. To split the non-test part to train and validation ones, we use four splitting strategies:

- Random Split, validation size = 0.2;
- K-Fold Cross-Validation (using 5 random folds);
- Temporal Global, validation size = 0.2;
- Temporal User, validation size = 0.2.

Bayesian Optimization is performed using Gaussian Process as a surrogate target function model and UCB as a point acquisition function. Hyperparameters optimized are the number of factors and the regularization constant. The number of initial randomly chosen points is 1, and the number of optimization iterations is 5.

Two kinds of target metrics are employed. The first one is RMSE that estimates rating prediction error. The second one is NDCG@K used for ranking evaluation:

$$NDCG@K = \frac{1}{|U_{valid}|} \sum_{j=1}^{|U_{valid}|} \frac{DCG@K}{IDCG@K}, \tag{1}$$

$$DCG@K = \sum_{k=1}^{K} \frac{2^{r^{true}(\pi^{-1}(k))} - 1}{log_2(k+1)}, \tag{2}$$

where U_{valid} is a set of users in $\mathcal{K}_{valid}$, $\pi^{-1}(k)$ is the element at kth position in the prediction for j-th user (if items are sorted by predicted ratings), and r^{true} is the true relevance function with values in [0, 1]. $IDCG@K$ is the maximum possible $DCG@K$ value, i.e., with $r^{true} = 1$ for each element. In experiments, K is set to 10, and r^{true} just returns 1 if an item is at its true position and 0 otherwise. If a user has rated less than 10 items in a validation set, K is lowered for them.

Consequently, for each dataset, 80 experiments are performed: 5 versions, 2 models, 2 metrics, and 4 splitting strategies. Both metric values and evaluation times are registered during the experiments.

3.3 Results Interpretation Methodology

After the experiments are performed, one has raw data containing metric and time values for each dataset version, each model, and each splitting strategy. Results must be interpreted in a way that answers the following questions:

Q1. How much metric values differ between different splitting strategies?
Q2. How much evaluation times differ between different splitting strategies?
Q3. Do metric values depend on what splitting strategy is used?

Q4. Does the proper choice of the strategy depend on the version of the dataset (i.e., on the dataset size and state in time)?

To compare metric values and evaluation times from different experiments, it is required for them to be normalized. NDCG@10 metric is already normalized, so its values will be used 'as is'. To compare RMSE values, a normalized version $NRMSE = RMSE/(r_{\max} - r_{\min})$ is used [1], where $r_{\max}$ and $r_{\min}$ are rating scale bounds of a dataset. To analyze how evaluation times differ between strategies, relative values are taken, i.e., $\delta_{time} = time/time_{\min}$, where $time_{\min}$ is the minimal evaluation time given for this dataset version, this model, and this metric kind.

Since most of the results reviewed in Sect. 2 show that the choice of the optimal splitting strategy highly depends on data in use, for an appropriate comparison one should analyze experimental results over not only all the datasets but also over some groups of them. According to the datasets' characteristics and domains, they have been divided into four groups:

- *Movies*: Movielens datasets;
- *Books*: LibraryThing and GoodRead Reviews dataset;
- *Drug*: Drug Recommendations dataset;
- *Customer*: Epinions and Amazon datasets.

It is assumed that models and strategies show roughly the same performance for all the datasets in the same group. When a larger grouping is needed for the analysis (i.e., to provide a sufficient number of points), 'Movies', 'Books', and 'Drug' groups are united to one group *'Non-Customer'*.

Next, to answer Q1 and Q2, it is enough to visualize results, e.g., at violin plots. Normalized metric values should be aggregated by the dataset group, model, metric kind, and splitting strategy. For relative evaluation times, the plotting is the same except that both times for NRMSE and NDCG@10 are taken and averaged.

To answer Q3, it is needed to perform a statistical test, where the independent variable is categorical (a splitting strategy), while the dependent one is continuous (a metric value). The null hypothesis is that normalized results do not differ between strategies (i.e., there is no difference in which strategy is used). Depending on the distributions of the values, either one-way ANOVA or Kruskall-Wallis test should be chosen. To perform the test, only results for $\mathcal{K}^{(1)}$ version of each dataset should be considered to make all observations independent within a single sample. To make a sample, we take observations for different splitting strategies given for the same dataset group ('All', 'Non-Customer', 'Customer'), model, and kind of a metric.

To answer Q4, it is needed to analyze how many times a strategy has been preferable over others when the same-numbered versions of datasets are used. To analyze this, we count how many times the best metric values are given for the strategy, per each version number within a dataset group (with separation of different models and metrics). After this, one has a couple of vectors of five integers representing the number of 'best hits' for 5 versions of datasets. Next, the Chi-Square test is applied to check the null hypothesis that there is no dependence between the dataset version and the number of times a strategy has been preferable over others for this version.

4 Results and Discussion

The source code of the experiments and raw results of all 1200 experiments are publicly available on GitHub (https://github.com/dapqa/dataset-split-experiments-public). For SVD and SVD++, we use our own fast implementations. For Bayesian Optimization, Nogueira's implementation [19] is used.

Figure 1 shows violin plots of measured metric values and evaluation times.

As the plots show, there is almost no difference between normalized metric values when different strategies are used (Q1). Absolute differences of confidence intervals' borders are around $1e-3$ (for the same model, metric, and dataset group), so they do not render on the plots. The only sample in which the confidence interval differs from others is NRMSE values for SVD models trained on customer datasets with the random splitting strategy. It shows higher NRMSE values, and this can be simply explained by poor random splits in particular experiments. At the same time, distribution medians differ slightly more. It is hard to state from the plots if these differences are important, but medians of NDCG@10 and RMSE values on movies and drugs datasets are slightly better when temporal splitting strategies are applied.

Evaluation times are obviously drastically higher in experiments with cross-validation. Their confidence intervals are high because the difference in evaluation times for different dataset versions multiplies when cross-validation is applied. It is important that violins for temporal splitting strategies are mostly the widest ones, with the maximal probability density around 1. Because the training procedure of

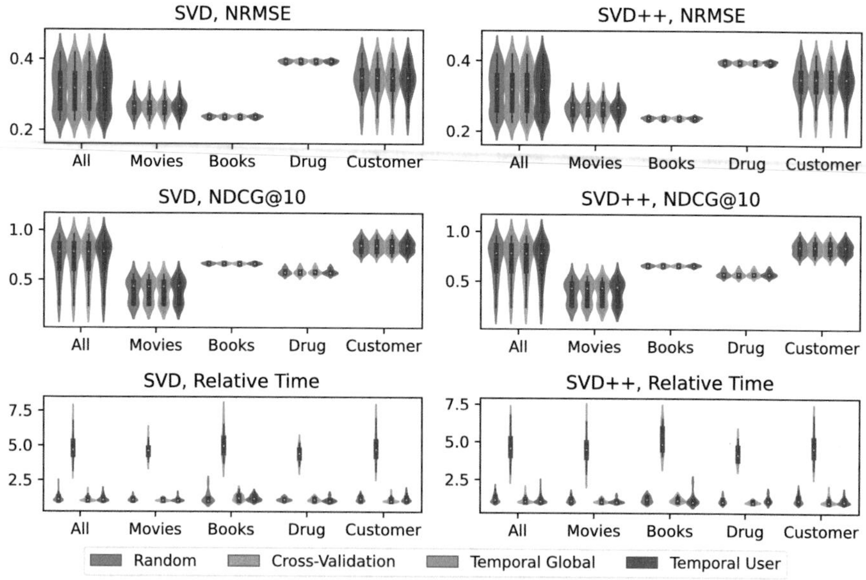

Fig. 1 Metric values and evaluation times distribution

Table 2 Kruskal-Wallis test p-values for the hypothesis that there is no difference between metric values when different splitting strategies are used

Dataset	SVD NRMSE	SVD NDCG@10	SVD++ NRMSE	SVD++ NDCG@10
All	0.998604	0.999234	0.997511	0.99443
Non-Customer	0.995512	0.976506	0.999856	0.980752
Customer	0.983305	0.999195	0.976745	0.975759

SVD and, especially, SVD++ algorithms heavily relies on how many items users have rated, temporal strategies may slightly decrease evaluation times, and the plots confirm it. So, the answer for Q2 is that the usage of temporal splitting strategies may lead to slightly lower evaluation times (but the difference is also insignificant if cross-validation is not used).

The distributions of metric values depicted in the plots are mostly non-normal. They are described more by their medians than by their means. Since the groups are enlarged for statistical tests to provide sufficient numbers of measurements, the distributions have been additionally analyzed for the 'Non-Customer' group (plots can be found on Github). The analysis confirms that these distributions are also non-normal. So, the non-parametric Kruskal-Wallis H-test is chosen to check the null hypotheses for Q3. The important characteristic of the Kruskal-Wallis test is that ranks of values are used instead of values themselves. Practically, if the difference of normalized metric values is low (i.e., around $1e-3$ or $1e-2$), these values may be considered equal ones. But, since the plots above show that most of the differences are of such low orders, metric values are not rounded and used 'as is'. Table 2 presents p-values computed in Kruskal-Wallis tests for the hypothesis that there is no difference between metric values when different splitting strategies are used.

All the p-values are very large and tend to 1. Consequently, the null hypothesis is accepted in all cases. The answer for Q3 is that there is no evidence that measured metric values depend on the splitting strategy choice when matrix factorization models are used.

When answering Q4, it is also important to take into account that, practically, small differences in metric values should not influence the splitting strategy choice. However, since all the differences analyzed above are of small orders of magnitude, metric values are not rounded as well as for Q3. Table 3 presents p-values computed in Chi-Square tests for the hypothesis that there is no dependence between the dataset version and the number of times a strategy has been preferable over others for this version.

The presented p-values demonstrate that almost in all cases there is no evidence that metric values given for a splitting strategy depend on which dataset version is used. The only values allowing reject the null hypothesis are NDCG@10 results for SVD++ models trained on customer datasets using temporal global splitting strategy ($p = 0.03$), NDCG@10 results for SVD models trained on non-customer datasets using temporal user one ($p = 0.09$), and NDCG@10 results for SVD++ models

Table 3 Chi-Square test p-values for the hypothesis that there is no dependence between the dataset version and the number of times a strategy has been preferable over others for this version.

Dataset group, strategy	SVD NRMSE	SVD NDCG@10	SVD++ NRMSE	SVD++ NDCG@10
All				
Random	0.218613	0.135888	0.542729	0.066298
Cross-validation	0.951864	0.608186	0.719043	0.788121
Temporal global	0.498165	0.584587	0.689886	0.597156
Temporal user	0.808792	0.774408	0.926149	0.147832
Non-customer				
Random	0.446052	0.516893	0.735759	0.342547
Cross-validation	0.945023	0.199148	0.735759	0.735759
Temporal global	0.446052	0.516893	0.406006	0.683371
Temporal user	0.735759	0.091578	0.930627	0.705136
Customer				
Random	0.406006	0.175599	0.406006	0.044583
Cross-validation	0.855695	0.472054	0.334162	0.702359
Temporal global	0.557825	0.865985	0.352012	0.030577
Temporal user	0.91858	0.816757	0.985344	0.213144

trained on customer (or all) datasets using random one ($p = 0.04$ or $p = 0.067$ respectively). If one takes a closer look into the examined numbers (which can be found on Github), they find that temporal strategies perform better on $\mathcal{K}^{(2)}$ version, while the random one on both $\mathcal{K}^{(1)}$ and $\mathcal{K}^{(5)}$. However, the absolute difference of metric values is still too low to assert that any strategy can be apriori preferable based on just the dataset size. In simple terms, the answer for Q4 is that if a strategy performs well on a smaller version of a dataset, it cannot be stated that this strategy is preferable for other smaller versions. Most likely, it will perform equally regardless of how large a part of a dataset is, and its performance will depend only on the peculiarities of the particular dataset.

The answers given to Q1, Q3, and Q4 are not obvious, since they contradict the intuition based on the fact that metric values have much higher differences on validation sets. However, the presented experimental results show the choice of the splitting strategy has a low impact on the actual latent factor model's performance. This cannot be explained by an assumption that examined models just do not fit these data, because metric values are acceptably good, and there are a plethora of existing results showing that matrix factorization models are accurate enough for some of these datasets. So, one can explain these findings by the fact that existing biases and interaction patterns are not heavily distorted in most splits, and the distinctions on validation sets exist only because of the peculiarities of the validation set sampling.

5 Conclusion

In this paper, we present the results of the research aimed at answering how much the data splitting strategy choice impacts the performance and evaluation time of latent factor recommender models. To reach this goal, 1200 experiments have been performed, using 15 datasets from four domains, their different states in time, four different splitting strategies (two general-purpose and two RS-specific ones), and two matrix factorization algorithms.

The numerical comparison shows absolute differences in prediction errors and ranking metric values are too low (no more than $1e-3$) to prefer one splitting strategy over others in terms of models' performance. There are few minor distinctions between measurements' medians and confidence intervals in favor of temporal strategies, but they hardly have any practical significance. The performed statistical tests confirm that there is no dependency between the splitting strategy choice and models' performance (all p-values tend to 1), as well as there is no one between the dataset state in time and which strategy should be chosen (only 4 out of 48 p-values are less than 0.1). The analysis of evaluation times given for different splitting strategies shows that temporal strategies slightly more often allow a model to be trained faster (while the usage of cross-validation obviously increases evaluation time 3 up to 6 times).

Consequently, according to the presented results, the choice of the data splitting strategy does not impact the performance of recommender models based on matrix factorization. Moreover, the usage of cross-validation for these models is not mandatory if one wants to avoid biases made by particular splits, so evaluation time may be greatly reduced by just applying a single-split strategy. When choosing the splitting strategy for a new business case, only domain-specific logic should be taken into account, but for research purposes, it is still important to use the same metric as in related works. In common, we recommend choosing temporal global or temporal user strategy. The performance of a model with this choice will be at least as good as with other ones, while the evaluation time may be lowered, and the problem of 'transaction leakage' will be avoided thus making estimations fairer.

References

1. Ricci F, Shapira B, Rokach L (2015) Recommender systems handbook, 2nd edn.
2. Rendle S, Zhang L, Koren Y (2019) On the difficulty of evaluating baselines: a study on recommender systems
3. Dacrema MF, Cremonesi P, Jannach D (2019) Are we really making much progress? A worrying analysis of recent neural recommendation approaches. In: Proceedings of the 13th ACM conference on recommender systems. ACM. https://doi.org/10.1145/3298689.3347058
4. Meng Z, McCreadie R, Macdonald C, Ounis I (2020) Exploring data splitting strategies for the evaluation of recommendation models. CoRR, arXiv:abs/2007.13237. https://arxiv.org/abs/2007.13237

5. Cañamares R, Castells P, Moffat A (2020) Offline evaluation options for recommender systems. Inf Retr J 23(4):387–410. https://doi.org/10.1007/s10791-020-09371-3
6. Kohavi R, et al (1995) A study of cross-validation and bootstrap for accuracy estimation and model selection. IJCAI 14:1137–1145; Montreal, Canada
7. Westerhuis JA, Hoefsloot HCJ, Smit S, Vis DJ, Smilde AK, van Velzen EJJ, van Duijnhoven JPM, van Dorsten FA (2008) Assessment of PLSDA cross validation. Metabolomics 4(1):81–89
8. Shi W, Wang L, Qin J (2020) User embedding for rating prediction in svd++-based collaborative filtering. Symmetry 12(1):121. https://doi.org/10.3390/sym12010121
9. Xu Y, Goodacre R (2018) On splitting training and validation set: a comparative study of cross-validation, bootstrap and systematic sampling for estimating the generalization performance of supervised learning. J Anal Test 2(3):249–262. https://doi.org/10.1007/s41664-018-0068-2
10. Joseph VR, Vakayil A (2021) SPlit: An optimal method for data splitting. Technometrics 1–23. https://doi.org/10.1080/00401706.2021.1921037
11. He X, Liao L, Zhang H, Nie L, Hu X, Chua TS (2017) Neural collaborative filtering. In: Proceedings of the 26th international conference on world wide web, pp 173–182
12. Harper FM, Konstan JA (2016) The MovieLens datasets. ACM Trans Interact Intell Syst 5(4):1–19. https://doi.org/10.1145/2827872
13. Ni J, Li J, McAuley J (2019) Justifying recommendations using distantly-labeled reviews and fine-grained aspects. In: Proceedings of the 2019 conference on empirical methods in natural language processing and the 9th international joint conference on natural language processing (EMNLP-IJCNLP). Association for Computational Linguistics. https://doi.org/10.18653/v1/d19-1018
14. Cai C, He R, McAuley J (2017) SPMC: socially-aware personalized Markov chains for sparse sequential recommendation. In: Proceedings of the twenty-sixth international joint conference on artificial intelligence. International joint conferences on artificial intelligence organization. https://doi.org/10.24963/ijcai.2017/204
15. Zhao T, McAuley J, King I (2015) Improving latent factor models via personalized feature projection for one class recommendation. In: Proceedings of the 24th ACM international on conference on information and knowledge management. ACM. https://doi.org/10.1145/2806416.2806511
16. Wan M, Misra R, Nakashole N, McAuley J (2019) Fine-grained spoiler detection from large-scale review corpora. arXiv:1905.13416
17. Chakraborty S (2021) Drug recommendations. https://www.kaggle.com/subhajournal/drug-recommendations
18. Nechaev A, Meltsov V, Strabykin D (2021) Development of a hyperparameter optimization method for recommendatory models based on matrix factorization. East-Eur J Enterp Technol 5(4 (113)):45–54. https://doi.org/10.15587/1729-4061.2021.239124
19. Nogueira F (2014) Bayesian optimization: open source constrained global optimization tool for Python. https://github.com/fmfn/BayesianOptimization

DBN_VGG19: Construction of Deep Belief Networks with VGG19 for Detecting the Risk of Cardiac Arrest in Internet of Things (IoT) Healthcare Application

Jyoti Mishra and Mahendra Tiwari

1 Introduction

A better healthcare system is the main problem for an increasing global population in the contemporary world. The Internet of Medical Things (IoMT) is a concept for a more comprehensive and widespread medical monitoring network [1]. IoMT refers to the use of Wi-Fi to connect medical equipment and allow for device-to-device (D2D) interaction. The most problematic question in recent days has been the amount of time it takes for internet services. By staying up with the current technological advances, three-dimensional (3D) footage can be retrieved at random intervals [2]. For reliable data measurement, the acquired extensive information is retrieved with minimum time. This will improve device allocation of resources and provide faster speeds for network infrastructures. IoMT, a subset of IoT, is a modern phenomenon in which numerous medical systems, such as intelligent heart rate monitors, smart glucose meters, intelligent bands, intelligent pacers, intelligent beat, etc., [3], be interrelated and communicate in order to display responsive health information that is being used by health care authorities, specialists, and healthcare facilities for greater support and medication [4]. Wi-Fi, Bluetooth, ZigBee, and other cellular platforms are among the heterogeneous networks that make up the IoMT. D2D communication is a critical component of the IoMT platform, as it is both efficient and reliable [5].

In the partitioning ensemble methods, meta-heuristic optimization procedures are used to divide a dataset into groups based on a certain criterion viewed as a fitness value [6]. This variable has a bigger influence on how these groupings are formed. When a suitable fitness function is chosen, the partitioning process is transformed into an optimal solution. Splitting is done in the N-dimensional area also with the goal of minimising range or maximising likeness among sequences, or the regularity is

J. Mishra (✉) · M. Tiwari
Department of Electronics and Communication, University of Allahabad, Allahabad, India
e-mail: Jyoti_jk@allduniv.ac.in

P. Singh et al. (eds.), *Machine Learning and Computational Intelligence Techniques for Data Engineering*, Lecture Notes in Electrical Engineering 998,
https://doi.org/10.1007/978-981-99-0047-3_64

optimized [7]. These techniques are widely used in various research projects because they are capable of clustering large datasets, such as signal/image processing for segmenting images, analysing homogeneous users to classify into groups, generating precise hidden equalisers, organising humans effectively in robotics based on actions, matching aftershocks in seismology from background conditions, obtaining high dimensional data reports, mining web text, and recognising image pattern [8].

The motivation of this work is as follows, Because of scientific innovations in Information and Communication Technology, the complete computational emphasis has shifted. Numerous modern communication channels are being created as a result of these improvements, with the Internet of Things (IoT) playing a crucial role. The Internet of Medical Things (IoMT) is a detachment of the Internet of Things in medical equipment exchange confidential information with one another. These progressions allow the medical diligence to remain close to someone while caring for its users [9]. The essential characteristics of a smart health service are low latency, data rate, and dependability, which are all critical for precise and successful assessment and treatment. For medical emergency systems, the crucial period assessment is the most useful characteristic to examine. Wearable devices powered by the Internet of Things can provide extremely dependable, latency connectivity, and data transmission.

The paper is organized as indicated here, In Sect. 2, the related research works are presented. Section 3 shows the proposed model for data optimization and data classification. Evaluation criteria showed in Sect. 4. The conclusion is finally presented in Sect. 5.

2 Related Works

The healthcare system involves deep learning (DL) approaches in the fields like diagnosing, predicting, and surveillance. It is believed by the health monitoring agents that by using DL techniques, life can be saved.

Tuli et al. [10], developed HealthFog, a new framework for combining ensemble supervised learning in Edge smart devices, and they used it as a real-life example of computerised heart disease analysis. HealthFog uses sensor devices to supply healthcare as an overhaul and reliably maintains the information of patients, that exists in the appearance of consumer queries. The density and strength of the Boltzmann deep belief neural network is used in [11] by Al-Makhadmeh et al., to evaluate periodically transmitted data to the health care centre (HOBDBNN). Deep knowledge of heart illness characteristics can get from the previous study and improves effectiveness through clever data manipulation. The HOBDBNN approach accurately detects disease with 99.03 percent accuracy and little time, effectively lowering heart disease by lowering diagnostic difficultyLong Short Term Memory (LSTM), an alternative of Recurrent Neural Networks, is a Deep Learning (DL) approach used in [12] by Meena et al. The suggested technique will aid in detecting harmful PM 2.5 levels and taking appropriate action. The information is split into one-hour segments using several air pollutants sensors. The error rate and the efficiency of estimating the PM 2.5 level are used to evaluate the efficiency. Ali et al. [13] offer an intelligent health center that forecasts cardiac disease by means of ensemble deep learning and feature fusion methods. To began, the feature-matching approach merges the generated features from numerous sensors with electronic records in gathering pertinent health data. Second, the retrieval quality technique reduces processing load and improves system performance by removing redundant and irrelevant data while focusing on the most important. Additionally, the likelihood model calculates a feature descriptor heaviness for every division, substantially improving system presentation. At last, the ensemble computational intelligence models are trained to forecast heart disease. AlZubi et al. [14] The composed in the sequence is processed by heuristic tube optimized sequence modular neural network (HTSMNN). The method examines gathered information continuously and the self-governing approach expects the transformation near at brain fruitfully. Developing the particle swarm optimization-oriented neural network for Parkinson sickness prediction is discussed in [15] by Haldar et al. The approach gathers patient information, which is then analyzed using a layer-by-layer system. The model parameters are changed on a regular basis based on the location and velocity of the features, which assist to recover the detection fee. The built organization is put into action, and its effectiveness is assessed. When contrasted with a multi-layer feed-forward system, the results show that the introduction system gives more precision but takes longer.

The existing works are related to the data optimization with machine learning and deep learning techniques which do not improve the predictive accuracy with their optimization level. Based on the comparison discussed is not enough for both optimizations with improving the accuracy. So, this research aims to propose a meta-heuristic algorithm for data optimization of data and the predictive analysis for cardiac attack risk prediction. Here, the Internet of-Medical-of Things (IoMT) module is used for data collection and uses the gravitational algorithm for data optimization, then the data classification is done using Deep Belief Networks (DBN) with VGG19 which are described as follows,

3 System Model

The dataset, obtained from public healthcare, contains more than 100,000 records comprising of 55 attributes. Few among them are age, gender, race, number of procedures, number of medications, number of diagnoses, readmissions, and so on.

The data has been initially collected using the IOT module, and this data has been preprocessed for improving the optimization of data. Here we use metaheuristic algorithms for data optimization in which gravitational search optimization algorithm for feature extraction. Then using this optimized data the cardiac attribute data has been classified for identifying the abnormal range. Here we establish Deep Belief Networks (DBN) with VGG19. By this classification the normal range and abnormal

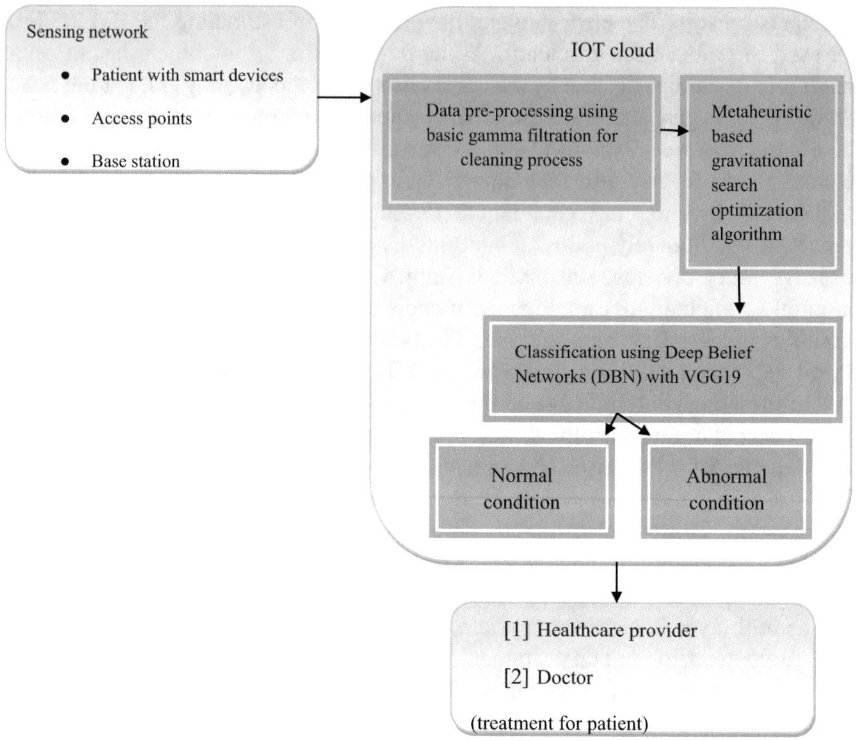

Fig. 1 Proposed design for cardiac arrest data analysis

range of diabetes have been classified for predictive analysis. The proposed design for cardiac arrest data analysis has been given in Fig. 1.

4 Data Forwarding from Sensing Network

The architecture of the IoT-based smart medical process is set by the sensing network. The model's major components are the information collector, IoT portal, backbone facilitator, and consumer applications. The smart health unit can be represented in mathematical form. Generally, for the Smart Healthcare Network (SHN):

$$SHN = E(G_c, J_{gate\_w}, B_{back\_faci}, A_a) \tag{1}$$

In (1), Smart Healthcare Network (SHN) variables are definite as here: E stands for function; G_c stands for data collector devices; J_{gate_w} stands for the IoT gateway in the intelligent medical center; B_{back_faci} stands for the backend facilitator, which offers the smart medical service's backend infrastructure; and A_a stands for the accessibility program. Equation (1) depicts the entire smart health unit, which is dependent on all of the proposed model's components:

$$SHN = (\mu + \Omega G_c + \alpha J_{gate_w} + B_{back_{faci}} + \pi^2) \quad (2)$$

anywhere, for every module,

$\Omega G_c =$ f (data gathering, data forward)

G_c is a method that provides a collection of data which is used to gather data before sending it to an IoT network equipment for analysis. Now specifically:

$$\text{devices} = \alpha J_{gate_w} \quad (3)$$

where,

$$\text{devices} = (bl_{tooth}, wifi, z - wave, WPAN, 3G, 4G, \ldots..n) \quad (4)$$

The module *devices* represents the communication technologies that are supported by the gateway device.

5 Pre-processing of Data

The collected information contains noise data that degrades the accuracy, thus the noisy data is extracted from the data. As during data maintenance process, the arriving information is examined in order to remove any corrupted or erroneous data from each database line, as a result, the noisy data is given by:

$$AY = \{ay_1, ay_2, ay_3, ay_4, \ldots, ay_n\}, \quad (5)$$

The Difference of Gaussian (DOG) technique can be used to recover the clarity of borders and previous details in electronic files. The DOG impulse response is defined as:

$$DOG(x, y) = \frac{1}{2\pi\sigma 1^2}.e^{\frac{x^2+y^2}{2\sigma 1^2}} - \frac{1}{2\pi\sigma 2^2}.e^{\frac{x^2+y^2}{2\sigma 2^2}} \quad (6)$$

In this case, the entries of τ 1 and τ 2 are chosen as 3.0 and 4.0 accordingly. As a result of this impact, the entire distinction created by the process is reduced, and the difference must be increased in later steps. Following the collection of data, it was assessed in terms of incomplete data in each row or column. If there is a null value or blank space in the information, it is substituted by finding the data's mean amount. The information's mean cost is computed to be:

$$Avg = \frac{\sum_{i=1}^{n} \text{na}y_1}{N} \tag{7}$$

where Avg indicates the mean price of the sorted list with an attribute data displayed. ay_1-N is the quantity of information in a certain point.

To make defibrillation prediction models easier, the generated dataset must be normalised to a number in the range of 0 to 1. The parameterised normalisation procedure is used for this reason since it is data-independent. As a result, while reducing noise from a database, the data quality is unaffected.

6 Feature Extraction Using Metaheuristic Based Gravitational Search Optimization Algorithm

The gravitational Search optimization Algorithm (GSOA) is a stochastic populace-based meta-heuristics approach that was developed to relay on Newton's laws of significance and action. Originally, the basic GSOA model was developed to find a solution for the continuous optimization problem. Agents/objects from preprocessed data were introduced in the search space within the dimension in order to determine an optimal solution where the principle of Newton's laws was followed. Here, the position of every agent describes a candidate solution Xi which is a vector in the search space. An agent whose performance is higher obtains more gravitational mass as heavier objects gain more attraction radius. In the lifespan of GSOA, Xi is adjusted successively by an agent with the positions of best agents in a Metaheuristic based gravitational search optimization algorithm (MGSOA) adapting Newton's laws. To explain in detail, a system with s agents is assumed where the position of the agent is given by:

$$X_i = \left(x_i^1, \ldots, x_i^d, \ldots, x_i^n\right); i = 1, 2, \ldots, s \tag{8}$$

where x_i^d presents the position of the agent in dimension d with search space dimension as n. For every agent, the gravitational mass after estimating the current data fitness is computed as given:

$$q_i(t) = \frac{fit_i(t) - worst(t)}{best(t) - worst(t)} \tag{9}$$

$$M_i(t) = \frac{q_i(t)}{\sum_{j=1}^{s} \text{n}q_j(t)} \tag{10}$$

here, Mi(t) and fiti(t) are the gravitational accumulation and fitness cost of *i*th mediator correspondingly at instance t. The best(t) and worst(t) are given by:

$$best(t) = \min_{j\in\{1,\ldots,s\}} fit_j(t) \tag{11}$$

$$worst(t) = \max_{j\in\{1,\ldots,s\}} f_{it_j}(t) \tag{12}$$

Agent acceleration of an is estimated by adding the forces of every agent in the set of KGSA based on the gravitational law using Eq. (13), agent acceleration estimated using motional law is given in Eq. (14):

$$F_i^d(t) = \sum\nolimits_{j\in Kbest, j\neq i}^{\mathrm{n}} \mathrm{n}rand_j G(t) \frac{M_j(t) M_i(t)}{R_{ij}(t) + \varepsilon} \left(x_j^d(t) - x_i^d(t)\right) \tag{13}$$

$$a_i^d(t) = \frac{F_i^d(t)}{M_i(t)} = \sum\nolimits_{j\in Kbest, j\neq i}^{\mathrm{n}} \mathrm{n}rand_j G(t) \frac{M_j(t)}{R_{ij}(t) + \varepsilon} \left(x_j^d(t) - x_i^d(t)\right) \tag{14}$$

where r and j is a random number distributed evenly ranging between [0,1],

- ε, a small value, helps to get rid of division by zero error when $R_{ij}(t)$ is zero,
- $R_{ij}(t)$ represents the Euclidean distance of agents i and j, defined as $kX_i(t), X_j(t)k_2$,
- K finest indicates the agents in KGSOA having superlative fitness rate as well as higher gravitational mass, where KGSOA is the time function which is initially assigned a K initial value and gets diminished with the period.
- G(t) is the gravitational constant that initially takes G initial value and decreases with time till Gend is reached as:

$$G(t) = G(G_{initial}, G_{end}, t) \tag{15}$$

Then, agent velocity and next position are estimated as:

$$v_i^d(t+1) = rand_i * v_i^d(t) + a_i^d(t) \tag{16}$$

$$x_i^d(t+1) = x_i^d(t) + v_i^d(t+1) \tag{17}$$

Various heart properties are obtained based on the preceding discussions, which are then processed using the described Metaheuristic-based gravitational search optimization technique (MGSOA). Throughout this procedure, a total of 25 vectors are calculated in order to classify the cardiac arrest prediction, which is detailed in the next part.

7 Construction of Deep Belief Networks (DBN) with VGG19

Here, the parameters are initialized during unsupervised pre-training so that the process of optimization ends with local minima of the cost function. The architecture of Deep Belief Networks (DBN) with VGG19 is shown in Fig. 2.

In DBM, the energy function E taking parameters v and h representing a pair of visible and hidden vectors respectively has the general form with weights matrix W as:

$$E(v, h) = -a^T v - b^T - v^T W \tag{18}$$

here a and b indicate the bias weights for visible and hidden units accordingly. With v and h in terms of E, the probability distribution P is given as:

$$P(v, h) = \frac{1}{Z} e^{-E(v,h)} \tag{19}$$

Here the normalizing constant Z is given by:

$$Z = \sum_{v',h'}^{\text{n}} \text{n} e^{-E(v',h')} \tag{20}$$

Moreover, the probability of v over hidden units is the sum of above-given equations and is given by:

$$P(v) = \frac{1}{Z} \sum\nolimits_h^{\text{n}} \text{n} e^{-E(v,h)} \tag{21}$$

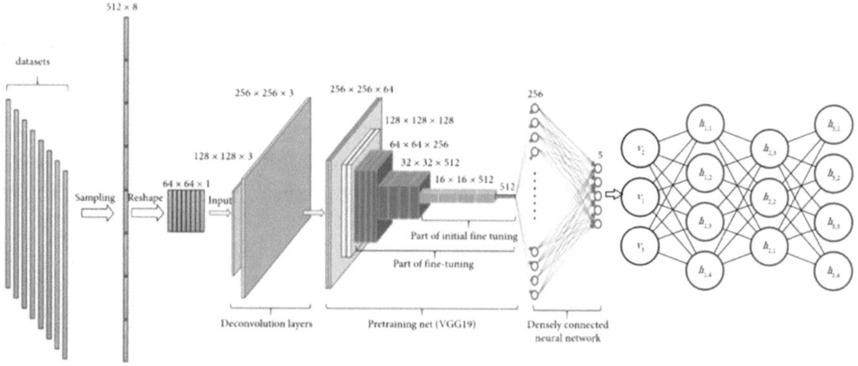

Fig. 2 Architecture of proposed DBN_VGG 19 based classification

log-likelihood difference of training data in terms of W is estimated as $\sum_{n=1}^{n=N} \mathrm{n} \frac{\partial log P(v^n)}{\partial W_{ij}} = \langle v_i h_j \rangle_{data} - \langle v_i h_j \rangle_{model}$ here $\langle v_i h_j \rangle_{data} and \langle v_i h_j \rangle_{model}$ represent the values expected for data and distribution model respectively. For log-likelihood-based training data, network weights are computed using the learning rate ε as $\Delta W_{ij} = \varepsilon\left(\langle v_i n_j \rangle_{data} - \langle v_i n_j \rangle_{model}\right)$. As neurons are not connected either at the hidden or visible layers, it is possible to obtain unbiased samples from $\langle v_i \mathrm{n} \rangle_{data}$. Further, the activation of hidden or visible units is conditionally independent for given h and c respectively. For given v, conditional property of is described as:

$$P(h \mid v) = \prod_{j}^{\mathrm{n}} \mathrm{n} P(\mathrm{n} \mid v) \tag{22}$$

where $n_j \in \{0, 1\}$ and the probability of $n_j = 1$ is given as:

$$P\left(n_j = 1 \mid v\right) = \sigma\left(b_j + \sum_{i}^{\mathrm{n}} \mathrm{n} v_i Wij\right) \tag{23}$$

Here the logistic function σ is specified as $\sigma(x) = \left(1 + e^{-x}\right)^{-1}$. Likewise, when $v_i = 1$, the conditional property is estimated by $P(v_i = 1 \mid v) = \sigma\left(a_i + \sum_j W_{ij} h_j\right)$ Generally, with $< v_i, n_j >$ the unbiased sampling is not straightforward, however is applicable for reconstructing the first sampling of v from h and then Gibbs sampling is used for multiple iterations. With this Gibbs sampling, every unit of the hidden and visible layer is updated in parallel. At last, with $< v_i, n_j >$, the proper sampling is computed by multiplying the expected and updated values of h and v.

Branch 1: The original features are sequentially passed through a convolution level of dimension 1 × 1 and a convolution level of size 2 × 2. This branch is not specially processed so that branch 1 can save the features in the original data as much as possible.

Branch 2: The original features are sequentially passed through a convolution level of dimension 1 × 1 and a standard pooling level of dimension 2 × 2, and finally a ReLU activation level is connected. Branch 2 uses the averaging pooling layer mainly to filter out the interference information in the original features.

Branch 3: The original features are sequentially passed through a convolution level of dimension 1 × 1 and a maximum pooling level of dimension 2 × 2, and finally a ReLU activation level is connected. Branch 3 adopts the maximum pooling level mainly to extract the features with higher brightness from the original features, so as to better find the defect area.

In order to allow the improved VGG19 backbone neural network to extract multi-level feature information, the 17th, 18th, and 19th layers were connected with a maximum and average feature extraction module and a convolution layer, respectively. Then, the multi-level features extracted from the three branch layers were combined to connect a global pooling layer and a full connection layer. Finally, the network was connected to the softmax classifier.

The labeling may well have distortion due to the physical resemblance of collected data. The asymmetrical cross-entropy error function helps limit noise's impact and reduce computational. The following is the explanation of the geometric cross-entropy error function:

$$I_{rce} = I_{ce} + I_{rce} \tag{24}$$

in the midst of them, l_{ce} is the cross-entropy defeat task and l_{rce} is the reverse cross-entropy task:

$$I_{rce} = -\sum\nolimits_{k=1}^{k} \mathrm{n} p(k) log log q(k) \tag{25}$$

As a result, the symmetric cross-entropy loss function is used as the model's loss function in this article to limit the impact of noise on the model's generalisation capacity. The optimised register is incessantly reorganized according to the aforementioned technique to obtain the optimised heaviness value, which is then repeated until the optimised requirements are met. The modular independent network generates information for each component independently, reducing the categorization process's complexity. The final information is evaluated by means of the train information to forecast variations in cardiac characteristics using a different working procedure that can manage a huge number of data.

8 Performance Analysis

The tentative result is approved using the constraint such as Accuracy, Precision, Recall, F1-score, Root Mean Square Error, and Mean Square Error. These constraints are evaluated with two baseline methods namely Heuristic Tubu Optimized Sequence Modular Neural Network (HTSMNN) and Particle Swarm optimized Radial basis Neural Networks (PSRNN) with the proposed Deep Belief Networks with VGG19 (DBN_VGG19). Training and development were performed on a single NVIDIA Tesla P40 GPU with a memory size of 24 GB. The CPU was an Intel Xeon Gold 6146 CPU @ 3.20 GHz, and the memory size was 16 GB. In such a system, the complete training of a U-Net network following the proposed methodology takes about 1.5 h. In the test phase, it takes less than 0.01 s to segment one image using the GPU, and about 10 s using the CPU alone.

Accuracy presents the ability of the overall prediction produced by the model. True positive (TP) and true negative (TN) provide the capability of predicting the nonappearance and existence of an attack. False positive (FP) and false negative (FN) presents the false predictions made by the used model. The formula for accuracy is given as in Eq. (26). Table 1 shows the accuracy between existing PSRNN, HTSMNN methods and proposed DBN_VGG19 method.

Table 1 Accuracy between existing PSRNN, HTSMNN methods, and proposed DBN_VGG19 method

Number of patients	PSRNN	HTSMNN	DBN_VGG19
1000	96.785	97.94	98.45
2000	96.885	97.845	98.3
3000	96.905	98.105	98.68
4000	97.055	98.235	99.2
5000	97.09	98.245	98.3

$$\text{Accuracy} = \frac{\text{True Positive} + \text{True Negative}}{\text{Total Instances}} * 100 \tag{26}$$

Figure 3 illustrates the comparison of accuracy between PSRNN, HTSMNN methods, and the proposed DBN_VGG19 method where the X axis shows the number of patients used for examination and the Y axis indicates the accuracy values. When compared, existing PSRNN and HTSMNN methods achieve 97.81% and 98.32% of accuracy respectively while the proposed DBN_VGG19 method achieves 99.2% of accuracy which is 2.61% better than PSRNN and 1.1% better than HTSMNN method.

Precision estimates the efficiency of the classification model. It is the probability of positive prediction if disease is identified and is also termed a True Positive Rate (TPR) which is estimated as

$$\text{Precision} = \frac{\text{True positive}}{\text{True positive} + \text{False Positive}} \tag{27}$$

Table 2 shows the comparison of precision between existing PSRNN, HTSMNN methods, and proposed DBN_VGG19 method.

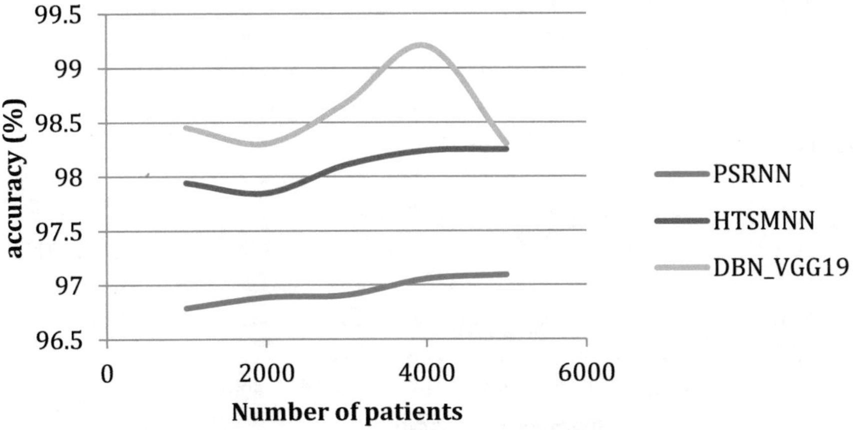

Fig. 3 Comparison of accuracy

Table 2 Comparison of precision between existing PSRNN, HTSMNN methods, and proposed DBN_VGG19 method

Number of patients	PSRNN	HTSMNN	DBN_VGG19
1000	97.82	98.12	99.1
2000	97.84	98.34	98.65
3000	97.72	98.23	98.3
4000	97.83	98.45	99.5
5000	97.92	98.52	99.3

Figure 4 illustrates the contrast of precision among PSRNN, HTSMNN methods and projected DBN_VGG19 method where X axis shows the number of patients used for examination and Y axis shows the precision values. When compared, existing PSRNN and HTSMNN methods achieve 97.2% and 97.61% of precision respectively while the proposed DBN_VGG19 method achieves 98.52% of precision which is 1.32% better than PSRNN and 1.2% better than HTSMNN method.

Recall is the probability of true negatives aptly identified and is also termed as True Negative Rate (TNR). The formula for specificity is given as

$$\text{Recall} = \frac{\text{True Positive}}{\text{True Positive} + \text{False negative}} \tag{28}$$

Table 3 shows the comparison of recall between existing PSRNN, HTSMNN methods, and proposed DBN_VGG19 method.

Figure 5 illustrates the comparison of recall between PSRNN, HTSMNN methods, and proposed DBN_VGG19 method where X axis shows the number of patients used for analysis and Y axis shows the recall values obtained in percentage. When

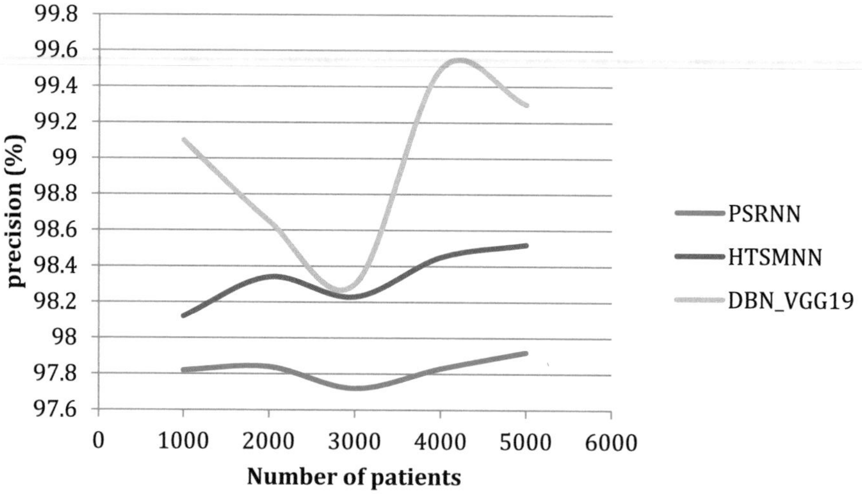

Fig. 4 Comparison of precision

Table 3 Comparison of recall between existing PSRNN, HTSMNN methods, and proposed DBN_VGG19 method

Number of patients	PSRNN	HTSMNN	DBN_VGG19
1000	97.23	97.76	98.1
2000	97.34	97.35	98.3
3000	97.14	97.98	98.43
4000	97.39	98.02	99.31
5000	97.35	97.97	98.31

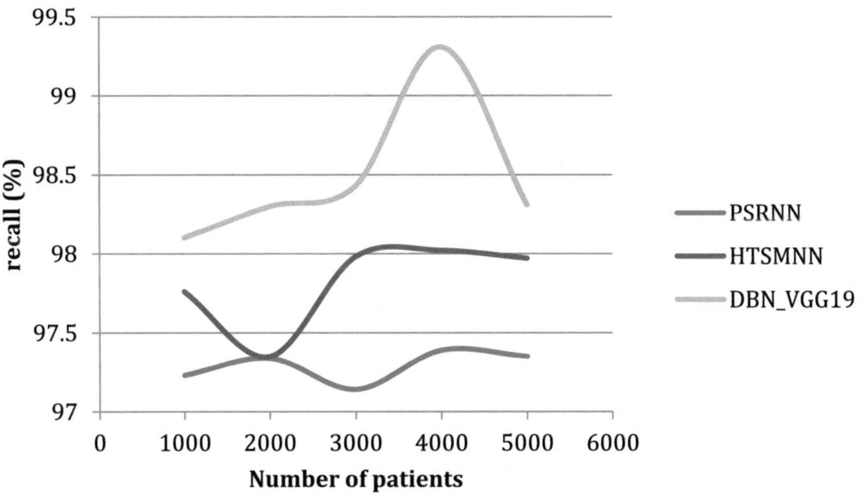

Fig. 5 Comparison of recall

compared, existing PSRNN and HTSMNN methods achieve 97.32% and 97.9% of recall respectively while the proposed DBN_VGG19 method achieves 98.46% of recall which is 1.14% better than PSRNN and 1.52% better than HTSMNN method.

- **Mean Square Error (MSE)**

The Mean Square Error (MSE) is uttered as a ratio, comparing a mean error (residual) to errors twisted by a small representation.

$$\text{MSE} = \frac{\sum_{i=1}^{n}(pi - Ai)^2}{\sum_{i=1}^{n}Ai} \frac{\sum_{i=1}^{n}(pi - Ai)^2}{\sum_{i=1}^{n}Ai} \tag{29}$$

Table 4 shows the contrast of MSE for existing PSRNN, HTSMNN methods and proposed DBN_VGG19method.

Figure 6 illustrates the comparison of MSE between PSRNN, HTSMNN methods, and proposed DBN_VGG19 method where X axis shows the number of patients used for analysis and Y axis shows the MSE values. When compared, existing PSRNN and HTSMNN methods achieve 0.28and 0.221 of MSE respectively while the proposed

Table 4 Contrast of MSE for existing PSRNN, HTSMNN methods and proposed DBN_VGG19 method

Number of patients	PSRNN	HTSMNN	DBN_VGG19
1000	0.256	0.234	0.20
2000	0.242	0.216	0.19
3000	0.263	0.213	0.17
4000	0.27	0.225	0.18
5000	0.26	0.205	0.13

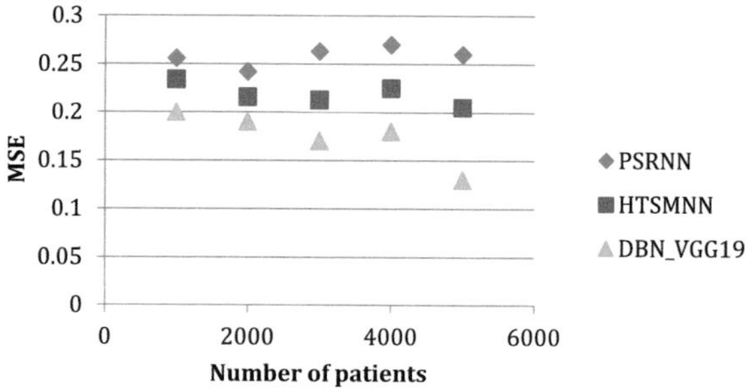

Fig. 6 Comparison of MSE

DBN_VGG19 method achieves 0.183 of MSE which is 0.103 better than PSRNN and 0.11 better than the HTSMNN method.

- **Mean Absolute Error (MAE)**

It is determine if errors among matching annotations express the similar occurrence. It includes the contrast of forecast versus experiential, following time versus initial time.

$$\text{MAE} = \sum_{i=1}^{n} (yi - xi) \tag{30}$$

Table 5 shows the comparison of MAE between existing PSRNN, HTSMNN methods, and proposed DBN_VGG19method.

Figure 7 illustrates the comparison of MSE between PSRNN, HTSMNN methods, and proposed DBN_VGG19 method where X axis shows the number of patients used

Table 5 Comparison of MAE between existing PSRNN, HTSMNN methods, and proposed DBN_VGG19 method

Number of patients	PSRNN	HTSMNN	DBN_VGG19
1000	0.335	0.283	0.20
2000	0.36	0.289	0.14
3000	0.32	0.256	0.12
4000	0.33	0.278	0.15
5000	0.34	0.267	0.14

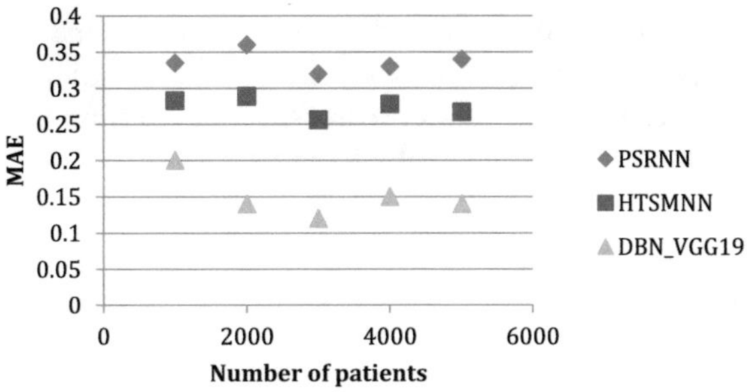

Fig. 7 Comparison of MAE

Table 6 Overall comparison between existing and proposed method

Parameters	PSRNN	HTSMNN	DBN_VGG19
Accuracy (%)	97.81	98.32	99.2
Precision (%)	97.2	97.61	98.52
Recall (%)	97.32	97.9	98.46
MSE	0.28	0.221	0.183
MAE	0.35	0.26	0.14

for analysis and Y axis shows the MSE values. When compared, existing PSRNN and HTSMNN methods achieve 0.35 and 0.26 of MAE respectively while the proposed DBN_VGG19 method achieves 0.14 of MSE which is 0.103 better than PSRNN and 0.11 better than the HTSMNN method. Table 6 shows the overall comparison between existing and proposed method.

9 Conclusion

This paper discusses the Metaheuristic based gravitational search optimization algorithm (MGSOA) based cardiac arrest prediction process. The work collects data features using a smart IoT sensor device by placing them on the patient's brain. The efficiency of the Meta-heuristic optimization algorithms is proved by solving several issues related to redundancy. Then, a Deep Belief Network with VGG19 (DBN_VGG19) is established where the classification is carried out. By this classification the normal range and abnormal range of data have been classified. Further, the method has also been compared with Heuristic Tubu Optimized Sequence Modular Neural Network (HTSMNN) and Particle Swarm optimized Radial basis Neural Networks (PSRNN) in terms of various parameters. As a result, the proposed classification network achieves 99.2% of accuracy, 98.52% of precision, 98.46% of recall, 0.183 of MSE, and 0.14 of MAE. The limitation is the problem of getting the networks to learn the high-level structure of the arteries and veins is still unsolved. Along with this issue, there is also room for improvement in the state of the art regarding another aspect: the image preprocessing. Although this technique demonstrated a positive impact on the A/V classification results, it does not improve the results. So, in the future, ensemble deep learning methods can be utilized to further improve the efficiency of the model.

References

1. RM, Maddikunta PKR, Parimala M, Koppu S, Gadekallu TR, Chowdhary CL, Alazab M (2020) An effective feature engineering for DNN using hybrid PCA-GWO for intrusion detection in IoMT architecture. Comput Commun 160:139–149
2. Maheswari GU, Sujatha R, Mareeswari V, Ephzibah EP (2020) The role of metaheuristic algorithms in healthcare. In: Machine learning for healthcare. Chapman and Hall/CRC, pp 25–40
3. Firdaus H, Hassan SI, Kaur H (2018) A comparative survey of machine learning and meta-heuristic optimization algorithms for sustainable and smart healthcare. Afr J Comput ICT Ref Format **11**(4):1–17
4. Murugan S, Jeyalaksshmi S, Mahalakshmi B, Suseendran G, Jabeen TN, Manikandan R (2020).Comparison of ACO and PSO algorithm using energy consumption and load balancing in emerging MANET and VANET infrastructure. J Crit Rev 7(9):2020
5. Abugabah A, AlZubi AA, Al-Obeidat F, Alarifi A, Alwadain A (2020) Data mining techniques for analyzing healthcare conditions of urban space-person lung using meta-heuristic optimized neural networks. Clust Comput 23:1781–1794
6. Saha A, Chowdhury C, Jana M, Biswas S (2021) IoT sensor data analysis and fusion applying machine learning and meta-heuristic approaches. Enabl AI Appl Data Sci 441–469
7. Suganya P, Sumathi CP (2015) A novel metaheuristic data mining algorithm for the detection and classification of Parkinson disease. Indian J Sci Technol 8(14):1
8. Salman I, Ucan ON, Bayat O, Shaker K (2018) Impact of metaheuristic iteration on artificial neural network structure in medical data. Processes 6(5):57
9. Li J, Liu LS, Fong S, Wong RK, Mohammed S, Fiaidhi J, Wong KK (2017) Adaptive swarm balancing algorithms for rare-event prediction in imbalanced healthcare data. PloS one 12(7):e0180830
10. Tuli S, Basumatary N, Gill SS, Kahani M, Arya RC, Wander GS, Buyya R (2020) HealthFog: an ensemble deep learning based smart healthcare system for automatic diagnosis of heart diseases in integrated IoT and fog computing environments. Futur Gener Comput Syst 104:187–200
11. Al-Makhadmeh Z, Tolba A (2019) Utilizing IoT wearable medical device for heart disease prediction using higher order Boltzmann model: a classification approach. Measurement 147:106815
12. Meena K, Mayuri AVR, Preetha V (2022) 5G narrow band-IoT based air contamination prediction using recurrent neural network. Sustain Comput Informat Syst 33:100619
13. Ali F, El-Sappagh S, Islam SR, Kwak D, Ali A, Imran M, Kwak KS (2020) A smart healthcare monitoring system for heart disease prediction based on ensemble deep learning and feature fusion. Inf Fusion 63:208–222
14. AlZubi AA, Alarifi A, Al-Maitah M (2020) Deep brain simulation wearable IoT sensor device based Parkinson brain disorder detection using heuristic tubu optimized sequence modular neural network. Measurement 161:107887
15. Haldar S (2013) 'Particle swarm optimization supported artificial neura network in detection of Parkinson's disease. Neurobiol Dis 58:242–248

Detection of Malignant Melanoma Using Hybrid Algorithm

Rashmi Patil, Aparna Mote, and Deepak Mane

1 Introduction

Melanoma can affect people of any age and can appear on any part of body, including places not exposed to sunlight. The face, scalp, neck, legs and arms are the most common sites for melanoma. However, under the fingernails or toenails; on the hands, soles, or tips of the toes and fingers; or on mucous membranes, such as skin covering the mouth, nose, vagina, and anus, melanoma may also develop. In previous decades, primary skin melanoma and skin lesions have been more common and have thus become a very serious public health problem.

There are multiple melanoma therapies implemented as chemotherapeutic drugs or combination therapies, such as neck dissection, chemotherapy, biochemotherapy, photodynamic therapy, immunotherapy, or targeted therapy. Selection of an appropriate therapeutic approach depends on clinical situation, cancer stage, and position of the patient. There are some tests present for early detection, and sometimes self-diagnosis will work for that we need to diagnose skin parts using full mirror. Another person's examination of the scalp and back of the neck is beneficial. Epiluminescence microscopy, often known as dermoscopy, is a painless medical method used to diagnose melanoma early. A portable gadget may be used by a clinician to detect the magnitude, shape, and pigmentation patterns in pigmented skin lesions.

Diagnosis of melanoma is possible using skin lesion images, if melanoma is diagnosed, then next it is essential to detect the cancer stage. If melanoma is detected at an early stage, then it is possible to recover, but if it is last stage melanoma, then

R. Patil (✉) · A. Mote
Computer Engineering, Savitribai Phule Pune University, Pune, India
e-mail: rashmiashtagi@gmail.com

D. Mane
Computer Engineering, Vishwakarma Institute of Technology, Pune, India

P. Singh et al. (eds.), *Machine Learning and Computational Intelligence Techniques for Data Engineering*, Lecture Notes in Electrical Engineering 998,
https://doi.org/10.1007/978-981-99-0047-3_65

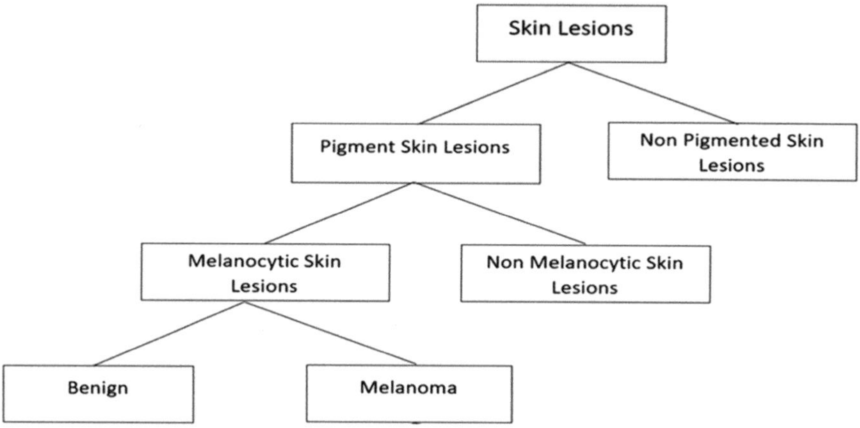

Fig. 1 Classification of skin lesions [1]

it will cause death to patients. So, melanoma present or not present is a big question and diagnosis if it is present at early stage is essential.

In the case of skin lesions, segmentation amounts are given for identification of the boundary that isolates the lesion region from the skin of a neighbour. It is possible to figure and use characteristics related to the asymmetry, boundary, colour, accessible structures in the lesion when the part of the lesion is defined, to advise arithmetical techniques to classify the analysis. Acquiring a precise segmentation of the lesion is therefore necessary particularly for the measurement of shape & border characteristics. A variety of dermoscopy image classification methods have been developed to indicate the difficulties. In general, these methods can be categorized as follows: RF, SVM and CNN.

As shown in Fig. 1, first skin lesion is categorized as pigmented and non-pigmented, if its pigmented then categorized as melanocytic or non-melanocytic. If its melanocytic then categorized as benign or melanoma.

The proposed CNN-SVM combined model is developed to classify melanoma from benign. SVM is used for classification, whereas CNN is utilized for training. CNN-SVM model architecture was built by replacing the CNN model's final output layer with an SVM classifier. In addition to making sense of CNN model, outcome values of the hidden layer are used as input features for other classifiers. It is assumed that such a combined model would incorporate the benefits of CNN and SVM.

In addition, the paper is organised as follows: In the subsequent segment, existing methodologies are portrayed; in the third area, framework design, and algorithm steps is given; in the fourth segment, dataset depiction, experimental setup, assessment metrics, and results examination are discussed; and finally, conclusion is mentioned.

2 Literature Review

Ganesan et al. [2] proposed research and showed how a combined hill climbing and FCM method may be used to identify and segment Melanoma skin cancer. In the segmentation process, they act as the first clusters. Clearly, the method's experimental results reveal that the procedure's success is dependent on the quantity of beginning seeds.

By using the illustration of solar lentigines, Bimastro et al.'s [3] goal is to aid diagnose melanoma. This software utilizes ABCD technique for feature extraction and for classification the decision tree is used. The ABCD approach is a medical technique utilized to diagnose cancer, they are asymmetry, border, colour, differential structure. Chen et al. [4] propose a U-Net multi-task model for effectively classifying melanoma lesion characteristics. The network has two tasks: classification and segmentation. The classification job determines if the lesion's properties are useable, while the segmentation task segments the attributes in the photos. Feng et al. [5] further identified the key reactive species affecting variations in feasibility between fibroblasts (L929) and melanoma cells (B16). Findings suggest that the CAP has a different cytotoxic effect on the viability of two cell lines. B16 cell growth is slowed.

Ganguly et al. [6] propose a CNN-based artificial eye melanoma diagnosis system. A typical database yields 170 pre-diagnosed samples, which are then pre-processed into lower resolution samples before being fed to CNN architecture. For detection of ocular melanoma, the proposed approach removes the need for independent feature extraction and classification. Utilizing an artificial neural network to identify ocular melanoma yields a 91.76 percent accuracy, despite fact that the recommended method necessitates a big estimate.

In the computer-aided diagnosis of skin cancer, automatic melanoma segmentation in image data is important. Current techniques can suffer from hole and shrink issues with minimizing potential of segmentation. Guo et al. [7] presented supplementary network of adaptive receptive filed learning to address these problems. In contrast, dependency between both foreground and background networks is applied to create use of a novel mutual loss, allowing the reciprocal effect between these two networks. Subsequently, this technique of cooperative training allows for semi-monitored learning and increases boundary sensitivity.

Hagerty et al. [8] suggested that the couple of approaches, with diverse error profiles, were synergistic. Their image processing work utilizes three handcrafted, clinical knowledge design and one biologically inspired image processing modules. Compared to clinical dermoscopy information, these identify various lesion characteristics in a standard pigment network, blood vessels and color distribution. Patient gender, age, lesion size, location and patient existing data submitted to the pathologist are part of the clinical module. The DL arm uses information transfer through a ResNet-50 algorithm that used for melanoma predict probability. In order to detect the overall risk of melanoma, each individual module's classification scores from all the processing arms are then assembled using logistic regression (LR).

Nandhini et al. [9] discussed classification of skin cancer by using Random forest classification technique. They applied random forest algorithm to classify the skin lesion. They classify a person's skin cancer into seven distinct categories based on dermoscopic photographs. Use the HAM10000 data-set to handle this problem. The finalized dataset contains 10,001 dermoscopic images that are published and publicly accessible via the ISIC archive as a readiness collection for academic ML purposes. Based on dermoscopic images, they classify a person's skin cancer into seven distinct forms. Through this study, a person will learn that if he/she does or does not suffer from some form of skin cancer, a person will have some certainty regarding skin cancer before heading to consult any doctor.

Rashmi and Bellary [10–12] presented different approaches to classify melanoma from benign, to find type and stage of melanoma. In [11] two strategies which detect stages of melanoma is presented. The first approach categorizes in 1 or 2 stages and second approach categorizes melanoma in 1, 2 or 3 stages. In [12] the cognitive techniques to identify type of melanoma is proposed. Gaikwad et al. [13] proposed the system to detect melanoma using deep neural networks.

3 System Architecture

CNN is often used in image identification and detection, as well as labelling and other tasks. Nonetheless, CNN has been demonstrated to be useful in detecting attacks. An input layer, convolutional layer, pooling layer, fully connected layer, and output layer make up a standard CNN. The convolutional filters, weights, and biases of the layers may be learned via forward- and back-propagation in the training process.

SVM is a nonlinear classification-capable supervised learning model. This model uses a method called the kernel to implicitly transform inputs into high-dimensional feature spaces. SVM may identifies intrusions with less training data.

Random Forest is a low-classification-error ensemble classifier that can handle enormous data sets, noise, and outliers. For better accuracy, each tree creates a classification. This strategy's use of several trees may result in slow real-time prediction. However, RF has excelled at identifying network threats.

Figure 2 provides a CNN-SVM combined model for melanoma classification. In combination with the sigmoid classifier, CNN is trained, where cross-entropy has been added with class separability data. The CNN-SVM model architecture was built by replacing the CNN model's final output layer with an SVM classifier. In addition to making sense of CNN model, outcome values of the hidden layer are used as input features for other classifiers. It is assumed that such a combined model would incorporate the benefits of CNN and SVM.

Melanoma and Benign Classification using CNN+SVM algorithm:

1. ISIC Images dataset of cancer-infected skin as input.
2. Perform preprocessing to resized images. (Image preprocessing is performing to make the data better suited to use in the model).

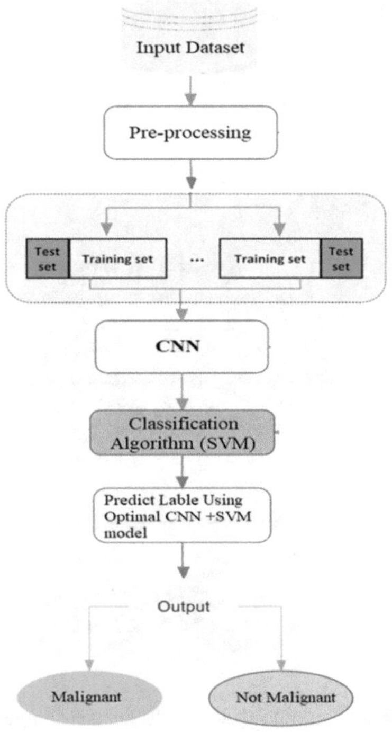

Fig. 2 System architecture

3. Perform feature extraction and segmentation of images using image thresholding techniques.
4. The CNN-SVM model architecture was built by replacing the CNN model's final output layer with an SVM classifier. CNN-SVM is implemented on last layer of CNN, in its place of traditional softmax function with the cross-entropy function (for computational loss).

$$\min \frac{1}{p}||w||_2^2 + \mathrm{C}\sum_{i=1}^{p} \max(0, 1 - y_i^{'}(w^T x_i + b))^2$$

where $||w||_2^2$ be the Euclidean norm.

C	be the penalty parameter
y′	be the actual label
w^Tx+ b	be the predictor function

5. The result is converted to case y ∈ {−1, +1}, and the loss is determined by the above Eq. Using Adam, the weight parameters are learned.

Table 1 Dataset information

	Training dataset	Testing dataset	Validation dataset
Total images	2000	600	150
Benign	1626	483	120
Melanoma	374	117	30

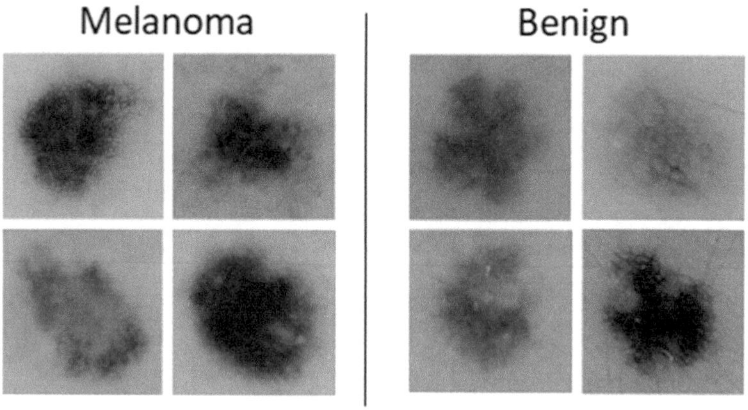

Fig. 3 Melanoma, Benign dataset sample

4 Result and Discussion

4.1 Dataset Description

We use ISIC 2017 dataset. This dataset includes training, validation, and a blind held-out test. We used the ISIC dataset that comprises of 2000 pictures for training. 600 photos were used for testing, and 150 photographs were used for validation. The dataset is described in Table 1 (Fig. 3).

4.2 Performance Parameters

Accuracy, recall, f1-score, and precision are used to evaluate proposed task performance. Measurements are derived from

$$\text{Accuracy} = \frac{T_n + T_p}{T_p + T_n + F_p + F_n}$$

$$\text{Precision} = \frac{T_p}{F_p + T_p}$$

Table 2 Fine tuning parameters for CNN and CNN-SVM algorithm

Parameters	CNN	CNN-SVM
Number of EPOCH	50	50
Batch size	8	16
Dropout Probability	1.0	1.0
Optimizer	Nadam	Adam
Activation function	Softmax	Softmax
Learning rate		

$$\text{Recall} = \frac{T_p}{T_p + T_n}$$

$$\text{F1} = 2\text{X}\frac{\text{Precision} \times \text{Recall}}{\text{Precision} + \text{Recall}}$$

where T_p, F_p, F_n, T_n are true positive, false positive, false negative, and true negative.

4.3 Experimental Setup

With an Intel i5 machine with two cores @2.3 GHz and 8 GB RAM, we did this research using the Notebook Jupiter platform and Python technology. This platform also includes NVIDIA GFORCE drivers, 500 GB SSD, and 1 TB HDD.

4.4 Tuning Parameter

Table 2 shows fine tuning parameters for CNN and CNN-SVM algorithm.

4.5 Result Analysis

Figure 4 presents the accuracy, recall, precision and f-measure comparison for RF, SVM, CNN, CNN+SVM algorithm. The CNN+SVM gives higher precision, recall F-measure and accuracy percentage than other algorithms (Fig. 5 and Table 3).

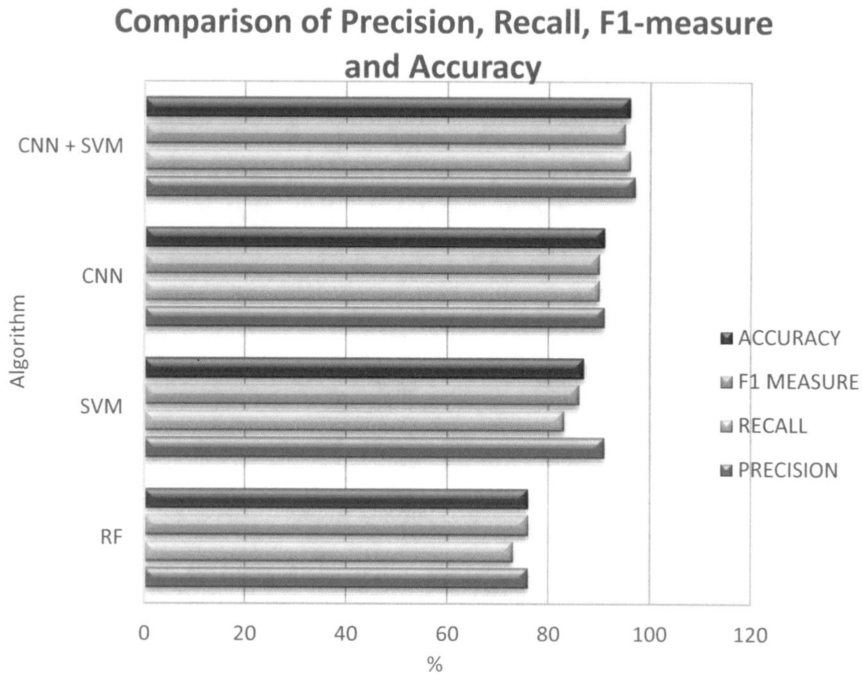

Fig. 4 Melanoma performance measure graph comparison

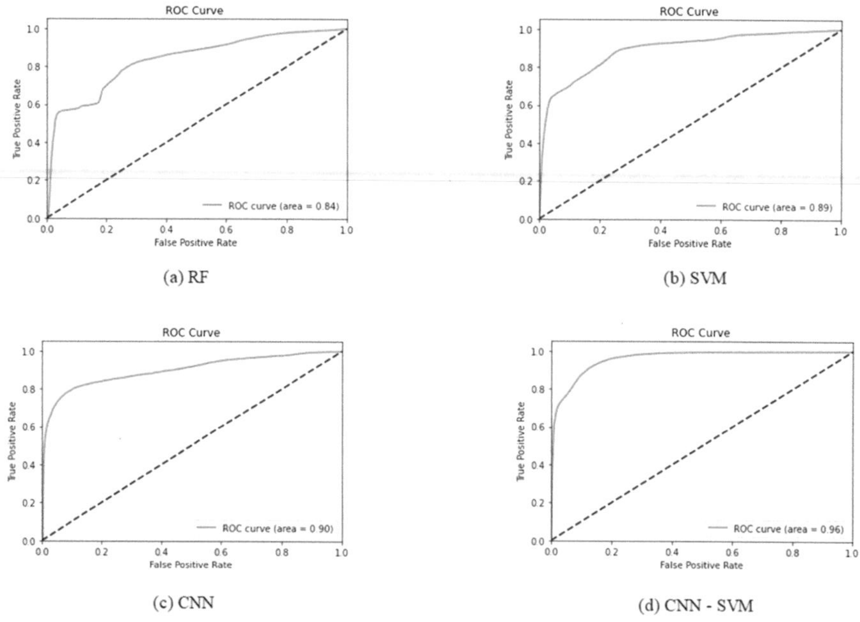

Fig. 5 ROC graph comparison

Table 3 Performance parameter comparison of algorithms

	Precision	Recall	F1 measure	Accuracy
RF	76	73	76	76
SVM	91	83	86	87
CNN	91	90	90	91
CNN+SVM	97	96	95	96

5 Conclusion

Many other forms of skin cancer are present. Since they arise from skin cells other than melanocytes. They appear to function very differently from melanomas and are therefore handled with various techniques. So, melanoma yes or no is big challenge and diagnosis of melanoma if it is present at an early stage is very essential. Here we present study related to melanoma present or not present. We studied some skin lesion classification techniques such as RF, CNN, and SVM. According to study we proposed a non-invasive method to differentiate melanoma from benign. CNN is used for training, SVM is used for classification. The advantages of CNN and SVM are used in proposed the CNN-SVM model. Further, the work can be carried out to find the stage and type of melanoma with more accurately.

References

1. Patil R, Sangve SM (2014) Competitive analysis for the detection of melanomas in dermoscopy images. Int. J. Eng. Res. Technol. (IJERT) 3(6)
2. Ganesan P, Vadivel M, Sivakumar VG, Vasanth (2020) Hill climbing optimization and fuzzy C-means clustering for melanoma skin cancer identification and segmentation. In: 2020 6th International conference on advanced computing and communication systems (ICACCS), Coimbatore, India, 2020, pp 357–361. https://doi.org/10.1109/ICACCS48705.2020.9074333
3. Bimastro KZ, Purboyo TW, Setianingsih C, Murti MA (2019) Potential detection of Lentigo Maligna melanoma on solar lentigines image based on android. In: 2019 4th International conference on information technology, information systems and electrical engineering (ICITISEE), Yogyakarta, Indonesia, pp 113–118. https://doi.org/10.1109/ICITISEE48480.2019.9003743
4. Chen EZ, Dong X, Li X, Jiang H, Rong R, Wu J (2019) Lesion attributes segmentation for melanoma detection with multi-task U-net. In: 2019 IEEE 16th International symposium on biomedical imaging (ISBI 2019), Venice, Italy, 2019, pp 485–488. https://doi.org/10.1109/ISBI.2019.8759483
5. Feng Z, Xu Z, Pu S, Shi X, Yang Y, Li X (2019) Cytotoxicity to melanoma and proliferation to fibroblasts of cold plasma treated solutions with removal of hydrogen peroxide and superoxide anion. IEEE Trans Plasma Sci 47(10):4664–4669. https://doi.org/10.1109/TPS.2019.2936055
6. Ganguly B, Biswas S, Ghosh S, Maiti S, Bodhak S (2019) A deep learning framework for eye melanoma detection employing convolutional neural network. In: 2019 International conference on computer, electrical & communication engineering (ICCECE), Kolkata, India, pp 1–4. https://doi.org/10.1109/ICCECE44727.2019.9001858

7. Guo X, Chen Z, Yuan Y (2020) Complementary network with adaptive receptive fields for melanoma segmentation. In: 2020 IEEE 17th International symposium on biomedical imaging (ISBI), Iowa City, IA, USA, pp 2010–2013. https://doi.org/10.1109/ISBI45749.2020.9098417
8. Hagerty JR et al (2019) Deep learning and handcrafted method fusion: higher diagnostic accuracy for melanoma dermoscopy images. IEEE J Biomed Health Inform 23(4):1385–1391. https://doi.org/10.1109/JBHI.2019.2891049. 2019 Jan 4 PMID: 30624234
9. Nandhini S et al (2019) Skin cancer classification using random forest. Int J Manage Human (IJMH) 4(3). ISSN: 2394-0913
10. Patil R (2021) Machine learning approach for malignant melanoma classification. Int J Sci Technol Eng Manage 3(1):40–46. http://ijesm.vtu.ac.in/index.php/IJESM/article/view/691
11. Patil R, Bellary S (2020) Machine learning approach in melanoma cancer stage detection. J King Saud Univ Comput Inf Sci. ISSN 1319-1578, https://doi.org/10.1016/j.jksuci.2020.09.002. http://www.sciencedirect.com/science/article/pii/S1319157820304572
12. Patil R, Bellary S (2021) Transfer learning based system for melanoma type detection. Revue d'Intelligence Artificielle 35(2):123–130. https://doi.org/10.18280/ria.350203
13. Gaikwad M et al (2020) Melanoma cancer detection using deep learning. Int J Sci Res Sci Eng Technol 7(3). https://doi.org/10.32628/IJSRSET

Shallow CNN Model for Recognition of Infant's Facial Expression

P. Uma Maheswari, S. Mohamed Mansoor Roomi, M. Senthilarasi, K. Priya, and G. Shankar Mahadevan

1 Introduction

Facial expressions, which are a vital aspect of communication, are one of the most important ways humans communicate facial expressions. There is a lot to comprehend about the messages we transmit and receive through nonverbal communication, even when nothing is said explicitly. Nonverbal indicators are vital in interpersonal relationships, and facial expressions communicate them. There are seven universal facial expressions that are employed as nonverbal indicators: laugh, cry, fear, disguise, anger, contempt, and surprise. Automatic facial expression recognition using these universal expressions could be a key component of natural human-machine interfaces, as well as in cognitive science and healthcare practice [1]. Despite the fact that humans understand facial expressions almost instantly and without effort, reliable expression identification by machines remains a challenge.

In that, infant facial expression recognition is developing as a significant and technically challenging computer vision problem as compared to adult facial expressions recognition. Since there is a scarcity of infant facial expression data, the recognition is mostly based on the building of a dataset. There are no datasets were publically available or created particularly to analyze the expression of infants. The creation of dataset for infant facial expression analysis is a big and challenging task. The ability to accurately interpret a baby's facial expressions is critical, as most of the expressions resemble those displayed in Fig. 1. This process leads to the development of identifying the action behind the scene. Although advancements in face detection,

P. U. Maheswari (✉)
Department of Information Technology, Velammal College of Engineering and Technology, Madurai, India
e-mail: umamahes.p@gmail.com

S. M. M. Roomi · M. Senthilarasi · K. Priya · G. S. Mahadevan
Department of ECE, Thiagarajar College of Engineering, Madurai, Tamil Nadu, India

P. Singh et al. (eds.), *Machine Learning and Computational Intelligence Techniques for Data Engineering*, Lecture Notes in Electrical Engineering 998,
https://doi.org/10.1007/978-981-99-0047-3_66

Fig. 1 Sample of infant facial expressions (neutral, laugh/neutral, cry/laugh, laugh) (difficult to categorize)

feature extraction processes, and expression categorization techniques have been made in recent years, developing an automated system that achieves this objective is challenging.

Automatic facial expressions classifiers have made significant progress, according to researchers. The facial Action Coding System (FACS) has been developed for classifying the facial movements by action units [2]. Traditional machine learning based classifiers such as Hidden Markov Model [3], Support Vector Machine (SVM) [4], and Bayesian network [5] are proposed for face facial expressions recognition. With the development of massive data and computing efficiency, deep learning-based face expression identification has become widely used. The face and Body of infants are detected using YOLOv3-tiny and achieved a classification accuracy of 94.46% for the face and 86.53% for the body of infants [6]. Two-stream CNNs model is employed for extracting local temporal and spatial features respectively [7].

Transfer learning-based models such as VGG16 [8], Resnet 18 & 50 [9] have been proposed for adult facial expressions recognition. Because infant facial expressions detection is required in parenting care, a deep neural network must be constructed to recognize the infant's facial expressions from images. Since there is an insufficient amount of data related to infant facial expressions, the transfer learning model based approaches have an issue of over fitting. To avoid this shallow network has been proposed for infant action recognition. The main contributions of the proposed work are.

- Introduction of a publically available facial expression dataset containing 5400 images.
- Proposal of reduced space and time constraint shallow network.
- Proposal of shallow convolutional neural network (CNN) based infant facial expression recognition.
- Performance comparison of the proposed network with the existing network.

Section 2 describes the dataset that has been created and the architecture of the shallow network model, and Sect. 3 presents the results and discussion and it is concluded in Sect. 4.

Table 1 Types of dataset

Actions	Count	Source
Cry	1800	Infant action database
Laugh	1200	Infant action database
Laugh	600	Website
Neutral	1800	Website

2 Methodology

The paper proposes a novel infant facial expressions recognition system. Due to the lack of neutral and laugh infant activity images in the dataset, the images are collected from the website and labeled them manually. Because the number of images in the collected dataset is relatively small, a shallow network rather than a deep convolutional network is created to increase the learning capacity thereby reducing the overfitting issue.

2.1 Dataset

The dataset contains 5400 images depicting three different types of infant face facial expressions: cry, laugh, and neutral. To appropriately describe additional universal facial expressions, these three expressions must first be recognized. For that process, initially the images of infant actions such as crying and laughing are gathered from the private database named infant action database [10]. The database contains footage of children performing various actions from which images of various actions have been taken. The images of neutral activity and some images of laughing were collected from the internet. Table 1 shows the number of images in each activity.

2.2 Proposed Shallow Network Architecture

For recognizing infant facial expressions, a shallow convolutional neural network with eleven simple layers was created. Figure 2 illustrates the proposed shallow network architecture. Two groups of convolutional layers, two maxpool layers, one fully-connected layer, one softmax layer, and one classification output layer form the network. The group contains one convolution layer followed by batch normalization and the relu activation function. With zero center normalisation, the input layer is fixed to a size of $224 \times 224 \times 3$. The input layer is then convolved with 64 filters with the size of 7×7. For independent learning the batch normalization is performed in the convolution filter output followed by relu activation layer to provide linearity.

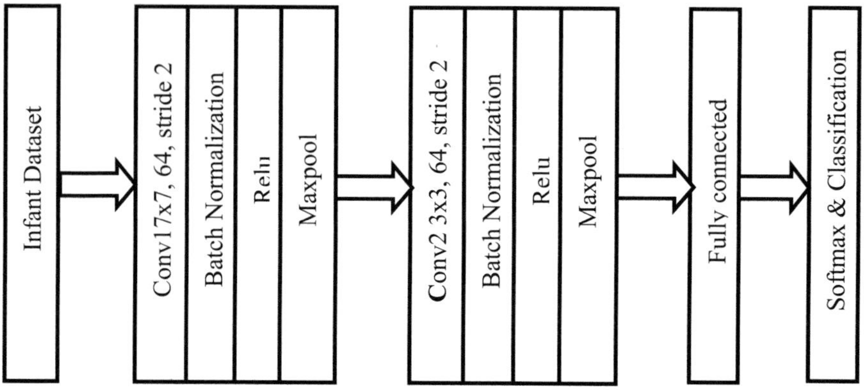

Fig. 2 The proposed shallow CNN model

After that the low-level features are extracted by performing the maxpool operation. This highlights the important features of an infant image and avoids fuzzy issues. The features are again convolved with 64 filters with the size of 3×3 with the batch normalization followed by an activation function and the maxpool layer. This structure reduces the learning parameters and improves the representation of features. At the end of the structure the fully connected layer is attached followed by a softmax and the classification output layer.

2.3 *Training*

Initially, the data is augmented in the training phase to keep the image size consistent. One of the most important adjustment hyperparameters is the learning rate, which is proportional to the gradient descent. To get the optimum feature learning of the infants' facial expressions, dynamic learning rate adjustment has been used. For that the initial learning rate is set to 0.01, and after every 100 iterations, the learning rate is multiplied by 0.1. The simplistic and generalized architecture with stochastic gradient descent with momentum has been designed, and it is well suited for the new infant images in order to avoid overfitting. After several adjustments made to the model the optimal training model is selected. That model matches the recognition properties of three facial expressions of infant actions for the limited number of dataset.

3 Results and Discussion

MATLAB 2021b in GPU processor is used to train and test the model. As illustrated in Fig. 3, the dataset mostly contains three different infant facial expressions: cry, laugh, and neutral. There are variances between each infant's facial expressions. They do, however, have certain striking features that make recognition tasks challenging. The number of images in each class is about 1800. Because the images collected in the dataset and on the website are of different sizes, data augmentation methods such as resizing are used to make the size distribution equal. The model's diversity and adaptability are improved as a result of this procedure. Precision, recall, and F-measures are introduced to the review process in addition to validation and testing accuracy.

The classical and standard networks for facial expressions recognition are Resnet 18, Resnet50, and VGG 16. As a result, these architectures are used in a comparison study with the suggested shallow network. The architectures of the proposed shallow network with Resnet50 and VGG16 are shown in Table 2.

The input image for the proposed shallow network is 224 × 224 pixels in size. The Conv1 layer has 64 filters with stride 2 and 7 × 7 convolution kernels. As a result, the output size is reduced to 112 × 112 pixels. After that, a batch normalization with the same scale and offset is created, with a relu activation layer and a max pool layer.

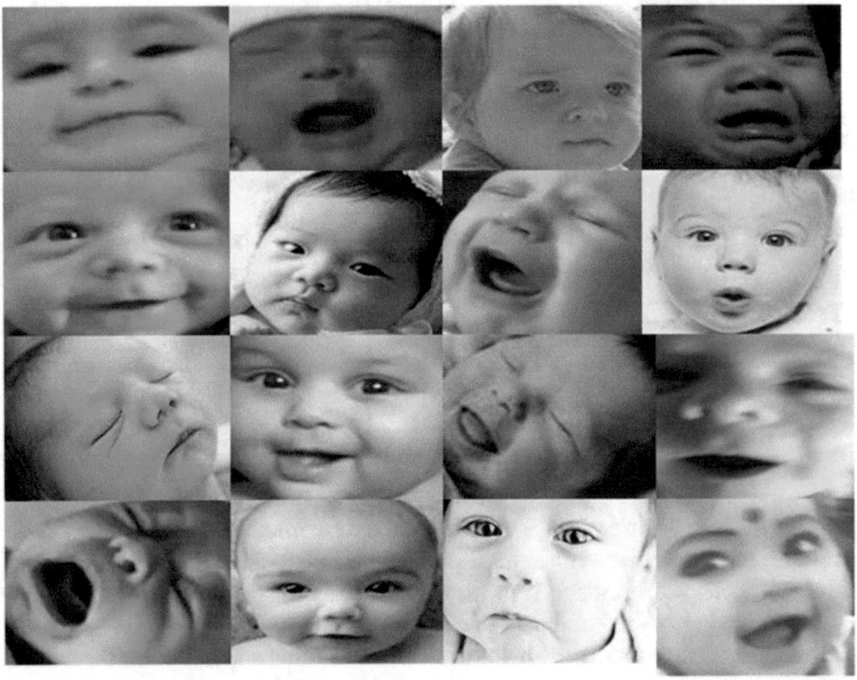

Fig. 3 Sample images of infant facial expressions

Table 2 Architectures of ResNet-18, ResNet-50, VGG16, and the proposed shallow CNN

Layer Name	Resnet-18	Resnet-50	VGG-16	Proposed
Conv1	7×7,64, stride 2	7×7,64, stride 2	3×3 max pool, stride 2 $\begin{bmatrix} 3x3, 64 \\ 3x3, 64 \end{bmatrix}$x2	7×7,64, stride 2 Batch normalization, Relu, 3×3 max pool, stride 2
Conv2	3×3 max pool, stride 2 $\begin{bmatrix} 3x3, 64 \\ 3x3, 64 \end{bmatrix}$x2	3×3 max pool, stride 2 $\begin{bmatrix} 1x1, 64 \\ 3x3, 64 \\ 1x1, 256 \end{bmatrix}$x3	3×3 max pool, stride 2 $\begin{bmatrix} 3x3, 128 \\ 3x3, 128 \end{bmatrix}$x2	3×3,64, stride 2 Batch normalization, Relu, 3×3 max pool, stride 2
Conv3	$\begin{bmatrix} 3x3, 128 \\ 3x3, 128 \end{bmatrix}$x2	$\begin{bmatrix} 1x1, 128 \\ 3x3, 128 \\ 1x1, 512 \end{bmatrix}$x3	3×3 max pool, stride 2 $\begin{bmatrix} 3x3, 256 \\ 3x3, 256 \\ 3x3, 256 \end{bmatrix}$x3	–
Conv4	$\begin{bmatrix} 3x3, 256 \\ 3x3, 256 \end{bmatrix}$x2	$\begin{bmatrix} 1x1, 256 \\ 3x3, 256 \\ 1x1, 1024 \end{bmatrix}$x3	3×3 max pool, stride 2 $\begin{bmatrix} 3x3, 512 \\ 3x3, 512 \\ 3x3, 512 \end{bmatrix}$x3	–
Conv5	$\begin{bmatrix} 3x3, 512 \\ 3x3, 512 \end{bmatrix}$x2	$\begin{bmatrix} 1x1, 512 \\ 3x3, 512 \\ 1x1, 2048 \end{bmatrix}$x3	3×3 max pool, stride 2 $\begin{bmatrix} 3x3, 512 \\ 3x3, 512 \\ 3x3, 512 \end{bmatrix}$x3	–
	Average pool, 1000-d fc	Average pool, 1000-d fc	fc with 4096 nodes	–
	6-d fc, softmax	6-d fc, softmax	fc with 4096 nodes, softmax with 1000 nods	3-d fc, softmax, classification

With only a 3×3 difference in convolution layer kernel size, the identical set of Conv2 layers is designed. This reduces the output size to 56×56. Finally, the fully connected layer with 3 categories with softmax and classification layer is added to determine the kind of infant facial expressions.

This entire design procedure helps to stabilize learning and minimize the number of epochs necessary to train the networks considerably. It prevents the computation required to learn the network from growing exponentially. Other networks are more complex in terms of space and thus take longer to execute and train. In comparison to the other networks in Table, the proposed network has a lower structural level of complexity. It also saves time and reduces the difficulty of the training process by minimizing the usage of hardware resources. As a result, the proposed method is more computationally efficient and produces superior performance results.

The images in the dataset are divided according to the pareto principle, with 70% of the data being used for training and 15% for validation. The remaining 15% is employed for testing purposes alone. So, out of 1800 images, 1260 are chosen for training, 270 for validation, and 270 for testing in each category. Figure 4 shows the proposed method's training curve, which includes training and validation accuracy as well as loss curve information.

The experiments were carried out on a local dataset, which was trained and validated using existing methods. In general, as the number of layers increases, the learning capacity increases as well. Nevertheless, if the learning capacity is large enough, overfitting issues may arise. It will perform excellently in the training phase but adversely in the test set. The proposed network's performance is compared to that of the existing network shown in Table 3. It shows the proposed method achieves better accuracy.

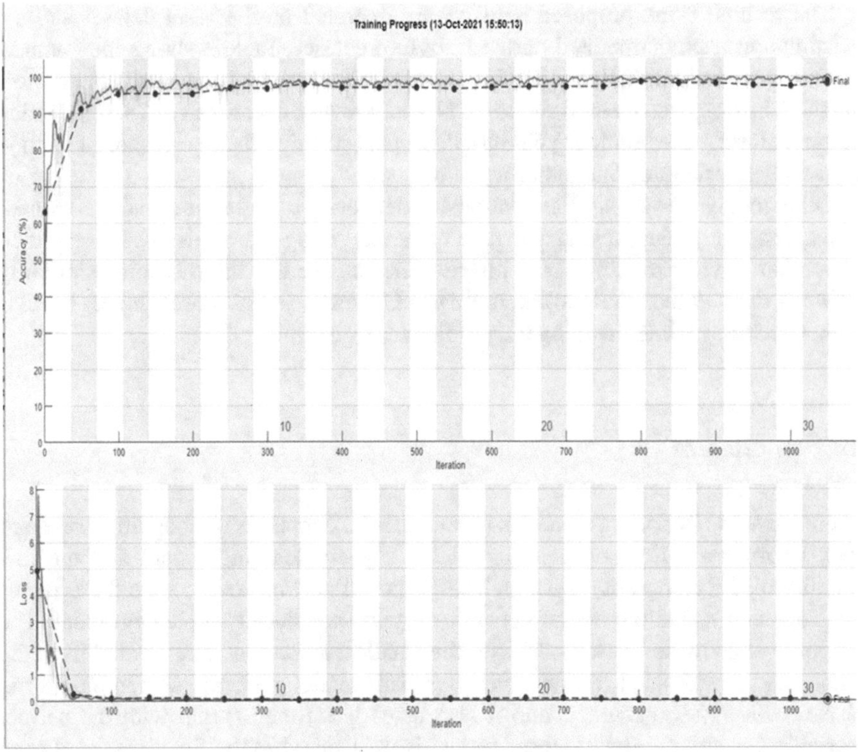

Fig. 4 Training and validation curve

Table 3 Performance comparison (Pretrained CNN vs. proposed shallow CNN)

	Laugh		Cry		Neutral		Overall	
	Precision	Recall	Precision	Recall	Precision	Recall	Testing accuracy	Average accuracy
Resnet-18	91.21	94.7	93	92.7	89	93.2	96	94
Resnet-50	94.3	95.6	91.74	90.6	94.3	95.6	92.6	90.26
VGG-16	92.1	94	89.91	90.3	94.26	94	95	93.6
Proposed method	97.8	98	96	97.2	98	97.8	97.8	98.82

Table 4 Performance comparison of proposed method vs VFESO-DLSE [11]

	Laugh		Cry		Neutral		Overall
	Precision	Recall	Precision	Recall	Precision	Recall	Average accuracy
VFESO-DLSE [11]	93.18	78.5	85.07	96.27	86.71	88.86	93.6
Proposed CNN	97.8	98	96	97.2	98	97.8	98.82

The results of the proposed network are operated on the local dataset and the existing approach is operated on the BabyExp dataset. Table 4 shows the comparisons of these two networks, and it shows the proposed method achieves better performance. The proposed method shallow CNN achieves a better result of 98.82%, which is about 10.92% greater than VEFSO-DLSE [11] showing that the proposed method extracts better features than other methods.

Other models show a similar relationship between real and expected facial expressions. The matrix demonstrates that some neutral images are mistakenly classified with others. When looking at the images, one can see that there is some similarity between them, which adds to the intricacy. However, the proposed strategy yields higher accuracy while avoiding the problem of overfitting.

4 Conclusion

There are some important concerns in infant facial expressions recognition research such as impaired learning capacity of the transfer learning model and the low stability of the recognition system. The proposed network overcomes these issues by proposing a shallow neural network as a space-reduced two-stage model. The proposed shallow network model uses the smallest amount of data created from the videos and downloaded from the websites. This model performs well in the testing phase, with 97.8% accuracy, and it also takes less time to train with the perfect learning capacity. So the proposed model is well suited to the field of surveillance parental care and provides better intelligent interpersonal interactions.

References

1. Altamura M, Padalino FA, Stella E (2016) Facial emotion recognition in bipolar disorder and healthy aging. J. Nerv. Mental Disease 204(3):188–193
2. Ekman P, Friesen W (1978) Facial action coding system: a technique for the measurement of facial movement. Consulting Psychologists Press
3. Cohen I, Sebe N, Sun Y, Lew M (2003) Evaluation of expression recognition techniques. In: Proceedings of international conference on image and video retrieval, pp 184–195
4. Ming Z, Bugeau A, Rouas J, Shochi T (2015) Facial action units intensity estimation by the fusion of features with multi-kernel support vector machine. In: 11th IEEE International conference and workshops on automatic face and gesture recognition (FG), pp 1–6
5. Padgett C, Cottrell G (1996) Representing face images for emotion classification. In: Conference on advances in neural information processing systems, 894900
6. Lee Y, Kim KK, Kim JH (2019) Prevention of safety accidents through artificial intelligence monitoring of infants in the home environment. In: International conference on information and communication technology convergence (ICTC), pp 474–477
7. Simonyan K, Zisserman A (2014) Two-stream convolutional networks for action recognition in videos. Adv Neural Inf Process Syst NIPS
8. Dhankar P (2019) ResNet-50 and VGG-16 for recognizing facial emotions. Int J Innov Eng Technol 13(4). ISSN :2319-1058
9. Li B, Lima D (2021) Facial expression recognition via ResNet-50. Int J Cogn Comput Eng 2:57–64. ISSN 2666-3074
10. Sujitha Balasathya S, Mohamed Mansoor Roomi S (2021) Infant action database: a benchmark for infant action recognition in uncontrolled condition. J Phys Conf Ser 1917:1742–6588
11. Lin Q, He R, Jiang P (2020) Feature guided CNN for baby's facial expression recognition, complexity, Hindawi, vol 2020, Article ID 8855885, 10 p

Local and Global Thresholding-Based Breast Cancer Detection Using Thermograms

Vartika Mishra, Subhendu Rath, and Santanu Kumar Rath

1 Introduction

Cancer is one of the second most causes of global death accounting 9.6 million of death as recorded by World Health organization [1]. One of the most common cancers among different cancers is breast cancer due to which 6,27,000 deaths have been reported. Hence, the early detection and treatment of the cancer can subsequently reduce the mortality rate. Breast cancer is usually detected in ducts, glands producing milk, tubes carrying milk to the lobules and nipples. Different modalities have been helpful in the detection of the breast cancer viz. Ultrasound, Mammography, Magnetic Resonance Imaging (MRI), thermography and many more.

Mammography is one commonly adopted methodologies in which the images are captured with X-ray exposures compressing the breasts. It has been observed that, difficulty arises when capturing the dense breast tissues thereby, the tumors of the respective area leaving undetected. Infrared Thermography is one of such techniques that has the potential to detect breast cancer even to the extent of eight to ten years earlier than diagnosis by application of Mammography [2]. IR is a non-ionizing, radiation free and non-contact technique. It helps in detecting the tumors based on the asymmetry analysis when both left and right breasts are compared for the tumor detection [3].

V. Mishra (✉) · S. K. Rath
NIT Rourkela, Rourkela 769008, Odisha, India
e-mail: vartikamishra151@gmail.com

S. K. Rath
e-mail: skrath@nitrkl.ac.in

S. Rath
VCU, Richmond, VA 23219, USA
e-mail: subhendu.rath@vcuhealth.org

P. Singh et al. (eds.), *Machine Learning and Computational Intelligence Techniques for Data Engineering*, Lecture Notes in Electrical Engineering 998,
https://doi.org/10.1007/978-981-99-0047-3_67

In this work, the red channel of the RGB image is taken for analysing the breast thermograms which helps to identify the abnormality of the breast thermograms. The red channel image is extracted from the RGB images. Further, these obtained red channel images are converted to the grey scale images. Adaptive mean thresholding, Adapting Gaussian thresholding and Otsu thresholding is applied for the breast thermograms for differentiating the background from the image and to get more precise edges. Further, the breast image is cropped into left and right part of the breast. The different statistical features are extracted from these breast thermogram images viz. Gray Level Co-Occurrence Matrix (GLCM), Gray Level Run Length Matrix (GLRLM), Neighbourhood Gray Tone Difference Matrix (NGTDM), Gray Level Size Zone Matrix (GLSZM) and Gray Level Dependence Matrix (GLDM). Relief-F methods is applied for feature selection method and further Decision Tree and Random Forest are applied for classifying among the healthy and unhealthy breast thermograms.

2 Literature Survey

This section presents a review of articles on breast thermogram classification based on different methods of feature extraction methods, feature selection methods and their analysis. Motta et al. proposed a method for tri-dimensional profile based on the temperature profile by applying the level set method for the automatic segmentation of the breast thermograms [4]. They applied Otsu thresholding for background segmentation. Ritam et al., proposed super pixel-based segmentation for automatic segmenting the breast thermograms [5].

Gogoi et al. have obtained a feature set of 24 features by applying the Mann Whitney Wilcoxon test. They applied six different classifiers among which ANN and SVM with RBF kernel gave the highest accuracy or 84.75% [6]. Sathish et al. have extracted the texture features in spatial domain. Further, by applying SVM classifier and ANN classifier, the later is giving a better accuracy of 80% [7]. De Santana et al., have extracted the geometry and texture features viz. Zernike and Haralick moments. Among different classifiers applied Extreme Machine Learning (ELM) and Multi-Layer Perceptron (MLP) obtained promising results. An accuracy of 76.01% is attained with Kappa index [8]. Sathish et al. have proposed a novel method based on temperature normalization of the breast thermograms and extracted the wavelet based local energy features. Further RSFS and GA are applied for selecting the best set of features among which RSFS is selecting the best set of features when classified with SVM classifier with Gaussian kernel function giving an accuracy of 91% [9]. Karthiga et al., have extracted the spatial domain and curvelet domain based statistical features and applied hypothesis testing. 16 best set of features are selected and classified with different classifiers viz. Logistic Regression, SVM, K-Nearest Neighbor, Naïve Bayes. Among these above-mentioned classifiers, cubic-SVM is giving a highest accuracy of 93%. [10]. A comparative analysis of the state of the art is represented in a tabular form in Table 1.

Table 1 State of the art comparison

Author's name	Dataset	Methodology applied
Motta et al. [4]	DMR database	Level set method
Ritam et al. [5]	147 thermograms	Super-pixel segmentation
Gogoi et al. [6]	80 thermograms	Mann Whitney test
Sathish et al. [9]	100 thermograms	Temperature based normalization
Karthiga et al. [10]	60 thermograms	Extracted spatial and curvelet domain features

It is observed from the above-mentioned literature survey that better preprocessing of images further helps in extracting better set of features. Thus, this research gap motivates towards getting a better vision of the images. The proposed work is based on the RGB image which corresponds to the information of different color channels for the respective image. The red channel images are extracted which indicates the higher probability of tumor for abnormal breast. Further, three thresholding methods are applied viz. Otsu thresholding, Adaptive Mean thresholding and Adaptive Gaussian thresholding. Texture features are extracted from these images and Relief-F method is applied for feature selection. The best set of features are classified with Decision Tree and Random Forest.

3 Dataset

For this study, a publicly available dataset comprising of a total number of 56 breast thermograms is considered. This dataset is available online at the Database for Mastology Research (DMR) repository of UFF, Brazil [11]. The breast thermograms, are captured with the FLIR SC-620 THERMAL camera having a specification of 640 $\times$ 480 spatial resolution which has 37 unhealthy and 19 healthy breast thermograms.

4 Breast Thermogram Analysis

In this study, breast thermograms are obtained by transforming the temperature matrix into images. On these obtained images, the red channel images are extracted from the three-channel RGB images. Further, they are converted to gray scale and thresholding-based techniques are applied for the images. Three different methods viz., Otsu thresholding, Adaptive Mean thresholding, and Adaptive Gaussian Thresholding are applied for obtaining the object from the background. Further, different texture features are extracted by the application of GLCM, GLSZM, GLRLM, NGTDM, and GLDM. After extracting the features from thermograms, the feature selection method Relief-F is applied for the selection of the most distinctive features.

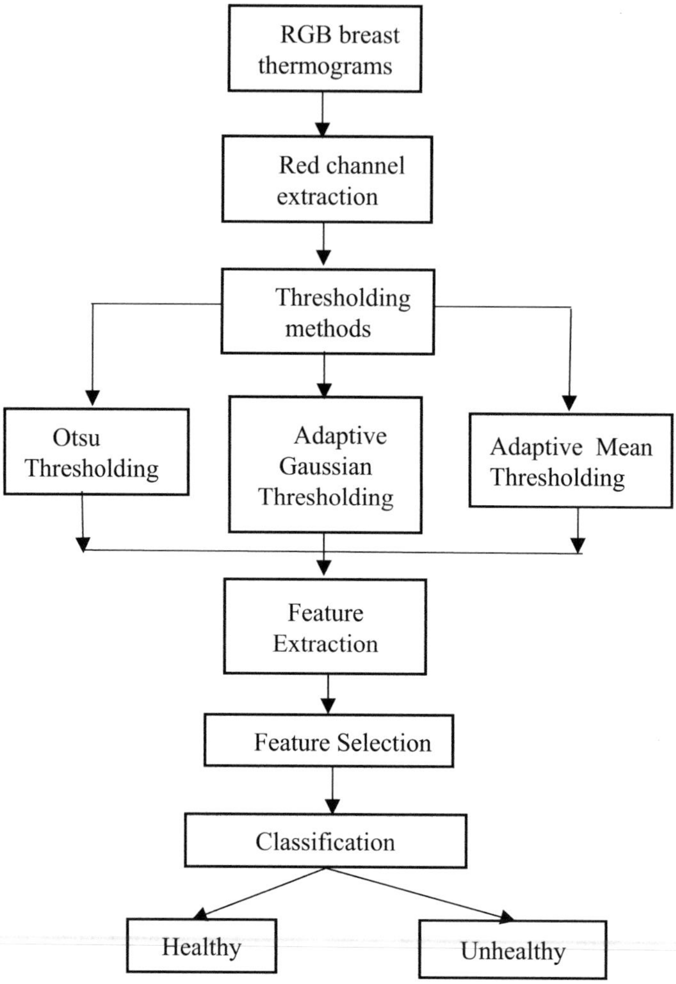

Fig. 1 Structure of the experiment executed

The feature subset obtained is then classified as healthy and unhealthy subjects. The respective flow of the work is explained in the form of a flow chart as shown in Fig. 1.

4.1 Pre-processing

The breast thermograms are obtained by transforming the temperature matrix. The images obtained are in RGB channel, where the red channel images are extracted for the analysis of the abnormality of the breast thermograms. Further, different thresholding methods, adaptive thresholding, and Otsu thresholding methods are applied to the images obtained.

Thresholding methods: The thresholding methods are divided into local thresholding and global thresholding. The traditional thresholding method was proposed wherein a suitable single threshold value was chosen to distinguish between the background and the object [12]. The pixel values above the threshold value are classified in one class and the pixel values below the threshold value are classified in another class. Selecting an optimal threshold value is a very important key for applying the thresholding technique. An Image I (a, b) of N levels of gray in an image gy (a, b) is defined with two gray levels as:

$$gy(a, b) = \begin{cases} 1, if I(a, b) \geq T \\ 0, if I(a, b) < T \end{cases} \ldots \quad (1)$$

where pixels corresponding to value 1 are objects and the pixels corresponding to 0 are background.

4.1.1 Adaptive Thresholding

The adaptive thresholding method is a local method where the threshold values of smaller regions are calculated. For finding the local threshold, the intensity values of each pixel are statistically examined for local neighborhood. It calculates different thresholds for different regions of the image and gives results which are better with varying illumination [13].

a. **Adaptive Mean thresholding**
   The value of the threshold is calculated by the mean of neighbouring (a, b) pixels.
b. **Adaptive Gaussian thresholding**
   The value of threshold is calculated as the gaussian weighted sum of the block size neighbourhood of (a, b) pixels.

4.1.2 Otsu Thresholding

It is a global thresholding technique and is applied on the bimodal images. It chooses the threshold value based on the minimization of the within-class variance by minimizing the weighted sum of class variance [14].

$$\sigma_w^2(t) = q_1(t)\sigma_1^2(t) + q_2(t)\sigma_2^2(t) \ldots \quad (2)$$

$$\sigma_b^2(t) = q_1(t)[1 - q_1(t)][\mu_1(t) - \mu_2(t)]^2 \ldots \quad (3)$$

$$\sigma^2 = \sigma_w^2(t) + \sigma_b^2(t) \ldots \quad (4)$$

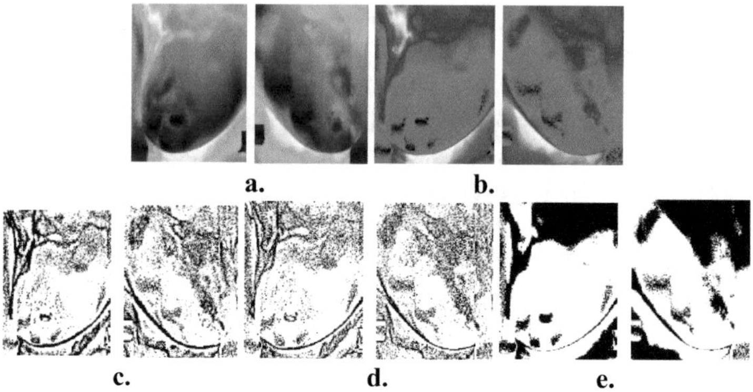

Fig. 2 The segmented left and right breast thermograms: **a** the breast thermograms; **b** the red channel breast thermograms; **c** the Adaptive Mean thresholding; **d** the Adaptive Gaussian thresholding; **e** the Otsu thresholding

where

$\sigma^2 : thetotalvariance,$

$\sigma_w^2(t) : thewithinclassvariance$

$\sigma_b^2(t) : the\ betweenclassvariance$

The images are further segmented between the left and right breast using the manual segmentation method as shown in Fig. 2.

4.2 Feature Extraction

The different techniques of feature extraction play a very significant role in detecting the breast cancer abnormalities. It is observed that the texture patterns are better reconginzed by the gray level intensity in the thermogram in the particular direction. The different intensity levels describe the mutual relationship among the neighbouring pixels of the image. In this work, from the segmented left and right breast, various features are extracted by applying techniques such as GLRLM, GLCM, GLSZM, GLDM, and NGTDM.

4.2.1 Gray Level Co-occurrence Matrix (GLCM)

GLCM is a second-order statistical feature extraction method which helps to analyze the textural properties of the image [15]. The different gray levels in the image mark the spatial orientations adjacent to the pixels from each other. Twenty-four different features are extracted from the breast thermograms viz. joint average (JA), autocorrelation (AC), cluster shade (CS), difference entropy (DEN), cluster tendency (CT), correlation (CR), inverse variance (IV), sum of squares (SOS), informational measure of correlation 2 (IMC2), contrast (C), inverse difference moment (IDM), maximum probability (MP) inverse difference normalized (IDN), inverse difference moment normalized (IDMN), joint entropy (JEN), sum entropy (SEN), inverse difference (ID), sum average (SA) difference average (DA), difference variance (DV), joint energy (JEY), maximal correlation coefficient (MCC), cluster prominence (CP), and informational measure of correlation 1 (IMC1),

4.2.2 Gray Level Run Length Matrix (GLRLM)

GLRLM statistical features are computed by the length of number of consecutive pixels which has the same gray level value. Here the occurrence of given gray colour is calculated on the basis of direction giving homogeneous pixels running for every gray-levels [16]. Total sixteen GLRLM features viz., long run emphasis (LRE), gray level non-uniformity normalized (GLNUN), short run emphasis (SRE), run length non-uniformity normalized (RLNUN), gray level variance (GLV), long run high gray level emphasis (LRHGLE), run percentage (RP), run entropy (REN), short run low gray level emphasis (SRLGLE), gray level non-uniformity (GLNU), run length non-uniformity (RLNU), high gray level run emphasis (HGLRE), low gray level run emphasis (LGLRE), short run high gray level emphasis (SRHGLE), run variance, and long run low gray level emphasis (LRLGLE) are extracted from the breast thermograms.

4.2.3 Gray Level Size Zone Matrix (GLSZM)

GLSZM feature extraction method quantifies the different gray level zones in an image [17]. Her, the connected pixels share the same gray level intensity. The features extracted from the matrix are large area emphasis (LAE), small area emphasis (SAE), size zone non-uniformity (SZNU), gray level non-uniformity normalized (GLNUN), zone variance (ZV), zone percentage (ZP), size zone non-uniformity normalized (SZNUN), low gray level zone emphasis (LGLZE), small area high gray level emphasis (SAHGLE), small area low gray level emphasis (SALGLZE), gray level non-uniformity (GLNU), large area high gray level zone emphasis (LAHGLE), large area low gray level emphasis (LALGLE), zone entropy (ZEN), high gray level zone emphasis (HGLZE), and gray level variance (GLV).

4.2.4 Gray Level Dependence Matrix (GLDM)

GLDM statistical texture features quantify gray-level dependencies [18]. Here, the features are calculated as the number of pixels which are connected within depending on the centre pixel. The different features extracted from the matrix are Large Dependence Emphasis (LDE), Dependence Non-Uniformity (DNU), gray Level

Non-Uniformity (GLNU),), Dependence Entropy (DEN), Dependence Variance (DV), Low gray Level Emphasis (LGLE), Dependence Non-Uniformity Normalized (DNUN), Large Dependence High gray Level Emphasis (LDHGLE), High gray Level Emphasis (HGLE), Small Dependence High gray Level Emphasis (SDHGLE), gray Level Variance (GLV), Small Dependence Emphasis (SDE), Small Dependence Low gray Level Emphasis (SDLGLE), and Large Dependence Low gray Level Emphasis (LDLGLE).

4.2.5 Neighbourhood Gray Tone Difference Matrix (NGTDM)

NGTDM texture features are calculated with quantifying differences based on the different gray level values and their average gray value which are within the distance [7]. The different features extracted are Coarseness (Co), Busyness (B), Contrast (C), Strength (SH), and Complexity (CL).

4.3 Feature Selection

The feature selection methods help in removal of the redundant and irrelevant features from the extracted set of features. This further assists in better result analysis with the selected subset of features. In this work, Relief-F is applied for selecting the best set of features.

The main idea behind implementing the Relief-F method is it helps in estimating the most qualitative features. It calculates on the basis of the distinguishes between different instances which are near to each other. The relief-F method finds the two nearest neighbors, each from a different class. This results in the adjustment of the feature weighting vector, where more weight is given to the features that are discriminating between the instances from the neighbors, belonging to different classes [19].

4.4 Classification

A classification model is used to draw a conclusion from the input instances given for training. It predicts the class for the new instance given for training. The classifier maps the given instance to that specific category. In this study, two different classifiers Decision Tree [20] and Random Forest [21] are applied for classifying the healthy and unhealthy breast thermograms.

5 Results Analysis

In this study, the red channel-based images are extracted from the RGB image. Further the three thresholding methods viz. Otsu thresholding, Adaptive Mean thresholding, and Adaptive Gaussian thresholding methods are applied for distinguishing the background from the object. Further, the thresholded images are applied for extracting the texture features viz. GLCM, GLRLM, GLDM, NGTDM and GLSZM. Further, two classifiers are applied for classifying the breast thermogram analysis among healthy and unhealthy breasts.

The PSNR (peak signal-to-noise ratio) value for the reconstructed image is calculated for the different three thresholding methods and it is observed that higher the value of the PSNR, the better is the quality of the image. The PSNR value differentiates the similarities between the original image and the reconstructed image. It is also defined as the root mean square of each pixel in the image. The different values for the image obtained are shown in Table 2, where the average values of the image are calculated. It is observed that the Adaptive Gaussian thresholding is giving the highest value thus depicting to be a better technique for reconstructing the image.

It is calculated as:

$$\text{PSNR} = 20 * log_{10}(maxvalue/RMSE)\ldots \tag{5}$$

$$\text{RMSE} = \sqrt{\frac{\sum_{i=1}^{r_m}\sum_{j=1}^{c_m}(I_o(a,b) - I_r(a,b))}{r_m * r_c}}\ldots \tag{6}$$

where

$r_m * r_c$ are the maximum number of rows and columns, $I_o istheoriginalimage$, and I_r is the reconstructed image respectively (Table 2).

The Otsu thresholding method is applied when the grey level intensities are clearly distinguished in the image. In this work, the Otsu thresholding method is applied for the red channel extracted images which depict the higher temperature of the surface of the breast thermograms. The features are extracted from these images and further applied for classification. Random Forest and Decision Tree are applied for classifying among the healthy and unhealthy breast thermograms. Among both classifiers, Random Forest is giving a better accuracy of 92.30% as compared to the decision Tree giving an accuracy of 85.90%. The parameters viz., precision, sensitivity, and specificity are also calculated and are shown in Table 4. The positive

Table 2 The average PSNR metrics calculated for different thresholding methods for healthy and unhealthy breast

Thresholding methods	Normal	Abnormal
Otsu thresholding	55.63	55.06
Adaptive mean thresholding	56.61	55.05
Adaptive Gaussian thresholding	58.61	58.05

predictive value for random Forest is high as the precision is giving a higher value of 93.39% as compared to the Decision Tree.

The Adaptive thresholding method is applied on the red channel of the breast thermograms which helps in analyzing the abnormality in the abnormal breast as the surface temperature of the unhealthy breast is higher due to the presence of the tumor in the cells. A block size of 11×11 is taken for the image for calculating the threshold value of every pixel block wise. Further, the features are extracted from these obtained images and two classifiers viz. Decision Tree and Random Forest are applied for classifying between the healthy and unhealthy breast thermograms. Among the two classifiers, Random Forest is giving a better accuracy of 94.67% by applying the Adaptive Gaussian thresholding method. Another method Adaptive Mean thresholding is giving an accuracy of 91. 12% which is comparatively less. The other performance parameter values are also calculated and are shown in Tables 5 and 6 for each thresholding method based on different classifiers.

However, among the three thresholding methods, images applied by Adaptive Gaussian thresholding are giving a better accuracy of 94.67% when Random Forest is applied for classifying the unhealthy and unhealthy breast as compared to the decision tree. The positive predictive value is higher for the random forest classifier than the decision tree as shown in Table 5. A box plot is calculated shown in Fig. 3 describing the accuracies obtained for different thresholding methods viz. Otsu thresholding, Adaptive Mean thresholding, and Adaptive Gaussian thresholding for two different classifiers viz. Random Forest and Decision Tree. A comparative observation of the proposed work with the state of the art is discussed in Table 3.

Table 3 Comparison of state-of-the-art with the proposed work

Author's name	Classifier	Accuracy (%)
Gogoi et al. [6]	ANN	84.29
Sathish et al. [7]	ANN	80
De Santanna et al. [8]	MLP	76.01
Sathish et al. [9]	SVM	91
Karthiga et al. [10]	SVM	93
Proposed work	Random forest	94.67

Table 4 The performance metrics in percentage obtained for the Otsu Thresholding images

Classifiers	Accuracy	Sensitivity	Specificity	Precision
Random forest	92.30	93.39	95.07	86.96
Decision tree	85.90	90.22	83.46	89.02

Table 5 The performance metrics in percentage obtained for the adaptive Gaussian thresholding images

Classifiers	Accuracy	Sensitivity	Specificity	Precision
Random forest	94.67	93.02	98.52	88.89
Decision tree	88.46	90.45	91.71	82.40

Table 6 The performance metrics in percentage obtained for the adaptive mean thresholding images

Classifiers	Accuracy	Sensitivity	Specificity	Precision
Random forest	91.12	92.96	91.56	88.00
Decision tree	89.35	93.52	90.18	87.72

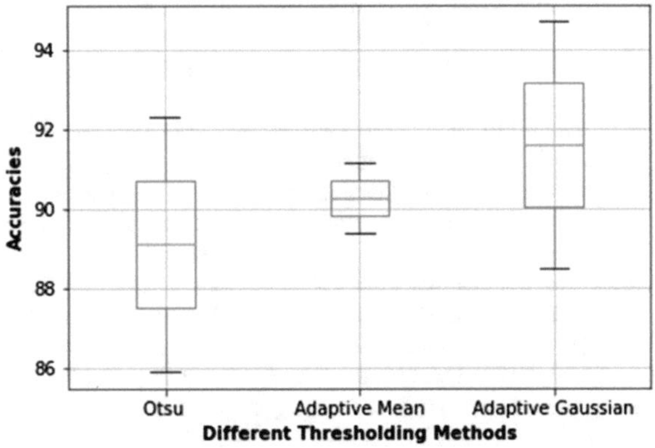

Fig. 3 Box plot of the accuracies for the different thresholding methods

6 Conclusion

In this study, the red channel-based images are extracted from the RGB images. Further, three thresholding methods viz. Otsu thresholding, Adaptive Mean thresholding, and Adaptive Gaussian thresholding methods are applied. Statistical features are extracted from the breast thermograms and classified among Random Forest and Decision tree. Among the three methods, Adaptive Gaussian thresholding is giving a better accuracy of 94.67% and the other performance parameters such as precision, specificity, and sensitivity are higher for the same as compared to the Decision Tree with the other two thresholding methods. For future work, automatic segmentation will be applied for obtaining more précised images which will improve the visual of the breast thermograms and assist in detecting the abnormality in the breast thermograms.

References

1. WHO, Cancer (2021). https://www.who.int/en/news-room/fact-sheets/detail/cancer
2. Tan TZ, Quek C, Ng GS, Ng E (2007) A novel cognitive interpretation of breast cancer thermography with complementary learning fuzzy neural memory structure. Expert Syst Appl 33(3):652–666
3. Nurhayati OD (2011) Principal component analysis combined with first order statistical method for breast thermal images classification 1
4. Motta LS, Conci A, Lima RCF, Diniz EM (2010) Automatic segmentation on thermograms in order to aid diagnosis and 2D modelling. In: Proceedings of the Tenth Workshop em Informatica Medica, Belo Horizonte, pp 1610–1619, MG, Brazil
5. Sharma R, Sharma JB, Maheshwari R, Baleanu D (2021) Early anomaly prediction in breast thermogram by hybrid model consisting of superpixel segmentation, sparse feature descriptors and extreme learning machine classifier. Biomed Signal Process Control 70:103011
6. Gogoi UR, Bhowmik MK, Ghosh AK, Bhattacharjee D, Majumdar G (2017) Discriminative feature selection for breast abnormality detection and accurate classification of thermograms. In: International conference on innovations in electronics, signal processing and communication (IESC), pp 39–44. IEEE
7. Sathish D, Kamath S, Prasad K, Kadavigere R (2018) Texture analysis of breast thermograms using neighbourhood grey tone difference matrix. Int J Bioinformat Res Appl 14(1–2):104–118
8. Santana MA de, Pereira JMS, da Silva FL, de Lima NM, de Sousa FN, de Arruda GMS, de Cássia Fernandes de Lima R, da Silva WWA, dos Santos WP (2018) Breast cancer diagnosis based on mammary thermography and extreme learning machines. Res Biomed Eng 34:45–53
9. Sathish D, Kamath S, Prasad K, Kadavigere R (2019) Role of normalization of breast thermogram images and automatic classification of breast cancer. Visual Comput 35(1):57–70
10. Karthiga R, Narasimhan K (2021) Medical imaging technique using curvelet transform and machine learning for the automated diagnosis of breast cancer from thermal image. Pattern Anal Appl 24(3):981–999
11. Conci A (2014) Breast thermograms data. http://visual.ic.uff.br/en/proeng/thiagoelias/
12. Sahoo PK, Soltani SAKC, Wong AKC (1998) A survey of thresholding techniques. Comput Vis Graph Image Process 41(2):233–260
13. Filipczuk P, Kowal M, Obuchowicz A (2011) Automatic breast cancer diagnosis based on k-means clustering and adaptive thresholding hybrid segmentation. In: Image processing and communications challenges, vol 3, pp 295–302. Springer, Berlin, Heidelberg
14. Otsu N (1979) A threshold selection method from gray-level histograms. IEEE Trans Syst Man Cybernet 9(1):62–66
15. Haralick RM, Shanmugam K, Dinstein IH (1973) Textural features for image classification. IEEE Trans Syst Man Cybernet 6:610–621
16. Tang X (1998) Texture information in run-length matrices. IEEE Trans Image Process 11(7):1602–1609
17. Thibault G, Fertil B, Navarro C, Pereira S, Cau P, Levy N, Sequeira J, Mari J-L (2013) Shape and texture indexes application to cell nuclei classification. Int J Pattern Recogn Artif Intell 27(1):1357002
18. Novitasari DCR, Lubab A, Sawiji A, Asyhar AH (2019) Application of feature extraction for breast cancer using one order statistic, GLCM, GLRLM, and GLDM. Adv Sci Technol Eng Syst J 4(4):115–120
19. Robnik-Sikonja M, Kononenko I (2009) Theoretical and empirical analysis of ReliefF and RReliefF. Mach Learn 53:23–69
20. Chao C-M, Yu Y-W, Cheng B-W, Kuo Y-L (2014) Construction the model on the breast cancer survival analysis use support vector machine, logistic regression and decision tree. J Med Syst 38(10):106–113
21. Nguyen C, Wang Y, Nguyen HN (2013) Random Forest classifier combined with feature selection for breast cancer diagnosis and prognostic. J Biomed Sci Eng 6(1):551–560

Multilevel Crop Image Segmentation Using Firefly Algorithm and Recursive Minimum Cross Entropy

Arun Kumar, A. Kumar, and Amit Vishwakarma

1 Introduction

Image segmentation is a pre-processing step in computer vision technology. During the processing step, the input image is analyzed with a signal processing technique, and the result is further explored to obtain desired information. In today's era image segmentation has been used in many applications such as medical imaging, remote sensing, the agriculture field, etc. In agriculture, image segmentation has been projected for various applications by providing useful information about the ripeness status of fruit, plant health, plant species identification, plant density population, and disease identification [1–3]. Mostly, the symptoms of the disease on crops are visible on the leaf, stem, and fruits. Hence, the disease can be identified by the color changes on leaf, stem, and fruits. However, diseases that affect the crop through its development phases such as recognition of banana leaf disease [4], analysis of rice leaf disease [5], and disease spot recognition from leaf [6]. The quality and quantity are harmed as a result of this disease.

Initially, crop-related problems can be resolved by agriculture experts, but currently by the advent of technology in agriculture farmers can prevent big losses. Similarly, they are now able to get a prediction about their products in many ways such as the condition of plant health, disease-related to crops, etc. The production of crops has a direct impact on the country's economy. The main problem of crop image segmentation is that the plant has a different intensity of color, weak local

A. Kumar (✉) · A. Kumar · A. Vishwakarma
PDPM Indian Institute of Information Technology, Design and Manufacturing, Jabalpur, India
e-mail: erarun15182@gmail.com

A. Kumar
e-mail: anilk@iiitdmj.ac.in

A. Vishwakarma
e-mail: amitv@iiitdmj.ac.in

P. Singh et al. (eds.), *Machine Learning and Computational Intelligence Techniques for Data Engineering*, Lecture Notes in Electrical Engineering 998,
https://doi.org/10.1007/978-981-99-0047-3_68

pixel correlation, and complex background [2, 7–9]. Due to this reason, crop image segmentation is a challenging task.

Several methods have been developed for the segmentation of an image [10, 11, 12–14]. It is divided into four parts (a) splitting and merging region-based; (b) histogram thresholding based; (c) clustering-based; and (d) texture analysis. Among all, the thresholding-based method provides the most hopeful results [15, 16, 17, 18]. Thresholding is categorized into two portions: bi-level and multilevel. An image can be divided into two regions for bi-level thresholding whereas, it can be divided into multiple regions for multilevel thresholding. Bilevel thresholding has an unfitting issue. Therefore multilevel thresholding has been widely used [19, 20, 21]. Researchers have done a lot of jobs in the last decades, and multilevel thresholding-based segmentation has been widely observed [22, 23].

A variety of efficient metaheuristic algorithms is utilized to solve the issue of multilevel thresholding for improvement in accuracy and fast convergence such as differential evolution (DE) [24], genetic algorithm (GA) [25], particle swarm optimization (PSO) [15], and wind-driven optimization [26, 27]. Kurban et al. [28] proposed a comparative analysis of swarm-based techniques for multilevel thresholding. Recently, the iterative algorithm has been used for the multilevel MCE for fast convergence [29]. Crop images have complex backgrounds, weak local pixel correlation, and different color intensities. Hence, multilevel thresholding becomes challenging. Therefore, an objective function (recursive minimum cross entropy) has been combined with a firefly (an efficient metaheuristic) algorithm to increase the accuracy.

In this paper, FFA has been combined with recursive minimum cross entropy to find the optimum threshold value. The proposed method has been tested on ten complex background crop images with high resolution and compared with an existing algorithm such as wind-driven optimization (WDO). Experimental results evidence that the proposed technique enhances the accuracy of the segmented image in terms of mean square error (MSE), peak signal-to-noise ratio (PSNR), structural similarity index (SSIM), and Feature similarity index (FSIM).

2 Proposed Methodology

2.1 Cross Entropy

Cross entropy is calculated as a function of two probability distribution functions $A = \{a_1, a_2, ..., a_Q\}$ and $B = \{b_1, b_2, ..., b_Q\}$ within the same set represented as:

$$C(A, B) = \sum_{i=1}^{Q} A_i^c \log \frac{A_i^c}{B_i^c}; c = \begin{cases} 1,2,3 for RGBimage \\ 1 for Grayimage \end{cases} \tag{1}$$

$$I_{th} = \left\{ \frac{\mu^c(1, th)I(x, y) < th}{\mu^c(th, L+1)I(x, y) \geq th} \right. \quad (2)$$

In Eq. (2), I is the original test image, th represents the threshold, $h^c(i)$ denotes the histogram calculated by amalgamating three color components to preserve the pixel information.

In Eq. (3), cross-entropy is evaluated as:

$$C(th) = \sum_{i=1}^{th-1} ih^c(i)\log\left(\frac{i}{\mu^c(1, th)}\right) + \sum_{i=th}^{L} ih^c(i)\log\left(\frac{i}{\mu^c(th, L+1)}\right) \quad (3)$$

The optimal threshold is calculated by minimizing Eq. (3) as:

$$th* = \arg\min_t C(th) \quad (4)$$

In general, the computational complexity for n level thresholding is $O(L^{n+1})$.

2.2 Recursive Minimum Cross Entropy

Recursive programming is used to decrease the computational computation of objective functions. Equation (4) can be simplified as:

$$C(th) = -\sum_{i=1}^{L} ih^c(i)\log(i) - \sum_{i=1}^{th-1} ih^c(i)\log\mu(1, th) - \sum_{i=th}^{L} ih^c(i)\log\mu(th, L+1)) \quad (5)$$

In Eq. (5), the first term is constant for a given image, the objective function can be reformulated as:

$$\lambda(th) = -\sum_{i=1}^{th-1} ih^c(i)\log\mu(1, th) - \sum_{i=th}^{L} ih^c(i)\log(\mu(th, L+1))$$
$$= -m^{c1}(1, th)\log\left(\frac{m^{c1}(1, th)}{m^{c0}(1, th)}\right) - m^{c1}(th, L+1)\log\left(\frac{m^{c1}(th, L+1)}{m^{c0}(th, L+1)}\right) \quad (6)$$

In Eq. (6), $m^{c0}(a, b)$ and $m^{c1}(a, b)$ denotes the zero-moment and first-moment of the histogram image. Assume gray image of L levels, N number of pixels, and r number of thresholds are required to partition the original crop image into $r + 1$ classes. For the convenience of calculation two dummy thresholds $th_0 = 0$ and $th_{r+1} = L$ are used. The objective function is then:

$$\lambda(th_c^1, th_c^2, \ldots\ldots, th_c^r) = m^{c1}(th_{i-1}, th_i) \log\left(\frac{m^{c1}(th_{i-1}, th_i)}{m^{c0}(th_{i-1}, th_i)}\right) \tag{7}$$

A gray image contains a single channel but a color image consists of three channels. The technique conferred in Eq. (7) is processed three times for RGB images. The computational complexity of multilevel thresholding reduces $O(nL^n)$ by using recursive minimum cross entropy [24], but still, this technique provides a good result for bi-level. Every threshold point increases the arithmetic formulation when it is prolonged for multilevel. Therefore, a firefly optimization algorithm has been incorporated with an objective function to search for the optimal threshold value more efficiently and increase the accuracy.

2.3 Multilevel Thresholding Using Firefly Algorithm

Firefly algorithm is inspired by fireflies because of their light-emitting behavior [30]. They produce chemical light form from their lower abdomen. Firefly bioluminescence, which comes in a variety of flashing structures, is used to communicate between two flies, to explore the prey, and discover mates. We make an effort to obtain r numbers of thresholds, $x = \{th_c^1, th_c^2, \ldots, th_c^r\}$, which can be attained by minimizing Eq. (7). The nature of the firefly algorithm is to resolve the maximization problem [30], and the objective function can be modified as the reciprocal of $\lambda(th_c^1, th_c^2, \ldots, th_c^r)$.

$$\varphi(th_c^1, \ldots, th_c^r) = 1/\lambda(th_c^1, \ldots, th_c^r) \tag{8}$$

The proposed method tries to obtain the $\varphi(th_c^1, \ldots, th_c^r)$ which maximize Eq. (8). For simplicity, the characteristic of fireflies was idealized, as (a) All fireflies are unisex and they would be captivated by another firefly irrespective of gender (b) The attractiveness of two fireflies is proportionate to their illumination, and hence lesser bright will move towards the brighter one. If there is not a single firefly that shines brighter than the others. It will start moving randomly. The Euclidian distance can be written as the distance between two fireflies i and j, which is:

$$g_{i,j} = \left\|x_i - x_j\right\| = \sqrt{\sum_{k=1}^{r} (x_{i,k} - x_{j,k})^2} \tag{9}$$

$$\beta \leftarrow \beta_0 e^{-\gamma g_{i,j}} \tag{10}$$

In Eq. (10), β_o and γ is the attractiveness and light absorption coefficient. The motion of the firefly is governed by:

$$x_{i,k} \leftarrow (1 - \beta)x_{i,k} + \beta x_{j,k} + u_{i,k} \tag{11}$$

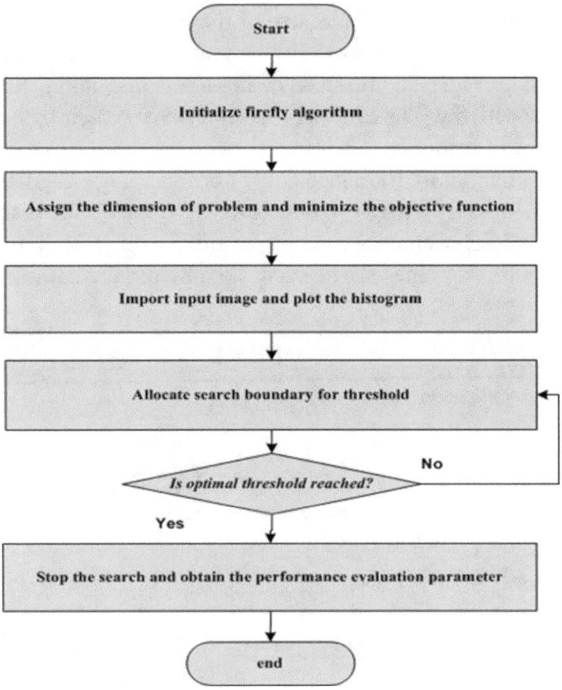

Fig. 1 The flow diagram of the proposed method

$$u_{i,k}^{\max} = \alpha(rand1 - 0.5) \tag{12}$$

If there is not a single firefly brighter than a specific firefly with optimum fitness, it will move freely as per the following equation:

$$x_{i^{\max},k} \leftarrow x_{i^{\max},k} + u_{i^{\max},k} \text{ for k = 1,2,...,r} \tag{13}$$

$$u_{i^{\max},k} = \alpha(rand2 - 0.5) \tag{14}$$

In Eq. (13),$rand1$ and $rand2$ are achieved from the uniform distribution; (c) The landscape of the fitness function $\varphi(x)$ affects the luminance of firefly. The details of the FFA-based recursive minimum cross entropy are summarised in the form of a flow diagram as shown in Fig. 1.

3 Results and Discussion

Ten different complex backgrounds with high-resolution crop images have been tested for the measurement of efficiency and accurateness of the proposed technique as shown in Fig. 2. The size of the image is 2048×1365 with gray level $L = 256$. The parameters used to implement the proposed technique are $\beta_o = 1$,$\alpha = 0.001$ and $\gamma = 1$. The number of iterations and initial firefly was taken as 300 and 20 respectively. The original crop images and their histogram are shown in Fig. 2. It is evident from Fig. 2 that multilevel thresholding of crop images is a challenging task.

Popular performance indicators in image segmentation such as PSNR, MSE, SSIM, and FSIM have been used to compare the results. The accuracy of the algorithm has been measured and justified using PSNR and MSE values. whereas, SSIM and FSIM values are utilized. To measure the similitude of the outcome image:

$$PSNR = 10 \log 10\left(\frac{(255)^2}{MSE}\right) \tag{15}$$

$$MSE = \frac{1}{M \cdot N} \sum_{i=1}^{M} \sum_{i=1}^{N} [I(i, j) - I_s(i, j)]^2 \tag{16}$$

In Eq. (16), I_s and I represents the outcome and original image respectively. The quality of an outcome image has been evaluated using SSIM [26].

$$SSIM(I, I_s) = \frac{(1\mu_I \mu_{I_s} + C_1)(2\sigma_{II_s} + C_2)}{(\mu_I^2 + \mu_{I_s}^2 + C_1)(\sigma_I^2 + \sigma_{I_s}^2 + C_2)} \tag{17}$$

In Eq. (17), $C_1 = (k_1 L)^2$ and $C_2 = (k_2 L)^2.\sigma_{I_s}, \sigma_I, \mu_I, \mu_{I_S}$ and σ_{II_s} represent the mean, standard deviation, and covariance of the outcome image and original crop image respectively. The value of k_1 and k_2 are taken as 0.01 and 0.03 by default. For RGB image, SSIM is defined as:

$$SSIM = \sum_c SSIM(I^c, I_s^c) \tag{18}$$

In Eq. (18), $c = 1$ and $c = 1, 2, 3$ for gray and RGB channel respectively.

FSIM is utilized for the evaluation of features of segmented and original test images [26].

$$FSIM(I, I_s) = \frac{\sum_{X \in \Omega} S_L(X) PC_m(X)}{\sum_{X \in \Omega} PC_m(X)} \tag{19}$$

In Eq. (19), PC_m and $S_L(X)$ are phase congruency and the similarity index. For RGB image FSIM is represented as:

$$FSIM = \sum_c FSIM(I^c, I_s^c) \tag{20}$$

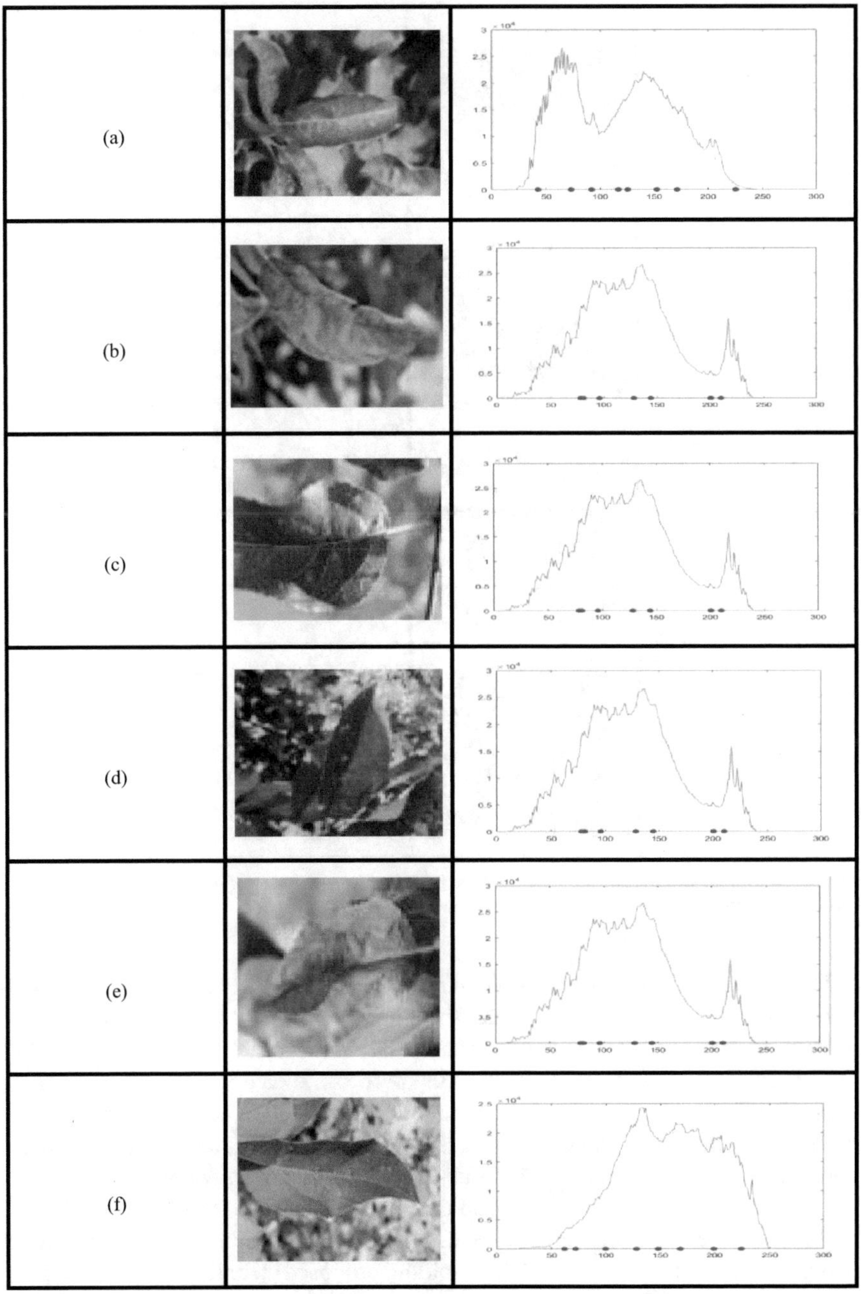

Fig. 2 Original crop images and corresponding histogram presented 8 levels of thresholding

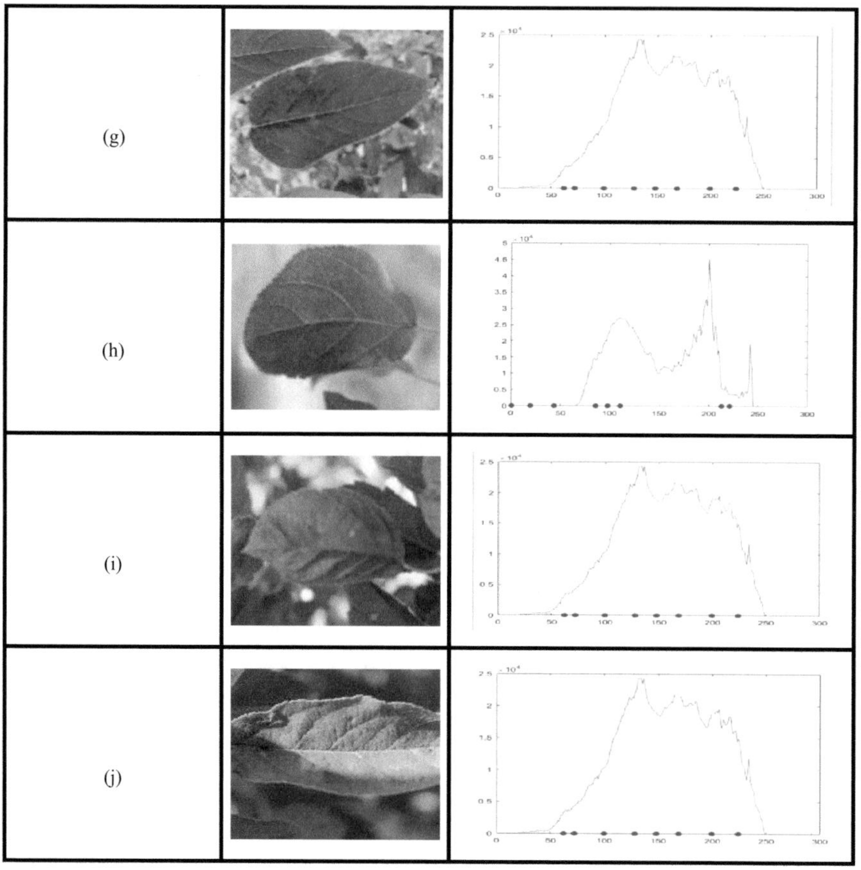

Fig. 2 (continued)

The histogram of the cropped image shown in Fig. 2 evidences that the segmentation is indeed a challenging task. Hence the proposed method automatically segments the crop images into different levels of thresholding (Threshold_L = 2, 5, 8, and 16). The experimental results evidence that the outcome image for the proposed technique visually looks better. In our experiment, 30 trials have been used to avoid the discrepancy. Among all the trails, the best results are selected and the visuals of the crop images are shown in Fig. 3 for a different level of thresholding.

The comparative analysis of the proposed method with WDO has shown in Table 1 for different level thresholding. The results are almost comparable for a lower level of threshold, but for a higher level of thresholding proposed technique is more accurate. The higher value of PSNR and lower value of MSE show the better accuracy of the proposed technique than WDO. SSIM and FSIM are used to measure the similarity between the outcome image and the original crop image. The outcome of the proposed

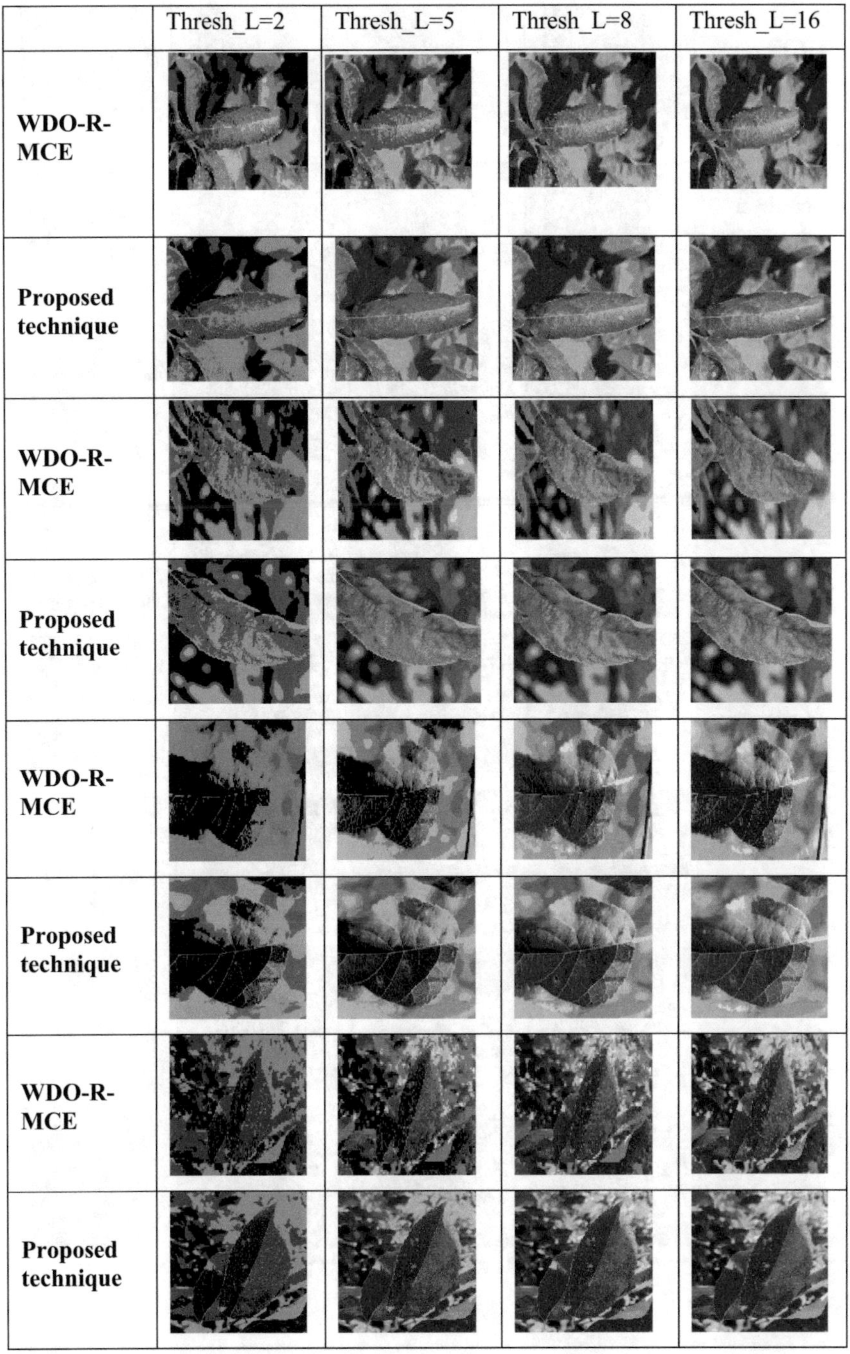

Fig. 3 Segmented image for a different level of thresholding (Th_level = 2, 5, 8, and 16) sequentially from a to j based on multilevel recursive MCE

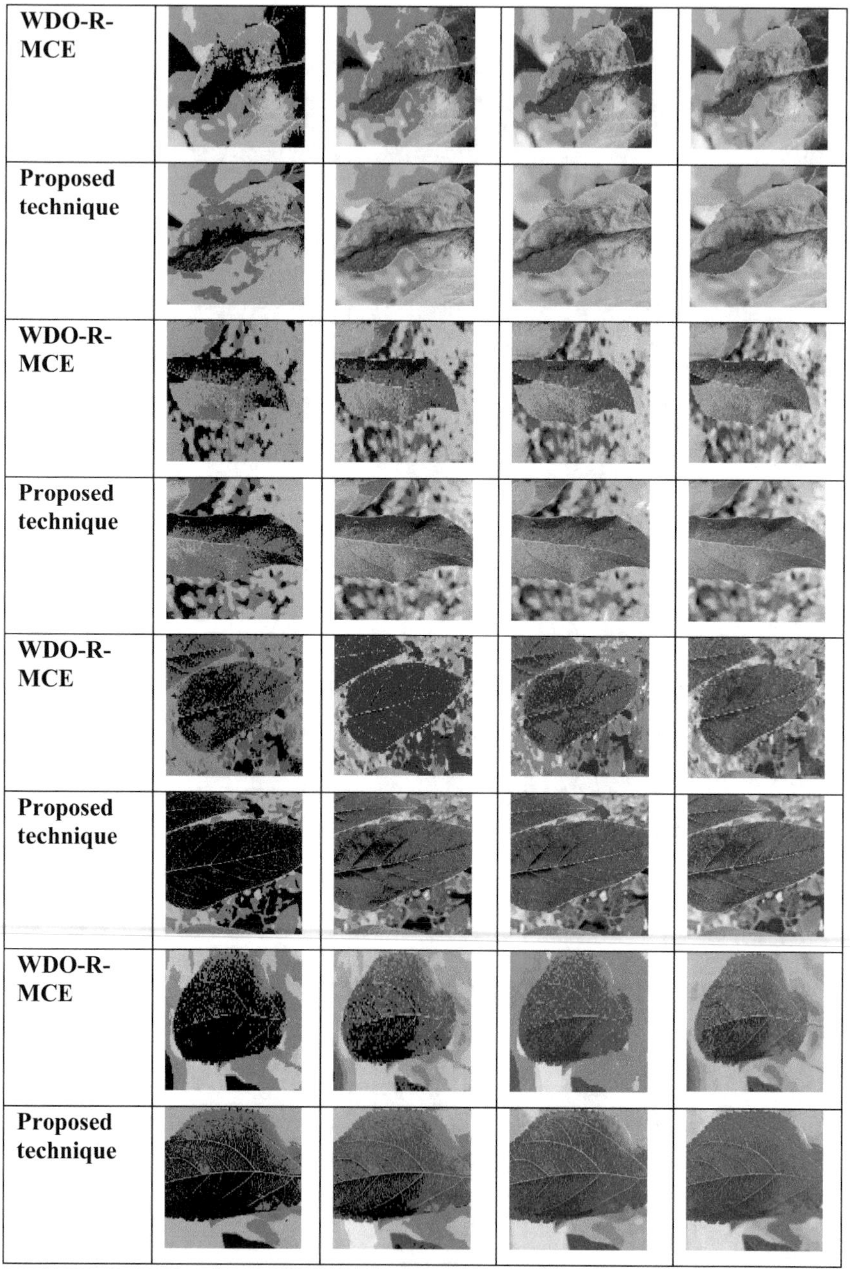

Fig. 3 (continued)

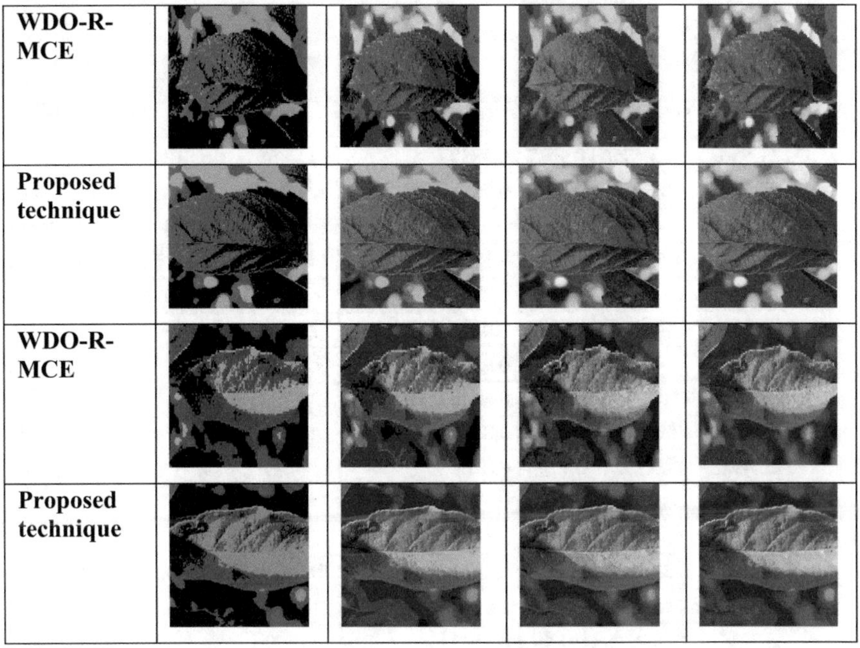

Fig. 3 (continued)

technique shows better results qualitatively and quantitatively as shown in Fig. 3 and Table 1 respectively.

Table 1 Comparison with WDO and proposed method using recursive MCE

Crop images	WDO-R-MCE				Threshold level	Proposed method			
	PSNR	MSE	SSIM	FSIM		PSNR	MSE	SSIM	FSIM
a	14.6523	2235.0012	0.9274	0.7217	2	14.6523	2235.0012	0.9274	0.7217
	19.3751	764.9858	0.9733	0.8319	5	19.3594	754.0665	0.9769	0.8122
	21.2907	517.6289	0.9810	0.882	8	22.5328	365.4070	0.9972	0.8502
	24.0584	309.5862	0.9881	0.9239	16	28.0300	104.5389	0.9142	0.9469
b	13.7601	2742.9771	0.9159	0.6698	2	13.7626	2797.2759	0.9142	0.6706
	19.3944	770.9065	0.9750	0.8103	5	20.6026	568.7240	0.9841	0.8088
	23.1074	334.1826	0.9892	0.8824	8	22.7008	352.7258	0.9916	0.8351
	27.8288	116.0870	0.9961	0.9455	16	24.9957	206.3199	0.9944	0.8723
c	13.9624	2616.9850	0.9233	0.6999	2	13.9093	2648.6268	0.9222	0.7001

(continued)

Table 1 (continued)

Crop images	WDO-R-MCE				Threshold level	Proposed method			
	PSNR	MSE	SSIM	FSIM		PSNR	MSE	SSIM	FSIM
	17.6563	1115.4456	0.9632	0.7889	5	17.9776	1036.8104	0.9669	0.7886
	20.5646	601.1270	0.9794	0.8513	8	20.8212	570.0206	0.9816	0.8347
	24.6453	260.2707	0.9908	0.9209	16	25.1789	199.6000	0.9947	0.9030
d	14.9487	2120.6359	0.9386	0.7130	2	14.9456	2123.3315	0.9386	0.7126
	19.9569	657.9991	0.9803	0.8572	5	20.1751	625.6480	0.9815	0.8623
	22.5712	364.6930	0.9880	0.9147	8	22.2491	387.9639	0.9878	0.9045
	26.0310	170.9247	0.9939	0.9525	16	26.5888	143.7545	0.9957	0.9485
e	13.7988	2718.3446	0.9238	0.6848	2	13.7988	2718.3446	0.9238	0.6848
	18.1834	1001.5439	0.9684	0.8118	5	19.7358	703.0113	0.9803	0.8100
	20.4988	609.3924	0.9795	0.8709	8	21.3437	489.9688	0.9856	0.8582
	23.5942	339.2449	0.9877	0.9273	16	24.5392	233.6570	0.9936	0.8888
f	12.5906	3635.9231	0.8968	0.6438	2	12.5906	3635.9231	0.8968	0.6438
	17.2589	1331.9862	0.9573	0.8022	5	19.0113	848.0393	0.9751	0.7987
	20.4926	675.4244	0.9774	0.8810	8	21.6121	449.9983	0.9886	0.8505
	25.7089	220.4210	0.9924	0.9508	16	27.0621	129.5270	0.9965	0.9327
g	10.7481	5575.9188	0.8335	0.6612	2	10.6843	5662.8597	0.8311	0.6602
	15.115	2174.8743	0.9276	0.7716	5	15.3305	2062.8139	0.9330	0.7390
	16.9345	1591.2679	0.9450	0.8235	8	18.0436	1206.7634	0.9594	0.8224
	20.7563	859.5986	0.9689	0.8871	16	23.3504	371.4321	0.9873	0.9008
h	12.4263	3725.7234	0.8923	0.7025	2	12.3393	3799.5820	0.8901	0.7019
	15.6997	1823.6140	0.9411	0.7684	5	15.9561	1725.6574	0.9478	0.7538
	17.4876	1279.3078	0.9567	0.8061	8	20.1577	638.2958	0.9832	0.8340
	22.9566	518.9836	0.9815	0.8919	16	23.4784	293.3884	0.9943	0.8725
i	12.1497	3972.2098	0.8783	0.6913	2	12.1545	3967.5816	0.8783	0.6910
	17.1928	1307.7845	0.9570	0.8095	5	17.3881	1245.6719	0.9599	0.8040
	20.8278	600.3443	0.9798	0.8666	8	21.2642	486.8999	0.9856	0.8342
	22.2282	478.8115	0.9826	0.8999	16	23.5261	290.3081	0.9913	0.8711
j	13.9598	2707.1156	0.9112	0.7274	2	13.9598	2707.1156	0.9112	0.7274
	17.7376	1128.9107	0.9599	0.7972	5	19.5878	752.8468	0.9741	0.8176
	20.5276	662.4808	0.9755	0.8583	8	20.8762	561.3922	0.9809	0.8367
	23.1149	420.9683	0.9838	0.9015	16	24.6043	231.8320	0.9938	0.8688

4 Conclusion

Crop images have different illuminations and complex backgrounds, and thus an efficient technique is required for multilevel thresholding. In this paper, the recursive minimum cross entropy (R-MCE) has been applied to reduce the complexity of the formulation. The R-MCE technique is combined with the firefly algorithm which searches the optimal threshold value more effectively and accurately than WDO. The result has been demonstrated qualitatively and qualitatively. It is evident from experimental results that FFA yields almost similar results for the lower level of thresholding. For a higher level of thresholding, FFA gives better results qualitatively and quantitively than WDO. The accuracy of the proposed technique has been confirmed using performance parameters such as MSE, PSNR, SSIM, and FSIM. The promising result of the proposed technique inspires us to use such objective functions in different fields of image processing.

References

1. Hemming J, Rath T (2001) PA—precision agriculture: computer-vision-based weed identification under field conditions using controlled lighting. J Agric Eng Res 78:233–243. https://doi.org/10.1006/JAER.2000.0639
2. Huang Y-P, Singh P, Kuo W-L, Chu H-C (2021) A type-2 fuzzy clustering and quantum optimization approach for crops image segmentation. Int J Fuzzy Syst 233(23):615–629. https://doi.org/10.1007/S40815-020-01009-2
3. Guerrero JM, Pajares G, Montalvo M, Romeo J, Guijarro M (2012) Support vector machines for crop/weeds identification in maize fields. Expert Syst Appl 39:11149–11155. https://doi.org/10.1016/J.ESWA.2012.03.040
4. Deenan S, Janakiraman S, Nagachandrabose S (2020) Image segmentation algorithms for banana leaf disease diagnosis. J Inst Eng Ser C 1015(101):807–820. https://doi.org/10.1007/S40032-020-00592-5
5. Pugoy RADL, Mariano VY (2011) Automated rice leaf disease detection using color image analysis 8009:93–99. https://doi.org/10.1117/12.896494
6. Pang J, Bai ZY, Lai JC, Li SK (2011) Automatic segmentation of crop leaf spot disease images by integrating local threshold and seeded region growing. N: Proceedings of 2011 international conference on image analysis and signal processing, IASP 2011, pp 590–594. https://doi.org/10.1109/IASP.2011.6109113
7. Haug S, Ostermann J (2014) A crop/weed field image dataset for the evaluation of computer vision based precision agriculture tasks. Lect Notes Comput Sci (including Subser Lect Notes Artif Intell Lect Notes Bioinf) 8928:105–116. https://doi.org/10.1007/978-3-319-16220-1_8
8. Gao R, Wu H (2015) Agricultural image target segmentation based on fuzzy set. Optik (Stuttg) 126:5320–5324. https://doi.org/10.1016/J.IJLEO.2015.09.006
9. Lu H, Cao Z, Xiao Y, Li Y, Zhu Y (2016) Joint crop and tassel segmentation in the wild. In: Proceedings—2015 Chinese automation congress, CAC 2015, pp 474–479. https://doi.org/10.1109/CAC.2015.7382547
10. Saɀ T, Çunkaş M (2015) Color image segmentation based on multiobjective artificial bee colony optimization. Appl Soft Comput 34:389–401. https://doi.org/10.1016/J.ASOC.2015.05.016
11. Wang C, Shi AY, Wang X, Wu FM, Huang FC, Xu LZ (2014) A novel multi-scale segmentation algorithm for high resolution remote sensing images based on wavelet transform and improved JSEG algorithm. Optik (Stuttg) 125:5588–5595. https://doi.org/10.1016/J.IJLEO.2014.07.002
12. Bhandari AK (2020) A novel beta differential evolution algorithm-based fast multilevel thresholding for color image segmentation. Neural Comput Appl 32:4583–4613. https://doi.org/10.1007/s00521-018-3771-z
13. Pare S, Bhandari AK, Kumar A, Singh GK (2019) Rényi's entropy and bat algorithm based color image multilevel thresholding. Adv Intell Syst Comput 748:71–84. https://doi.org/10.1007/978-981-13-0923-6_7

14. Pare S, Bhandari AK, Kumar A, Bajaj V (2017) Backtracking search algorithm for color image multilevel thresholding. Signal Image Video Process 122(12):385–392.
15. Harnrnouche K, Diaf M, Siarry P (2010) A comparative study of various meta-heuristic techniques applied to the multilevel thresholding problem. Eng Appl Artif Intell 23:676–688. https://doi.org/10.1016/J.ENGAPPAI.2009.09.011
16. Cheng HD, Jiang XH, Sun Y, Wang J (2001) Color image segmentation: advances and prospects. Pattern Recognit 34:2259–2281. https://doi.org/10.1016/S0031-3203(00)00149-7
17. Bhandari AK, Srinivas K, Kumar A (2021) Optimized histogram computation model using cuckoo search for color image contrast distortion. Digit Signal Process 118: 103203. https://doi.org/10.1016/j.dsp.2021.103203
18. Pare S, Kumar A, Singh GK, Bajaj V (2020) Image segmentation using multilevel thresholding: a research review. Iran J Sci Technol Trans Electr Eng 44
19. Mala C, Sridevi M (2016) Multilevel threshold selection for image segmentation using soft computing techniques. Soft Comput 5:1793–1810. https://doi.org/10.1007/S00500-015-1677-6
20. Bhandari AK, Kumar A, Singh GK (2015) Modified artificial bee colony based computationally efficient multilevel thresholding for satellite image segmentation using Kapur's, Otsu and Tsallis functions. Expert Syst Appl 42:1573–1601. https://doi.org/10.1016/j.eswa.2014.09.049
21. Pare S, Kumar A, Bajaj V, Singh,GK (2016) A multilevel color image segmentation technique based on cuckoo search algorithm and energy curve. Appl Soft Comput 47: 76–102. https://doi.org/10.1016/j.asoc.2016.05.040
22. Tellaeche A, Burgos-Artizzu XP, Pajares G, Ribeiro A (2008) A vision-based method for weeds identification through the Bayesian decision theory. Pattern Recognit 41:521–530. https://doi.org/10.1016/J.PATCOG.2007.07.007
23. Nandhini S, Parthasarathy S, Bharadwaj A, Harsha Vardhan K (2021) Analysis on classification and prediction of leaf disease using deep neural network and image segmentation technique. annalsofrscb.ro. 25:9035–9041
24. Pare S, Kumar A, Bajaj V, Singh GK (2017) An efficient method for multilevel color image thresholding using cuckoo search algorithm based on minimum cross entropy. Appl Soft Comput 61:570–592. https://doi.org/10.1016/J.ASOC.2017.08.039
25. Pare S, Bhandari AK, Kumar A, Singh GK, Khare S (2015) Satellite image segmentation based on different objective functions using genetic algorithm: a comparative study. In: IEEE international conference on digital signal processing (DSP), pp 730–734
26. Bhandari AK, Singh VK, Kumar A, Singh GK (2014) Cuckoo search algorithm and wind driven optimization based study of satellite image segmentation for multilevel thresholding using Kapur's entropy. Expert Syst Appl 41:3538–3560. https://doi.org/10.1016/J.ESWA.2013.10.059
27. Kumar A, Kumar A, Vishwakarma A, Lee GKS (2022) Multilevel thresholding for crop image segmentation based on recursive minimum cross entropy using a swarm-based technique. Compt Electron Agric 16:630–649. https://doi.org/10.1049/sil2.12148
28. Kurban T, Civicioglu P, Kurban R, Besdok E (2014) Comparison of evolutionary and swarm based computational techniques for multilevel color image thresholding. Appl Soft Comput 23:128–143. https://doi.org/10.1016/J.ASOC.2014.05.037
29. Lei B, Fan J (2020) Multilevel minimum cross entropy thresholding: a comparative study. Appl Soft Comput 96:106588. https://doi.org/10.1016/J.ASOC.2020.106588
30. Horng MH, Liou RJ (2011) Multilevel minimum cross entropy threshold selection based on the firefly algorithm. Expert Syst Appl 38:14805–14811. https://doi.org/10.1016/J.ESWA.2011.05.069

Deep Learning-Based Pipeline for the Detection of Multiple Ocular Diseases

Ananya Angadi, Aneesh N. Bhat, P. Ankitha, Parul S. Kumar, and Gowri Srinivasa

1 Introduction

Visual impairment is a serious condition that plagues close to 2.2 billion individuals globally [1]. Timely detection and prevention of ocular diseases can help stall or prevent visual impairment in up to half of these cases. However, the world faces considerable challenges and inequity in terms of availability and access to eyecare [2]. The lack of affordable and easily accessible healthcare is the greatest risk factor for blindness. The impact of being blind extends much further than just losing the ability to see, it is a disability which exacerbates poverty and isolation. South Asia currently has the highest percentage of visually impaired people [3]. Thus, a diagnostic system for the pre-screening of ocular diseases or continuous monitoring that is accessible to people is a need of the hour, to identify those at-risk as early as possible.

Computer-aided diagnostic (CAD) systems have greatly contributed to making eyecare more equitable [4, 5]. The benefits are twofold: CAD overcomes the physical distance between patients and ophthalmologists, while simultaneously reducing the burden on doctors by aiding diagnosis and monitoring, allowing them to attend to a larger group of patients [6]. Considerable progress has been made in building diagnostic tools for relatively ocular common diseases like diabetic retinopathy and glaucoma [7–9]. However, tools that detect individual diseases necessitate multi-fold testing for different diseases or, when used in isolation, result in other potential diseases remaining undiagnosed. Thus, there is a need to create a unified system capable of enabling multi-disease detection.

Retinal image analysis is a non-invasive approach to detecting diseases of the eye. Recent advancements in fundus imaging have enabled ophthalmologists to diagnose diseases by visual inspection of retinal images [10]. However, the diagnostic procedure differs from disease to disease. For instance, Diabetic Retinopathy is identified by the presence of exudates, microaneurysms, and hemorrhages [11], whereas Retinitis Pigmentosa is characterized by pigment deposits [12]. Detecting these diseases in

A. Angadi · A. N. Bhat · P. Ankitha · P. S. Kumar · G. Srinivasa (✉)
PES Center for Pattern Recognition, Department of Computer Science and Engineering, PES University, Bengaluru, India
e-mail: gsrinivasa@pes.edu

P. Singh et al. (eds.), *Machine Learning and Computational Intelligence Techniques for Data Engineering*, Lecture Notes in Electrical Engineering 998,
https://doi.org/10.1007/978-981-99-0047-3_69

isolation will require distinct screenings. We aim to build a computer vision model which can enable holistic screening of retinal images which can aid ophthalmologists in the diagnostic process.

Deep learning methods are increasingly being used in medical image analysis due to the improvement in performance over traditional image processing methods [13, 14]. They work particularly well when data availability is high, and require minimal domain expertise [15, 16]. We propose a deep learning-based training pipeline for the detection of seven eye diseases, namely, Diabetic Retinopathy, Media Haze, Drusen, Myopia, Age-Related Macular Degeneration, Optic Disc Edema, and Retinitis. For every one of these diseases, we train three models and compare the relative performance of the models to find the most suitable model for each disease.

2 Exploratory Data Analysis

We have used the Retinal Fundus Multi-Disease Image Dataset (RFMiD) for training and testing our model [17]. Published in November 2020, the data was made available to researchers via the Retinal Image Analysis for Multi-Disease Detection (RIADD) Grand Challenge that was a part of the International Symposium on Biomedical Imaging (ISBI) 2021 [17]. With a training set of 1920 images and an evaluation set of 640 images, this dataset covers several ocular diseases that appear frequently in clinical situations. Images are annotated by ophthalmologists. Two of the sample images depicting a healthy retina and one with Diabetic Retinopathy are shown in Fig. 1. Out of the 1920 images comprising the training dataset, 1519 (79.1%) are classified as at-risk. Within the at-risk samples, the most frequent diseases detected were Diabetic Retinopathy (found in 24.75% of the at-risk patients), Media Haze (20.86%), Optical Disc Coloboma (18.56%), and Tessellated Fundus (12.24%). On the other hand, the diseases with the least occurrence were Shunt (0.3%), Retinitis Pigmentosa (0.3%), Tortuous Vessels (0.3%), Macular Hole (0.7%), and Parafoveal Telangiectasia (0.7%). The occurrence of a few diseases can be found in Table 1.

Since the data associates each image with the occurrence of 28 diseases, the dataset is exceedingly sparse. 93.56% of the values in the dataset were found to be 0. Chi-squared tests to check for correlation between diseases yielded no significant

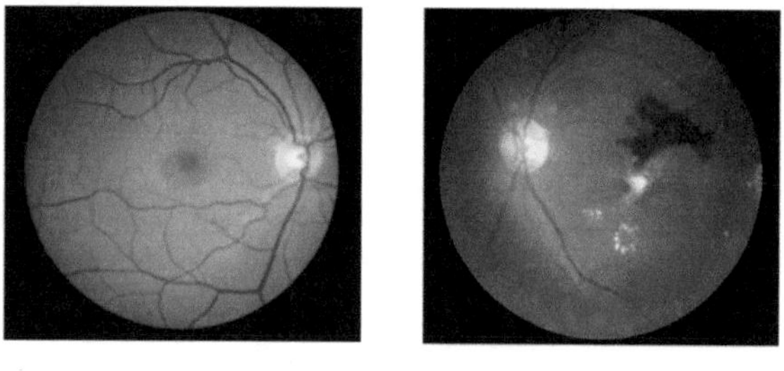

Fig. 1 Image depicting a healthy retina on the left and a retina afflicted by Diabetic Retinopathy on the right

Table 1 Disease Occurrence

Name of the disease	Label	No.of positive samples
Diabetic retinopathy	DR	376
Media Haze	MH	317
Drusen	DN	138
Myopia	MYA	101
Age-related macular degeneration	ARMD	100
Optic disc edema	ODE	58
Retinitis	RS	43

co-occurrence between any pair of diseases. Hence, for all intents and purposes, the diseases can be considered independent with respect to this data. The data also suffers from class imbalance, as can be seen in Fig. 2 The number of at-risk samples (close to 80%) outweighs the number of not-at-risk samples (20%). For individual diseases, the positive samples available for training are very limited in comparison with the availability of negative samples. For example, considering Age-Related Macular Degeneration (ARMD), there are 100 samples (5%) classified as positive while the remaining 1820 samples (95%) are classified as negative. Due to this imbalance, it is difficult to train models that can correctly classify disease occurrence [18]. Thus, there is a need to improve the representation of various classes in the dataset, either by augmenting positive samples or creating a representative, stratified sample from the data.

There are three considerations that point to an imbalance in the data: degree of concept complexity, size of the training set, and the level of imbalance among the classes [19]. Since disease prediction is an image classification task, the degree of

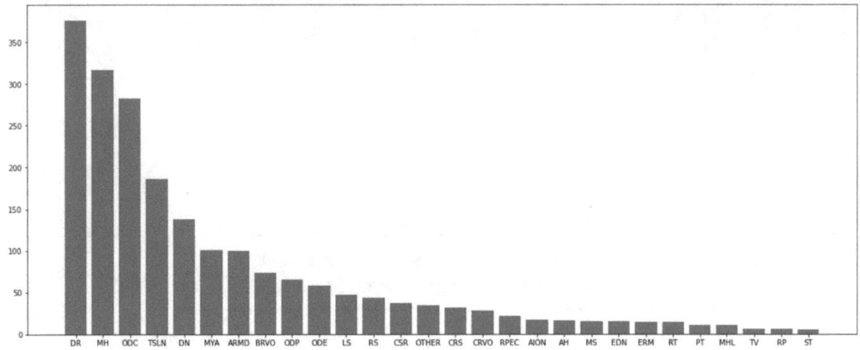

Fig. 2 Sharp decrease in no. of positive samples as we move away from common diseases, indicative of class imbalance

concept complexity is high. The training set is fairly limited in size, and there is a considerable imbalance between the classes (approximately 25:1 negative:positive samples ratio).

3 Proposed Methodology

We have selected 7 diseases for consideration, belonging to the high occurrence, mid-range, and low occurrence categories. This ensures that the results can be extended to other diseases of the same group, while minimizing duplication of effort. This section details the approach used for model building. A schematic diagram of the workflow and components used in the solution is presented in Fig. 3.

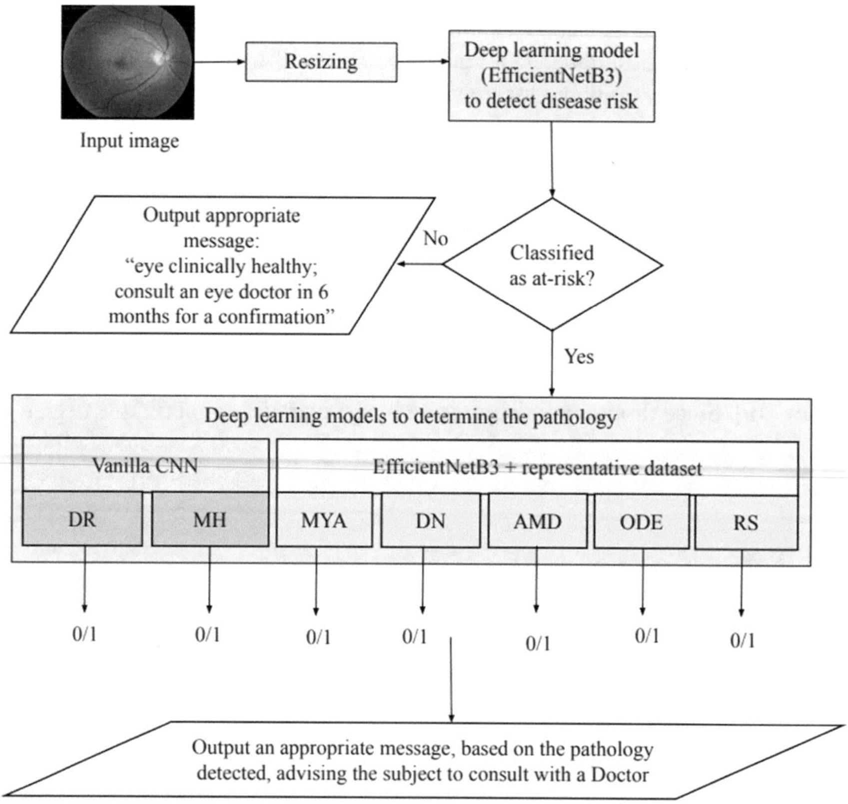

Fig. 3 Sequence of steps in the proposed pipeline

3.1 Preprocessing

A large portion of the original image is constituted by pixels that do not provide any useful information to the model. These background regions were cropped out in order to focus on the fundus. The images were then resized to 32 × 32 to reduce the dimension of the data, while ensuring the features of the disease remain intact. The input size of the network is matched with the resized data. Since the total number of samples in the dataset is limited, real-time data augmentation was performed in order to increase the amount of unique data seen by the model in each epoch of training. This ensures the model has the power to generalize and facilitates better performance [20]. Transformations such as random rotations, flips, and changes in brightness were employed for this purpose.

3.2 Detection of Presence of a Disease

After the pre-processing step, we train a deep learning model, the EfficientNetB3 pre-trained on ImageNet, to detect the presence of a disease as a binary classification problem; the output of this stage results in a tag of 0 for the absence of any diseases and 1 for the presence of a disease (the disease is not yet determined at this stage). EfficientNet has been found to perform better and produce greater accuracy than other pre-trained models such as ResNet and InceptionNet, while also being more lightweight [21, 22]. These features of EfficientNet lend itself to be the natural choice for model building. Within the various EfficientNet architectures, we make use of the B3 model for 2 reasons: (1) B3 was the first variant to support Transfer Learning, and (2) It is lightweight as compared to higher versions [21]. The model is pre-trained on ImageNet and then fine-tuned by subjecting it to an equal number of positive and negative samples from our data. The positive samples comprise a subset of images of various pathologies in the same proportion they occur in the original dataset. The negative samples constitute the images labeled as clinically healthy.

3.3 Training to Detect the Type of Disease

Images that are determined to have one more of the seven diseases under study are then checked for each individual disease. In order to find the best model for each disease, we compared three different Machine Learning models based on various performance parameters.

Convolutional Neural Network (CNN) Our first approach consists of a Vanilla CNN. The original images with the background cropped out are resized to 360 × 360 and input to the CNN, since the neural network applies downsampling as a part of the processing. This network comprises six sets of a convolutional layer, followed by a

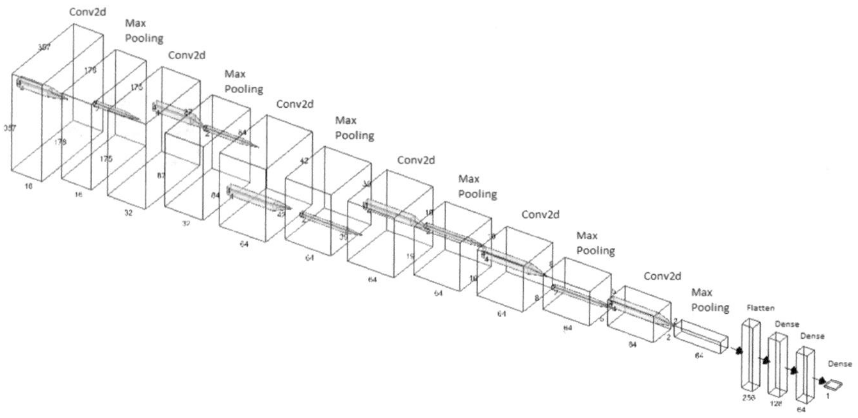

Fig. 4 Vanilla CNN Architecture (Trained from scratch)

pooling layer and a dropout layer for feature extraction. There are two fully connected layers at the end for classification. The last layer uses the sigmoid activation function, so as to produce a value between 0 and 1 as the output. A train-test split of 70:30 has been used, with Adam as the optimizer and binary cross entropy as the loss function. A schematic diagram of the network is presented in Fig. 4. The network is trained for 20 epochs. Real-time data augmentation was performed to ensure diversity in data for training.

Transfer Learning While the CNN is well suited for images with sufficient representation, it tends to overfit when sufficient samples are not presented to this classification model. As a rule of thumb, a minimum of 1000 examples per class are desired in order to train a neural network from scratch. This ensures a robust model free from overfitting. However, with the use of pre-trained models, this number can be reduced significantly [23]. Since the RFMiD data has 73 samples per class on average, we expect transfer learning to perform better than Vanilla CNN. Hence, we employed transfer learning using EfficientNetB3 pre-trained on ImageNet to train the model. Using a pre-trained model avoids overfitting and reduces the time and resources required for training. Three fully connected layers were added to the EfficientNet model in order to fine-tune it to our dataset. Real-time data augmentation was performed for this model as well. The architecture for the same is shown in Fig. 5.

Transfer Learning on a Representative Dataset. From our exploratory data analysis, we conclude that our dataset is imbalanced. The presence of imbalance leads to bias in the models trained over this data. This results in poor prediction performance for the minority class (in our case, disease presence) as compared to the majority class (disease absence) [19]. Some of the methods used to solve class imbalance found in literature, are resampling, boosting, feature selection, etc. [24]. There are two varieties of resampling data, oversampling involves replication of minority class samples

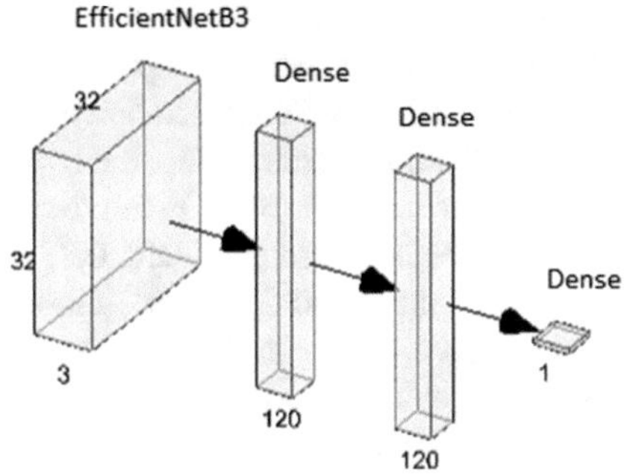

Fig. 5 Transfer Learning Model (EfficientNetB3 pre-trained on ImageNet, with 3 additional layers for fine-tuning)

to match the number of majority class samples, however, this may lead to overfitting. The other method, undersampling, involves removing samples of the majority class therefore it has a risk of loss of valuable information. Synthetic Minority Oversampling Technique (SMOTE) involves selectively oversampling the minority class and is a resource intensive process [24]. Keeping in mind the respective disadvantages of each method we experimented and found stratified random downsampling followed by real-time upsampling to be the most suitable method for solving class imbalance in our dataset. We therefore trained our Efficient Net model on this new representative dataset we created. The following steps were followed to create this dataset'-

- Collect all positive samples for the disease.
- Collect an equal number of negative samples, with each disease appearing in the relative proportion (approximately) in which it appears in the original dataset. All the remaining diseases need not be part of the dataset. We include a subset of diseases in decreasing order of frequency until the requirement is met.

3.4 Evaluation

Once the model is trained, we determine its accuracy, sensitivity, specificity, and F1-score with respect to the test data set, as a measure of its performance.

4 Experimental Results

Each of the above-described models was trained for 7 diseases, namely, Diabetic Retinopathy (DR), Media Haze (MH), Drusen (DN), Myopia (MYA), Age-Related Macular Degeneration (ARMD), Optic Disc Edema (ODE), and Retinitis (RS), as well as for disease risk. 70% of the training data was used for training, while the remaining 30% was used for validation. For every disease, the accuracy, specificity, sensitivity, and F1-score produced by the 3 models on the validation data were compared. The results are summarized in Table 2, with the green font emphasizing the best result obtained after comparison between the results from all 3 of the models, i.e., CNN, EfficientNet-Based Transfer Learning, and EfficientNet-Based Transfer Learning with representative dataset. From the results, we see that for DR and MH, both high occurrence (300+ positive samples) diseases, the Vanilla CNN model produced the best accuracies {DR: 0.78, MH: 0.73} and F1-scores {DR: 0.75, MH: 0.61} as compared to the other 2 models. For mid-range diseases (100-300 positive samples) including DN, ARMD, and MYA, the EfficientNet model trained on a representative dataset produced better F1-scores {DN: 0.53, MYA: 0.83, ARMD: 0.71} compared to the other 2 models. The same pattern was observed for low occurrence diseases, namely ODE and RS, the EfficientNet with representative dataset model gave the best accuracies {ODE: 0.71, RS: 0.88} and F1-scores {ODE: 0.62, RS: 0.88} as compared to the other 2 models. For disease risk, of the 3 models, the EfficientNet model using the complete dataset was found to work best, with an accuracy of 0.77 and an F1-score of 0.77. Figure 6 shows an example of results obtained from the Vanilla CNN model for Diabetic Retinopathy.

Post selecting the best-performing model for each disease, we evaluated the overall model on the validation dataset provided by the challenge organizers. The dataset consisted of 640 images, of which we have used 185 images to test our model, so as to achieve a 90:10 train:test ratio. This test subset was created by preserving the relative proportion of positive samples in the validation dataset, thus keeping the test dataset representative. Each image was first checked for disease risk. If classified as at-risk, it was checked for each of the 7 diseases under consideration. Table 3 summarizes the results obtained.

5 Discussion

The success of the Vanilla CNN method with respect to high occurrence diseases (Diabetic Retinopathy and Media Haze) can be attributed to higher amounts of data available for training. There are approximately 4 negative samples for each positive sample of DR in the dataset. This allows the model to have high prediction accuracy for both positive and negative classes. However, as the number of positive samples decreases and the level of class imbalance increases, as observed in the cases of low occurrence diseases, the Vanilla CNN model fails to deliver the same level of performance. Since the model sees very few images belonging to the positive class during training (ARMD: 18 negative samples for each positive sample), the model produces good accuracy, but very poor recall.

We expect that the EfficientNet model will work better than Vanilla CNN for such diseases since transfer learning is known to produce better results when data available for training is limited. However, the results show that it does not produce a significant improvement in performance. While the EfficientNet model works well for disease risk prediction, the same effect is not observed in other cases. This is because the problem of class imbalance still persists in the data. When we train the

Table 2 Comparison of ML models

Disease type	CNN				EfficientNet -based transfer learning				EfficientNet-based transfer learning with representative dataset			
	Acc	Sens	Spec	F1	Acc	Sens	Spec	F1	Acc	Sens	Spec	F1
Disease Risk	0.79	1	0	0	0.77	0.76	0.79	0.77	0.73	0.5	0.96	0.66
DR	0.78	0.7	0.8	0.75	0.85	0.4	0.95	0.6	0.6	0.87	0.3	0.45
MH	0.73	0.5	0.78	0.61	0.85	0.12	0.99	0.22	0.63	0.34	1	0.30
DN	0.93	0	1	0	0.93	0	1	0	0.54	0.47	0.61	0.53
MYA	0.94	0	1	0	0.97	0.43	0.99	0.6	0.86	1	0.72	0.83
ARMD	0.95	0	1	0	0.92	0.4	0.96	0.56	0.78	0.96	0.55	0.71
ODE	0.97	0	1	0	0.97	0	1	0	0.71	0.46	0.95	0.62
RS	0.98	0	1	0	0.98	0	1	0	0.88	0.85	0.92	0.88

Table 3 Results on evaluation set: expected versus predicted

	DR	MH	DN	MYA	ARMD	ODE	RS	Not At Risk
DR	0.57	0	0.06	0.04	0.04	0	0.02	0.27
MH	0.125	0	0.25	0.16	0.09	0.09	0.16	0.125
DN	0	0	0.31	0.15	0	0	0	0.54
MYA	0	0	0.06	0.64	0.06	0	0.06	0.12
ARMD	0	0	0	0	0.77	0	0	0.23
ODE	0.17	0	0	0	0.17	0.33	0	0.33
RS	0	0	0	0	0	0	0.75	0.25
Not At Risk	0	0	0	0	0	0	0	1

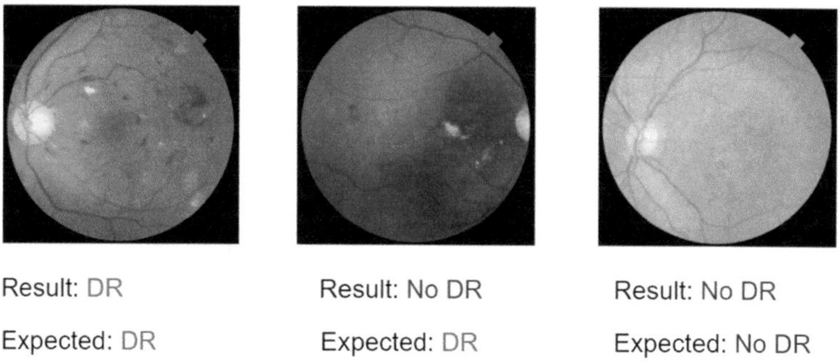

Fig. 6 Example results of Vanilla CNN model for Diabetic Retinopathy

EfficientNet model on a representative dataset to reduce the class imbalance, we see a marked improvement in results. The model produces good prediction accuracy for both positive and negative classes, on account of balanced data (approximately 1:1 negative: positive samples).

6 Reproducible Research

In the spirit of reproducible research, we have made the entire code and models used to obtain the results presented in this paper available in Reproducible Research Repository online.

7 Conclusion

We have thus built a deep learning-based pipeline for detecting multiple ocular diseases, with a special focus on class imbalance in small datasets. Transfer learning was seen to surpass traditional deep learning methods, particularly in diseases with very few positive training samples (< 50). Stratified random downsampling brought about a marked increase in both recall and F1-score. A fresh image input to the system will be first checked for disease risk, followed by each of the diseases. The methodology used can be extended to more rare as well as common diseases such as Central Retinal Vein Occlusion (CRVO), Optic Disc Cupping (ODC), Anterior Ischemic Optic Neuropathy (AION), etc.

References

1. World Health Organization (2019) World report on vision
2. Trimmel J (2016) Inequality and inequity in eye health. Community Eye Health 29(93):1
3. Murthy GV (2017) Eye care in South Asia, 1988–2018: developments, achievements and future challenges. Community Eye Health 30(100):99
4. Doi K (2007) Computer-aided diagnosis in medical imaging: historical review, current status and future potential. Comput Med Imaging Graph 31(4–5):198–211
5. Maksoud EAA, Ramadan M, Barakat S, Elmogy M (2019) A computer-aided diagnoses system for detecting multiple ocular diseases using color retinal fundus images. In: Machine learning in bio-signal analysis and diagnostic imaging, pp 19–52. Academic Press
6. Zhang Z, Srivastava R, Liu H, Chen X, Duan L, Wong DWK, Kwoh CK, Wong TY, Liu J (2014) A survey on computer aided diagnosis for ocular diseases. BMC Med Inform Decis Mak 14(1):1–29
7. Alyoubi WL, Shalash WM, Abulkhair MF (2020) Diabetic retinopathy detection through deep learning techniques: a review. Inform Med Unlocked 20:100377
8. Mookiah MRK, Acharya UR, Chua CK, Lim CM, Ng EYK, Laude A (2013) Computer-aided diagnosis of diabetic retinopathy: a review. Comput Biol Med 43(12):2136–2155
9. Gnanaselvi JA, Kalavathy GM (2021) Detecting disorders in retinal images using machine learning techniques. J Ambient Intell Humanized Comput 12(5):4593–4602
10. Patton N, Aslam TM, MacGillivray T, Deary IJ, Dhillon B, Eikelboom RH, Yogesan K, Constable IJ (2006) Retinal image analysis: concepts, applications and potential. Prog Retin Eye Res 25(1):99–127
11. Pratt H, Coenen F, Broadbent DM, Harding SP, Zheng Y (2016) Convolutional neural networks for diabetic retinopathy. Procedia Comput Sci 90:200–205
12. Brancati N, Frucci M, Gragnaniello D, Riccio D, Di Iorio V, Di Perna L (2018) Automatic segmentation of pigment deposits in retinal fundus images of Retinitis Pigmentosa. Comput Med Imaging Graph 66:73–81
13. Xu K, Feng D, Mi H (2017) Deep convolutional neural network-based early automated detection of diabetic retinopathy using fundus image. Molecules 22(12):2054
14. Abràmoff MD, Garvin MK, Sonka M (2010) Retinal imaging and image analysis. IEEE Rev Biomed Eng 3:169–208
15. Hussain M, Bird JJ, Faria DR (2018) A study on CNN transfer learning for image classification. In: UK workshop on computational intelligence, pp 191–202. Springer, Cham
16. Litjens G, Kooi T, Bejnordi BE, Setio AAA, Ciompi F, Ghafoorian M, Van Der Laak JA, Van Ginneken B, Sánchez CI (2017) A survey on deep learning in medical image analysis. Med Image Anal 42:60–88

17. Pachade S, Porwal P, Thulkar D, Kokare M, Deshmukh G, Sahasrabuddhe V, Giancardo L, Quellec G, Mériaudeau F (2021) Retinal fundus multi-disease image dataset (RFMiD): a dataset for multi-disease detection research. Data 6(2):14
18. Krawczyk B (2016) Learning from imbalanced data: open challenges and future directions. Prog Artif Intell 5(4):221–232
19. Japkowicz N, Stephen S (2002) The class imbalance problem: a systematic study. Intell Data Anal 6(5):429–449
20. Chlap P, Min H, Vandenberg N, Dowling J, Holloway L, Haworth A (2021) A review of medical image data augmentation techniques for deep learning applications. J Med Imaging Radiat Oncol
21. Tan M, Le Q (2019) Efficientnet: rethinking model scaling for convolutional neural networks. In: International conference on machine learning, pp 6105–6114. PMLR
22. Wang J, Yang L, Huo Z, He W, Luo J (2020) Multi-label classification of fundus images with efficientnet. IEEE Access 8:212499–212508
23. Weiss K, Khoshgoftaar TM, Wang D (2016) A survey of transfer learning. J Big Data 3(1):1–40
24. Zheng W, Jin M (2020) The effects of class imbalance and training data size on classifier learning: an empirical study. SN Comput Sci 1(2):1–13

Development of a Short Term Solar Power Forecaster Using Artificial Neural Network and Particle Swarm Optimization Techniques (ANN-PSO)

Temitope M. Adeyemi-Kayode, Hope E. Orovwode, Chibuzor T. Williams, Anthony U. Adoghe, Virendra Singh Chouhan, and Sanjay Misra

1 Introduction

Energy is vital to our society today, as it ensures our quality of life and supports our economy. It is the most often used item nowadays. It is vital to our socio-economic progress. Today, we witness the detrimental consequences of relying on fossil fuels for energy. Fossil fuel combustion produces toxins that damage the environment and public health. We utilize energy in many ways every day. With the rising expense of fossil fuels and the diminishing availability of fossil fuels, the globe has begun to harness renewable energy [1].

Renewable energy is derived from inexhaustible natural resources. Renewable energy technologies enable us to harness clean energy from the sun, wind, biomass, and other sources. It may also rapidly exceed global energy consumption. Almost every location has an exploitable amount of renewable energy resources [2].

T. M. Adeyemi-Kayode · H. E. Orovwode · C. T. Williams · A. U. Adoghe
Covenant University, Ota, Nigeria
e-mail: mercy.john@covenantuniversity.edu.ng

H. E. Orovwode
e-mail: hope.orovwode@covenantuniversity.edu.ng

C. T. Williams
e-mail: chibuzor.williams@stu.cu.edu.ng

A. U. Adoghe
e-mail: anthony.adoghe@covenantuniversity.edu.ng

V. S. Chouhan (✉)
Manipal University Jaipur, Jaipur, India
e-mail: darbarvsingh@yahoo.com

S. Misra
Ostfold University College, Halden, Norway
e-mail: sanjay.misra@hiof.no

P. Singh et al. (eds.), *Machine Learning and Computational Intelligence Techniques for Data Engineering*, Lecture Notes in Electrical Engineering 998,
https://doi.org/10.1007/978-981-99-0047-3_70

This paper will use MATLAB to predict solar energy characteristics in Nigeria. In addition to understanding the supply side of load balancing, electricity suppliers and grid operators have studied the relative cost of non-renewable energy production. Also, the implications of successfully incorporating renewable energy sources into the grid must be investigated. As forecasting improves, solar will be better positioned to expand and integrate into the global energy mix [3].

Finding techniques and technology for accurate solar forecasting will be crucial in the future of solar energy. Inaccurate solar predictions and undeveloped technology may be expensive. While solar forecasting is a newer technique, researchers are working on new methodologies to predict solar irradiance. It allows grid management to anticipate and balance energy output and demand. A suitable solar forecasting methodology allows grid operators to deploy their controlled units more effectively.

Nigeria's solar potentials are enormous. Hypothetically, it would be possible to generate 1850×10^3 GWh of solar electricity per year if just 1% of Nigeria's landmass is covered with solar modules [4–6]. With a solar intensity ranging between 3.5 and 7.0 kW/m/day, this resource could potentially increase the current energy consumption of Nigeria over a hundred times [7, 8]. According to the Nigerian Bulk Electricity Trading Company (NBET), as of the 10th of November 2015, there were around 8 Power Purchase Agreements (PPAs) that had indicated interests in developing Solar power projects [9].

2 Methodology

2.1 Data Collection

Data were obtained from (Solcast, 2021) for the different power distribution companies areas (DISCOs) chosen for this study. This data includes parameters such as air temperature, dew point, global horizontal irradiance (GHI), direct normal irradiance (DNI), diffuse horizontal irradiance (DHI), relative humidity, surface pressure, azimuth, precipitable water, and wind speed.

The data dates from 30th December 2014 to 31st December 2020. In order to correspond with the load demand data earlier collected, the author ensured that the solar and wind data derived from Solcast was fully representative of the DISCOs Area Zones earlier stated. These locations are listed in Table 1.

2.2 Nigerian Solar Data

When sunlight passes through the atmosphere, it interacts with atmospheric molecules, resulting in absorption and scattering. The quantity of radiation scattered in the atmosphere that hits a particular place on the ground is known as diffuse

Table 1 Location and coordinates for the DISCO locations

Locations	Coordinates
Abuja	8.8941° N, 7.186° E
Benin	6.335° N, 5.6037° E
Eko	6.4549° N, 3.4246° E
Enugu	6.3849° N, 7.5139° E
Ibadan	8.9669° N, 4.3874° E
Ikeja	6.6018° N, 3.3515° E
Jos	9.2182° N, 9.5179° E
Kaduna	12.1222° N, 6.2236° E
Kano	12.0022° N,8.592° E
Port-Harcourt	4.8396° N, 6.9112° E
Yola	9.3265° N, 12.3984° E

horizontal irradiance (DHI). The number of molecules, contaminants, and clouds in the sky can alter the quantity of diffuse irradiance on any particular day. As cloud cover rises, diffuse irradiance becomes a higher proportion of total irradiance. Global horizontal irradiance (GHI) is radiation measurement on a horizontal surface. Direct normal irradiance (DNI) refers to the direct sunbeam that passes through the atmosphere [10].

In this study, GHI would be considered the forecaster's output. The average GHI value across all the DISCO locations in 2017–2020 is given in Table 2. As seen in Table 2, the northern regions have the highest GHI values. The values are also highlighted in the GHI map in Fig. 1.

The average monthly accumulated irradiances (GHI) from 2017 to 2020 calculated (W/m^2/month) at the eleven locations (DISCOs) are summarized in Table 2. March records the highest GHI values across several DISCO states like Eko, Ikeja, Ibadan,

Table 2 Average global horizontal irradiance (W/m^2) 2017–2020

Locations	GHI values
Abuja	219.59
Benin	195.65
Eko	194.73
Enugu	207.87
Ibadan	221.58
Ikeja	194.73
Jos	233.94
Kaduna	241.37
Kano	246.68
Port-Harcourt	180.56
Yola	238.54

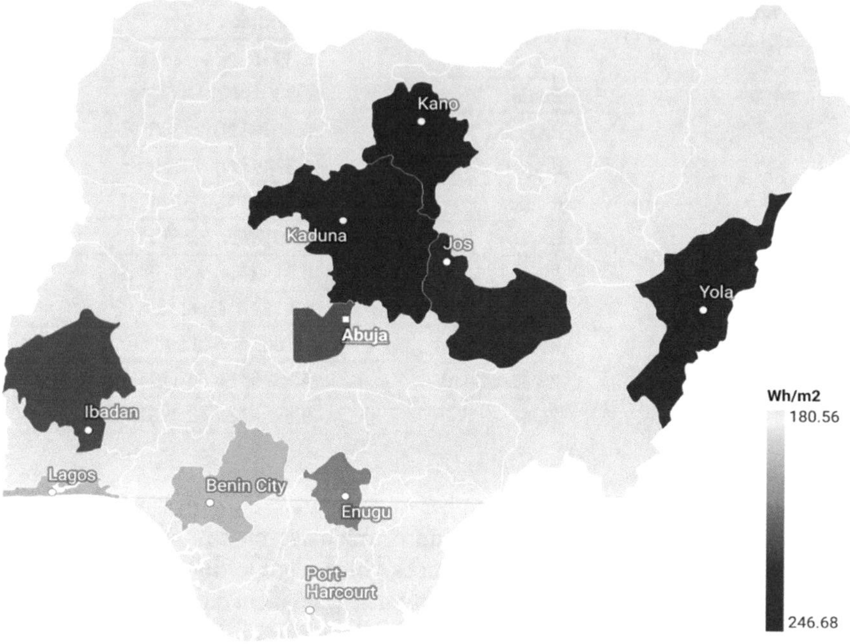

Fig. 1 Map showing GHI values for DISCOs locations in Nigeria

Jos, Yola, Kano, Kaduna, and Abuja from the data. The lowest GHI record for most DISCO is in August as seen in Benin, Enugu, Eko, Ikeja, Ibadan, Jos, Yola, and Abuja. Historically, these months correspond to the hottest and wettest months in Nigeria [11]. The lowest and highest GHI values for each DISCO are highlighted in Appendix 1.

2.3 *Solar Forecasting Using Artificial Neural Networks and Particle Swarm Optimization*

In various applications, the combination of neural networks with PSO has proven to be effective. This innovative method combines the Adaline weight vector updating rule that Widrow–Hoff introduced with the PSO updating rules for adaptive harmonic estimation. To improve its performance, the time factor $\text{iter}_{\max} - \text{iter}/\text{iter}_{\max}$ is paired with Adaline's updating rule [12].

The neural network is used to train the newly constructed updating rule. The suggested ANN–PSO technique for harmonic estimation is depicted in Fig. 2 as a block diagram.

Seven steps were taken to train an ANN using PSO: data collection, network creation, network configuration, setting weights and biases to their initial values,

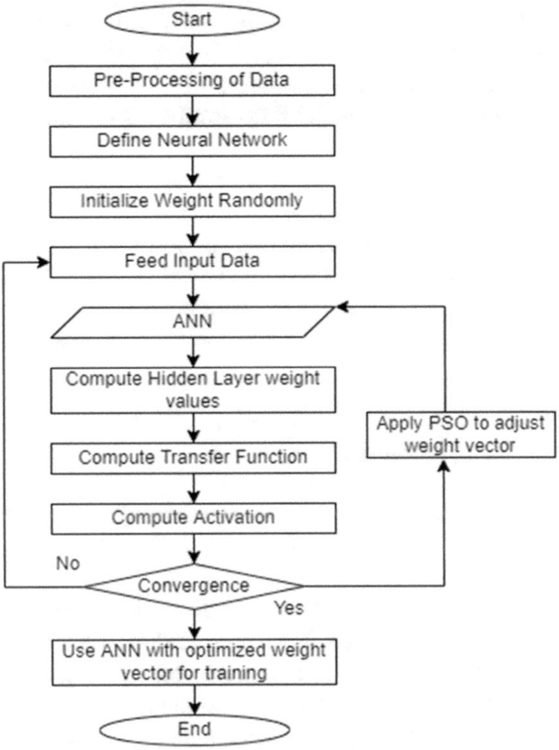

Fig. 2 Flowchart of neural network training using PSO

training the network with PSO, confirming network validity, and using the developed network.

The classical PSO version is used for this study. To initiate the PSO, the population value, the tolerance value used in this work are defined, and the maximum iteration value will be varied later in work to show the effect of a higher or lower number of iterations. The inertia weight constant (w) in this code is given in Eq. 1

$$w = 0.1 + r * 0.4 \tag{1}$$

The value of w could vary depending on the desired output result. The personal best (pbest) and global best (gbest) are also initialized.

Each particle's fundamental behavior gravitates toward two attractors: the particle's best position and the particle's neighbors' best position. The best location is the pbest, while the best location inside a neighborhood is the neighborhood best (nbest) [13]. The velocity, $\boldsymbol{Vi}$ of the i-th particle, is computed using Eq. 2

$$\boldsymbol{Vi}(t+1) = w\boldsymbol{V}_i(t) + c_1\boldsymbol{r}_1(t)\big(\boldsymbol{y}_i(t) - \boldsymbol{x}_i(t)\big) + c_2\boldsymbol{r}_2(t)\big(\hat{\boldsymbol{y}} - x_i(t)\big) \tag{2}$$

where w is the inertia weight, C_1 and C_2 are the acceleration coefficients, $\boldsymbol{r}_1(t), \boldsymbol{r}_2(t) \sim U(0, 1))^{n_x}$, n_x is the dimension of the search space, $x_i(t)$ is the current position of the i-th particle, $\boldsymbol{y}_i(t)$ is the particle's pbest position, and $\widehat{\boldsymbol{y}}(t)$ is the gbest position. Particle positions are updated using Eq. 3

$$\boldsymbol{x}_i(t+1) = \boldsymbol{x}_1(t) + \boldsymbol{v}_i(t) \tag{3}$$

Also, brief experimentation for best tuning of the cognitive acceleration coefficients C_1 and C_2 was performed by altering the C_1 and C_2 until the best coefficient is derived. The acceleration coefficients control the movement of particles. C_1 and C_2 are the cognitive and social components, respectively.

3 Results and Discussion

3.1 *Results Showing Average GHI Across the Year*

From the results of GHI gotten from the 11 selected locations (DISCOs) across Nigeria, it is observed that Kano, Kaduna, and Yola have the highest accumulated GHI value. This means that more solar energy can be generated from these locations in the same period compared to the other 8 locations chosen for this research. The results show that the accumulated average monthly irradiance measured in these locations was highest in March, April, and November across all the locations over four years (2017–2020). Kano has the highest average GHI annually of 246.69 kWh/m^2/year, Kaduna with 241.1 kWh/m^2/year, and Yola with 238.5569 kWh/m^2/year to make up the top three locations with the highest GHI out of the 11 selected locations. The lowest accumulated average GHI annually was Port-Harcourt, with 180.56 kWh/m^2/year.

3.2 *Results Showing GHI Change Across Seasons*

From these results, the highest average irradiance is from the northern part of the country, and this is because the rainy season in the south lasts longer than in the north (March–September) while the north lasts from June to September, and this means more solar energy can be generated from the northern locations. The least GHI accumulated from the southern locations chosen is usually between June and August. Using Ikeja as a case study, the average sum of GHI accumulated from 2017 to 2020 is 827.92 kWh/m^2/year in March. In June, it is 669.209 kWh/m^2/year, in July 645.75 kWh/m^2/year, and in August 724.29 kWh/m^2/year, the drastic drop in the GHI accumulated is as a result of the beginning of the rainy season in that part of the country. While in the northern parts, the least average GHI accumulated is usually from July to early September, using Yola as a case study, in May the average sum of GHI accumulated from 2017 to 2020 is 965.17 kWh/m^2/year, in July 870.66 kWh/m^2/year, in August 845.24 kWh/m^2/year. It starts increasing again in September 900.89 kWh/m^2/year.

3.3 Effect of Climate Change on Global Horizontal Irradiance (GHI)

It is observed that in the locations chosen, the average GHI increased across the years (2017–2019) in most months and decreased in 2020. These changes can be attributed to the climate change that has been exponentially increasing in the world for some time now. It is also observed that 2019 accumulated the most GHI across the 11 locations, using Kaduna as a case study. In 2017 it accumulated an average GHI of 240.98 kWh/m^2/year. In 2018 an average GHI of 240.23 kWh/m^2/year, in 2019, an average GHI 248.03 kWh/m^2/year, and in 2020, an average GHI of 236.26 kWh/m^2/year.

Figures 3 and 4 compare the changing regression patterns for the dataset chosen to be trained and tested by the ANN-PSO algorithm. These changing patterns result from varying the cognitive acceleration coefficients (C_1 and C_2) and the number of iterations during training. Error calculations and other comparative analyses were also performed in this study. Figure 3 is the result when the cognitive acceleration coefficient $C_1 = 1.5$ and $C_2 = 2.5$ while in Fig. 4, $C_1 = 1.0$ and $C_2 = 2.5$.

From the results in Figs. 3 and 4, it was observed through experimentation that by varying the C_1, C_2, and the number of iterations, the best value for C_1 was 1.0, C_2 was 2.5, and the best value for the number of maximum iterations to achieve the best regression coefficient (R) was 1200.

The nRMSE, MSE, MAPE, and R comparing the forecasted and actual 2017–2020 raw dataset for each array has been analyzed using the ANN-PSO algorithm for the simulation result.

Tables 3 and 4 show the result of the MAPE, MSE, nRMSE, and R values for both variations of C_1 and C_2. Columns with 1 represent when $C_1 = 1.0$, $C_2 = 2.5$ and max iteration = 1000 while those with 2 represent when $C_1 = 1.5$, $C_2 = 2.5$ and max iteration = 1200.

From Ref. [14], the model considers satellite and weather-based measurements and employs an intensive structure for neural networks that can be generalized across locations derived. The model produced an nRMSE value of 31.21%. Also, from Ref. [15], a zoning scenario validated a framework that enables the models to predict solar radiation accurately in areas where such data are not accessible. For all situations tested, the results demonstrate good generalization abilities with NRMSE ranging from 7.59 to 12.49% as the best performance for solar irradiance projections, respectively, for 1–6 h ahead. In this study, the nRMSE ranged from 0.7813 to 13.8948%.

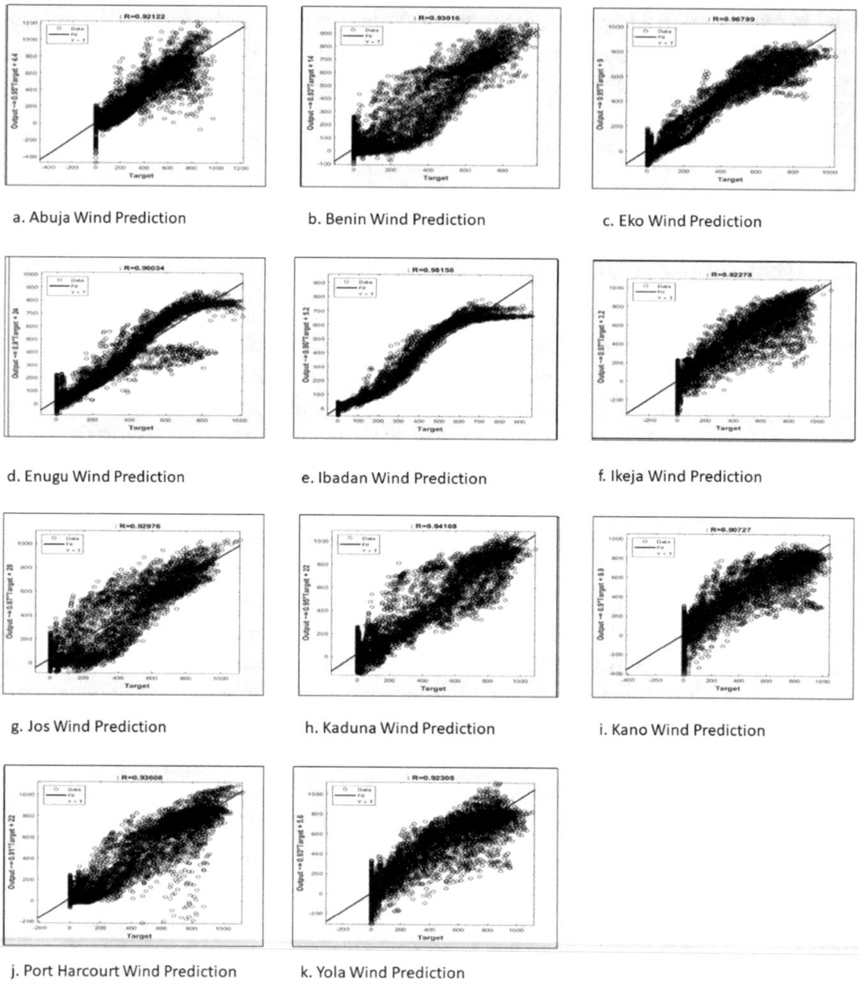

Fig. 3 Regression coefficient for DISCOs when $C_1 = 1.5$ and $C_2 = 2.5$

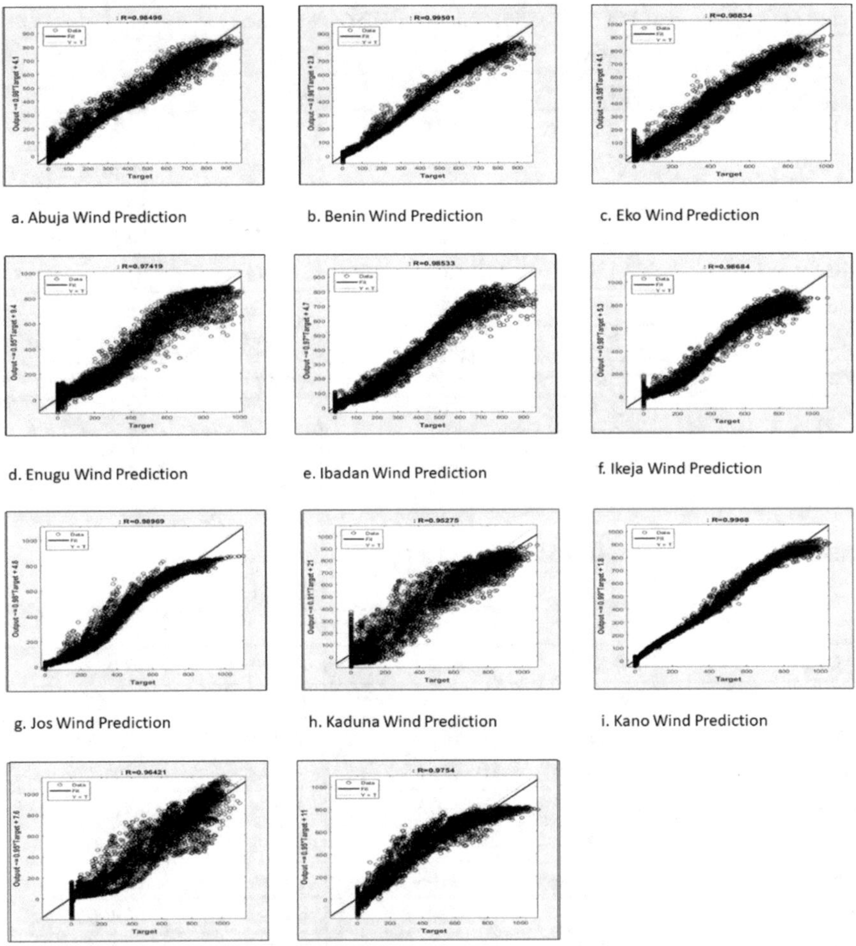

a. Abuja Wind Prediction
b. Benin Wind Prediction
c. Eko Wind Prediction
d. Enugu Wind Prediction
e. Ibadan Wind Prediction
f. Ikeja Wind Prediction
g. Jos Wind Prediction
h. Kaduna Wind Prediction
i. Kano Wind Prediction
j. Port Harcourt Wind Prediction
k. Yola Wind Prediction

Fig. 4 Regression coefficient for DISCOs when $C_1 = 1.0$ and $C_2 = 2.5$

Table 3 Comparison of MAPE and MSE values for DISCO locations

	MAPE		MSE	
	1	2	1	2
Abuja	2.21	6.85	2.29	1.45
Benin	7.48	4.04	713.44	9.88
Eko	2.50	2.65	3.91	6.02
Enugu	1.20	1.03	1.78	4.84
Ibadan	1.03	4.02	1.80	1.19
Ikeja	4.40	5.98	1.83	2.30
Jos	5.67	5.34	9.15	1.20
Kaduna	3.80	3.65	5.03	1.62
Kano	1.70	5.96	1.33	7.53
Port-Harcourt	6.39	5.98	1.90	1.04
Yola	3.07	3.59	6.41	1.87

Table 4 Comparison of R and nRMSE values for DISCO locations

	R		nRMSE	
	1	2	1	2
Abuja	0.98684	0.92278	2.7675	15.9551
Benin	0.99501	0.93016	0.9476	11.4606
Eko	0.97419	0.96034	7.8919	5.0487
Enugu	0.98834	0.96789	2.2142	6.2012
Ibadan	0.98969	0.92976	2.361	13.0901
Ikeja	0.99533	0.98158	2.4522	3.3391
Jos	0.95275	0.94108	11.2862	12.6755
Kaduna	0.9754	0.92305	6.8495	16.4899
Kano	0.96421	0.933608	13.8948	7.4029
Port-Harcourt	0.98496	0.92122	2.485	9.7309
Yola	0.9968	0.90727	0.7813	439.067

Reference [16] estimated solar potential using a neural network and geographic and meteorological data as inputs (latitude, longitude, altitude, month, mean sunshine duration, and mean temperature) and got an R-value of 99.89%. Reference [17] used MLP to forecast GHI and obtained a value of 0.99; similarly, Ref. [18] obtained a range of 0.981–0.998 by using a model comprising of Particle Swarm Optimization, Genetic Algorithm and Artificial Neural Network (PSO-GA-ANN). This model is similar to this study's model, which produced its best R-value of 0.9968. In Ref. [19], the study proposed a Long Short Term Memory (LSTM) algorithm and obtained a value of 0.2478 MAPE and 6.7207 RMSE for the solar data analyzed. This study obtained a MAPE of 3.07% in Yola and 5.67% in Jos. This result outperforms Ref. [19].

Also, the best value for C_1 and C_2 is given as 1 and 2.5; respectively, as evidenced by the results, the forecaster performed comparably to other existing solar forecasters in the literature.

4 Conclusion

This study was carried out to attain the peak hours and locations for effective and efficient solar power generation in Nigeria. The input parameters used in this study are hours, days, month, year, DNI, DHI, and air temperature, while the GHI value is used as the output parameter.

The estimation was done by using ANN and the PSO (ANN-PSO) algorithm to improve the performance. It was observed that the performance with the PSO algorithm was better than the performance of just the ANN. Error calculations such as nRMSE, MSE, R, and MAPE were also carried out to get the most accurate values.

Appendix 1: Average Monthly Solar Irradiance

Month	A	B	C	D	E	F	G	H	I	J	K
January	188.48	200.74	210.1	200.1	194.1	*226.67*	239.6	237.6	235.3	239.743	228.21
February	**200.9**	197.76	**216.86**	195.1	187.6	225.09	239.8	232.1	241.8	245.414	224.19
March	184.42	209.96	218.7	**236.2**	**210.8**	**255.19**	**266.9**	**266.1**	**273.7**	**274.763**	**247.62**
April	197.63	207.55	203.54	226.5	203.1	235.16	243.2	247.2	272.9	266.275	234.87
May	175.75	**211.95**	216.53	208.2	195.2	231.72	236.2	242.4	258.6	257.976	237.94
June	158.97	191.4	200.26	166.9	162.9	220.28	220.4	233.9	246.2	237.313	212.89
July	156.5	157.38	182.96	158.2	147.3	205.72	202.7	207.7	236.8	245.696	194.97
August	154.31	**168.53**	**169.18**	**170.9**	**152.5**	**176.09**	**187.2**	**191.1**	229.4	200.977	**165.92**
September	164.05	182.47	181.98	197.3	168.6	196.32	229.5	218.2	259.7	242.074	199.08
October	**151.76**	190.84	202.09	190.3	168.3	217.99	240.5	236.4	245.6	243.905	226.7
November	177.52	200.12	223.59	206.2	190.2	233.48	251.7	254.1	237.9	241.642	240.61
December	177.97	188.12	200.67	179.06	175.84	194.84	221.2	220.8	**200.85**	**192.833**	203.95

Key: A—Port harcourt, B—Benin, C—Enuga, D—Eko, E—Ikeja, F—Ibadan, G—Jos, H—Yola, I—Kano. J—Kaduna, K—Abuia

References

1. Ellabban O, Abu-Rub H, Blaabjerg F (2014) Renewable energy resources: current status, future prospects and their enabling technology. Renew Sustain Energy Rev 39:748–764
2. Bull SR (2001) Renewable energy today and tomorrow. Proc IEEE 89(8):1216–1226
3. Lerner J, Grundmeyer M, Garvert M (2009) The importance of wind forecasting. Renew Energy Focus 10(2):64–66
4. Etukudor C et al (2021) Yield assessment of off-grid PV systems in Nigeria. In: 2021 IEEE PES/IAS PowerAfrica. IEEE
5. Etukudor C et al (2018) Optimum tilt and azimuth angles for solar photovoltaic systems in South-West Nigeria. In: 2018 IEEE PES/IAS PowerAfrica. IEEE
6. Gbenga A et al (2019) The influence of meteorological features on the performance characteristics of solar photovoltaic storage system. In: Journal of Physics: Conference Series. IOP Publishing
7. Charles A (2014) How is 100% renewable energy possible for Nigeria? Global Energy Network Institute
8. Ohunakin OS, Adaramola MS, Oyewola OM, Fagbenle RO (2014) Solar energy application and development in Nigeria: drivers and barriers. Renew Sustain Energy Rev 32:294–301
9. Isoken G, Idemudia DBN (2016) Nigeria power sector: opportunities and challenges for investment in 2016. In: Client alert white paper, pp 1–15
10. Mohanty S et al (2017) Forecasting of solar energy with application for a growing economy like India: survey and implication. Renew Sustain Energy Rev 78:539–553
11. Obibineche C, Igbojionu DO, Igbojionu JN Design, development and evaluation of a bucket drip irrigation system for dry season vegetable production in South-Eastern Nigeria. Turk J Agric Eng Res 2(1):183–192
12. Vasumathi B, Moorthi S (2012) Implementation of hybrid ANN–PSO algorithm on FPGA for harmonic estimation. Eng Appl Artif Intell 25(3):476–483
13. Engelbrecht A (2012) Particle swarm optimization: velocity initialization. In: 2012 IEEE congress on evolutionary computation. IEEE
14. Lago J et al (2018) Short-term forecasting of solar irradiance without local telemetry: a generalized model using satellite data. Sol Energy 173:566–577
15. Marzouq M et al (2020) Short term solar irradiance forecasting via a novel evolutionary multi-model framework and performance assessment for sites with no solar irradiance data. Renew Energy 157:214–231
16. Sözen A, Arcaklioğlu E, Özalp M (2004) Estimation of solar potential in Turkey by artificial neural networks using meteorological and geographical data. Energy Convers Manage 45(18–19):3033–3052
17. El Alani O, Ghennioui H, Ghennioui A (2019) Short term solar irradiance forecasting using artificial neural network for a semi-arid climate in Morocco. In: 2019 International conference on wireless networks and mobile communications (WINCOM). IEEE
18. Jamali B et al (2019) Using PSO-GA algorithm for training artificial neural network to forecast solar space heating system parameters. Appl Therm Eng 147:647–660
19. Gundu V, Simon SP (2021) Short term solar power and temperature forecast using recurrent neural networks. Neural Process Lett 53(6):4407–4418

A Rule-Based Deep Learning Method for Predicting Price of Used Cars

Femi Emmanuel Ayo, Joseph Bamidele Awotunde, Sanjay Misra, Sunday Adeola Ajagbe, and Nishchol Mishra

1 Introduction

Studies have shown that price prediction for used cars is an essential task. Recently, demand for used automobiles has increased, while demand for new cars has decreased drastically [1]. As a result, many car buyers are seeking better alternatives to buying new cars. Some people prefer to buy cars through lease rather than buy them outright due to income and expenditure of such individuals, and other factors to take an informed decision by the buyer. Under a lease contract, buyers pay an agreed sum of money in installments for the item purchased for a pre-determined period of time. These lease installments depend on the car's projected price. As a result, sellers are interested in the fair estimated price of these cars. This necessitated the need for the development of an accurate price prediction mechanism for the used cars. The

F. E. Ayo
Department of Computer Science, McPherson University, Seriki-Sotayo, Abeokuta, Nigeria

J. B. Awotunde (✉)
Department of Computer Science, Faculty of Information and Communication Sciences, University of Ilorin, Ilorin 240003, Nigeria
e-mail: awotunde.jb@unilorin.edu.ng

S. Misra
Department of Computer Science and Communication, Ostfold University College, Halden, Norway
e-mail: Sanjay.misra@hiof.no

S. A. Ajagbe
Department of Computer Engineering, Ladoke Akintola University of Technology, Ogbomoso, Nigeria
e-mail: saajagbe@pgschool.lautech.edu.ng

N. Mishra
Rajiv Gandhi Technical University, Bhopal, India

P. Singh et al. (eds.), *Machine Learning and Computational Intelligence Techniques for Data Engineering*, Lecture Notes in Electrical Engineering 998,
https://doi.org/10.1007/978-981-99-0047-3_71

application of the prediction power of machine learning (ML) algorithms can be useful in this regard [2].

In this study, a rule-based feature selection method based on deep learning for used car price prediction was developed. The developed method deployed a hybrid feature selection technique rooted in a rule-based engine for the selection of essential features and a neural network model for the prediction task. The motivation of this study is to (1) improve prediction accuracy through a robust feature selection rooted in a rule-based engine. (2) investigate used car price prediction using both numeric and Boolean attributes.

The remaining parts of this research are structured as follows: The second section contains various literature reviews. The methodology is presented in Sect. 3. The implementation and outcomes are presented in Sect. 4. Finally, Sect. 5 wraps up and introduces future study.

2 Literature Review

The estimated prices of used cars are difficult and almost impossible to determine. The prevailing increase in demand for used cars and decrease in demand for new cars has influenced the purchase decision of prospective car owners, prompting them to search for other alternatives to the purchase of new cars. Hence, the need for an accurate price prediction mechanism for the used cars. The use of machine learning prediction power is advisable for this purpose. Recently, authors in [3] performed a comparative study on the performance of regression-based ML models. The developed machine learning models were trained with a dataset collected from German e-commerce website on the used car shopping. The results showed that gradient boosted regression trees performed better with a MAE of 0.28, in contrast to random forest regression and multiple linear regression with a mean absolute error of 0.35 and 0.55 respectively. Authors in [4] developed an ensemble model that includes ANN, Support Vector Machine (SVM) and Random Forest (RF) ML techniques. The dataset helped in the developed ensemble model was collected from the web portal autopijaca.ba using a coded web scraper. The individual predictive ability of the machine learning algorithms in the ensemble model was compared using the same dataset to find the final prediction model. The model was integrated into a Java application and the results showed an accuracy of 87.38%. Pudaruth [5] developed a predictive model to forecast the price of used automobiles in Mauritius using historical data machine learning techniques.

The developed predictive model was based on data gathered from daily newspapers in the past. The predictive model includes k-Nearest Neighbour (KNN), naïve Bayes, and decision trees. The predictions from the individual model were compared in order to find the best, and they were assessed for the prediction results. The linear regression model and the KNN showed better results than the other machine learning models. Pandey et al. [6] developed an RF model to predict the selling price of used cars on historical datasets. The results showed that the developed model is accurate for the prediction of the selling price of used cars on both large and small datasets. Yang et al. [7] developed a Convolutional Neural Network (CNN) model on bicycle and car datasets to forecast a product's pricing based on its image and see the features that contributed to the forecasted price. The results of the developed model were compared with linear regression on the histogram of oriented gradients (HOG) and a multiclass SVM. The performance of the developed CNN model showed better results on both datasets, compared with HOG and SVM models. Arefin [8] developed a second-hand Tesla car prediction model using a machine learning technique. The prediction model includes decision trees, SVM, random forest, and deep learning models. The boosted decision tree regression when compared to other prediction models, produced superior outcomes. Samruddhi and Kumar [9] developed a KNN regression algorithm to forecast the cost of secondhand cars. The developed model was trained on used automobile data gathered from the Kaggle website. The developed model was evaluated on the collected data using different testing and training ratios. The results showed a high accuracy of 85% with the developed model.

Several datasets have few samples but with a large number of features. Therefore, feature selection approaches are needed to choose the most important characteristics to improve the accuracy of the classifiers. There are two types of strategies for selecting features: scalar and vector. The scalar technique selects a subset of individual attributes, while the vector method can be divided into three categories: Filtering, wrapping, and embedding are three approaches to consider [10]. The filter method selects based on the dataset statistics, the fittest features, and wrapper approach selects the fittest features based on the model [11]. Similarly, the embedded method makes use of a basic function in order to raise the performance level of the classifier in the search for the fittest features. The task of price prediction for used cars can be addressed with ML methods [12].

Deep learning (DL) and ML are tools in artificial intelligence (AI), and they make use of data to make an informed decision. They use the hidden knowledge to fetch the information for an informed decision that revealed in the output. They use highly interconnected processing units to perform supervised and unsupervised learning [13]. The DL methods helps to learn data directly from data thereby avoiding erroneous handcrafted features. It has techniques such as Artificial Neural Network (ANN), Fully Convolutional Network (FCN), among others, which are used in diverse studies especially for prediction, like forecasting; it is a strategy that uses previous data as inputs to make well-informed predictions about the direction of future trends. Segmentation is a technique for separating an image into several pieces in order to transform the image's representation into a more meaningful one that can be easily examined [11]. Classification is an approach that aids the assist radiologists in image interpretation by analyzing images and providing a second opinion for diagnosis among other tasks [14]. The ANN is a simple example of the supervised learning method which requires huge amount of labeled data for training through weight adjustments. Therefore, there are needs for features extraction technique to select important features from the huge amount of data to increase accuracy. The car price prediction datasets include many features and thus needs a feature selection technique.

Genetic search uses the simple Genetic Algorithm (GA) to perform the search for the best solutions. GA is an evolutionary algorithm that uses the computer to imitate the natural selection process. GA was originally suggested by Folorunso et al. [15] and has become a popular area in machine learning [15] The GA begins with a population of randomly generated individual programs. The evaluation of the fitness of an individual in the population is a function of the different types of fitness measures. In each iteration of the current solution, a computer-simulated gene is recombined and mixed to produce new better solutions. The Correlation-based Feature Selection Subset Evaluator (CfsSubsetEval) evaluates the weight of a subset of features based on the fitness of the individual feature to the classifier as well as the degree of intercorrelation that exists between them [16]. The characteristics chosen are those which are substantially correlated with the class yet have low intercorrelation. A rule-based engine uses its inference rules and the rule base for correct decisions. It is a widely used tool in AI applications, and an inference engine with a function to evaluate based on various relationship with its rule base. The inference engine resolves uncertainty by matching input values with its rule base in order to select the best option for execution. The implementation stage is used to select an alternative method as a result of the conflict-resolution stage for the selection of best option.

3 Material and Method

The ML models can be used in the prediction of car use, and in this study a car price prediction model was proposed using the DL model with hybrid feature selection models based on Genetic search, Correlation-based, Subset Evaluator and the rule-based engine (CfsSubsetEval).

3.1 Data

The dataset used for this study can be downloaded from the link: https://archive.ics.uci.edu/ml/datasets/Automobile. It comprises of three types of entities: (i) the specification of a car in terms of different features (features 3–25), (ii) its assigned insurance risk score (feature 1), (iii) when compared to comparable cars, its normalized losses in use (feature 2). The second entity relates to the degree to which the car is morerisky than its indicated price. Cars are given a risk factor symbol when they are first purchased. Then, if it is more risky, this symbol is either increased (or decreased) by adjusting the scale. This is referred to as "symboling" by actuaries. A value of −3 suggests that the vehicle is likely to be generally safe, while a value of +3 indicates that it is possibly harmful. The relative average loss payment per insured vehicle year is the third entity. This figure is standardized for all cars in a given size category (compact two-door cars, station wagons, sports/specialty cars, and so on) and indicates the average annual loss per vehicle. The dataset comprises

205 instances and 26 attributes. The collected dataset was formatted and preprocessed into the Attribute-Relation File Format (ARFF). The ARFF is an ASCII text file that describes a list of instances with a set of attributes. Table 1 shows a detailed description of the dataset.

Table 1 Car price prediction dataset

S/N	Attribute	Data type	Attribute range
1	Symbolling	Integer	−3, −2, −1, 0, 1, 2, 3
2	Normalized-losses	Continuous	Continuous from 65 to 256
3	Make	Nominal	Audi, bmw, chevrolet, dodge, honda, isuzu, jaguar, mazda, mercedes-benz, mercury, mitsubishi, nissan, peugeot, plymouth, porsche, renault, saab, toyota, volkswagen, Volvo
4	Fuel-type	Nominal	Diesel, gas
5	Aspiration	Nominal	Std, turbo
6	Number-of-doors	Nominal	Four, two
7	Body-style	Nominal	Hardtop, wagon, sedan, hatchback, convertible
8	Drive-wheels	Nominal	4wd, fwd, rwd
9	Engine-location	Nominal	Front, rear
10	Wheel-base	Continuous	Continuous from 86.6 120.9
11	Length	Continuous	Continuous from 141.1 to 208.1
12	Width	Continuous	Continuous from 60.3 to 72.3
13	Height	Continuous	Continuous from 47.8 to 59.8
14	Curb-weight	Continuous	Continuous from 1488 to 4066
15	Engine-type	Nominal	Ohcf, ohcv, rotor, dohc, dohcv, l, ohc
16	Number-of-cylinders	Nominal	Six, three, twelve, two eight, five, four
17	Engine-size	Continuous	Continuous from 61 to 326
18	Fuel-system	Nominal	Idi, mfi, mpfi, spdi, 1bbl, 2bbl, 4bbl, spfi
19	Bore	Continuous	From 2.54 to 3.94, there is no break
20	Stroke	Continuous	Continuous between 2.07 and 4.17
21	Compression-ratio	Continuous	From 7 to 23 h nonstop
22	Horsepower	Continuous	From 48 to 288 in a row
23	Peak-rpm	Continuous	Continuous between 4150 and 6600
24	Cty-mpg	Continuous	From 13 to 49 continuously
25	Highway-mpg	Continuous	From 16 to 54 continuously
26	Price	Continuous	Continuous between 5118 and 45,400

3.2 Proposed Methodology

This study proposed an enhanced hybrid feature selection model for the selection relevant features and ANN was used in dataset classification for the purpose of car price prediction. The CfsSubsetEval was used to determine the best feature subsets and GA was used as the search strategy for the best feature subset. The rule-based engine checks for similar feature subsets with the least number of features using the feature subset. CfsSubsetEval and GA help in improving the prediction accuracy of the developed model by selecting the most essential features from the dataset. CfsSubsetEval considers each feature's specific prediction ability, as well as the degree of redundancy between them, to determine the value of a subset of features [17]. The correlation coefficient is used to calculate the correlation between a subset of attributes and the target class label, as well as the intercorrelations between features. The importance of a group of features increases as the correlation between features and classes develops, but reduces as the intercorrelation grows. The selected features are then used as input for the ANN to perform the car price prediction. Figure 1 shows the architecture for the developed model for use in car price prediction. It comprises the car dataset, data preprocessing, hybrid feature selection phase, rule-based engine phase, selected feature, ANN, and prediction stage [18].

The proposed methodology is defined by the following five steps in Eqs. 1–10.

Step 1: A feature V_i is said to be relevant iff there exists some v_i and c for which $p(V_i = v_i) > 0$ such that

$$p(C = c|V_i = v_i) \neq p(C = c) \tag{1}$$

Step 2: It is possible to predict the correlation between a composite test made up of the accumulated elements and the outside variable. If the correlation between each test component and the external variable is accumulated, as well as the inter-correlation

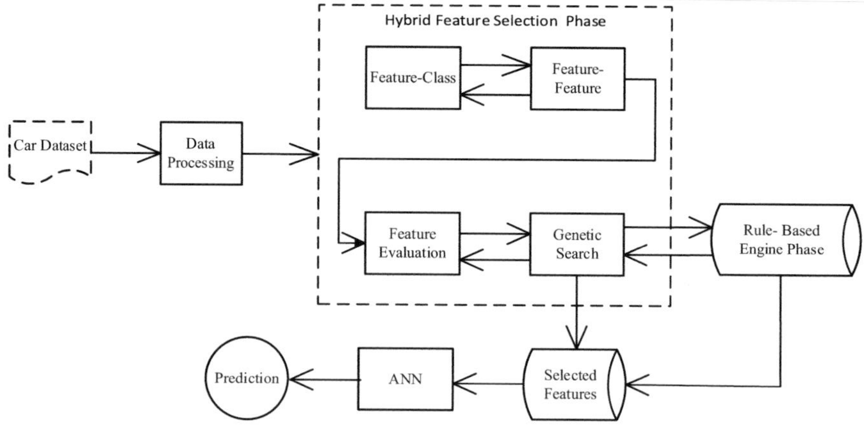

Fig. 1 Proposed Architecture for car price prediction

between each pair of components, this information is available.

$$r_{zc} = \frac{k\overline{r_{zi}}}{\sqrt{k + k(k-1)\overline{r_{ii}}}} \tag{2}$$

where r_{zc} is the relationship between the totaled components and the external variable, k is the number of components, r_{zi} is the average of the correlations between the components and the outside variable, and r_{ii} is the average inter-correlation between components.

Step 3: In this genetic search, the fitness function is a linear mixture of an accuracy term and a simplicity term:

$$Fitness(X) = \frac{3}{4}A + \frac{1}{4} = \left(1 - \frac{S+F}{2}\right), \tag{3}$$

where X represents a feature subset, A represents ANN's average cross-validation accuracy, the number of examples or training samples is denoted by S, while the number of subset features is denoted by F.

Step 4: The suggested network intrusion detection using ANN has the following detection model define in (4).

$$\widehat{Y}_{(t+p)} = f_{ANN}\left(Y_t^{(d)}, X_t^{(d)}\right) \tag{4a}$$

where $Y_t^{(d)}$ and $X_t^{(d)}$ are the observed value and training dataset of the identified class label and $\widehat{Y}_{(t+p)}$ is the calculated class label for the targeted sample data at time t, as well as during the detection period p

$$Y_t^{(d)} = \{Y_t, Y_{t-1}, Y_{t-2}, \ldots, Y_{t-d}\} \tag{4b}$$

$$X_t^{(d)} = \{X_t, X_{t-1}, X_{t-2}, \ldots, X_{t-d}\} \tag{4c}$$

Step 5: If there are multiple feature subsets ($F_>$) with a similar optimal solution, the rule-based engine provides (V_i) the set of features (X_F) with the best fitness (F_{hi}) to the basic classifier as in (5).

$$R = \begin{cases} V_i, if\, V_i \in F_> \bigcap X_F \\ V_i, if\, F_{hi} \bigcap \varnothing \end{cases} \tag{5}$$

The inaccuracy of the network is determined using (6), and the weight change at a single neuron input is computed using (7).

$$\delta_k = o_k(1 - o_k)(t_k - o_k) \tag{6}$$

The output term $o_k(1 - o_k)$ is added because of the sigmoid function, where δ_k represents the error of the connected output node o_k, t_k represents the target output presented by the training data, and $o_k(1 - o_k)$ represents the output term presented by the training data.

$$\Delta w_{ki}(n) = \eta.\delta_k.x_i \tag{7}$$

where $\Delta w_{ki}(n)$ is the weight change for the kth neuron activated by the input x_i at time step η required for the weight of the adjustment process, η is the learning rate, which is always 1, and δ is the output node error. After calculating the weight change, the new weight is obtained by multiplying the old weight by the new weight $\Delta w_{ki}(n)$ with the change in weight value of (7) as shown in (8).

$$\Delta w_{ki}(n + 1) = w_{ki}(n) + \Delta w_{ki}(n) \tag{8}$$

By combining all the weights together $w_{kh}(n)$ exacerbated by the associated output's error (δ_k), the error of the concealed nodes can be determined in the same way as the error of the linking output node, and as revealed by multiplying the result by the output of the hth hidden node (o_h) (9).

$$\delta h = o_h(1 - o_h)\sum w_{kh}.\delta_k \tag{9}$$

With the help of the new weight of the buried layer may be computed (10).

$$\Delta w_{kh}(n + 1) = x_h + (\delta_h \cdot \Delta w_{kh}) \tag{10}$$

3.3 Evaluation Metrics

3.3.1 Correlation Coefficient (CC)

The CC calculates the strength of the relationship between two variables. The value close to 1 is strongly and positively correlated while −1 is weakly and negatively correlated. The CC is depicted mathematically by Eq. 11.

$$r = \frac{\sum(x - m_x)(y - m_y)}{\sqrt{\sum (x - m_x)^2 \sum (y - m_y)^2}} \tag{11}$$

where m_x and m_y are the means of x and y variables.

3.3.2 Mean Absolute Error (MAE)

The MAE is used to measure the closeness of a predicted value to the actual value. The MAE is depicted by Eq. 12. A smaller error depicts that the predicted value is close to the actual value.

$$MAE = \frac{1}{n}\sum_{i}^{n}|\hat{y}_i - y_i| \tag{12}$$

where $\hat{y}_i$ = prediction, y_i= true value, n = total number of instances

3.3.3 Root Mean Square Error (RMSE)

The RMSE is the square root of the differences between the predicted values and the actual values. A lower value of RMSE is better. In all cases, RMSE is always larger or equal to the MAE and both can range between the value [0, ∞]. The RMSE can be depicted mathematically by Eq. 13.

$$RMSE = \sqrt{\frac{\sum_{i=1}^{n}(\hat{y}_i - y_i)^2}{n}} \tag{13}$$

where $\hat{y}_i$ = prediction, y_i = true value, n = total number of instances.

Steps in the model

1. Collection of car datasets.
2. Conversion of car datasets into arff format.
3. Calculate the correlation between feature subsets and classification.
4. Select feature categories that are highly correlated.
5. Determine the fitness value for the feature subsets you selected.
6. The subset of features with the highest fitness value is returned.
7. If two feature subsets have the same fitness rating, return the feature subset with the fewest subset features.
8. For used car price prediction, apply a NN to the selected feature subsets.

4 Implementation and Results

4.1 Implementation

The implementation was done using JAVA programming language on a Windows 10 PC with an Intel Pentium CPU operating at 2.40 GHz and 4.00 GB of memory.

Table 2 Comparison of algorithms for feature selection

S/N	Algorithm	Features	Features selected	Acc (%)
1	BestFirst + CfsSubsetEval	5	f5, f8, f14, f15, f24	0.9744
2	Ranker + ClassifierAttributeEval	24	f24, f23, f8, f9, f10, f7, f6, f5, f2, f3, f4, f11, f12, f13, f21, f22, f19, f20, f18, f14, f15, f16, f17, f1	–
3	Ranker + LatentSemanticAnalysis	1	f1	–
4	SubsetSearchForwardSelection + CfsSubsetEval	5	f5, f8, f14, f15, f24	0.9730
5	ScatterSearchV1 + CfsSubsetEval	5	f5, f8, f14, f15, f24	0.9748
6	Ranker + CorrelationAttributeEval	24	f24, f8, f16, f18, f15, f21, f5, f1, f4, f3, f9, f7, f6, f20, f23, f13, f2, f10, f19, f12, f17, f11, f22, f14	0.9723
7	GreedyStepwise + CfsSubsetEval	5	f5, f8, f14, f15, f24	0.9750
8	RandomSearch + CfsSubsetEval	4	f5, f14, f15, f24	0.9740
9	CfsSubsetEval + GeneticSearch + Rule-Based-Engine (CfsGSRBE)	8	f3, f4, f8, f9, f12, f16, f19, f24	0.9843

The developed application includes tools such as GA library, CfsSubsetEval library and ANN Weka API. The resources for the dataset used can be downloaded from https://archive.ics.uci.edu/ml/datasets/Automobile. The improvement in the simulation results by balancing the data in the classification were managed by putting k-fold (k = 5) cross-validation. At random, the training data was partitioned into five sections of equal size. The generated model was tested on a single subset, while the data from the remaining four subsets was used for training purposes. The implementation setup for the developed model are: no of Epochs = 500, learning rate = 0.3 and momentum = 0.2. In order to demonstrate the importance of feature selection on the base classifier, Table 2 shows the comparison of different feature selection algorithms.

4.2 Results

The results indicates that the developed rule-based hybrid model selected 8 features with superior accuracy of 0.9843 compared to related feature selection algorithms. The superior accuracy value was due to the rule-based hybrid feature selection method used with the ANN as the base classifier. The program interface envelops the system modules and shows the results output for the developed model for price prediction for used cars. Table 3 shows the comparison of the developed model in bold with related classifiers for price prediction for used cars. The developed rule-based hybrid feature selection and ANN showed better results having correlation coefficient, mean absolute error accuracy and root mean squared error of 0.9818, 1.1809 and 1.5643 respectively. Table 4 shows the sample prediction results for the developed rule-based hybrid feature selection and ANN for price prediction for used cars. The results show close prediction to the actual values and therefore validated that the developed model is efficient for used car price prediction. Figure 2 shows the graphical comparison of the developed model with other related models.

Table 3 Comparison of classifiers

S/N	Evaluation metric	CfsSubsetEval + GeneticSearch + Rule-Based-Engine + ANN (CfsGSRBEANN)	M5P	KNN	SVM
1	Correlation coeffficient	0.9818	0.9762	0.9162	0.8831
2	Mean absolute error	1.1809	1.0904	1.2293	3.0681
3	Root mean squared error	1.5643	1.4929	2.768	3.0681

Table 4 Sample prediction results for CfsGSRBEANN

S/N	Actual value	Predicted value
1	27	26.069635773582682
2	26	24.36254857283255
3	30	29.181857783808944
4	22	22.378622466069267
5	25	27.118482086415703
6	20	20.681270216071
7	29	28.581076933938537
8	28	27.825520081453888
9	53	53.10425758926064
10	43	42.91052148620654

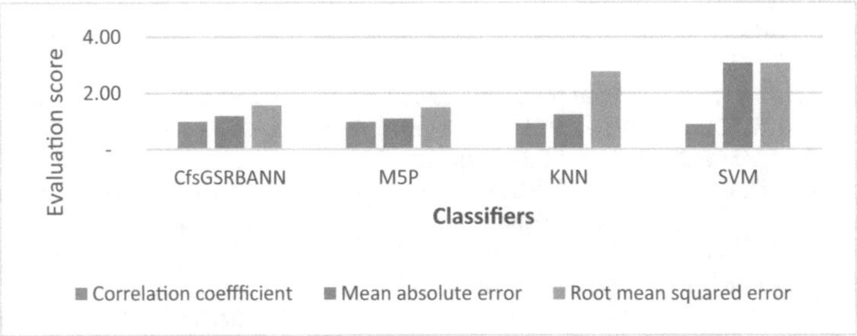

Fig. 2 Results of different classifiers for prediction of the used car price

5 Conclusions and Future Directions

In this research, the hybrid feature selection based on rules and ANN model was used for used car price prediction. The focus of the paper is to build a robust architecture for accurate price prediction through feature selection. The developed model includes CfsSubsetEval, GA, a rule-based engine and ANN. The CfsSubsetEval returns the attribute-class relationship with the highest correlation as the selected features. Based on the selected features, the GA searches and returns the features that have the highest fitness value. Rule-based engine check for similar feature subsets and the feature subset with the fewest features is returned. The ANN was then used for the final price prediction on the selected features. The research was evaluated using the CC, MAE and RMSE with 0.9818, 1.1809 and 1.5643 respectively. The results of the evaluation conducted showed that the developed model outperform the other related and existing models. In the future, a more robust feature selection algorithm will be used to increase accuracy for used car price prediction.

References

1. Awotunde JB, Chakraborty C, Adeniyi AE (2021) Intrusion detection in industrial internet of things network-based on deep learning model with rule-based feature selection. Wirel Commun Mobile Comput 7154587
2. Shehadeh A, Alshboul O, AI-Mamlook RE, Hamedat O (2021) Machine learning models for predicting the residual value of heavy construction equipment: An evaluation of modified decision tree, LightGBM, and XGBoost regression. Autom Constr 129:103827
3. Monburinon N, Chertchom P, Kaewkiriya T, Rungpheung S, Buya S, Boonpou P (2018) Prediction of prices for used car by using regression models. In: 2018, 5th international conference on business and industrial research (ICBIR), pp 115–119
4. Gegic E, Isakovic B, Keco D, Masetic Z, Kevric J (2019) Car price prediction using machine learning techniques. TEM J 8(1):113
5. Pudaruth S (2014) Predicting the price of used cars using machine learning techniques. Int J Inf Comput Technol 7(4):753–764
6. Pandey A, Rastogi V, Singh S (2020) Car's selling price prediction using random forest machine learning algorithm. In: 5th international conference on next generation computing technologies (NGCT-2019)
7. Yang RR, Chen S, Chou E (2018) AI blue book: vehicle price prediction using visual features. arXiv:1803.11227
8. Arefin SE (2021) Second hand price prediction for tesla vehicles. arXiv:2101.03788
9. Samruddhi K, Kumar RA (2020) Used car price prediction using K-nearest neighbor based model. Int J Innov Res Appl Sci Eng 4(2):629–632
10. Kabir MR, Onik AR, Samad T (2017) A network intrusion detection framework based on bayesian network using wrapper approach. Int J Comput Appl 166(4):13–17
11. Awotunde JB, Ogundokun RO, Jimoh RG, Misra S, Aro TO (2021) Machine learning algorithm for cryptocurrencies price prediction. Stud Comput Intell 2021(972):421–447
12. Ayo FE, Folorunso SO, Abayomi-Alli AA, Adekunle AO, Awotunde JB (2020) Network intrusion detection based on deep learning model optimized with rule-based hybrid feature selection. Inf Secur J 29(6):267–283

13. Thanh DNH, Hai NH, Hieu LTP, Prasath VBS (2021) Skin lesion segmentation method for dermoscopic images with convolutional neural networks and semantic segmentation. Comput Opt 45(1):122–129
14. Ajagbe SA, Idowu IR, Oladosu JB, Adesina AO (2020) Accuracy of machine learning models for mortality rate prediction in a crime dataset. Int J Inf Process Commun 10(1 & 2):150–160
15. Folorunso SO, Awotunde JB, Ayo FE, Abdullah KKA (2021) RADIoT: the unifying framework for IoT, radiomics and deep learning modeling. Intell Syst Ref Libr 209:109–128
16. Fernandez A, Lopez V, del Jesus MJ, Herrera F (2015) Revisiting evolutionary fuzzy systems: Taxonomy, applications, new trends and challenges. Knowl-Based Syst 80:109–121
17. Ogundokun RO, Awotunde JB, Misra S, Abikoye OC, Folarin O (2021) Application of machine learning for Ransomware detection in IoT devices. Stud Comput Intell 972:393–420
18. Afolabi AO, Oluwatobi S, Emebo O, Misra S, Garg L, Evaluation of the merits and demerits associated with a diy web-based platform for e-commerce entrepreneurs. In: International conference on information systems and management science. Springer, Cham, pp 214–227

Classification of Fundus Images Based on Severity Utilizing SURF Features from the Enhanced Green and Value Planes

Minal Hardas, Sumit Mathur, and Anand Bhaskar

1 Introduction

DR is a complication of diabetes that affects the eyes. It is caused by high blood sugar because of diabetes, which damages the retina [1]. The retina is provided with a supply of blood vessels and nerves. These blood vessels to the retina develop tiny swellings called microaneurysms, which are prone to hemorrhage [2]. The interruption to the supply of nutrients and oxygen triggers the formation of new blood vessels across the eye [3]. The new blood vessels formed are brittle and prone to breakage. Both the microaneurysms and the newly formed blood vessels may rupture the path. This causes the leakage of blood into the retina and blurriness of vision. DR may even lead to blindness [4]. A few specks or spots floating in the visual field of the eye are because of the presence of DR. DR can be classified into two main stages as shown in Fig. 1, NPDR (Non-proliferative diabetic retinopathy) and PDR (Proliferative diabetic retinopathy). NPDR is the primary stage of diabetic eye disease [5]. In NPDR, the tiny blood vessels leak and cause swelling of the retina [6]. People with NPDR have blurry vision. NPDR can be further classified as mild, moderate, and severe. PDR is the more advanced stage of diabetic eye disease [7]. In this stage, the retina starts growing new blood vessels called neovascularization [8]. These fragile new vessels often bleed into the vitreous. PDR is very serious as it causes loss of central and peripheral vision of a person.

The workflow diagram of the proposed system is shown in Fig. 2.

The fundus camera captures the images of the retina. This image is pre-processed and the enhanced green and value color planes are extracted from it. The two-color planes are combined and the AGVE algorithm is applied to it. Then the SURF algorithm is used to extract the strongest feature points. The red score feature is

M. Hardas (✉) · S. Mathur · A. Bhaskar
Sir Padampat Singhania University, Udaipur 313601, India
e-mail: minal.sudarshan@spsu.ac.in

P. Singh et al. (eds.), *Machine Learning and Computational Intelligence Techniques for Data Engineering*, Lecture Notes in Electrical Engineering 998,
https://doi.org/10.1007/978-981-99-0047-3_72

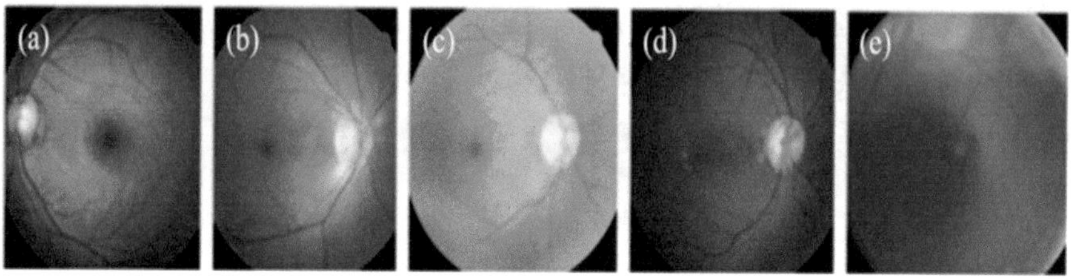

Fig. 1 Classification of DR: NPDR and PDR. NPDR is the early stage of DR, which comprises edema and hard exudates. Later stages comprise vascular occlusion, a restriction of blood supply to the retina, and an increase in macular edema, known as PDR. Images of retina with **a** No DR **b** Mild NPDR **c** Moderate NPDR **d** Severe NPDR **e** PDR

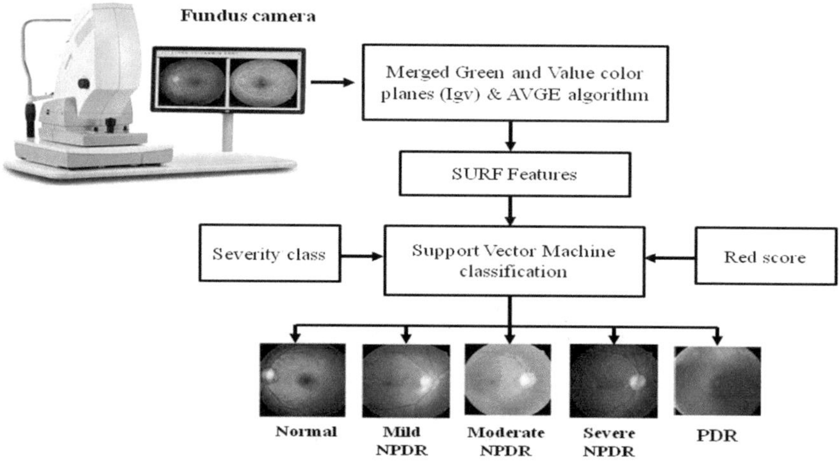

Fig. 2 Workflow diagram of the proposed system. The retinal fundus image is passed through the combined enhanced green and value color plane. The extracted SURF features, along with the red score and the DR severity level, are fed to the SVM classifier for predicting the level of severity

computed and normalized to a total number of pixels. DR severity level is obtained by calculating the ground truths of each of the abnormalities. Red score feature with the DR severity level is then added to the previous SURF feature vector to classify various stages of DR.

The remaining sections of the study are organized as follows. Section 2 presents the literature review. Section 3 presents the methodology of the proposed system. Results and discussions of the proposed work take place in Sects. 4 and 5 respectively. Finally, Sect. 6 concludes the study.

2 Literature Review

Several automated algorithms have been over viewed by the authors for detecting and grading the severity of DR.

Pires and his group [9] introduced an effective method for directly assessing the refer-ability of patients without preliminary DR lesion detection. The accuracy of the proposed method was improved using SVM. Lachure and his team [10] developed an automatic screening of DR by detecting red and bright lesions using GLCM as a feature extraction technique in digital fundus images. The SVM outperformed better over KNN classifier in detecting the severity level of DR. Abbas et al. [11] developed a novel automatic recognition system for recognizing the five severity levels of diabetic retinopathy (SLDR) using visual features and a deep-learning neural network (DLNN) model.

A method for detecting DR was presented by Costa and his group [12] using a Bag-of-Visual Words (BoVW) method. Their work comprised extracting dense and sparse features using SURF and CNN from the image. The sparse feature SURF model outperformed well as compared with the dense feature CNN model. A fully automated algorithm using deep learning methods for DR detection was developed by Gargeya et al. [13]. Their work focused on preprocessing the fundus images and training the deep learning network with the data-driven features from the data set. The tree-based classification model classified the fundus image into grade 0 (no retinopathy) or grade 1 (severity from mild, moderate, severe, or proliferative DR). A hybrid machine learning system was proposed by Roy and his group [14] for DR severity prediction from retinal fundus images. Their work comprised CNNs with dictionary-based approaches trained for DR prediction. The resulting feature vectors were concatenated and a Random forest classifier was trained to predict DR severity.

Islam and his team [15] described an automated method using a bag of words model with SURF for DR detection in retinal fundus images. Their work comprised detecting the interesting SURF points, and the classification was performed using SVM. A computer-assisted diagnosis was introduced by Carrera and his team [16] to classify the fundus image into one of the NPDR grades. The SVM and a decision tree classifier were used to figure out the retinopathy grade of each retinal image. Koh and his group [17] presented an automated retinal health screening system to differentiate normal image from abnormal (AMD, DR, and glaucoma) fundus images. They extracted the highly correlated features from the PHOG and SURF descriptors using the canonical correlation analysis approach. The system was evaluated using a tenfold cross-validation strategy using the k-nearest neighbor (k-NN) classifier.

A dictionary-based approach for severity detection of diabetic retinopathy was introduced by Leeza and her team [18]. Their work comprises creating the dictionary of visual features and detecting the points of interest using the SURF algorithm. The images were further classified into five classes: normal, mild, moderate, and severe NPDR and PDR, using the radial basis kernel SVM and Neural Network. Gayathri and her team [19] have focused on the detection of diabetic retinopathy using binary and multiclass SVM. Their work comprised extracting Haralick and Anisotropic Dual

Tree Complex Wavelet Transform (ADTCWT) features for reliable DR classification from retinal fundus images. Gadekallu et al. [20] developed a hybrid principal component analysis (PCA) firefly-based deep neural network model for the classification of diabetic retinopathy. Their work employed scalar technique for normalizing the dataset, PCA fo7r feature selection, Firefly algorithm for dimensionality reduction, and machine learning classifiers and DNN for DR detection.

3 Methodology

The proposed method for estimating the severity levels of DR comprises diverse processes, such as preprocessing, segmentation, feature extraction, and classification. First, the original input image is transformed into a grayscale image. Then, the RGB and HSV color planes are extracted. The optic disk and the blood vessels are segmented. Further, the enhanced green and value sub color planes are merged and the AGVE algorithm is used to obtain the average grey value in a fundus image. The ground truth sum of each abnormality, such as microaneurysms, hemorrhages, hard exudates, and soft exudates, is used to calculate the severity level of a fundus image. The red score is generated and the SURF algorithm is used to extract the most important feature points from the fundus image. An SVM is trained for these features and the DR severity level is predicted. Figure 3 shows the overall process of the proposed work.

The original image was converted to a grayscale image and the areas around the fundus image were masked using a threshold of 10 (Eqs. 1 and 2).

$$g(x, y) = 0 \forall f(x, y) < 10 \quad (1)$$

$$g(x, y) = f(x, y) \forall f(x, y) \geq 10 \quad (2)$$

where g (x, y) is the mask image and f (x, y) is the gray scale image.

The original input color image is divided into three color planes R, G, and B. The green color plane is then selected and a maximum pixel intensity value was determined.

A green channel enhancement was done in two steps. First, the masked pixels found from the masked image are set to 255 and second, the green channel color pixels are normalized (Eq. 3), and contrast stretch between 0 and 255.

$$\mathrm{h(x,\ y)} = \mathrm{Ig(x,\ y)} - \min(\mathrm{Ig(x,\ y)}) / \max(\mathrm{Ig(x,\ y)}) - \min(\mathrm{Ig(x,\ y)}) \quad (3)$$

where, Ig (x, y) is the green channel color image, h (x, y) is the enhanced green channel color image, min (x, y) and max (x, y) are the minimum and maximum values of Ig (x, y) and (x, y) is the current co-ordinates under consideration.

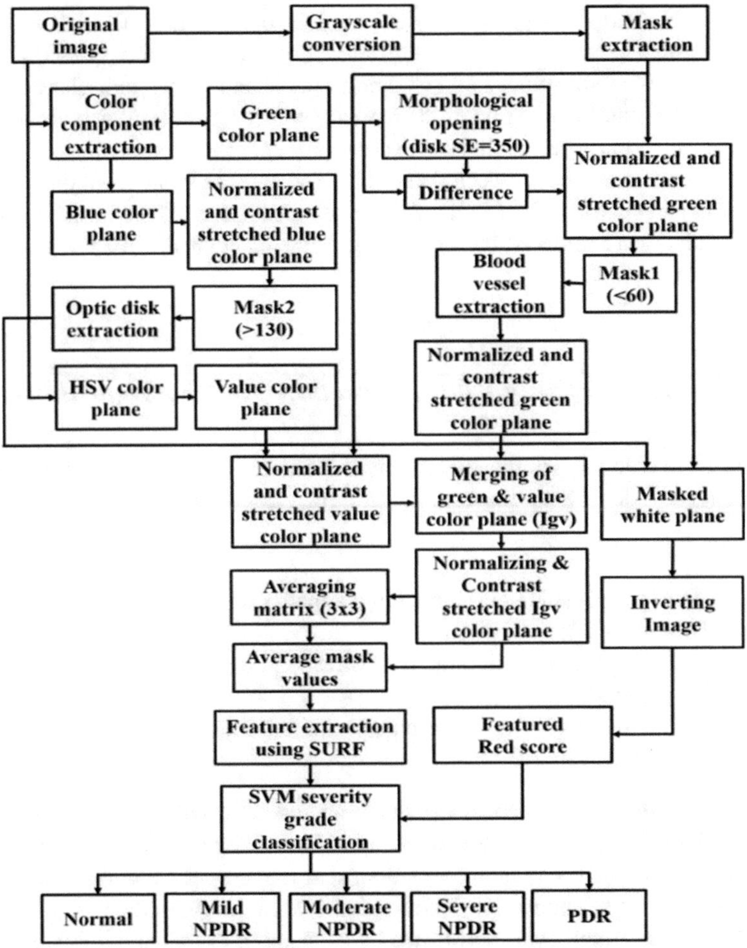

Fig. 3 The proposed method for predicting the severity levels of DR using the AGVE algorithm and SVM. An original input image is converted into a grayscale image. RGB and HSV color planes are extracted. The green and value sub color planes are merged and the AVGE algorithm is applied to extract the average gray value in the image. Then, the strongest feature points are extracted using the SURF algorithm. The SURF features, along with the red score feature and severity level, are used to train the SVM and the DR severity level is predicted

The blue color plane image was enhanced in the same way as that of the green color planes. The vessels visible in the green channel are enhanced by creating a mask with a threshold value of 60. The optic disk visible in the blue channel was masked with the thresholding level of 130. The original image was then converted into an HSV color plane and the value color plane and green color plane were merged with equal magnitude (Eq. 4).

$$\mathrm{Igv}\,(x,\ y) = 0.5 * \mathrm{Ig}\,(x,\ y) + 0.5 * \mathrm{Iv}\,(x,\ y) \tag{4}$$

where, Ig and Iv are the enhanced green color plane and value color planes.

The minima and the maxima of the newly obtained Igv color plane are extracted and for masking purposes, all the max values are replaced by min values. Igv image is then contrast stretched between 0 and 255. The average value from the entire Igv fundus image is extracted using the AGVE algorithm and all the masked pixels of the original Igv are replaced by the average value calculated. This image is called I_{AGVE} image.

3.1 The Average Gray Value Extraction (AGVE) Algorithm

The AGVE is a novel technique to extract the average grey level value of a fundus image. Figure 4 depicts an illustration of the AGVE algorithm

Algorithm

Step1. The Igv image is transformed into a matrix of 3 × 3, given that the fundus image comprises a round border.
Step2. All the rows and columns that correspond to an edge are ignored for the averaging purpose.
Step3. Select the central pixel to ensure the maximum accuracy with the grey scale value extraction.

The image was then given to the mapping function that saturates top 1% and bottom 1% of all the pixel values to remove any spike noise to get the maximum information. This saturated image was then sent to the SURF algorithm for feature extraction and the strongest 320 feature points were considered as feature matrix. The DR severity grading was performed by calculating the ground truths of each of the abnormalities and generating a red score for each image.

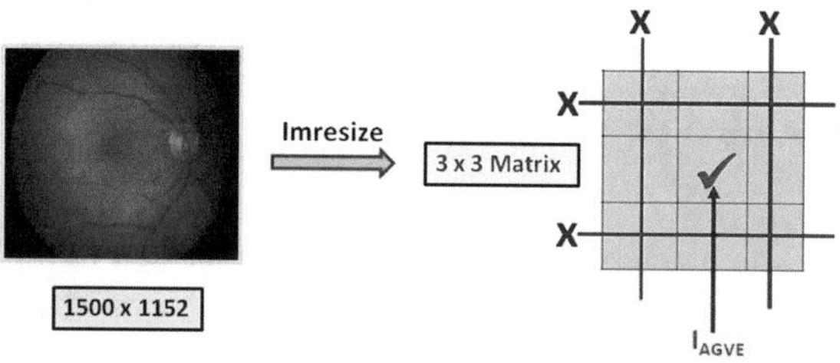

Fig. 4 Illustration of the AGVE algorithm

3.2 Red Score Calculation

The green channel image extracted from the original image was morphologically opened with a structural element (disk) of a size of 350 pixels to form the complete background picture. It was then subtracted from the green channel image to enhance the blood vessels and reddish parts in the eye. The outside image was then masked with the white value and all the values below 50 are preserved while the remaining values are converted as 255. This image was then normalized (Eqs. 3 and 4) and scaled up to 0 and 255. The image negative was then computed at a grayscale level. Finally, the red score feature was computed and normalized to a total number of pixels. This red feature score was then added to the previous 320 feature vector for severity classification.

3.3 Severity Level Generation

To gain the supervisory severity level the ground truths of four types of abnormalities such as hard exudates, soft exudates, microaneurysms, and hemorrhages from the DIARETDB1 dataset were used. All the four types of output images were binarized using Otsu's method and their corresponding sum was calculated. The binarized ground truth image of the hemorrhage was used and all the four connected component objects were then separated and the total number of objects was computed for that image.

The severity level was then defined depending upon the abnormality and count in the images.

Level 1: Ground truth score of all abnormalities is Zero. Indicates no DR and the person is normal.
Level 2: Ground truth score of the microaneurysms is greater than zero and all other scores are zero. Indicates mild NPDR.
Level 3: Ground truth score of the hemorrhages or microaneurysms greater than zero then the flag is set and the exudates score is checked to be positive. Indicates moderate NPDR.
Level 4: Hemorrhage count greater than 20. Indicates severe NPDR.
Level 5: None of the severity maps matched. Indicates PDR.

Finally, the extracted SURF features, red score count and severity class were fed to the SVM classifier and the severity classification model was trained and tested.

4 Results

DIARETDB1 dataset was used for the experimentation purposes. Images were captured with the 50-degree field-of-view digital fundus camera with varying imaging settings controlled by the system and the analysis was performed in MATLAB 2019a. The dataset comprises ground truths of four types of abnormalities, such as hard exudates, soft exudates, microaneurysms, and hemorrhages.

The performance of the proposed system has been validated using 76 images from the dataset, out of which 70% randomized images were used for training and the remaining 30% were reserved for testing. This process was sequentially repeated 5 times to optimize the final SVM model. The outputs of all the 76 images were classified into 5 different severity levels. It was seen that only first 4 levels were assigned to the fundus images while the last level was ignored throughout the calculations. As it was assumed that the occurrence frequency of this level is very less in the DIARETDB1 dataset.

The 8% randomly selected images were then tested, and the accuracy was computed. To compute accuracy, the number of correctly classified DR images was divided by the total number of DR images in the dataset. Hence, if the model classified 8 DR images accurately out of 10, then the accuracy of the model was 80%.

Result analysis

Figure 5 visualizes the output of the proposed system at various phases of DR severity level detection.

The confusion matrix of the proposed system is shown in Table 1. Out of total 76 images, only 1 image was found to be inaccurately classified. The Level 2 image was placed under categories as Level 3 image. There were no images of severity level 5 hence, they are not shown in the confusion matrix.

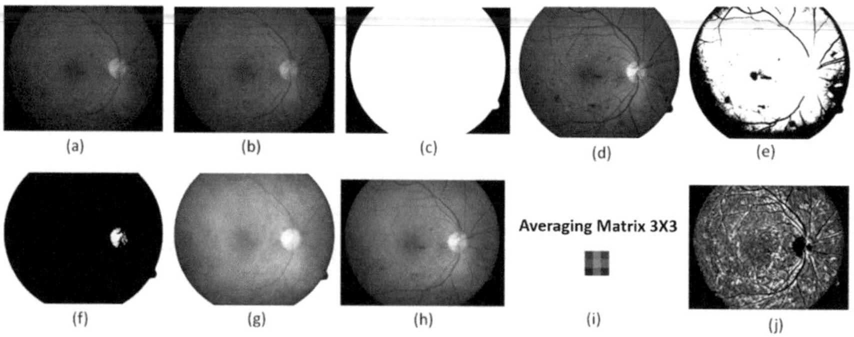

Fig. 5 Output at various stages for the proposed method during DR severity detection. **a** Original image of the retina. **b** Grayscale image. **c** Mask image. **d** Enhanced green color plane. **e** Extracted blood vessels. **f** Optic disc extraction. **g** Enhanced value color plane. **h** Enhanced combined green and value color planes. **i** Averaging matrix 3X3. **j** SURF feature points

Table 1 Confusion matrix

	Actual					
	Severity Level 1	Level 1	Level 2	Level 3	Level 4	Image count
Predicted	Level 1	5	0	0	0	5
	Level 2	0	18	0	0	18
	Level 3	0	1	40	0	41
	Level 4	0	0	0	12	12
	Image count	5	19	40	12	76

Table 2 and Fig. 6 shows a detailed classifier outcome of the proposed model in terms of four accuracy measures.

It is observed that the DR images under level 1 (normal) category are classified by the accuracy of 100%, sensitivity of 100%, specificity of 100%, and F1 score of 1. Next, the DR images under level 2 categories are classified by the accuracy of 98.68%, sensitivity of 95%, specificity of 100%, and F1 score of 0.97. Similarly, the DR images under level 3 categories are classified by the accuracy of 98.68%, sensitivity of 100%, specificity of 98%, and F1 score of 0.99. In the same way, the DR images under level 4 are classified by accuracy of 100%, sensitivity of 100%, specificity of 100%, and F1 score of 1. These values shows that the proposed system effectively grades the DR images under respective classes.

The accuracy obtained for the different number of folds in five test runs of the proposed system is shown in Table 3. During the 5 different runs, the average accuracy

Table 2 Performance measures of DR severity levels

Severity Level	Accuracy (%)	Specificity (%)	Sensitivity (%)	F1 score
Level 1	100	100	100	1
Level 2	98.68	100	95	0.97
Level 3	98.68	98	100	0.99
Level 4	100	100	100	1

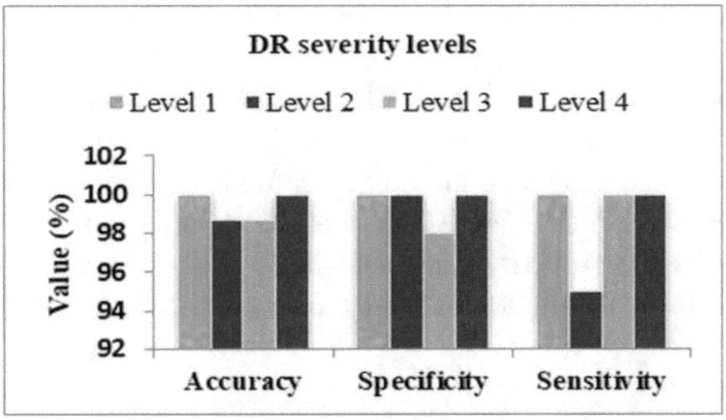

Fig. 6 Comparison of DR severity levels based on performance metrics

Table 3 Accuracy values for varying number of folds for five test runs

Number of folds	Accuracy (%)						Standard deviation
	1st run	2nd run	3rd run	4th run	5th run	Average	
1	54.5	52.6	54.5	56.4	55.2	54.64	0.89
2	73.3	72.1	72.4	74.2	73.3	73.06	0.51
3	94.9	93.2	91.2	92.8	95.4	93.5	1.51
4	96.7	95.5	94.5	96.5	97.4	96.12	0.9
5	99.3	98.2	98.6	97.4	100	98.7	0.45

gradually increases from 54.64% to 98.68%. The standard deviation at the lowest accuracy was around 0.89 and at fivefold validation it could reach up to 0.45. The deviation for the intermediate runs was found to be 1.51. For single fold cross-validation, the average accuracy obtained was 54.64%, which was the lowest among all the folds tested. The average accuracy increases with an increasing number of folds, with the highest average accuracy of 98.68% being obtained at the five-fold cross-validation. Throughout the five test runs the accuracy values obtained are fairly consistent for each number of folds. Overall, the trend of accuracy was seen improving till the five-fold validation and then got saturated. Hence, we decided to select five folds as the optimal number of folds.

5 Discussions

The comparison of the proposed method with the existing methods in the literature is shown in Table 4. Out of all the systems reported in the literature, our system could achieve one of the highest accuracy of 98.68% and sensitivity of 98.75%. Similarly, the proposed method could deliver the highest specificity of 99.5% which was similar to as reported by Lachure et al. [10]. The Area Under Curve (AUC) was not computed by most of the reported literature, but amongst the papers that reported it, the proposed method had the maximum AUC of 1. The average F1 score for all the severity levels was found out to be 0.99. The proposed method uses a unique combination of the green sub-color plane from RGB and the value sub-color plane from the HSV plane to enhance the overall features of the fundus image. An AGVE algorithm was applied on the merged plane to extract the important information from the eye. The red score generated for each image enabled to highlight the reddishness of the eye. Due to the unique combination of the color planes and the selection of vital features using AGVE and SURF, the proposed method was able to accurately detect the severity grade level of DR from a fundus image. Hence, our method could achieve the highest accuracy among the other methods reported literature.

Table 4 Performance comparison of other algorithms for DR detection

Author	Technique	Accuracy (%)	S ens itivity (%)	S pecificity (%)	Area Under Curve (AUC)
Pires et al. [9]	Extraction of mid level features-BossaNova and Fisher Vectorand Fisher Vector	–	–	96.4	–
Lachure et al. [10]	Morphological operations for detecting red and bright lesions and GLCM	–	90	100	–
Abbas et al. [11]	Vis ual features and a deep-learning neural network (DLNN) model	–	92.18	94.5	0.924
Costa et al. [12]	Dense and sparse feature extraction using CNN and SURF	–	–	–	0.93
Gargeya et al. [13]	Deep learning network with data driven features	–	94	98	0.97
Ro y et al. [14]	CNNs with generative and discriminative bag-of-word dictionary	61	92	–	–
Islam et al. [15]	SURF features with Kmeans clustering	94.4	94	94	0.95
Carerra et al. [16]	Quantitative feature extraction	85	94.6	–	–
Koh et al [17]	PHOGand SURF descriptors using the canonical correlation analysis approach	96.21	95	97.42	–
Leeza et al. [18]	Computation of descriptive features through SURF and histogram of oriented gradients	98.3	95.92	98.9	–

(continued)

Table 4 (continued)

Author	Technique	Accuracy (%)	S ens itivity (%)	S pecificity (%)	Area Under Curve (AUC)
Gayathri et al. [19]	Haralick and Anis otropic Dual-Tree ComplexWavelet Transform (ADTCWT) features	97.42	97.5	0.994	–
Gadekallu et al. [20]	Principal Component Analys is (PCA) and firefly based deep neural network model	73.8	83.6	83.6	–
Alyoubi [21]	Deep learning based models for DR detection and localization of DR lesions	89	97.3	89	–
Proposed method	Severity classification using SURF features from the enhanced green and value plane and AGVE algorithm	98.68	98.75	99.5	1

6 Conclusion

Diabetic retinopathy (DR) is a vision-related consequence of long-term diabetes. If not treated on time it can lead to complete loss of vision. As a result, an early diagnosis and a concise approach for identifying and rating the severity level of DR is required. Thus, our suggested study comprises four levels of severity classification approach employing an SVM classifier that predicts different levels of DR severity. The system performed pre-processing on the input fundus images to extract the enhanced green and value plane, which were then merged. The AGVE algorithm was used to highlight important details in the merged green and value plane image. Finally, the SURF and our novel red score method were used to extract the features of the fundus images, which were further used to train an SVM classifier. The unique method of feature extraction enabled the classifier to detect the severity accurately in the fundus images. Our system achieved an average accuracy of 98.68% and an average F1 score of 0.99. The proposed system can be easily deployed on the currently available equipment at hospitals and laboratories, which would enable the doctors to detect DR accurately and efficiently.

References

1. Nentwich MM, Ulbig MW (2015) Diabetic retinopathy-ocular complications of diabetes mellitus. World J Diabetes 6(3):489
2. Junior SB, Welfer D (2013) Automatic detection of microaneurysms and hemorrhages in color eye fundus images. Int J Comput Sci Inf Technol 5(5):21
3. Fruttiger M (2007) Development of the retinal vasculature. Angiogenesis 10(2):77–88
4. Dwyer MS, Melton LJ, Ballard DJ, Palumbo PJ, Trautmann JC, Chu C-P (1985) Incidence of diabetic retinopathy and blindness: a population-based study in rochester. Minnesota Diabetes Care 8(4):316–322
5. Mesquida M, Drawnel F, Fauser S (2019) The role of inflammation in diabetic eye disease in Seminars in immunopathology. Springer 41(4):427–445
6. Maher RS, Kayte SN, Meldhe ST, Dhopeshwarkar M (2015) Automated diagnosis non-proliferative diabetic retinopathy in fundus images using support vector machine. Int J Comput Appl 125(15)
7. Davidson JA, Ciulla TA, McGill JB, Kles KA, Anderson PW (2007) How the diabetic eye loses vision. Endocrine 32(1):107–116
8. Chang J-H, Gabison EE, Kato T, Azar DT (2001) Corneal neovascularization. Curr Opin Ophthalmol 12(4):242–249
9. Pires R, Avila S, Jelinek HF, Wainer J, Valle E, Rocha A (2015) Beyond lesion-based diabetic retinopathy: a direct approach for referral. IEEE J Biomed Health Inform 21(1):193–200
10. Lachure J, Deorankar A, Lachure S, Gupta S, Jadhav R (2015) Diabetic retinopathy using morphological operations and machine learning. In: IEEE international advance computing conference (IACC). IEEE, India, pp 617–622
11. Abbas Q, Fondon I, Sarmiento A, Jimenez S, Alemany P (2017) Automatic recognition of severity level for diagnosis of diabetic retinopathy using deep visual features. Med Biol Eng Compu 55(11):1959–1974
12. Costa P, Campilho A (2017) Convolutional bag of words for diabetic retinopathy detection from eye fundus images. IPSJ Trans Comput Vis Appl 9(1):1–6
13. Gargeya R, Leng T (2017) Automated identification of diabetic retinopathy using deep learning. Ophthalmology 124(7):962–969
14. Roy P, Tennakoon R, Cao K, Sedai S, Mahapatra D, Maetschke S, Garnavi R (2017) A novel hybrid approach for severity assessment of diabetic retinopathy in colour fundus images. In: IEEE 14th international symposium on biomedical imaging. IEEE, Australia, pp 1078–1082
15. Islam M, Dinh AV, Wahid KA (2017) Automated diabetic retinopathy detection using bag of words approach. J Biomed Sci Eng 10(5):86–96
16. Carrera EV, Gonzalez A, Carrera R (2017) Automated detection of diabetic retinopathy using svm. In: IEEE XXIV international conference on electronics, electrical engineering and computing (INTERCON). IEEE, Peru, pp 1–4
17. Koh JE, Ng EY, Bhandary SV, Laude A, Acharya UR (2018) Automated detection of retinal health using phog and surf features extracted from fundus images. Appl Intell 48(5):1379–1393
18. Leeza M, Farooq H (2019) Detection of severity level of diabetic retinopathy using bag of features model. IET Comput Vision 13(5):523–530
19. Gayathri S, Krishna AK, Gopi VP, Palanisamy P (2020) Automated binary and multiclass classification of diabetic retinopathy using haralick and multiresolution features. IEEE Access 8(504):497–557
20. Gadekallu TR, Khare N, Bhattacharya S, Singh S, Reddy Maddikunta PK, Ra I-H, Alazab M (2020) Early detection of diabetic retinopathy using pca-firefly based deep learning model. Electronics 9(2):274

Hybrid Error Detection Based Spectrum Sharing Protocol for Cognitive Radio Networks with BER Analysis

Anjali Gupta and Brijendra Kumar Joshi

1 Introduction

Cognitive Radio (CR) is a prominent methodology to conquer the resource shortages of the radio spectrum. To maximize the spectrum usage efficiency, an unlicensed or SU is permitted to acquire the licensed bands issued to a PU in CR networks [1]. Through the similar spectrum band, the simultaneous transmissions of PU and SU are permitted by the spectrum-sharing method. The corresponding approach can be categorized in to two types including underlay and overlay models [2]. In case of underlay model, the signal power of SU is inadequate and so the interference to PU is less than the standard level of noise. The licensed spectrum can be accessed by SU even if PU is in operating condition. In case of overlay model, SU collaborates with PU by sharing the power of SU and improves the QoS (Quality of service) of PU [3, 4]. This process is demonstrated as the cooperative spectrum sharing. PU permits SU to exchange the licensed spectrum for compensation purpose and hence CN accomplishes spectrum sensing, resource optimization, allocation and mobility [5].

Sensing of spectrum holes is the initial step involved in sharing of spectrum. The respective spectrum portions are determined by the SU or cognitive user which can be utilized by PU with respect to space, frequency domain and time. Once, if the accurate determination of spectrum holes are established, the further step to be pursued is distribution and allocation of the obtainable band to the corresponding SUs [6]. In order to undergo coordination between SUs and PUs, accessing of enhanced spectrum is considered to be the third major requirement. The fourth step is spectrum

A. Gupta (✉)
Shri Vaishnav Vidyapeeth Vishwavidyalaya, Indore, MP, India
e-mail: chitraanjali@yahoo.com

B. K. Joshi
Military College of Telecommunication Engineering, MHoW, MP, India

P. Singh et al. (eds.), *Machine Learning and Computational Intelligence Techniques for Data Engineering*, Lecture Notes in Electrical Engineering 998,
https://doi.org/10.1007/978-981-99-0047-3_73

handoff which administrates the PU's working, switching and to neglect the delay or collision between the users, for this the mobility of SUs are considered. Different approaches of cooperative spectrum-sharing have been implemented by depending on Amplify-and-Forward (AF) and Decode-and-Forward (DF) protocols [7, 8]. Thus, it can be stated that, Maximum Ratio Combining (MRC) policy does not offer overall diversity gain to PU in CR networks. Also, several approaches have been established to diminish the propagation of error and to attain a diversity gain in case of practical systems. The major contributions of this research are given below:

- To propose a hybrid error detection based spectrum sharing model by the intrusion detection scheme for error mitigation in the cognitive radio networks.
- Design a Neumann series based minimum mean square error assisted detector (NS-MMSEAD) to reduce the computational complexities and enhance the performance of the CRN system.
- Evaluation of the developed model is analyzed in terms of BER under PU and SU, throughput, end to end delay, average power consumption and average total utility.

2 Related Work

Jain et al. [9] had developed interference cancellation techniques for spectrum sharing in CRNs. To attain a desired quality of service (QoS) for both the primary and secondary systems, the author introduced a three phase cooperative decode and forward relaying. In addition to this the space time block coding (STBC) was developed to cancel the interference at both the primary and secondary receiver. Finally, the BER and outage probability under both the primary and secondary system is analyzed.

Lakshmi et al. [10] had proposed a simplified swarm based spectrum sharing model in CRN technology. The maximum utilization of the spectrum is an essential concern of this research. Here, the performance analysis of Throughput, Latency, End-To-End Delay, Average Power Consumption, Average Adaptation Time and Average Total Utility are analyzed.

Zhang et al. [11] had developed BER analysis of chaotic CRN over slow fading channels. The performance of BER is analyzed to attain high security and high flexibility. In this cognitive network, the chaotic sequence had been generated instead of time and frequency domain. Here, BER had been computed over the slow fading channels such as Additive White Gaussian Noise (AWGN) channel, slow flat Rayleigh, Rician, and Nakagami fading channels.

Bhandari et al. [12] had developed energy detection based spectrum sharing technique in the CRN technology. In order to characterize the spectrum sharing, the channel allocation time, probability of false alarm detection and spectral efficiency are the major performance measures. Receiver Operating Characteristics (ROC) curve is made to analyze the detector performance. Here, the allocation of the spectrum to the SU has been done by a coalition based cooperative game. Vickrey–Clarke–Groves (VCG) auction mechanism was introduced for spectrum allocation to each cognitive user.

Kim et al. [13] developed an improved spectrum sharing protocol for CRNs with multiuser cooperation. Here, the authors have proposed a cooperative maximal ratio combining scheme for error reduction and attain gain diversity. An optimization problem was considered to optimize the BER of the SU. The performance of BER and spectral efficiency are analyzed. The major challenge of CRN is spectrum sharing. However, the existing methods suffer due to some major drawbacks such as spectrum scarcity, computational complexity and error in the primary and secondary systems. To overcome this, the author introduced error detection based spectrum sharing the model to mitigate the error and improve the system performance.

3 Proposed Methodology

The proposal introduces spectral sharing approach for the CRNs. Effective utilization of spectrum resources is an important factor in wireless communication which reduces spectrum scarcity. Sharing of available spectrum to users is a prominent issue. So, the hybrid error detection based spectrum sharing model is proposed for efficient spectrum sharing. In CRNs, a SU can access the licensed bands of a PU as a compensation for transmission. During transmission, the detection error at the SU degrades the performance of the system. So, the minimum mean square error assisted detector is introduced to mitigate the error propagation. However, the computational complexity of Minimum Mean Square Error (MMSE) is high. In order to reduce the computational complexity, the Neumann Series (NS) is jointly contributed with MMSE to be called as Neumann Series-Minimum Mean Square Error Assisted Detector (NS-MMSEAD). To evaluate the hybrid scheme, the BER and diversity order of PU and SU are analyzed and compared with existing methods.

3.1 Spectrum Sharing System Model

Let us consider the cognitive radio network with the primary and secondary system. The primary and secondary systems are intended with primary transmitter P_T, secondary transmitter P_R, secondary transmitter S_T and secondary receiver S_R respectively. Several factors affect the spectrum sharing and the factors includes BER, available free carriers, transmission power etc. This paper aims at sharing the available spectrum to different cognitive users. For perfect spectrum sharing, the transmission power of each subcarrier is evaluated. Along with this, the Multiple Quadrature Amplitude Modulation (MQAM) is used and the MQAM order is utilized for spectrum sharing. Let us consider the specific allocation as:

$$X = \begin{bmatrix} x_{11} & x_{12} & \cdots & x_{1Q} \\ x_{21} & \vdots & \ddots & x_{2Q} \\ \vdots & \vdots & \ddots & \vdots \\ x_{k1} & x_{k2} & \cdots & x_{kQ} \end{bmatrix} \tag{1}$$

here, $x_{kq} = \left(M_{kq}, TP_{kq}\right)$. . The Total Transfer Rate (TTR) of the cognitive network system is given in Eq. (2).

$$TTR = \mathrm{B} * \sum_{q=1}^{Q} \sum_{k=1}^{K} V_{kq} \tag{2}$$

Here, V_{kq} is represented as $V_{kq} \cong 0.5 * \log_2 M_{kq}$. B signifies the subcarrier bandwidth. K and Q specify the cognitive users and subcarriers. M_{kq} resembles the modulation order of QAM and TP_{kq} denotes the transmission power of users and subcarriers.

$$\begin{aligned} &\left(\mathrm{B} * \sum_{q=1}^{Q} V_{kq}\right) \geq TR_{\min} \\ &E \leq E_{\max}, \\ &TTP \leq TP_{\max} \end{aligned} \tag{3}$$

Here, $TR_{\min}$ specifies the minimum transfer rate of cognitive users, $E_{\max}$ describes the maximum permissible BER, $TP_{\max}$ denotes the maximum usage of total power of the system. E resembles the average BER over the system and TTP specifies the total transmission power of the system [14].

$$E = avg\left(E_{kq}\right) \tag{4}$$

$$E_{kq} = \left(\frac{4}{\log_2 M_{kq}}\right)\left(1 - \frac{1}{\sqrt{\log_2 M_{kq}}}\right) erfc\left(\sqrt{\frac{3 * \log_2 M_{kq} * TP_{kq}}{\left(M_{kq} - 1\right) * Q}}\right) \tag{5}$$

Here, $erfc$ denotes the complementary error function. Therefore, the Total Transmit Power (TTP) is represented as:

$$TTP = \sum_{q=1}^{Q} \sum_{k=1}^{K} TP_{kq} \tag{6}$$

3.2 Error Detection Based Spectrum Sharing Protocol

The improved spectrum sharing model based error detection approach is proposed for an efficient spectrum sharing. Here, in this paper the MMSEAD is utilized to overcome or detect the errors. But, due to the modulation order the computational complexity burden grows up rapidly. To progress the computational issues and mitigate error propagation, the NS is incorporated with the MMSE detector. In the first time slot T_s, P_T transmit their data to P_R in sequence. After the data transmission, the received signals of P_R, S_T and S_R are concurrently expressed below:

$$\begin{aligned} r\,P_R,\, j &= G\,P_{T_j},\, P_R x_j + n\,P_R,\, j \\ r\,S_T,\, j &= G\,P_{T_j},\, S_T x_j + n S_T,\, j \\ r\,S_R,\, j &= G\,P_{T_j},\, S_R x_j + n S_R,\, j \quad j = 1, 2, \ldots.., T_s \end{aligned} \tag{7}$$

Here, G specifies the fading gain between the transmitter and receiver. x_j specifies the transmitted data. In the $(T_s + 1)$th time slot, the received signals of P_R and S_R are represented as:

$$\begin{aligned} r\,P_R,\, T_s + 1 &= G\,S_T,\, P_R\,x\,S_T + n\,P_R,\, T_s + 1 \\ r\,S_R,\, T_s + 1 &= G\,S_T,\, S_R\,x\,S_T + n\,S_R,\, T_s + 1 \end{aligned} \tag{8}$$

The MMSE detection is employed to attain an optimal BER at P_R and S_R. Based on MMSE equalization, the estimate of the transmitted signal is defined in Eq. (9).

$$x = \left(H^H H + \frac{\sigma_n^2}{\mathrm{P}} I_k \right)^{-1} H^H y = W^{-1} a \tag{9}$$

Here, $a = H^H y$ defines the matched filter output of y, I defines the identity matrix and P specifies the average transmit power per user. Therefore, the MMSE weight matrix W can be defined as:

$$W = G + \frac{\sigma_n^2}{\mathrm{P}} I_l \tag{10}$$

G defines the gain matrix and $H^H H$ describes the Hermitian positive definite matrix [15]. The estimated symbol is represented as:

$$\hat{x}_j = \mu_j x_j + \mathrm{Z}_j \tag{11}$$

Here, Z_j represents the noise plus interference and μ_j represents the equalized channel gain. When compared to all detectors, the MMSE detector minimizes the mean square error and it solves the given problem:

$$\hat{x} = \arg\min \mathrm{P}_{x,k} = \left[\|\delta - x\|^2 \right] \tag{12}$$

Here, $\delta = Ey + a$ and E specify the $U \times V$ matrix. $a \in C^K$ and C^K resembles the complex constellation for the modulation order. The detector attains worst BER performance but the computational complexity of MMSE is high. To reduce the computational complexity, NS is employed by the matrix polynomial with the MMSE detector [16].

The posteriori mean $\Im^p$ and variance υ^p can be defined as:

$$\Im^p = \left(\Im + \frac{1}{\sigma^2} \upsilon^p \left(H^H y \right) - H^H H \Im \right) \tag{13}$$

$$\upsilon^p = \left(\upsilon^{-1} + \frac{1}{\sigma^2} \left(H^H H \right)^{-1} \right) \tag{14}$$

Here, $\Im$ and υ signify the priori mean and variance respectively. The gain matrix becomes diagonal and it results in NS expansion. The decomposition of regularized gain matrix is defined as $F = \upsilon^{-1} + \frac{1}{\sigma^2} G$. F^{-1} is approximated in the NS and it is expressed as:

$$F^{-1} = \sum_{j=0}^{L} \left(I_l - D^{-1} V^{-1} - \frac{1}{\sigma^2} D^{-1} G \right) D^{-1} \tag{15}$$

Here, l represents the transmission link. D and V specify the diagonal and vector element respectively. The extrinsic mean and variance are expressed as given in Eqs. (16) and (17) respectively [17].

$$\Im_k^e = \upsilon_K^e \left(\frac{\Im_k^p}{\upsilon_k^p} - \frac{\Im_k}{\upsilon_k} \right) \tag{16}$$

$$\upsilon_k^e = \left(\frac{1}{\upsilon_k^p} - \frac{1}{\upsilon_k} \right)^{-1} \tag{17}$$

Based on this scheme, the error gets mitigated and the computational complexity gets diminished. Along with this, the developed model computes the BER of both the PUs and SUs.

4 Result and Discussion

This study implements the error detection based cognitive network model for an efficient spectrum sharing. In order to solve the shortage of spectrum usage the spectrum sharing is an essential concern. The implementation of the proposed model is evaluated in Matlab platform to analyze the effectiveness of the spectrum sharing model. The number of users assumed is 50 and the available subcarriers is considered as 2000. The modulation technique used is MQAM. Both the modulation order and the transmission power are generated randomly. In addition to this, the minimum transfer rate is set as 64 Kbits/s. The maximum usage of total power is considered as 0.0005%. The proposed model is evaluated in terms of BER, throughput, delay and utilization. The analysis of BER is computed under both the PUs and SUs in terms of detection and modulation schemes.

4.1 Performance Analysis

(a) BER analysis under PUs and SUs. BER is calculated by comparing the total number of bit errors in transmitted and received sequence bits. The BER is analyzed under both the PUs and SUs. The BER of PU [13] is expressed in Eq. (18).

$$E_{PU} = \frac{1}{\log_2 M_{PU}} \frac{1}{|A_x|} \sum_{x \in A_x} \mu_{PU} \left(x_{ST}, \hat{x}_{ST}\right) E_{PU} \left(x \rightarrow \hat{x}\right) \tag{18}$$

The BER of SU is expressed in Eq. (19).

$$E_{SU} = \frac{1}{\log_2 M_{SU}} \frac{1}{|A_x|} \sum_{x \in A_x} \mu_{PU} \left(x_{ST}, \hat{x}_{ST}\right) E_{SU} \left(x \rightarrow \hat{x}\right) \tag{19}$$

Figure 1 illustrates the BER of detection schemes under PU and SU respectively. From Fig. 1 it is seen that the proposed detection scheme attains better performance compared to the existing methods. By varying SNR, the BER under both the PUs and Sus are evaluated. The existing detection schemes are computationally complex and degrade the system performance due to the detection errors. The existing schemes are limited to estimate the mean and standard deviation. Under SUs, the proposed method attains a BER of 10^{-3} and 10^{-4} when SNR = 20 and 35 dB respectively. Under PUs, the proposed scheme grasps a BER of 10^{-4} and 10^{-8} when SNR = 20 and 35 dB consecutively.

Figure 2 illustrates the BER of modulation schemes under PU and SU. Here, the modulation techniques of Quadrature Phase Shift Keying (QPSK), Binary Phase Shift Keying (BPSK) and Quadrature Amplitude Modulation (QAM) [13] are compared with MQAM. However, MQAM attains a low BER as compared to QPSK, BPSK and QAM. For PU, the MQAM modulation technique yields a BER of 10^{-6} and 10^{-8} when SNR = 35 and 45 dB. For SU, the MQAM modulation technique yields a BER of 10^{-5} and 10^{-6} when SNR = 30 and 50 dB.

(b) Throughput. The network parameter throughput is defined as the amount of data transmitted from the sender to the receiver in a particular time period. The expression of the throughput is given below:

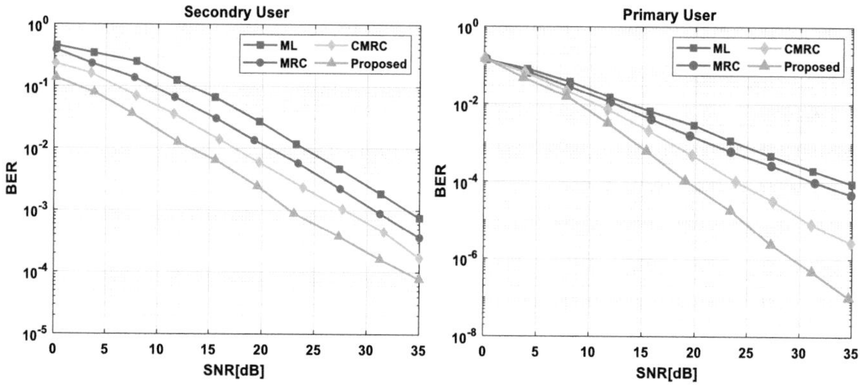

Fig. 1 BER of Detection Schemes. **a** BER for SU, **b** BER for PU

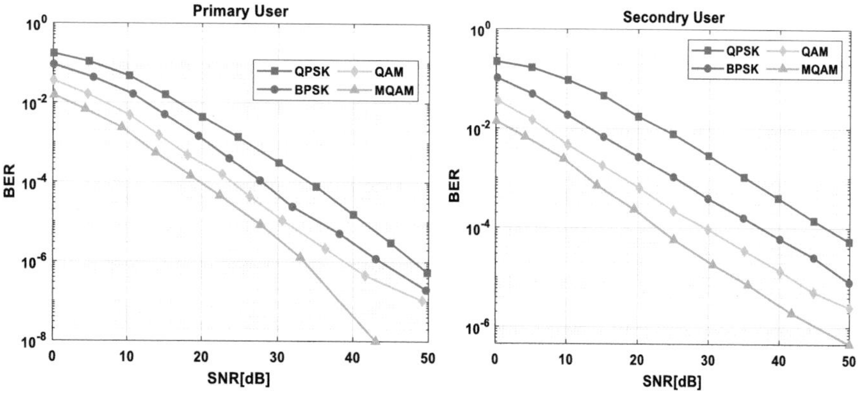

Fig. 2 BER of Modulation Techniques. **a** BER for SU, **b** BER for PU

$$Throughput = \frac{Transmitted\ data}{Time\ Taken} \tag{20}$$

The term throughput is an essential parameter of the network and it improves the network quality. Figure 3 illustrates the throughput of the proposed model and that of existing schemes. The proposed spectrum sharing design is compared with the existing methods of Simplified Swarm Optimization (SSO), Genetic Algorithm (GA) and Particle Swarm Optimization (PSO) [10]. From Fig. 3 it is seen that, the proposed model attains high throughput compared to existing techniques.

(c) End to End delay. The term delay is defined as the time needed for the destination to receive the data generated from the source application. The mathematical expression of the End to End delay is defined below:

$$Delay = (D + (N - 1)) * {}^{L}/_{T_R} \tag{21}$$

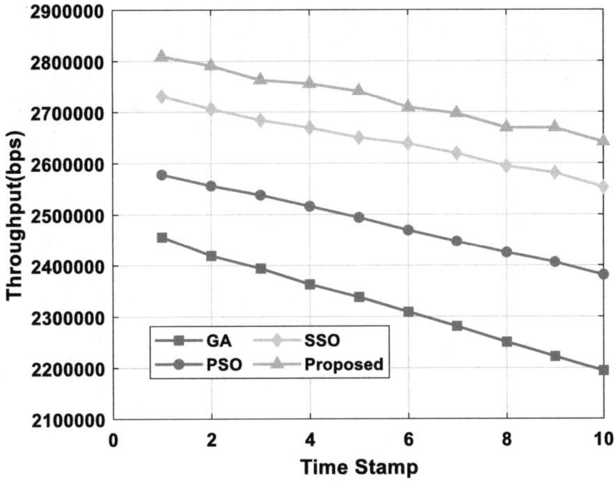

Fig. 3 Throughput

Here, D defines the transmitted data, N resembles the number of links, L denotes the data length and T_R specifies the transmission rate. The network parameter end to end delay is insisted with two phases namely, end point application delay and network delay. Based on end point applications, the end point delay is emphasized. The network delay is termed as the time difference between the first bit and the last bit at the receiver. Figure 4 illustrates the End to End delay parameter. From Fig. 4 it is sensed that the delay of proposed scheme is very low when compared to existing approaches. The lower value of delay reflects that the proposed model is highly efficient.

(d) Average Power Consumption. The average power consumption is defined as the measure of average power utilized by the nodes and this is evaluated by the simulation process. The average power consumption is measured in terms of mW. The power is stated as $P = V \times I$. V and I refers to the voltage and current. n defines

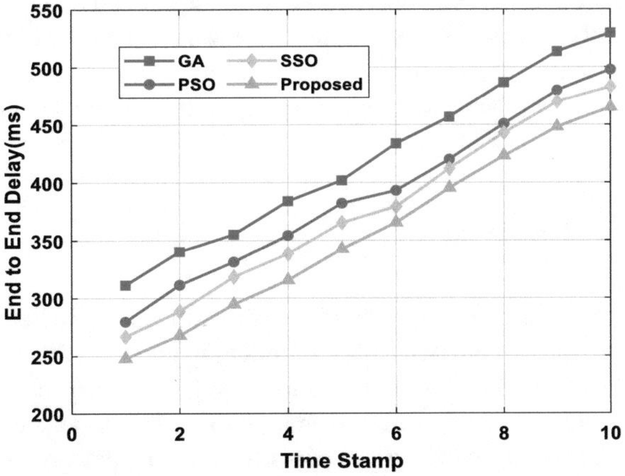

Fig. 4 End to End delay

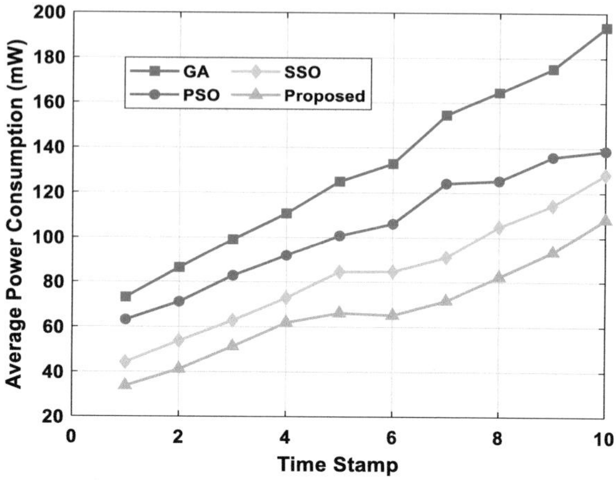

Fig. 5 Average Power Consumption

the number of nodes. The mathematical expression for average power consumption is defined below:

$$A_p = \frac{1}{n} \sum_{i=1}^{n} (V_i \times I_i) \quad (22)$$

Figure 5 illustrates the average power consumption metrics. When compared to existing methods, the major difference is the proposed method yields low power consumption and the low value reflects that the proposed method is highly accurate. From the simulation results it is proved that the proposed model is highly prominent.

(e) Average Total Utility. The parameter of average total utility is computed by the total utility of spectrum throughout. Figure 6 illustrates the average total utility by varying the number of nodes. The utilization of the proposed model is high when compared with the existing methods and the high value reflects that the proposed model is better. If the number of nodes is 100 and 400, the proposed model attends utilization of 89% and 83% whereas the existing methods of GA, PSO and SSO [10] reach an utility of 80% and 75%, 83% and 78%, 86% and 80% respectively. The numerical values of the network parameters are shown in Table 1.

5 Conclusion

The spectrum that is not effectively utilized by the PU or licensed user is accessed by the SU or CR user without producing any disturbances and this is the fundamental principle of CRN. In this paper the hybrid error detection based spectrum sharing model is presented. The design of hybrid error detection scheme mitigates the errors and enhances the system performance. The performance of BER, throughput, delay and the spectrum utility are computed and compared with existing models. The work

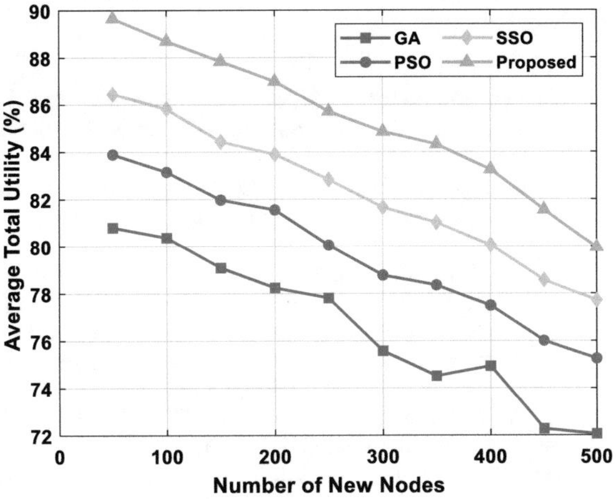

Fig. 6 Average total Utility

Table 1 Network parameters

Time stamp	Throughput (bps)			
	GA	PSO	SSO	Proposed
2	2,403,172	2,544,284	2,704,200	28,095 (bps)
4	2,343,530	2,497,877	2,664,740	279,800 (bps)
6	2,284,000	2,451,635	2,625,395	27,000 (bps)
8	2,224,380	2,405,396	2,586,034	269,879 (bps)
10	2,165,184	2,358,778	2,546,276	265,941 (bps)
2	333	303	284	260 (mS)
4	378	348	335	320 (mS)
6	429	387	378	360 (mS)
8	482	445	441	420 (mS)
10	527	491	482	460 (mS)
2	91	77	61	40 (mW)
4	115	96	79	50 (mW)
6	136	110	91	70 (mW)
8	166	129	109	80 (mW)
10	192	141	131	110 (mW)

can be extended with the construction of neural network architectures concentrating on spectrum sharing with various applications.

References

1. Sarala B, Rukmani Devi S, Joselin Jeya Sheela J (2020) Spectrum energy detection in cognitive radio networks based on a novel adaptive threshold energy detection method. Comput Commun 152:1–7
2. Haldorai A, Kandaswamy U (2019) Cooperative spectrum handovers in cognitive radio networks. In: Intelligent spectrum handovers in cognitive radio networks. Springer, Cham, pp 1–18
3. Afzal H, Rafiq Mufti M, Raza A, Hassan A (2021) Performance analysis of QoS in IoT based cognitive radio Ad Hoc network. Concurr Comput: Pract Exp 33(23):e5853
4. Guda S, Rao Duggirala S (2021) A Survey on cognitive radio network models for optimizing secondary user transmission. In: 2021 2nd international conference on smart electronics and communication (ICOSEC). IEEE, pp 230–237
5. Saradhi DV, Katragadda S, Valiveti HB (2021) Hybrid filter detection network model for secondary user transmission in cognitive radio networks. Int J Intell Unmanned Syst
6. Saraç S, Aygölü Ü (2019) ARQ-based cooperative spectrum sharing protocols for cognitive radio networks. Wirel Netw 25(5):2573–2585
7. Pandeeswari G, Suganthi M, Asokan R (2021) Performance of single hop and multi hop relaying protocols in cognitive radio networks over Weibull fading channel. J Ambient Intell Humaniz Comput 12(3):3921–3927
8. Perumal B, Deny J, Sudharsan R, Muthukumaran E, Subramanian R (2021) Analysis of amplify forward, decode and amplify forward, and compression forward relay for single and multi-node cognitive radio networks
9. Jain N, Vashistha A, Ashok Bohara V (2016) Bit error rate and outage analysis of an interference cancellation technique for cooperative spectrum sharing cognitive radio systems. IET Commun 10(12):1436–1443
10. Rajalakshmi, Sumathy P (2020) Spectrum allocation in cognitive radio–simplified swarm optimization based method. Int J Eng Adv Technol (IJEAT) 9(3):2249–8958
11. Zhang L, Lu H, Wu Z, Jiang M (2015) Bit error rate analysis of chaotic cognitive radio system over slow fading channels. Ann Telecommun-Annales Des Télécommunications 70(11):513–521
12. Bhandari S, Joshi S (2021) A modified energy detection based dynamic spectrum sharing technique and its real time implementation on wireless platform for cognitive radio networks. Indian J Eng Mater Sci (IJEMS) 27(5):1043–1052
13. Kim T-K, Kim H-M, Song M-G, Im G-H (2015) Improved spectrum-sharing protocol for cognitive radio networks with multiuser cooperation. IEEE Trans Commun 63(4):1121–1135
14. Mishra S, Sagnika S, Sekhar Singh S, Shankar Prasad Mishra B (2019) Spectrum allocation in cognitive radio: A PSO-based approach. Periodica Polytechnica Electri Eng Comput Sci 63(1):23–29
15. Gao X, Dai L, Ma Y, Wang Z (2014) Low-complexity near-optimal signal detection for uplink large-scale MIMO systems. Electron Lett 50(18):1326–1328
16. Wang F, Cheung G, Wang Y (2019) Low-complexity graph sampling with noise and signal reconstruction via Neumann series. IEEE Trans Signal Process 67(21):5511–5526
17. Khurshid K, Imran M, Ahmed Khan A, Rashid I, Siddiqui H (2021) Efficient hybrid Neumann series based MMSE assisted detection for 5G and beyond massive MIMO systems. IET Commun 14(22):4142–4151

Lie Detection with the SMOTE Technique and Supervised Machine Learning Algorithms

M. Ramesh and Damodar Reddy Edla

1 Introduction

The term "Brain-Computer Interface (BCI)" [1] refers to a procedure for mapping the electrical potential produced from the brain to a device such as a wheelchair, prosthetic, or computer. Invasive and non-invasive methods are used to measure brain potentials. Non-invasive data collection methods are most commonly employed in this area of brain computing research. Earlier, polygraphy [2] tests to decide whether a person was lying or not. These tests depend on the autonomic nervous system activity such as rapid heartbeat, sweating rate, and muttering. If these polygraph examinations are conducted secretly, they might fail to identify the guilty party. Using an autonomic nervous activity to identify the guilty is insufficient for deciding their guilt. Brain-Computer Interface (BCI) technology can assist by recording the guilty person's brain activity in various ways. Electroencephalography is one of the various BCI procedures routinely utilized (EEG). Non-invasively [3], EEG detects brain movement from the brain scalp. It is important to note that various brain reactions produce distinct EEG signals. One of the tests applies the "Concealed Information Test (CIT)" to detect lying in these current situations. Subjects are required to take on the role of either a criminal or an innocent victim prior to performing the CIT. The subject to be shown to be a criminal must perform various criminal-related tasks. Subjects are tested on their understanding of the crime scene during the CIT and correctly answer questions. An EEG device captures the subject's cognitive activity as people answer a series of questions based on the answers. A piece of acquisition equipment is used to acquire and evaluate these EEG signals. Event-

M. Ramesh (✉) · D. R. Edla
Department of Computer Science and Engineering, National Institute of Technology Goa, Farmagudi, Ponda, 403401 Goa, India
e-mail: m.ramesh@nitgoa.ac.in

D. R. Edla
e-mail: dr.reddy@nitgoa.ac.in

P. Singh et al. (eds.), *Machine Learning and Computational Intelligence Techniques for Data Engineering*, Lecture Notes in Electrical Engineering 998,
https://doi.org/10.1007/978-981-99-0047-3_74

Related Potentials (ERPs) are the scientific name for these brain responses (ERP). When participating in any cognitive exercises and being asked about a crime-related subject. A positive peak is evoked 300 milliseconds after the probe is asked. In fundamental terms, this positive effect leads to an ERP component known as P300. These are different; crime-related questions are a stimulus for the topic. As a result, human intents do not show these brain neuronal activities and cannot differentiate between innocent and guilty. Different researchers have used various methodologies to categorize the EEG signals produced by individuals' brains. The participant is typically exhibited with three categories of stimuli: Target, Probe, and Irrelevant while doing CIT. The questions known as both guilty and innocent participants are the target stimuli. Stimuli that are unknown to any of the individuals are irrelevant. The probing is the sporadic appearance of crime-related stimuli. Many writers reported a variety of CITs and processes for EEG data classification using this set of stimuli.

Farwell et al. [4] developed a CIT and used EEG equipment to record reactions. They used the bootstrapping technique to capture the reactions using three EEG electrodes, Cz, Pz, and Fz. The P300 response is concerned with amplitude difference and correlation. Bootstrapping is a technique for measuring distributions by randomly picking a limited number of samples. Other authors' work is based on bootstrapping analyses using various CITs. Another study used a bootstrapping approach to investigate three different forms of oddball stimulus. In most circumstances, an unusual stimulation is applied in subject-based psychological research. A one-of-a-kind stimulation causes the subject's brain to react considerably. Statistical analysis [5–7] was also employed in other investigations to detect deceit. Because EEG data is biological, it contains a variety of artifacts. Because the EEG program typically overlaps over time, the P300 [6] component of ERP is used to detect deception behavior that cannot be split into a single trial. Using multiple pattern recognition approaches, non-P300 and P300 were separated.

Gao et al. [6], for example, used a template matching technique based on Independent Component Analysis (ICA). Using ERP data, this approach was denoised and separated into non-P300 and P300 components. P300 components were retrieved and rebuilt from decomposed data using a template matching method at the Pz location. Arasteh et al. [8] used three EEG channels, Cz, Pz, and Fz, to test the Empirical Mode Decomposition (EMD) approach. The EMD decomposes the signal into multiple Intrinsic Mode Functions (IMFs) and offers temporal and frequency domain characteristics. Using Empirical Mode Decomposition, the authors demonstrated greater accuracy when compared to a single-channel electrode. Other researchers developed unique EEG data categorization processes and feature extraction approaches. To avoid highly favorable results, a genetic SVM employs a stringent validation framework [9]. As a result, the authors obtained numerous ideal solutions and chose the most common outcome to arrive at one optimal solution.

Above, researchers drawback the handling to the imbalance EEG signal data. The 16-channel electrode for the human brain to capture EEG channels data from the concealed information test. This research proposes handling the EEG signal imbalanced data dealt with the SMOTE technique, and machine learning methods will aid in the analysis of large amounts of data to lie detection. This system used the

SMOTE method to remove imbalanced EEG channel data sets and different classifiers, such as the KKN, DT, LR, RM, and SVM methods, to improve the system's lie detection accuracy. The rest of the paper is as follows: The following Methodology is explained in Sect. 2. The experimental data and analysis are presented in Sect. 3. The conclusion is described in Sect. 4 and Feature work is described in Sect. 5.

2 Methodology

2.1 Supervised Machine Learning Algorithms

This proposed work uses the SMOTE technique to deal with the imbalanced EEG channel data set. This paper provides a strategy for lie detection using classification methods and enhancing classification accuracy using an ensemble of classifiers. The EEG channel data is split into a 90% training set and a 10% test set. Individual classifiers are trained using a train set and test data at each classifier stage. The performance of the classifiers is evaluated using test data.

2.2 K-Nearest Neighbor (KNN)

KNN is a simple way to classify data and help non-linearly [10] or statistical training approaches. K is the number of adjacent neighbors used, directly defined in the object creator or assessed using the declared value's upper limit [11–13]. All problems are categorized in the same way, and each new case is classified by calculating its resemblance to all previous examples. When an unrecognized selection is received, the closest neighbor method searches the pattern space for the k training samples immediately adjacent to the unidentified model. The test instance can predict many neighbors based on their distance, and two separate approaches for translating the space into a weight are introduced [14–16]. The approach has various advantages, including that it is logically tractable and easy to apply. Because it only deals with a single instance, the classifier does well in lie detection. The best-fit parameters for the data set in this investigation were n neighbors two and leaf size 50.

2.3 Decision Tree (DT)

One of the most well-known and widely used machine learning algorithms is the decision tree. A DT produces decision logic to analyze and coordinate decisions for organizing data objects into a tree-like structure [10]. DT often the multiple nodes, with the top-level as the parent or root node and the others as child nodes. Every inner node with at least one child node assesses the input variables or aspects. The evaluation result directs components to the appropriate child node, and the evaluation and branching procedure is repeated until the leaf node is reached. The terminal or leaf nodes describe the outcomes of the decisions. DT is easy to learn and faster, yet it is necessary for many approaches [17]. The decisions' results are described based on the terminal or leaf nodes. While DT is simple to learn and master, it is essential for many techniques [18]. The classifier in this experiment produced the most suitable classification results for the applied EEG channel data set by selecting the maximum depth value of 7.

2.4 Logistic Regression (LR)

This is a LR technique that can be used for the traditional statistics as well as machine learning algorithms. It expands the general regression modeling to a EEG channels data set, denoting a particular instance's probability of appearance or non-occurrence [19]. It predicts the subsequent variables. We must consider two factors: weight and size. To utilize the line to forecast lie detection given the EEG data, consider the x-axis the guilty class and the y-axis the innocent class. LR is used to determine the chances of a new observation belonging to a specific class, such as 0 and 1. To implement the LR as binary classification, a threshold value is assigned that specifies the separation into two groups. For example, a probability value more than 0.5 is defined as "Guilty class," whereas a probability value less than 0.5 is designated as "Innocent class." The LR model is generalizable as a multinomial logistic regression model [20].

2.5 Random Forest

Random Forest is a DT-based [21] ensemble learning-based data categorization approach. It grows many trees during the training stage and a forest of decision trees [22]. Every tree, or forest member, forecasts the class label for every instance during testing. A majority vote is used to determine the conclusion for each test data when each tree accurately denotes a class label [23]. The class label with the most significant votes should be used on the test data. This process is done for each data set in the collection.

2.6 Support Vector Machine (SVM)

SVM is a classification-based technique [16] that employs kernels to calculate the distance between two values. It then constructs a border or hyper-plane whose primary function is to maximize the distance between the nearest value points of two separate classes, therefore separating them. In our situation, the two types are guilty and innocent class. The maximum margin hyper-plane is the name of the hyper-plane. The dot product of vectors is used to calculate the distance between two points, and the function [16] is defined as:

$$A(x) = L_0 + \sum_{i=1}^{n} (p_i \times (a \times b_i)) \quad (1)$$

$A(x)$ is SVM distance function
L_0,p_i are coefficients defined while training of data
a is new input vector
b_i is previous support vectors.

2.7 *Synthetic Minority Oversampling Technique (SMOTE)*

The SMOTE [24] approach is an oversampling strategy used in lie detection for a considerable time to deal with imbalanced class EEG data. SMOTE improves the number of data samples by producing random synthetic data from the minority class equivalent to the majority class using Euclidean distance [25] and the nearest neighbors. Because new samples are created based on original characteristics, they become identical to the original data. Because it introduces extra noise, SMOTE is not required for high-dimensional data. The SMOTE technique is used in this study to create a new training data set. SMOTE increased the number of data samples from 1200 to 2500 for each class.

2.8 *Performance Metrics*

There are some ways to assess the performance of machine learning models. An analytical study is likely to benefit from various examination instruments [26]. Figure 1 depicts the suggested lie detection flowchart. The metrics accuracy, precision, sensitivity, specificity, and F1-Score assess the various types of metrics in machine learning-based algorithms. The confusion matrix [27] assists the calculating all four metrics after learning the train data and providing test data for each classifier and step evaluation. True Positive (TP), True Negative (TN), False Positive (FP), and False-Negative (FN) are associated with the confusion matrix. In this search, we can consider the positive class as nothing but the innocent class and the negative class as nothing but the guilty class. The performance metrics indicators are listed below.

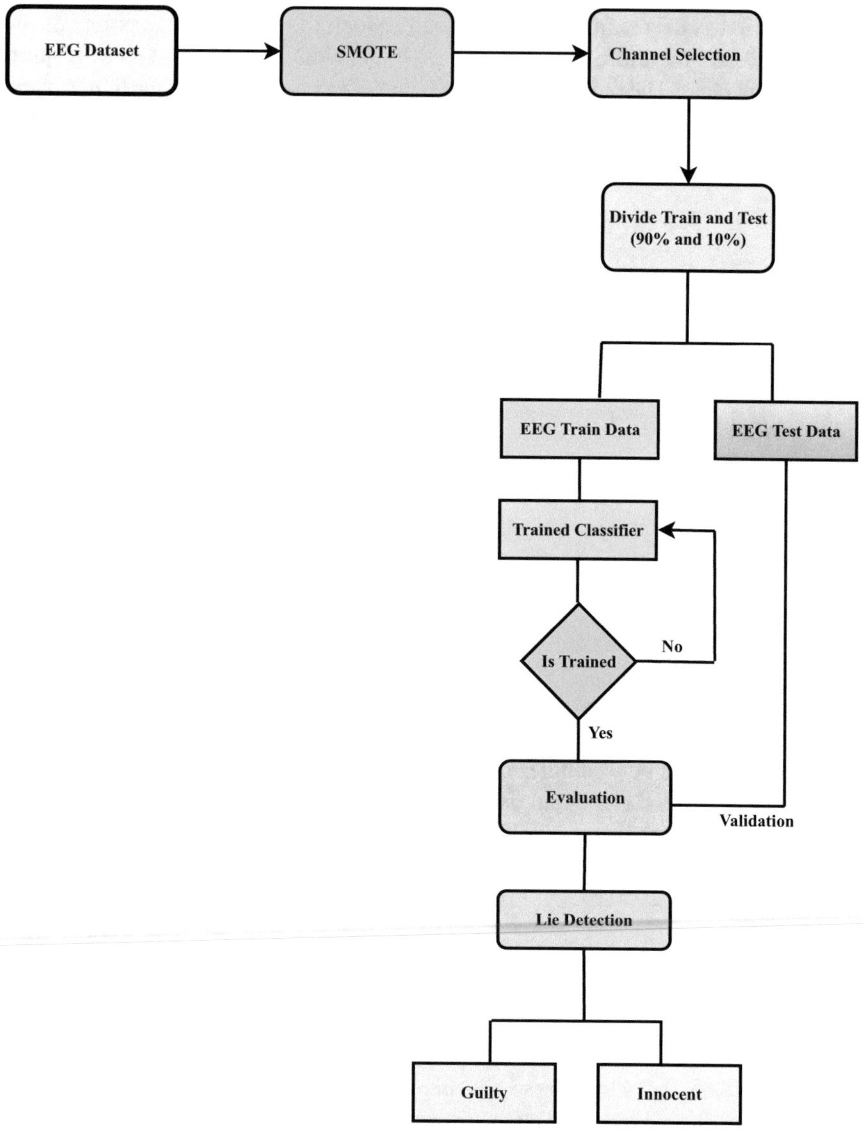

Fig. 1 Flowchart for lie detection using EEG data

$$Precision(\%) = \frac{TP}{TP + FP} \times 100 \quad (2)$$

$$Accuracy(\%) = \frac{TP + TN}{TP + TN + FP + FN} \times 100 \quad (3)$$

$$Sensitivity(\%) = \frac{TP}{TP + FN} \times 100 \quad (4)$$

$$Specificity(\%) = \frac{TN}{TN + FP} \times 100 \quad (5)$$

$$F_1score(\%) = 2 \times \frac{Recall \times Precision}{Recall + Precision} \times 100 \quad (6)$$

3 Experimental and Analysis

This section discusses all experiments' experimental designs and outcomes for lie detection. Implement the EEG data set, which was produced in CIT utilizing a 16-channel electrode brain cap, in this study. The SMOTE approach was used to elevate the minority of the class data from an inconsistent data collection. SMOTE was established to unbalance the data set balance data set. On the balanced data set, machine learning models were trained and assessed in accuracy, sensitivity, specificity, and F-Score. In this case, the outcomes with imbalanced data were compared to the results with balanced data.

3.1 Data Acquisition

We used EasyCap set [4] (EEG 32 channels Cap Set), sixteen electrodes, a brain vision recorder [24], and V-amplifier for the signal acquisition. In this case, EEG data recording for the placing Ag/AgCl electrodes (10–20 international system) at CP2, CP6, CP5, CP1, O1, O2, Oz, C3, Cz, C4, P4, Pz, P3, and Fz, FC2, FC1 sites as well as an electrode on the forehead as ground. These subjects are tested using three different stimuli types: target, irrelevant, and probe stimuli. A 15.4-inch display screen to be used to show images to the subjects. One is a probe stimulus, two are target stimuli, and the rest are irrelevant stimuli. These images of famous people will act as a probe, eliciting a P300 [28] response from the brain. Celebrities images elicit a P300 response, even though some of the random unknown elicits a non-P300 response. A brain vision analyzer is used to record and study the output of the answers. Subjects are trained to reply "no" or "yes" to probe stimuli while reading the stimulus. Subjects who answer "no" are assumed to be lying. This probe image is only shown once and generates P300 because it is associated with crime. Subjects' need causes them to say "no" to irrelevant stimuli, implying that they are telling the truth. The P300 will not produce a response because irrelevant stimuli are more likely to occur and are unrelated to the crime. When trained subjects recognize the image as a target stimulus, they respond with "yes". P300 response is unusual and has nothing to do with a crime with which the subject is still familiar.

3.2 *Feature Extraction*

The EEG data for the ten subjects (S1, S2,..., S10) were recorded from 16 channels, but subject-6 data could not be considered throughout the entire study process due to numerous artifacts. As a result, we used 16-channel data from 9 subjects (16*9) for 30 trials (lie session and truth session) for both sessions. Before evaluating CIT data, we pre-processed the EEG signal for artifact reduction using the HEOG and VEOG approaches [29]. Several statistical methods, such as Fourier transform, frequency-time domain features, and wavelet transformation methods, would be employed to extract features from the signal. After that, specified classifiers differentiate between the innocent and guilty classes. The entire experimental approach used in previous work is presented in detail [30].

3.3 *EEG Data Set*

The SMOTE approach is used to manage the unbalanced EEG data set in this research effort. An acquisition device is used to gather this information. A Concealed Information Test (CIT) will be carried out to investigate human lying behavior. The goal is to detect lies as "guilty" or "innocent" classes for binary classification. The EEG channel data collection contains 5930 recordings. 585 test records and 5345 train records are available. Each one has sixteen channels and only one class attribute: guilty or innocent. A data collection's categorization attribute can be set to "0" for guilty class records and "1" for innocent class records. This study uses imbalanced data sets, such as 3600 guilty and 1745 innocent records.

3.4 *Experimental Environment*

The Python programming language implements the proposed SMOTE methodology with machine learning technologies. The Intel i5 processor runs the PC on Windows ten operating system and has 8 GB of RAM.

Table 1 Results without SMOTE model on EEG data

Classifier	Accuracy	Precision (%)	Specificity (%)	Sensitivity (%)	F1 score (%)
KNN	78.4	76.67	75.50	81.35	78.94
DT	73.90	68.25	68.79	80	73.66
LR	72.91	67.11	67.11	80	72.99
Random forest	73.35	70.57	67.11	79.66	74.84
SVM	80.1	77.39	75.50	84.74	80.90

3.5 Experimental Results Without Smote

The supervised machine learning classifiers will be applied on a complete EEG channel imbalanced data set. Some classifiers performed well on evaluation metrics, whereas others did not. KNN, Decision Tree, LR, RF, and SVM classifiers were used in this study to lie detection using EEG channel data. Tabulated 1 shows the performance of machine learning models on the entire set of channel data with an imbalanced EEG data set. According to the data in Tabulated 1, The KNN classifier scored 78.4% accuracy, 76.67% precision, 75.50% specificity, 81.35% sensitivity, and 78.94% F1-Score. The DT classifier achieved 73.90% accuracy, 68.25% precision, 68.79% specificity, 80% sensitivity, and 76.33% F1-Score. The LR classifier achieved 72.99% accuracy, 67.11% precision, 67.11% specificity, 80% sensitivity, and 72.99% F1-Score. The RM classifier achieved 73.35% accuracy, 70.57% precision, 67.11% specificity, 79.66% sensitivity, and 74.84% F1-Score. The SVM classifier achieved 80.1% accuracy, 77.39% precision, 75.50% specificity, 84.74% sensitivity, and 80.90% F1-Score. In this proposed method using SVM classifier achieved the highest accuracy compared with other classifiers (Table 1).

3.6 Experimental Results with Smote

A similar analysis of supervised machine learning classifiers has been accomplished on a complete set of lie detection EEG channel imbalanced data sets. These imbalanced data sets handle the SMOTE technique and apply the machine learning algorithms to improve system performance. Following the improvement of the imbalance, data such as minority class is raised, followed by the use of the classifier for training data and evaluation of performance using test data for each classifier. Some of the classifiers exhibited favorable evaluation metrics results, and some exhibited poor performance. This work has applied KNN, Decision Tree, LR, RF, and SVM classifiers to lie detection using EEG channel data. Tabulated 2 demonstrates the performance evaluation of machine learning models on the complete set of channels data with an imbalanced EEG data set. According to the results in Tabulated 2,

Table 2 Results with SMOTE model on EEG data

Classifier	Accuracy	Precision (%)	Specificity (%)	Sensitivity (%)	F1 score (%)
KNN	93.97	89.74	90.60	98	93.69
DT	92.15	89.35	**96.60**	94	91.61
LR	94.89	91.41	92.28	98	94.59
Random forest	90.60	89.55	96.0	96	92.66
SVM	**95.64**	**93.15**	93.25	**98**	**95.51**

the KNN classifier performed 93.97% accuracy, 89.74% precision, 90.60% specificity, 98% sensitivity, and 93.69% F1-Score. The DT classifier performed well with 92.15% accuracy, 89.35% precision, 96.60% specificity, 94% sensitivity, and 91.61% F1-Score. The LR classifier performed 97.89% accuracy, 91.41% precision, 92.28% specificity, 98% sensitivity, and 94.59% F1-Score. The RM classifier performed 93.06% accuracy, 89.55% precision, 90.60% specificity, 96% sensitivity, and 92.66% F1-Score. The SVM classifier performed with 95.64% accuracy, 93.15% precision, 93.525% specificity, 98% sensitivity, and 95.51% F1-Score. Using the SVM classifier achieves the highest accuracy in this proposed method compared with other classifiers because it handles high-dimensional data, but the other classifiers do not.

3.7 Comparison Between the SMOTE and Without SMOTE

Without SMOTE, it is necessary to use an unbalanced data set and classifiers to lie detection and system performance. The SMOTE approach was used to elevate the minority of the class data from an inconsistent data collection. SMOTE was established to unbalance the data set balance data set. Machine learning models were trained and evaluated in accuracy, precision, recall, and F-Score on the balanced data set. Compared to the performance results with unbalanced data, the SMOTE technique produced the best results, such as accuracy of **95.64**%, precision **93.15**%, the sensitivity of **98**%, specificity of **93.52**%, and F-Score of **93.5**% (Table 2).

4 Conclusion

Using machine learning algorithms to process raw EEG channel data of lie information will aid human lie detection in real-time systems. This research proposes an effective and efficient SMOTE technique and a machine learning-based technique for lie detection. EEG channel data is an imbalance data set recorded using 16-channel electrodes in the CIT test using a brain cap. These skewed data are processed using the SMOTE methodology to improve the minority of the class data and machine learning methods such as a KKN, DT, LR, RM, and SVM to forecast lie detection and improve system performance. The performance of machine learning models is compared across all EEG channel data and SMOTE feed data. In this study's experiments, SVM classifiers with the SMOTE technique achieved the highest accuracy of 95.64%, the precision of 93.15%, and the F1 score of 95.51%, with the decision tree classifier achieving the highest specificity of 96.6%.

5 Feature Work

The SMOTE technique was proposed in this study to remove the imbalanced data. Increase the minority class until it equals the majority class. This method has the disadvantage of producing redundant data in the data set. The duplication of data is hampering the system's performance. More research is needed to use the various techniques for removing data duplication.

References

1. Bablani A, Edla DR, Tripathi D, Cheruku R (2019) Survey on brain-computer interface: an emerging computational intelligence paradigm. ACM Comput Surv (CSUR) 52(1):1–32
2. Farwell LA, Donchin E (1991) The truth will out: Interrogative polygraphy ('lie detection') with event-related brain potentials. Psychophysiology 28(5):531–547
3. Ramadan RA, Vasilakos AV (2017) Brain computer interface: control signals review. Neurocomputing 223:26–44
4. Farwell LA, Donchin E (1991) The truth will out: interrogative polygraphy ('lie detection') with event-related brain potentials. Psychophysiology 28(5):531–547
5. Rosenfeld JP, Soskins M, Bosh G, Ryan A (2004) Simple, effective countermeasures to P300-based tests of detection of concealed information. Psychophysiology 41:205–219
6. Rosenfeld JP, Labkovsky E, Winograd M, Lui MA, Vandenboom C, Chedid E (2008) The Complex Trial Protocol (CTP): a new, countermeasure-resistant, accurate, P300-based method for detection of concealed information. Psychophysiology 45(6):906–919
7. Bablani A, Edla DR, Tripathi D, Kuppili V (2019) An efficient concealed information test: EEG feature extraction and ensemble classification for lie identification. Mach Vis Appl 30(5):813–32
8. Moradi AMH, Janghorbani A (2016) A novel method based on empirical mode decomposition for P300-based detection of deception. IEEE Trans Inf Forensics Secur 11(11):2584–2593
9. Farahani ED, Moradi MH (2017) Multimodal detection of concealed information using genetic-SVM classifier with strict validation structure. Inform Med Unlocked 9:58–67
10. Luo X, Lin F, Chen Y, Zhu S, Xu Z, Huo Z, Yu M, Peng J (2019) Coupling logistic model tree and random subspace to predict the landslide susceptibility areas with considering the uncertainty of environmental features. Sci Rep 9(1):1–13
11. Cover T, Hart P (1967) Nearest neighbor pattern classification. IEEE Trans Inf Theor 13(1):21–27
12. Dasarathy BV (1991) Nearest neighbor (NN) norms: NN pattern classification techniques. IEEE Comput Soc Tutorial 10012834200

13. Raviya KH, Gajjar B (2013) Performance Evaluation of different data mining classification algorithm using WEKA. Indian J Res 2(1):19–21
14. Kotsiantis SB, Zaharakis I, Pintelas P (2007) Supervised machine learning: a review of classification techniques. Emerg Artif Intell Appl Comput Eng 160:3–24
15. De Mantaras RL, Armengol E (1998) Machine learning from examples: inductive and Lazy methods. Data Knowl Eng 25(1–2):99–123
16. Jain H, Yadav G, Manoov R (2021) Churn prediction and retention in banking, telecom and IT sectors using machine learning techniques. In: Advances in machine learning and computational intelligence, pp 137–156. Springer, Singapore
17. Quinlan JR (1986) Induction of decision trees. Mach Learn 81–106
18. Cruz JA, Wishart DS (2006) Applications of machine learning in cancer prediction and prognosis. Cancer Inform 2:117693510600200030
19. Hosmer Jr DW, Lemeshow S, Sturdivant RX (2013) Applied logistic regression, vol 398. Wiley
20. Dreiseitl S, Ohno-Machado L (2002) Logistic regression and artificial neural network classification models: a methodology review. J Biomed Inf 35(5–6):352–359
21. Breiman L (2001) Random forests. Mach Learn 45(1):5–32
22. Hasan SMM, Mamun MA, Uddin MP, Hossain MA (2018) Comparative analysis of classification approaches for heart disease prediction. In: 2018 international conference on computer, communication, chemical, material and electronic engineering (IC4ME2), IEEE, pp 1–4
23. Quinlan JR (1986) Induction of decision trees. Mach Learn 81–106
24. Blagus R, Lusa L (2015) Joint use of over- and under-sampling techniques and cross-validation for the development and assessment of prediction models. BMC Bioinf 16(1):1–10
25. Chawla NV (2009) Data mining for imbalanced datasets: an overview. In: Data mining and knowledge discovery handbook. Springer, pp 875–886
26. Lim T-S, Loh W-Y, Shih Y-S (2000) A comparison of prediction accuracy, complexity, and training time of thirty-three old and new classification algorithms. Mach Learn 40(3):203–228
27. Hay AM (1988) The derivation of global estimates from a confusion matrix. Int J Remote Sens 9(8):1395–1398
28. Abootalebi V, Moradi MH, Khalilzadeh MA (2009) A new approach for EEG feature extraction in P300-based lie detection. Comput Methods Programs Biomed 94(1):48–57
29. Svojanovsky (2017) Brain products. Accessed: 15, 2017. http://www.brainproducts.com/
30. Dodia S, Edla DR, Bablani A, Cheruku R (2020) Lie detection using extreme learning machine: a concealed information test based on short-time Fourier transform and binary bat optimization using a novel fitness function. Comput Intell 36(2):637–658

Printed in the United States
by Baker & Taylor Publisher Services